2016 International SoC Design Conference (ISOCC 2016)

Jeju, South Korea
23 – 26 October 2016

IEEE Catalog Number: CFP1669E-POD
ISBN: 978-1-5090-3220-4

Copyright © 2016 by the Institute of Electrical and Electronics Engineers, Inc
All Rights Reserved

Copyright and Reprint Permissions: Abstracting is permitted with credit to the source. Libraries are permitted to photocopy beyond the limit of U.S. copyright law for private use of patrons those articles in this volume that carry a code at the bottom of the first page, provided the per-copy fee indicated in the code is paid through Copyright Clearance Center, 222 Rosewood Drive, Danvers, MA 01923.

For other copying, reprint or republication permission, write to IEEE Copyrights Manager, IEEE Service Center, 445 Hoes Lane, Piscataway, NJ 08854. All rights reserved.

***This publication is a representation of what appears in the IEEE Digital Libraries. Some format issues inherent in the e-media version may also appear in this print version.**

IEEE Catalog Number: CFP1669E-POD
ISBN (Print-On-Demand):   978-1-5090-3220-4
ISBN (Online):   978-1-5090-3219-8
ISSN: 2163-9612

Additional Copies of This Publication Are Available From:

Curran Associates, Inc
57 Morehouse Lane
Red Hook, NY 12571 USA
Phone: (845) 758-0400
Fax: (845) 758-2633
E-mail: curran@proceedings.com
Web: www.proceedings.com

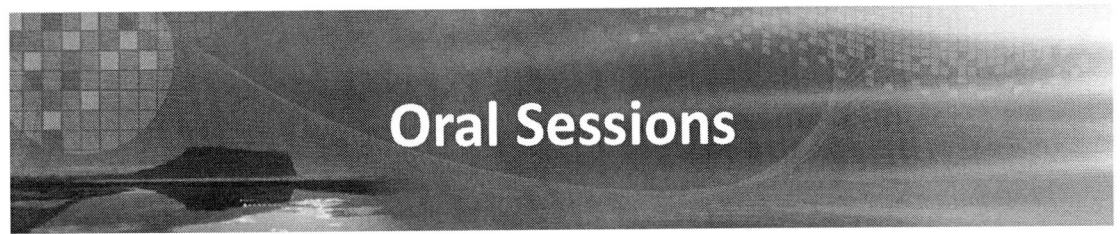

Oral Sessions

Monday, October 24, 2016

PEC
Power and Energy Circuits and Systems

13:45~15:15 Ballroom 2
Chair: Sukho Lee (*ETRI, Korea*)

[PEC-1]
A 200-kHz/6.78-MHz Wireless Power Transmitter Featuring Concurrent Dual-Band Operation.....1
Dukju Ahn[1], Jiwoong Park[2], and Patrick P. Mercier[2]
[1] *Incheon National University, Korea*
[2] *University of California San Diego (UCSD), USA*

[PEC-2]
Invited Paper: **Electrical Performance Analysis of Biogas Fuelled Generator with Purifier.....3**
Edison E. Mojica, and Arnold C. Paglinawan
Mapua Institute of Technology, Philippines

[PEC-3]
Power and Area Calibration of Switch Arbiter for High Speed Switch Control and Scheduling in Network-on-Chip.....5
Sangeeta Singh[1], JVR Ravindra[1], and B Rajendra Naik[2]
[1] *Vardhaman College of Engineering, India*
[2] *Osmania University, India*

[PEC-4]
An Efficient On-Chip Network with Packet Compression Capability.....7
Malihe Vafaiee[1], Majid Jalili[2], Reza Sabbaghi-Nadooshan[1], and Hamid Sarbazi-Azad[2]
[1] *Islamic Azad University, Iran*
[2] *Institute for Research in Fundamental Science (IPM), Iran*

CASE 1

Emerging Technologies 1

13:45~15:15 Ballroom 3
Chair: Sohmyung Ha(*New York University, Abu Dhabi*)

[CASE1-1]
Prediction-Based Latency Compensation Technique for Head Mounted Display.....9
Song-Woo Choi, Min-Woo Seo, and Suk-Ju Kang
Sogang University, Korea

[CASE1-2]
Invited Paper: **Review of Low Power Image Sensors for Always-On Imaging.....11**
Jaehyuk Choi
Sungkyunkwan University, Korea

[CASE1-3]
MEMS Resonator Based Thermometer SoC Design in CMOS 0.18 μm Standard Process.....13
Chong-Yang Lin, and Kuei-Ann Wen
National Chiao Tung University, Taiwan

[CASE1-4]
A Decouple structured Gyroscope with Integrated Readout Circuit on Standard 0.18μm 1P6M CMOS Technology.....15
Chun-Lin Chien, and Kuei-Ann Wen
National Chiao Tung University, Taiwan

[CASE1-5]
Monolithic MEMS Resonator Based Pressure Sensor and Readout Design.....17
Po-Chun Chuang, and Kuei-Ann Wen
National Chiao Tung University, Taiwan

ARM 1

Wireless and RF ICs

16:00~17:30 Ballroom 2

Chair: Hung-Wen Lin(*YuanZe University, Taiwan*)

[ARM1-1]
Implementation of RF Frequency Synthesizer for Smart Utility Network System.....19
Dong-Shik Kim, Won-Sang Yoon, and Sang-Hoon Chai
Hoseo University, Korea

[ARM1-2]
Invited Paper: **LNA Topologies for RX Carrier Aggregation.....21**
Jusung Kim[1], Keunkwan Ryu[1], Sungchan Kim[1], and Sanghun Lee[2]
[1] *Hanbat National University, Korea*
[2] *Wavepia Corporation, Korea*

[ARM1-3]
A design of Dual-band Smart Tag.....23
Jin-ho Kim, and Yong Moon
Soongsil University, Korea

[ARM1-4]
Design of 28GHz CMOS Phased Array T/R Circuits for 3-Dimensional Beamforming Applications.....25
Sungjin Shin, and Hyunchol Shin
Kwangwoon University, Korea

[ARM1-5]
A study of META-Voltage Controlled Oscillator and Prescaler using 65nm CMOS Process.....27
No yong Kwon, Bo ra Kim, and Yong Moon
Soongsil University, Korea

[ARM1-6]
Invited Paper: **Recent Advances in TSV Inductors for 3D IC Technology.....29**
Bruce Kim[1], and Sang-Bock Cho[2]
[1] *City University of New York, USA*
[2] *Ulsan University, Korea*

DIGITAL 1

Processor, Embedded Systems & Applications

16:00~17:30 Ballroom 3
Chair: Suk-Ju Kang(*Sogang University, Korea*)

[DIGITAL1-1]

Approximate Stochastic Computing (ASC) for Image Processing Applications.....31

Ramu Seva[1], Prashanthi Metku[1], Kyung Ki Kim[2], Yong-Bin Kim[3], and Minsu Choi[1]
[1]*Missouri University, USA*
[2]*Daegu University, Korea*
[3]*Northeastern University, USA*

[DIGITAL1-2]

Design and Implementation of Multi-Mode Block Adaptive Quantizer for Synthetic Aperture Radar.....33

Yu-Liang Tsai[1], Pei-Yun Tsai[1], Ching-Horng Lee[2], Li-Mei Chen[2], and Sz-Yuan Lee[2]
[1] *National Central University, Taiwan*
[2] *National Space Program Office (NSPO), Taiwan*

[DIGITAL1-3]

Mapping Table-based Fisheye Image Correction for Low Computational Complexity.....35

Yong Deok Ahn, and Suk-Ju Kang
Sogang University, Korea

[DIGITAL1-4]

Cryptographic Coprocessor Design for IoT Sensor Nodes.....37

Weizhen Wang, Jun Han, Zhicheng Xie, Shan Huang, and Xiaoyang Zeng
Fudan University, China

[DIGITAL1-5]

Software-Based Embedded Core Test Using Multi-Polynomial for Test Data Reduction.....39

Soyeon Kang, Inhyuk Choi, Hyeonchan Lim, Sungyoul Seo, and Sungho Kang
Yonsei University, Korea

DIGITAL 2
Digital Signal Processing Systems & Applications

16:00~17:30 Ballroom 4
Chair: Chip Hong Chang(*Nanyang Technological University, Singapore*)

[DIGITAL2-1]
Motion Vector Smoothing of Boundary of Moving Object for Frame Rate Up-Conversion.....41
Ho Sub Lee[1], Suk-Ju Kang[2], and Young Hwan Kim[1]
[1] *Pohang University of Science and Technology(POSTECH), Korea*
[2] *Sogang University, Korea*

[DIGITAL2-2]
Invited Paper: **Low-power and Real-time Computer Vision On-chip.....43**
Wei Pang[1], Hantao Huang[2], Fengwei An[3], and Hao Yu[1,2]
[1] *Southeast University, China*
[2] *Nanyang Technological University, Singapore*
[3] *Hiroshima University, Japan*

[DIGITAL2-3]
Image Interpolation Based on Hessian Analysis.....45
Sangho Yoon, and Young Hwan Kim
Pohang University of Science and Technology(POSTECH), Korea

[DIGITAL2-4]
Sharpness-aware Real-time Haze Removal for Advanced Driver Assistance Systems.....47
Joonggeun Ahn, Jihoon Kim, and Youngjoo Lee
Kwangwoon University, Korea

[DIGITAL2-5]
A New Scheme for Secret-Hiding in DSP Circuits.....49
Sumedh Dhabu, and Chip-Hong Chang
Nanyang Technological University, Singapore

Tuesday, October 25, 2016

ARM 2

Digital Signal Processing Systems & Applications

08:00~09:00 Ballroom 2

Chair: Jusung Kim(*Hanbat National University, Korea*)

[ARM2-1]

A Programmable ΔΣ SAR-ADC with Charge Shuttling Technique.....51

Kohei Yamada, Yosuke Toyama, and Hiroki Ishikuro

Keio University, Japan

[ARM2-2]

A 1-V 8-Bit 0.84uW SAR ADC for Biomedical Applications.....N/A

Tasnim B. Nazzal, and Soliman A. Mahmoud

University of Sharjah, United Arab Emirates

[ARM2-3]

11-bit 1.8uW 40KS/s Segmented SAR ADC for Sensor Applications.....55

Behnam Samadpoor Rikan, Sang-Yun Kim, and Kang-Yoon Lee

Sungkyunkwan University, Korea

[ARM2-4]

A Pipelined Time Stretching for High Throughput Counter-based Time-to-Digital Converters.....57

Seongheon Shin, and Hyung-Joun Yoo

Korea Advanced Institute of Science and Technology (KAIST), Korea

ARM 3

High-Speed Interface and Wireline ICs

08:00~09:00 Ballroom 3

Chair: Hyouk-Kyu Cha(*Seoul National University of Science and Technology, Korea*)

[ARM3-1]

A Low-Jitter Self-Biased Phase-Locked Loop for SerDes.....59

Heng-zhou Yuan, Yang Guo, Yao Liu, Bin Liang, Qian-cheng Guo, and Jia-wei Tan
National University of Defense Technology, China

[ARM3-2]

36-Gb/s CDR IC using simple passive loop filter combined with passive load in phase detector.....61

Keiji Kishine[1], Hiroshi Inoue[1], Kosuke Furuichi[1], Natsuyuki Koda[1], Hiromu Uemura[1], Hiromi Inaba[1], Makoto Nakamura[2], and Akira Tsuchiya[3]
[1]University of Shiga Prefecture, Japan
[2]Gifu University, Japan
[3]Kyoto University, Japan

[ARM3-3]

All-Synthesizable Transmitter Driver and Data Recovery Circuit for USB2.0 Interface.....63

Kihwan Seong, Won-Cheol Lee, Byungsub Kim, Jae-Yoon Sim, and Hong-June Park
Pohang University of Science and Technology (POSTECH),

DIGITAL 3

Circuits & Systems for Communications

08:00~09:00 Ballroom 4
Chair: Youngjoo Lee(*Kwangwoon University, Korea*)

[DIGITAL3-1]
Power-Efficient Partially-Adaptive Routing in On-chip Mesh Networks.....65
Majid Jalili[1], Julien Bourgeois[2], Hamid Sarbazi-Azad[1]
[1]Institute for Research in Fundamental Sciences (IPM), Iran
[2]UFC/FEMTO-ST Institute, France

[DIGITAL3-2]
Hash-Table and Balanced-Tree Based FIB Architecture for CCN Routers.....67
Kenta Shimazaki[1], Takashi Aoki[2], Takahiro Hatano[2], Takuya Otsuka[2], Akihiko Miyazaki[2], Toshitaka Tsuda[1], and Nozomu Togawa[1]
[1]WASEDA University, Japan
[2]NTT Corporation, Japan

[DIGITAL3-3]
Low Latency IFFT Design for 3GPP LTE.....69
Yeon-Jin Kim[1], Zheyan Piao[1], In-Gul Jang[2], Kyung-Ju Cho[3], and Jin-Gyun Chung[1]
[1]Chonbuk National University, Korea
[2]Electronics and Telecommunications Research Institute (ETRI), Korea
[3]Wonkwang University, Korea

[DIGITAL3-4]
A 0.5V/22μW Low Power Transceiver IC for Use in ESC Intra-body Communication System.....71
Yuhwai Tseng, Tinyou Lin, Songwen Yau, Yingchieh Ho, and Chauchin Su
National Chiao Tung University, Taiwan

DIGITAL 4

Digital IC and VLSI Architectures 1

13:45~15:30 Ballroom 2
Chair: Kristofor Gibson(*UC San Diego, USA*)

[DIGITAL4-1]
Area Efficient Neuromorphic Circuit Based on Stochastic Computation.....73
Kiwon Yoon, Suhyeong Choi, and Youngsoo Shin
Korea Advanced Institute of Science and Technology (KAIST), Korea

[DIGITAL4-2]
Invited Paper: **A 4.1mA Adaptive Duty-Cycle Corrector Loop with Background Calibration in 45nm CMOS Process.....75**
Esther Kim, Deokgwan Jeong, and Taehyoun Oh
Kwangwoon University, Korea

[DIGITAL4-3]
A New Approach to Binarizing Neural Networks.....77
Jungwoo Seo[1], Joonsang Yu[1], Jongeun Lee[2], and Kiyoung Choi[1]
[1]*Seoul National University, Korea*
[2]*UNIST, Korea*

[DIGITAL4-4]
Customized SRAM design for low power video code applications.....79
Sangkyu Lee, Hoyoung Tang, Kyungrak Choi, and Jongsun Park
Korea University, Korea

[DIGITAL4-5]
ISFET with Built-in Calibration Registers through Segmented Eight-bit Binary Search in Three-Point Algorithm Using FPGA.....81
Cyrel Ontimare Manlises[1], Febus Reidj G. Cruz[1,2], and Wen-Yaw Chung[2]
[1]*Mapua Institute of Technology, Philippines*
[2]*Chung Yuan Christian University, Taiwan*

[DIGITAL4-6]
Power-efficient Partitioning and Cluster Generation Design for Application-Specific Network-on-Chip.....83
Jiayi Ma, Cong Hao, Wencan Zhang and Takeshi Yoshimura
Waseda University, Japan

DIGITAL 5

Digital IC and VLSI Architectures 2

13:45~15:30 Ballroom 3
Chair: Youngmin Kim(*Kwangwoon University, Korea*)

[DIGITAL5-1]
Computation of Modular Multiplicative Inverses Using Residue Signed-Digit Additions.....85
Shugang Wei
Gunma University, Japan

[DIGITAL5-2]
Invited Paper: **Design Techniques for Ultra-Efficient Computing.....87**
Dongsuk Jeon
Seoul National University, Korea

[DIGITAL5-3]
An Ultra-Low Power AES Encryption Core in 65nm SOTB CMOS Process.....89
Van-Phuc Hoang[1], Van-Lan Dao[1], and Cong-Kha Pham[2]
[1]*Le Quy Don Technical University, Vietnam*
[2]*The University of Electro-Communications, Japan*

[DIGITAL5-4]
Cell-Based Delay Locked Loop Compiler.....91
Pei-Ching Huang, and Shi-Yu Huang
National Tsing Hua University, Taiwan

[DIGITAL5-5]
Hybrid GDI-NCL for Area/Power Reduction.....93
Prashanthi Metku[1], Ramu Seva[1], Kyung Ki Kim[2], Yong-Bin Kim[3] and Minsu Choi[1]
[1]*Missouri University of Science & Technology, USA*
[2]*Daegu University, Korea*
[3]*Northeastern University, USA*

[DIGITAL5-6]
A High-performance Circuit Design Algorithm using Data Dependent Approximation.....95
Kazushi Kawamura, Masao Yanagisawa, and Nozomu Togawa
Waseda University, Japan

DIGITAL 6

Memory Circuits & Systems

13:45~15:30 Ballroom 4
Chair: Seokhyeong Kang(*UNIST, Korea*)

[DIGITAL6-1]
Discussion of Cost-effective Redundancy Architectures.....97
Keewon Cho, Jooyoung Kim, Hayoung Lee, and Sungho Kang
Yonsei University, Korea

[DIGITAL6-2]
Equalization Scheme Analysis for High-Density Spin Transfer Torque Random Access Memory.....99
Beomsang Yoo[1], Taehui Na[1], Byungkyu Song[1], Jung Pill Kim[2], Seung H. Kang[2], and Seong-Ook Jung[1]
[1]*Yonsei University, Korea*
[2]*Qualcomm Incorporated, USA*

[DIGITAL6-3]
Variation-Tolerant and Low Power Look-Up Table (LUT) Using Spin-Torque Transfer Magnetic RAM for Non-volatile Field Programmable Gate Array (FPGA).....101
Kangwook Jo, Kyungseon Cho, and Hongil Yoon
Yonsei University, Korea

[DIGITAL6-4]
Disturb-free 5T Loadless SRAM Cell Design with Multi-Vth Transistors Using 28 nm CMOS Process.....103
Chua-Chin Wang, and Chia-Lung Hsieh
National Sun Yat-Sen University, Taiwan

[DIGITAL6-5]
Single-Flux-Quantum Cache Memory Architecture.....105
Koki Ishida[1], Masamitsu Tanaka[2], Takatsugu Ono[1], and Koji Inoue[1]
[1]*Kyushu University, Japan*
[2]*Nagoya University, Japan*

[DIGITAL6-6]
Parallel Decoding for Multi-Stage BCH decoder.....107
Prashanthi Metku[1], Ramu Seva[1], Kyung Ki Kim[2], Yong-Bin Kim[3] and Minsu Choi[1]
[1]*Missouri Univ of Science & Technology, USA*
[2]*Daegu University, Korea*
[3]*Northeastern University, USA*

[DIGITAL6-7]
A RAM cache approach using Host Memory Buffer of the NVMe interface.....109
JuHyung Hong, SangWoo Han, and Eui-Young Chung
Yonsei University, Korea

ARM 4
Analog Circuits

15:45~17:15 Ballroom 2
Chair: Dong-Woo Jee(*Ajou University, Korea*)

[ARM4-1]
A Passband Lock Loop Circuit System for Band Pass Filter.....111
Hung-Wen Lin[1], and Jin-Yi Lin[2]
[1]*Yuanze University, Taiwan*
[2]*Wpisil Technologies Inc., Taiwan*

[ARM4-2]
Invited Paper: **A 11mV Single Stage Thermal Energy Harvesting Regulator with Effective Control Scheme for Extended Peak Load.....113**
Priya.V, Murali. K. Rajendran, Shourya Kansal, and Ashudeb Dutta
Indian Institute of Technology, India

[ARM4-3]
Buffer With Neuron MOSFETs for Class-G Headphone Driver.....115
Yuki Matsuda[1], Sumio Fukai[2], Akio Shimizu[1], and Yohei Ishikawa[1]
[1]*Ariake College, Japan*
[2]*Saga University, Korea*
Kangwook Jo, Kyungseon Cho, and Hongil Yoon
Yonsei University, Korea

[ARM4-4]
Energy-efficient spread second capacitor capacitive-DAC for SAR ADCs.....117
Sung-min Lee, Ju Eon Kim, Dong-Hyun Yoon, and Kwang-Hyun Baek
Chung-Ang University, Korea

[ARM4-5]
A design of new voltage to current converter with high linearity and wide tuning.....119
Yui-Hwan Sa, Pyo-Hoon Son, Ki-Hong Kim, Hi-Seok Kim, and Hyeong-Woo Cha
Cheongju University, Korea

CASE 2

Emerging Technologies 2

15:45~17:15 Ballroom 3

Chair: Tso-Bing Juang(*National Pingtung University, Taiwan*)

[CASE2-1]

A Fully Integrated High-efficiency Step-up DC-DC Converter for Energy Harvesting Applications.....121

Seyed Mohammad Noghabaei, and Mohamad Sawan
Polytechnique Montreal, Canada

[CASE2-2]

Invited Paper: **An Integrated Optical Parallel Adder as a First Step Towards Light Speed Data Processing.....123**

Tohru Ishihara[1], Akihiko Shinya[2], Koji Inoue[3], Kengo Nozaki[2], and Masaya Notomi[2]
[1]*Kyoto University, Japan*
[2]*NTT Basic Research Laboratories, Japan*
[3]*Kyushu University, Japan*

[CASE2-3]

Invited Paper: **Integrated Circuits Design Using Carbon Nanotube Field Effect Transistor.....125**

Yong-Bin Kim
Northeastern University, USA

[CASE2-4]

Memory Efficient Hardware Accelerator for Kernel Support Vector Machine Based Pedestrian Detection.....127

Asim Khan and Chong-Min Kyung
Korea Advanced Institute of Science and Technology (KAIST), Korea

CASE 3

Emerging Technologies 3

15:45~17:15 Ballroom 4
Chair: Bruce Kim(*City University of New York, USA*)

[CASE3-1]

A TSV Test Structure for Simultaneously Detecting Resistive Open and Bridge Defects in 3D-ICs.....129

Young-woo Lee, Junghwan Kim, Inhyuk Choi and Sungho Kang
Yonsei University, Korea

[CASE3-2]

Novel Pixel Calibration Circuit for Bolometer-Type Uncooled Infrared Image Sensor.....131

Sang-Hwan Kim[1], Byoung-Soo Choi[1], Jang-Kyoo Shin[1], Jae-Hyoun Park[2], and Kyoung-Il Lee[2]
[1]*Kyungpook National University, Korea*
[2]*Korea Electronics Technology Institute, Korea*

[CASE3-3]

A Novel Frequency-shift Readout System for CEA Concentration Detection Application.....133

Deng-Shian Wang, Yun-Shen Liu, and Chua-Chin Wang
National Sun Yat-Sen University, Taiwan

[CASE3-4]

Fifth-Order Powerline Interference Rejection Filter Tailored Towards EEG Detection System.....N/A

Aisha Abdallah and Soliman Mahmoud
University of Sharjah, United Arab Emirates

[CASE3-5]

Low Power DPOTA Based Instrumentation Amplifier used in EEG Systems.....N/A

Aisha A. Alhammadi, Soliman A. Mahmoud, and Ahmed S. Elwakil
University of Sharjah, United Arab Emirates

[CASE3-6]

Design of a Configurable Bit-Resolution CMOS Image Sensor for the Image Depth Extraction.....139

Seongjoo Lee, and Minkyu Song
Dongguk University, Korea

SoC

SoC Design Methodology

15:45~17:15 Udo

Chair: Taegeun Yoo(*Nanyang Technological University, Singapore*)

[SoC-1]

P-Backtracking: A New Scan Chain Diagnosis Method with Probability.....141

Tae Hyun Kim, Hyun Yul Lim, Sungho Kang
Yonsei University, Korea

[SoC-2]

Design-Time Energy Optimization for Asymmetric Multiprocessor System-on-Chip.....143

Yonghee Yun, and Young Hwan Kim
Pohang University of Science and Technology(POSTECH) Korea

[SoC-3]

eFuse based IC Authentication Architecture.....145

Seung-Yeob Lee and Joon-Sung Yang
Sungkyunkwan University, Korea

[SoC-4]

Process Variation-aware Bridge Fault Analysis.....147

Heetae Kim, Inhyuk Choi, Jaeil Lim, Hyunggoy Oh, and Sungho Kang
Yonsei University, Korea

[SoC-5]

A Test Methodology to Screen Scan-Path Failures.....149

Junghwan Kim, Young-woo Lee, Minho Cheong, Sungyoul Seo, and Sungho Kang
Yonsei University, Korea

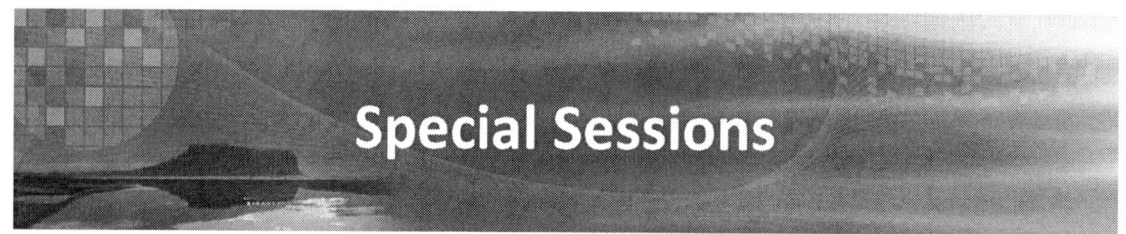

Special Sessions

Monday, October 24, 2016

SS 1

Energy-Efficient Radios and Cognitive Radios for Emerging Wireless

13:45~15:15 Ballroom 4
Chair: Woogeun Rhee(*Tsinghua University, China*)

[SS1-1]
CMOS Energy Efficient Integrated Radios for Emerging Low Power Standards.....151
Mustafijur Rahman, and Ramesh Harjani
University of Minnesota, USA

[SS1-2]
A 400MHz 3-10Mbps Transceiver IC with ~0.3 nJ/bit TX/RX Energy Efficiency for Body Area Applications.....153
Zhaoyang Weng, Jingjing Dong, Hanjun Jiang, and Zhihua Wang
Tsinghua University, China

[SS1-3]
Time-Varying Circuit Approaches for Software Defined and Cognitive Radio Applications.....155
Sudhakar Pamarti[1], N. Sinha[1], S. Hameed[1], and M. Rachid[2]
[1]*University of California, USA*
[2]*Silvus Technologies, USA*

[SS1-4]
A 0.5-V Sub-mW Energy-Efficient Receiver in 0.18-μm CMOS for IoT Applications.....157
Tse-Wei Wang, Yi-Lin Tsai, Chong-Rong Lee, Fu-Lian Hung, and Tsung-Hsien Lin
National Taiwan University, Taiwan

SS 2

Video Processing Circuits and Systems for Smart Vehicles

13:45~15:15 & 16:00~17:30 Mara
Chair: Hyuk-Jae Lee(*Seoul National University, Korea*)

[SS2-1]
Automatic Image Deviation Detection for AVM Auto-Calibration.....159
Jiwon Bang, Junghwan Pyo, and Yongjin Jeong
Kwangwoon University, Korea

[SS2-2]
Hardware implementation of fast traffic sign recognition for intelligent vehicle system.....161
Eunchong Lee, Sang-Seol Lee, Youngbae Hwang and Sung-Joon Jang
Korea Electronics Technology Institute, Korea

[SS2-3]
Front Collision Warning based on Vehicle Detection using CNN.....163
Jiwon Bang, Junghwan Pyo, and Yongjin Jeong
Kwangwoon University, Korea

[SS2-4]
Development of Burst Error Effect Reduction Algorithm for CAN using Interleaver Method.....165
Ronnie O. Serfa Juan[1,2], Min-Woo Jeong[1], and Hi-Seok Kim[1]
[1]Cheongju University, Korea
[2]TUP, Philippines

[SS2-5]
Hardware Implementation of aggregated channel features for ADAS.....167
Hohyon Song, Bosun Jeong, Hyunkyu Choi, Taeho Cho, and Hweihn Chung
Nextchip Co. Ltd., Korea

[SS2-6]
Improvements in Parallel SIMD Implementation of Single Image Defogging.....169
Truong Q. Nguyen[1], Hannoh Yoon[2], and Kristofor B. Gibson[1]
[1]University of California, USA
[2]Mtek Vision Co., Ltd., Korea

[SS2-7]
Dehazing in Color Filter Array Domain.....171
Yeejin Lee, Truog Q. Nguyen, and Changyoung Han
University of California, USA

[SS2-8]

Nighttime Image Enhancement Applying Dark Channel Prior to Raw Data From Camera.....173

Yan Gong, Yeejin Lee, and Truong Q. Nguyen
University of California, USA

[SS2-9]

Pedestrian Detection Aided by Temporal Prior.....175

Zhaowei Cai, Matthew Jacobsen and Nuno Vasconcelos
University of California, USA

[SS2-10]

Moving Objects Detection using Classifying Object Proposals for Driver Assistance System.....177

Kunyao Chen[1], Subarna Tripathi[1], Youngbae Hwang[2], and Truong Nguyen[1]
[1] *University of California, USA*
[2] *Korea Electronics Technology Institute, Korea*

[SS2-11]

Dense Stereo-based Real-time ROI Generation for On-road Obstacle Detection.....179

Soon Kwon[1,2], and Hyuk-Jae Lee[1]
[1]*Seoul National University, Korea*
[2]*DGIST, Korea*

SS 3

Challenges and opportunities for future memory circuits and systems of Internet of Things

13:45~15:15 Udo

Chair: Kyeong-Sik Min(*Kookmin University, Korea*)

[SS3-1]
High Bandwidth Memory(HBM) with TSV Technique.....181

Jong Chern Lee, Jihwan Kim, Kyung Whan Kim, Young Jun Ku, Dae Suk Kim, Chunseok Jeong, Tae Sik Yun, Hongjung Kim, Ho Sung Cho, Sangmuk Oh, Hyun Sung Lee, Ki Hun Kwon, Dong Beom Lee, Young Jae Choi, Jaejin Lee, Hyeon Gon Kim, Jun Hyun Chun, Jonghoon Oh, and Seok Hee Lee
SK Hynix Inc., Korea

[SS3-2]
Emulation of Processing in Memory Architecture for Application Development.....183

Jin-San Kwon, Tae-ho Hwang, and Dong-Sun Kim
Korea Electronics Technology Institute, Korea

[SS3-3]
Implementation of a Low-Overhead Processing-in-Memory Architecture.....185

Young-Jong Jang, Byung-Soo Kim, Dong-Sun Kim, and Tae-ho Hwang
Korea Electronics Technology Institute, Korea

[SS3-4]
High Density PCM(Phase Change Memory) Technology.....187

Hongsik Jeong
Tsinghua University, China

SS 4

Sensor and sensor system for IoT

16:00~17:30 Udo
Chair: Toru Shimizu(*Keio University, Japan*)

[SS4-1]

Robust Optical Fingerprint Sensor to Moisture Fingerprints.....189

Young-Hyun Baek
UnionCommunity Co., Ltd., Korea

[SS4-2]

Deep Learning Application Trial to Lung Cancer Diagnosis for Medical Sensor Systems.....191

Ryota Shimizu, Shusuke Yanagawa, Yasutaka Monde, Hiroki Yamagishi,
Mototsugu Hamada, Toru Shimizu, and Tadahiro Kuroda
Keio University, Japan

[SS4-3]

Normally-off Power Management for Sensor Nodes of Global Navigation Satellite System.....193

Takashi Nakada[1], Toshifumi Nakamoto[2], Toru Shimizu[3], and Hiroshi Nakamura[1]
[1]*The University of Tokyo, Japan*
[2]*Core Corporation, Japan*
[3]*Keio University, Japan*

[SS4-4]

Low-Power Multi-Sensor System with Normally-off Sensing Technology for IoT Applications.....195

Masanori Hayashikoshi[1], Hideyuki Noda[1], Hiroyuki Kawai[2], and Hiroyuki Kondo[3]
[1]*Renesas Electronics Corporation, Kanazawa University, Japan*
[2]*Tokushima Bunri University, Japan*
[3]*Renesas Electronics Corporation, Japan*

[SS4-5]

A large scale access-control list for IoT security comprising embedded IP-core and DDR DRAM.....197

Kazunari Inoue[1], and Yuji Yano[2]
[1]*National Institute of Technology, Japan,*
[2]*Osaka-city University, Japan,*

[SS4-6]

$3D^2$ Processing Architecture
- High Reliability and Low Power Computing for Novel Nano Tactile Sensor Array -.....199

Kiyotaka Komoku[1], Kazutami Arimoto[1], Tomoyuki Yokogawa[1], Hitoshi Yamauchi[1], Yoichiro Sato[1], and Hidekuni Takao[2]

[1]*Okayama Prefectural University, Japan*
[2]*Kagawa University, Japan*

Tuesday, October 25, 2016

SS 5
Hardware Security for Intelligent Things

08:00~09:00 Mara
Chair: Noriyuki Miura(*Kobe University, Japan*)

[SS5-1]

Attack Sensing against EM Leakage and Injection.....201

Noriyuki Miura[1], and Shivam Bhasin[2]

[1]*Kobe University, Japan*
[2]*Nanyang Technological University, Singapore*

[SS5-2]

How to Design Hardware Prime Field Multipliers for Bilinear Pairing.....203

Daisuke Fujimoto, Yusuke Nagahama, and Tsutomu Matsumoto

Yokohama National University, Japan

SS 6

Computational Devices, Circuits and Systems

08:00~09:00 Udo

Chair: Letian Huang(*University of Electronic Science and Technology of China, China*)

[SS6-1]

A Lightweight Metric for The Evaluation of Network Congestion in NoC-based MPSoC.....205

Yang Huang[1], Letian Huang[1], and Xiaohang Wang[2]

[1]*University of Electronic Science and Technology of China, China*

[2]*South China University of Technology, China*

[SS6-2]

An Efficient FPGA Implementation for odd-even sort based KNN algorithm using OpenCL.....207

Hai Peng[1], Letian Huang[1], and John Chen[2]

[1]*University of Electronic Science and Technology of China, China*

[2]*Intel, China*

[SS6-3]

An Address Remapping Algorithm to Reduce Power Consumption in NoC-based Chip-Multiprocessors.....209

Shuyu Chen[1], Letian Huang[1], and Song Li[2]

[1]*University of Electronic Science and Technology of China, China*

[2]*Inspur Group, China*

[SS6-4]

Neural Network based Seizure Detection System using Raw EEG Data.....211

Tianchan Guan[1], Letian Huang[2], Xiaoyang Zeng[1], Tianchan Guan[3], and Mingoo Seok[3]

[1]*Fudan University, China*

[2]*University of Electronic Science and Technology of China, China*

[3] *Columbia University, USA*

SS 7

Design, Analysis and Tools for Integrated Circuits and Systems (DATICS)

13:45~15:30 Mara

Chair: Ka Lok Man(*Xi'an Jiaotong-Liverpool University in Suzhou, China*)

[SS7-1]

Low-Cost Concurrent Error Detection Schemes for Logarithmic Converters.....213

Tso-Bing Juang[1], Ying-Ren Lee[1], and Chin-Chieh Chiu[2]

[1]*National Pingtung University (NPTU), Taiwan*

[2]*National Chao Tung University (NCTU), Taiwan*

[SS7-2]

Digital Image Preprocessing and Hair Artifact Removal by using Gabor Wavelet.....215

Uzma Jamil[1], Shehzad Khalid[1], and M.Usman Akram[2]

[1]*Bahria University,　Pakistan*

[2]*National University of Sciences & Technology, Pakistan*

[SS7-3]

A Flexible Software Defined Radio-based UHF RFID Reader Based on the USRP and LabView.....217

Wang Yuechun[1], Ka Lok Man[1], Robert G. Maunder[2], Jin Kyung Lee[3] and Kyung Ki Kim[3]

[1]*Xi'an Jiaotong-Liverpool University, China*

[2]*University of Southampton, United Kingdom*

[3]*Daegu University, Korea*

[SS7-4]

Radio Frequency Energy Harvesting Technology.....219

Lanxiang Wang, Menglong He, Zhao Wang, Mark Leach, Jingchen Wang, Kalok Man, and Eng Gee Lim

Xi'an Jiaotong-Liverpool University, China

[SS7-5]

Skew Control Methodology for Useful-Skew Implementation.....221

SangGi Do, Seungwon Kim, and Seokhyeong Kang

UNIST, Korea

[SS7-6]

A scheme for interference avoidance in Cognitive Radio channel allocation and transmission.....N/A

Aamir Nadeem, Murad Khan, Bhagya Nathali Silva, and Kijun Han

Kyungpook National University, Korea

[SS7-7]

μPnP-WAN: Wide Area Plug and Play Sensing and Actuation with LoRa.....225

Fan Yang[1], Gowri Sankar Ramachandran[1], Piers Lawrence[1], Sam Michiels[1], Wouter Joosen[1], and Danny Hughes[1,2]

[1] KU Leuven, Belgium
[2] VersaSense NV, Belgium

SS 8

Embedded software for Internet of Things

13:45~15:30 Udo
Chair: Sang Yub Lee(*KETI, Korea*)

[SS8-1]

CAN FD Controller for In-Vehicle System.....227

Jung Woo Shin, Jung Hwan Oh, Sang Muk Lee, and Seung Eun Lee
Seoul National University of Science and Technology,Korea

[SS8-2]

Design of An Area-Efficient Hardware Filter for Embedded System.....229

Ji Kwang Kim, Oh Seong Gwon, and Seung Eun Lee
Seoul National University of Science and Technology,Korea

[SS8-3]

A Network Architecture Design of Embedded System for Media Service in Bus.....231

Sang Yub Lee, Duck Keun Park , Jae Jin Ko, Jae Kyu Lee, and Choul Jun Kang
Korea Electronics Technology Institute, Korea

[SS8-4]

A Study on Improvement of Recognition Accuracy by Applying Machine Learning Algorithms to the Vision-Based Traffic Condition Analysis System.....233

Keonhee Lee, Hyuntae Ju, Yong Mu Jeong, and Soo-Young Min
Korea Electronics Technology Institute, Korea

[SS8-5]

A Study on Improvement of Vision-Based Traffic Condition Analysis System by Comparing Feature Data of Images.....235

Eunae Park, Hyuntae Ju, Yong Mu Jeong, and Soo-Young Min
Korea Electronics Technology Institute, Korea

[SS8-6]

A Study on river water level monitoring method in a debris barrier.....237

Hyo Sub Choi, and Deepak Ghimire
Korea Electronics Technology Institute, Korea

[SS8-7]
Software Design for GUI Display in the Wearable Device.....239
Gyutae Oh, Inhye Park, Sang-Yub Lee, and Jaejin Ko
Korea Electronics Technology Institute, Korea

2016 International SoC Design Conference

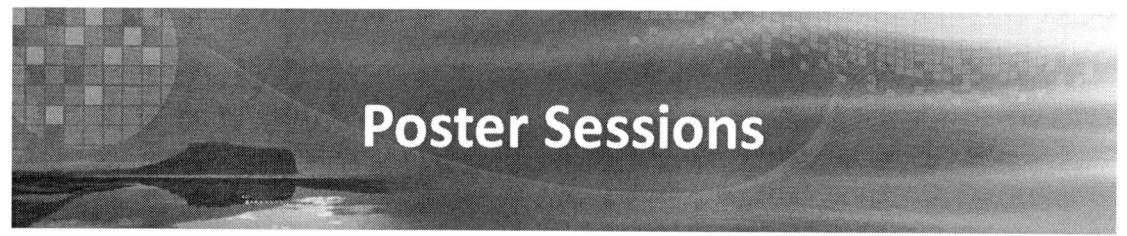

Poster Sessions

Tuesday, October 25, 2016

Poster Session

15:45~17:15 Lobby
Chair: Youngmin Kim (*Kwangwoon University, Korea*)
Youngjoo Lee (*Kwangwoon University, Korea*)

Analog/RF/Mixed-Signal Circuits

[PS-1]
Resolution Tunable Ring Oscillator type TDC.....241
Himchan Park, Zhang-Zhi Yu, Jinwoo Kim, and Jinwook Burm
Sogang University, Korea

[PS-2]
Novel 8-T CNFET SRAM Cell Design for the Future Ultra-low Power Microelectronics.....243
YoungBae Kim[1], Qiang Tong[1], Ken Choi[1], and Yunsik Lee[2]
[1]*Illinois Institute of Technology, USA*
[2]*UNIST, Korea*

[PS-3]
Low Area 10-Bit DAC Based on Programmable Current Output Buffer for AMOLED Column Driver.....N/A
Tao Huang, and Hong Ge Li
BeiHang University, China

[PS-4]
A MDLL-based Multi-Phase Clock Multiplier.....247
Junsub Yoon, and Jongsun Kim
Hongik University, Korea

[PS-5]
Proposal for sensitive frequency demodulator for 10-Gb/s transmission labeling signal system.....249
Natsuyuki Koda, Kosuke Furuichi, Hiromu Uemura, Hiromi Inaba, and Keiji Kishine
The University of Shiga Prefecture, Japan

[PS-6]
A Transient Enhanced External Capacitor-Less LDO With A CMOS Only Sub-Bandgap Voltage Reference.....251
Chang-Bum Park, Chan-Kyeong Jung, and Shin-Il Lim
Seokyeong University, Korea

[PS-7]
A Fast-Locking Clock Multiplying DLL.....253
Jongsun Kim, and Bongho Bae
Hongik University, Korea

[PS-8]
A CMOS 10-bit SAR ADC with Threshold Configuring Comparator for 5 MSBs.....255
Sang Heon Lee, Seong Jae Hyeon, Kim Jong Gu, and Kwang Sub Yoon
Inha University, Korea

[PS-9]
A Low-Power 10-bit Single-Slope ADC Using Power Gating and Multi-Clocks for CMOS Image Sensors.....257
Byoung-Kwan Jeon, Seong-Kwan Hong, and Oh-Kyong Kwon
Hanyang University, Korea

[PS-10]
A 200-Mb/s to 3-Gb/s Wide-band Referenceless CDR Using Bidirectional Frequency Detector.....259
Nguyen Huu Tho, Kyung-Sub Son, Kyongsu Lee, and Jin-Ku Kang
Inha University, Korea

[PS-11]
Design of High-Linearity Delay Detection Circuit for 10-Gb/s Communication System in 65-nm CMOS.....261
Kosuke Furuichi, Hiromu Uemura, Natsuyuki Koda, Hiromi Inaba, and Keiji Kishine
University of Shiga Prefecture, Japan

[PS-12]
Design of Pseudo-Random Bit Sequence Generator with Adjustable Sinusoidal Jitter.....263
Hong-Jhih Chen, Jau-Ji Jou, and Tien-Tsorng Shih
National Kaohsiung University of Applied Sciences, Taiwan

[PS-13]

A study of the referenceless CDR based on PLL.....265

JiHoon Kim, YoungJu Hwang, and Yong Moon
Soongsil University, Korea

[PS-14]

A design of NFC Analog Front-End with the Frequency Selector.....267

Jin-ho Kim, and Yong Moon
Soongsil University, Korea

Circuits and Systems for Emerging Technologies

[PS-15]

Speed-Adaptive Ratio-Based Lane Detection Algorithm for Self-Driving Vehicles.....269

Seongrae Kim, Junhee Lee, and Youngmin Kim
Kwangwoon University, Korea

[PS-16]

A Design of Tunable Component for Font End Module.....271

Suk-Hui Lee[1], Ki-Jin Kim[1], K.H. Ahn[1], Sung-Il Bang[2]
[1]*Korea Electronics Technology Institute, Korea*
[2]*Dankook University, Korea*

[PS-17]

Efficient and Real-time Stereo Matching Hardware Architecture for High-resolution Image.....273

Haengson Son, Seonyoung Lee, and Kyoungwon Min
Korea Electronics Technology Institute, Korea

[PS-18]

A Low-Power, Low-Noise Neural Recording Amplifier for Implantable Biomedical Devices.....275

Hyung Seok Kim, and Hyouk-Kyu Cha
Seoul National University of Science and Technology, Korea

[PS-19]

Design of Emotion Lighting Control System on the Power Spectrum Algorithm.....277

Su-Jeong Yun, and Chi-Ho Lin
Semyung University, Korea

[PS-20]

A Low-Power Capacitive-Feedback CMOS Neural Recording Amplifier for Biomedical Applications.....279

Hyung Seok Kim, and Hyouk-Kyu Cha
Seoul National University of Science and Technology, Korea

[PS-21]

Development of an IoT-based Visitor Detection System.....281

Hyoung-Ro Lee[1,2], Chi-Ho Lin[1], and Won-Jong Kim[2]

[1]*Semyung University, Korea*

[2]*Electronics and Telecommunications Research Institute(ETRI), Korea*

[PS-22]

Current Mode Four-Quadrant Multiplier Design Using CNTFET.....283

Gyunam Jeon[1], Minsu Choi[2], Kyung Ki Kim[3], and Yong-Bin Kim[1]

[1]*Northeastern University, USA*

[2]*Missouri Univ of Science & Technology,USA*

[3]*Daegu University, Korea*

Digital VLSI Circuits and Systems

[PS-23]

A Flexible MCMC Detector ASIC.....285

Dominik Auras, Sebastian Birke, Tobias Piwczyk, Rainer Leupers, and Gerd Ascheid

RWTH Aachen University,Germany

[PS-24]

Throughput Enhancemnet with Optimal Fragmented MSDU Size for Fragmentation and Aggregation Scheme in WLANs.....287

Eunbi Ku, Chulho Chung, Byungcheol Kang, and Jaeseok Kim

Yonsei Universitiy, Korea

[PS-25]

Design of NFC transceiver for automotive applications.....289

Yeong-Gyo Gim, and Shiho Kim

Yonsei Universitiy, Korea

[PS-26]

Possibility Verification of Drone Detection Radar based on Pseudo Random Binary Sequence.....291

Sung Jun Lee, Jae Ho Jung, and Bonghyuk Park

Electronics and Telecommunications Research Institute (ETRI), Korea

[PS-27]

Design of Low Latency Successive Cancellation Decoder for Polar Codes.....293

Zheyan Piao, and Jin-Gyun Chung

Chonbuk National University, Korea

[PS-28]

Adaptive Approximate Adder (A^3) to Reduce Error Distance for Image Processor.....295

Sunghyun Kim, and Youngmin Kim
Kwangwoon University, Korea

[PS-29]

Artificial Neural Network Implementation in FPGA: A Case Study.....297

Shuai Li[1], Ken Choi[1], and Yunsik Lee[2]
[1]*Illinois Institute of Technology, USA*
[2]*UNIST, Korea*

[PS-30]

Resource-Efficient FPGA Architecture of Canny Edge Detector.....299

Yunseok Jang , Junwon Mun, and Jaeseok Kim
Yonsei University, Korea

[PS-31]

Standing Wave Oscillator Based Clock Distribution.....301

Wei Zhang, Youde Hu, Keji Cui, Dongxuan Bao, Dashan Pan, Lebo Wang, and Lirong Zheng
Fudan University, China

[PS-32]

Area-efficient and High-speed Binary Divider Architecture for Bit-Serial Interfaces.....303

Yunho Park, Jonghyuk Kwon, and Youngjoo Lee
Kwangwoon University, Korea

[PS-33]

Hardware Design Exploration of Fully-Connected Deep Neural Network with Binary Parameters.....305

Jinkyu Kim, Juyeob Kim, Byungjo Kim, Miyoung Lee, and Joohyun Lee
Electronics and Telecommunications Research Institute(ETRI), Korea

[PS-34]

A Pre-characterization Method for Multiple Single-Event Transient Analysis in Cell-based Designs.....307

Jong Kang Park, Jun-Sung Go, and Jong Tae Kim
Sungkyunkwan University, Korea

[PS-35]

Hardware implementation of fast high dynamic range processor for real-time 4K UHD video.....309

Sang-Seol Lee, Eunchong Lee, Youngbae Hwang, and Sung-Joon Jang
Korea Electronics Technology Institute, Korea

[PS-36]

A low power, high speed FinFET based 6T SRAM cell with enhanced write ability and read stability.....311

Rahaprian Mudiarasan[1], Qiang Tong and Ken Choi[1], and Yunsik Lee[2]
[1] *Illinois Institute of Technology, USA*
[2] *UNIST, Korea*

[PS-37]

A Dual-Retention Time Architecture towards Secure and High Performance STT-RAM Main Memory Subsystem.....313

Taemin Lee, and Sungjoo Yoo
Seoul National University, Korea

[PS-38]

Selective Refresh to Avoid Read Disturb Errors in STT-RAM Main Memory.....315

Taemin Lee, and Sungjoo Yoo
Seoul National University, Korea

[PS-39]

Design of eMMC Controller with Multiple Channels.....317

Chulhoon Kim, and Chanho Lee
Soongsil University, Korea

[PS-40]

Energy-Based Iterative Cost Aggregation in Depth Estimation with a Stereo Camera.....319

Nguyen Xuan Truong, and Huyk-Jae Lee
Seoul National University, Korea

[PS-41]

Implementation of Low Complexity Inter Prediction for IoT Systems.....321

Jaehyuk So, Junwon Mun, Kyungmook Oh, and Jaeseok Kim
Yonsei University, Korea

[PS-42]

Halo Effect Suppression for Single Image Haze Removal Method.....323

Geun-Jun Kim, and Bongsoon Kang
Dong-A University, Korea

[PS-43]

A Design of Real Time Detection IP with Color Detection for Surveillance.....325

Chang-Hee Park, Hyun-Tae Kim, Young-Min Jang, and Sang-Bock Cho
University of Ulsan, Korea

[PS-44]

Non-Photorealistic Rendering from Real Video Sequences with Discontinuity Reduction Using Fast Video Segmentation..... 327

Lu Xiao, Xiao-Xuan Huang, and Yi-Chang Lu
National Taiwan University, Taiwan

[PS-45]

An H.265/HEVC 4K UHD Slim Codec Design with Shared Prediction Unit Architecture.....329

Sukho Lee, and Hyunmi Kim
ETRI, Korea

[PS-46]

Fast CU Size Decision Method for HEVC Using CU Split Information of Adjacent Frames.....331

Young Ho Kim[1], Tae Sun Kim[1], Myung Hoon Sunwoo[1], and Jae Heon Jeong[2]
[1] Ajou University, Korea
[2] Hyundai Mobis, Korea

[PS-47]

The parallelization of convolution on a CNN using a SIMT based GPGPU.....333

Heekyeong Jeon, Kwanho Lee, Seonghyung Han, and Kwangyeob Lee
Seokyeong university, Korea

[PS-48]

Transmission Timing Configuraiton for Control and Non-Payload Communication of Unmanned Aerial Vehicle.....335

Tae Chul Hong, Kunseok Kang, Kwangjae Lim, and Jae Young Ahn
Electronics and Telecommunications Research Instituten, Korea

[PS-49]

A System-level Design of MapReduce-based Embedded Multiprocessor System-on-Chips.....337

Huajuan Zhang, Hao Xiao, and Ning Wu
Nanjing University of Aeronautics and Astronautics, China

Power and Energy Circuits

[PS-50]

Radio-Frequency Energy-Harvesting IC with DC-DC Converter.....339

Donghoon Seong, Kichang Jang, Wonjoon Hwang, Hyeondeok Jeon, and Joongho Choi
University of Seoul, Korea

[PS-51]

Design and Verification of sensorless BLDC motor start-up Logic with FPGA.....341

Won-Ki Park, Hyun-Young Lee, Byeong-Chan Jeon, and Sung-Chul Lee
KETI, Korea

[PS-52]

A Dimmable and Power-compensated AC Direct LED Driver with High Efficiency.....343

Donglie Gu[1], Shengpeng Tang[1], Jianxiong Xi[1], Lenian He[1], and Kexu Sun[2]
[1] *Zhejiang University, China*
[2] *Southern Methodist University, USA*

[PS-53]

HV Switch Using Differential Voltage Shaping Driver for 13 Series Li-ion Battery Cells BMS.....345

Tzung-Je Lee
Cheng Shiu University, Taiwan

[PS-54]

A CMOS buck converter with PFM / Hysteretic mode.....347

Tae-Heon Lee, Jong-Gu Kim, and Kwang-Sub Yoon
Inha University, Korea

[PS-55]

A Single Inductor Multiple Output(SIMO) Buck/Boost DC-DC Converter with Output Error-Driven Random Control.....349

Hyunbin Park, and Shiho Kim
Yonsei University, Korea

[PS-56]

A Synchronous Boost Converter with High Speed and High Accuracy Peak Current Control Unit.....351

Shengpeng Tang[1], Xianzhi Meng[1], Donglie Gu[1], Jianxiong xi[1], Lenian He[1], and Kexu Sun[2]
[1] *Zhejiang University, China*
[2] *Southern Methodist University, USA*

[PS-57]

A Digital Low-Dropout(DLDO) Regulator with 14dB Power Supply Rejection Enhancement.....353

Byung Gun Joung, Yangho Seo, and Chulwoo Kim
Korea University

SoC Design Methodology

[PS-58]

FPGA Power Estimation Simulator for Dynamic Input Data.....355

Taehee You, Jeongbin Kim, Minyoung Im, and Eui-Young Chung
Yonsei University, Korea

[PS-59]

Full System Verification of Compatible Microprocessors with a Dual Physical Core Verification Platform.....357

Jyun-Yan Li, and Ing-Jer Huang
National Sun Yat-sen University, Taiwan

[PS-60]

Memory ECC Architecutre Utilizing Memory Column Spares.....359

Jong Hyuk Park[1], and Joon-Sung Yang[2]
[1] *Samsung Electronics, Korea*
[2] *Sungkyunkwan University, Korea*

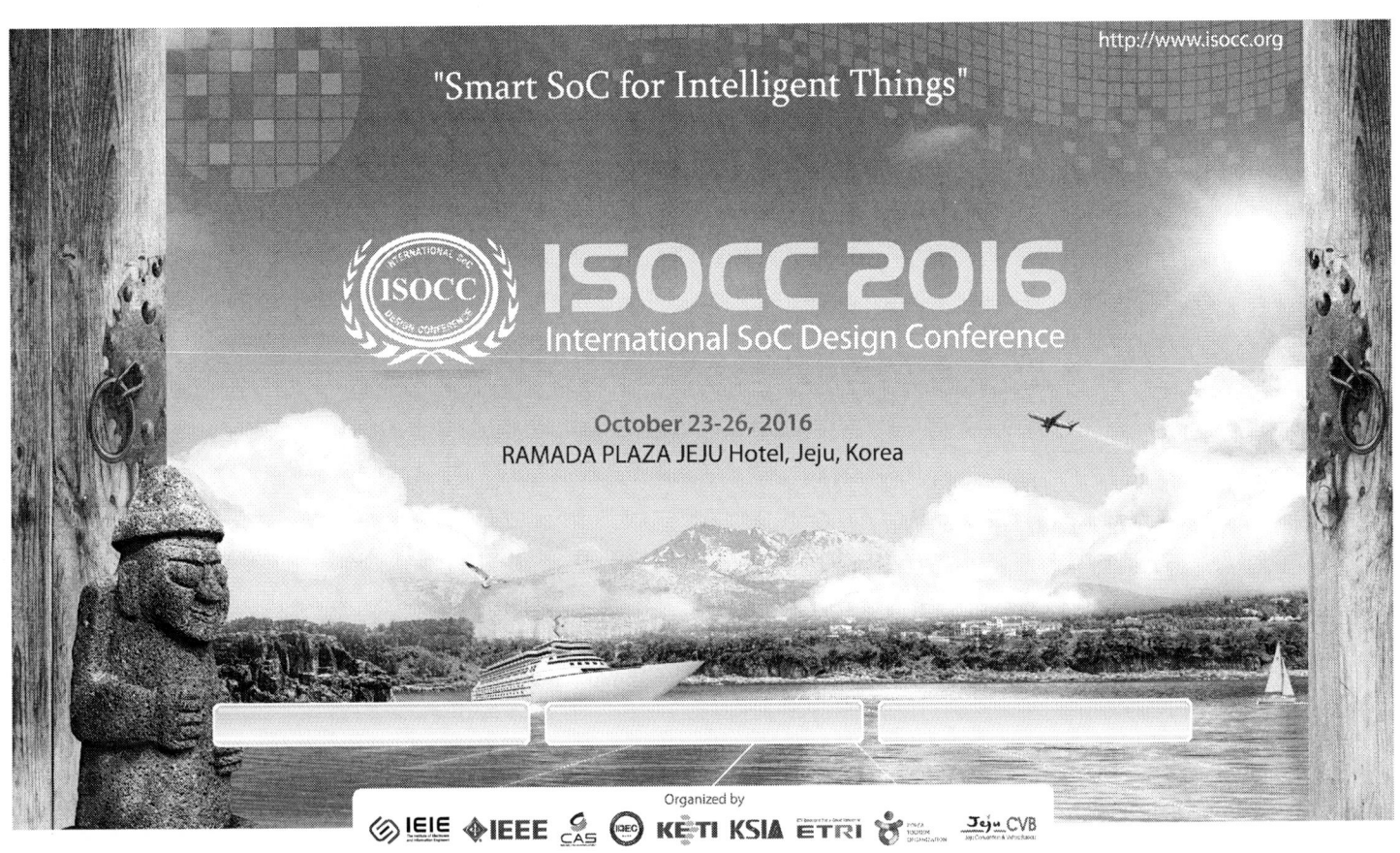

2016 International SoC Design Conference

Foreword

Welcome to ISOCC 2016, the 13th International SoC Design Conference!

On behalf of the Organizing Committee, I wish to invite all of the experts on the SoC, embedded SW, design automation area to the 13th International SoC Design Conference (ISOCC 2016). ISOCC 2016 will be held in Jeju Island, Korea, October 23- 26, 2016.

ISOCC has established a long tradition as an annual conference providing the world's premier SoC design forum for leading researchers from academia and industries. It is no arguing that SoC were continuously evolving to meet the industries needs. And super-deep nano technology and new emerging technology provide lots of possibilities for SoC's evolution. Embedded SW is already deeply integrated with SoC and reveals its superior ability. ISOCC 2016 might be the chance to see and experience the SoC with SW, super-deep nano technology, and emerging technology for the future.

The island of Jeju, Korea, warmly referred to as "Volcanic Island full of Allure" offering an expansive variety of things to see and do. The island's mixture of volcanic rock, beach and sands, and warm temperate climate, make it very popular islands for foreigners around the world. The island offers visitors a wide range of activities: hiking, trailing, catching sunrises and sunsets over the ocean, viewing majestic waterfalls, riding horses, or just lying around on the sandy beaches.

It is my great pleasure to invite you to ISOCC 2016. The technical sessions of ISOCC 2016 will provide a timely chance to discuss the recent progress in diverse fields of SoC design. The keynote and plenary speeches delivered by the leading experts will provide the deepest insights of SoC design.

During your stay in Korea, I hope you will enjoy not only the technical diversity of topics but also the natural beauty and culture of fantastic island, Jeju.

I look forward to seeing you at ISOCC 2016.

Yun Sik Lee
General Chair
ISOCC 2016

Message from TPC Chair

On behalf of the Technical Program Committee (TPC), we would like to welcome you all to the 13th International SoC Conference (ISOCC 2016) and we are glad to introduce the technical program covering a wide range of topics on the SOC designs. Theme of ISOCC 2016 is "Smart SoC for Intelligent Things".

The TPC received 216 regular papers from 17 countries this year. All submitted papers have been carefully reviewed in 5 technical tracks, "Analog/RF/Mixed-Signal Circuits", "Digital VLSI Circuits and Embedded Systems", "SoC Design Methodology", "Power and Energy Circuits", and "Circuits and Systems for Emerging Technologies". TPC selected 65 and 60 papers for oral and poster presentations, respectively. As a result, the overall acceptance ratio was 57.87%. Because of the limited time and sessions available, it was impossible to select all excellent papers for inclusion in the technical program.

In addition to the contributed papers, 8 keynote speeches, 10 invited papers, 48 special session papers, 6 tutorials, industrial demos, and chip design contest demos (CDC) are also presented in ISOCC 2016.

We would like to express our sincere gratitude to all those who have contributed to the technical program, including authors, reviewers, special session organizers, organizing committee members, and technical program committee members. Without their devotion and efforts, it would be impossible to hold the successful ISOCC 2016.

We hope that ISOCC 2016 technical program will satisfy your interests.

Kwang-Hyun Baek
Technical Program Committee Chair
ISOCC2016

Invitation to ISOCC 2017

It is my great pleasure to invite you to the 14th International SoC Conference (ISOCC 2017), which will be held from November 5th to 7th, 2017 at the Grand Hilton Hotel, Seoul, Korea. This conference is one of the most highly acclaimed annual conferences in the field of SoC. This year's conference will be held in conjunction with IEEE Asian Solid-State Circuit Conference (ASSCC 2017). We are truly glad to host such a prestigious event in the capital city of Korea. The conference will mainly focus on the latest developments and trends, as well as current and future outlook of the design and applications of System on a Chip.

The organizing committee is gearing up for an exciting and informative conference program including plenary lectures, symposia, workshops, tutorials on a variety of topics, poster presentations and various social programs for over 1,000 participants from around the world.

We invite you to join us at the ISOCC'2017, where you will be sure to have a meaningful experience with industry peoples and scholars from around the world. All members of the ISOCC'2017 organizing committee look forward to meeting you in Seoul Korea.

Sincerely,

Shiho Kim
Yonsei University,
General Chair of ISOCC 2017

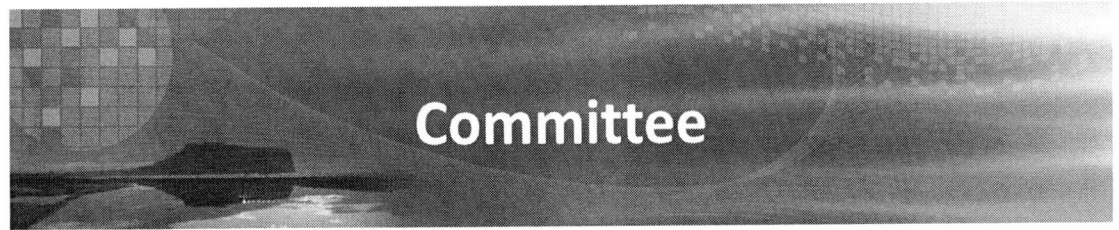

Committee

Organizing Committee

General Chair
Yun Sik Lee (*UNIST, Korea*)

General Co-Chairs
Chan-Ho Lee (*Soongsil University, Korea*)
Jun Jin Kong (*Samsung Electronics, Korea*)
Zhihua Wang (*Tsinghua University, China*)

General Vice Chairs
Shiho Kim (*Yonsei University, Korea*)
Sung Weon Kang (*ETRI, Korea*)
Chip Hong Chang (*Nanyang Technological University, Singapore*)

Steering Committee Chairs
Jinwook Burm (*Sogang University, Korea*)
Jun Rim Choi (*Kyungpook National University, Korea*)
Kwang Sub Yoon (*Inha University, Korea*)
Kyeongsoon Cho (*Hankuk University of Foreign Studies, Korea*)
Shin Il Lim (*Seokyeong University, Korea*)

Conference Secretary
Kyung Ki Kim (*Daegu University, Korea*)

Special Session Chairs
Byeong-Gyu Nam (*Chungnam National University, Korea*)
Min-Kyu Song (*Kookmin University, Korea*)
Minkyu Je (*KAIST, Korea*)
Kuk Tae Hong (*LG Electronics, Korea*)
Byeongho Choi (*KETI, Korea*)
Toru Shimizu (*Keio University, Japan*)

Finance Chairs
Chi Ho In (*Semyung University, Korea*)
Kyeong-Sik Min (*Dongguk University, Korea*)
Dong Kyue Kim (*Hanyang University, Korea*)
Hyuk-Jae Lee (*Seoul National University, Korea*)
Seo Kyu Lee (*Pixelplus, Korea*)
Kyung Ki Kim (*Daegu University, Korea*)

IEEE Liaison Chairs
Myung Hoon Sunwoo (*Ajou University, Korea*)
Kwang Sub Yoon (*Inha University, Korea*)
Jin Sang Kim (*Kyung Hee University, Korea*)

Publication Chairs
Yong Ho Song (*Hanyang University, Korea*)
Kwang-Yeob Lee (*Seokyeong University, Korea*)
Suk-Ju Kang (*Sogang University, Korea*)

Publicity Chairs
Changsik Yoo (*Hanyang University, Korea*)
Hyung Tak Kim (*Hongik University, Korea*)
Jong Sun Kim (*Hongik University, Korea*)
Ji-Hoon Kim (*Seoul National University of Science & Technology, Korea*)

Local Arrangement Chairs
Jae Yun Lim (*Jeju National University, Korea*)
Byungin Moon (*Kyungpook National University, Korea*)
Seokhyeong Kang (*UNIST, Korea*)

Poster Session Chairs
Youngmin Kim (*Kwangwoon University, Korea*)
Youngjoo Lee (*Kwangwoon University, Korea*)

Chip Design Contest Chairs
Kyoung Rok Cho (*Chungbuk National University, Korea*)
Tae Wook Kim (*Yonsei University, Korea*)

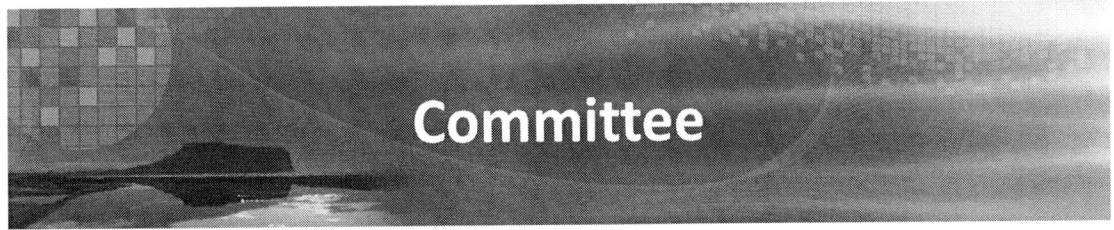

Committee

Technical Program Committee

Technical Program Chair
Kwang Hyun Baek (*Chung Ang University, Korea*)

Technical Program Co-Chairs
Yong-Bin Kim (*Northeastern University, USA*)
Yong Moon (*Soongsil University, Korea*)
Kyuwon(Ken) Choi (*Illinois Institute of Technology, USA*)
Jongsun Park (*Korea University, Korea*)

Track Chairs

Analog/RF/Mixed-Signal Circuits
Kang-Yoon Lee (Sungkyunkwan University, Korea)
Dong-Woo Jee (Ajou University, Korea)
Marvin Onabajo (Northeastern University, USA)
Josef Dobes (Czech technical University, Czech Republic)
Sanquan Song (Samsung Display America Lab, USA)
Hanlim Lee (Chung-Ang University, Korea)

Power and Energy Circuits
Baris Taskin (*Drexel University, USA*)
Franklin Bien (*UNIST, Korea*)
Yong Sin Kim (*Korea University, Korea*)

Digital VLSI Circuits and Systems
Jae-Sun Seo (*Arizona State University, USA*)
Minsu Choi (*Missouri University of Science and Technology, USA*)
Ji-Hoon Kim (*Seoul National University of Science & Technology, Korea*)
Youngjoo Lee (*Kwangwoon University, Korea*)

SoC Design Methodology
Youngmin Kim (*Kwangwoon University, Korea*)
Jaeyong Chung (*Incheon University, Korea*)
Yan Li (*NanoSemi Inc., USA*)

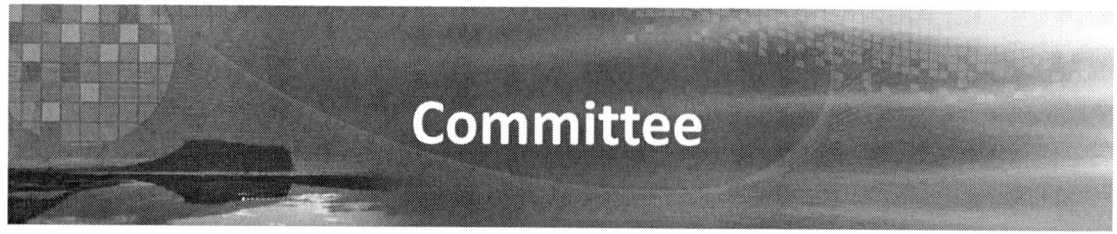

Circuits and System for Emerging Technologies
Youngcheol Chae (*Yonsei University, Korea*)
Hyouk-Kyu Cha (*Seoul National University of Science and Technology University, Korea*)
Bruce Kim (*The City College of New York, USA*)

Advisory Committee

Young Hwan Kim (*POSTECH, Korea*)
Hang Geun Jeong (*Chonbuk National University, Korea*)
Hi Seok Kim (*Cheongju University, Korea*)
Shin Il Lim (*Seokyeong University, Korea*)
Kyeongsoon Cho (*Hankuk University of Foreign Studies, Korea*)
Yeon Mo Chung (*Kyung Hee University, Korea*)
Sang Bock Cho (*University of Ulsan, Korea*)
Jin Gu Kang (*Inha University, Korea*)
Joongho Choi (*University of Seoul, Korea*)

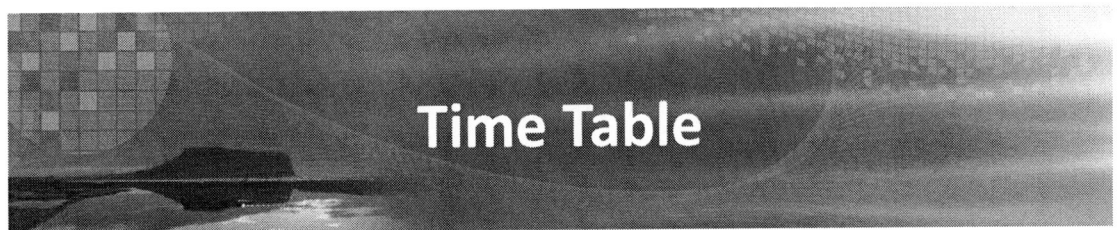

Time Table

Sunday ~ Monday, October 23-24, 2016

Oct. 23, Sunday — Venue: 2nd Floor

From	Till	Lobby	Ballroom2	Ballroom3
12:30	13:00	REGISTRATION		
13:00	14:30	REGISTRATION	Tutorial 1-1	Tutorial 2-1
14:30	14:45		Break	
14:45	16:15		Tutorial 1-2	Tutorial 2-2
16:15	16:30		Break	
16:30	18:00		Tutorial 1-3	Tutorial 2-3
18:00	18:30		Break	
18:30	20:30		Welcom Reception	

Oct. 24, Monday — Venue: 2nd Floor / 8th Floor

Morning

From	Till	Lobby	Ballroom1	Ballroom2	Ballroom3	Chuja	Mara	Udo	Tamra
8:00	8:15					CDC-1	CDC-2	CDC-3	
8:15	8:30					CDC-1	CDC-2	CDC-3	
8:30	8:45					CDC-1	CDC-2	CDC-3	
8:45	9:00			Break					
9:00	9:15	CDC Panel 1	Opening Ceremony						
9:15	10:00	CDC Panel 1	Keynote 1-1						
10:00	10:45		Keynote 1-2						
10:45	11:00	REGISTRATION	Break						
11:00	11:45	REGISTRATION	Keynote 1-3						
11:45	12:30	REGISTRATION	Keynote 1-4						
12:30	13:45	REGISTRATION	Lunch						

Afternoon

From	Till	Lobby	Ballroom1	Ballroom2	Ballroom3	Ballroom4	Mara	Udo	Tamra
13:45	14:00	CDC Panel 2		PEC: 213, 169, 110, 124	CASE1: 173, 228, 029, 030, 032	SS1: 052, 019, 053, 059	SS2: 050, 116, 122, 133, 210, 257	SS3: 195, 208, 222, 289, 288	Intelligent IC Workshop (13:00 ~ 17:30)
14:00	14:15								
14:15	14:30								
14:30	14:45								
14:45	15:00								
15:00	15:15								
15:15	15:30	Chip Design Contest (CDC) Poster Exhibition							
15:30	15:45	Chip Design Contest (CDC) Poster Exhibition							
15:45	16:00	Chip Design Contest (CDC) Poster Exhibition							
16:00	16:15		ARM1	007, 033, 152, 193, 232, 291	DIGITAL 1: 260, 094, 182, 041, 246	DIGITAL 2: 127, 074, 129, 150, 027	SS2: 258, 259, 274, 280, 281	SS4: 201, 207, 236, 253, 282, 283	
16:15	16:30								
16:30	16:45								
16:45	17:00								
17:00	17:15								
17:15	17:30								
17:30	18:00	Break							
18:00	20:00	Banquet							

ARM1	Wireless and RF ICs
CASE	Emerging Technologies
CDC	Chip Design Contest
Digital1	Processor, Embedded Systems & Applications
Digital2	Digital Signal Processing Systems & Applications
PEC	Power and Energy Circuits and Systems
SS1	Energy-Efficient Radios and Cognitive Radios for Emerging Wireless
SS2	Video Processing Circuits and Systems for Smart Vehicles
SS3	Challenges and opportunities for future memory circuits and systems of Internet of Things
SS4	Sensor and sensor system for IoT SIC2015 Project - Mobile/Automotive SoC

2016 International SoC Design Conference

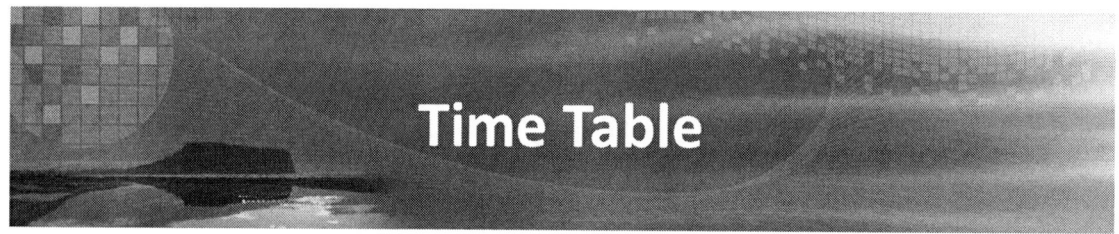

Time Table

Tuesday ~ Wednesday, October 25-26, 2016

Time		Oct. 25 Tuesday									Oct. 26 Wednesday
		2nd Floor								8th Floor	8th Floor
From	Till	Lobby	Ballroom1	Ballroom2	Ballroom3	Ballroom4	Mara	Udo	Tamra	Ora
8:00	8:15			ARM2 043	ARM3 058	DIGITAL3 120	SS5 025	SS6 174		
8:15	8:30			081	098	031	114	175		
8:30	8:45			103	186	249	147	176		
8:45	9:00			153		051	171	177		
9:00	9:15			Break						
9:15	10:00			Keynote 2-1						
10:00	10:45			Keynote 2-2						Discussion
10:45	11:00	R		Break						
11:00	11:45	E		Keynote 2-3						
11:45	12:30	G		Keynote 2-4						
12:30	13:45	I		Lunch						
13:45	14:00	S		DIGITAL4 042	DIGITAL5 106	DIGITAL6 151	SS7 073	SS8 044		
14:00	14:15	T		145	108	229	090	045		
14:15	14:30	R		183	131	237	119	048		
14:30	14:45	A		247	179	112	121	055		
14:45	15:00	T		146	262	166	178	060		
15:00	15:15	I		022	248	263	180	123		
15:15	15:30	O				036	184	017		
15:30	15:45	N		Break						
15:45	16:00			ARM4 256	CASE2 140	CASE3 159	SoC 233			
16:00	16:15			255	142	082				
16:15	16:30	Poster Session		255	134	115	135			
16:30	16:45			214	244	002	190			
16:45	17:00			095		021	198			
17:00	17:15			148	279	080				
17:15	17:30			Break						
17:30	18:00			Closing Ceremony						

ARM2   Data Converters
ARM3   High-Speed Interface and Wireline ICs
ARM4   Analog Circuits
CASE   Emerging Technologies
Digital3 Circuits & Systems for Communications
Digital4 Digital IC and VLSI Architectures 1
Digital5 Digital IC and VLSI Architectures 2
Digital6 Memory Circuits & Systems
SoC   SoC Design Methodology
SS5   Hardware Security for Intelligent Things
SS6   Computational Devices, Circuits and Systems
SS7   Design, Analysis and Tools for Integrated Circuits and Systems (DATICS)
SS8   Embedded software for Internet of Things

2016 International SoC Design Conference

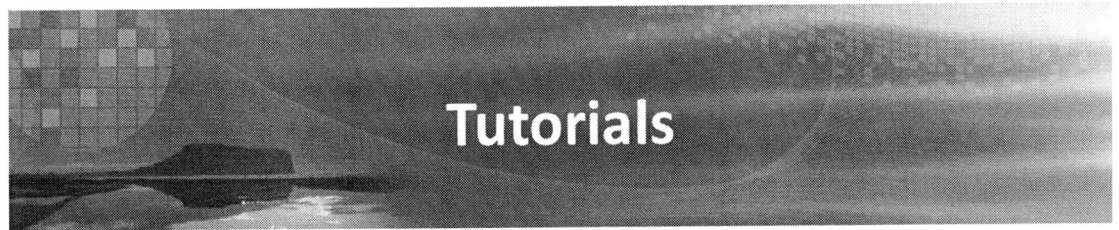

Tutorials

13:00~18:00 Sunday, October 23, 2016

[Tutorial 1-1(13:00~14:30 Ballroom 2)]
Chair: Youngmin Kim (*Kwangwoon University, Korea*)

Deep Learning for Computer Vision

Junmo Kim
Ph.D, Professor, Korea Advanced Institute of Science and Technology (KAIST), Korea

Biography
Junmo Kim received the B.S. degree from Seoul National University, Seoul, Korea, in 1998, and the M.S. and Ph.D. degrees from the Massachusetts Institute of Technology (MIT), Cambridge, in 2000 and 2005, respectively. From 2005 to 2009, he was with the Samsung Advanced Institute of Technology (SAIT), Korea, as a Research Staff Member. He joined the faculty of KAIST in 2009, where he is currently an Assistant Professor of electrical engineering. His research interests are in image processing, computer vision, statistical signal processing, and information theory.

Abstract
Recently deep learning has become one of the most powerful and popular machine learning techniques due to its record-breaking performances in a variety of recognition tasks including speech recognition and image classification. Deep learning also changed the paradigm of pattern recognition in that it allows us to automatically discover hierarchical features from data instead of relying on hand-crafted features. In this tutorial, I will provide an overview of deep learning discussing what have been the main difficulties of training deep neural networks and how these difficulties have been overcome by recent breakthroughs. I will also introduce several deep learning techniques such as restricted Boltzmann machine (RBM), deep belief network (DBN), deep neural network (DNN), and convolutional neural network (CNN) and talk about how they are applied to computer vision problems such as a large scale image classification.

2016 International SoC Design Conference

[Tutorial 1-2(14:45~16:15 Ballroom 2)]
Chair: Youngmin Kim (*Kwangwoon University, Korea*)

Design strategies for wearable sensor interface circuits for patient-specific monitoring

Jerald Yoo
Ph.D, Professor, Masdar Institute of Science and Technology, United Arab Emirates

Biography

Jerald Yoo received the B.S., M.S., and Ph.D. degrees in Department of Electrical Engineering from the Korea Advanced Institute of Science and Technology (KAIST), Daejeon, Korea, in 2002, 2007, and 2010, respectively. Since 2010, he has been with Electrical Engineering and Computer Science, Masdar Institute, Abu Dhabi, United Arab Emirates, where he is currently an associate professor. He developed low-energy Body Area Network (BAN) transceivers and wearable body sensor network using Planar-Fashionable Circuit Board (P-FCB) for continuous health monitoring system. His research focuses on low energy circuit technology for wearable bio signal sensors, BAN transceivers, ASIC for piezoelectric Micromachined Ultrasonic Transducers (pMUT) and SoC design to system realization for wearable healthcare applications. He is an author of a book chapter in Biomedical CMOS ICs (Springer, 2010).

Dr. Yoo is the recipient or a co-recipient of several awards: IEEE International Circuits and Systems (ISCAS) 2015 Best Paper Award (BioCAS Track), ISCAS 2015 Runner-Up Best Student Paper Award, the Masdar Institute Best Research Award (2015) and the Asian Solid-State Circuits Conference (A-SSCC) Outstanding Design Awards (2005). He is the Vice Chair of IEEE Solid-State Circuits Society (SSCS) United Arab Emirates (UAE) Chapter. Currently he serves as a Technical Program Committee member in IEEE Asian Solid-State Circuits Conference (A-SSCC), IEEE Custom Integrated Circuits Conference (CICC) and in IEEE International Solid-State Circuits Conference (ISSCC) Student Research Preview (SRP). He is also an Analog Signal Processing Technical Committee (ASPTC) member of IEEE Circuits and Systems Society (CASS). He is a senior member of IEEE.

Abstract

Wearable healthcare sensor provides attractive opportunity for semiconductor sector. The target here is to mitigate the impact of chronic diseases by providing continuous yet adequate low noise monitoring and analysis of physiological signals. Wearable environment is challenging for circuit designers due to its unstable skin-electrode interface to begin with. Wet and dry electrodes have very different electrical characteristic that needs to be addressed. Also, in wearable environment, trade-off between available resource and performance among the components (analog front-end and digital back-end) is of crucial.

This short course will cover the design strategies of bio interface circuits for such wearable sensors. We will first explorer the difficulties, limitations and potential pitfalls in wearable interface and strategies to overcome such issues. After that, system level considerations for better key metrics such as energy efficiency will be introduced. Several state-of-the-art instrumentation amplifiers that emphasize on different parameters will also be discussed. We will then see how the signal analysis part impacts the analog interface circuit design; we will also cover how we can achieve patient-specific monitoring. The talk will conclude with interesting aspects and opportunities that lie ahead.

2016 International SoC Design Conference

[Tutorial 1-3(16:30~18:00 Ballroom 2)]
Chair: Youngmin Kim (Kwangwoon University, Korea)

Wearable and Patchable Integrated Sensors for Real-time Continuous Monitoring of Vital Physiological Signals

Zheng Yuanjin
Ph.D, Professor, Nanyang Technological University, Singapore

Biography

Dr. Zheng Yuanjin received his B.Eng. from Xi'an Jiaotong University, P. R. China in 1993, M. Eng. from Xi'an Jiaotong University, P. R. China in 1996, and Ph.D. from Nanyang Technological University, Singapore in 2001. From July 1996 to April 1998, he worked at the National Key Lab of Optical Communication Technology, University of Electronic Science and Technology of China. He joined the Institute of Microelectronics, A*SATAR on 2001 as a senior research engineer, and then promoted to a principle investigator and group leader for wideband RFIC design group. Here, he has leaded and developed various CMOS RF transceivers and baseband SoC for WLAN, WCDMA, Ultra-wideband, and low power medical radio etc. Since July, 2009, he joined Nanyang Technological University as an assistant professor. He has been working on electromagnetic and acoustics physics and devices, biomedical imaging especially photoacoustics / thermoacoustics imaging and 3D imaging, energy harvesting circuits and systems etc.

Dr. Zheng has published more than 250 journal and conference papers, 22 patents filed/granted and 5 book chapters. He served as session chairs and TPC chairs/members for several international conferences. He has successfully leaded and contributed numerous public funded research and industry projects.

Abstract

Continuous health monitoring in hospital and/or home conditions has been of interest to doctors and healthcare practitioners for a long time. Recording of physiological and psychological variables in real-life conditions could be especially useful in management of accurate and chronic disorders or health problems, e.g., for stoke, shock, high blood pressure, diabetes, neural disorder, chronic pain, or severe obesity etc.. Physiological signals have been used as important indicator of the vital signs of human kinds, and real time monitoring of various physiological signals can predicate and be preventive to many serious life attacks. Furthermore, real-life long-term monitoring of health could be good measurement of treatment effects at home care, in situations where the subjects live their daily life. In this tutorial, firstly we will introduce and cover different types of physiological signals such as electrocardiogram (ECG), blood oxygen saturation (SO2), neural spike, blood core temperature, and Glucose etc. Secondly, the typical and vibrant integrated wearable or patchable sensors to acquire and measure the physiological signals are introduced and presented. Furthermore, the newly developed novel physiological sensors will be illustrated and demonstrated. The vital important applications and future development aspects are briefly envisioned.

2016 International SoC Design Conference

[Tutorial 2-1(13:00~14:30 Ballroom 3)]
Chair: Youngjoo Lee (*Kwangwoon University, Korea*)

CMOS Image Sensor and Its Noise

Manlyun Ha
Ph. D., Dongbu Hitek Co. Ltd., Korea

Biography
- Bachelor of Electronics, Kyoungpook Nat. Univ. (1991~1997)
- Master of Electrical Engineering & Computer Science, KAIST (1998~1999)
- Ph.D of Electrical Engineering & Computer Science, KAIST (2000~2004)
- MagnaChip Semiconductor (2005~2008): 0.13~ 0.11um tech. CIS Tech. Development
- Dongbu Hitek (2009~Present):0.13um~0.11um CIS Tech. Development (Mobile & Nonmobile),
Under 90nm CIS Tech. Development (FSI & BSI)

Abstract
In this lecture, as a first, I will review the recent CMOS image sensor's application and technology development trend. And then, as a next, I will introduce the understanding and the developing status on the Image Noise. Recently the application area of the CMOS Image Sensor become wide continuously and, mainly, it can be divided as Mobile Phone, High-end Camera (DSLR/DSC), Time-of-Flight, Automotive, Security, Medial, and Science & Space area. The market demand for the CIS technology increases even yet and nowadays, thanks to appearance of the Dual camera system in mobile phone, its demand expects to become twice simply. Automotive Sensor is also very impressive application market but needs higher reliability and frame rate than other applications. Security Sensor receives the attention with increasing needs on the social safety request. The main medical sensor is the capsule endoscope and X-ray sensor.

Even though many application areas there are, the main technology stem can be listed as two types. Fist one is the FSI (Frontside Illumination) and second one is the BSI (Backside Illumination). As well-known, the FSI technology is based on the traditional semiconductor processing technology and many important technologies have been developed, but due to the process limitation, now this is used for making the large pixel products and some special aim sensors.
The BSI technology shows so impressive performance but have so complicate process and high cost at the same time. Nevertheless, recently the BSI technology applied for the high-end camera and become general trend and this expand its application area to the large pixel products continuously.

These application conditions and process technology developments in CMOS Image Sensor is always running with the understanding of the Image Noise. This understanding of the Image Noise acts as one of the big motivation of technology development. To understand the Image noise, I will review the operation of the unit pixel at first and then design and layout related noise will be reviewed, too. List up the noise from the image and then also summarize and understand the physical basis of these noises, we can imagine the technical method to solve these noises. I check the direction of the technology development from the review of recent results to improve the image quality in the process and design area. This will give us the opportunity to develop more improved products.

2016 International SoC Design Conference

[Tutorial 2-2(14:45~16:15 Ballroom 3)]
Chair: Youngjoo Lee (*Kwangwoon University, Korea*)

Designing High-Performance Circuits Using Inverter Based Amplifiers

Ramesh Harjani
Ph.D, Professor, University of Minnesota, USA

Biography

Ramesh Harjani is the E.F. Johnson Professor of Electronic Communications in the Department of Electrical & Computer Engineering at the University of Minnesota. He is a Fellow of the IEEE. He received his Ph.D. in Electrical Engineering from Carnegie Mellon University in 1989. He was at Mentor Graphics, San Jose before joining the University of Minnesota and has been a visiting professor at Lucent Bell Labs, Allentown, PA and the Army Research Labs, Adelphi, MD. He co-founded Bermai, Inc, a startup company developing CMOS chips for wireless multi-media applications in 2001. His research interests include analog/RF circuits for wired and wireless communication systems.

Dr. Harjani received the National Science Foundation Research Initiation Award in 1991 and Best Paper Awards at the 1987 IEEE/ACM Design Automation Conference, the 1989 International Conference on Computer-Aided Design, and the 1998 GOMAC. His research group was the winner of the SRC Copper Design Challenge in 2000 and the winner of the SRC SiGe challenge in 2003. He is an author/editor of eight books. He was an Associate Editor for IEEE Transactions on Circuits and Systems Part II, 1995-1997, Guest Editors for the International Journal of High-Speed Electronics and Systems and for Analog Integrated Circuits and Signal Processing in 2004 and a Guest Editor for the IEEE Journal of Solid-State Circuits, 2009-2011. He was a Senior Editor for the IEEE Journal on Emerging & Selected Topics in Circuits & Systems (JETCAS), 2011-2013. He was the Technical Program Chair for the IEEE Custom Integrated Circuits Conference 2012-2013, the Chair of the IEEE Circuits and Systems Society technical committee on Analog Signal Processing from 1999 to 2000 and a Distinguished Lecturer of the IEEE Circuits and Systems Society for 2001-2002.

Abstract

Amplifiers form the core of many analog and mixed-signal circuits. However, the design of differential pair based OTAs is becoming increasingly difficult in finer geometries due to lower supply voltages. Inverter based designs have proven to have better transconductance efficiency, higher swing and better linearity but have degraded CMRR, worse PSRR and limited PVT tolerance. In this tutorial, we discuss traditional amplifiers and why inverter based amplifiers are better suited for lower supplies. We then describe the design procedure for inverter based OTA designs with an emphasis on improving their performance, including PVT tolerance, CMRR and PSRR. In particular, we introduce new biasing techniques for inverters to improve their PVT tolerance. We will finally validate our designs using measurement results from a number of fabricated designs.

[Tutorial 2-3(16:30~18:00 Ballroom 3)]

Chair: Youngjoo Lee (*Kwangwoon University, Korea*)

Low-Power High-Resolution ADCs for Sensor Applications

Youngcheol Chae
Ph.D., Professor, Yonsei University, Korea

Biography

Youngcheol Chae received the Ph.D degree in electrical and electronic engineering from Yonsei University, Seoul, Korea in 2009. From 2009 to 2011, he was a post-doctoral researcher with Delft University of Technology in the Netherlands. Dr. Chae is currently an Assistant Professor at Electrical and Electronic Engineering Department, Yonsei University. After joining Yonsei since 2012, he leads the Mixed-Signal IC Laboratory working on low-power data converters, high performance sensors and interface circuits. He has authored or co-authored 20 patents and over 50 technical papers. Dr. Chae received the Outstanding Teaching Award from Yonsei University in 2013 and 2014, respectively. He received a VENI grant from Dutch Technology Foundation STW in 2011. In 2008, He received a Gold Prize (1st) at Human-Tech Thesis Award (14th) from Samsung Electronics. Dr. Chae is a member of several technical program committees.

Abstract

Low-power, high-resolution analog-to-digital converters (ADCs) are highly required in many sensor applications. Such ADCs often result in poor energy-efficiency and high-power consumption, thus making them unsuitable for the use in battery-powered sensor systems. This tutorial covers the design of low-power high-resolution ADCs for smart sensor interfaces, which include both architectural- and circuit- level techniques to maintain their power-efficiency. It also describes practical design aspects from several design examples.

2016 International SoC Design Conference

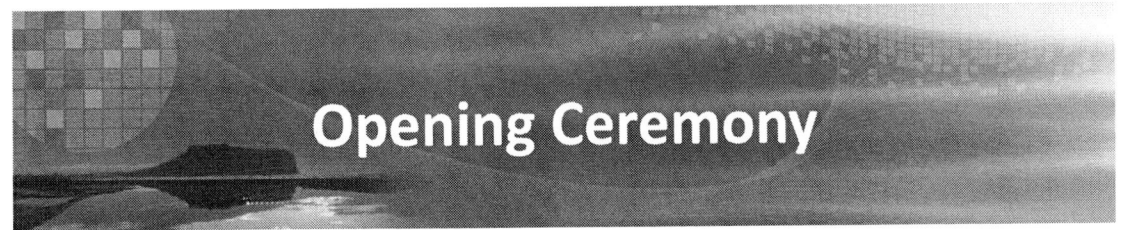

Opening Ceremony

09:00~09:15 Monday, October 24, 2016
Ballroom 1

Chair : Kwang Hyun Baek (*Chung Ang University, Korea*)

Welcome Address

Yun Sik Lee, General Chair (*UNIST, Korea*)

Conference Statistics

Kwang Hyun Baek, TPC Chair (*Chung Ang University, Korea*)

Announcements

Kwang Hyun Baek, TPC Chair (*Chung Ang University, Korea*)

2016 International SoC Design Conference

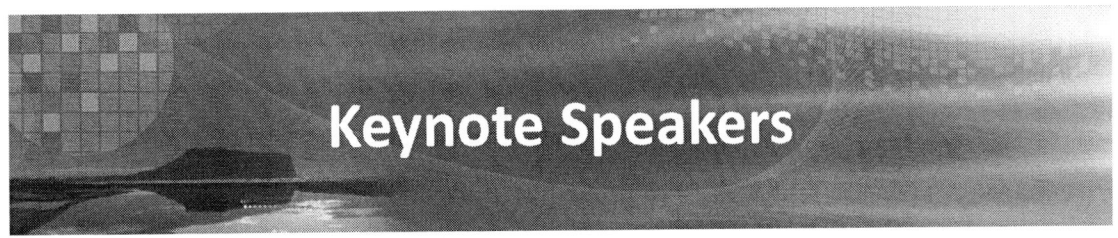

[Keynote #1-1]

Oct. 24, 9:15 ~10:00 Ballroom 1

Deirdre Hanford
Executive Vice President, Customer Engagement, Synopsys, USA

Innovation in the Age of Smart Everything

Biography

Deirdre Hanford is Executive Vice President, Customer Engagement. In her current role, her organization's mission is to ensure the successful adoption of the company's technology into customers' demanding environments. In addition, Ms. Hanford oversees Synopsys' engagements with its key customers and ecosystem partners.

Ms. Hanford has held a variety of positions, including leadership roles in applications engineering, sales and marketing. In 2001, Ms. Hanford was a recipient of the YWCA Tribute to Women and Industry (TWIN) Award and the Marie R. Pistilli Women in EDA Achievement Award. Ms. Hanford served as the Chairman of American Electronics Association in 2008. She currently chairs Brown University's Engineering Advisory Committee and serves on the Engineering Advisory Board for UC Berkeley's College of Engineering.
Ms. Hanford earned a BSEE from Brown University and an MSEE from UC Berkeley.

Abstract

Smart Everything is here, driving the next wave of innovation forward. Virtual reality and computer vision are transforming work and entertainment. Autonomous driving and advanced driver assist advancements are revolutionizing and disrupting the automobile industry. Machine learning is transforming how professionals leverage compute power.

Hardware and software designers face challenges as they create the SoCs to enable these innovations. Additionally, the adoption and deployment of smart everything solutions is complicated by security concerns. Ms. Hanford's presentation will explore how the demand for secure and innovative solutions is driving engineers to create ever faster, smarter, and safer SoCs.

2016 International SoC Design Conference

[Keynote #1-2]

Oct. 24, 10:00~10:45   Ballroom 1

Kou-Hung Lawrence Loh, Ph.D.
Corporate Senior Vice President, MediaTek Inc., President, MediaTek USA Inc., USA

IC Design Challenges for Big Digital Semiconductor Companies

Biography

Dr. Kou-Hung Lawrence Loh is a Corporate Senior Vice President of MediaTek Inc. He oversees the company's Central Engineering Group, responsible for engineering the company's SOCs and chipsets design, development and implementation activities for all MediaTek's product lines including mobile communication, application processors, wireless connectivity, IOT, automotive, home entertainment, optical storage and broadband/networking business. He is also serving as President of MediaTek USA responsible for the company's global operations in Europe and America. Dr. Loh started his first circuit design position at IMP and later he joined Cirrus Logic, where his last position was Director of Analog IC Engineering. In 1998, Dr. Loh founded Silicon Bridge Inc., where he successfully led a number of analog/mixed-signal IC development projects with major semiconductor companies including MediaTek and Altera Corporation. Before joining MediaTek in 2004, Dr. Loh had contributed to IC design industry in areas of read/write channels for magnetic and optical storage, high-performance analog filters, solid-state fingerprint sensors, high-speed SERDES and wireline transceivers for various business applications.

Dr. Loh received his Ph.D. degree in Electrical Engineering from Texas A&M University, College Station, Texas. He has authored/co-authored dozens of technical papers/patents in areas of analog and mixed-signal integrated circuits/systems design and has contributed many panel talks and invited keynote speeches at numerous international conferences and professional communities. Dr. Loh had served on ISSCC International Technical Program Committee for 5 consecutive years since 2005. He is currently serving on Steering Committee of A-SSCC and also on Board of Directors for Global Semiconductor Alliance (GSA).

Abstract

Recent consolidations in semiconductor industry have led to fewer mega-sized companies.   IC product companies have been generally categorized into two distinctive groups: 'big digital' companies to continue to invest in most advanced IC fabrication technologies to chase after Moore's Law, and 'non big digital' companies to squeeze out more juices from more matured processes. MediaTek, as one of the most representative big digital companies, would need to provide multiple ICs to enable its increasingly complicated and demanding 'integrated platforms' including digital-intensive SOCs and companion chips to support various less-to-non-digital functions such as power management, analog/mixed-signal and RF functions. IC designers of big digital companies therefore face unprecedented challenges to deal with increasing complexities and non-idealities caused by increasingly sophisticated processes, more demanding system performances and cost pressures. In this presentation we will share issues and challenges which IC design engineers of big digital companies have been facing and dealing with their solid technical skills, experiences and innovations/creativities.

2016 International SoC Design Conference

[Keynote #1-3]

Oct. 24, 11:00~11:45 Ballroom 1

Seo-Kyu Lee, Ph.D.
CEO, Founder, PIXELPLUS, KOREA

Sensorization in automotive IoT

Biography

Dr. Seo-Kyu Lee is the founder and CEO of PIXELPUS. PIXELPLUS is a fabless semiconductor company that researches, develops and manufactures image sensors and imaging solutions. PIXELPLUS has been focusing on security and monitoring camera solutions and automotive camera solutions, and offers total solutions for image sensor-based camera applications by penetrating medical and home application markets in the future.

Chairman Lee since April 2000. Prior to Pixelplus, he served as CMOS image sensor project leader as well as CCD research and development team leader at LG Semiconductor Co., Ltd. In addition, he also served as a design team leader in CCD development division and held various engineer positions at LG Semiconductor Co., Ltd. Dr. Lee received a Ph.D. in electronic and electrical engineering from Pohang University of Science & Technology in 1997, a master's degree in electrical engineering from Yonsei University in 1993 and a bachelor's degree in physics and electronic engineering from Sogang University in 1985.

Abstract

Understanding IoT, what does it mean for automotive business in IoT and what is the trend for automotive tech, apply sensors in autonomous vehicle, ADAS system and challenges. Automotive market player partnership status and trend of sensorization.

[Keynote #1-4]

Oct. 24, 11:45~12:30 Ballroom 1

Tom Beckley
Senior Vice President & General Manager of the Custom IC & PCB Group, Cadence, USA

Smart Products: Complexity, Security & Collaboration

Biography

Tom Beckley is Senior Vice President and General Manager of the Custom IC & PCB Group. His product responsibilities include the Virtuoso® design environment, physical design and routing, and simulation product lines for full-custom digital and analog design; infrastructure such as the OpenAccess database and the process design kits (PDKs) that are essential for physical IC design; the Allegro® and OrCAD® design and routing solutions; and Sigrity™ high-speed analysis solutions for PCBs and IC packaging.

In addition, Beckley is executive sponsor of the Cadence Quality Initiative, a sustained corporate focus on developing and deploying processes enabling design, implementation, and delivery of high-quality, full-featured products.

Beckley joined Cadence in 2004 via the acquisition of Neolinear, where he served as President and CEO. Neolinear developed innovative auto-interactive and automated analog/RF tools and solutions for mixed-signal design. Prior to Neolinear, Beckley was head of the Systems Division at Avant! Corporation. He came to Avant! through the acquisition of Xynetix Design Systems, the market leader in advanced IC packaging and systems-level virtual prototyping, where he was President and CEO. Prior to Xynetix, Beckley held engineering and management positions at Harris Corporation and General Motors.

Beckley received his B.S. in mathematics and physics from Kalamazoo College and an MBA from Vanderbilt University.

Abstract

We are at an early stage in the world's most significant electronics/software-driven transformation. The rapid unfolding of the autonomous vehicle is but one example. From Tesla to Google to Uber to Nvidia to Samsung, practically every semiconductor, software and systems company is looking to be part of the driverless car opportunity. Mobile devices are similarly transforming. Today, they comprise high performance computers, radios, sensors, and more enabling a vast software stack and thousands of software apps. Smart phones are quickly becoming personal computing hubs that access and enable IoT devices for health, gaming, home management and more. While the driverless car includes robotics and AI, advanced node processor scaling is providing unprecedented cost effective, high performance compute power. When coupled with ultra-fast, high capacity memory from leading suppliers including SK Hynix and Samsung, along with Cloud access, new opportunities unfold for semis and systems across industrial and home robotics, data mining, and more.

This presentation will focus on new methods and solutions for designing, verifying and assessing functional safety/security for our unfolding world of smart, interconnected products. These Smart Product systems include multiple complex hardware technologies – SoCs, radios, MEMs, high speed boards – and an ever growing software stack.

2016 International SoC Design Conference

[Keynote #2-1]

Oct. 25, 09:15~10:00 Ballroom 1

Walden C. Rhines, Ph.D.
Chief Executive Officer and Chairman of the Board of Directors, Mentor Graphics, USA

SMART SoC's for INTELLIGENT THINGS

Biography

WALDEN C. RHINES is Chairman and Chief Executive Officer of Mentor Graphics, a leader in worldwide electronic design automation with revenue of about $1.2 billion in 2015. During his tenure at Mentor Graphics, revenue has nearly quadrupled and Mentor has grown the industry's number one market share solutions in four of the ten largest product segments of the EDA industry.

Prior to joining Mentor Graphics, Rhines was Executive Vice President of Texas Instruments' Semiconductor Group, sharing responsibility for TI's Components Sector, and having direct responsibility for the entire semiconductor business with more than $5 billion of revenue and over 30,000 people.

During his 21 years at TI, Rhines managed TI's thrust into digital signal processing and supervised that business from inception with the TMS 320 family of DSP's through growth to become the cornerstone of TI's semiconductor technology. He also supervised the development of the first TI speech synthesis devices (used in "Speak & Spell") and is co-inventor of the GaN blue-violet light emitting diode (now important for DVD players and low energy lighting). He was President of TI's Data Systems Group and held numerous other semiconductor executive management positions.

Rhines has served five terms as Chairman of the Electronic Design Automation Consortium and is currently serving as a director. He is also a board member of the Semiconductor Research Corporation and First Growth Children & Family Charities. He has previously served as chairman of the Semiconductor Technical Advisory Committee of the Department of Commerce and as a board member of the Computer and Business Equipment Manufacturers' Association (CBEMA), SEMI-Sematech/SISA, Electronic Design Automation Consortium (EDAC), University of Michigan National Advisory Council, Lewis and Clark College and SEMATECH.

Dr. Rhines holds a Bachelor of Science degree in metallurgical engineering from the University of Michigan, a Master of Science and Ph.D. in materials science and engineering from Stanford University, a master of business administration from Southern Methodist University and an Honorary Doctor of Technology degree from Nottingham Trent University.

Abstract

Since 1995, the number of bits of FLASH memory produced each year has been growing at an 85% compound average growth rate while the cost per bit decreased more than 45% per year. While the balance between logic and memory transistors was approximately equal in the early 1990's, memory bits constitute 99.7% of all transistors produced today. This change in the memory-to-logic transistor ratio is a major step toward filling the need for improved pattern recognition, a requirement to intelligently process, store, retrieve and make use of the rapidly growing petabytes of photographic, video and audio data. Further advances in memory and processor technology will result in a discontinuity in computer architectures. Dr. Rhines will analyze this evolution and demonstrate how brain-like computer architecture features will make the IOT-generated base of sensor data much more valuable.

2016 International SoC Design Conference

[Keynote #2-2]

Oct. 25, 10:00~10:45 Ballroom 1

Amara Amara, Ph.D.
*Professor, Director of Research, ISEP(Institu Supérieur d'Electronique de Paris), France IEEE
CASS ExCom, Vice-President for Conferences, President IEEE France Section*

Extremely Low-Leakage memory and logic circuits based on Tunneling-FET device

Biography

Prof. Amara AMARA obtained Ph.D. in computer science in 1989 and a Master in 1984 in microelectronics and computer science from Pierre and Marie Curie University (Paris VI). In 1988 he joined IBM research and development laboratory at Corbeil-Essonnes as invited researcher where he was involved in SRAM memory design with advanced CMOS technologies. From 1989 to 1992, he was associate professor developing microelectronics academic programs for CEMIP (Microelectronic Center of Paris Iles-de-France) and took part actively in many European Research Projects. In 1992, he joined ISEP (Paris Institute for Electronics) in charge of the microelectronics laboratory where he headed a joint team (Paris VI and ISEP) involved in High Speed GaAs VLSI circuit design and developed curricula in Microelectronics. He was involved in education management and research management for almost 34 years. Currently he is Deputy Managing Director of ISEP in charge of Research and International Cooperation. His research interests are mainly focusing on Low Power and Low Voltage circuit design techniques and on Circuit and Device Co-design for advanced technologies (SOI, DGates FD SOI, Ultra Thin Body SOI, T-FET...). In 1999, he spent his sabbatical at Stanford University working as a visiting researcher in Professor De Micheli's group.

He launched a well-established laboratory called LISITE, which is now composed of 21 researchers, 18 PhD students split over 3 teams in addition to visiting scholars and visiting researchers. The laboratory is mostly funded by industry through collaborative research projects or bilateral research agreements.

Amara has a strong activity within IEEE since he joined this prestigious organization in 1992. He has been Vice President of the French IEEE Section since January 2004. He has been elected President of the IEEE France Section for a 3 years term starting from January 2014. From 2000 to 2004 he was Chairman of the IEEE-CAS France Chapter (Recipient of the 2004 Best Chapter of the Year Award and a Certificate of Appreciation from IEEE Regional Activities), he was member of the IEEE CASS Board of Governors (2008-2014) and was elected in November 2013 member of CASS Executive Committee as Vice President for Conferences (term 2014-2015) and reelected in November 2015 in the same position for the term 2016-2017. As VP for conferences he is the Chair of ISCAS (IEEE CASS Flagship Conference) Steering Committee. As President of the IEEE France Section, he represents France in the IEEE Region 8 board.

Amara is member of numerous Conference Technical Program Committees and Conference organizing Committees, member of the Editorial Board of Microelectronics Journal (ELSEVIER). He chaired IEEE ICICDT 2008 in Grenoble and IEEE ISCAS 2010 in Paris for which he has been awarded the Bronze Medals respectively of the cities of Grenoble and Paris.

Amara published a book on Molecular Electronics (more than 3000 chapters downloads from Elsevier web site) and co-edited two books: "Double-Gate FD SOI devices and circuits" and "Emerging Technologies and Circuits". He is author of 6 patents and co-author of numerous papers published in IEEE conferences and journals. He supervised 20 PhD students on subjects related to his research interests.

Abstract

Nowadays CMOS transistors have become too leaky and have weak performance at low voltage. CMOS leakage is increasing with technology scaling and is of critical concerns for circuit designers. In advance technology nodes, it has been the limiting factor in improving the performance, especially at low voltages. There are various techniques that minimize leakage currents but never eliminate them systematically. In this presentation, we will introduce a new Tunneling FET (T-FET) structure developed in our laboratory and show how to use it to obtain innovative and efficient systems in terms of power consumption. New memory architectures and a new paradigm for designing logic circuits will be presented.

[Keynote #2-3]

Oct. 25, 11:00~11:45 Ballroom 1

Yong Lian, Ph.D.
Professor, York University, Canada
Fellow of IEEE and Fellow of Academy of Engineering Singapore

Ultra Low Power SoC for Healthcare: Challenges and Future

Biography

Dr. Yong Lian received the B.Sc degree from College of Economics & Management of Shanghai Jiao Tong University in 1984 and the Ph.D degree from the Department of Electrical Engineering of National University of Singapore (NUS) in 1994. He worked in industry for more than 9 years before joining NUS in 1996. He was appointed as the first Provost's Chair Professor in the Department of Electrical and Computer Engineering in 2011. Currently, he is a professor in the Department of Electrical Engineering and Computer Science of York University. His research interests include low power techniques, continuous-time signal processing, biomedical circuits and systems, and computationally efficient signal processing algorithms. He has received more than US$25 million in research funds from various sources. He is the Founder of Clearbridge VitalSigns Pte Ltd, a start-up for commercializing wireless biomedical sensor technologies.

Dr. Lian received more than 20 awards for his research including the 1996 IEEE Circuits and Systems Society's Guillemin-Cauer Award, the 2008 Multimedia Communications Best Paper Award from the IEEE Communications Society, 2011 IES Prestigious Engineering Achievement Award, 2012 Faculty Research Award, 2013 Outstanding Contribution Award from Hua Yuan Association and Tan Kah Kee International Society, 2014 Chen-Ning Yang Award in Science and Technology for New Immigrant, and the latest 2015 Design Contest Award in 20th International Symposium on Low Power Electronics and Design. He is also the recipient of the National University of Singapore Annual Teaching Excellence Awards in 2009 and 2010, respectively.

Dr. Lian is the President-Elect of the IEEE Circuits and Systems (CAS) Society, Steering Committee member of the IEEE Transactions on Biomedical Circuits and Systems. He was the Editor-in-Chief of the IEEE Transactions on Circuits and Systems II for two terms from 2010 to 2013. He served many positions in the IEEE CAS Society including Vice President for Publications, Vice President for Asia Pacific Region, Chair of the Biomedical Circuits and Systems Technical Committee, Chair of DSP Technical Committee, Distinguished Lecturer, etc. He is the founder of several conferences including BioCAS, ICGCS, and PrimeAsia. Dr. Lian is a Fellow of IEEE and Fellow of Academy of Engineering Singapore.

Abstract

According to the World Health Organization's 2008 report, the top three leading causes of death worldwide are coronary heart disease, cerebrovascular diseases, and lower respiratory infections. Governments and biomedical companies are pouring millions of dollars into research and development to find solutions for these diseases, and technological platforms to support disease management. Wireless biosensors are one such platform, with their ability to measure vital signs that indicate the presence, or onset, of pathology. According to the report from Wearable World, the wearable market for mobile health applications and associated devices will grow at a compound annual growth rate of 61% to reaching $26 billion in revenue by 2017. The market and economic forces drive the development of ultra-low power wearable wireless biomedical sensors. SoC solutions that integrate sensors, analog, digital logic, memory, RF, energy harvesters and power management will become the norm in the IoT era. This talk will discuss the challenges in the design of self-powered wearable wireless biomedical sensor chip including regulatory requirements, skin-electrode interface, design considerations of analog frontend, ADC, signal processing, and wireless transceiver. The focus is on the low power system architecture that utilizes the continuous-in-time and discrete-in-time (CTDA) signal flow to maximize energy efficiency.

[Keynote #2-4]

Oct. 25, 11:45~12:30 Ballroom 1

Gabriel A. Rincón-Mora, Ph.D.
Professor, Georgia Institute of Technology, USA

Powering Intelligent IoT Microsensors

Biography

Prof. Gabriel A. Rincón-Mora worked for Texas Instruments in 1994-2003, was an Adjunct Professor at Georgia Tech in 1999-2001, and has been a Professor at Georgia Tech since 2001 and a Visiting Professor at National Cheng Kung University in Taiwan since 2011. He is a Fellow of the IEEE and a Fellow of the IET, and his scholarly products include 9 books, 4 book chapters, 38 patents issued, over 170 publications, over 26 commercial power-chip designs, and over 110 invited talks. Awards include the National Hispanic in Technology Award from the Society of Professional Hispanic Engineers, the Charles E. Perry Visionary Award from Florida International University, a Commendation Certificate from the Lieutenant Governor of California, the IEEE Service Award from IEEE CASS, the Orgullo Hispano and the Hispanic Heritage awards from Robins Air Force Base, and two "Thank a Teacher" certificates from Georgia Tech. Georgia Tech inducted him into the Council of Outstanding Young Engineering Alumni in 2000 and Hispanic Business magazine named him one of "The 100 Most Influential Hispanics" in 2000.

Abstract

Wireless microsensors can not only monitor and manage power consumption in small- and large-scale applications for space, military, medical, agricultural, and consumer markets but also add energy-saving and life-saving intelligence to large infrastructures and tiny devices in remote and difficult-to-reach places. Ultra-small systems, however, cannot store sufficient energy to sustain monitoring, interface, processing, and telemetry functions for extended periods. And replacing or recharging the batteries of hundreds of networked nodes is prohibitive, and often impossible. This is why alternate sources are the subject of ardent research today. Except, power densities are low, and in some cases, intermittent, so supplying functional blocks is challenging. Plus, tiny lithium-ion batteries and super capacitors, while power dense, cannot sustain life for long. This keynote illustrates emerging charger-supply systems that draw power from tiny energy-harnessing transducers, inductively coupled coils, fuel cells, and batteries to sustain microsystems for extended periods.

A 200-kHz/6.78-MHz Wireless Power Transmitter Featuring Concurrent Dual-Band Operation

Dukju Ahn
Department of Electrical Engineering,
Incheon National University
Incheon, Korea
dahn@inu.ac.kr

Jiwoong Park and Patrick P. Mercier
Department of Electrical and Computer Engineering
University of California San Diego (UCSD)
La Jolla, CA, USA
pmercier@ucsd.edu

Abstract— **A wireless power transfer (WPT) transmitter is presented that can simultaneously power devices operating at 200 kHz and 6.78 MHz, enabling a true multi-standard WPT transmitter. To achieve this, the proposed design utilizes two coils, each optimized for efficient power transfer when operating alone. By placing the coils co-axially on a single charging stand, concurrent power transfer is possible. However, nominally doing this invokes significant eddy current losses, degrading WPT efficiency. To combat this, an eddy-current filter is included in the 200 kHz path. The proposed design is built into a 12.5 x 8.9 cm² charging pad area, and at 25 mm separation, the system is able to concurrently power two smartphone-sized receiving devices at 25 mm separation. The system achieves a total power delivery of 9 W and 7.4 W with efficiencies of 78% and 70.6% at 6.78 MHz and 200 kHz, respectively.**

Keywords; Wireless power transfer, dual-band, multi-standard, resonant power transfer, wireless charging

I. INTRODUCTION

There are currently three standards utilized for wireless power transfer in the industry today: the Alliance for Wireless Power (A4WP), the Wireless Power Consortium (WPC), and the Power Matters Alliance (PMA). Unfortunately, the operating frequency of each standard differs, making these standards incompatible with one another. For example, the A4WP standard utilizes a 6.78 MHz ± 15 kHz carrier frequency, while the WPC and PMA standards use frequency ranges of 110~205 kHz and 110~300 kHz, respectively. With different frequency standards for each receiver device, a single power transmitter cannot conventionally charge incompatible receivers, never mind doing so simultaneously. To enable a universal charger, there is a need to develop a single transmitter which can accommodate multiple receivers operating with different standards, and therefore at different frequencies, while ideally also doing so concurrently.

In this work, based on [1], we propose to use two separate Tx coils which can be individually designed and optimized for each frequency: 200 kHz and 6.78 MHz. This enables dual-frequency operation with high efficiency. To minimize the increase in coil volume, the 200 kHz coil is embedded within

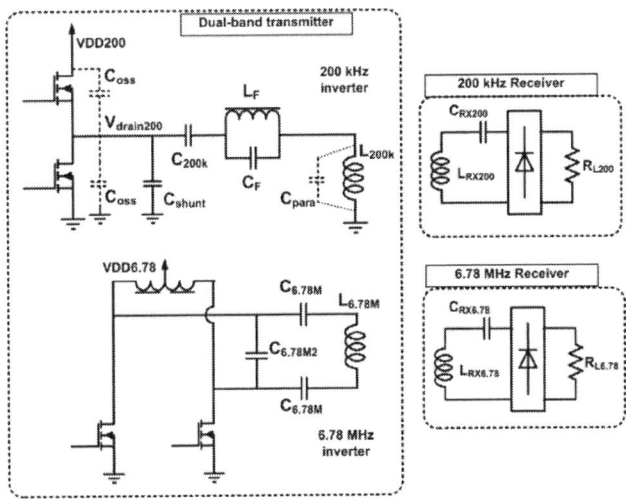

Fig. 1. Proposed dual-band wireless power transmitter which can simultaneously power two receivers of different frequency standard.

the geometry of the 6.78 MHz coil (i.e., co-axially). The receiver design is not altered by the concurrent operation, and thus existing designs can still operate correctly with the proposed transmitter.

II. PROPOSED TRANSMITTER

To enable *concurrent* dual-band frequency operation with high efficiency, the proposed transmitter is implemented with two separate coils, $L_{6.78M}$ and L_{200k}, appropriately sized for maximal efficiency at 6.78 MHz and 200 kHz, respectively. As illustrated in Fig. 1, each coil features shunting and/or blocking filters to reduce coupled losses. Here, an auxiliary resonant tank, L_F-C_F, is added in series with the 200 kHz stage, and is tuned to a resonant frequency of 6.78 MHz. This tank then acts as a filter whose impedance is high at 6.78 MHz, which helps to minimize both the undesired crosstalk from the 6.78 MHz power carrier to 200 kHz power transistors, and the eddy current losses of the 6.78 MHz transmitter through the 200 kHz path. The C_{200k} is then tuned to compensate for the sum of L_{200k} and L_F. Similarly, capacitors $C_{6.78M}$, which are nominally required in the 6.78 MHz path for resonant operation, act as a high-impedance filter to the 200 kHz inverter, minimizing losses during 200 kHz operation.

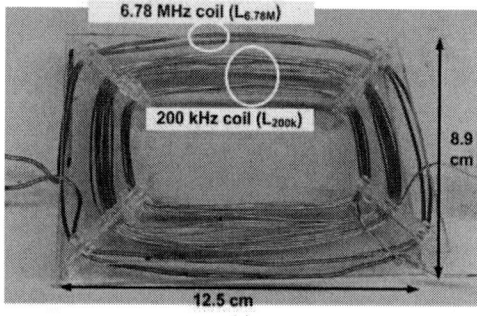

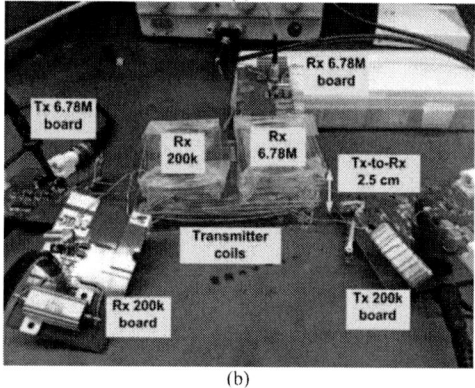

Fig. 2. Fabricated coils and measurement setup.

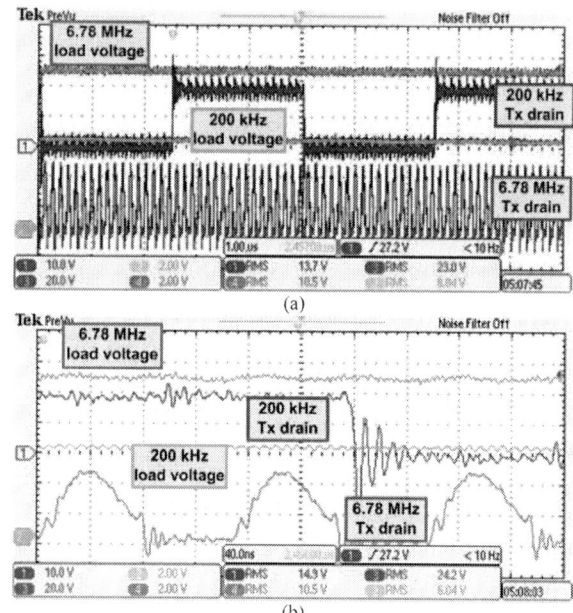

Fig. 3. Measured waveforms showing simultaneous operation. One frequency operation does not affect the other frequency mode. (a) 1 μsec/div (b) 40 nsec/div

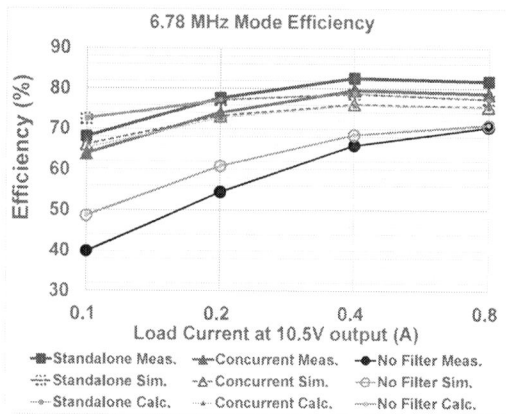

Fig. 4. 6.78 MHz efficiencies. The efficiency degradation from standalone mode due to concurrent operation is less than 4.2%. The efficiency degradations are severe when eddy current is not blocked by L_F-C_F filter.

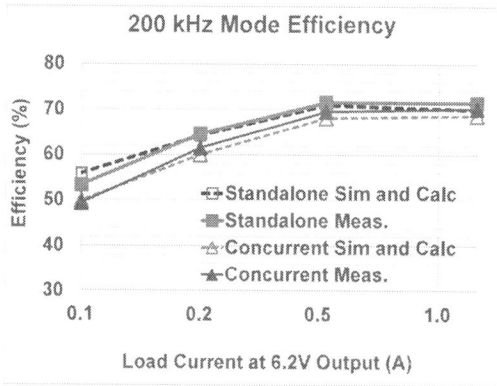

Fig. 5. 200 kHz efficiencies. The efficiency degradations due to concurrent operation are less than 4%.

III. MEASUREMENT RESULT

Fig. 2 shows the fabricated coils and measurement setup. Fig. 3 shows the measured waveforms. Note that one frequency does not affect the other frequency, enabling *simultaneous* power transmission at two different frequencies. Fig. 4 and Fig. 5 show the Tx-to-Rx power transmission efficiencies. It can be seen that the efficiency degradation due to multi-band operation is minimal.

ACKNOWLEDGEMENTS

This work is supported by the Technology Development Program for Commercializing System Semiconductor funded by the Ministry Of Trade, Industry and Energy (MOTIE, Korea). (No 10041126, Title: International Collaborative R&BD Project for System Semiconductor)

REFERENCES

[1] D. Ahn and P. Mercier, "Wireless Power Transfer with Concurrent 200 kHz and 6.78 MHz Operation in a Single Transmitter Device," IEEE Trans. Power Electron., vol. 31., no. 7, Jul. 2016.

Electrical Performance Analysis of Biogas Fuelled Generator with Purifier

Edison E. Mojica
COE- University of Perpetual Help System Dalta
GS- Mapua Institute of Technology
Manila, Philippines
edisonmojica@gmail.com

Arnold C. Paglinawan
School of Electrical, Electronics, and Computer Engineering
Mapua Institute of Technology
Manila, Philippines
acpaglinawan@gmail.com

Abstract— **The paper focuses on the evaluation of locally-made biogas-powered generators to determine its electrical performance efficiency using the purified fuel. Since the demand of energy in the world is greatly increasing, the efficiency and financial concerns of the system is a major aspect. Locally-made biogas generator was chosen for the study due to its availability and abundance in the Philippines. The data on technical specifications of the generator, the biogas consumption of the system (m³/hr), the biogas fuel at the initial and purified percentage concentration, and the electrical output (kW) were collected in operating unit. For the locally-made biogas generator, the average flow rate was found to be 6.0 m³/hr with the biogas initial concentration of 66% to 4.9 m³/hr with the biogas initial concentration of 84% with the purifier. The overall filtration efficiencies were found to increase the biogas concentration by 18%. The efficiency of the operating unit was influenced by the system efficiency and engine efficiency. The filtered biogas in a locally-made generator was found viable with an annual operating cost of PhP 1.75 per kWh as compared to the PhP 2.02 per kWh of the unfiltered fuel.**

Keywords; biogas; biogas generator; renewable energy; purifier

I. INTRODUCTION

The utilisation of renewable energy nowadays as source of energy is inevitable. The use of the energy has proven different benefits including the energy security of a given region due to its diverse supply. This also limits the import of the fossil fuels resulting in economic and social benefits. But, what highlights most is the ecological effects due to the elimination of destructive substances and greenhouse gas emissions that are harmful to the environment [1].

Biogas is a biological organic-rich material which occurs during the digestion and decomposition process of a biodegradable materials. The harmful and flammable gaseous mixture of methane (CH_4) and carbon dioxide (CO_2) are released but only the methane gas is utilised to generate electricity [2].

II. COMPOSITION OF THE BIOGAS FUEL

In an existing locally-made biogas fuelled generator unit, the impurities were identified and measured using the methane gas analyzer. The impurities were determined through a sampling of 1200 ppm biogas fuel. There were 0.5% Hydrogen Sulphide (H_2S), 5% water vapor, 36% Carbon Dioxide (CO_2), and 1% Nitrogen (N). The methane gas (CH_4) was only 57.5% of the total content which makes the electricity generation less efficient.

H_2S is present in biogas resulting to the anaerobic digestion of organic material containing sulphur. This is the most corrosive of several trace gases present in biogas, while gas quality varies greatly depending on the source of the gas supply. Without the removal of H_2S, production efficiency is reduced and repair costs are greatly increased [3] [4].

Biogas from digesters is normally collected from headspace above a liquid surface or very moist substrate, the gas is usually saturated with water vapor. The combination of the H_2S and water vapor reacts to form sulfuric acid (H_2SO_4) which can result in severe corrosion in pipes and other equipment that will come in contact with the biogas. Likewise, even if the H_2S is removed from the biogas, the water vapor can still react with CO_2 to form carbonic acid (H_2CO_3), which is also corrosive. The removal of the water vapor will eliminate the possibilities of clogging or freezing the piping system and nozzles, corrosion, [3].

Carbon dioxide has the highest percentage in the biogas component among all the impurities. This causes corrosion to the whole system, lowers the calorific value, and damages the alkali fuel cells. On the other hand, the effect of nitrogen on the final output is a dilution of the energy content of the biogas. The most common way of dealing with nitrogen in biogas is to limit introduction of it before or during biogas cleaning and to simply accept whatever the resulting nitrogen levels are in the final bio - methane product [3].

III. BIOGAS FUEL PURIFIER SYSTEM

The first stage of the purification was the condenser. This was designed to remove the water vapor in the methane gas. The next stage of filtration was the iron filling filter that aimed to remove the hydrogen sulphide. This was placed after the water vapor filtration to prevent the rusting of iron fillings. The last stage of filtration was an active carbon that aimed to filter the carbon dioxide. It was placed at the last stage of filtration since this should not be used in wet condition and in a high hydrogen sulfide concentration. Figure 1 shows the biogas fuel purifier system.

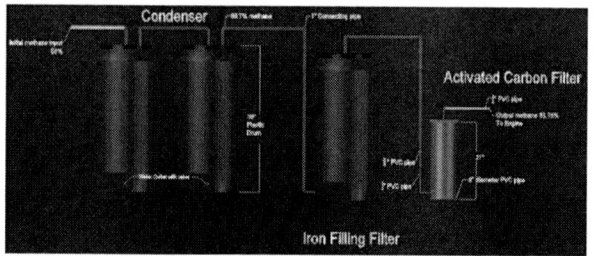

Fig. 1. Biogas fuel purifier system

IV. EVALUATION OF THE EFFICIENCY OF THE SYSTEM

The evaluation was made to compare the unfiltered biogas to the filtered biogas. Table 1 shows that there was an increased in concentration of methane from the unfiltered methane with 66% concentration to filtered methane with 84% concentration. It was found to filter 28% of the impurities upon the process of filtrations. It also showed that upon filtration, the process average flow rate decreased with an increased in the biogas flow rate content. This means that with the lesser flow rate, more methane are contained to the flow. The filtered biogas utilization increased the current output resulting to the increase in the power input to the engine and the power output of the generator.

TABLE I
EVALUATION EFFICIENCY OF THE FILTERED BIOGAS

BIOGAS	AVE. FLOW RATE (m³/hr)	AVERAGE METHANE CONTENT (%)	BIOGAS FLOW RATE CONTENT (m³/hr)	ENGINE POWER INPUT (kW)	GENERATOR POWER OUTPUT (kW)
Unfiltered	6.0	58	3.94	17.406	16.8
Condenser	5.9	66.7	3.94	19.94	16.9
Filtered	4.9	83.75	4.1	20.7	17.12

Table 2 shows the engine and system efficiency. Engine efficiency is the relationship between the total energy contained in the fuel and the amount of energy used to perform useful work [5]. The system efficiency is the measurement of the power output of the engine over the power output of the generator. The lowered impurities resulted to the increase in the calorific value of the system and the higher concentration of methane was easier to combust.

TABLE II
ENGINE AND SYSTEM EFFICIENCY

BIOGAS	Engine Efficiency (%)	System Efficiency (%)
Unfiltered	83.10	84
Condenser	84.76	84.5
Filtered	85.27	86.15

V. COST BENEFIT ANALYSIS

Cost per power output (in Philippine Peso per kilowatt-hour) of each generator was assessed to determine which generator cost the least in generating power. Table 3 shows the cost and return analysis of each biogas generator. Life span of the system was found to be 7 years for the locally-made biogas generators using filtered biogas and about 5 years for the generators using unfiltered biogas. The cost per power output of unfiltered biogas has dropped from PhP2.02/kWh to the filtered biogas of PhP1.75/kWh.

TABLE IIII
COST AND RETURN ANALYSIS OF EACH BIOGAS GENERATOR SYSTEM

PARAMETERS	Generator	
	Unfiltered Biogas	Filtered Biogas
Initial Cost of Biogas Generator System, (in PhP)	800,000.00	800,000.00
Life Span, (Years)	5	7
Salvage Value (10 % of I.C), (in Php)	80,000.00	80,000.00
Depreciation, (in Php)	144 000	102 857.14
Interest of Investment, (in Php)	35,992.00	35,992.00
Taxes and Insurances ,(in Php)	8,000.00	8,000.00
Total Fixed Cost, (in Php)	187,992.00	146,849.14
Labor Cost, (in Php)	56,940.00	56,940.00
Total Operating Cost, (in Php)	203,789.14	167,076.86
Annual Electricity Production(kWh)	122,640	125,779
Cost per Power Output, (Php / kWh)	2.02	1.75

ACKNOWLEDGMENT

The author would like to acknowledge the dean and the faculty members of the College of Engineering of the UPHSD Las Pinas Campus and Molino Campus, and the faculty of the Graduate School of Mapua Institute of Technology.

REFERENCES

[1]   M. M. A. Manzurul Alam, Md.; Sumaiya, Sharaf, "Cost Analysis and Simulation of Stand-alon Hybrid Renewable Energy System (HRES)," *Military Institute of Technology* 2014.

[2]   H. Y. Zhiyi Wang, Jinqing Peng, Lin LU, "Analysis of Energy Utilization on Digestion Biogas Tri-Generation in Sewage Treatment Works," *Springer-Verlag* 2014.

[3]   M. r. Maikel Fernandez, Rosa Maria Perez, Jose Manuel Gomez, Domingo Cantero, , "Hydrogen Sulphide Removal From Biogas by an Anoxic Biotrickling Filter Packed with Pall Rings," *Chemical Engineering Journal,* 2014.

[4]   M. r. Maikel Fernandez, Rosa Maria Perez, Jose Manuel Gomez, Domingo Cantero,, "Biogas Biodesulfurization in an Anoxic Biotrickling Filter Packed with Open-Pore Polyurethane Foam," *Journal of Hazardous Materials,* 2014.

[5]   R. H. Y. Manabe, H. Kita, K. Takitani, "Cooperative Control of Energy Storage Systems and Biogas Generator for Multiple Renewable Energy Power Plants," *Power Systems Computation Conference,* 2014.

Power and Area Calibration of Switch Arbiter for High Speed Switch Control and Scheduling in Network-on-Chip

Sangeeta Singh, JVR Ravindra
Center for Adv. Computing Research Lab (C-ACRL)
Dept. of Elec. and Comm. Engg
Vardhaman College of Engineering, Shamshabad,
Kacharam, Hyderabad, Telangana, India
email:sangeetasingh@ieee.org, jayanthi@ieee.org

B Rajendra Naik
Dept. of Elec. and Comm. Engg
University College of Engineering,
Osmania University
Hyderabad, Telangana, India
email:rajendranaikb@gmail.com

Abstract— **Network-on-chip (NoC) is being considered as a promising model to overcome the communication bottleneck of future multicore systems. It plays an important role in determining the area and power of the entire chip. As a basic component in on-chip router, arbiter has a large impact on the performance of router. A centralized arbiter called switch arbiter resolves conflicts between the input ports to get the access for the same output port. The design of such power and area efficient switch arbiter for Network on Chip is a major challenge in DSM technology. This paper proposes a novel switch arbiter that reduces power dissipation and consumes less area in comparison to conventional switch arbiter. The novel switch arbiter has been designed using twisted ring counter instead of ring counter to increase its performance. Simulation results have been carried out using Cadence Design Framework with 180nm technology.**
Index Terms —**Switch Arbiter, Twisted Ring Counter, NoC**

I. INTRODUCTION

Due to technology scaling over several generations, it is possible to integrate diverse functional blocks on a chip. These blocks can be connected by a communication network to form a heterogeneous system [1]. As the number of components continues to increase, the communication architecture has to be modified to achieve low area and high performance [2]. Traditionally, the system designs were based on critical paths and clock trees which presented an increased amount of delay and power consumption. Network on chip (NoC) is an interconnect fabric used to connect sub system blocks on a chip. The advantages of NoC on performance, scalability, and reusability make it more suitable design approach to solve these interconnection problems when compared to traditional bus based architecture [3]. NoC architecture basically contains three main components: 1) switches (routers) 2) inter-switch links and 3) repeaters [4]. Among them, the design and implementation of a router plays crucial role in high performance NoCs. The router consists of an arbiter which provides connection between input and output ports. Switch arbiter being a centralized arbiter is one of the important components of the router in network on chip that impacts the performance of the router [5]. It is used in router to resolve the conflicts between the multiple input ports, when both the input

ports access for the same output port. Most of the designs used for arbiters were based only on achieving the performance and scalability issues [6].

II. LITERATURE SURVEY

There are many arbitration schemes developed for efficient implementation of an on chip router. The different types of arbiters available are round robin arbiter, fixed priority arbiter, lottery arbiter, and token ring arbiter. Among all, round robin token scheme is the standard scheme which is used in Network on Chip for bus arbitration [7]. This arbitration scheme offers low complexity and provides fairness in scheduling. An NXN order switch arbiter consists of a bus arbiter, 'N' no. of AND gates and an OR gate [8]. Traditional 4-bit switch arbiter is shown in Fig.1.

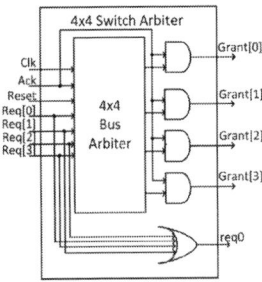

Fig.1. 4x4 Switch Arbiter

In general an NXN switch arbiter comprises of 'N' Request signals and 'N' Grant signals. Request signals of the current level are combined to generate only one request, so as to trigger the next higher level. The internal block diagram of 4x4 bus arbiter is presented in [8]. It consists of a D flip-flop, priority encoder (logic) blocks, an N-bit ring counter and 'N' N-input OR gates. The output of the ring counter acts as the enable signals to the priority logic blocks. Thus, only one enabled priority logic block can activate a grant signal. A conventional bus arbiter uses a ring counter which requires 'n' flip flops for 'n' states that increases with increase in number of requests. This increases power dissipation and cell area.

III. PROPOSED SWITCH ARBITER

The proposed switch arbiter uses a twisted ring counter instead of ring counter to generate '2n' states using 'n' flipflops for a mod-N ring counter and thus enhances its performance. Fig.2 shows the block diagram of the proposed 2-bit twisted ring counter. The priority of inputs are arranged in descending order from in[0] to in[3] in the priority logic blocks.

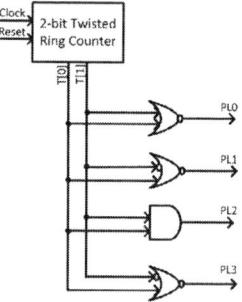

Fig.2. 2-Bit Twisted Ring Counter

The block diagram of modified 4X4 bus arbiter is shown in Fig.3. The ring counter in traditional 4X4 bit bus arbiter is replaced with the proposed 2-bit twisted ring counter. The modified bus arbiter is used in switch arbiter to improve its characteristics.

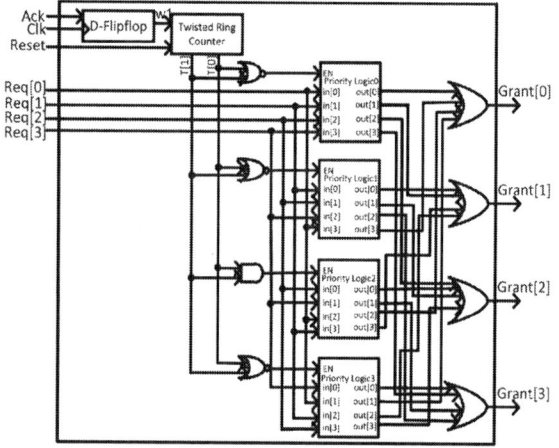

Fig.3. Proposed 4X4 Bus Arbiter

IV. EXPERIMENTAL RESULTS

This section presents the comparison between power and cell area results of the 4-bit bus arbiter and switch arbiter with existing circuits [5]. These circuits are simulated on Cadence NC Launch platform and synthesized using Cadence RTL compiler with UMC 0.18 micron technology.

TABLE I
POWER MEASUREMENT OF 4X4 SWITCH ARBITER

Parameter	Conventional 4X4 Switch Arbiter	Proposed 4X4 Switch Arbiter
Leakage Power (mw)	0.29	0.28
Dynamic Power(mw)	15.673	10.618
Total Power(mw)	15.703	10.646

The outputs in Table I depicts that for 4X4 switch arbiter, the leakage power is decreased by 2.5%, dynamic power by 32.25% and total power is reduced by 32.19 %.

TABLE II
NO. OF CELLS AND CELL AREA MEASUREMENT OF 4X4 SWITCH ARBITER

Parameter	Conventional 4X4 Switch Arbiter	Proposed 4X4 Switch Arbiter
Cells	36	38
Area(Sq.micron)	768	709

The reduction in the area of 4X4 switch arbiter cell is 7.68% when compared to conventional 4X4 switch arbiter as shown in Table II. The comparison between power outputs of traditional and proposed switched arbiter is shown in Fig.4.

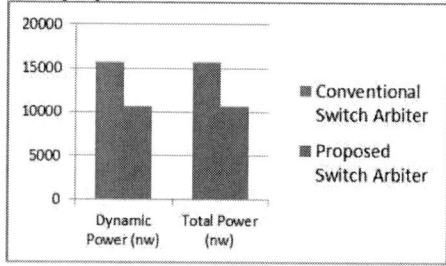

Fig.4. Power comparison between Conventional and Proposed Switch Arbiter

V. CONCLUSION

In this paper, a novel switch arbiter is designed and implemented by using a twisted ring counter instead of a traditional ring counter. The proposed switch arbiter reduced leakage power by 17% and dynamic power by 32.25%. Also the cell area in the proposed switch arbiter is reduced by 7.68% when compared to conventional design. The comparison results show that the proposed switch arbiter is more efficient than conventional switch arbiter in terms of power and area.

REFERENCES

[1] Yuho Jin, et al., "Communication-Aware Globally-Coordinated On-Chip Networks", IEEE Trans. Parallel and Distrib. Syst., Vol. 23, no. 2, pp. 242-253, February 2012.

[2] Radu Marculescu, et al., "Outstanding Research Problems in NoC Design: System, Microarchitecture, and Circuit Perspectives", IEEE Trans. VLSI Syst., Vol. 28, pp.12-21, 2009.

[3] Daniel Schinkel, et al., "Low-Power, High-S peed Transceivers for Network-on-Chip Communication", IEEE Trans. VLSI Syst. Vol.17, pp.12-21, 2009.

[4] Gursharan Reehal, et al., "A Systematic Design Methodology for Low-Power NoCs", IEEE Trans. VLSI Syst., pp.1-11, 2015.

[5] Yanhua Liu, et al., "A dynamic Adaptive Arbiter for Network-on-Chip", Journal of Microelectronics in Electronic Components and Materials. Vol.43, No.2, pp.111-118, 2013.

[6] Si Qing Zheng, et al., "Algorithm-Hardware Co-design of Fast Parallel Round-Robin Arbiters", IEEE Trans. Parallel Distrib. Syst., Vol.18, No.1, pp.84-95, 2007.

[7] H. Jonathan, et al., "A Fast Arbitration Scheme for Terabit Packet Switches", Globecom, 1999.

[8] E.S. Shin, et al., "Round-robin Arbiter Design and Generation", in Proc. of ACM/ISSS, pp.243-248, 2002.

An Efficient On-Chip Network with Packet Compression Capability

M. Vafaiee[1], M. Jalili[2], R. Sabbaghi-Nadooshan[1], H. Sarbazi-Azad[2]

[1] Islamic Azad University, Central Tehran Branch, Tehran, Iran

[2] Institute for Research in Fundamental Science (IPM), Tehran, Iran

m.vafaiee@yahoo.com, majalili@ipm.ir, r_sabbaghi@iauctb.ac.ir, azad@ipm.ir

Abstract— **In order to reduce inter-core data communication load, increase effective bandwidth, reduce storage space and power consumption, various solutions have been suggested for on-chip networks. The aim of this study is to employ data compression for the packets delivered between cores over the inter-core on-chip network. The packets transferring the cache lines between cores are compressed before transmission. We use a full-system simulator to evaluate and compare different systems that employ different compression methods.**

Keywords: *NoC, Packet Compression, Cache, Simulation.*

I. INTRODUCTION

According to Moore's Law increasing the number of cores per processor continues. On the other hand, the memory speed growth is not comparable to that of the core [1]. Compression is a proper way for reducing this gap in multi-core systems. Adopting compression of data packets in the network-on-chip (NoC) has several advantages and some disadvantages. As a matter of fact, compression increases the effective capacity and bandwidth while reduces the total power consumption. Disadvantages include fabrication of the on-chip compression and decompression logics. By using data compression, data size is reduced so that more space will be available for packet storage and transmission. One of the main benefits is the bandwidth increase across the main memory and cache levels. Besides, compression in a latency-critical context (e.g. caches) must be used with much more attention, since the compression/decompression logics add extra latency to the read/write path. Additionally, due to the small size of the data blocks in the cache, the compression ratio is limited [2][9].

In 2000, Zhang et al introduced FVC method for data compression in the cache. In this method, several specific patterns were investigated by studying the behavior of different benchmarks and recording the most frequent data in the memory [3]. Alameldeen and Wood proposed FPC method by compressing the cache line in a word-by-word manner [4]. Since 70% of stored data in the cache are zero bytes, Villa et al took advantage of this feature to reduce the length of stored data in the cache and used it to reduce energy consumption [5]. Chen et al, by using two methods based on the dictionary and repeated patterns, have compressed data blocks in the cache (C-pack) [6]. The proposed scheme of Pekhimenko [7] is called BΔI that was based on the fact that most cache lines experience a small dynamic range of changes.

In this paper, we evaluate the effect of compressing data blocks in a multi-core system before transmitting them over the on-chip network. Our evaluation shows that using compression can considerably improve both the performance and power consumption in a multi-core system.

II. COMPRESSION METHODS

In this section, the most important and popular compression methods are reviewed.

A. FVC method

In the FVC method, a dictionary of frequent words is formed and in a cache sorted. When the cache size is large enough, the chance of finding a word in the dictionary will be high. In FVC, after constructing the dictionary, the words of the cache data block are searched in the dictionary and replaced by their equivalent codes. Assuming a dictionary of n frequent words of d bits, a k-word cache line can be compressed to $k \log n$ bits [3].

B. FPC method

This algorithm is based on the observation that some data patterns are repeated most. So, instead of using a cache to tore frequent word, a small meta-data buffer associated to each cache line stores the 7 most frequent words [4]. FPC, compared to FVC method, works faster as it does not require accessing the dictionary cache.

C. DZC method

In this method, a byte with a value of zero is displayed with a zero bit. Experiments show that when $k=8$, k-bit groups of binary data in cache lines exhibit the highest level of locality. DZC detects 8-bit all-zero groups and replace them with a single zero-bit. The advantage of this method over previous methods is that it uses no dictionary and since only zero bytes are desired, the compression hardware (or algorithm) is simple [5].

D. C-pack method

In C-pack both dynamic and static compression techniques are used. At the time of data compression, firstly, a static table of patterns (static dictionary) is examined to find a given data word and in case of hit, it is replaced by its corresponding code word. If the considered data word does not match a pattern in the static dictionary, it is entered into a dynamic dictionary [6] which is formed with respect to workload's behavior. Due to the existence of a dynamic dictionary and also the need for searching each data word in 2 dictionaries, this method has a lower performance with respect to simple compression techniques and consumes more power.

E. BΔI method

In this method, a packet is divided into smaller units and the data unit that has the minimum distance from the rest of data units is selected as the base and stored at the beginning of the compressed line. The data of other units are not stored and only their differences from the base data (Δ) are stored. Compared to other methods, this method is simple and works very well in real applications. This method does not require a dictionary and can be implemented in hardware with low cost [7].

III. EVALUATION RESULTS

To evaluate the effect of different compression methods on the performance and power consumption of a mesh-like interconnected multi-core system, we use gem5 simulator [8]. The specifications of the evaluated system and workloads are listed in Table 1. We conduct the simulation for 1 billion instructions while skipping the first 500 million instructions as warm-up. We load Ruby memory module to enable the NoC in our simulation experiments.

Fig. 1 and Fig. 2 depict the simulation results for PARSEC benchmarks using different compression methods. Fig. 1 reports the compression ratio of the cache lines for different compression methods. As can be seen in this figure, C-pack with an average compression ratio of 6.13 among the evaluated workloads is the best.

Fig. 2 reports the performance and power consumption values of the NoC using C-pack method normalized to the results taken for the baseline system without compression. As can be seen in the figure, the average packet latency is improved by about 20% when C-pack compression scheme is used while power consumption is also reduced by 24.3%, on average.

TABLE I.   DETAILS OF THE SIMULATED SYSTEM WITH 4 CORES

Evaluated System	Description
Processor Cores	16 processors, 4x4 mesh NoC
Private L1 Cache	L1 Size = 32KB, 64-byte lines, LRU
L2 Cache	L2 Size = 256KB per core, 64-byte lines
Evaluated workloads (from PARSEC Benchmark suit)	*blackscholes, bodytrack, dedup, facesim, ferret, fluidanimate, streamcluster, vips, x264*

VI. CONCLUSION

Data compression in the cache reduces the size of needed memory and more space remain for the data storage. This reduction in data provides the opportunity for reducing energy dissipation. The aim of this study was to evaluate and compare different methods of compression on data in NoC. After reviewing the some common and superior methods in data compression by performing simulations through *gem5*, the C-

pack procedure regardless of disadvantages and other factors shows the more compression ratio and storage space than other methods.

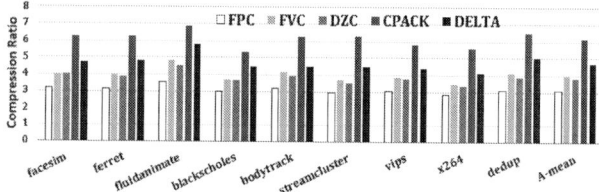

Figure 1 Data Compression ratio of different methods and workloads.

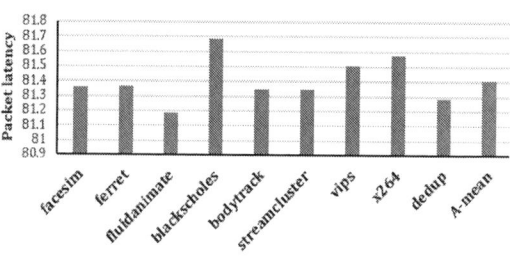

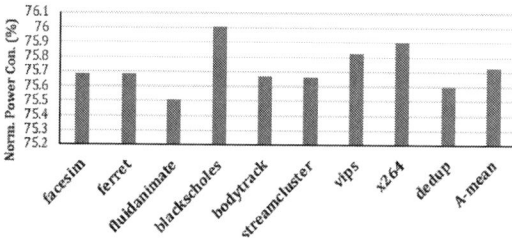

Figure 2 Average packet latency (top), and power consumption (bottom) of C-pack normalized to the baseline system without compression.

REFERENCES

[1] G. E. Moore, "Cramming More Components onto Integrated Circuits," *Electronics*, pp. 114–117, April 1965.

[2] Y. Zhang, J. Yang and R. Gupta, "Frequent Value Locality and Value-Centric Data Cache Design," ACM SIGOPS Operating Systems Review, pp. 150-159, 2000.

[3] J. Yang, Y. Zhang and R. Gupta, "Frequent Value Compression in Data Caches," Technical Report, Univ. of Arizona, Dept. of CS,Tucson, AZ, June 2000.

[4] A. R. Alameldeen and D. A. Wood, "Frequent Pattern Compression: A Significance-Based Compression Scheme for L2 Caches," Technical Report 1500, Computer Sciences Department, 2004.

[5] L. Villa, M. Zhang and K. Asanovi´c, "Dynamic Zero Compression for Cache Energy Reduction," Proceedings of the 33rd annual ACM/IEEE international symposium on Microarchitecture, pp. 214-220, 2000.

[6] X. Chen, L. Yang, R. P. Dick, L. Shang and H. Lekatsas, "C-Pack: A High-Performance Microprocessor Cache Compression Algorithm," VLSI Systems, IEEE, pp. 1196-1208, 2010.

[7] G. Pekhimenko, V. Seshadri and O. Mutlu, "Base-Delta-Immediate Compression: Practical Data Compression for On-Chip Caches," international conference on Parallel architectures and compilation techniques, pp. 377–388, 2012.

[8] N. Binkert et al., "The gem5 simulator," SIGARCH Comput. Archit. News, 2011.

[9] M. Jalili, H. Sarbazi-Azad, "Tolerating More Hard Errors in MLC PCMs Using Compression", ICCD, 2016.

Prediction-Based Latency Compensation Technique for Head Mounted Display

Song-Woo Choi, Min-Woo Seo and Suk-Ju Kang

Dept. of Electronic Engineering, Sogang University
Republic of Korea
E-mail: songwoo602@gmail.com, yoynok08@gmail.com, sjkang@sogang.ac.kr

Abstract—In this paper, we propose a prediction-based latency compensation system for head mounted display. Specifically, the proposed system uses a linear extrapolation of head orientations for prediction based on biological data of body. The experimental results show that the proposed system compensates a latency up to 53 milliseconds with 1.083 degrees of a minimum average error.

Keywords; Virtual reality, head mounted display, prediction

I. INTRODUCTION

Virtual Reality (VR) technologies have more potentiality over the other display technologies in market because of their immersive user experience. In particular, the display industry starts to pay attention to head mounted display (HMD) devices due to their outstanding immersive experience by wide field-of-view (FOV) and low prices. However, many users feel gaps on their sense between the reality and virtual reality because of a motion-to-photon latency [1]. If there is a motion-to-photon latency on the display, incorrect images are outputted in front of the user's eyes. Because the user cannot see the correct image correspond to the motion, the motion sickness occurs [2]. In this paper, we propose a system that outputs the predicted images by *k*-steps ahead prediction of head orientation in order to compensate a motion-to-photon latency in HMD device.

II. PROPOSED SYSTEM

The whole system is expressed in the block diagram of Fig. 1. Current head orientations can be acquired from an inertia measurement unit (IMU) of an HMD device [3]. After that, an equation, which is obtained by a curve fitting, predicts orientations at the next step using the previous values by a linear extrapolation technique. The output image is rendered by a game engine based on the predicted orientation data. The detailed operations of the proposed method are described below.

A. Prediction of Head Orientations Based on Linear Extrapolation

A linear extrapolation, a simpler and faster techniques than polynomial extrapolation, is used to minimize a latency problem without the large load of computation [4]. The equation for linear extrapolation is as follows:

$$y(x_{t+k}) = y_{t-1} + \frac{x_{t+k} - x_{t-1}}{x_t - x_{t-1}}(y_t - y_{t-1}) \qquad (1)$$

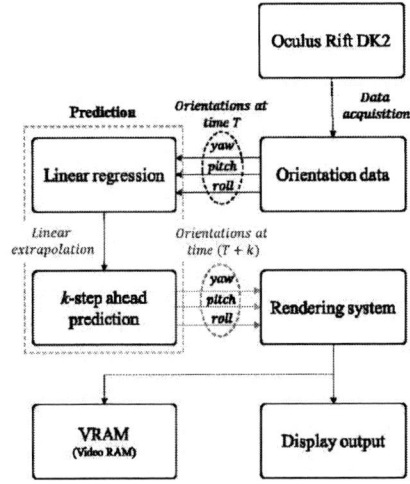

Figure 1. Overall architecture of proposed method

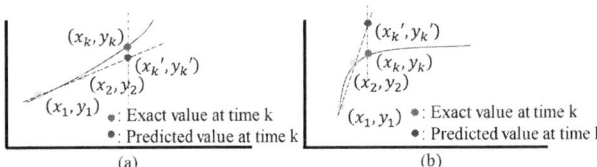

Figure 2. Limitations of the linear extrapolation (a) gradual change of orientations (b) sudden change of orientations

where $y(x_{t-k})$ denotes predicted orientations at *k* steps ahead from the current state. x_t, x_{t-1}, y_t, and y_{t-1} denote a frame number and the head orientations at time *t* and time *t-1*, respectively.

B. Limiting Divergent Value

After the prediction by linear extrapolation, predicted values fit with the exact curve well at linear region where the angular acceleration is constant as shown in Fig. 2 (a). However, as shown in the Fig. 2 (b), if the current value differs rapidly to the previous value, the error between an exact value and the predicted value is higher than the value of (a). This error gets higher when increasing steps of prediction.

In order to compensate the error, we use a maximum rotation speed of a neck measured from biological experiments, which is widely used in many documents [5]. Then, the maximum rotation speed decides a maximum or

978-1-5090-3220-4/16 $31.00 © 2016 IEEE

TABLE I. PREDICTION ERROR OF HEAD ORIENTATIONS AT EACH STEP

Prediction steps (Frames)	Yaw rotation			Pitch rotation			Roll rotation		
	Predicted angles (°)	Exact angles (°)	Error(°)	Predicted angles (°)	Exact angles (°)	Error(°)	Predicted angles (°)	Exact angles (°)	Error(°)
	$T = 100$			$T = 200$			$T = 230$		
1	29.967	29.242	0.724	-6.070	-6.160	0.090	-4.900	-4.600	0.300
2	30.305		1.063	-6.040		0.120	-5.200		0.600
3	30.643		1.401	-6.010		0.150	-5.500		0.900
4	30.981		1.739	-5.980		0.180	-5.800		1.200

TABLE II. AVERAGE PREDICTION ERROR OF HEAD ORIENTATIONS

Prediction steps (Frames)	Yaw rotation			Pitch rotation			Roll rotation		
	Average error (°)	Compensated latency (s)	Compensation rate (%)	Average error (°)	Compensated latency (s)	Compensation rate (%)	Average error (°)	Compensated latency (s)	Compensation rate (%)
1	0.069	0.013	26.667	0.768	0.013	26.667	0.208	0.013	26.667
2	0.271	0.027	53.333	1.461	0.027	53.333	0.833	0.027	53.333
3	0.609	0.040	80.000	2.670	0.040	80.000	1.839	0.040	80.000
4	1.083	0.053	106.667	4.497	0.053	106.667	3.248	0.053	106.667

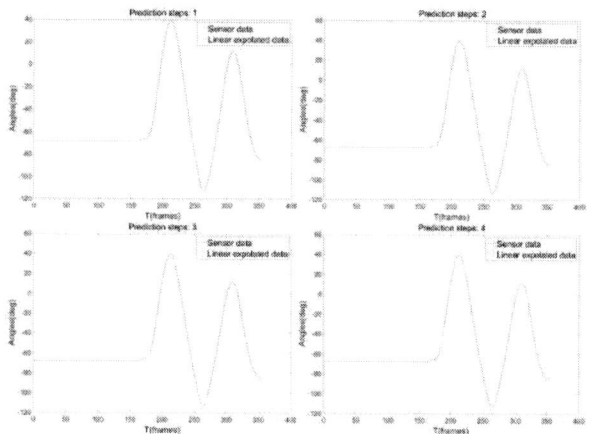

Figure 3. Prediction based on linear extrapolation (top-left: 1 step, top-right: 2 steps, bottom-left: 3 steps, bottom right: 4 steps).

minimum limit of changes for each direction and they are given with constant values. The limits prevent predicted values from increasing/decreasing excessively over the current value. The limits are calculated as follows:

$$c_{limit} = \begin{cases} T \cdot v_{neck} & , \text{if } \Delta y > T \cdot v_{neck} \\ 0 & , \text{otherwise} \end{cases} \quad (2)$$

where c_{limit} denotes limit of orientation changes, T and V_{neck} denote period of refresh rate and maximum speed of neck, respectively. Δy denotes difference of orientations at the current time and previous time. Finally, the predicted orientations are calculated with following equations:

$$y(x_{t+k})' = \begin{cases} y(x_{t+k}) + c_{limit} & , \text{if } y(x_{t+k}) > 0 \\ y(x_{t+k}) - c_{limit} & , \text{otherwise} \end{cases} \quad (3)$$

where $y(x_{t-k})'$ denotes the final result of prediction.

III. EXPERIMENTAL RESULTS

In the experimental results, we set 50ms of latency as a target compensation. Then, we compared predicted orientations and the exact orientations from sensor. A graphical representation of predicted orientations and the exact orientations are expressed in Fig. 3. Errors were computed between the predicted values and the exact values at every points. Average errors were computed for each axial directions by averaging the calculated errors. Also, a compensated latency and compensation rate were calculated with the target latency. The Table I, II show the results.

IV. CONCLUSION

In this paper, we proposed a system using prediction technique that compensated the latency on display for an HMD device. In the proposed method, a prediction method based on linear extrapolation head motion data was applied. Then, a scene was rendered corresponding to the motion of a user with the predicted orientations. Finally, the predicted orientations were compared to the exact orientations after k steps. As a result, the proposed method showed 106.667% of compensation rate with 1.083 degrees of error in yaw direction.

ACKNOWLEDGMENT

This research was supported by Basic Science Research Program through the National Research Foundation of Korea (NRF) funded by the Ministry of Science, ICT & Future Planning (2014R1A1A1004746), a grant (16CTAP-C114672-01) from infrastructure and transportation technology promotion research Program funded by Ministry of Land, Infrastructure and Transport of Korean government, and LG Display.

REFERENCES

[1] D. Kanter, "Graphics Processing Requirements for Enabling Immersive VR," AMD White Paper, 2015.

[2] F. Zheng, T. Whitted, A. Lastra, P. Lincoln, "Minimizing Latency for Augmented Reality Displays: Frames Considered Harmful," Mixed and Augmented Reality (ISMAR), 2014 IEEE International Symposium on. IEEE, pp. 195-200, September 2014.

[3] S. M. LaValle, A. Yershova, M. Katsev, M. Antonov, "Head Tracking for the Oculus Rift," 2014 IEEE International Conference on Robotics and Automation (ICRA). IEEE, 2014. pp. 187-194, May 2014.

[4] D. C. Montgomery, P. A. Elizabeth, and G. G. Vining. "Introduction to linear regression analysis". John Wiley & Sons, 2015.

[5] G. E. Grossman, R. J. Leigh, L. A. Abel, D.J. Lanska, S. E. Thurston, "Frequency and Velocity of Rotational Head Perturbations During Locomotion," Experimental brain research, vol. 70, no. 3, pp. 470-476, May 1988.

Review of Low Power Image Sensors for Always-On Imaging

Jaehyuk Choi

Dept. of Semiconductor Systems Engineering
Sungkyunkwan University
Suwon, Korea
choix215@skku.edu

Abstract— **Energy-efficient design approaches for always-on imaging will be reviewed. Circuit design techniques including dynamic voltage scaling (DVS), dynamic current scaling (DCS), and dynamic frequency scaling (DFS) will be described. In addition, energy-efficient architecture for image signal readout will be described. Finally, power reduction techniques by adaptively suppressing spatial and temporal bandwidth of image signals will be introduced. These energy-efficient design approaches will be illustrated with design examples including 1.36 µW/Frame adaptive CIS for wireless sensor node, 3.4 µW/Frame CIS with integrated feature extractor for object-detection, and 45.5 µW (@ 15fps) always-on CIS for mobile and wearable devices.**

Keywords; CMOS image sensor; CIS; low power; always-on

I. INTRODUCTION

Traditionally, researches on CMOS image sensors (CIS) focused on noise-suppression. Owing to application of pinned photodiodes and low-noise readout circuit, noise in CIS was reduced to few electrons, which overwhelms the performance of CCD image sensors in these days. This noise suppression enabled to achieve extremely low noise with small pixels (i.e., high-resolution) even under extremely low illumination. This research trend is quite reasonable because the main purpose of CIS is to capture high-quality images. Users always want to get a high-quality picture from their DSCs or camcorders. Beyond DSCs, CISs are integrated in most mobile devices, which are enabled by low-cost CMOS image sensors. Even in mobile devices, the main goal is same as the case of DSCs. The power consumption was not so an important issue compared to the other main parameters such as resolution, SNR, and dynamic range because the power consumption can be sacrificed for better image quality. Another main reason is that a CIS needs to be turned on only when users take a picture, which does not require severe power consumption.

Now, new applications of the CIS are emerging: from traditional CIS for picture-taking to the CIS as a new type of sensor for augmented reality (AR), virtual reality (VR) and internet-of-things (IoT). In the last few years, many wearable devices have been introduced into the market. A watch-type device has a camera module for video communication. A glass-type device also integrates a camera module in order to achieve AR by providing digital information interpreted from captured images. Not only as a cellphone in traditional applications, the mobile device works as a platform for AR, VR, and IoT where integrated CISs collect images from real world, recognize, and analyze objects and environment from the scene, and provide virtual but useful information to users. These new applications in mobile and wearable devices have strong potential to change our daily life by application to gaming, shopping, travelling, education, and so on. The new applications of CIS are not limited in mobile and wearable devices. Drones and tiny wireless sensor nodes (that autonomously operate in remote sites) enable acquiring information even in unreachable area, which significantly extends user-experience. By connecting all of them though network, we can achieve true IoT based on images captured from real world.

As noted, a CIS integrated in mobile and wearable devices has been dedicated to taking pictures at specific time and is turned on at the user's request. For the application of AR, VR, and IoT, a CIS should continuously capture images in order to deliver information without missing events. In other words, a CIS should be "always-on". In addition to avoid missing events, one important benefit of always-on imaging is enhanced usability from instantaneous response. The examples of the enhanced usability by the always-on operation can be found in traditional devices such as electronic watch, network-attached storage (NAS), and always-on display in mobile devices..

In order to implement always-on imaging in mobile and wearable devices, we can consider using an existing CIS in the camera modules. This is a cost-effective solution because an existing camera module would be used, i.e., no extra circuits in the CIS or extra chip packages in the module would be required for always-on imaging. However, an existing CIS consumes high power, greater than 50 mW, because it is dedicated and optimized to capture high-resolution, low-noise images. Because of the limited energy supplied by a battery, low power consumption is a crucial factor in always-on imaging. Even though many power reduction techniques such as voltage scaling reduces power consumption, SNR is easily traded off from the power reduction.

The power budget of always-on imaging in wireless sensor nodes is even more limited than in mobile and wearable devices. Among many networked sensor devices that consist of IoT, consider wireless sensor nodes which collect information from the wide area covered by wireless sensor networks. The distributed sensor nodes autonomously operate at extremely low power consumption from energy harvesting to monitor wide and unreachable areas for military surveillance, environmental monitoring, and biomedical systems such as endoscopy, retinal prosthesis, etc. In applications for these wireless sensor nodes, extremely low power consumption under one microwatt is required because tiny sensor nodes should operate from the limited energy sources by energy harvesting from the ambient. Because most state-of-the-art image sensors require large power consumption over 50 mW, they are not applicable to wireless sensor nodes. As noted earlier, the power consumption of CMOS image sensors is difficult to scale down because voltage scaling directly suppresses SNR and degrades image quality.

In the design of always-on image sensor, main goal is to reduce power consumption for continuous operation with limited energy source, while avoiding significant reduction of SNR for post image processing (such as object detection and recognition) required for smart sensing. In this paper, energy-efficient design approaches for always-on imaging will be reviewed. Circuit design techniques including dynamic voltage scaling (DVS), dynamic current scaling (DCS), and dynamic frequency scaling (DFS) will be described. In addition, energy-efficient architecture for image signal readout will be described.

Finally, power reduction techniques by adaptively suppressing spatial and temporal bandwidth of image signals will be introduced. These energy-efficient design approaches will be illustrated with design examples including 1.36 µW/Frame adaptive CIS for wireless sensor node [1], 3.4 µW/Frame CIS with integrated feature extractor for object-detection [2], and 45.5 µW (@ 15fps) always-on CIS for mobile and wearable devices [3].

REFERENCES

[1] J. Choi, S. Park, J. Cho, and E. Yoon, "An Energy/Illumination-Adaptive CMOS Image Sensor With Reconfigurable Modes of Operations," *IEEE J. Solid-State Circuits*, vol. 50, no. 6, pp. 1438–1450, Jun. 2015.

[2] J. Choi, S. Park, J. Cho, and E. Yoon, "A 3.4 µW Object-Adaptive CMOS Image Sensor With Embedded Feature Extraction Algorithm for Motion-Triggered Object-of-Interest Imaging," *IEEE J. Solid-State Circuits*, vol. 49, no. 1, pp. 289–300, Jan. 2014.

[3] J. Choi, J. Shin, D. Kang, and D. –S. Park, "Always-On CMOS Image Sensor for Mobile and Wearable Devices," *IEEE J. Solid-State Circuits*, vol. 51, no. 1, pp. 130–140, Jan. 2016.

MEMS Resonator Based Thermometer SoC Design in CMOS 0.18 μm Standard Process

Chong-Yang Lin
Department of Electronic Engineering
National Chiao Tung University
Hsinchu 300, Taiwan
E-mail: tyu070707@gmail.com

Kuei-Ann Wen
Department of Electronic Engineering
National Chiao Tung University
Hsinchu 300, Taiwan
E-mail: stellawen@mail.nctu.edu.tw

Abstract—**This paper presents a thermometer SoC design that can be manufactured and monolithically integrated in the ASIC standard CMOS process. A high gain low power trans- impedance amplifier(TIA) circuit and phase locked loop(PLL) are designed for readout. The resonator based thermometer has been demonstrated with a sensitivity -5.7Hz/°C(-139ppm) in a temperature range from -40°C to 120°C. Power consumption of the readout circuit is only 190.4μW.**

Keywords—ASIC/MEMS process; thermometer; resonator; TIA; PLL

I. INTRODUCTION

Due to growing demand in the sensor applications, there has been a push to create combo sensors and reduce overall chip size. So, trying to scale down devices and finding out fabrication and packaging techniques to co-fabricate many sensors on a single wafer is an important issue [1]. Resonant sensors have become a great solution for sensing many physical parameters such as pressure, temperature, viscosity, mass, etc. [2]. Therefore, we choose the resonator as the sensing structure for environmental temperature detecting and maybe extend to other physical parameters.

The proposed MEMS resonator was fabricated under UMC 0.18μm 1-poly-6-metal standard CMOS-MEMS process [3] and could be monolithically integrated with CMOS readout circuitry without the need of any post-processing after fabrication from foundry. This fabricated way can save a lot of chip area and highly integrate in single wafer for mass production.

In this design, the interdigital transducer method is adopted to complete the resonator for environment temperature sensing. Unlike other designs [4], it can easily combine other sensing functionalities in single structure, such as magnetic field sensors, accelerometers, etc.

II. STRUCTURE OF THE CMOS MEMS RESONATOR

Fig. 1 (a) shows the SEM of the resonator based sensing structure. The mechanical resonator is composed of two comb finger structures for driving and sensing, and a moveable shutter structure to form mechanical resonance. The shutter is suspended above the substrate by four symmetric springs. Etching holes were added in the center of the shutter to ensure it can be released completely. For monolithic design, four

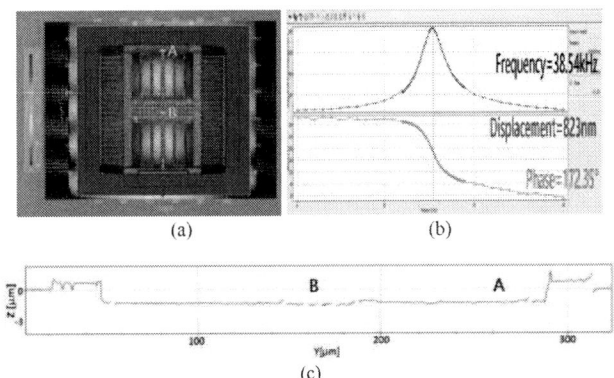

(a)   (b)

(c)

Fig. 1. (a) The SEM of the resonator based sensing structure, (b) the in-plane vibration analyzer measurement result of frequency response of the proposed MEMS resonator under P = 1atm, and (c) the white light interferometer measurement result of the resonator based sensing structure.

anchors were placed at the corners to avoid wire bond from resonator electrode to CMOS circuitry.

Fig. 1 (b) shows frequency response of resonator based sensing structure under pressure is 1atm and the voltage of piezoelectric sheet is 10 volts. The measured resonant frequency *f0* was 38.54 kHz compared to 41.11 kHz in FEA simulation and the Q factor was ranging from 167 to 201 compared to 360 in FEA simulation.

Fig. 1 (a) and (c) shows the white light interferometer measurement result of resonator based sensing structure. The light blue region is the suspension region, so it must sink slightly. The height difference of the surface of the red light A to B is less the 250nm, so the resonator manufactured in this process is very stable. The measured result of the overlap of the

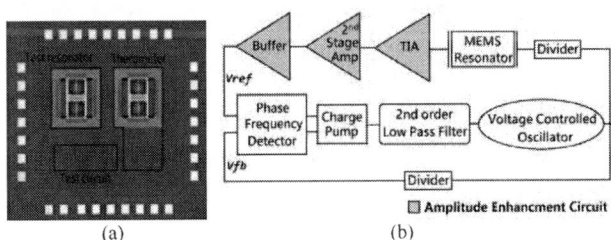

(a)   (b)

Fig. 2. (a) SEM of resonant based thermometer, and (b) System Archiecture of the proposed temperature sensor

finger is larger than 407.6μm², which is represent more than 92% initial capacitance.

III. READOUT CIRCUITY OF THE TEMPERATURE SENSOR

Fig. 2 (b) shows the system architecture of the proposed environmental thermometer. The PLL circuitry provides a stable clock to drive the resonator, and it can track the resonant frequency shifts due to the environmental temperature changes. When the temperature changes, the PLL reference frequency (*Vref*) that gets from sensing node of the resonator shifts, and then the PLL will track the *Vref*. So, we can get the temperature information in real-time from the controlled voltages of the voltage controlled oscillator.

The CMOS PLL clock synthesizer targets for sensor applications that require low power consumption. Based on the integer-N architecture, the PLL clock synthesizer produces 40 kHz output signal to drive the MEMS resonator. The loop has 101.06 Hz bandwidth with phase margin 53.7°. The charge pump output current is only 4.8nA, and the gain of the voltage controlled oscillator is 255kHz/V. Fig. 3 shows the simulation result of resonant frequency and controlled voltage variations with different temperatures. The proposed thermometer has been demonstrated with a sensitivity -5.7Hz/°C (-139ppm) or 0.224mV/°C in a temperature range from -40°C to 120°C.

Assuming quality factor is equal to 183, extracted from the measurement result of the previous design, the output current induced by the MEMS resonator can be calculated as in (1).

$$i_o \cong V_{dc} \frac{\partial c_{sense}}{\partial t} = V_{dc} \frac{\partial c_{sense}}{\partial x} f_0 \Delta x = 0.39 \, nA \qquad (1)$$

For around 200mV output swing, the total gain of the amplitude enhancement circuit should be as high as 173.98dBΩ. The first stage is resistance-feedback TIA which can convert capacitance change of the comb finger to voltage variation. The second stage is capacitive-feedback amplifier that completes the total gain of amplitude enhancement. The third stage is inverter string which shortens the rising time of the sensing signal for ensuring the phase frequency detector proper functioning.

The resonator based sensing structure is replaced by the equivalent RLC model which has the capacitance variations due to motional and environmental parameter changes for co-simulation of the temperature sensing system. It takes 65ms for

TABLE I. PERFORMANCE COMPARSION OF TEMPERATURE SENSOR

	[2][a]	[4][a]	[5]	This work
Technology	N/A	0.22μm CMOS	0.18μm/ 65nm CMOS	0.18μm CMOS
Method	Freq. shift	Freq. shift	Freq. shift	Freq. shift
Range (°C)	20, 90	20, 120	-40, 120	-40, 120
Resolution (°C)	0.1	1	0.01(est)	0.69
Power (μW)	N/A	N/A	61.5	190.4
Area (mm²)	N/A	N/A	7.26[b]	0.74

a. Without readout circuit
b. Two chips are wire bonded. (Estimated)

the PLL to lock in from the initial condition of VCO controlled voltage is 515mV (equivalent to 120 °C) to the stable controlled voltage 551mV (equivalent to -40 °C), which represents the overall sensing range. The ripple is less than 154μV, equivalent to 0.69 degrees Celsius variation. The power consumption of PLL is 17.13μW, and the overall circuit is 190.4μW.

IV. CONCLUSION

This paper provides an effective solution for single chip environmental thermometer SoC design. The presented thermometer has been demonstrated with a sensitivity -5.7Hz/°C (-139ppm) or 0.224mV/°C in a temperature range from -40°C to 120°C. TABLE I. shows comparison of co-simulation result the proposed system with previous works.

V. ACKNOWLEDGMENT

This research was sponsored in part by the National Science Council of Taiwan under grant of MOST 104-3115-E-009-022. The authors appreciate the UMC and the National Chip Implementation Center (CIC), Taiwan, for supporting the CMOS chip manufacturing.

VI. ACKNOWLEDGMENT

[1] V. A. Hong et al., "Capacitive sensor fusion: Co-fabricated X/Y and Z-axis accelerometers, pressure sensor, thermometer," 2015 Transducers - 2015 18th International Conference on Solid-State Sensors, Actuators and Microsystems (TRANSDUCERS), Anchorage, AK, 2015, pp. 295-298.

[2] H. Fatemi, M. J. Modarres-Zadeh and R. Abdolvand, "Passive wireless temperature sensing with piezoelectric MEMS resonators," 2015 28th IEEE International Conference on Micro Electro Mechanical Systems (MEMS), Estoril, 2015, pp. 909-912.

[3] S. H. Tseng, Y. T. Hsieh, C. C. Lin, H. H. Tsai and Y. Z. Juang, "CMOS MEMS resonator oscillator with an on-chip boost DC/DC converter," 2015 Transducers - 2015 18th International Conference on Solid-State Sensors, Actuators and Microsystems (TRANSDUCERS), Anchorage, AK, 2015, pp. 1981-1984.

[4] R. Mahameed et al., "Fully monolithic MEMS based thermal sensor in 22 nm CMOS technology for SoC thermal managemet," 2013 Transducers & Eurosensors XXVII: The 17th International Conference on Solid-State Sensors, Actuators and Microsystems (TRANSDUCERS & EUROSENSORS XXVII), Barcelona, 2013, pp. 734-737.

[5] Y. Zhu et al., "An Energy Autonomous 400 MHz Active Wireless SAW Temperature Sensor Powered by Vibration Energy Harvesting," in IEEE Transactions on Circuits and Systems I: Regular Papers, vol. 62, no. 4, pp. 976-985, April 2015

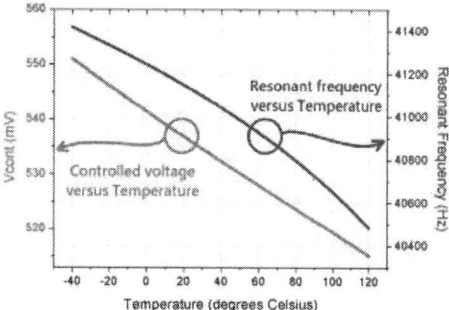

Fig. 3. The simulation result of resonant frequency and controlled voltage variations with different temperatures.

A Decouple structured Gyroscope with Integrated Readout Circuit on Standard 0.18μm 1P6M CMOS Technology

Chun-Lin Chien
Department of Electronic Engineering
National Chiao Tung University
Hsinchu 300, Taiwan
E-mail:chunlin1011@gmail.com

Kuei-Ann Wen
Department of Electronic Engineering
National Chiao Tung University
Hsinchu 300, Taiwan
Email:stellawen@mail.nctu.edu.tw

Abstract— **A dual proof-mass structured gyroscope integrated with readout circuit have been proposed. The C to V stage is achieved by the differential difference amplifier (DDA) which has advantages of high gain, low temperature and process dependence. Chopper Stabilization (CHS) and Corrected Double Sampling (CDS) is used to suppress low-frequency noise and compensate DC offset. The gain of DDA is 25dB and power consumption of total readout circuit is 791μW. The mems part is fabricated with standard 1P6M 0.18um CMOS process and the mechanical sensitivity is 11.26aF/°/s. Chip area is 1.9x1.7mm².**

Keywords; CMOS-MEMS; Gyroscope; readout circuit, capacitive sensing

I. INTRODUCTION

MEMS gyroscopes and accelerometers are two essential sensors for most modern applications. Highly integrated design is always the design challenge especially for IOT, and wearable devices. Among various MEMS gyroscopes and to consider integration, CMOS-MEMS gyroscopes [1] have the advantages of small size, lower parasitic, and potentially lower cost over the two-chip hybrid solutions. The DDA circuit can be chosen as the readout interface to get both high gain and low-temperature dependence [2]. In gyroscope design, one of the major challenges is the external acceleration acting on the sensing-axis [3] and it will cause interferences. To reduce the external acceleration, a compensation method uses the differential way has been introduced [4]. This work proposed a solution to use the differential reading of the multiple sensed signals to successfully integrate multi-sensors as well as read out circuit monolithically.

II. STRUCTURE OF THE CMOS MEMS GYROSCOPE

The proposed MEMS part was fabricated with UMC 0.18 μm 1-poly-6-metal standard CMOS-MEMS process [5] provided by CIC (National Chip Implementation Center) and could be monolithically integrated with CMOS readout circuitry without the need of any post-processing after fabrication from the foundry. By both the anisotropic silicon oxide etching (DRIE) and the isotropic silicon substrate etching, the desired microstructure will release after the isotropic silicon substrate etching and become movable for generating capacitance

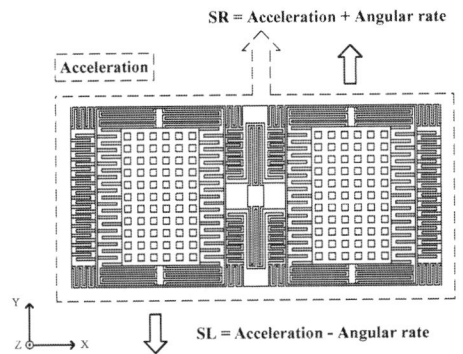

Fig. 1. Principle of dual function structure

variation.

The schematic view of the gyroscope is shown in Fig. 1. An outer decouple ring is used to completely decouples the Coriolis sense mode from vibration driving mode. To eliminate low-frequency proof mass motion due to external accelerations, the dual proof masses were chosen. The dual masses are driven along x-axis in an anti-phase mode and the Coriolis force will push the proof masses along the y-axis when the system rotate along the z-axis. Assuming the system was accelerating, the displacement will be generated by the acceleration and Coriolis force. The angular rate can be calculated by subtracting SR and SL; meanwhile, the acceleration can be generated by add SR and SL. To reduce the curl rate in the 1P6M process, we chose METAL156 as the main structure and METAL234 as the routing layer. The total capacitance is 206fF and the sensitivity is 11.26aF/°/s.

III. SYSTEM ARCHITECTURE

The system architecture is shown in Fig. 2. The driving module will generate ac signal which is used to drive the proof mass in the resonator frequency and detect the output current to ensure the gyroscope is working in the requirement frequency. The external offset

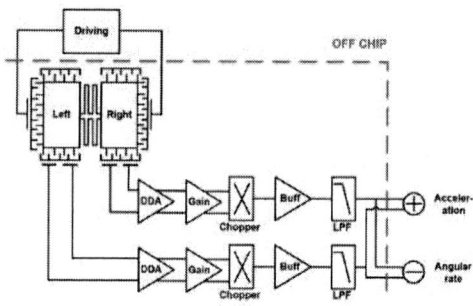

Fig. 2. System architecture

will be calculated by add and subtract two anti-phase signals it will be generated by the control MCU.

The fully differential capacitive bridge represents the sensing capacitors. Consider the temperature independence and the SNR in 100f order sensing capacitance, differential difference amplified [6] was chosen as the C-to-V stage. The output of the DDA can be derived as

$$v_{out+} - v_{out-} = -\frac{2\Delta C_s \cdot (C_{fix} + C_{fb})}{(2C_{s0} + C_p) \cdot C_{fb}} \cdot V_P$$

where Vp is the bias voltage, ΔC_s are the change of sensing capacitance due to Coriolis force. Since the gain of DDA is based on capacitance ratio, it is inherently independent of temperature and the gain was achieve 25dB. To reduce the noise signal and to eliminate offset, Chopper Stabilization (CHS) and Corrected Double Sampling (CDS) was adapted. In this paper, the sensing signal has shifted to the odd harmonic frequencies of the chopper frequency. The simulation result of the input noise is 66.55nV/√Hz at 500kHz.

IV. CONCLUSION

The SEM of Decouple Gyroscope is shown in Fig. 3(a). Fig.3 (b) shows measurement of resonator frequency, the driving mode and sensing mode is about 6.85 kHz and 7 kHz. A Decouple Gyroscope with Integrated Readout Circuit on Standard 0.18μm 1P6M CMOS Technology has been integrated on the ASIC. Modulating the capacitance signal to a chopping frequency reduces the low-frequency noise and canceling the offset with CDS technique. The chip area

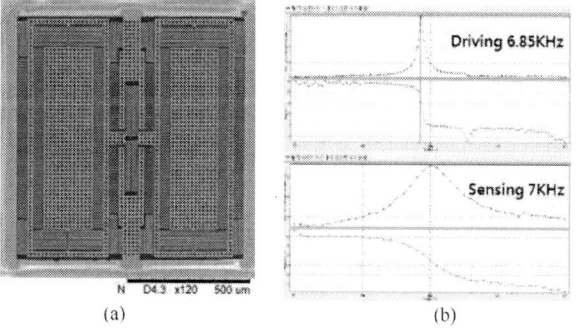

Fig. 3. (a)The photograph taken by SEM (b)Resonator frequency

is 1.9x1.7mm². The Digital system and the driving circuit will be integrated in the future.

V. ACKNOWLEDGEMENT

This research was sponsored in part by the National Science Council of Taiwan under a grant of MOST 104-3115-E-009-022. The authors appreciate the UMC and the National Chip Implementation Center (CIC), Taiwan, for supporting the CMOS chip manufacturing.

TABLE I. SYSTEM PARAMETER

Param Eters	CMOS-MEMS GYROSCOPE			
	[1]	[7]	[8]	This Work
Detected Axis	Z	Z	Z	Z
Area(mm²)	N/A	6.25	2.25	3.41
Power(mW)	N/A	3.8	0.4	0.791
Mechanical Sensitivity (aF/°/s)	0.14	N/A	80	11.26
MEMS Process	TSMC 2P4M 0.35μm	SOG-bulk microma chining	40um-thick SOI	UMC 0.18μm CMOS MEMS

REFERENCES

[1] Hung-Yao Hung; Dou-Ru Chang; Wen-Pin Shih, "Design and simulation of a CMOS-MEMS gyroscope with a low-noise sensing circuit," in Computer Communication Control and Automation (3CA), 2010 International Symposium on , vol.2, no., pp.253-256, 5-7 May 2010

[2] Hongzhi Sun; Kemiao Jia; Xuesong Liu; Guizhen Yan; Yu-Wen Hsu; Fox, R.M.; Huikai Xie, "A CMOS-MEMS Gyroscope Interface Circuit Design With High Gain and Low Temperature Dependence," in Sensors Journal, IEEE , vol.11, no.11, pp.2740-2748, Nov. 2011

[3] K. Azgin, "High Performance MEMS Gyroscopes," M.Sc. Thesis, Middle East Technical University Feb. 2007.

[4] Sonmezoglu, S.; Gavcar, H.D.; Azgin, K.; Alper, S.E.; Akin, T., "Simultaneous detection of linear and coriolis accelerations on a mode-matched MEMS gyroscope," in Micro Electro Mechanical Systems (MEMS), 2014 IEEE 27th International Conference on , vol., no., pp.32-35, 26-30 Jan. 2014

[5] S. H. Tseng, Y. T. Hsieh, C. C. Lin, H. H. Tsai and Y. Z. Juang, "CMOS MEMS resonator oscillator with an on-chip boost DC/DC converter," 2015 Transducers - 2015 18th International Conference on Solid-State Sensors, Actuators and Microsystems (TRANSDUCERS), Anchorage, AK, 2015, pp. 1981-1984.

[6] H. Sun et al., "A CMOS-MEMS Gyroscope Interface Circuit Design With High Gain and Low Temperature Dependence," in IEEE Sensors Journal, vol. 11, no. 11, pp. 2740-2748, Nov. 2011.

[7] S. R. Chiu, C. Y. Sue, L. P. Liao, L. T. Teng, Y. W. Hsu and Y. K. Su, "A fully integrated circuit for MEMS vibrating gyroscope using standard 0.25um CMOS process," 2011 6th International Microsystems, Packaging, Assembly and Circuits Technology Conference (IMPACT), Taipei, 2011, pp. 315-318.

[8] A. Sharma, M. F. Zaman and F. Ayazi, "A 104-dB Dynamic Range Transimpedance-Based CMOS ASIC for Tuning Fork Microgyroscopes," in IEEE Journal of Solid-State Circuits, vol. 42, no. 8, pp. 1790-1802, Aug. 2007.

Monolithic MEMS Resonator Based Pressure Sensor and Readout Design

Po-Chun Chuang
Department of Electronic Engineering
National Chiao Tung University
Hsinchu 300, Taiwan
E-mail: zak.chuang@gmail.com

Kuei-Ann Wen
Department of Electronic Engineering
National Chiao Tung University
Hsinchu 300, Taiwan
E-mail: stellawen@mail.nctu.edu.tw

Abstract— A monolithic MEMS resonator based pressure sensor and monolithically integrated with TIA (trans-impedance amplifier) readout circuitry has been fabricated in standard 1p6m AISC process. Dependence of the quality factor and ambient pressures are well known to resonator designers and it will be feasible to integrate a quality factor readout circuitry to detect ambient pressure. By measuring the sample resonator, the air pressure changes from 100 Pa to 1600 Pa, the Q factor will change from 2566 to 452 with resonant frequency in 15.4 k Hz and the readout circuit is designed accordingly. The system power consumption is 332.82 µW with 1.8 V power supply and sensitivity is 0.0203 mV per Q.

Keywords—ASIC/MEMS process; pressure sensor; resonator; TIA;

I. INTRODUCTION

Using resonator to act a pressure sensory exhibit better long-term stability than piezo-resistive and capacitive sensors [1][2], both of which need diaphragms that are susceptible to wear and tear. Furthermore, resonant sensors have low susceptibility to hysteresis effects, which can adversely affect the behavior of other sensor types [3]. In this paper, we implemented the resonator based pressure sensor, which can be fabricated in the UMC 0.18 µm 1P6M CMOS-MEMS standard process without any post-processing and monolithically integrated with CMOS readout circuitry. In this fabrication, we can achieve lower cost, lower area and high reliability. Then to design the quality factor detector to reform the resonator to be pressure sensor. It not only increased the functionality of the resonator but to make multiple sensor fabricated with the same process being feasible.

II. STRUCTURE OF THE CMOS MEMS RESONATOR

The dynamic performance of MEMS structure can be described as a spring-mass damped vibrating system [4]. The quality factor of the MEMS sensor can be inferred as

$$Q = \frac{\sqrt{m_e k_e}}{b_e} \qquad (1)$$

where m_e, k_e and b_e are an effective mass, effective spring constant, damping coefficient. The source of damping in MEMS structures can be categorized into two types of squeeze film damping and slide film damping, and both type of damping will contribute to the damping coefficient b and quality factor Q. The formula of squeeze film damping coefficient and slide film damping coefficient are given as

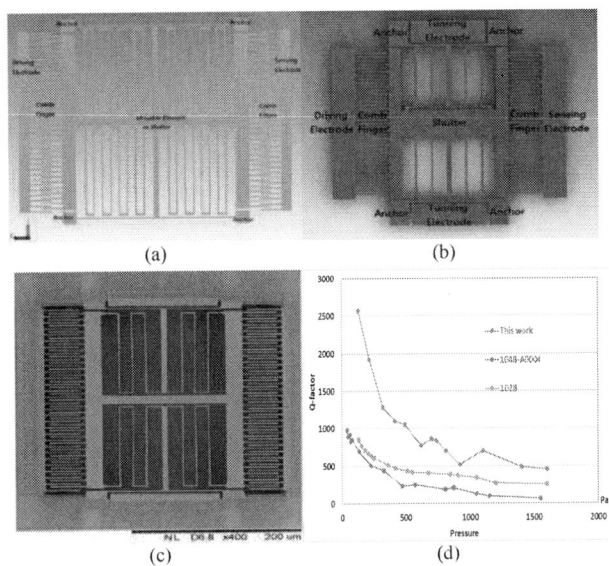

Fig. 1. (a)102B resonator (b)104B resonator (c)This work (d)The measurement result of (a)(b)(c).

$$b_{squeeze} = 7.2 N_{gn} \mu l_{ov} \frac{h_r^3}{d_0^3} \qquad (2)$$

$$b_{slide} = N_{gn} \mu \frac{A}{d_0^3} \qquad (3)$$

where N_{gn}, μ, l_{ov}, h_r, d_0, A are the number of air gaps, the viscosity of air, parallel-plate overlap, thickness, initial gap between a rotor and stator, and the overlap area between parallel-plate structures. Since the viscosity changes mainly according to the pressure of the environment [3][5], we can use the relationship of Q factor and pressure to sense the air pressure.

To verify the dependence of the Q value vs. pressure, we designed different structure of resonator. The sensing structure consists of two comb-finger structures for driving and sensing electrodes with movable shutter structure for mechanical resonance, as shown in Fig. 1 (a) (b) (c). The measurement result of quality factor versus pressure shows on Fig. 1 (d). It reveals that Fig. 1 (c) has higher quality factor and sensitivity with large mass, spring constant and lower number of gaps, parallel-plate overlap.

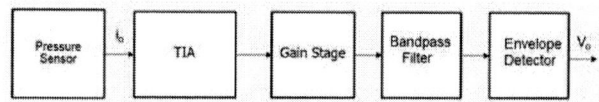

Fig. 2. System Archiecture of the proposed pressure sensor.

With Fig.1 (c), the air pressure was from 100 to 1600 Pa, Q factor was from 2566 to 452 in the resonant frequency 15.4 kHz. With the transducer designed, the displacement of the sensing figures are proportional to the Q factor and thus are proportional to the air pressure. Hence, we can measure the capacitance variation caused by the displacement and thus to measure the ambient pressure.

III. READOUT CIRCUITY OF THE PRESSURE SENSOR

Fig. 2 shows the system architecture of the proposed pressure sensor. Since the output current was generated on the sensing electrodes when the resonator was driving, the TIA readout circuitry was designed to detect the capacitance variation due to quality factor change of the resonator structure caused by ambient pressure variation. The output current signal of MEMS resonator can be derived as

$$i_o \cong \frac{V_{dc}^2 \xi^2 N_{gn}^2 \varepsilon_0^2 h_r^2 f_0 V_{ac} Q}{d_0^2 k_e} \qquad (4)$$

where ξ is the constant that models additional capacitance due to fringing electric fields, ε_0 is the permittivity constant, f_0 is the resonant frequency.

To obtain the output voltage in the order of several tens mili-volts, the total gain of the readout circuit should be large than 100MΩ. The first stage is resistance-feedback TIA that can convert and enhance input signal. The second stage is capacitive-feedback amplifier that can complete the total gain of amplitude enhancement. The band-pass filter can reduce the noise of forward stages and sensor. The output signal is converted from resonating waveform into voltage level by the last stage envelope detector for ambient pressure estimation.

Fig. 3 (a) shows the co-simulation result of sensor and readout circuit. The resonator based pressure sensor is replaced by the equivalent RLC model of which the quality factor is 2566, then the output voltage after Envelope detector is about 700 mV.

Fig. 3 (b) shows the co-simulation result of output voltage in different quality factor. The output voltage is proportional to quality factor and its slope is 0.0203 mV per Q.

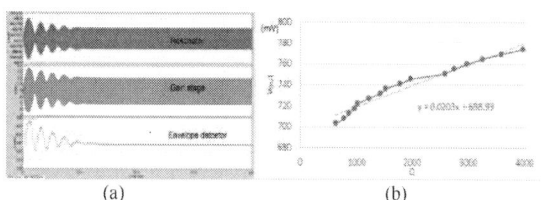

(a)   (b)

Fig. 3. (a) co-simulation result of sensor and readout circuit, (b) different Q with output voltage.

TABLE I.   PERFORMANCE COMPARSION OF PRESSURE SENSOR

	[1]	[2]	[6]	This work
Technology	SiGe CMOS MEMS	MetalMUMPs	amorphous silicon carbide (a-SiC)	1P6M 0.18μm ASIC CMOS MEMS
Sensor type	Diaphragm	Diaphragm	Beam Resonator	Comb-finger Resonator
Range (kPa)	10~100	0~30	1~120	0~1.6
Area (mm²)	0.56	4.1	1	0.119
Power consumption(μW)	N/A	16.56	520	332.82

IV. CONCLUSION

Table I shows comparison of different types of pressure sensor. In this paper, we designed a monolithic MEMS resonator based pressure sensor, which can be fabricated on the standard CMOS-MEMS process and monolithically integrated with readout circuitry. The measurement pressure is from 100 to 1600 Pa, and Q factor is from 2566 to 452 in the resonant frequency 15.4 kHz. The sensor area is 0.119 mm². The power consumption of overall system is 332.82 μW with 1.8 V power supply and sensitivity is 0.0203 mV per Q.

V. ACKNOWLEDGMENT

This research was sponsored in part by the National Science Council of Taiwan under grant of MOST 104-3115-E-009-022. The authors appreciate the UMC and the National Chip Implementation Center (CIC), Taiwan, for supporting the CMOS chip manufacturing.

REFERENCES

[1] Sundararajan, A.D.; Rezaul Hasan, S.M., "Elliptic Diaphragm Capacitive Pressure Sensor and Signal Conditioning Circuit Fabricated in SiGe CMOS Integrated MEMS," in *Sensors Journal, IEEE*, vol.15, no.3, pp.1825-1837, March 2015.

[2] J. A. Montiel-Nelson, J. Sosa, R. Pulido, A. Beriain, H. Solar, and R. Berenguer, "Digital output MEMS pressure sensor using capacitance-to-time converter," in Design of Circuits and Integrated Circuits (DCIS), 2014 Conference on, 2014, pp. 1-4.

[3] Banerji, Saoni, et al. "CMOS-MEMS resonant pressure sensors: optimization and validation through comparative analysis." Microsystem Technologies (2016): 1-17.

[4] F. Y. Kuo, C. F. Chang and K. A. Wen, "CMOS 0.18 μm standard process capacitive MEMS high-Q oscillator with ultra low-power TIA readout system," IEEE SENSORS 2014 Proceedings, Valencia, 2014, pp. 911-914.

[5] F. R. Blom, S. Bouwstra, M. Elwenspoek, J. H. J. Fluitman, "Dependence of the quality factor of micromachined silicon beam resonators on pressure and geometry", J. Vac. Sci. Technol. B, vol. 10, Issue 1, pp. 19-26, Jan 1992.

[6] K. Allidina, M. A. Taghvaei, F. Nabki, P. V. Cicek, and M. N. El-Gamal, "A MEMS-based vacuum sensor with a PLL frequency-to-voltage converter," in *Electronics, Circuits, and Systems, 2009. ICECS 2009. 16th IEEE International Conference on*, 2009, pp. 583-586.

Implementation of RF Frequency Synthesizer for Smart Utility Network System

Dong-Shik Kim, Won-Sang Yoon, Sang-Hoon Chai
Dept. Electronic Engineering
Hoseo University
Asan, Korea
dongshik.kim1205@gmail.com

Abstract— **A wideband frequency synthesizer has been designed and fabricated to generate 780MHz(China), 868MHz(Europe), and 915MHz(Korea, North America) of three bands at the same time. It will be applied in the IEEE 802.15.4g SUN system. Measurement results show the frequency synthesizer has wide bandwidth of 1527~2020MHz(about 27.8% of center frequency) and low phase noise characteristics of -98.63dBc/Hz at 100KHz offset, -122.05dBc/Hz at 1MHz offset.**

Keywords; IEEE802.15.4g; SUN; IoT; Frequency Synthesizer; PLL; Circuit Design

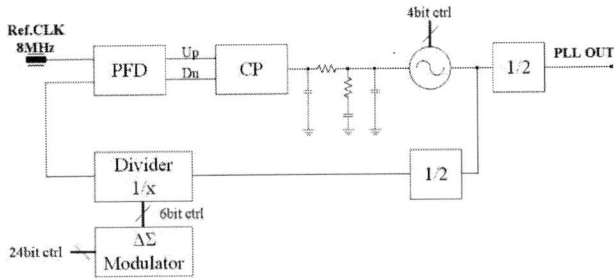

Figure 1. Structure of the proposed frequency synthesizer.

I. INTRODUCTION

In recent research, while increasing the energy efficiency AMI(Advanced Metering Infrastructure) to provide a comfortable living environment based energy management devices of significance has been emphasized. Vigorous research on the AMR(Automatic Meter Reading) has been conducted based on PLC and Zigbee technology. However, it is difficult for Zigbee to obtain high link margin characteristics and it has limits of mesh routing technology especially in long range outdoor environments. So development of a novel international standards, wireless transmission technology has become necessary in connection with a smart grid [1]. Also, unlicensed frequency band including 700MHz~1GHz and 2.4GHz has been considered as frequency band for IEEE 802.15.4g SUN system that can support a variety of network application in along with a smart gird [2]. In this study, wideband frequency synthesizer was designed and fabricated to support three bands at the same time, which are 780MHz(China), 868MHz(Europe), and 915MHz(Korea, North America) band. After that, applicability of the IEEE 802.15.4g SUN system was examined. The structure of frequency synthesizer generating LO signal in transceiver is as shown Figure 1. As a frequency synthesizer has a significant effect on the characteristics of the transceiver input and output signals, each block was optimized so that phase noise and spurious performance get excellent values. The VCO oscillation frequency to generate the I-Q signals having a phase difference of 90 degrees was designed the 1558~1856MHz of bandwidth to contain twice of 779~982MHz using frequency. In addition, using the fractional-N divider structure in order to make the selection of a large number of the channels.

II. FREQUENCY SYNTHESIZER DESIGN

A. VCO

Transmission channels used for SUN system is the five channels of Chinese band(779~787MHz) with 2MHz spacing, the ten channels of North America band(902~928MHz) with 2MHz spacing, and the two channels of Europe band (868.3MHz) with 600KHz spacing relative to the center frequency. Phase noise and spurious components should be excellent at the point of 1MHz which is adjacent channel spacing. The structure of the LC tank having excellent Q characteristic are applied to the VCO in order to obtain an excellent phase noise characteristics, and the NP core topology with excellent power consumption and waveform symmetry in the using frequency 1.5~2.0GHz was selected [3-4]. The 1.6V regulator which is tolerant to voltage variation were used instead of bias circuit that induces flicker noise [5]. As the proposed VCO in this study requires wideband frequency characteristics, cap-bank that allows an adjustment of the wideband with forth steps combined was used. Simulation result shows that output frequency range of the designed VCO corresponding to a control voltage is 1526~2077MHz(about 30.6% of center frequency). It is margin of 32MHz at lower frequency, 221MHz at higher frequency compared to the target value of 1558~1856MHz range.

B. Other Block and Layout

It is difficult the integer divider to divide the output frequency of the VCO as it operate at the fastest frequency in the entire frequency synthesizer block. Thus, a prescaler with fast operating speeds up to 5GHz[6] was used for the integer divider to divide 1/64~1/127 via 6bit control signal with D-F/F and logic gate. In order to obtain fractional division ratio, the delta-sigma modulator was designed in RTL. The RC 3rd order filter with a bandwidth of 40KHz to charge pump subsequent stage was used in order to apply a stable control voltage of the VCO, and designed in tuning possible way off-chip type from the outside. The proposed frequency synthesizer was implemented using 0.18 μm 2-poly-6metal CMOS RF analog process. The overall size of the chip is 1.0 x 0.4 mm², designed layout as shown in Figure 2. The VCO, PFD, charge pump, divider, delta-sigma modulator and bias circuit is located from the left side of the layout. For noise characteristics improvement and reduce interference between each block, a differential structure was used for the signal input stage and the output stage, and a guard-ring was used in the main block. Furthermore, the power supplies of the analog and digital blocks are separated, and also the power supplies of the digital blocks are separated according to the power consumption [5-6].

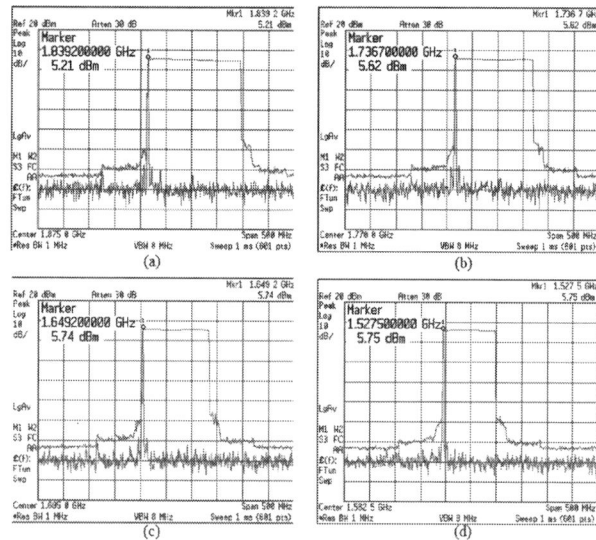

Figure 2. Layout of the designed frequency synthesizer

III. MEASUREMENT AND CONCLUSION

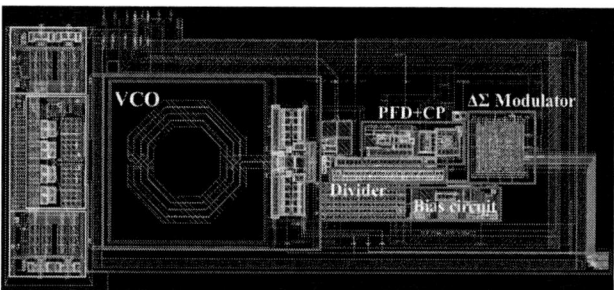

Figure 3. Measured spectrum of the proposed frequency synthesizer.

Figure 3 shows measurement result of the output signal of the frequency synthesizer via another output terminal in the transceiver chip. The output frequency is 1527~2020MHz (about 27.8% of center frequency), there is a margin of 30.5MHz at lower frequency, 164MHz at higher frequency compared to design target value 1558~1856MHz. Figure 4 shows measured result of 100Hz~1MHz away offset frequency at transmit frequency 1848MHz. At 100KHz offset was measured -98.63dBc/Hz and at 1MHz offset was measured -122.05dBc/Hz. This indicates excellent characteristics than -90dBc/Hz at 100KHz offset, -110dBc/Hz at 1MHz offset required in a typical wideband system. According to the measurement results, if complement the reference clock and the chip peripheral circuits of the test board, it is should be used widely as a transceiver with the excellent quality in the SUN system.

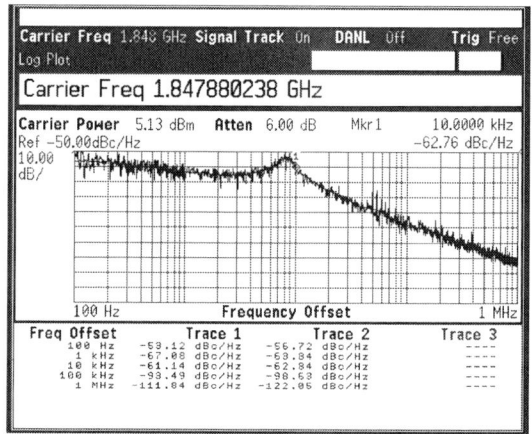

Figure 4. Measured phase noise of the proposed frequency synthesizer.

ACKNOWLEDGMENT

This work was supported by the National Research Foundation of Korea(NRF) grant funded by the Korea government(MSIP) (No.2016R1C1B1013862). Also, partially supported by IDEC and ETRI.

REFERENCES

[1] Sang-Seong Choi, Cheol-Ho Shin, and Mi-Gyeong Oh, "Wireless Transmission Technology Standardiztion Trend for Smart Utility Network," TTA Journal, vol. 113, pp. 122–131, Jan 2011.

[2] Cheol-Ho Shin, Mi-Gyeong Oh, and Sang-Seong Choi, "IEEE 802.15.4g SUN Standardization Technology Trend," National IT Industry Promotion Agency Weekly Technology Trends, vol. 1483, pp. 1–13, Feb 2011.

[3] M. Haase, V. Subramanian, T. Zhang, and A. Hamidian, "Comparison of CMOS VCO technologies," in Proc. Conference on PRIME, pp. 1–4, July 2010.

[4] B. Muer, M. Borremans, M. Steyaert, and G. Puma, "A 2GHz Low-phase-noise integrated LC-VCO set with fliker-noise upconversion minimization," IEEE Journal of Solid State Circuits, vol. 35, no. 7, pp. 1034-1038, Jul 2000.

[5] Ho-Yong Kang, Nae-Soo Kim, and Sang-Hoon Chai, "Implementation of 1.9GHz RF Frequency Synthesizer for USN Sensor Nodes," The Institute of Electronics Engineers of Korea, vol. 46, SD, no. 5, pp. 49-54, May 2009.

[6] Ho-Yong Kang, Se-Han Kim, Cheol-Sig Pyo, and Sang-Hoon Chai, "Implementation of 5.0GHz Wide Band RF Frequency Synthesizer for USN Sensor Nodes," The Institute of Electronics Engineers of Korea, vol. 48, SD, no. 4, pp. 32-38, Apr 2011.

LNA Topologies for RX Carrier Aggregation

Jusung Kim, Keunkwan Ryu, Sungchan Kim

Electronics and Controls Engineering
Hanbat National University
Daejeon, Korea
jusungkim@hanbat.ac.kr, kkryu@hanbat.ac.kr,
sckim@hanbat.ac.kr

Sanghun Lee

R & D Center
Wavepia Corporation
Gyeonggi-do, Korea
platune@wavepia.com

Abstract— **Performance of the low-noise amplifier (LNA) determines the sensitivity, impedance matching (reflection), and other critical parameters of the receiver. Carrier aggregation (CA) in LTE-Advanced and upcoming 5G requires the LNA to support multiple-outputs without degrading its dynamic range (DR) performance and thus requires the architectural changes. In this paper, several LNA topologies are presented which can support RX carrier aggregation. Each topology is meticulously characterized in its performance. Simulation results reveal that the cascode divert-switch and cascode-shutoff in common-source configuration shows comparable performance to the legacy operation.**

Keywords- Low Noise Amplifier (LNA), receiver front-end, carrier aggregation (CA), single-input multi-ouput (SIMO), multi-input multi-output (MIMO), LTE-Advanced

I. INTRODUCTION

Ever increasing data crunch in wireless communication necessitates the means to increase the data rate with a spectral efficiency. Carrier aggregation (CA) is one of the key features to increase the data rate given scarce bandwidth spectrum. CA raises the bandwidth and the capacity by combining two or more bands (inter-band CA), or several channels within the single band (intra-band CA) [1]. CA is also the technical solution to overcome the spectrum fragmentation and can be applied to a wide variety of spectrum scenarios.

Several receiver architectures at both the system and the circuit level have been proposed to support carrier aggregation [1, 2, 3]. Indispensable block in carrier aggregation receiver as well as single-carrier receiver is high performance LNA. Inductor-degenerated LNA in common-source configuration is the dominant topology for conventional receivers in 2G/3G/4G era. Inductor-degenerated LNA is widely adopted due to its simultaneous noise and power match by shifting the optimum noise impedance Z_{opt} to the desired value, but only in a narrowband around a single frequency [4]. Narrowband operation of inductor-degenerated LNA is not a problem in CA operation *per se*. However, to support inter-band as well as intra-band CA, LNA needs to have single-input multi-output (SIMO) characteristic with good impedance match, noise-figure (NF), and linearity.

In this paper, several LNA topologies suitable for RX carrier aggregation is presented. Performance of the presented LNAs is evaluated with the receiver architecture based on a single trans-conductance (gm) driving a current-mode passive mixer

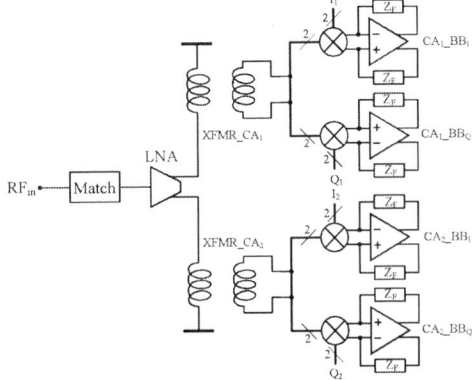

Figure 1. Receiver front-end architecture with 2 CA support.

terminated with a low-impedance trans-impedance amplifier (TIA) [5].

II. RECEIVER TEST-BENCH

Fig 1. shows the simulated receiver front-end architecture to evaluate different LNA topologies under 2 CA scenario. LTE Band 7 (2.62 - 2.69 GHz) was simulated for the comparison. L-section off-chip matching lumped elements are configured with RLC equivalent circuit parameters adopted from Murata Chip S-parameters & Impedance Library [6]. CA capable LNA is used as a single-gm of the receiver front-end. Major noise and nonlinearity (distortion) are generated from the LNA in the single-gm architecture and thus performance of different LNA topologies directly impacts the receiver performance. Conventional single-input single-output (SISO) LNA is designed for band 7 and is simulated with the test-bench without CA. Gain of the single-carrier receiver front-end as a reference is set at 56.6dB. NF and current consumption of the single-carrier receiver is 1.33dB and 7.4mA, respectively.

III. LNA TOPOLOGIES AND COMPARISON

A. 3-coil BALUN LNA

3-coil LNA in Fig. 2(a) enables CA operation with magnetically coupled inductors. LNA tank impedance is lowered by a factor of 2 due to 2 inductive coupled loads in 2 CA. Gain of LNA is reduced and NF is increased accordingly. Moreover, isolation between different CA path is limited by coupling coefficient ($k_{CA1(2) \rightarrow CA2(1)}$).

978-1-5090-3220-4/16 $31.00 © 2016 IEEE

Table 1
RECEIVER PERFORMANCE SUMMARY AND COMPARISON

	Reference (SISO)	Split-cascode		Cascode divert-switch		Cascode shut-off	
Mode	non-CA	CA	non-CA	CA	non-CA	CA	non-CA
Gain (dB)	56.6	54.2	56.7	56.1	59.5	57.1	58.9
NF (dB)	1.33	2.09	1.46	1.82	1.45	1.76	1.52
IIP3 (dBm)	-6.2	-4.2	-6.7	-4.7	-8.5	-6.7	-7.7
Current (mA)	7.4	19.8	7.4	21.3	10.8	21.3	11.1

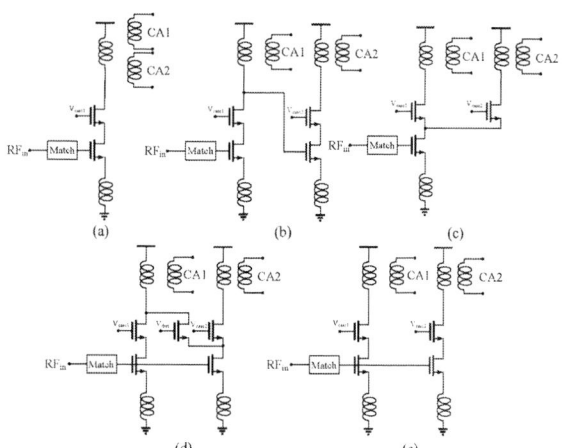

Figure 2. LNA topologies for CA. (a) 3 coil BALUN LNA. (b) 2-stage LNA. (c) Split cascode LNA. (d) Cascode divert switch LNA. (e) Cascode shut-off LNA.

B. 2-stage LNA

2-stage LNA is shown in Fig. 2(b). 2nd stage is sized smaller not to load the 1st stage of the LNA. CA1 is a single-stage design and its performance is comparable to non-CA case. On the other hand, CA2 path has 2-stage LNA and, even with large degeneration in 2nd stage, linearity is poor due to cascaded design approach [7]. NF for CA2 is primarily set by that of 1st stage design.

C. Split cascode LNA

Split cascade LNA shown in Fig. 2(c) realizes multi-outputs by signal splitting at the cascode node. Gm is shared between CA paths. Therefore, gain is reduced by $20 \cdot \log N$ (dB) without current scaling. Even with current scaling by $N \cdot X$ for N CA, gain reduction of around $10 \cdot \log N$ (dB) is inevitable due to square-root dependence of gm with respect to the current consumption in CMOS device (short channel effects are ignored). Shared gm dictates small variation in Z_{in} for different operation modes (non-CA and CA). NF is degraded due to significant noise contribution from the cascode devices and larger contribution from TIA due to smaller LNA gain.

D. Cascode divert switch LNA

Fig. 2(d) shows the cascode divert switch LNA. Gm is separated between different CA paths. However, both gm are ON for non-CA and CA modes in order to minimize the change in impedance matching (S_{11}). Noise contribution due to the cascode device is the same as a conventional LNA without sharing gm as in the split-cascode LNA. Divert-switch is utilized to improve gm in non-CA mode (CA1 ON, CA2 OFF). Asymmetric tank loading due to the divert-switch incur separate optimization process for the transformer loading.

E. Cascode shut-off LNA

From the circuit configuration, cascode shut-off LNA in Fig. 2(e) is similar to the cascade divert switch in Fig. 2(d) except the divert switch transistor. In CA mode, their operation is also

similar with slightly better isolation with cascode shut-off LNA due to the void of divert switch. In non-CA operation, gate bias of 2nd gm transistor is properly ON. At the same time, cascode bias of 2nd CA is shut-off to avoid throwing additional current. With this bias scheme, 2nd gm transistor operates in saturation and triode regions in CA and non-CA, respectively. Impedance matching (S_{11}) is slightly disturbed due to the change in C_{gs} for different CA modes.

IV. SIMULATION RESULTS & CONLUSION

The circuit was designed and simulated with TSMC 65nm CMOS technology. Simulated performance of several LNA topologies within the context of a single-gm receiver front-end is summarized in Table I and compared with conventional SISO receiver. Both the cascode divert-switch and the cascode shut-off topologies shows promising results with the support of SIMO (CA) receiver. Simulation results exhibit < 0.5dB noise penalty in CA operation. Linearity performance (IIP_3) is comparable to the legacy operation.

ACKNOWLEDGMENT

This work was supported by the National Research Foundation of Korea (NRF) grant funded by the Korea Government (MSIP) (No. 2016R1C1B1012042)

REFERENCES

[1] S. C. Hwu and B. Razavi, "An RF Receiver for Intra-Band Carrier Aggregation," *IEEE Journal of Solid-State Circuits*, vol. 50, no. 4, pp. 946–961, April 2015.

[2] C. Wu, Y. Wang, B. Nikolic, and C. Hull, "A Passive-Mixer-First Receiver with LO Leakage Suppression, 2.6dB NF, >15dBm Wide-Band IIP3, 66dB IRR Supporting Non-contiguous Carrier Aggregation," *IEEE RFIC Symp.*, May 2015, pp. 155-158.

[3] L. Sundstrom *et al.*, "A receiver for LTE rel-11 and beyond supporting non-contiguous carrier aggregation," *IEEE ISSCC.*, Feb 2013, pp. 336-337.

[4] J. Kim, S. Hoyos, and J. Silva-Martinez, "Wideband Common-Gate CMOS LNA Employing Dual Negative Feedback With Simultaneous Noise, Gain, and Bandwidth Optimization," *IEEE Microwave Theory and Techniques*, vol. 58, no. 9, pp. 2340–2351, September 2010.

[5] J. Kim and J. Silva-Martinez, "Low-Power, Low-Cost CMOS Direct-Conversion Receiver Front-End for Multistandard Applications," *IEEE Journal of Solid-State Circuits*, vol. 48, no. 9, pp. 2090–2103, September 2013.

[6] Murata, "Murata Chip S-Parameter & Impedance Library," [Online]. Available: http://www.murata.com/tool/download/mcsil.

[7] B. Razavi, *RF Microelectronics*. Englewood Cliffs, NF: Prentice-Hall, 1997.

A design of Dual-band Smart Tag

Jin-ho Kim and Yong Moon

Department of Electronic Engineering, Soongsil University, Korea

Email : jh4747h@naver.com, moony@ssu.ac.kr

Abstract— The dual-band smart tag is designed and is fabricated using a 0.18 μm 1-Poly 4-Metal CMOS Process, and the area is 5mm × 5mm. The dual-band smart tag can recognize and demodulate the frequency bands of both UHF band (868 ~ 956 MHz) and HF band (13.56 MHz). The Digital block for verification is programmed in Arduino Uno board. Consequently, the dual-band smart tag communicates between the HF/UHF reader and the tag.

Keywords—dual-band smart tag; RFID; NFC;

I. INTRODUCTION

RFID(Radio-Frequency Identification), the technology to recognize information in the radio-wave of near / far distance, uses various frequency bands. HFID(High-Frequency Identification) named NFC(Near Field Communication) uses the frequency band of 13.56 MHz, and it has been applied to the access control systems, electronic cash, transportation card, and smart card. UHFID(Ultra High-Frequency Identification) uses the frequency band of between 868 MHz and 956 MHz, and it has been applied to the distribution, logistics, SCM(Supply chain management), and automatic toll collection. The current distribution and logistics system uses HF and UHF systems concurrently. It is very ineffective, because current system uses different tags in the same radio system. In this paper, we have designed the Smart Tag for the transmission of the two bands and the verification of the Digital block has been completed by using the Arduino Uno board.

II. THE DESIGN OF DUAL-BAND SMART TAG

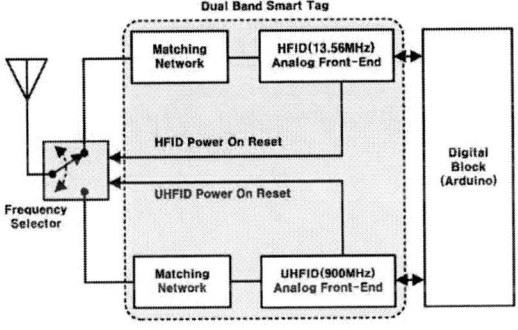

Figure 1. Block diagram of the proposed dual-band smart tag

Figure 1. shows the block diagram of the proposed dual-band smart tag. It consists of the analog front-end and the frequency selector. The analog front-end transmits and receives individual data of both HF and UHF bands. The frequency selector distinguishes between HF and UHF signals, each the analog front-end is operated according to the output voltage of the frequency selector.

A. The analog front-end

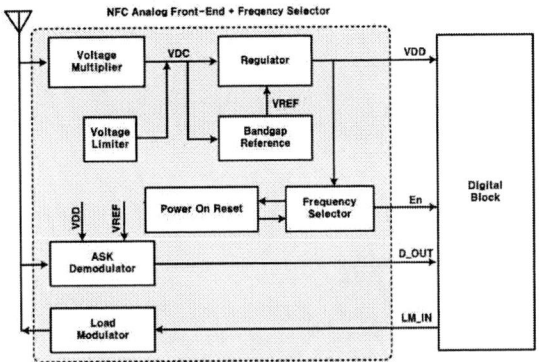

Figure 2. Block diagram of the proposed analog front-end

Figure 2. shows the block diagram of the proposed analog front-end. The analog front-end consists of the power supply part, the ASK demodulator and the load modulator. Using the signal received from the reader, the power supply part generates the energy for supplying the current to the internal circuit. The voltage multiplier generates the large DC voltage due to the multi-stage structure. The voltage limiter is required because large DC voltage may cause damage to the internal circuitry. The bandgap reference and the regulator generate stable DC voltage to supply the VDD for other circuits. The ASK demodulator generates digital data and transmits them to the digital block, and the load modulator sends the reader response signal from the digital block. The power supply part composed of the DC rectifier to convert the AC signal to DC signal, the voltage multiplier, the bandgap reference, the regulator, the voltage limiter to prevent the damage of internal circuit, and the power-on reset circuit to reset digital block after data transmission.

B. The frequency selector

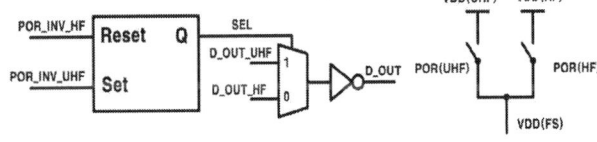

Figure 3. The proposed frequency selector circuit

Figure 3. shows the proposed frequency selector circuit. Two signals from the dual-band antenna enter the frequency selector. It distinguishes between HF and UHF signals. If the frequency selector decides the signal as the HF signal, the HF analog front-end is turned on. If the frequency selector decides the signal as the UHF signal, the UHF analog front-end is turned on.

III. MEASUREMENT RESULT

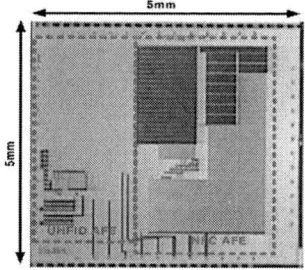

Figure 4. Photograph of the dual-band smart-tag chip

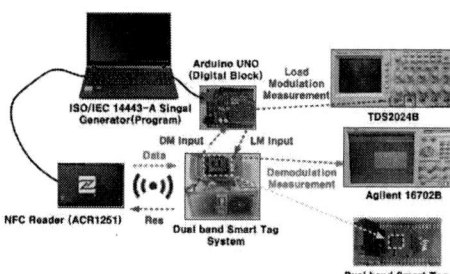

Figure 5. Measurement Environment

Figure 4. shows the photograph of the dual-band smart tag chip. We used a 0.18μm 1-Poly 4-Metal CMOS Process, and the area is 5mm×5mm. Figure 5. shows the measurement environment to measure the Test board. The NFC reader (ACR1251) generates the input signal of the ASK 100% satisfied the ISO/IEC-14443A standard. After receiving the input signal, we measured the test board output. The amplitude changes according to the response signal and the demodulated signal is measured using the TDS2024B (Oscilloscope).

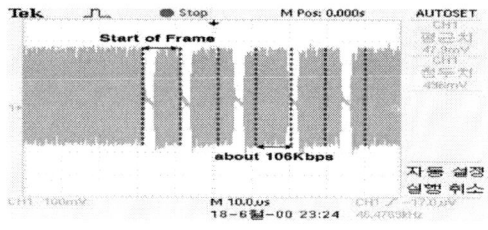

Figure 6. Measurement of the reader signal

Figure 6. shows the result of measurement of the ALL_REQ signal requesting the response to the digital block. Figure 7. shows the demodulation signal after receiving the ALL_REQ. The demodulation has been completed, the digital signal enter into the input of the Arduino Uno, and the program distinguishes both High("1") and Low("0"), and sends the response signal of 847KHz bit rate with Manchester coding.

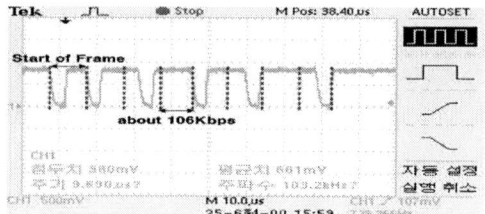

Figure 7. Measurement of the demodulation signal

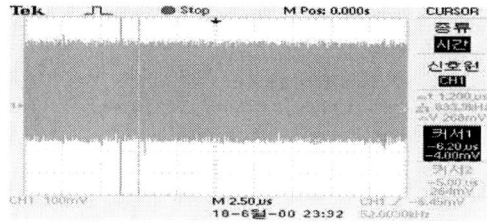

Figure 8. The signal prior to sending the response signal

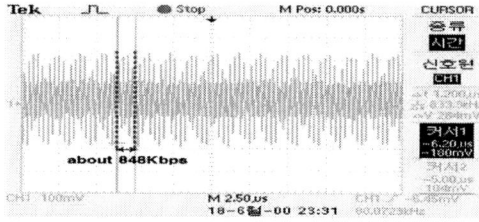

Figure 9. The signal after sending the response signal

Figure 8. shows the signal between the reader and the antenna prior to sending the response signal. Figure 9. shows the signal between the read and the antenna after sending the response signal. The amplitude change of signal can be seen from the measurement when the reader and the tag communicate.

IV. CONCLUSION

We have designed the dual-band smart tag. According to the induced signal, the test board produces the demodulated signals. We have integrated the Analog Front-End of UHF band and HF band in a single chip. The dual-band smart tag will replace the previous two tags and would be used widely in distribution and logistics systems.

ACKNOWLEDGMENT

This work was supported by the Human Resources Development program (No.20144030200600) of the Korea Institute of Energy Technology Evaluation and Planning (KETEP) grant funded by the Korea government Ministry of Trade, Industry and Energy.

REFERENCES

[1] Junghyun Cho and Shiho Kim, "Design of single-chip NFC transceiver,"Journal of The Institute of Electronics Engineers of Korea (IEEK), Vol.44, No.1, pp.68-75, Jan. 2007.

Design of 28GHz CMOS Phased Array T/R Circuits for 3-Dimensional Beamforming Applications

Sungjin Shin and Hyunchol Shin

High-Speed Integrated Circuits and Systems Lab.
Kwangwoon University, Seoul, Korea

Abstract— **A 28GHz CMOS phased array T/R circuit is designed for 3-dimensional beamforming system in 5G millimeter-wave communication applications. The T/R circuit integrates a phase shifter, a digital step attenuator and a SP8T switch. It provides 64-state attenuations in 0.5dB step via 6-bit attenuator circuit, and 32-state phase shifts in 11.25° step via 5-bit phase shift circuit, and overall a 14dB insertion loss, 7.6dBm input-referred P1dB.**

Keywords-component; 5G, Phase shifter, Attenuator, Beamforming System

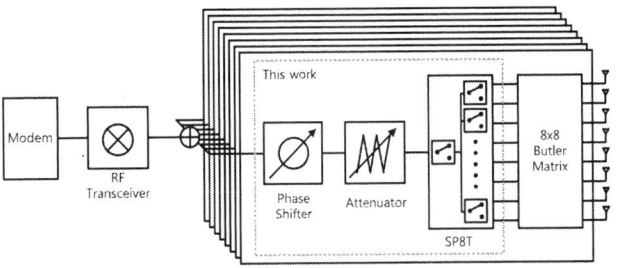

Fig. 1 Three dimensional beamforming system architecture

I. INTRODUCTION

Among many enabling technologies for 5G wireless communications, millimeter-wave 3-dimensional beamforming array antenna system is considered essential because it can compensate the high path loss at the high frequency and support small-cell operation [1]. Beamforming can be categorized in two types, one is the phase array continuous type, and the other is the switching type. In order to support both types in a single system, we need a 28GHz integrated T/R circuit that consists of phase shifter, attenuator, and switch, preferably in CMOS. However, previous works at this frequency region did not provide sufficient integration level, that is only supporting attenuator [2-4] or phase shifter [5], or operate in much lower frequency region [6-8].

II. CIRCUIT DESIGN AND RESULTS

This work presents a CMOS T/R core circuit element for 28-GHz 3-D beamforming in 5G millimeter-wave radio communications. It gives high integration level by including a phase shifter, a digital step attenuator, and SP8T switch, all in 65nm CMOS and operating in 28GHz. By combining eight of this element circuit and 8x8 Butler Matrix, this configuration can support a phased-array continuous beamforming in azimuth direction and a switched beamforming in elevation direction, which is suited for 3-D beamforming CMOS radio for 5G mobile applications.

Fig. 1 is the designed 3D beamforming system architecture for millimeter-wave 5G transceiver. The baseband signal from MODEM is up-converted to RF via the 28GHz RF transceiver, and splits into eight phased array T/R elements. In each element, the RF signal passes through a phase shifter and a attenuator, and subsequently splits again into eight path by using SP8T switch. Finally, the signal is fed to the 8x8 Butler matrix. By combining eight of these elements in parallel, a phased array type continuous beamforming can be done by controlling the phase and amplitude, and a switched beamforming can be done by controlling the SP8T-Butler pair.

Fig. 2 is the phase shifter circuit schematic, in which the FET widths and the resistance values are marked together. The phase shifter is based on Pi-type circuit for 11.25°, 22.5°, 45°, and 90°. On the other hand, the 180° shift is obtained by switching between the high-pass and low-pass LC circuits, which helps lessen the insertion loss otherwise caused by the series FET in the Pi-type circuit.

Digital step attenuator provides 6-bit 64-state attenuations in 0.5dB step. Fig. 3 shows the schematic. T-type circuit is chosen for small attenuation levels of 0.5, 1, 2, and 4 dB. It is because, compared to other conventional Pi-type or bridged-T type circuit, the T-type circuit can be realized with a smaller resistance value for a given attenuation level. On the other hand, Pi-type circuit is chosen for the higher attenuation levels of 8 and 16 dB, because the rather high on-resistance of the shunt FET can be tolerated here.

SP8T switch is designed by connecting a SPDT and two SP4T switches in series, as shown in Fig. 4. Compared to the conventional way of a direct SP8T implementation, this hybrid configuration is found to give better isolation characteristics because of the additional isolation given by the first SPDT switch.

The top circuit is designed by connecting the phase shifter, attenuator, and SP8T switch. The full-path performances are characterized via RF circuit simulations and shown in Fig. 5. Fig. 5(a) is the total attenuation level and Fig. 5(b) is the total phase shift. At 28GHz, the insertion loss is 14 dB, port-to-port isolation is at least 27 dB, and the return loss is 26 dB. Table I compares the performance of this work with other works. As shown, this work provides higher integration level including phase shifter, attenuator, and switch in CMOS at the higher operation frequency of 28GHz.

978-1-5090-3220-4/16 $31.00 © 2016 IEEE

TABLE I. PERFORMANCE COMPARISON

	Frequency (GHz)	Technology	Integration Level	Insertion Loss (dB)	Return Loss (dB)	Atten. Step (dB)	Number of Atten. States	Phase Step (degree)	Number of Phase States	Input P1dB (dBm)
This Work	28	65nm CMOS	Phase Shifter+Attenuator+SP8T	14	26	0.5	64(6bit)	11.25	32(5bit)	7.6
[2]	22-29/57-64	0.18um Bi-CMOS	Attenuator	7.9/11.1	10	1	16(4bit)	N.A.	N.A.	14/11
[3]	10-67	0.18um Bi-CMOS	Attenuator	8.4-15.2	8.7	2.9±0.1	16(4bit)	N.A.	N.A.	14
[4]	DC-2.5	0.13um CMOS	Attenuator	2.6	8.2	N.A.	N.A.	N.A.	N.A.	2.5
[5]	75-85	65nm CMOS	Phase Shifter	25±2.1	8	N.A.	N.A.	22.5	16(4bit)	N.A.
[6]	8.5-10	0.18um CMOS	Phase Shifter +Attenuator+Amp	11	10	0.5	64(6bit)	5.625	64(6bit)	11.5
[7]	8.5-10.5	0.13um CMOS	BDGA+Phase Shifter+Attenuator	N.A.	N.A.	1	32(5bit)	5.625	64(6bit)	6.5
[8]	0.8-1.8	0.7um GaAs	Phase Shifter+Attenuator+SPDT	14.5	15	0.5	64(6bit)	5.625	64(6bit)	N.A.

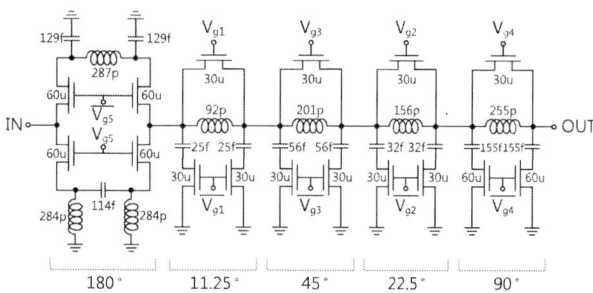

Fig. 2 5-bit phase shifter schematic

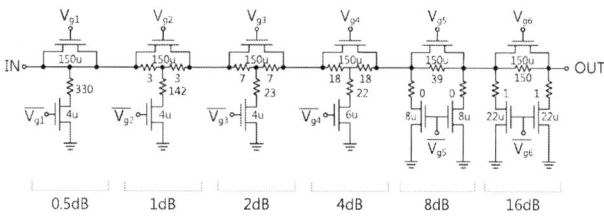

Fig. 3 6-bit attenuator schematic

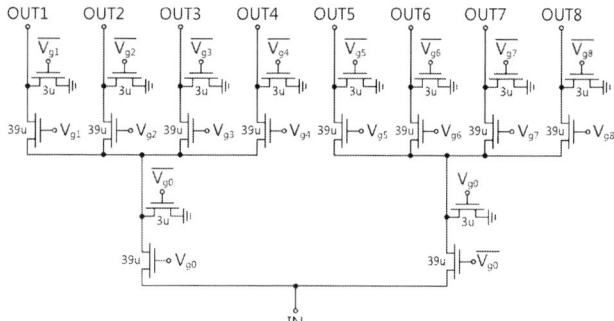

Fig. 4 SP8T switch schematic

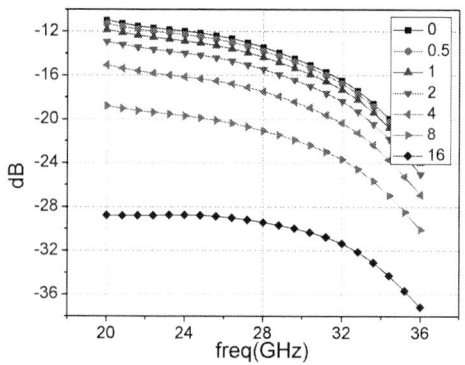

(a)

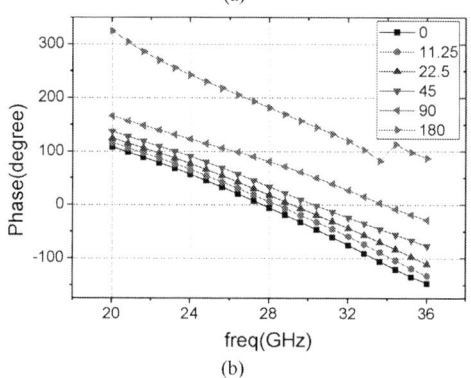

(b)

Fig. 5 Simulation results. (a) Attenuation level. (b) Phase shift

ACKNOWLEDGMENT

This work was supported by ICT R&D Program of MSIP/IITP [14-911-01-001, Development of quasi-millimeter-wave channel-adaptive antennas and transceivers]

REFERENCES

[1] W. Roh *et al.*, IEEE Communications Magazine, pp.106-113, Feb. 2014.

[2] J. Bae *et al.*, IEEE Trans. Microw. Theory Tech., vol. 64, no.6, pp.1867-1875, Jun. 2016.

[3] J. Bae *et al.*, IEEE Trans. Microw. Theory Tech., vol. 61, no.12, pp. 4118-4129, Dec. 2013.

[4] H. Dogan *et al.*, in Symp. VLSI circuits Dig. Tech. Papers, pp. 90-93, Jun. 2005.

[5] H. Lee *et al.*, IEEE Trans. Circuits Syst. II: Exp. Briefs, vol. 62, no.1, pp. 1-5, Jan. 2015.

[6] K. Gharibdoust *et al.*, IEEE Trans. Microw. Theory Tech., vol. 60, no.7, pp. 2192-2202, Jul. 2012.

[7] S. Sim *et al.*, IEEE Trans. Microw. Theory Tech., vol.61, no.1, pp.562-569, Jan. 2013.

[8] N. D. Doddamani *et al.*, in Proc. of ICSCN 2007, pp. 302-307

A study of META-Voltage Controlled Oscillator and Prescaler using 65nm CMOS Process

META-VCO and Prescaler using 65nm CMOS Precess

No yong Kwon, Bo ra Kim, Yong Moon
School of Electronic Engineering, Soongsil University, Seoul, Korea
e-mail : kny0572@naver.com, qnfm10@naver.com, moony@ssu.ac.kr

Abstract— The VCO (Voltage Controlled Oscillator) and the high speed prescaler are designed using 65nm CMOS technology with the frequency of 28.5GHz 5G mobile communication system. The simulation result show that the VCO has 28.4~28.8GHz tuning range and the prescaler divides the VCO output. The phase noise of the VCO is -173.75dBc/Hz at 1MHz and -181.43dBc/Hz at 10MHz offset frequency.

VCO; Prescaler; 5G; phase noise; (key words)

I. INTRODUCTION

Due to the development of the mobile communication device, the wireless data usage is increasing. The data usage is beyond the range that can be accommodated in the existing 4G networks (LTE / WiBro) after 2020. So, 5G-network are being developed as next-generation mobile communications. 5G network transmits the data fast by using the ultra-high frequency band of 20GHz ~ 40GHz. The PLL is used to generate ultra-high frequency stably, the block diagram is shown in Figure 1. The voltage controlled oscillator to generate high frequency exists in the PLL. The frequency divider block is used to lower ultra-high-frequency to low-frequency. In this paper, the voltage-controlled oscillator and prescaler were designed using 65nm CMOS process.

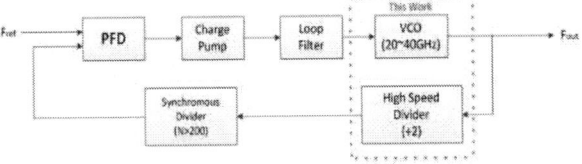

Figure1. 29 GHz PLL for the wireless communication system.

II. EASE OF USE

A. Meta-structure for the proposed VCO

Meta-material is used to have special electrical properties that can not be found in nature. In this paper, we use the SRR structure by controlling the variables of the SRR structure to obtain the LC resonance characteristics. Figure 2 shows the proposed meta-structure resonator array.

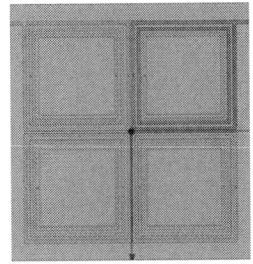

Figure 2. LC resonator structure of the SRR array (top view)

B. META-VCO

Figure 3 show the META-VCO circuit using the meta-material designed using HFSS. The META-VCO is composed of meta-material for the inductor role, NMOS cross coupled pair having negative resistance, and two varactors (C_{VAR1} and C_{VAR2}). It was designed to control the oscillation frequency by adjusting V_{CTRL}. Because meta-material has the parasitic capacitance, the parasitic capacitance of META-VCO is larger than the conventional LC-VCO. So we have designed the meta-material having high operating frequency to obtain operating frequency of 28.5GHz. R_{ISOL} is very small resistance less than 15Ω. It prevents short circuit of the OSCP and OSCM node to the VDD.

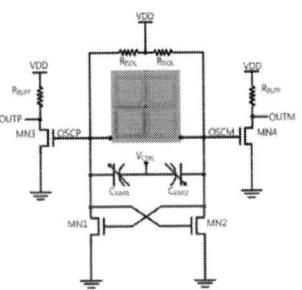

Figure 3. META-VCO circuit

C. Prescaler

The VCO frequency (F_{VCO}) is divided to several tens of MHz in PLL.

The CML-divider consists of two latches. The sample pair in the master latch and the hold pair in the slave latch are operated when CLK is 'high'. The hold pair in the master

---

Identify applicable sponsor/s here. If no sponsors, delete this text box.
(sponsors)

978-1-5090-3220-4/16 $31.00 © 2016 IEEE   ISOCC 2016

latch and the sample pair in the slave latch are operated when CLK is 'low'. Since the output is connected to the input as negative feedback, CLK frequency should be in the operating frequency range of the frequency divider, so CML-divider outputs the frequency divided signal from the CLK.

In this paper, we proposed the structure as shown in Figure 4, which replaces the resistance error in the layout of PMOS. The proposed architecture is composed of sample pair and hold pair. The sample pair is connected PMOS but hold pair is not connected. The sample pair was designed with twice W / L ratio of the hold-pair to send larger current to the sample pair.

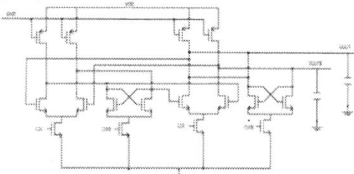

Figure 4. The proposed structure of CML divider

III. IMPLEMENTATION

Figure 5 shows the resonance frequency characteristic. S_{11} is -26.4dB and S_{21} is -12.1dB. Figure 6 shows The META-VCO simulation results. The META-VCO has the operating range of 28.4 ~ 28.8GHz by controlling V_{CTRL}. The phase noise at 1MHz offset frequency is -173.75dBc / Hz, 10MHz offset frequency is -181.43dBc / Hz.

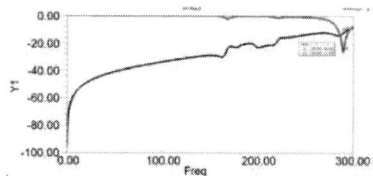

Figure 5. HFSS resonant characteristics graph of the proposed META structure

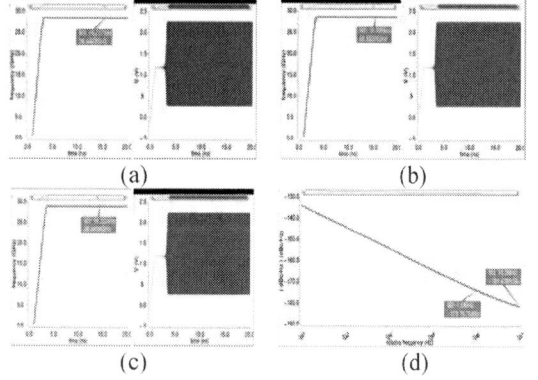

Figure 6. (a) V_{CTRL} = 0V (b) V_{CTRL} = 0.6V (c) V_{CTRL} = 1.2V (d) Phase noise

The output of the VCO is 28.8GHz when V_{CTRL} = 1.2V. Figure 7 shows the waveform of the VCO and the frequency divider. Output waveform after passing the divider is shown in figure 6, and the frequency is about 14GHz. The final output waveform of the divider has the output swing of 440mV.

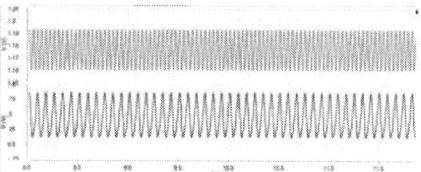

Figure 7. Simulation results prescaler

IV. CONCLUSIONS AND FUTURE RESEARCH DIRECTIONS

We designed the VCO using meta-structure based on the CMOS process in this study and the high-speed divider for dividing the frequencies using with CMOS technology. It was confirmed by the simulation for high-speed data transmission at high frequency. In this study META-VCO and CML are applicable for PLL which is operating at mm-wave frequency range.

ACKNOWLEDGMENTS

This research was supported by the MSIP(Ministry of Science, ICT and Future Planning), Korea, under the ITRC(Information Technology Research Center) support program (IITP-2016-H8501-16-1010) supervised by the IITP(Institute for Information & communications Technology Promotion)

Reference

[1] Jongsuk Lee and Yong Moon, "A study on Voltage Controlled Oscillator and High Speed 1/4 Frequency Divider using 65nm CMOS Process", Journal of The Institute of Electronics and Information Engineers, Vol.51, No.11, pp. 107-113, Nov, 2014

[2] Takayuki Sekiguchi , Shuhei Amakawa, Noboru Ishihara, and Kazuya Masu, "An 8.9mW 25Gb/s Inductorless 1:4 DEMUX in 90nm CMOS", SoC Design Conference (ISOCC), pp. 404-407, 2009.

[3] Muhammad Usama and Tad. A. Kwasniewski , "A 40-GHz Frequency Divider in 90-nm CMOS Technology", IEEE North-East Workshop on Circuits and Systems, pp. 41-43, 2006.

[4] Jung-Woong Park, Se-Hyuk Ahn, Hye-Im Jeong, and Nam-Soo Kim, "High-speed CMOS Frequency Divider with Inductive Peaking Technique", Transactions on Electrical and Electronic Materials, Vol. 15, No. 6, pp. 299-314, Dec, 2014

[5] Hyoungjun Kim and Chulhun Seo, "Resonant Wireless Power Transfer System with High Efficiency using Metamaterial Cover," Journal of The Institute of Electronics and Information Engineers, Vol.51, No.1, pp47-51, Jan. 2014.

[6] Choi Jaewon and Seo Chulhun, "Low Phase Noise Push-Push VCO using Microstrip Square Open Loop Multiple Split Ring Resonator and Rat Race Coupler," IEEE, Asia-Pacific Microwave Conference(APMC), pp.394-397, Dec. 2010.

Recent Advances in TSV Inductors for 3D IC Technology

Bruce Kim
Department of Electrical Engineering
City University of New York, CCNY
New York, U.S.A.
Bruce.Kim@ieee.org

Sang-Bock Cho
Department of Electrical Engineering
Ulsan University
Ulsan, South Korea
sbcho@ulsan.ac.kr

Abstract— **In this paper, we present a summary of 3D TSV inductors and their advances in RF applications. We describe a new inductor structure that uses fewer ground planes with the same functionality. Results were derived from a 3D full-wave simulation performed up to 2 GHz.**

Keywords; 3D inductor, TSV, 3D full-wave simulation.

I. INTRODUCTION

The current trend towards Internet of Things (IOT), System in Package (SiP) and Package-on-Package (PoP) requires meeting the power requirements of heterogeneous technologies while maintaining minimal package size. 3D chip stacking has emerged as a potential solution due to its high-density integration in a 3D power electronics packaging regime. As an integral part of many power electronics applications, TSV-based inductors are becoming a popular choice because of their high inductance density due to the reduced on-chip footprint compared to conventional planar inductors. Depending on the requirement, the values of these inductors could range from a few nanohenries to hundreds of microhenries. Small inductors with a high quality factor are mainly used for RF filter applications, whereas large inductors are used in power electronics packaging.

For high inductance it is necessary to use ferromagnetic materials. A conventional ferromagnetic metal core like nickel could offer high permeability, which can help to boost inductance. However, the magnetic field lines within a metal core induce eddy current, which can have multiple adverse effects in power electronics packaging. For example, it has long been known that this current can increase the resistance in transformer winding [1]. Eddy current can also heat up the core of the inductor, which impedes the heat sink process in 3D packaging. One way to decrease the eddy current is to pattern and laminate the core block into multiple segments orthogonal to the direction of the magnetic field line. Another method is to increase the resistivity of the core material so that the eddy current is limited to a very small magnitude.

II. AVAILABLE 3D INDUCTORS

Researchers have created various 3D inductors. 3D Glass Solutions in New Mexico, USA, manufactures 3D inductors in glass substrate, illustrated in Fig. 1.

High-Q inductors from 3D Glass Solutions substantially reduce form factors and offer improved performance over other competitive technologies. The technology provides unique inductor solutions for both high-power and high-quality factor designs for microwave and RF frequencies. [2]

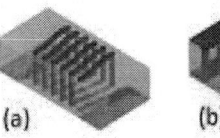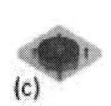

Fig. 1. 3D inductors in glass substrates from 3D Glass Solutions in New Mexico, USA for inductance values up to (a) 16nH, (b) 8nH, and (c) 12nH (www.3dglasssolutions.com).

Yousef et al. have produced 3D inductors using TSV. They propose 3D radio frequency (RF) integrated inductors with high inductance (L) values and enhanced quality factors (Q). The 3D inductors were designed in TSMC 0.18-µm technology and simulated using Momentum [3]. Fig. 2 shows their 3D monolithic inductor.

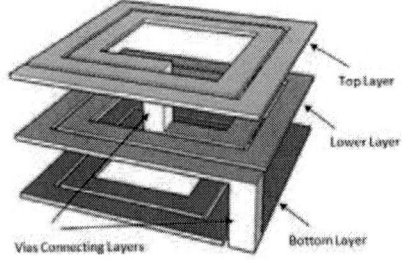

Fig. 2. 3D Monolithic inductor for RF applications [3].

Kim et al. provide TSV 3D inductors that are wound, as shown in Fig. 3 [4]. They describe the design and analysis of 3D TSV inductors for integrated sensor applications.

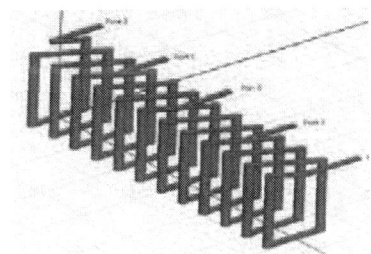

Fig. 3. 3D TSV inductor structure for RF applications [4].

Fig. 4 depicts an improved 3D inductor structure using magnetic materials [5]. In that paper, an inductance value of

978-1-5090-3220-4/16 $31.00 © 2016 IEEE

200nH is achieved up to 100MHz. The 3D inductor is designed for lower frequency applications.

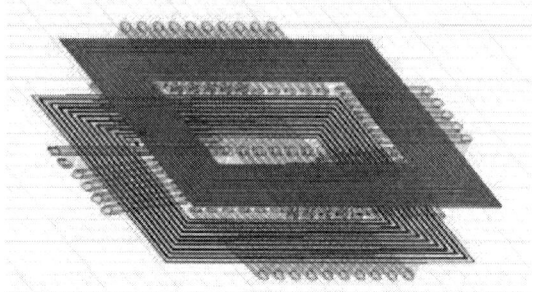

Fig. 4. Inductor with patterned nickel core [5].

A single TSV can be presented as a lumped parameter model with respect to its reference ground plane as in Fig. 5. There are two kinds of substrate coupling for the TSVs: (1) coupling of each TSV with its reference ground plane; (2) self-coupling between the two nearest TSVs of the inductor. However, apart from primary coupling, there is also secondary coupling among the distant TSVs, which can be ignored depending upon the relative contributions of the primary couplings [6].

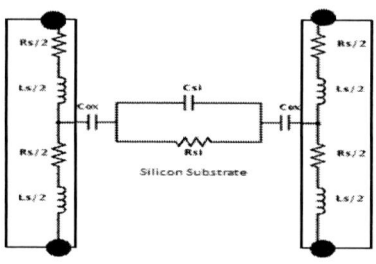

Fig. 5. Lumped parameter model of a TSV with respect to the ground plane [6].

For the different TSV structures in [6], the dimensions considered are shown in Table 1. The conductance for silicon substrate was considered as 2.5 S/m. The entire structure was confined within a 1.5 mm x 1.5 mm x 0.8 mm air radiation boundary. The dimension of the boundary was confirmed upon the convergence of the simulation results using a commercial EM simulation tool.

TABLE 1.
DIMENSIONS OF VARIOUS DESIGN PARAMETERS [6].

Design Parameter	Dimension (µm)
Radius of TSV	7.5
Dielectric liner thickness	1
Length of TSV	200
Interconnect length	300
Interconnect width	10
Interconnect thickness	8
Pitch (p) between two loops	100
Ground TSV distance for A& C	80
Ground TSV distance for B	50

Fig. 6 illustrates simulated inductor values for differently structured ground planes [6].

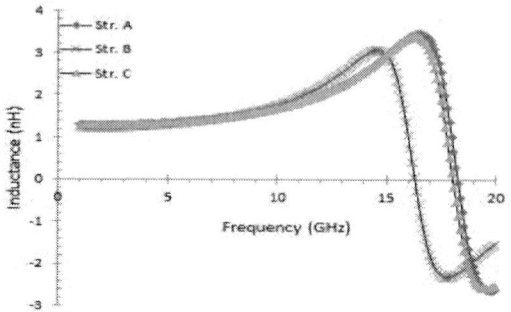

Fig. 6. Inductance of different structures with 2 loops.

SUMMARY

There have been several different 3D inductors for RF and millimeter applications. We have worked on TSV inductors that provide inductance of 4.52 nH at 2 GHz and a quality factor of 9.1 at 2 GHz. This could be achieved with an inductance density of about 65 nH/mm². It can be concluded that with decreased pitch and increased ground plane distance, high inductance values can be realized at the cost of inductor area. Furthermore, using specific geometry, the number of TSVs can be reduced with no change in functionality.

ACKNOWLEDGMENT

We would like to thank S. Mondal and J. Gamboa for providing simulation results.

REFERENCES

[1] M.S. Kim, R.P. Markondeya, Z. Wu, V. Sundaram, and R. Tummala, "Innovative electrical thermal co-design of ultra-high Q TPV-based 3D inductors in glass packages," in *2016 IEEE Electronic Components and Technology Conf.*, pp. 2384-2388.

[2] J. Kim, R. Shenoy, K. Lai, and J. Kim, "High-Q 3D RF solenoid inductors in glass," in *2014 IEEE Radio Frequency Integrated Circuits Symp.*, pp. 199-200.

[3] K. Yousef, H. Jia, A. Allam, R. Pokharel, M. Ragab, and K. Yoshida, "Design of 3D integrated inductors for RFICs," in *2012 Japan-Egypt Conf on Electronics, Communications and Computers*, pp. 22-25.

[4] B. Kim, S. Mondal, and S.H. Noh, "Tunable 3D TSV-based inductor for integrated sensors," in *2014 IEEE Electronics Packaging Technology Conf.*, pp. 322-325.

[5] S. Mondal, J. Gamboa, and B. Kim, "Development of TSV-based inductors in power electronics packaging," presented at the IEEE Int. Conf. on Electronics Packaging, Chnagsha China, Aug. 2015.

[6] B. Kim, S. Mondal, S-B. Cho, and J. Gamboa, "A novel TSV inductor structure for RF applications," presented at IEEE ECTC, May 2015.

Approximate Stochastic Computing (ASC) for Image Processing Applications

Ramu Seva[1], Prashanthi Metku[1], Kyung Ki Kim[2], Yong-Bin Kim[3] and Minsu Choi[1]

[1]Dept of ECE, Missouri Univ of Science & Technology, Rolla, MO, USA, {rs2k6,pmcmc,choim}@mst.edu
[2]Dept of Electronic Eng., Daegu University, Gyeongsan, Korea, kkkim@daegu.ac.kr
[3]Dept of ECE, Northeastern University, Boston, MA, USA, ybk@ece.neu.edu

Abstract—SC (stochastic computation) has been found to be very advantageous in image processing applications because of its lower area consumption and low-power operation. However, one of the major issues with the SC is its long run-time requirement for accurate results. In this paper, a new technique called the approximate stochastic computing (ASC) approach called the approximate stochastic computing (ASC) focusing on image processing applications is proposed to reduce the computation time of a SC by a factor of 16 at a trade-off of an error percentage of 3.13% in the absolute stochastic value ($[0, 1)$) computed. The proposed technique considers only the first four MSBs of the image pixel value for SC, which introduce a maximum error of 6.25% in the stochastic output. Attempts have been made to reduce this error to 3.13% by linearly increasing the clock cycles from 16 to 17 rather than exponentially (ex: $32, 64, 128, 256...$). Experimental results from SC edge detection circuit indicate that this technique is a promising approach for efficient approximate image processing.

I. INTRODUCTION

SC has its roots from 1960's and it is used for probability representation using digital bit streams [1, 2]. SC has been successfully applied to many applications like image processing, neural networks, LDPC coders, and factor graphs [3]. However, the extensive use of stochastic computation is limited because of it's long run-time requirement and inherant inaccuracy. Recent improvements have mainly focused on improving the accuracy and performance of the stochastic circuits by sharing consecutive bit streams, sharing the stochastic number generators, exploiting the correlation, and using the spectral transform approach for stochastic circuit synthesis [4, 5, 6, 7]. In this paper, a new approach called Approximate Stochastic Computing (ASC) to decrease the computation time has been proposed and analyzed. The proposed ASC is motivated by certain applications, such as audio, video and image processing, where an approximate or less than optimal solution is acceptable in lieu of smaller hardware circuit and faster operation. The proposed ASC technique has been validated by a specific image processing application called edge detection in this paper, although it can be used for various applications.

This paper is organized as follows. Section II gives the background of the proposed design approach and presents analytic observations supporting the proposed approach. Section III discusses about the image processing application implemented in this paper. Then, the validity of the proposed approach is demonstrated by using edge detection image processing application. Finally, Section IV makes the conclusion.

II. APPROXIMATE STOCHASTIC COMPUTING FOR EDGE DETECTION

Image processing belongs to the class of applications which demonstrate inherent error resilience, where approx-imate computing techniques can be used to design efficient digital systems [8]. Approximate computing (AC) is different from SC in a way that it does not involve assumptions and circuits involve deterministic designs rather than probabilistic designs implemented in SC. AC uses statistical properties of data and algorithms to trade quality for energy reduction and/or faster operation [9].

In this work, we combine both the AC and SC to build an ASC edge detection circuit which reduces the computation of a SC and provide area/speed efficient design with an error bound of 3.13%. As an input to the proposed ASC edge detection circuit, a grayscale bitmap image has been used where each pixel's value is represented in $8b$ binary number. Then, an approximation of pixel value is initially done where the pixel value of an image (ranges from 0 to 255) represented by 8 bit length is truncated to 4 bits (ranges from 0 to 16) by considering the first four MSBs of the binary value.

The maximum weight of a binary value is represented by always the MSBs in a binary representation. The LSBs always have a smaller percentage of the weight. This paper states that what ever may be the length of the binary value, the first four most significant bits always represent more than 90% of the total value of the number in a stochastic implementation.

By ignoring the first four LSBs an error percentage of 100% is recorded for smaller binary values with 4 MSBs are 0000_2, but when the absolute error after converting it to a stochastic bit stream, it does not exceed beyond 6.25% for a stochastic bit length upto $16, 384$ bits. Fig. 1 shows the saturation of the absolute error as the length of the stochastic bit stream increases. By this, one can conclude that though an error of 100% for certain binary input values is introduced the overall error in the final output is not affected much. This absolute error in the stochastic value is reduced further in this paper by considering the $5th$ MSB bit of the 8 bit binary input. Here a decision block is used where if the $5th$ bit of binary input is set, the counter used in the stochastic to binary conversion unit starts counting from 1 other wise it starts counting from 0. This approach helps in reducing the absolute stochastic value error from 6.25% to 3.125% almost equal to 50% reduction without much of a hardware and time overhead unlike the traditional SC where the clock cycles have to be increased exponentially for error reduction in the output value. The experimental results are shown in Fig. 1 where the absolute error in the stochastic value is reduced to 50% by increase in just one clock cycle.

Fig. 2 shows the proposed circuit which decreases the error percentage by increasing the clock cycles linearly rather than exponentially. Initially a delay of one clock cycle is added where the $5th$ bit in the binary bit length is checked to the initiate the counter from a value 0 or a

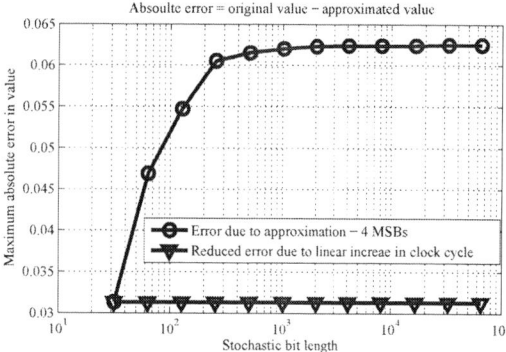

Fig. 1: Figure showing the absolute value of the error in between the original and approximated value.

1. Hence, a linear increase in clock cycle to reduce the error. The proposed approach is used for converting the binary to stochastic input values and the stochastic to binary output values. For pixel value in the range of 0 to 255 the output from this circuit has to be scaled accordingly and the absolute error introduced in the output by this does not exceed more than 6%.

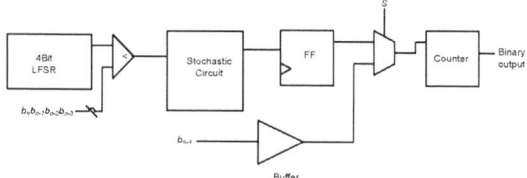

Fig. 2: Figure showing the proposed circuit for stochastic bit generation.

The next section contains the experimental results of the above mentioned approach being implemented on an edge detection circuit, which is an important image processing application.

III. SIMULATION RESULTS

There are a large number of edge detection operators available, each designed to be sensitive to certain types of edges. Most of these operators can be efficiently implemented by the stochastic circuits. Roberts cross operator based edge detection circuit described in paper [10] has been used for this work. Fig. 3 shows the difference between the original gray scale image when a 8-bit length (ranges from 0 to 255) and a 4-bit length (ranges from 0 to 16) is considered. A slightly more noise in the input image in later case is seen because of the error introduced but still the output image is in tolerable limits and shows all the edges as compared to the 8-bit length gray scale image.

From Fig. 3, it can be observed that the fixed 4 bit truncation yields a visually acceptable result with merely 6.25% error and 16 times reduction in the total number of clock cycles. Furthermore, adding one more clock cycle by checking the weight of the 5th MSB, even more accurate result with only 3.13% error can be generated.

Fig. 3: Top: Original grayscale image with 8-bit binary pixel value and its edge detection resulte; Bottom: Same grayscale image considering first 4 MSBs and its edge detection result.

IV. CONCLUSION

In this paper, an approximate computing technique has been used to reduce the stochastic computation time drastically by approximately 16 times and still achieve acceptable results. The edge detection results suggest that this approach can be beneficial to design efficient circuits with smaller circuit size and faster operation for image processing applications, where 100% accuracy is not needed. Future work would be to implement the same design techniques to various image processing applications as well as arithmetic circuits to reduce the error percentage further by increasing the clock cycles linearly rather than exponentially.

REFERENCES

[1] B. R. Gaines, "Stochastic computing," in *Proceedings of the April 18-20, 1967, spring joint computer conference.* ACM, 1967, pp. 149–156.

[2] B. Gaines, "Stochastic computing systems," in *Advances in information systems science.* Springer, 1969, pp. 37–172.

[3] A. Alaghi and J. P. Hayes, "Survey of stochastic computing," *ACM Transactions on Embedded computing systems (TECS)*, vol. 12, no. 2s, p. 92, 2013.

[4] P. Li and D. J. Lilja, "Accelerating the performance of stochastic encoding-based computations by sharing bits in consecutive bit streams," in *Application-Specific Systems, Architectures and Processors (ASAP), 2013 IEEE 24th International Conference on.* IEEE, 2013, pp. 257–260.

[5] H. Ichihara, S. Ishii, D. Sunamori, T. Iwagaki, and T. Inoue, "Compact and accurate stochastic circuits with shared random number sources," in *Computer Design (ICCD), 2014 32nd IEEE International Conference on.* IEEE, 2014, pp. 361–366.

[6] A. Alaghi and J. P. Hayes, "A spectral transform approach to stochastic circuits," in *Computer Design (ICCD), 2012 IEEE 30th International Conference on.* IEEE, 2012, pp. 315–321.

[7] A. Alaghi and J. Hayes, "STRAUSS: Spectral Transform Use in Stochastic Circuit Synthesis," 2012.

[8] J. Han and M. Orshansky, "Approximate computing: An emerging paradigm for energy-efficient design," in *2013 18th IEEE European Test Symposium (ETS).* IEEE, 2013, pp. 1–6.

[9] R. Venkatesan, A. Agarwal, K. Roy, and A. Raghunathan, "Macaco: Modeling and analysis of circuits for approximate computing," in *Proceedings of the International Conference on Computer-Aided Design.* IEEE Press, 2011, pp. 667–673.

[10] A. Alaghi, C. Li, and J. P. Hayes, "Stochastic circuits for real-time image-processing applications," in *Proceedings of the 50th Annual Design Automation Conference.* ACM, 2013, p. 136.

Design and Implementation of Multi-Mode Block Adaptive Quantizer for Synthetic Aperture Radar

Yu-Liang Tsai and Pei-Yun Tsai
Department of Electrical Engineering,
National Central University,
Taoyuan, Taiwan

Ching-Horng Lee, Li-Mei Chen, Sz-Yuan Lee
National Applied Research Laboratory (NARL)
National Space Program Office (NSPO),
Hsinchu, Taiwan

Abstract— **A high-speed and high-resolution block adaptive quantizer is designed and implemented for current synthetic aperture radar imaging systems. To solve the problem of exponential growth in complexity due to the requirements of large BAQ output wordlengths, input scaling and hybrid comparison architectures are proposed. Significant saving in the threshold memory size is achieved and a good balance of path delay and arithmetic complexity is attained by two techniques. The proposed design supports 291MHz and 12-bit ADC with output wordlengths of 2, 3, 4, 6 bits.**

Keywords; Block Adaptive Quantizer (BAQ); Synthetic Aperture Radar (SAR)

I. INTRODUCTION

Synthetic aperture radar (SAR) is one of the essential instruments in the satellite payloads for remote-sensing. Because of the advantages for providing all-weather day-and-night imagery, SAR systems develop rapidly during these two decades to support the earth observation missions. However, onboard storage and transmission bandwidth are limited in the satellite. Thus, SAR raw data are usually compressed before they are received by the ground station.

The block adaptive quantizer (BAQ) algorithm was first adopted in 1989 [1] by NASA to generate 2-bit compressed outputs from 8-bit analog-to-digital converter (ADC) results. As the technology evolves, high-resolution SAR images are desired. The flexible BAQ (FBAQ), used in the Environmental Satellite (Envisat) from 2002 to 2012, then offers various compression ratios to support 2-bit, 3-bit, and 4-bit outputs [2] from 8-bit I/Q signals sampled by about 20-MHz frequency. Sentinel-1, launched in April 2014, implements flexible dynamic BAQ (FDBAQ), producing 3, 4, 5 output bits for samples generated by 10-bit 260-MHz ADC with the Lloyd-Max quantizer [3]. The signal-to-thermal-noise ratio is taken into consideration to set the proper compression ratio. It is thus clear that recent BAQ design must be capable of dealing with higher ADC sampling rates as well as larger ADC word-lengths, and supporting more flexible BAQ output word-length settings.

In this paper, a multi-mode BAQ is designed and implemented. Given 12-bit ADC resolution, we aim to provide 2, 3, 4, 6 BAQ output bits. The number of segments for classifying reflection signal strength can be set to 128 or 256. Three block sizes for statistics estimation are offered, 128, 256 or 512. Generally speaking, the complexity of the BAQ grows exponentially as the output word-length increases. Thus, we propose an alternative architecture that can significantly reduce

the memory size of the look-up table. In addition, the hybrid comparison architecture is proposed to further reduce arithmetic complexity. Finally, the simulation and implementation results of the proposed multi-mode BAQ are given. We also demonstrate its capability for high-speed operation.

II. MULTI-MODE BLOCK ADAPTIVE QUANTIZER

The conventional BAQ architecture [1] is shown in Fig. 1(a). The magnitude of I/Q signals are accumulated for B samples. The lookup table stores the mapping from the average magnitude to the standard deviation of Gaussian distributed input signals. The standard deviation, reflecting the signal strength, then is used as the input to the threshold memory so that the threshold can be adapted properly according to the signal strength [1][4]. If the range of the standard deviation is partitioned into S segments and the output word-length of BAQ is set to N, then the size of the threshold memory will become

$$\sum_{N \in \Omega} S(2^{N-1} - 1)W_{TH} \qquad (1)$$

where W_{TH} is the wordlength of the threshold and Ω denotes the set containing supported BAQ output wordlengths. Therefore, as the value of N increases, the size of the threshold memory increases exponentially. Besides, more segments to support fine resolution also entail high hardware complexity.

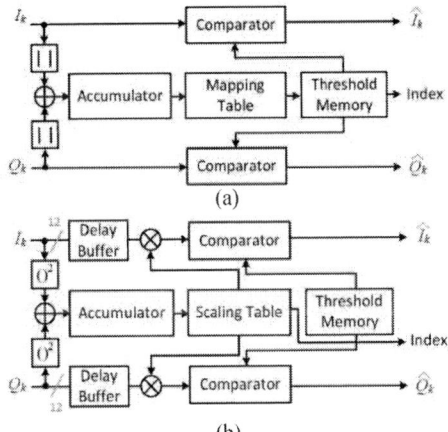

Fig. 1 (a) Conventional BAQ architecture. (b) Proposed BAQ architecture

We then propose a new BAQ architecture, which is capable to handle large N values, as shown in Fig. 1(b). The variance of one block with B samples is generated at the output of the accumulator. However, a scaling table and a multiplier replace the mapping table in the conventional BAQ. Denote the variance of full-scale output signals as σ_F^2. The scaling table saves the scaling factor $\sqrt{\sigma_F^2/\sigma^2}$. Then, the I/Q signal is scaled up by the multiplier with the scaling factor. Originally, for signal $\bar{I}_k, \bar{Q}_k$ with variance σ^2, vector $[\sigma d_1\ \sigma d_2\ ...\ \sigma d_{2^{N-1}-1}]$ is generated from the threshold table to the comparator, where d_i is the threshold of Lloyd-Max quantizer for Gaussian distributed data with zero mean and variance σ_F^2. In our design, signal signals $\bar{I}_k$ and $\bar{Q}_k$ are first scaled to $\sigma_F \bar{I}_k/\sigma$ and $\sigma_F \bar{Q}_k/\sigma$. Consequently, only one threshold vector $[d_1\ d_2\ ...\ d_{2^{N-1}-1}]$ is required for each N value in the threshold memory. Compared to (1), the size of the threshold memory is reduced to

$$\sum_{N\in\Omega}(2^{N-1}-1)W_{TH} \tag{2}$$

and the size of the scaling table can be expressed as SW_{SC}, where W_{SC} is the wordlength of the scaling table.

The number of comparisons for generating BAQ outputs also grows rapidly as the N value increases. If full degree of parallelism is used, the number of comparators will be $(2^{N-1}-1)$ and a 2^{N-1}-to-$(N-1)$ encoder is required. On the other hand, if serial binary comparison is used, the number of comparators becomes $(N-1)$. However, the path delay of the serial comparison architecture is longer than that of the parallel comparison architecture. To tradeoff path delay and complexity for satisfying high-throughput and high-resolution requirements, a hybrid architecture is designed. As shown in Fig. 2, serial R-stage comparison is implemented. For each stage r, parallel comparison is done, and thus $(2^{N_r-1}-1)$ comparators and a 2^{N_r-1}-to-(N_r-1) encoder are realized. Note that $\sum_{r=1}^{R}N_r = N-1$. Hence, when R is set to 2, uniform partition can achieve much more complexity reduction, namely $N_1 = \lfloor (N-1)/2 \rfloor$ and $N_2 = \lceil (N-1)/2 \rceil$. Also, if N_2 is one of the settings for the BAQ output wordlength, hardware sharing can be performed. Consequently, the hybrid comparison architecture not only balances the path delay and the complexity but also achieves hardware sharing among different operation modes of the output wordlength.

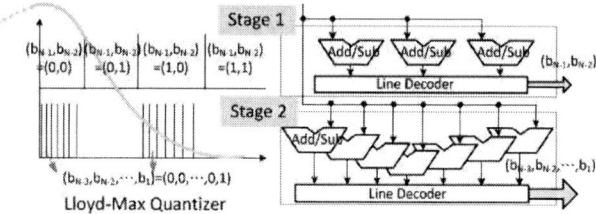

Fig. 2 Hybrid architecture of comparators.

III. IMPLEMENTATION RESULTS

The proposed design is then implemented on the FPGA platform. Assume that 12-bit ADC is employed. Fig. 3 shows the fixed-point simulation results for BAQ output wordlengths equal to 2, 3, 4, 6. The block size B is set to 512 and the variance range is partitioned into 128 segments ($S = 128$). The implementation loss is less than 0.5dB. The wordlength (W_{TH}) of the threshold in the lookup table is 12 bits and the wordlength (W_{SC}) of the scaling table are 14 bits and 15 bits

for $S = 128$ and $S = 256$, respectively. Thus, the memory reduction can be summarized in Table I. We can see that if multiple output wordlengths are considered in the implementation, which is a typical case in current BAQ systems, then significant memory saving can be attained with the proposed architecture. From the synthesis report of using Virtex-7 FPGA, the operating frequency achieves 291MHz. The number of slice LUTs is 1630, and 266 slice registers are used. Therefore, a multi-mode BAQ, which can support high-speed and high-resolution requirements with low complexity, is realized.

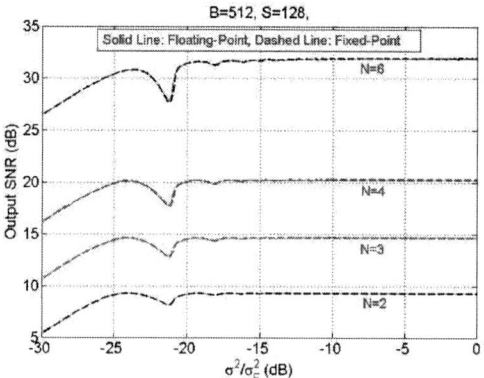

Fig. 3 Performance of the multi-mode BAQ.

Table I Saving in memory sizes.

		N	2,3	2,3,4	2,3,4,6
Conv. BAQ	Threshold Table	S=128	6144	16896	64512
		S=256	12288	33792	129024
Proposed BAQ	Threshold Table		48	132	504
	Scaling Table	S=128	1792	1792	1792
		S=256	3840	3840	3840
	Total (S=256)		3888	3972	4344
Saving	S=256		68.4%	88.2%	96.6%

IV. CONCLUSION

A BAQ suitable for high-speed and high-resolution SAR imaging systems is implemented. An input-scaling strategy is proposed to reduce the threshold memory size. A hybrid comparison architecture is also adopted to balance the path delay and arithmetic complexity. Both techniques are effective to achieve the goal especially when larger BAQ output wordlengths are desired.

REFERENCES

[1] R. Kwok, and W. T. K. Johnson, "Block adaptive quantization of Magellan SAR data" in IEEE Transactions on Geoscience and remote sensing, vol. 27, no. 4, pp. 375-383, July 1989.

[2] ESR, "BEST- Basic Envisat SAR Toolbox" User manual, Version 4.0.2., March 2005.

[3] P. Guccione, et al. "Sentinel-1A: analysis of FDBAQ performance on real data" in IEEE Transactions on Geoscience and Remote Sensing, vol. 53, No. 12, pp. 6804-6812, Dec. 2015.

[4] W. C. Fang, "VLSI processor design of real-time data compression for high resolution imaging radar," IEEE International ASIC Conference and Exhibit, 1994, pp. 441-444

Mapping Table-based Fisheye Image Correction for Low Computational Complexity

Yong Deok Ahn and Suk-Ju Kang

Dept. of Electronic Engineering, Sogang University
Republic of Korea
E-mail: ahnyd09@gmail.com, sjkang@sogang.ac.kr

Abstract— In this paper, we proposed a mapping table-based fisheye image correction to reduce computation time. Specifically, the proposed algorithm uses the field of view correction model and camera coordinate conversion when performing an image interpolation for generating an image with the target image size. The experimental results show that the proposed algorithm reduces the computation time up to 15.85% while improving perceptual image quality, compared with the benchmark algorithm.

Keywords; fisheye image correction, image interpolation, low computation time

I. INTRODUCTION

A fisheye lens is used in various types of cameras [1]. Especially, these pictures have wide field of view (FOV) up to 180 degrees. Although the wide FOV gives the user many advantages for capturing wide region, images have strong radial distortion. Therefore, radial distorted images should be corrected by adjusting the curved line based on the vision processing [2]. Radial distorted images can be corrected using a FOV correction model [3]. Using this model, image size and resolution should be increased to generate an undistorted image with original size. In this case, the complexity of the total process could be very high in image sequences requiring the high frame rate. In this paper, the mapping table-based fisheye image correction model is proposed to reduce the computation complexity. The proposed algorithm significantly reduces the complexity by concurrently performing the interpolation and radial correction based on the FOV correction model.

II. PROPOSED METHOD

Fig. 1 shows the overall block diagram of the proposed algorithm. The detailed processes are as follows.

A. Parameter Calculation and Coordinate Conversion

Intrinsic parameters are composed of the focal length of f and g, principle point of u_x and u_y, and interpolation scale factor of s. These parameters are calculated by the general camera calibration [4].

Next, Euclidean coordinate for an input image is converted into the affine coordinate to correct distortion using the FOV correction model as follows:

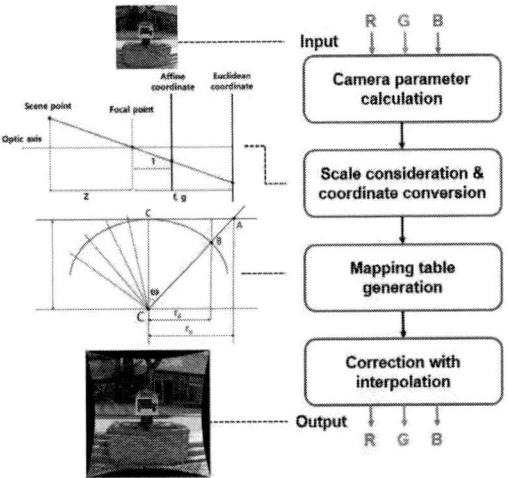

Figure 1. Overall block diagram of the proposed algorithm

$$\left(sf, sg, -su_x, -su_y\right) = s \times \left(f, g, -u_x, -u_y\right) \quad (1)$$

$$\begin{bmatrix} x_{d\_n} \\ y_{d\_n} \\ 1 \end{bmatrix} = \begin{bmatrix} sf & 0 & -su_x \\ 0 & sg & -su_y \\ 0 & 0 & 1 \end{bmatrix}^{-1} \begin{bmatrix} x_d \\ y_d \\ 1 \end{bmatrix} \quad (2)$$

where x_d and y_d denote indexes of x and y in an Euclidean coordinate, x_{d_n} and y_{d_n} denote indexes of x and y in an affine coordinate.

Then, the distorted parameter (r_d) and undistorted parameter (r_u), which is matched with r_d, are calculated in (3) and (4). Next, the coordinate conversion is performed in (5).

$$r_d = \sqrt{x_{d\_n}^2 + y_{d\_n}^2} \quad (3)$$

$$r_u = \frac{\tan\left(r_d \omega\right)}{2 \tan \frac{\omega}{2}} \quad (4)$$

$$\begin{bmatrix} x_{c\_r} \\ y_{c\_r} \\ 1 \end{bmatrix} = \begin{bmatrix} sf & 0 & -su_x \\ 0 & sg & -su_y \\ 0 & 0 & 1 \end{bmatrix} r_u \begin{bmatrix} x_{d\_n} \\ y_{d\_n} \\ 1 \end{bmatrix} \quad (5)$$

B. Mapping Table Generation and Image Interpolation

In the proposed algorithm, the coordinate mapping table, which assign the representative value for multiple values in the same position, is generated. Then, the proposed algorithm performs image interpolation considering the target image size with correcting distortion. In case of the interpolation technique, the nearest neighborhood and bilinear [5] interpolations are used with correction for adjustable image size. In this process, the representative pixel is selected based on the corresponding binary number, where 0 is the selected signal as follows:

$$I_f\left(x_{c\_r}, y_{c\_r}\right) = I_{in}\left(x_t, y_t\right), \; if \; I_{lt}\left(x_t, y_t\right) = 0,$$
$$I_f\left(x_{c\_r}, y_{c\_r}\right) = 0, \; else, \quad (6)$$

where I_f, I_{in}, and I_{lt} denote a final result image, an interpolated image, and the mapping table. x_t and y_t denote indexes of x and y in an interpolated coordinates.

III. EXPERIMENTAL RESULT

The performance of the proposed algorithm was evaluated from the perspective of computation time and image quality. The method, which does not consider overlapping pixels, was used for a benchmark. For the test image sets, landscape test image (960 × 960) had three different viewing angle parameter of ω (2.2, 2.5, 2.8) with 40 frames. The interpolation technique increased horizontal and vertical resolution by two and four times. Table 1 shows the average computation time for the benchmark and the proposed algorithms. The performance of the proposed algorithm was evaluated by using two interpolation techniques. The proposed algorithm reduced the average computation time by 2.17% and 7.02% when ω = 2.2. In case that ω = 2.5, the proposed algorithm reduced the average computation time by 5.31% and 8.29%. In case that ω = 2.8, it reduces the average computation time by 7.71% and 15.85%. Fig. 2 shows the image quality for the benchmark and the proposed algorithms. The benchmark lost original data when correcting the radial distortion while the proposed algorithm maintains the original image quality because of considering target image size when correcting radial distortion.

IV. CONCLUSION

In this paper, the mapping table-based fisheye image correction was proposed. The proposed algorithm used the FOV correction model with the combination of an image interpolation and distortion for reducing computation complexity and perceptual image quality. In the experimental result, it reduced the computation time by up to 15.85% while enhancing perceptual image quality.

TABLE I. AVERAGE COMPUTATION TIME WITH FISHEYE IMAGES

Test sequence (number of frames)	Increase	Nearest neighborhood interpolation			Bilinear interpolation		
		Benchmark [s]	Proposed [s]	Diff. [%]	Benchmark [s]	Proposed [s]	Diff. [%]
Test image set ω= 2.2 (40)	2	5.52	5.40	2.17	13.04	12.33	5.46
	4	39.13	36.56	6.56	88.16	81.97	7.02
Test image set ω= 2.5 (40)	2	4.06	3.85	5.31	9.50	8.73	8.18
	4	23.08	21.39	7.31	51.88	47.58	8.29
Test image set ω= 2.8 (40)	2	3.02	2.79	7.71	6.77	5.83	13.95
	4	11.00	9.81	10.80	22.51	18.94	15.85

(a)

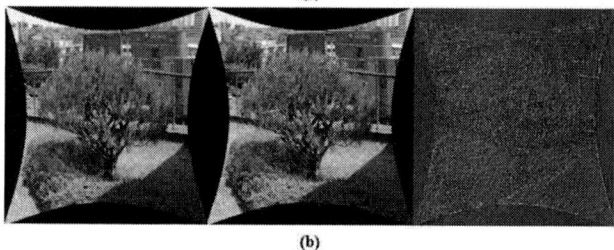

(b)

Figure 2. Image quality comparison for test set (a) bilinear interpolation. ω =2.2 (left:, benchmark, center: proposed algorithm, right: difference between left and right), (b) bilinear interpolation, ω =2.5 (left:, benchmark, center: proposed algorithm, right: difference between left and right).

ACKNOWLEDGMENT

This research was supported by a grant (16CTAP-C114672-01) from infrastructure and transportation technology promotion research Program funded by Ministry of Land, Infrastructure and Transport of Korean government and supported by Korea Electric Power Corporation through Korea Electrical Engineering & Science Research Institute (R15XA03-08)

REFERENCES

[1] Z. Huaibin, and J. K. Aggarwal, "3D reconstruction of an urban scene from synthetic fish-eye images," Image Analysis and Interpretation, Proceedings. 4th IEEE Southwest Symposium. IEEE, pp. 219-223, April 2000.

[2] Wei. Jin, L. Chen-Feng, H. Shi-Min, M. Ralph R, T. Chiew-Lan "Fisheye video correction," IEEE Transactions on Visualization and Computer Graphics, vol. 18, no. 10, pp.1771-1783, Octorber 2012.

[3] F. Devernay, and O. Faugeras, "Straight lines have to be straight," Machine vision and applications, vol. 13, no. 1, pp. 14-24, August 2001.

[4] S. Shishir, and J. K. Aggarwal, "Intrinsic parameter calibration procedure for a (high-distortion) fish-eye lens camera with distortion model and accuracy estimation," Pattern Recognition, vol. 29, no. 11, pp. 1775-1788, February 1996.

[5] T. M. Lehmann, C. Gonner, and K. Spitzer, "Survey: Interpolation methods in medical image processing," IEEE Trans. Medical Imaging, vol. 18, no. 11, pp. 1049-1075, November 1999.

Cryptographic Coprocessor Design for IoT Sensor Nodes

Weizhen Wang, Jun Han*, Zhicheng Xie, Shan Huang and Xiaoyang Zeng

State Key Laboratory of ASIC and System, Fudan University

Shanghai, China

*Email: junhan@fudan.edu.cn

Abstract—**Together with the popularity of internet of things (IoT), the information security in IoT is becoming an urgent issue. In this paper, a cryptographic coprocessor is designed to provide security for IoT sensor nodes. This design incorporates the application-specific instruction set to support popular cryptography algorithms like advanced encryption standard (AES) and elliptic curve cryptography (ECC). The tailored instruction provides good flexibility, high energy efficiency and low latency. Local instruction and data ram is integrated into our coprocessor to reduce the instruction transfer between embedded processor and the coprocessor. Our work was synthesized under TSMC 65 nm technology. The measurement results show that our design consumes 32.7 uJ for each ECC point multiplication and 3 nJ for each AES encryption and decryption when working at 10 MHz.**

Keywords: IoT, Cryptography, ECC, AES, Application-specific instruction set processor (ASIP)

I. INTRODUCTION

The advancement in IoT is not only making our life more convenient but also leading to some serious problems. Amongst them, the potential risk of sensitive data leakage is one of the most concerned problems. The conventional solution is to utilize the public-key cryptography (PKC) for secure key exchange along with the symmetric-key cryptography (SKC) like AES for secure data transfer. However, due to the strict power and resources constraints in IoT sensor nodes, software implementation of encryption algorithms with high efficiency is proved to be hard. The SoC implementation with hardware accelerator or dedicated hardware for both PKC and SKC has gained a lot of attention for the reason of energy efficiency and acceptable performance.

Since modular multiplication is the kernel operation in PKC and consumes a lot of time and energy, various kinds of modular multipliers have been proposed and integrated into SoC as accelerators to reduce the ECC runtime. However, previous works show that the approach of integrating dedicated hardware into SoC results in massive data and instructions transferred between the embedded processor and the dedicated hardware [1] and becomes the performance bottleneck of the whole system. In this paper, we propose and implement a cryptographic ASIP with local instruction and data storage to speed up the cryptographic tasks with little data and instruction

transfer.

ECC and RSA are the two popular PKC algorithms. Both of them can be used to provide efficient solution for authentication in communication systems. ECC compared with RSA uses much smaller key size when provides the same level of security [2]. What's more, the implementation of ECC in F_2^m consumes less power when compared with that in F_P with the same key size [3]. Thus the ECC over F_2^m is an proper choice for applications like IoT sensor nodes where power and resources are limited.

Since AES is one of the most popular encryption algorithm for data transfer, instructions for AES key expansion, encryption and decryption are also included in our customized instruction set. As a result, our coprocessor could be programmed to support AES encryption and decryption with different operation modes like cipher block chaining (CBC), electronic code book (ECB) and offset code book (OCB).

This paper is organized as follows. In section II, we introduce the implementation of the proposed cryptographic coprocessor. In section III, the SoC platform which our coprocessor is integrated with is introduced. Moreover, the performance evaluation of the proposed coprocessor is also provided in this section. Finally, the conclusion is obtained in section VI.

II. CRYPTOGRAPHIC COPROCESSOR ARCHITECTURE

The overall architecture of the coprocessor with a 5-stage pipeline is illustrated in Fig. 1.

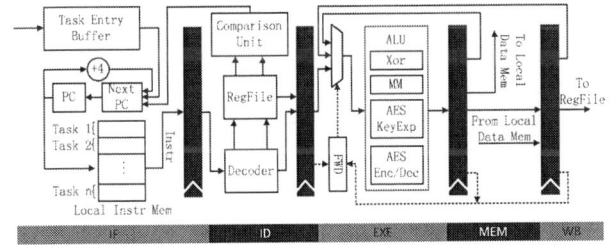

Figure 1.   Hardware Architecture for the Cryptographic Coprocessor

Several tasks like point doubling, point addition, point multiplication, AES key expansion, AES encryption and decryption are stored in the local instruction memory. Entries of those tasks are stored in a lookup table, namely, the task

---

This work was supported by the National Natural Science Foundation of China (61574040, 61234002, 61525401), and the Project of the State Key Laboratory of ASIC and System (2015ZD003).

978-1-5090-3220-4/16 $31.00 © 2016 IEEE

entry buffer in Fig. 1. When the coprocessor start working, the program counter (PC) is set to the entry of a certain task so as to control coprocessor to perform the required task. Since the instruction memory is pre-programmed, the coprocessor is able to work continually.

The FWD unit in Fig. 1 is a forwarding control unit, which is utilized to resolve the data hazard and improve the throughput. The arithmetic logic unit (ALU) mainly consists of a Montgomery modular multiplier (MM), XOR, AES key expansion and AES encryption and decryption unit. The digit serial Montgomery modular multiplier over F_2^m proposed in [4] is implemented with digit size w=16. Moreover, instead of using irreducible polynomial with fixed degree, our design could support irreducible polynomial with different degrees. The maximum degree supported is 256. The irreducible polynomials $P(x)=x^{163}+x^7+x^6+x^3+1$ and $P(x)=x^{233}+x^{74}+1$ ecommended by National Institute of Standards and Technology (NIST) are supported by our design. The implementation of Sbox transformations with lookup table results in high throughput but consumes a lot of area, which is not capable for applications like IoT sensor nodes. In our design, the Sbox transformations are implemented based on the composite-field to reduce the area consumption [5]. The register file is used to store the subkeys generated by key expansion unit, and the generated subkeys are used by AES encryption and decryption unit, which fulfills one AES round.

III. SoC Verification and Performance Evaluation

The proposed coprocessor is integrated with SoC platform depicted in Fig. 2. The DataIn, DataOut and TaskReg registers are mapped to memory address values so that they could be accessed by the embedded processor in the same way that it accesses memory. The TaskReg registers receive commands, named macro instructions, from the embedded processor and the macro instructions are decoded to index the task entry in task entry buffer shown in Fig. 1. Since the local data and instruction memory is deployed in our coprocessor, only indispensable data is transferred between embedded processor (OpenRISC 1200) and the coprocessor.

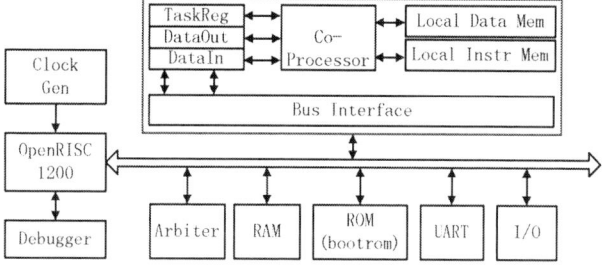

Figure 2.   SoC Verification

Our work was synthesized under TSMC 65 nm technology. The synthesis results show that the coprocessor requires 162 kGates and 36% of the total area is consumed by the local storage. When it comes to the runtime of each ECC point multiplication, compared with previous works listed in Table I, our work outperforms previous works with speedup of 34, 1.61 and 1.40, respectively. It should be noted that our design could

be configured for irreducible polynomials with different degrees and the supported key size is larger than previous works, which means our design offers security of higher level.

TABLE I.   RUNTIME FOR EACH POINT MULTIPLICATION

Works	Key Size	Cycles	Comment
Wenger *et al.* [6]	163	7216905	Software, openMSP 430
Wenger *et al.* [6][a]	163	341835	Dedicated hardware
Hein *et al.* [7]	163	296299	Dedicated hardware
This Work	233	212270	ASIP

a. The result without squarer unit is selected for a fair comparison

TABLE II.   ENERGY CONSUMED BY THE COPROCESSOR[a]

Task	Cycles	Power (mW)	Energy (nJ)
AES key expansion	12	2.11	2.53
AES encryption	14	2.13	2.98
AES decryption	14	2.23	3.12
ECC point multiplication[b]	212270	1.54	32689.58

a. Clocked @ 10MHz, Power supply @ 1.2V

b. The NIST recommended curve B-233 [8] is selected for measurement

The energy consumption when the proposed coprocessor performs certain tasks is measured with Synopsys PrimeTime PX and the results are shown in Table II. As shown in Table II, the coprocessor consumes approximately 3 nJ for each AES encryption and decryption

IV. Conclusion

In this paper, a cryptographic coprocessor for IoT applications is proposed. The coprocessor incorporates local data and instruction memory to save the bandwidth when it is integrated with SoC. The tailored instruction was capable of supporting ECC over F_2^m and AES algorithms. The simulation results show that our design achieves small latency compared with related works. The power estimation results show good energy efficiency for cryptographic tasks.

[1]   Xu Guo and Patrick Schaumont. Optimizing the hw/sw boundary of an ecc soc design using control hierarchy and distributed storage. In Proceedings of the Conference on Design, Automation and Test in Europe, pages 454–459. European Design and Automation Association, 2009.

[2]   Neal Koblitz. Towards a quarter-century of public key cryptography. Springer, 2000.

[3]   Erich Wenger and Michael Hutter. Exploring the design space of prime field vs. binary field ecc-hardware implementations. In Nordic Conference on Secure IT Systems, pages 256–271. Springer, 2011.

[4]   Miroslav Knezevic, Frederik Vercauteren, and Ingrid Verbauwhede. Speeding up barrett and montgomery modular multiplications.

[5]   Atri Rudra, Pradeep K Dubey, Charanjit S Jutla, Vijay Kumar, Josyula R Rao, and Pankaj Rohatgi. Efficient rijndael encryption implementation with composite field arithmetic. In International Workshop on Cryptographic Hardware and Embedded Systems, pages 171–184. Springer, 2001.

[6]   Erich Wenger. Hardware architectures for msp430-based wireless sensor nodes performing elliptic curve cryptography. In International Conference on Applied Cryptography and Network Security, pages 290–306. Springer, 2013.

[7]   Daniel Hein, Johannes Wolkerstorfer, and Norbert Felber. Ecc is ready for rfid–a proof in silicon. In International Workshop on Selected Areas in Cryptography, pages 401–413. Springer, 2008.

[8]   PUB FIPS. 186-4. Digital Signature Standard (DSS), 2013

Software-Based Embedded Core Test Using Multi-Polynomial for Test Data Reduction

Soyeon Kang, Inhyuk Choi, Hyeonchan Lim, Sungyoul Seo and Sungho Kang

Department of Electrical & Electronic Engineering
Yonsei University
Seoul, Korea
{adhoner, ihchoi, lhcy92, sungyoul}@soc.yonsei.ac.kr and shkang@yonsei.ac.kr

Abstract— **Software-based self-test (SBST) is a self-test where processors and intellectual property (IP) cores test itself using an embedded memory. However, an environment-limited memory size is one of the biggest challenges. In this paper, we present a new SBST solution using multiple polynomials. For reducing the required test data, the polynomials consist of a primitive polynomial and (BM)-algorithm based polynomials and each polynomial generates pseudo random patterns and deterministic patterns respectively. Experimental results show that this SBST method reduces the size of the test program without a reduction of the fault coverage.**

Keywords; SBST; IP-core test; Polynomial

I. INTRODUCTION

Increased number of transistors and smaller size of system-on-chip (SoC) made the testing more difficult. The main reason is that the accessibility of the embedded cores is limited and the amount of the data for testing is increased. Built-in self-test (BIST) and scan architecture are regarded as a solution to improve testability. BIST and scan architecture can achieve high fault coverage in return for introducing some modifications of the design. This modification increases the area overhead, deteriorates the performance, and generates excessive power during testing.

To overcome these problems, software-based self-test (SBST) is getting attention. Instead of utilizing an external tester, SBST uses on-chip resources. It downloads a test program to the memory, and tests processor and intellectual property (IP) cores. For this reason, SBST can test at-speed without any modification of the circuit which may result in the degradation of the circuit function [1].

The proposed SBST is depicted in Figure 1. The proposed method generates both pseudo random and deterministic patterns to test IP cores. Polynomials and seeds are downloaded from ATE to the on-chip memory. Based on the polynomials, the processor generates software-based linear-feedback shift-register (LFSR) and produces test vectors using each corresponding seed. The test vectors test the IP cores and the responses are transmitted back to the memory after compressed by multiple-input signature register (MISR). Experimental results show that the suggested method reduces

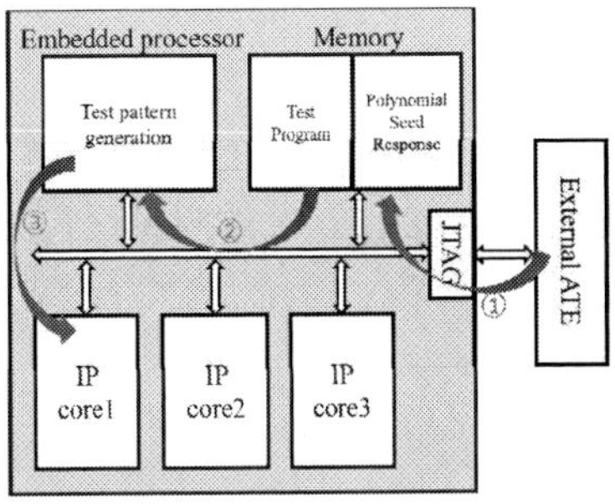

Figure 1. SoC architecture for testability

the software size while the fault coverage remains same to the original test vectors.

II. PROPOSED IDEA

Since the size of a memory inside a chip is limited, it is impossible to store all deterministic patterns for all IP cores in the memory. Therefore, the effort to reduce the test software size is inevitable. The proposed SBST uses the pseudo random patterns in advance the deterministic patterns to reduce the amount of data which needs to be stored in the memory. The rest of the faults which are random pattern resistant, are detected by the deterministic patterns.

Both pseudo random and deterministic patterns are stored as a form of polynomials and seeds. Pseudo random patterns are generated from the primitive polynomials and deterministic patterns are generated from the BM-algorithm based polynomials [2,3]. Procedure 1 shows sequences generating the polynomials, *Gen_Poly*. To generate the polynomials, original test vectors should be generated by automatic test pattern generator (ATPG). Then a primitive polynomial and a seed is generated based on the length *l* which is user defined parameter (line1-2). Using this primitive polynomials, pseudo random patterns are generated and the original test vectors collapsing with the pseudo random patterns are removed (line3-10). The remainder patterns are used as inputs to generate BM-based polynomials (line11-14).

In the procedure 2, test vector generating sequences inside the chip processor are described. The polynomials, seeds, the

---

This work was supported by Institute for Information & communications Technology Promotion(IITP) grant funded by the Korea government(MSIP) (No.R-20160229-002930, Development of Application Program Optimization Tools for High Performance Computing Systems)

Procedure 1. *Gen_Poly(l, v)*

input: primitive polynomial length l, original test
vector v

output: polynomial set p, seed set s

```
1:    p_0 ← Gen_Primitive_Poly(l)
2:    s_0 ← Gen_Primitive_Seed(p_0)
3:    r_n ← Gen_Pseudo_Random_vec(p_0, s_0)
4:    for i = 1 to M do
5:        for j = 1 to N do
6:            if r_i = v_j
7:                Remove(v_j, v)
8:            end if
9:        end for
10:   end for
11:   for n = 1 to N' do
12:       p_n ← Gen_BM_Poly(v_n)
13:       s_n ← Gen_BM_Seed(v_n, p_n)
14:   end for
```

Procedure 2. *Gen_Test_Vec(p, s, M, N)*

input: polynomial set p, seed set s, the number of
random pattern M, and deterministic pattern
N

output: random test vector r, deterministic test
vector d

```
1:    for n = 1 to M do
2:        r_n ← Gen_Pseudo_Random_Vec(p_0, s_0)
3:    end for
4:    for n = 1 to N do
5:        d_n ← Gen_Determin_Vec(p_n, s_n)
6:    end for
```

number of random patterns and deterministic patterns are inputs to the procedure *Gen_Test_Vec*. First, the pseudo random patterns are generated by first polynomial and stored in vector r (line1-3). The processor generates software-LFSR based on the primitive polynomial which generates the test vectors. Then the deterministic patterns are generated by the other polynomials and saved in vector d in similar ways (line 4-6). Each of the BM-algorithm based polynomial generates a deterministic test vectors. The vectors r and d are the final outputs of the processor to test the IP cores.

III. EXPERIMENTAL RESULTS

The experiments are performed based on some ISCAS89 benchmark circuits assuming the circuits are implemented on a single SoC which needs to be tested by SBST. The necessary polynomials and seeds for each core are compared with the test patterns of [4] in Table 1 row 1 to 4. The patterns used in [4] are deterministic test patterns which guarantee high fault coverage in return for high test data size. Contrastively, the proposed idea use polynomials to reduce the test data size without the loss of the fault coverage.

For the proposed method, 32-bit software-based LFSR is used to generate 100 different pseudo random test patterns. The length of the polynomials for the deterministic patterns are all summed and shown in column 8. The sizes of both polynomials are then combined and shown in column 9. It shows that in all cores, the amount of memory needs to store the test data are reduced.

In row 5, the size of the test programs is compared. The size for [4] is smaller than the proposed one. Though, the total size for both test pattern and test program shown in row 6 shows that the proposed method is about 50% smaller than the size of [4].

IV. CONCLUSION

In this paper, a new SBST method of low memory size is proposed. The proposed idea uses primitive and BM-algorithm based polynomials. The processor loads the polynomials and generates software-based LFSR which produces test patterns. The experiment results show that the proposed method reduced the size of data necessary to be stored in the on-chip memory.

REFERENCES

[1] M. Grosso, W.J. Pérez H, D. Ravotto, E. Sanchez, M. Sonza Reorda, and J. Velasco Medina, "A software-based self-test methodology for system peripherals," 15th IEEE European Test Symposium, 2010, pp 195–200.

[2] J. Rajski, and J. Tyszer, "Primitive polynomials over GF(2) of degree up to 660 with uniformly distributed coefficients," Journal of Electronic testing, bol. 19, no. 6, 2003, pp. 645.

[3] V. R. Sidorenko, G. Richter, and M. Bossert, "Linearized shift-register synthesis," IEEE Trans. Inf. Theory, vol. 57, no. 9, Sep. 2011, pp. 6025–603.

[4] L. Kuen-Jong, C. Chia-Yi, and H. Yu-Ting, "An embedded processor based SOC test platform," in Proceeding IEEE Internation Symposium on Circuit and Systems, 2005.

TABLE I. COMPARISON OF THE SIZE OF THE TEST PATTERNS AND THE TEST PROGRAMS

Core	[4]			Proposed				
	Number	Length	Total (bytes)	Primitive Poly.		BM Poly.		Total (bytes)
				Number	Length	Number	Length	
s38417	387	1,664	80.5 K	245	32	142	121,650	30.4 K
s15850	121	611	9.2 K	17	32	104	33,448	8.4 K
s13207	147	700	12.9 K	18	32	129	49,620	12.4 K
Test Pattern	-		102.6 K	-				51.2 K
Test Program	-		0.4 K	-				3.3 K
Total	-		103 K	-				54.5 K

Motion Vector Smoothing of Boundary of Moving Object for Frame Rate Up-Conversion

Ho Sub Lee
Electrical Engineering Department
POSTECH
Pohang, Republic of Korea
hslee5210@postech.ac.kr

Suk-Ju Kang
Electrical Engineering Department
Sogang University
Seoul, Republic of Korea
sjkang@sogang.ac.kr

Young Hwan Kim
Electrical Engineering Department
POSTECH
Pohang, Republic of Korea
youngk@postech.ac.kr

Abstract—**This paper presents a motion vector smoothing method to increase the accuracy of the motion vector in the boundary of moving object. Existing motion vector smoothing methods cannot generate the accurate motion vectors in the boundary of moving object and it induces the block artifacts. To reduce these block artifacts, the proposed method detects the object boundary and extracts the moving region using the temporal information. Then, the boundary of moving object is extracted using the object boundary and the moving region information. Finally, the motion vector smoothing is performed to correct the falsely detected motion vector in the boundary of moving object. Experimental results demonstrate that the proposed method successfully increased the peak signal-to-noise ratio and the structural similarity index by up to 4.12 dB and 0.0626, respectively, compared to the benchmark methods.**

Keyword—Moving object boundary, motion vector smoothing, frame rate up-conversion

I. INTRODUCTION

Frame rate up-conversion (FRUC) is a technique that increases the frame rate of video sequences by inserting interpolated frames between original frames. Recently, it has been used for various applications such as film-to-video conversion, television standard conversion, and liquid crystal displays (LCDs) to reduce motion blur [1].

The typical process of FRUC comprises two components: motion estimation (ME) and motion compensated interpolation (MCI). ME calculates the motion vectors (MVs) of object between consecutive frames. MCI generates the interpolated frame using the MVs calculated in the ME step. ME process is the important because the performance of FRUC mostly affected by the accuracy of MVs [2]. When the FRUC is applied, various exceptional situations of FRUC such as block artifacts in letter box, caption, boundary of moving object, and scrolling text region must be handled to enhance the performance of FRUC [2]. Among them, block artifacts in the boundary of moving object caused by the different object motions in its neighboring occur frequently in the FRUC process. In addition, they induce visually uncomfortable watching experience. Therefore, the detection boundary of moving object is necessary when the MV smoothing is applied to the FRUC.

In this paper, we propose a MV smoothing method to improve the quality of the interpolated frame in the boundary

This research was supported by the LG Display and the MSIP (Ministry of Science, ICT and Future Planning), Korea, under the "ICT Consilience Creative Program" (IITP-R0346-16-1007) supervised by the IITP (Institute for Information & communication Technology Promotion).

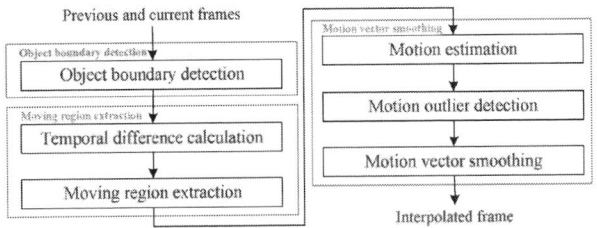

Fig. 1. Overview of the proposed method

of moving object. This paper is organized as follows. Section II describes the proposed MV smoothing of the boundary of moving object. Section III provides the experimental results. Finally, section IV concludes the paper.

II. PROPOSED METHOD

Fig. 1 shows the overview of the proposed method. First, the proposed method detects the object boundary using the relative total variation (*RTV*) [3]. After that, it extracts the moving region (*MR*) using the temporal information between previous and current frames. And, it extracts the boundary of moving object using the boundary detection and the *MR* information. Finally, it performs the MV smoothing using the boundary of moving object information.

A. Object boundary detection

To detect the object boundary, we calculate the *RTV* of the image as follows:

$$RTV(i, j) = \frac{D_x}{|D_x|+1} + \frac{D_y}{|D_y|+1} + \frac{50}{D_x^2 + D_y^2 + 1}, \quad (1)$$

where D_x and D_y denote the total variation of horizontal and vertical direction respectively.

If the *RTV* is less than or equal to threshold T_1, the block of the image is regarded the region which has the moderate gradient variation and can be detected as the object boundary region. Otherwise, the block of the image is regarded the region which has the large or small gradient variation and can be detected as the texture or smooth region. The threshold T_1 for detection of the object boundary was set to 5 empirically.

B. Moving region extraction

To extract the *MR* of the continuous frames, we calculate the temporal difference as follow:

$$diff(i,j) = \left| p_t(i,j) - p_{t-1}(i,j) \right|,$$

$$MR(i,j) = \begin{cases} 1, & if \ diff(i,j) > T_2 \\ 0, & otherwise \end{cases}, \qquad (2)$$

$$T_2 = \sqrt{\frac{1}{N-1}\sum_{i=1}^{N}(p-\bar{p})^2},$$

where $p_t(i,j)$ and $p_{t-1}(i,j)$ denote the current and previous frame at location (i,j) pixel, N denotes the number of total pixels in a frame, $\bar{p}$ denotes the mean of total pixel value in a frame, and T_2 denotes the threshold value for extraction of the *MR*. This process is based on the assumption that the temporal difference exists when the objects are moved between the consecutive frames.

C. MV smoothing

To reduce the block artifacts in the boundary of moving object (*BMO*), MV smoothing is performed based on the information of object boundary and *MR*. This step is based on the assumption that the MVs of *BMO* are similar to its *MR*. In MV smoothing, we refine the corrected MV by calculating the mean value of MVs. For the extraction of the *BMO*, if the *RTV* is less than T_1 and *MR* is 1, the region is designated as the *BMO*. If we detect the *BMO*, we refine the mean of its neighboring MVs in the search window. Otherwise, we refine the mean of its inlier MVs in the search window. The block and the search window size was set to 5 × 5 blocks and 8 × 8 pixels, respectively. These sizes provide a good trade-off between the performance and computational complexity of the FRUC [4].

III. EXPERIMENTAL RESULTS

For the performance evaluation of the proposed and the benchmark methods, peak signal to noise ratio (PSNR) and structural similarity index (SSIM) [5] were used to measure the quality of the interpolated frames. For the test sequences, we used *padam-padam*, *alone in love*, and *descendants of the sun* with a HD resolution. For the benchmark methods, we used two benchmark methods: method 1 [6], and method 2 [7].

As shown in Table I, the proposed method showed higher PSNR and SSIM values for each test sequence when compared to the benchmark methods. Fig. 2 shows the interpolated frames generated by the proposed and the benchmark methods. As shown in Fig. 2, the quality of the interpolated frame generated by the proposed method visually outperformed the quality of the interpolated frames generated by the benchmark methods especially in the boundary of moving object.

IV. CONCLUSTION

In this paper, we proposed a motion vector smoothing method of the boundary of moving object. In the proposed method, the object boundary detection is performed and the moving region is extracted using the temporal information. And then, the boundary of moving object is extracted using the

TABLE I. AVERAGE PSNRs[dB] AND SSIMs OF THE PROPOSED AND BENCHMARK METHODS

Test sequences (# of frames)	Methods	PSNR	SSIM
Padam-padam (99)	Benchmark 1	23.77	0.8249
	Benchmark 2	25.42	0.8767
	Proposed	25.84	0.8875
Alone in love (40)	Benchmark 1	33.37	0.9329
	Benchmark 2	37.15	0.9511
	Proposed	37.49	0.9525
Decendants of the sun (118)	Benchmark 1	26.68	0.8836
	Benchmark 2	30.42	0.9038
	Proposed	30.75	0.9051

(a) (b)

(c) (d)

Fig. 2. Interpolated frame results of the benchmark and proposed methods (a) original frame, (b) benchmark 1, (c) benchmark 2, and (d) proposed methods

object boundary and the moving region information. Finally, the motion vector smoothing is performed based on the boundary of moving object. In the experimental results, the proposed method showed that the average PSNR and SSIM up to 4.12 dB and 0.0626 higher than those of the benchmark methods.

REFERENCES

[1] K. Sekiya and H. Nakamura, "Eye-trace integration effect on the perception of moving pictures and a new possibility for reducing blur on hold-type displays," *SID Symp. Digest of Tech. Papers*, vol. 33, no. 1, pp. 930-933, May 2002.

[2] C. Bartels, CN Cordes, B Riemens, and G. D. Haan, "A system approach to high-quality picture-rate conversion." *Journal of the Society for Information Display*, vol. 18, no. 11, pp. 922-930, Nov 2010.

[3] L. Xi, Q. Yan, Y. Xia, and J. Jia, "Structure Extraction from Texture via Relative Total Variation," *ACM Transaction on Graphic*, vol. 31, no. 6, Nov 2012.

[4] J. Zhai, J. L. K. Yu, and S. Li, "A low complexity motion compensated frame interpolation method," *Proc. IEEE International Symposium on Circuits and Systems*, vol. 5, pp. 4927-4930, Sept. 2005.

[5] Z. Wang, A.C. Bovik, H. R. Sheikh, and E. P. Simoncelli, "Image quality assessment: from error visibility to structural similarity," *IEEE Trans. Image Process.*, vol. 13, no. 4, pp. 600-612, Apr 2004.

[6] Y. Ling, J. Wang, Y. Liu, and W. Zhang, "A novel spatial and temporal correlation integrated based motion-compensated interpolation for fraem rate up-conversion." *IEEE Trans. Consumer Electron*, vol. 54, no. 2, pp. 863-869, May 2008.

[7] S.-J. Kang, D.-G. Yoo, S.-K Lee, and Y. H. Kim, "Median filter-based adaptive motion vector smoothing." *International Symposium on Consumer Elcetronics*, pp. 745-748, May 2009.

Low-power and Real-time Computer Vision On-chip (Invited)

Wei Pang[1], Hantao Huang[2], Fengwei An[3], Hao Yu[1,2]

[1] National ASIC System Engineering Research Center, Southeast University, Nanjing 210096, Jiangsu, China
[2] School of Electrical and Electronic Engineering, Nanyang Technological University, Singapore, 639798
[3] Hiroshima University, Higashi-hiroshima City, Japan
Email: [1]220141209@seu.edu.cn, [2]{hhuang013, haoyu}@ntu.edu.sg, [3]anfengwei@hiroshima-u.ac.jp

Abstract— Computer vision on chip is critical for many emerging applications such as advanced driver assistance system (ADAS), which requires a low-power and real-time image data analytics. Therefore, designing a computer-vision accelerator on-chip to achieve high throughput as well as low power is greatly needed. This paper reviews how to have ASIC realization of standard computer vision algorithms such as SIFT/SURF. The first work is a feature-based recognition co-processor with peak power consumption of 31.5mW for real-time recognition of VGA images. The second work is a face recognition accelerator with 23mW for 5.5 frame/s HD images.

Keywords; computer vision, recognition coprocessor, face recognition accelerator, real-time

I. INTRODUCTION

Computer vision has attracted significant attention for developing the recent advanced driver assistance system (ADAS). With the increasing resolution of CMOS image sensors, the computational requirement and energy consumption are also growing rapidly. It hence requires to design a computer vision accelerator on chip to achieve low-power and high throughput performance.

Many object detection algorithms are based on either Scale-Invariant Feature Transform (SIFT) [1,2], Speeded Up Robust Feature (SURF) [3] or Viola-Jones [4] method. SIFT is robust to detect local features in images, such as changes in rotation, scale and brightness, and is hence widely used in object detection and image tracking. However, due to the abundant information in SIFT features, it suffers from high memory requirement and slow processing rate. To speed up SIFT, SURF has been proposed, using Haar features and integral images. SURF is stable while calculating local extremums of the image, but it still needs a great amount of memory space to store integral image. The Viola-Jones based object detection algorithm is one of the fastest method to face detection still with the need of feature memory accesses. As such, all three algorithms need cost excessive amount of power consumption, and are difficult to be realized in real-time processing. One needs specially designed hardware such as on-chip computer-vision accelerator for high throughput yet low power. In this paper, we introduce two recent works: a feature-based recognition coprocessor and a face recognition accelerator on this regard.

This work is sponsored by grants from Singapore MOE Tier-2 (MOE2015-T2-2-013), NRF-ENIC-SERTD-SMES-NTUJTCI3C-2016 (WP4) and NRF-ENIC-SERTD-SMES-NTUJTCI3C-2016 (WP5).

II. FEATURE-BASED RECOGNITION COPROCESSOR

A. Hardware Implementation for Cell-Based SURF Descripter

The SURF descriptor with fixed scale which contains the Haar-wavelet responses in x and y directions, and the polarity of their intensity changes can be used in object detection, as shown in Fig. 1. In this work, the hardware implementation for the SURF descriptor extraction with scalable input-image size is developed with a pixel-based pipeline architecture. It can be used for capturing the target objects among complex backgrounds in real time. Here, the processing speed is only limited by the pixel-transfer frequency from the image sensor. Additionally, the pixel-based pipeline architecture avoids the need of frame buffers since each input pixel is immediately consumed.

Figure 1. Object recognition with cell-based SURF descriptor.

Here the SURF descriptor feature vector (SURF-FV) stores only intermediate values of 4-dimensional local feature vectors for one-row of cells instead of an entire frame or integral image. A sliding-window (SW) scheme with cell-based features for simultaneous partial nearest-neighbor-search (PNNS) recognition in all related windows can benefit from the sequential extraction of only local cell-descriptor vectors in this architecture. As a result, construction of SURF-FV and recognition processing are completed for each SW nearly at the same time. Furthermore, by operating with cell-based rather than window-based FVs, it can easily take advantage of multiple usage of each cell in the same SW and in different SWs.

B. Experimental Results

The prototype chip of the feature-based recognition coprocessor is fabricated in 65nm CMOS technology as shown in Fig. 2. The average power consumption is 515 µW at 0.5 V supply voltage and 12.5 MHz, which is sufficient for synchronization with real-time pixel-transfer speed at 30 fps from QQVGA-image sensors. The peak power consumption is 31.5mW at 1 V supply and 200MHz, which enables real-time recognition processing of VGA images.

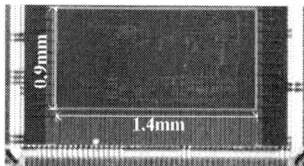

Chip size	1.4 mm×0.9 mm
Voltage	0.5 V-1 V
Clock	12.5-200 MHz
Power	0.515-31.5 mW
SRAM	28 KB

Figure 2. Die photo of the feature-based recognition coprocessor.

III. FACE RECOGNITION ACCELERATOR

A. Accelerator Architecture

In this accelerator design, we develop a Viola-Jones based object detection. The faces in an input image are firstly detected and then recognized in the following second stage (Fig. 3). In order to perform recognition consistently, the faces are normalized in size. PCA is used to extract Eigen-face from the candidate images, and SVM is utilized for classification [5].

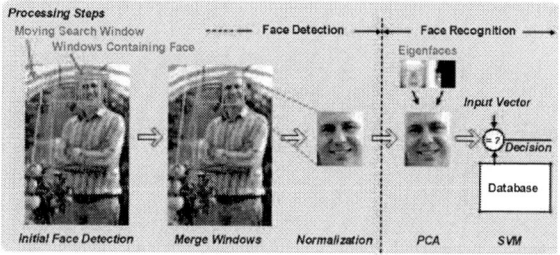

Figure 3. Processing steps of proposed accelerator.

The number of search windows tends to increase linearly as the image size grows, especial for HD images. As a result, detection speed is the bottleneck and feature memory accesses dominate energy consumption. Since some feature memory is accessed more frequently than others, the feature memory is separated into two parts; a small 1.5kB latch-based memory with lower read energy contains stages 1~5 while a large SRAM memory stores all remaining stages. Experiment on HD images shows that this scheme reduces read energy by 20% for 99.4% of rejected windows are discarded in the first 5 stages.

A hybrid-search scheme is proposed for further reducing the feature memory accesses with low energy consumption of the detection block. Instead of detecting windows around a face and then averaging to obtain an accurate center location, larger search steps are employed to coarsely detect a face. Smaller search steps are used around the point for better localization once a face is detected. This strategy reduces the number of search windows by 12× and achieves 93% accuracy. Recognition accuracy is 81% for 32 classes under the LFW face database [6].

The accelerator requires 492KB memory space to store coefficients such as support vectors ad features. The designed 5T-SRAM bitcell consists of an inverter pair and decoupled read transistor, which can provide aggressive low power operation and voltage scalability. With an L-shape layout, the logic-rule 5T bitcell occupies 7.2% less area than a conventional 6T bitcell; the read energy is 38% less compared to the 6T design.

B. Experimental Results

The face recognition accelerator is fabricated in 40nm COMS as shown in Fig. 4. The minimum voltage of a 12kB 5T SRAM array is 450mV. While processing HD images at 5.5fps throughput, the system consumes 23mW from 600mV with 4.5nJ/pixel, which enables continuous real-time face detection and recognition even in mobile applications.

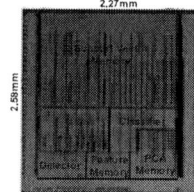

Technology	40nm
Vdd	600mV
Clock Freq.	100MHz
Core Area	2.58mm·2.27mm
Input Video	1280×720 (HD)
Power	23mW
Efficiency	4.5 nJ/pixel

Figure 4. Die photo of the face recognition accelerator.

IV. CONCLUSIONS

Computer vision algorithm is difficult to be implemented for real-time and low-power applications. Specially designed hardware accelerator can be a good solution to achieve high throughput and low power consumption. In this paper, we have reviewed a feature-based recognition coprocessor and a face recognition accelerator. In future, CNN-based common image signal processing will be further developed to compare with the works discussed in this paper.

REFERENCES

[1] G. Lowe, "Object recognition from local scale-invariant features," In International Conference on Computer Vision, Corfu, Greece, 1999, pp.1150–1157.

[2] G. Lowe, "Distinctive image features from scale-invariant keypoints", International journal of computer vision, 2004, pp.91-110.

[3] H. Bay, T. Tuytelaars, and V. Gool, "SURF: Speeded Up Robust Features," Computer Vision & Image Understanding, vol.110, no.3, pp.404-417, 2006.

[4] P. Viola, and J. Michael, "Robust real-time object detection". International Journal of Computer Vision, vol.4, 2001.

[5] D.S. Jeon, Q. Dong, Y.J. Kim, X.L. Wang, S. Chen, H. Yu, D. Blaauw, and D. Sylvester, "A 47mW Face Recognition Accelerator in 40nm CMOS with Mostly-Read 5T Memory", IEEE Symposium on VLSI Circuits, June 2015.

[6] G. Huang, M. Mattar, H. Lee, el., "Learning to align from scratch," Advances in Neural Information Processing Systems, 2012, pp. 764-772.

Image Interpolation Based on Hessian Analysis

Sangho Yoon
Department of Electrical Engineering
Pohang University of Science and Technology
Pohang, Republic of Korea
ysh4568@postech.ac.kr

Young Hwan Kim
Department of Electrical Engineering
Pohang University of Science and Technology
Pohang, Republic of Korea
youngk@postech.ac.kr

Abstract— **This paper proposes a novel interpolation method using edge orientation vector calculated by Hessian matrix. Existing polynomial-based interpolation methods cause blurring effects on edge. In addition, existing edge-based interpolation methods are suffered from edge aliasing and color distortion. To compensate for these problems, we propose an improved edge-based interpolation method considering edge direction on neighbor pixels. To interpolate image, unknown pixels are interpolated by internal division. To determine internal division ratio, the proposed method calculates the edge orientations for diagonal pixels and norm of pixel value differences for horizontal and vertical pixels. The proposed method improves the quality of interpolated images by increasing the average peak signal-to-ratio by 3.33 dB compared to benchmark method.**

Keywords; Hessian matrix; eigenvector orientation; internal division;

I. INTRODUCTION

Image interpolation is a technique to construct the unknown pixels among the known pixels. The purpose of image interpolation is to produce a high-resolution (HR) image from a low-resolution (LR) image.

Image interpolation methods are categorized into two types: polynomial-based and edge-based interpolation. Polynomial-based interpolation uses mathematically modeled polynomials such as bilinear or bicubic interpolation. However, the results of polynomial-based interpolation methods generate blurring effects on edge.

To compensate for edge blurring effect, edge-based interpolation consider edge pixels as outliers of interpolation [1]-[5]. New edge-directed interpolation (NEDI) minimizes linear mean square error of block-based weighted sum to approximately estimate pixel weight of interpolation [1]. Directional filtering and data fusion (DFDF) products two observation set which has orthogonal directions to consider two diagonal directions of edges [2]. Fast curvature-based interpolation (FCBI) selects interpolating direction by calculation of 2nd derivatives of two diagonal directions [3]. Soft-decision adaptive interpolation (SAI) uses maximum a posterior of HR-LR image block pairs based on least square approximation [4]. Robust SAI (RSAI) uses weighted least square to be robust to outliers [5]. However, those interpolation methods cause aliasing on edges because various edge orientations are not considered. In addition, color distortion

This research was supported by the LG Display and the MSIP (Ministry of Science, ICT and Future Planning), Korea, under the "ICT Consilience Creative Program" (IITP-R0346-16-1007) supervised by the IITP (Institute for Information & communication Technology Promotion).

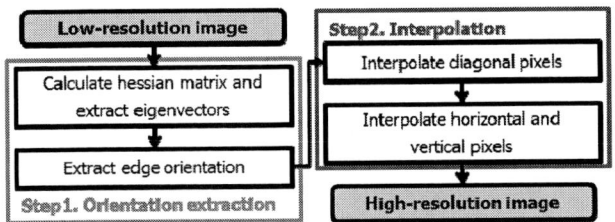

Figure 1. Flowchart of the proposed method

occurs nearby edges because of the large interpolating region and unstably assigned weight in interpolating process.

In this paper, we propose interpolation method by using edge orientation calculated by eigenvector of Hessian matrix to avoid aliasing on edge line. In addition, the proposed method uses internal division of pixel values for removal of color distortion. Section II presents proposed interpolation method considering orientations. Experimental results are presented in Section III. Finally, a conclusion is drawn in Section IV.

II. PROPOSED METHOD

The proposed image interpolation is composed of two steps (Fig. 1). First, orientation extraction is performed using eigenvectors of Hessian matrix to consider edge direction. Edge orientation is calculated by eigenvector corresponding to the largest eigenvalue of Hessian matrix. Each of variables is defined as follow:

$$v_1 = [x, y]^T, \theta = \tan^{-1}(y/x), a = |x|, b = |y|. \quad (1)$$

where v_1 is an eigenvector corresponding to the largest eigenvalue, θ is an orientation, and a and b are used as absolute values of (x, y).

Second, the unknown pixels are interpolated by internal division of neighbor pixels as follow:

$$X = \begin{cases} \dfrac{b-a}{2b}D + \dfrac{a+b}{2b}A, & \text{if } \dfrac{\pi}{4} \le \theta < \dfrac{\pi}{2} \\[2mm] \dfrac{a+b}{2a}B + \dfrac{a-b}{2a}A, & \text{if } 0 \le \theta < \dfrac{\pi}{4} \\[2mm] \dfrac{a+b}{2a}A + \dfrac{a-b}{2a}B, & \text{if } -\dfrac{\pi}{4} \le \theta < 0 \\[2mm] \dfrac{b-a}{2b}A + \dfrac{a+b}{2b}D, & \text{otherwise} \end{cases} \quad (2)$$

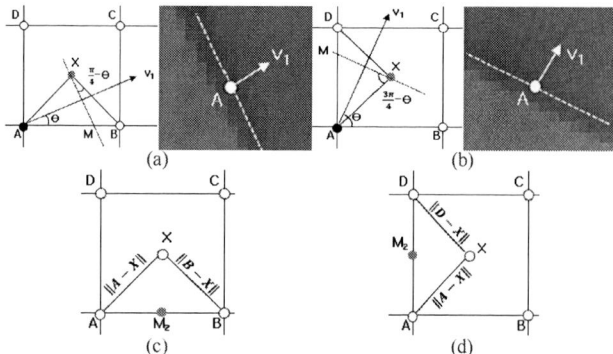

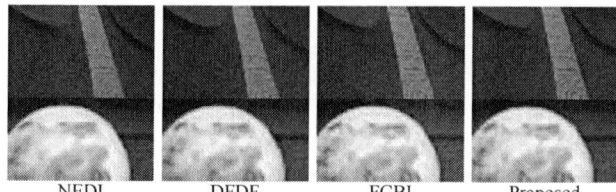

Figure 3. Interpolated image result of the benchmark methods and the proposed method

Figure 2. Interpolation of diagonal pixels with (a) acute angle and (b) obtuse angle, interpolation of (c) horizontal pixels, and (d) vertical pixels

where each parameter represents as follow: X is an unknown diagonal pixel, and A, B, C, D are relatively corresponding to the lower left pixel, the lower right pixel, the upper right pixel, and the upper left pixel as shown in Fig. 2. Diagonal pixels are interpolated by internal division using orientation vectors. Along the edge direction which is perpendicular to v_1, similar pixel values are distributed. Thus, if the unknown diagonal pixel X moves along the edge direction, it meets line AB or line AD. Its intersection point is denoted as M. Then, pixel X can be assumed to have same pixel value of M. If the case of acute angle, M is internal pixel between line AB and it can be calculated by internal division between A and B (Fig. 2a). Its division ratio can be expressed with respect to the absolute value of elements of v_1. In the case of obtuse angle, M is internal pixel between line AD and it can be calculated by internal division through the same process (Fig. 2b).

After interpolation of diagonal pixel, horizontal and vertical pixels are interpolated as follow:

$$M_2 = \begin{cases} \dfrac{\|A-X\|\cdot B + \|B-X\|\cdot A}{\|A-X\| + \|B-X\|}, & \text{if } M_2 \text{ is a horizontal pixel} \\ \dfrac{\|A-X\|\cdot D + \|D-X\|\cdot A}{\|A-X\| + \|D-X\|}, & \text{if } M_2 \text{ is a vertical pixel} \end{cases} \quad (3)$$

where M_2 is an unknown horizontal or vertical pixel (Fig. 2c, 2d). To estimate horizontal and vertical pixels, we assume that unknown pixel M_2 has the same pixel value between the interpolated diagonal pixel X and two neighbor pixels. In the case of horizontal pixel, neighbor pixels are A and B (Fig. 2c); otherwise, neighbor pixels are A and D (Fig. 2d). Then M_2 is an internal division of neighbor pixels and its division ratio is expressed with respect to norm of pixel value difference.

III. EXPERIMENTAL RESULTS

The proposed interpolation method was tested on 24 natural images from Kodak [6]. Benchmark interpolation methods are as follow: bicubic interpolation, new edge-directed interpolation (NEDI) [1], directional filtering and data fusion (DFDF) [2], fast curvature-based interpolation (FCBI) [3], soft-decision adaptive interpolation (SAI) [4], and robust soft-decision adaptive interpolation (RSAI) [5]. Quality of images generated by each interpolation method was evaluated by peak signal-to-ratio (PSNR) [7] with the original images. The proposed method increases the average of PSNR by 31.16 dB compared to NEDI (TABLE I).

In addition, the proposed method generates visually improved interpolated images (Fig. 3). Aliasing on edge line are alleviated by considering various edge orientation. In addition, color distortion is removed because the proposed method uses internal division to interpolate unknown pixels.

IV. CONCLUSION

This study proposed a new image interpolation method that uses internal division of neighbor pixels. To estimate division ratio, eigenvectors of image Hessian matrix are calculated. Diagonal pixels are estimated by the internal division with the orientation of eigenvector corresponding to the largest eigenvalue. Horizontal and vertical pixels are estimated by the internal division of neighbor pixels with pixel differences. In the experiment results, aliasing and color distortion is alleviated. As a result, the average PSNR was improved by 31.16 dB.

REFERENCES

[1] X. Li and M. T. Orchard, "New edge-directed interpolation," IEEE Trans. Image Process., vol. 10, no. 10, pp. 1521-1527, 2001.

[2] L. Jhang and X. Wu, "An edge-guided image interpolation algorithm via directional filtering and data fusion," IEEE Trans. Image Process., vol. 15, pp. 2226-2238, 2006.

[3] A. Giachetti and N. Asuni, "Fast artifact free image interpolation," Proc. Brit. Machine Vis. Conf. (BMVC), pp. 123-132, 2008.

[4] K.-W. Hung and W.-C. Siu, "Fast image interpolation using bilateral filter," IET Image Processing, vol. 6, no. 7, pp. 1-14, 2012.

[5] K.-W. Hung and W.-C. Siu, "Robust soft-decision interpolation using weighted least squares," IEEE Trans. on Image Process., vol. 21, no. 3, pp. 1061-1069, 2012.

[6] The Kodak color image dataset, http://r0k.us/graphics/kodak/.

[7] Z. wnag, A. C. Bovik, H. R. Sheikh, and E. P. Simoncelli, "Image quality assessment: from error visibility to structural similarity," IEEE Trans. Image Process., pp. 600-612, 2004.

TABLE I. PSNR(dB) OF BENCHMARK METHODS AND THE PROPOSED METHOD

	Bicubic	NEDI	DFDF	FCBI	SAI	RSAI	Proposed
Avg.	28.15	27.83	28.02	28.84	29.93	30.05	**31.16**
Max.	32.09	32.13	32.80	34.67	35.91	**36.24**	34.69
Min.	22.13	21.70	22.07	22.14	22.69	22.69	**25.02**
Std.	2.78	2.84	2.84	3.43	3.77	3.77	**2.69**

Sharpness-aware Real-time Haze Removal for Advanced Driver Assistance Systems

Joonggeun Ahn, Jihoon Kim and Youngjoo Lee

Department of Electronic Engineering
Kwangwoon University
20, Gwangun-ro, Nowon-gu, Seoul 01897, Republic of Korea
Email: jgahn.ims@gmail.com, yjlee@kw.ac.kr

Abstract— *This paper proposes a new haze-removal scheme for automotive applications. In contrast that the previous works focus on enhancing the quality of haze-free images, the proposed algorithm sharpens the edges of lanes by adopting the weighting filter. Moreover, the previous internal steps, which cannot improve the sharpness of lanes, are removed to reduce the computing overheads. As a result, the proposed haze-removal algorithm improves the sharpness of lanes remarkably, while providing the real-time image processing for advanced driver assistance systems.*

Keywords; Advanced driver assistance systems; Lane detection; Haze removal algorithm; Real-time image processing

I. INTRODUCTION

The camera-based lane detection is one of the key technology of the advanced driver assistance system (ADAS) [1]. However, the previous works cannot recognize lanes when the input images are unclear due to the poor weather conditions including rainy, dull, foggy and misty weathers. In automotive systems, therefore, the haze removal is an essential preprocessing for ensuring the safety of camera-based detection solutions.

This paper presents a new haze removal algorithm, which is suitable for the real-time ADAS. Unlike to the conventional complex removal schemes that only focus on improving the general image qualities [2]–[4], the proposed work concentrates on sharpening the edges of lanes with acceptable complexity. Due to the clear edges from the proposed scheme, therefore, the following detection step can recognize more lanes effectively. Targeting the real-time solution, in addition, we remove several operations in the conventional haze removal, which are not related to the sharpness of lanes. Compared to the previous work, simulation results show that the proposed algorithm enhances the sharpness of lanes by 51%. Due to the simplified operations, the processing speed is also boosted up to 34 fps, supporting the real-time automotive applications successfully.

II. PREVIOUS WORK

The effects of haze on the image can be model as

$$I(x) = J(x)T(x) + A(1-T(x)), \qquad (1)$$

where $I(x)$ is the captured image that suffers from haze, $J(x)$ is the haze-free image, $T(x)$ is the transmission of colored image, and A is the airlight coefficient [4]. In order to recover the haze-free image $J(x)$, the previous haze removal algorithm calculates

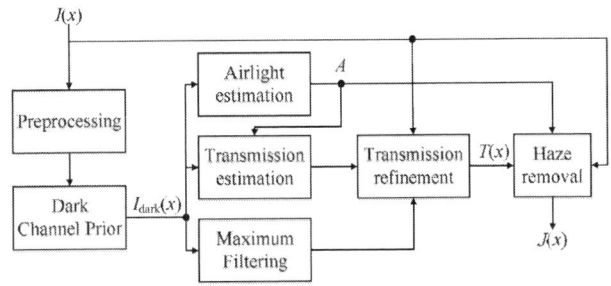

Figure 1. The conventional haze removal algorithm [4].

$T(x)$ and A by adopting the concept of dark channel prior (DCP) [2]. Fig. 1 shows the processing steps of the previous DCP-based haze removal [4]. In the previous algorithm, the input image $I(x)$ is firstly moved to the preprocessing step to adjust the illumination intensity. Then, the amount of haze intensity $I_{dark}(x)$ is obtained by using DCP step followed by the maximum filter. Based on $I_{dark}(x)$, the transmission and airlight values are estimated as depicted in Fig. 1. Note that the estimated $T(x)$ is refined by using the guided filter for better image quality. Finally, the haze-free image is recovered by using the input image, the refined transmission, and the airlight information.

Although the previous algorithm reconstructs the haze-free image in effect, each processing step is proposed to improve the quality of haze-free images. In the lane detection for ADAS, however, the edge information should be highlighted rather than the similarity between the original and the reconstructed images. To improve the detection accuracy of lanes, therefore, we newly propose the sharpness-aware haze-removal technique.

III. PROPOSED HAZE REMOVAL ALGORITHM

Fig. 2 illustrates the proposed haze-removal algorithm. To increase the sharpness of lanes, the weighting filter is newly addressed after performing the haze removal. Basically, the proposed algorithm is based on the hypothesis that the road is generally darker than lanes. To validate this concept, we make a histogram showing the distributions of grey-level pixels in roads as depicted in Fig. 3(a). Note that most of values are bounded by a certain threshold as the area of bright lanes is negligible by nature. The filter widens a gap in pixel values between road and non-road regions, leading to the sharp edges. When the pixel value is less than the threshold α, as shown in Fig. 3(b), the proposed filter multiplies the corresponding pixel

---

This work was supported by the National Research Foundation (NRF) grant funded by the Korea government (MSIP) (2016R1C1B1007593).

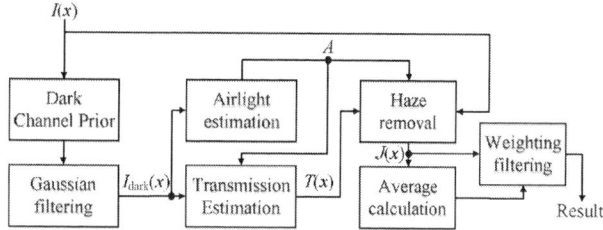

Figure 2. The proposed sharpness-aware haze-removal algorithm.

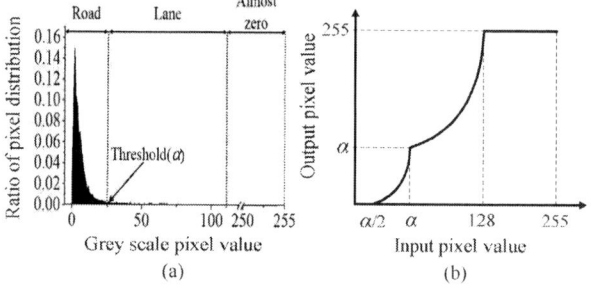

Figure 3. (a) The distribution of pixel values of the road image and (b) the transfer function of the proposed weighting filter.

Figure. 4 (a) The original haze image, (b) the haze-free image from the previous work [4], and (c) the haze-free image using the proposed method.

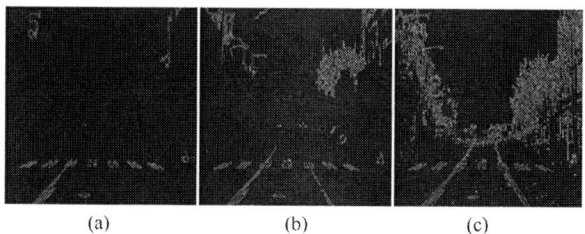

Figure. 5 The edge maps from (a) the original image, (b) the haze-free image using the previous work [4], and (c) the haze-free image associated with the proposed algorithm.

TABLE I.   SIMULATION RESULTS

	MTF50 $(10^{-3}$ c/p)	Processing speed[a] (fps)
Original haze image	62	N. A.
Conventional algorithm [4]	150	25.31
Proposed algorithm	227	34.02

a. Using Open CV on PC with 3.6GHz intel Core i7 4770 processor

value by less than one to further darken the road region. On the other hand, the pixels are gradually brightened by adjusting the weighting factor properly. Note that the maximum weighting factor is limited by two so that the shape of lanes is preserved.

To reduce the computational complexity, the haze removal itself is also modified as shown in Fig. 2. As we are focusing on the sharpness of lanes not the quality of the haze-free image, the maximum filter for adjusting DCP results is eliminated. Instead, the Gaussian filter is introduced for eliminating noises in $I_{dark}(x)$ that may cause unwanted false edges. Due to the smoothing effect of the Gaussian filter, the time-consuming guided filter is also removed in our work. Therefore, the proposed algorithm relaxes the processing complexity even it adds the weighting filter after obtaining the haze-free image.

IV. EXPERIMENTAL RESULTS

In this section, we compare the proposed sharpness-aware haze removal to the previous work in [4]. For the same sample haze image shown in Fig. 4(a), the haze-free images from the previous method and the proposed algorithm are depicted in Fig. 4(b) and Fig. 4(c), respectively. Note that the proposed work produces more clear lanes even it degrades the overall image quality. To show the impact of the proposed work more precisely, the edge detection process is applied to each image as shown in Fig. 5 [5]. It is noticeable that the edge map based on the proposed work in Fig. 5(c) is superior to the other maps from the original haze image and the one using the previous method as depicted in Fig. 5(a) and Fig. 5(b), respectively. For quantitative comparisons, as summarized in Table 1, we use the concept of modulation transfer function (MTF) that measures the sharpness of images [6]. Compared to the previous work, the proposed method enhances the MTF50 metric by 51%. By optimizing the internal operations, in addition, our algorithm processes 34 frames per second, which is 1.3 times faster than

the previous work. Targeting the automotive system, therefore, the proposed haze-removal technique provides more reliable edge information even boosting the processing rates remarkably.

V. CONCLUSION

In this paper, we have presented a new haze removal scheme sharpening the lane edges for the following lane detection in automotive solutions. The proposed weighting filter effectively highlights the lanes and the optimized haze removal steps shorten the processing delay. Hence, the proposed approach is suitable for ensuring the safety of the real-time ADAS by providing the clear lanes even in the unclear weather conditions.

REFERENCES

[1] M. Park, K. Yoo, and Y. Lee, "Diagonally-reinforced lane detection scheme for high-performance advanced driver assistance systems," *IEIE J. Semicond. Technol. Sci.*, in press.
[2] V. Toka *et al.*, "A fast method of fog and haze removal," in *Proc. IEEE Int. Conf. Acoust. Speech. Signal Process. (ICASSP)*, 2016, pp. 1224–1228.
[3] Y.-H. Shiau, H.-Y. Yang, P.-Y. Chen, and Y.-Z. Chuang, "Hardware implementation of a fast and efficient haze removal method," *IEEE Trans. Circuits Syst. Video Technol.*, vol. 23, no. 8, pp. 1369–1374, Aug. 2013.
[4] W.-T. Kim, H.-W. Bae, and T.-H. Kim, "Fast and efficient haze removal," in *Proc. IEEE Int. Conf. Consum. Electron. (ICCE)*, 2015, pp. 360–361.
[5] J. Canny, "A computational approach to edge detection," *IEEE Trans. Pattern Anal. Mach. Intell.*, vol. PAMI-8, no. 6, pp. 679–695, June 1986.
[6] G. D. Boreman, *Modulation transfer function in optical and electro-optical systems*, Bellingham, WA: SPIE Press, 2001.

A New Scheme for Secret-Hiding in DSP Circuits

Sumedh Dhabu and Chip-Hong Chang

School of Electrical and Electronic Engineering, Nanyang Technological University, Singapore.

Abstract— **This paper presents a new secret-hiding method for watermarking or fingerprinting the digital signal processing circuits. It uses the characteristic noise introduced by the truncation or rounding off of the results of arithmetic operations to distinguish between the otherwise identical circuits. Watermark is physically and functionally integrated with the circuit and can be extracted dynamically. This unique advantage is exemplified by a practical scenario of wireless sensor node authentication.**

I. INTRODUCTION

The increasing complexity of digital IC designs has led to a paradigm shift in the design strategies, where the designer uses off-the-shelf intellectual property (IP) cores to increase the design productivity. The growing market for IP reuse has necessitated the use of watermarking or fingerprinting to deter IP theft and unauthorized IP distribution. Digital signal processing (DSP) circuits such as digital filters can be found in virtually all kinds of intelligent devices (e.g., in biomedical signal processing systems, communication systems etc.) and thus are an important class of IP cores (and a potential location for secret-hiding within an integrated system). In this paper, we propose to embed the watermark (or any secret) into the fixed-point implementation of a DSP circuit. The truncation and rounding off errors, which have been traditionally considered as noise, are used advantageously for watermark obfuscation. The watermark can be easily extracted on-the-fly even for a packaged chip by legitimate users without disturbing the normal operation - a pragmatic feature that is not found in existing watermarking schemes in signal processing circuits.

II. PROPOSED SCHEME

Consider an arithmetic operation (e.g., multiplication) in the fixed-point implementation of a DSP circuit, where the precision of the output can be different from the precision of the input(s) and none of the inputs are zero. Let FB_1 and FB_2 be the number of bits used for the fractional part of two multiplicands. Provided that the error introduced by this change of precision is tolerable, it suffices to use FB_3 ($< FB_1 + FB_2$) bits in the fractional part of the output. In this case, the designer has a choice of using either truncation or rounding off for all such nodes throughout the designed circuit. In the proposed method, an M-bit secret (key) is embedded in the circuit by the choice of truncation vs. rounding off at M such nodes, e.g., a zero bit of the key dictates the use of truncation for a corresponding node and a one bit dictates rounding off. For a circuit with M nodes, there are 2^M possible circuits corresponding to 2^M key combinations. All the 2^M circuits are essentially identical from functionality point of view (e.g., frequency response specifications of a digital filter). The IP designer has in principle $C = 2^M - 2$ potential choices to embed his watermark, excluding the two obvious choices of truncation at all nodes and rounding off at all nodes. Therefore, the probability of coincidence is $P_c = 1/C$.

As the output of each circuit is characterized by the choice of truncation vs. rounding off at each of its M nodes determined by the M-bit key, the input signals that result in the output signals for all the circuits being different can be used to differentiate between these otherwise identical circuits to detect the embedded key, watermark or fingerprint. The database of input signals (and the corresponding output signals) to be used for detecting the embedded key, can be created offline before the actual circuit implementation as follows. A random input signal consisting of L samples is applied to all the circuits. The corresponding output signals, each consisting of L (or more) samples, are compared. If the output signals of any two circuits are exactly same (i.e. all the samples in the two output signals are identical), it is discarded. If the output signals of all the circuits are different, then this input signal and its corresponding output signals are stored in the database. A database of such I input signals (I can be arbitrarily large) and their corresponding output signals is established. This database is created at HDL and simulation level and can therefore be automated, without physically implementing all the circuits.

For detecting the embedded key or extracting the watermark / fingerprint, one of the stored input signals from the database is applied to the circuit in question. By comparing with the corresponding stored output signals, the embedded key can be identified. For every input signal from the stored I signals, the output signal of each of the circuits is different from the output signal of the rest. Therefore, ideally the embedded key can be detected with 100% accuracy. To avoid false rejection due to random signal processing errors, multiple input signals can be used.

This method is beneficial in light-weight hardware authentication in sensor networks, wireless body area networks, etc. Consider a sensor network with a master node and any slave sensor nodes within its periphery can connect to and communicate with it. It is important to verify the identity of each and every sensor node connecting in the network. First, M nodes in the signal processing circuit of each slave sensor are identified and the corresponding database is created. The master sensor node in the sensor network maintains this database of input and output signals and keys. Each of the slave sensor nodes is given a unique M-bit identification key and it contains the circuit designed with combination of truncation vs. rounding off at selected nodes determined by its M-bit identification key. To authenticate a slave node, the master node selects and sends a random input signal(s) from the database to the slave sensor node. By comparing the received

---

This research is supported by the Singapore Ministry of Education AcRF Tier I grant MOE 2014-T1-002-141 (RG186/14)

TABLE I. QUALITATIVE COMPARISON OF PROPOSED SCHEME WITH EXISTING WATERMARKING METHODS FOR SIGNAL PROCESSING CIRCUITS

Proposed scheme	Comparison with other methods
Number of allowable watermark bits (M) depends on order of the filter (N). For FIR filter, $M = N/2 + 1$ (only multipliers are marked) or $M = N/2 + 1 + N$ (adders can also be marked). N is preserved upon watermarking. Impulse response of all the filters is exactly the same.	In [1, 2], number of allowable watermark bits is limited by the frequency response discretization. Order of the watermarked filter needs to be increased in [1], e.g., $N = 31$ for an unmarked filter becomes $N = 40$ with decipherable watermark of $M = 7$. In our method, for $N = 40$, $M = 21$ (or $M = 41$ if adders are also marked). In [1, 2], watermark is not stealthy, as watermarked filters can be distinguished by observing the impulse response of various filters.
Suitable for higher order filters. NOT dependent on implementation platform or technology.	For higher order filters, the choices of memory locations for storing coefficient bits will be limited and therefore area and timing overheads may become significant for method in [4]. Also, it is dedicated to FPGA based IPs.
NO additional information required for extraction of watermark. It can be done in a non-invasive way.	In [1, 3, 4], there is no clear way of tracking and identifying a watermark. Additional netlist-level or layout-level information about the chip in question is required to identify watermark, which may not be available.
NO additional circuitry is needed for sensing the watermark extraction sequence or routing output, etc.	Method in [5] needs additional circuitry to detect the presence of signature extraction sequence at the input, to generate proper signals for the application of signature input patterns, and to route the signature bit blocks to the system output.

TABLE II. BEST AND WORST CASE AREA AND DELAY FOR FPGA AND ASIC IMPLEMENTATION OF FIR FILTERS WITH $N = 128$ AND 256

		$N_1 = 128$, $M_1 = 65$, $P_c = \sim 2.7 \times 10^{-20}$		$N_2 = 256$, $M_2 = 129$, $P_c = \sim 1.5 \times 10^{-39}$	
		All truncation	All rounding off	All truncation	All rounding off
FPGA	Area (number of slices)	2346	2769 (+18%)	4279	5078 (+19%)
	Post-PAR minimum period (in ns)	3.0	2.7	2.7	2.9
ASIC	Area (in μm^2)	180466.5	194659.4 (+8%)	307408.9	339872.0 (+11%)
	Delay (in ns)	2.2	2.4	2.3	2.5

output signal returned by the slave node against the database, the master node can validate the identity of the sensor node. This is more secure than relying on the key stored in the memory of the slave sensor node for identification, which can be easily copied or tampered with. Also, whenever a node is being authenticated, the same identity message is sent every time, making it vulnerable to eavesdropping and playback. On the contrary, with our proposed method, a different input signal can be selected for interrogation by the master node every time and therefore the output from the slave node will be different.

III. COMPARISON AND IMPLEMENTATION RESULTS

A qualitative comparison of the proposed scheme with existing works targeting DSP circuits (mainly the digital filters) is provided in Table I. To appreciate the variation in hiding different secret, the upper and lower bounds on the area of a watermarked circuit can be approximated by rounding off at all the nodes and truncation at all the nodes, respectively. Consider two symmetric coefficient FIR filters with $N_1 = 128$ and $N_2 = 256$, implemented in transposed direct form. When only the multipliers in these filters are considered for watermark embedding by our proposed scheme, the maximum allowable number of watermarking bits are $M_1 = 65$ (with $P_c = \sim 2.7 \times 10^{-20}$) and $M_2 = 129$ (with $P_c = \sim 1.5 \times 10^{-39}$). The two bounds of embedding an M-bit key for these filters are presented in Table II by implementing them in Xilinx Virtex 6 xc6vlx760-1ff1760 FPGA and in ASIC based on STM CMOS 65 nm LPGP 0.9 V standard cell library. The mapping to FPGA is performed using Xilinx ISE whereas the ASIC circuits are synthesized by Synopsys Design Compiler. The percentage increase in the area of each filter with rounding off at all the nodes over that with truncation at all the nodes is indicated in parenthesis. It should be noted that for the same filter order, the embedding capacity can be increased to $M =$ $N/2 + 1 + N$ by including truncation vs. rounding off of adders for watermarking. While the bounds of embedding cost are demonstrated with two FIR filter examples due to the page constraint, the scheme is applicable to other DSP circuits wherein truncation or rounding off of arithmetic results in a DSP circuit to match the final output precision will not violate the overall system specifications.

IV. CONCLUSION

This paper proposed a novel idea of embedding the watermark bits in terms of choice of truncation vs. rounding off at various nodes in a signal processing circuit and using these characteristic errors to detect the watermark. This scheme is also suitable for IP fingerprinting and hardware authentication.

REFERENCES

[1] A. Rashid, J. Asher, W. H. Mangione-Smith, and M. Potkonjak, "Hierarchical watermarking for protection of DSP filter cores," in *Proc. 1999 IEEE Custom Integrated Circuits*, San Diego, California, USA, May 1999, pp. 39-42.

[2] R. Chapman and T. S. Durrani, "IP protection of DSP algorithms for system on chip implementation," in *IEEE Trans. Signal Processing*, vol. 48, no. 3, pp. 854-861, Mar 2000.

[3] J. L. Wong, J. Q. Ya, and M. Potkonjak, "Watermarking multiple constant multiplications solutions," in *Proc. 38th Asilomar Conf. on Signals, Syst. and Computers*, Pacific Grove, California, USA, Nov. 2004, vol. 1., pp. 67-71.

[4] Wei Dai, H. K. Kwan, and Huapeng Wu, "IP protection for FPGA implementation of DSP algorithms," in *Proc. 48th Midwest Symp. Cir. and Syst.*, Covington, Kentucky, USA, Aug. 2005, vol. 2, pp. 1418-1421.

[5] E. Castillo, U. Meyer-Baese, A. Garcia, L. Parrilla, and A. Lloris, "IPP@HDL: Efficient Intellectual Property Protection Scheme for IP Cores," *IEEE Trans. Very Large Scale Integration (VLSI) Syst.*, vol. 15, no. 5, pp. 578-591, May 2007.

A Programmable ΔΣ SAR-ADC with Charge Shuttling Technique

Kohei Yamada, Yosuke Toyama, and Hiroki Ishikuro

Keio University, Yokohama, Japan

E-mail: yamada@iskr.elec.keio.ac.jp

Abstract— **This paper presents an ADC with programmability between SAR-only mode and delta-sigma (ΔΣ) assisted mode. The ΔΣ assisted mode brings 1st order noise shaping for resolution enhancement. Proposed charge shuttling technique makes it possible to share a charge re-distribution capacitor array for DAC in SAR, feedback DAC, and integrator capacitor in ΔΣ loop and improve the accuracy. The prototype ADC fabricated in 65-nm CMOS achieved SNDR of 44.35 dB at sampling rate of 32 MHz and power consumption of 0.55mW. The SNDR is improved to 62.9dB by ΔΣ assisted mode when the signal bandwidth is 60 kHz.**

Keywords; programmable, SAR, delta-sigma, charge shuttling

I. INTRODUCTION

Currently, Internet of Things (IoT) and M2M are attracting much attention, which are based on the wireless sensor network. Since there is vast variety of sensing signal, required circuit performance for sensing interface strongly depends on the application. From the view point the reduction of cost and turn-around-time for development, the function and performance of the interface circuits should be programmable.

Charge re-distribution type SAR-ADCs have achieved low power consumption and high-energy efficiency. Resolution variable SAR architectures with split-capacitor or MUX have been reported [1-3]. However, the influences of parasitic capacitance of bridge capacitor and component mismatches make it difficult to obtain high resolution. To improve the resolution, architectures with noise shaping by ΔΣ modulation in SAR-ADC has been reported [4,5]. Ref. [5] has achieved SNR higher than 120dB, but OpAmps for integrator have to operate N times to obtain resolution improvement, which increase the conversion time.

In this work, programmable ΔΣ SAR-ADC has been developed. The proposed architecture introduces charge shuttling technique which makes it possible to share the capacitor array for DAC in SAR, feedback DAC and integrator capacitor in ΔΣ loop and reduce the operation time of OpAmp.

II. PROPOSED ΔΣ SAR-ADC WITH CHARGE SHUTTLING TECHNIQUE

First, Fig.1 shows the proposed circuit schematic diagram and a timing chart in the ΔΣ assisted mode. The ADC consists of 8-bit binary weighted capacitor array, inverter based OpAmp for integrator, comparator, SAR logic, and mode control logic. The nonlinearity of the ADC can be calibrated by post signal processing using INL data of 8-bit capacitor array. Although, the schematic is described as a single-ended form for simplicity, the actual circuit was implemented in differential form. The comparator has one differential pair which detects the differential voltage of the capacitor array.

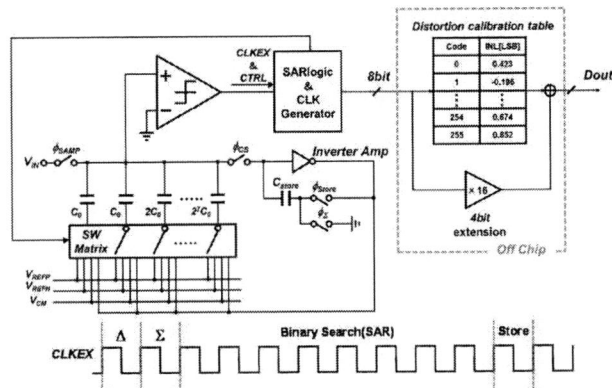

Fig. 1. The proposed circuit diagram and the timing chart

Fig.2 shows the key operation of the proposed charge shuttling technique. At first, input signal is sampled at the top plate of capacitor array while each bottom plates are set at V_{REFP} or V_{REFN} depending on the AD conversion results of previous cycle. This corresponds to the subtraction (Δ operation). At Σ operation phase (Fig.2 (a)), the bottom plate of the capacitor array are connected to the output of the OpAmp and the bottom plate of a store capacitor (C_{store}) is connected to common mode voltage. As a result, the charge stored in C_{store} during the previous conversion cycle is transferred into the capacitor array and signal integration is carried out. At quantization phase, since the integrated signal is in the capacitor array, a two-input comparator (one differential pair comparator) can be used to convert the signal to digital code by SA procedure. After the quantization phase, the bottom plates of the capacitor array are connected to common mode voltage and bottom plate of C_{store} is connected to the output of OpAmp. As a result, the integrated signal charge in the capacitor array returns to the C_{store} (Fig.2 (b)). In the proposed technique the signal charge shuttles between the capacitor array and C_{store} in one conversion cycle.

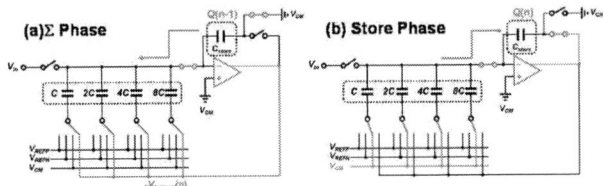

Fig. 2. The proposed charge shuttling operation

III. Measurement Results

The prototype ADC was fabricated in 65nm CMOS process and occupied circuit area is 150μm x 200μm. The total capacitance of 8-bit binary-weighted capacitor array is 512fF.

Fig.3 shows the relation between sampling frequency versus SNDR of the 8bit SAR-only mode, the DNL and INL of the capacitor array. In the 8bit SAR-only mode, flat frequency response is obtained from low frequency to about 30MHz. The achieved ENOB is 7.735bit. Although the SAR-only mode has power scalable characteristic, the power consumption saturates at 130μW at low sampling frequency because of leakage power. The leakage power can be suppressed by power gating technique [7]. The mismatch of the 8-bit capacitor array can be detected by using a technique in ref. [7.8]. The ΔΣ assisted mode can be used for precise detection of the mismatch of capacitor array. Once the table data of the INL is obtained, it is used for nonlinearity calibration by digital post signal processing. The circuit overhead is smaller than the usually used technique such as dynamic element matching. Fig.4 shows the measured spectrum of ΔΣ assisted mode when the input signal level and frequency are -9 dBFS and 12kHz, respectively. SNDR of 50.5dB is obtained when the signal bandwidth is 60kHz. SNDR was improved by 12.4 dB when the digital post-signal process is carried out.

The chip photograph is shown in Fig.5 and measured ADC performance is summarized in TABLE I.

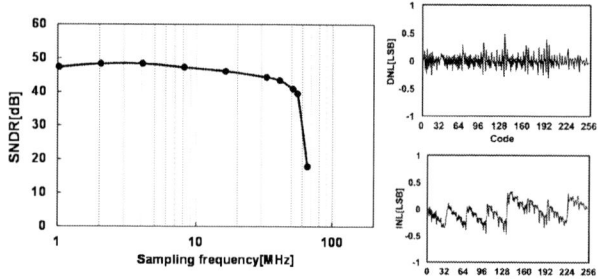

Fig. 3. Sampling frequency versus SNDR of SAR mode and DNL/INL of the 8bit capacitor array

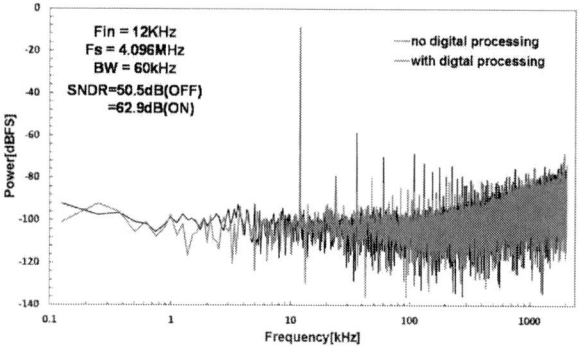

Fig. 4. FFT spectrum at ΔΣ assisted mode

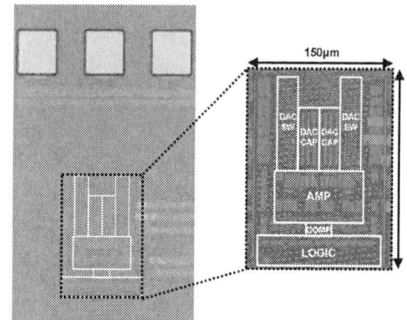

Fig. 5. Chip microphotograph

TABLE I
Measured Performance Summary

	SAR-Only mode	*ΔΣ-Assisted mode*
V_{DD}	1.2V(analog & digital)	
CMOS Process	65nm	
Sampling Frequency	32MHz	4.096MHz
Bandwidth	Nyquist	60kHz
SNDR	44.35dB	62.9dB
Power	550.4μW	264μW
FoM	124fJ	1.93pJ

IV. Conclusion

A programmable ΔΣ SAR-ADC is presented. The proposed charge shuttling technique which enabled the time-sharing of capacitor array between DAC in SAR, feedback DAC and integrator capacitor in ΔΣ loop. The prototype chip fabricated in 65nm CMOS has achieved 32MS/s and 44.35dB SNDR at 8-bit SAR-only mode. By the combination of ΔΣ assisted mode and post-signal processing, SNDR of 62.9dB is obtained within the signal bandwidth of 60kHz.

References

[1] Pieter Harpe, et al., "A 7-to-10b 0-to-4MS/s Flexible SAR ADC with 6.5-to-16fJ/conversion-step" *ISSCC* pp. 472-474, 2012.

[2] Marcus Yip, et al., "A Resolution-Reconfigurable 5-to-10-Bit 0.4-to-1 V Power Scalable SAR ADC for Sensor Applications" *IEEE JSSC*, Vol. 48, No. 6, pp. 1453-1464, 2013.

[3] Zhangming Zhu, et al., "A 6-to-10-Bit 0.5 V-to-0.9 V Reconfigurable 2 MS/s Power Scalable SAR ADC in 0.18 CMOS" *IEEE TCAS*, Vol. 62, No. 3, 2015.

[4] Jeffrey A. Fredenburg, et al., "A 90-MS/s 11-MHz-Bandwidth 62-dB SNDR Noise-Shaping SAR ADC" *IEEE JSSC.*, Vol. 47, No. 12, 2012.

[5] Youngcheol Chae, et al., "A 6.3 μW 20 bit Incremental Zoom-ADC with 6 ppm INL and 1mV Offset" *IEEE JSSC*, Vol. 44, No. 2, 2012.

[6] Sekimoto.R, et.al., "A 0.5-V 5.2-fJ/Conversion-Step Full Asynchronous SAR ADC With Leakage Power Reduction Down to 650 pW by Boosted Self-Power Gating in 40-nm CMOS" *IEEE JSSC.*, Vol. 48, No.11, pp. 2628-2636, 2013.

[7] James.L.McCreary,et al., "Precision Capacitor Ratio Measurement Technique for Integrated Circuit Capacitor Arrays," IEEE Trans. Instrument and Measurement, vol. IM-28, no. 1, pp.11-17, 1979.

[8] A. Shikata, et al., "A 0.5 V 1.1 MS/sec 6.3fJ/Conversion-Step SAR-ADC With Tri-Level Comparator in 40 nm CMOS," *IEEE JSSC*, Vol.47, No.4, pp. 1022-1030, 2012.

Gap in pagination due to withheld paper.

Pages 53-54

11-bit 1.8uW 40KS/s Segmented SAR ADC for Sensor Applications

Behnam Samadpoor Rikan, Sang-Yun Kim and Kang-Yoon Lee

College of Information and Communication Engineering Sungkyunkwan University

Suwon, South Korea

Email: [behnam, ksy0501, klee]@skku.edu

Abstract— **This paper proposes an 11-b 40KS/s Successive Approximation Register (SAR) Analog-to-Digital Converter (ADC) structure for sensor applications. Segmented structure is adopted in capacitive DAC to improve the linearity and decrease the power consumption. 500 aF custom-designed unit capacitors are applied in CDAC to reduce the area and to keep the INL and DNL within 1 LSB of an 11 bit ADC. A prototype ADC was implemented in CMOS 0.18μm technology. This structure consumed 1.8μW and achieved 67.27-dB SNDR and 83.7-dB SFDR at 40KS/s under a 1.8-V supply. The figure of merit (FOM) was 37fJ/conversion-step.**

Keywords- SAR ADC; segmented DAC; linearity; thermometer codes

I. INTRODUCTION

Low-power sensor applications applied in environmental monitoring, bio-potential recording and wireless autonomous sensor networks, require highly power-efficient ADCs, typically with resolutions over 10-b. SAR ADC requires mainly digital circuits and simple analog circuits. Nonetheless, a SAR ADC traditionally requires good capacitor matching to achieve the resolution of 10-b or higher, which leads to large ADC size. The aim of this work is to increase the accuracy of highly efficient SAR ADCs beyond 10-b while improving the power efficiency and to keep the area in reasonable size. For this purpose segmented SAR ADC is applied in which 4 MSB bits are decided with thermometer codes [1]. 500aF custom-designed unit capacitors are applied in CDAC to reduce the area and to keep the INL and DNL within 1 LSB for an 11-b ADC.

II. PROPOSED SAR ADC STRUCTURE

The advantage of a binary-weighted DAC is its simplicity because no decoding logic is required but there are several major drawbacks. These drawbacks are all associated with major bit transitions. At the mid-code transition, the most significant capacitor value needs to be matched to the sum of all the other capacitor values within 0.5 LSB's. Because of statistical spread, it is difficult to achieve and such matching can never be guaranteed. Therefore the monotonicity is not guaranteed which can result in a typical differential nonlinearity (DNL) plot. In addition, the errors caused by the dynamic behavior of the switches result in glitches in the output signal. Such glitches contain highly

nonlinear signal components, even for small output signals and will manifest itself as spurs in the frequency domain.

In order to solve the above-mentioned drawbacks, segmented SAR ADC is designed here in which the 4 MSB bits are decided by thermometer codes. Fig. 1 presents the proposed segmented SAR ADC structure. For simplicity, the single-ended structure is shown here while the real design is differential. 4 MSB bits are decided as thermometer codes ($T<14:0>$) while 7 LSB bits decisions are binary ($B<6:0>$). Simple inverters with minimum sizes are applied as switches in the CDAC to reduce the size of the CDAC and its complexity. To keep the comparison in the middle of input range (common-mode level) half of the capacitors in CDAC ($128c \times 8$) are connected to V_{REFT} (top reference voltage) and the other half ($128c \times 7 + (64c + ... + c + c)$) are connected to V_{REFB} (bottom reference voltage).

In sampling cycle, the input is sampled at the top plates of the capacitors. The first comparison is done at the end of holding cycle (Fig. 2a). There are two cases for the first comparison. If the first comparison is decided to be low, then the sequence continues to switch $128c$ capacitors from V_{REFB} to V_{REFT} one by one. The switching for thermometer codes decision will continue until the first step in which the comparison is decided to be high (Fig. 2b, B1-B7). As soon as the comparator decides high, 4 MSB bits decision is done, one of the $128c$ capacitors switches back to V_{REFB}, $64c$ switches to V_{REFT} and comparison continues to decide the 7 LSB binary codes.

Otherwise, if the first comparison is decided to be high, one of the $128c$ capacitors will switch from V_{REFT} to V_{REFB}.

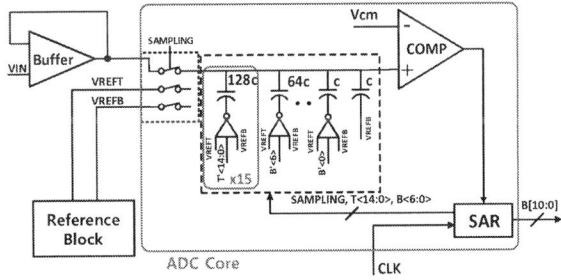

Figure 1. Block diagram of the proposed SAR ADC

"This work was supported by "the Technology Innovation Program" (10052624, Development of SoC for Positioning Service using BLE v4.2 IP) funded By the Ministry of Trade, industry & Energy (MI, Korea)"

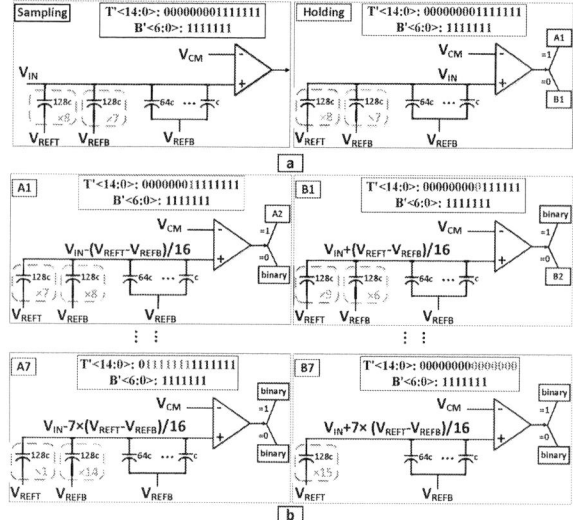

Fig. 2. Switching for a) sample and hold b) thermometer codes decision

Thermometer codes comparison and switching the capacitors from V_{REFT} to V_{REFB} will continue until the first step in which the comparison is decided to be low (Fig. 2b, A1-A7). As soon as "low" is detected in the comparator, the 4 MSB bits decision is done and comparison continues to decide the 7 LSB binary codes. In A7, if the comparison decides to be high again, the comparison enters binary while the last 128c switches to V_{REFB} and 64c switches to V_{REFT}. T'<14:0> and B'<6:0> on Fig. 2 are the control bits of the inverter type switches.

According to this switching scheme for thermometer codes decision, if the input value is around V_{REFT}, there will be switching in 128c×8 of the thermometer capacitors. This is the maximum number of switching in thermometer capacitors. If the input is around common mode voltage where second comparison level is different than first comparison, there will be switching only in 128c of the thermometer capacitors. Therefore according to input signal level, there will be switching in at least 128c and at most 128c×8 of the thermometer capacitors.

III. EXPERIMENTAL RESULTS

The ADC is implemented in 0.18μm 1.8V CMOS process. The active area of the ADC is 0.195 mm² as shown in Fig. 3. Test buffers area also included in this size.

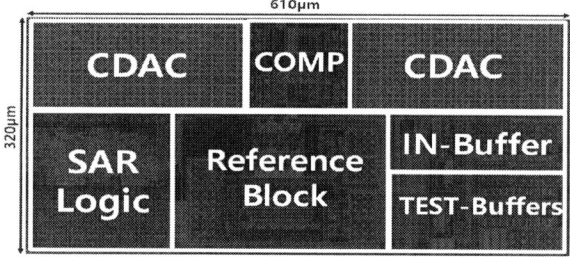

Fig. 3. Proposed ADC micrograph

The static performance is characterized through differential nonlinearity (DNL) and integral nonlinearity (INL) measurement. The measured DNL and INL are -0.49/0.48 LSB and -0.76/0.72 LSB respectively as shown in Fig. 4. The 2048-point FFT spectrum of the proposed ADC is implemented in Fig. 5. The input signal here is a 10 KHz sinusoidal wave sampled at 40KS/s. For this input frequency the effective number of bits (ENOB) is 10.57 bit. The SNDR, and SFDR are 65.73dB, 76.08dB respectively. This ADC consumes approximately 1μA from 1.8V source so the power consumption of this structure is 1.8μW. The ENOB around Nyquist frequency is 10.23 bits so the figure of merit is calculated to be 37 fJ/Conv-step. Finally Table I summarizes the performance of the ADC.

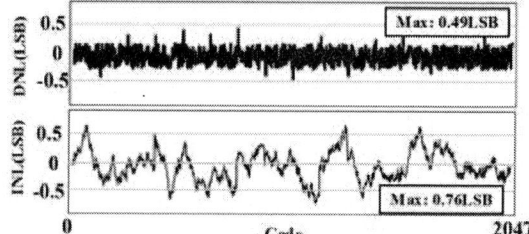

Fig. 4. DNL/INL of the proposed ADC

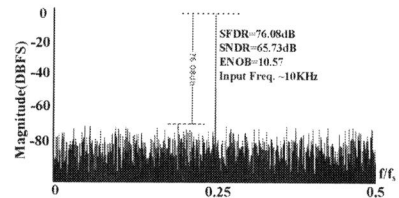

Fig. 5. FFT spectrum for a 10 KHz sin wave sampled at 40KS/s

TABLE I. PERFORMANCE SUMMARY AND COMPARISON

Parameter	This work	[2]	[3]
Number of bits	11	12	10
Power(μW)	1.8	66.7	330/470
ENOB	10.23~10.88	9.2	9.15~9.49
SFDR(dB)	71.2~83.7	57.3	62.2~66.5
INL(LSB)	-0.76/0.72	-	-0.72/0.56
DNL(LSB)	-0.49/0.48	-	-1/0.998
Sampling(KS/s)	40	40	10000
FOM(fJ/Conv-step)	37	2800	58

REFERENCES

[1] C.-H. Lin and K. Bult, "A 10-b, 500-MSample/s CMOS DAC in 0.6 mm² ," IEEE J. Solid-State Circuits, vol. 33, no. 12, pp. 1948–1958, Dec. 1998.

[2] D. J. Kim, Y. O. Kim and G. C. Ahn, "A 12-bit 40-kS/s VCM-Based Switching C-C SAR ADC," International SOC Design Conference (ISOCC), pp. 83-84, 2015.

[3] B. S. Rikan, D. H. Seo and K. Y. Lee, "A 10-b 10MS/s SAR ADC with Power and Accuracy Control of the Comparator" International SOC Design Conference (ISOCC), pp. 225-226, 2015.

A Pipelined Time Stretching for High Throughput Counter-based Time-to-Digital Converters

Seongheon Shin and Hyung-Joun Yoo

School of Electrical Engineering
Korea Advanced Institute of Science and Technology (KAIST)
Daejeon, Republic of Korea
ssh9120@kaist.ac.kr, hjyoo53@kaist.ac.kr

Abstract— **This paper proposes a pipelined time stretching technique for high throughput counter-based time-to-digital converters (TDC). Time stretching technique is used to increase the resolution of counter-based TDCs, yet it carries an inherent weakness of having a long conversion time due to the stretching phase. Without significant increment of chip area, the proposed pipelined time stretching method is realized with addition of a time splitter, a switching circuit and a fine capacitor. Pipelining of sampling and stretching efficiently improves the conversion rate by 29% according to the simulation results with a TSMC 0.25-μm CMOS process.**

Keywords; conversion rate; conversion time; counter-based time-to-digital converter (TDC); high throughput; pipelining; sampling rate; time interpolation; time stretching

I. INTRODUCTION

With growing demands on data converters having wide range, free from the voltage headroom issue and of low power consumption, counter-based time-to-digital converters (TDC) are widely studied in today's research field [1–4]. To enhance resolution of a counter-based TDC, which is solely dependent on the reference clock frequency, it is commonly accompanied with time stretching technique first suggested in 1978 [1].

Although the resolution of today's counter-based TDCs with time stretching are advanced up to a competitive level to other state-of-the-art TDCs while consuming lower power [2–4], time stretching technique carries an inherent issue: low conversion rate or a low throughput. Since the conversion includes the stretching phase, this type of TDC cannot avoid suffering from low conversion rate which must be overcome to be utilized in communication applications [2]. This paper proposes pipelined time stretching technique for counter-based TDCs where the conversion time can be reduced at the cost of minor increment of chip components and power.

II. PIPELINED TIME STRETCHING TECHNIQUE

The structure of the conventional counter-based TDC with time stretching is described in Fig. 1 (without the dotted box) and its operation is briefly illustrated in Fig. 2 (stretch factor = 4). A fractional time-residue (T_F) after the reference-clock counting is sampled as a voltage at a fine capacitor (C_{F1}) with a DC current (I_F). Then, T_F is stretched during the stretching

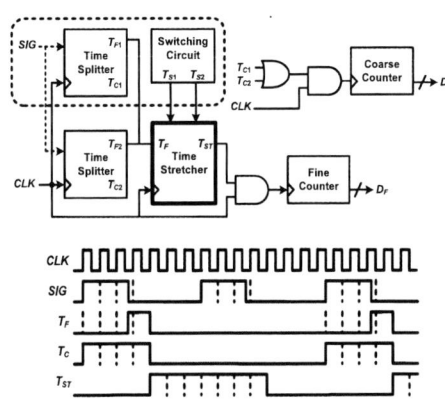

Fig. 1. Block diagram of a counter-based TDC with time stretching and timing diagram of operation

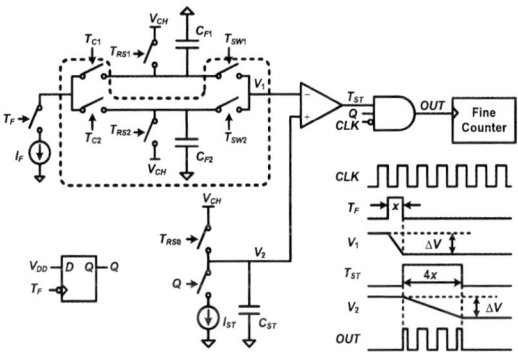

Fig. 2. Operation of a time stretching circuit

phase (T_{ST}) until the voltages, V_1 and V_2 change the state of the comparator. The stretch factor (T_{ST}/T_F) is determined by the product of current ratio (I_F/I_{ST}) and capacitance ratio (C_{ST}/C_F), and resolution of the TDC is improved by size of stretch factor.

As depicted in Fig. 1, the input signal (*SIG*) is digitized with the course counter during the coarse time, T_C, whereas the fractional time, T_F is stretched into T_{ST} and counted with the fine counter. It is clear that only after the signal stretching is finished and the system is reset, the next sampling begins.

This work was supported in part by Mobile Sensor and IT Convergence Center (MOSAIC) and IC Design Education Center (IDEC).

978-1-5090-3220-4/16 $31.00 © 2016 IEEE

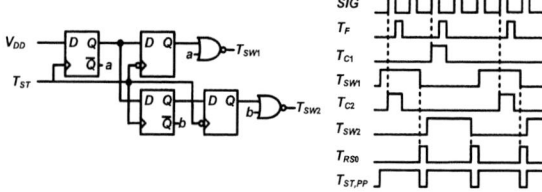

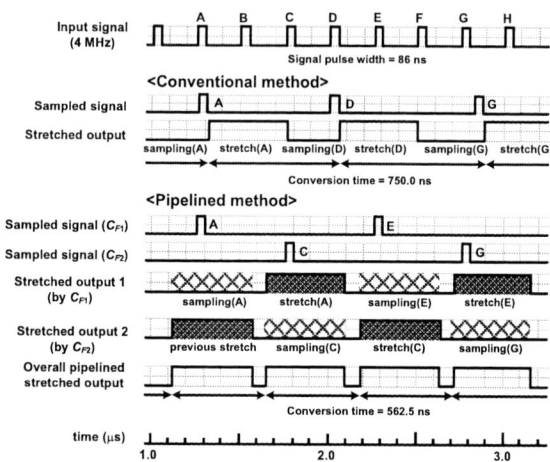

Fig. 3. Block diagram and timing diagram of the switching circuit

As discussed in [2], the size of the system is dominantly occupied by the time stretcher that contains a large-sized C_{ST}, the stretching capacitor. The dotted box in Fig. 1 highlights the additional parts to the conventional system. Figure 2 presents the circuit of the proposed time stretcher, where the dotted box indicates modification to the conventional time stretcher. Four additional switches and an additional fine capacitor, C_{F2} are the newly introduced components.

Four signals, T_{C1}, T_{C2}, T_{SW1}, and T_{SW2}, fed into the switches, are the key elements that implement pipelined operation of the proposed system. Figure 3 partially describes the switching circuit and illustrates the concept of how the signals are generated. The switching circuit generates T_{SW1} which is HIGH while T_F that was sampled as a voltage across C_{F1} is stretched, and T_{SW2} that is HIGH while T_F at C_{F2} is stretched. The same principle can be used to generate the reset signal at each falling edge of T_{ST} to provide sufficient time (4 clock periods in this case) for the large C_{ST} to be charged to the initial voltage (V_{CH}).

The switching scheme enables the proposed time stretching circuit to operate in a pipelined manner as follows. When T_{SW1} is HIGH, SIG is sampled as a voltage across C_{F2} by charging it with a DC current, I_F. At this phase, time stretching is done with the sampled voltage across C_{F1} and the discharging voltage across C_{ST}. When the stretching ends, T_{SW1} is set to LOW and reset phase, T_{RS0}, is HIGH. Then T_{SW2} becomes HIGH and SIG is sampled as a voltage across C_{F1} while time stretching goes on with sampled voltage in C_{F2} and discharging voltage at C_{ST} with a DC current, I_{ST}. When the stretching ends, a reset phase (T_{RS0}) follows. The resulting pipelined stretched time pulse, $T_{ST,PP}$ has no wasting phase except the reset phase as in Fig 3. Overall, the conversion time is effectively reduced.

III. RESULTS OF SIMULATION

Such simple variation of the conventional time stretching system is powerful for reducing the conversion time, theoretically down to 50% of the original when the period of sampling is about the same with of the stretched pulsewidth. Figure 4 shows the timing diagrams of the simulation results of the conventional and the proposed time stretching circuits. To provide a reasonable comparison, some commonly observed design parameters are chosen [2–4]. The stretch factor is set to 16, the reference clock frequency is 32 MHz, input signal frequency is 4 MHz, I_F/I_{ST} is 20/40 μA, and C_{ST}/C_F is set to 3.6/28.8 pF, identically applied to both the conventional and the proposed system.

Simulation in a TSMC 0.25-μm CMOS process shows the stretch factor (T_{ST}/T_F) of 16.0–16.2 in both time stretching systems. Table I summarizes and compares power, conversion

Fig. 4. Simulation results of counter-based TDC with time stretching of the conventional and the proposed pipelined method

TABLE I.   PERFORMANCE COMPARISON

Method	Power	Area	Conversion rate	FoM
Conventional	1.322 mW	0.137 mm²	750 ns 1.33 MS/s	7.746
Pipelined	1.507 mW	0.177 mm²	562.5 ns 1.78 MS/s	5.810

FoM = (Power*T_S)/2^{ENOB} [pJ/conv.-step]

rate, area and figure-of-merit (FoM). In the simulation, the period of input signal is set to 0.5 μs whereas the stretched pulsewidth is set to 0.39 μs. While the conventional method shows the average measurement time of 750.0 ns, the proposed method has the average measurement time of 562.5 ns, which means approximately 25% reduction.

IV. CONCLUSION

A pipelined time stretching technique for high throughput counter-based TDCs is proposed to alleviate the inherent problem of long conversion time of time stretching. Using the pipelined time stretching, the conversion time is reduced by up to 50% theoretically at the cost of minor increment in power consumption and area. When the proposed system is designed with sufficient design margin, simulation results demonstrate the conversion rate improvement by 25% from 1.33 MS/s to 1.78 MS/s with additional 0.185 mW of power and 0.04 mm² of the chip size with overall enhancement of the system's FoM.

REFERENCES

[1] B. Turko, "A picosecond resolution time digitizer for laser ranging," *IEEE Trans. Nucl. Sci.*, vol. NS-25, no. 1, pp. 75–80, Feb. 1978.

[2] M. Kim and S. Kim, "A low-cost and low-power time-to-digital converter using triple-slope time stretching," *IEEE Trans. Circuits Syst. II, Exp. Briefs*, vol. 58, no. 3, pp. 169–173, Mar. 2011.

[3] E. Raisanen-Ruosalainen, T. Rahkonen, and J. Kostamovaara, "An integrated time-to-digital converter with 30-ps single-shot precision," *IEEE J. Solid-State Circuits*, vol. 35, pp. 1507–1510, Oct. 2000.

[4] P. Chen and Y Shen, "A low-cost low-power CMOS time-to-digital converter based on pulse stretching," *IEEE Nuclear Science*, vol.53, pp. 2215–2220, Aug. 2006.

A Low-Jitter Self-Biased Phase-Locked Loop for SerDes

Heng-zhou Yuan, Yang Guo, Yao Liu, Bin Liang, Qian-cheng Guo, Jia-wei Tan

College of Computer, National University of Defense Technology

Changsha 410073, China

Abstract—**The paper presents a fully integrated multiphase output low-jitter CMOS phase-locked loop for 1.25 Gb/s to 6.25Gb/s wireline SerDes transmitter clocking. The self-biased bandwidth technology with simplified structure is applied to reduce the sensitivity to process variations. A differential charge pump which has property of low mismatch is proposed. The self-biased technology is used to make the bandwidth track the division ratio, which will improve suppression of VCO noise at higher output frequency. The simulation results under 65nm show good jitter performance.**

Keywords-PLL, self-biased, low-jitter, serializer-deserializer.

I. INTRODUCTION

To meet the increasing demand of protocols cost-effectively, SerDes(SERializer-DESerializer) transceivers must frequently span multiple data rates to enable new link rates while supporting compatibility with legacy rates[1]. The modulation of input reference clock and division ratios of feedback frequency divider will influence the bandwidth and stability, which will cause issues of poor jitter performance and even leading to unstable. Self-biased phase-locked loops (PLLs)[4,2], which can chooses the optimal operating bias levels adaptively, can solve the problem mentioned above. In the high date rate application of SerDes system, higher bandwidth seems better to reduce jitter, but the traditional self-biased PLLs have removed the effect of division ratio N, the bandwidth is unchanged when the input reference clock is stable.

To better adapt to SerDes system, the output frequency is designed to be range from 1.25GHz(clock-data-recovery circuits operate at full rate) to 3.125GHz(clock-data-recovery circuits operates at half rate). The modified self-biased PLL have established a relationship between bandwidth and division ratio. The transfer functions are derived to analyze the performance of PLL. A load balanced Voltage-Controlled-Oscillator(VCO) will provide 8 phase outputs. A high matched differential charge pump(CP) is designed to lower supply voltage and adapt to process migration.

II. CIRCUIT DESCRIPTION

As will be demonstrated in Figure 1, the dual-path loop filter offers simple control of closed-loop bandwidth and damping through dialing the gains of the two paths.

This flexibility extends the PLL applicability across a wide spectrum of networking protocols for a given design effort.The simplified open loop transfer function can be written as :

$$H(s)_{open} = \frac{\dfrac{kI_{cp1}K_{vco}}{2\pi C_2 N}s + \dfrac{kI_{cp1}R_1 K_{vco}}{2\pi C_1 C_2 N}}{s^2} = \frac{K_P s + K_I}{s^2} \quad (1)$$

The bandwidth and damping factor is modeled by the following equation:

$$\omega_n = \sqrt{K_I} = \sqrt{\frac{kI_{cp1}R_1 K_{vco}}{2\pi C_1 C_2 N}} \propto \sqrt{N}\omega_{ref} \quad (2)$$

$$\zeta = \frac{K_P}{2\sqrt{K_I}} = \frac{1}{2}\sqrt{\frac{kI_{cp1}C_1 R_1 K_{vco}}{2\pi C_2 N}} \propto \sqrt{N}\omega_{ref} \quad (3)$$

Wherein, ω_{ref} is the reference radian frequency. Equation (2) shows that ,when N is unchanged, the bandwidth is proportional to input reference clock. When N is increased, the bandwidth will increase to suppress more VCO noise to achieve better jitter performance. That will establish a relationship between N and bandwidth. Equation (3) mean the PLL will be more stable at high frequency.

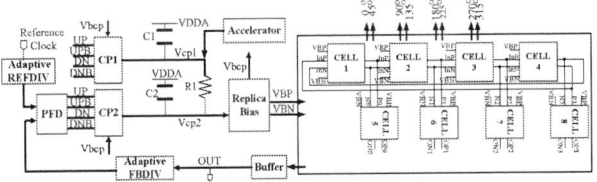

Figure 1. Block Diagram of PLL

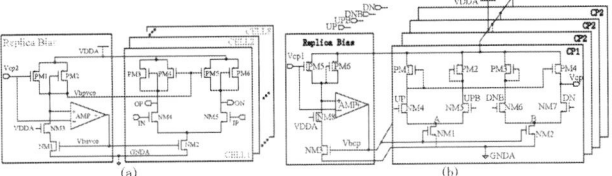

Figure 2. Circuits of VCO and Charge Pump

Figure 2 show the circuit of VCO and CP. The diffential delay cells are designed for high swing and fast switching to achieve better noise performance by suppressing power supply common noise. Great area of replica-biased circuits and VCO cells are designed to redece intrinsic noise. Broad bandwidth is used to suppress VCO noise.

Charge pump is the second major PLL noise contributor. Flicker noise is main noise source, while thermal noise is mostly suppressed by broad bandwidth. The differential structure of CP is essential. Better RMS jitter performance is

978-1-5090-3220-4/16 $31.00 © 2016 IEEE

achieved by suppressing power supplycommon noise. Two pairs of charge pump are designed to alleviate the charge sharing. Trading off between noise performance and current mismatch is sophisticated. Figure 3 show the transient current mismatch and DC mismatch of the proposed CP.

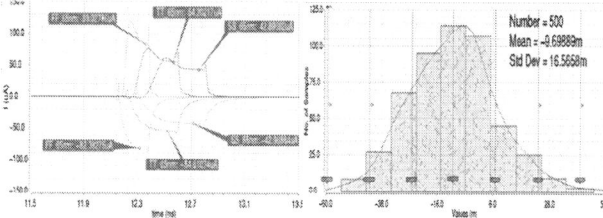

Figure 3. Mismatch of Charge Pump

Figure 4. Common-Centroid Layoutsof SerDes and PLL

SerDes lanes which consist of RX and TX are aligned with the PLL. Noise from other circuits of SerDes will be isolated by the decoupling capacitors. The layout has an active area of 0.384mm².

III. SIMULATION RESULTS

The post-layout simulation is in the state of 125MHz reference signal, the REFDIV sets the division ration to be 1, and the FBDIV sets the division ratio to be 25(output 3.125GHz). The RMS jitter and stability are calculated by Matlab program in Figure 5(a). The output spur and eye diagram are shown in Figure 5(b), the deterministic jitter is about 2.4ps.

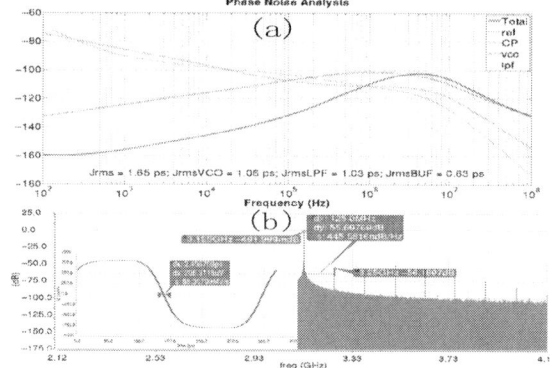

Figure 5. Common-Centroid Layoutsof SerDes and PLL

We compare the proposed PLL with some recently published papers in TABLE I. The proposed PLL can provide wider operating range. The jitter performance is good among all the references. A good FoM of -221.1db@3.125GHz is relative good, which demonstrating the good power efficiency.

TABLE I.   OUTPUT NOISE AND PHASE NOISE

Ref. Year	[3] 2014	[4] 2010	[5] 2006	This Work
Tech. (nm)	65	45	180	65
Power (V)	0.8	2.5	1.8	1.2
Power (mW)	0.78	70	25	28.8
RMS Jitter (ps)	1.7@0.9 GHz	0.99@2. 5GHz	2.36@1 GHz	1.65@3. 125 GHz
Output [MHz]	390-1410	2500	1000	1250-3125
Input [MHz]	40-350	100	100	100-300
FoM	-236.5	-221.6	-218.6	-221.1

IV. CONCLUSION

This work has improved the self-biased PLL state of the art designs in three ways: first, VCO noise has been suppressed by introducing a relationship between division ratio and bandwidth; and second, by presenting a high matched CP, deterministic jitter has been reduced. The simulation under 65nm testify the good jitter performance.

ACKNOWLEDGMENT

This research was supported by National Natural Science Foundation of China Program (No.61133007).

REFERENCES

[1] Alvin L. S. Loke , et al., "A Versatile 90-nm CMOS Charge-Pump PLL for SerDes Transmitter Clocking," IEEE J. Solid-State Circuits , vol. 41,pp. 1894 - 1907, No. 8, Aug 2006.J. Clerk Maxwell, A Treatise on Electricity and Magnetism, 3rd ed., vol. 2. Oxford: Clarendon, 1892, pp.68–73.

[2] J. Maneatis, et al., "Low-jitter process-independent DLL and PLL based on self-biased techniques," IEEE J. Solid-State Circuits, vol. 31, pp. 1723–1732, Nov. 1996.

[3] W. Deng, et al., "A 0.0066mm2780µW fully synthesizable PLL with a current-output DAC and an interpolative phase-coupled oscillator using edge-injection technique," ISSCC Dig.Tech. Papers, pp. 266-267, Feb., 2014.

[4] D. Fischette, et al., "A 45nm SOI-CMOS dual-PLL processor clock system for multi-protocol I/O," ISSCC Dig. Tech. Papers, pp. 246-247, Feb., 2010.

[5] M. Brownlee, et al., "A 0.5 to 2.5GHz PLL with fully differential supplyregulated tuning," IEEE J. Solid State Circuits, vol. 41, pp. 2720-2728, Dec. 2006.

36-Gb/s CDR IC using simple passive loop filter combined with passive load in phase detector

Keiji Kishine, Hiroshi Inoue, Kosuke Furuichi
Natsuyuki Koda, Hiromu Uemura and Hiromi Inaba
University of Shiga Prefecture
2500, Hikone Hassaka,
Shiga, 522-8533, Japan
Email: kishine.k@e.usp.ac.jp

Makoto Nakamura
Gifu University
1-1 Yanagido, Gifu City
Gifu, 501-1193, Japan
Email: m_naka@gifu-u.ac.jp

Akira Tsuchiya
Kyoto University
Yoshida Honmachi, Sakyoku
Kyoto, 606-8501, Japan
Email: A. Tsuchiya@vlsi.kuee.kyoto-u.ac.jp

Abstract—A 36-Gb/s clock and data recovery (CDR) circuit with a simple passive loop filter is presented. By combining the passive load for output magnitude generation in the phase detector with a passive loop filter, the cutoff frequency and high-frequency response in the loop filter can be controlled independently. The CDR consists of a half-rate decision circuit that provides higher speed operation. To confirm the validity of the proposed topology, we fabricated a 36-Gb/s CDR IC with the 65-nm MOSFET process. It provides an 18-GHz extracted clock (half rate) signal and 18-Gb/s recovered date (1:2DEMUX output). The measured jitter was lower than 1.15 ps rms. The area of chip, including an I/O buffer circuit, was 1 mm^2, and the power consumption was 290 mW.

I. INTRODUCTION

A clock and data recovery(CDR) circuit is one of the most important elements in transceivers for optical communication systems. For CDR circuits used in the high-speed networks, an analog phase-locked loop(PLL) is often used. The key issues in designing the CDR IC are compactness, high-speed, low-jitter operation and ease of design. An active filter is most popularly used as a loop filter in PLL[1]. However, it sometimes leads to larger area than that for a passive filter. A passive lag-lead filter is simple and small, but the cutoff frequency and high frequency response, in which the output jitter and the acquisition characteristics depend on them[2][3], can not be set independently. In this paper, we propose a half-rate CDR circuit with a simple passive low-pass filter that is combined with the passive load in a phase detector. The half-rate CDR can provide higher speed operation. Furthermore, the cutoff frequency and high frequency response of the filter, which are strongly dependent on the filter characteristics, can be set independently by adjusting only a capacitor and the resistance. To verify the the proposed method, we fabricated a 36-Gb/s CDR IC with the 65-nm MOSFET process and investigated its operation characteristics. It shows an acquisition characteristics that can be controlled by setting the filter parameter from outside the chip. It also provides low jitter characteristics. This indicates that the proposed method can be adaptable to analog CDR ICs.

II. HALF-RATE CDR CIRCUIT WITH COMPACT FILTER

A. Conventional half-rate CDR

Figure 1 shows a conventional half-rate CDR that consists of a half-rate decision circuit. Figure 2 shows the most popular active and passive lag-lead filters. The transfer functions for

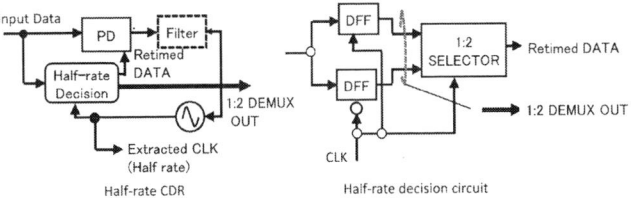

Fig. 1. Conventional half-rate CDR.

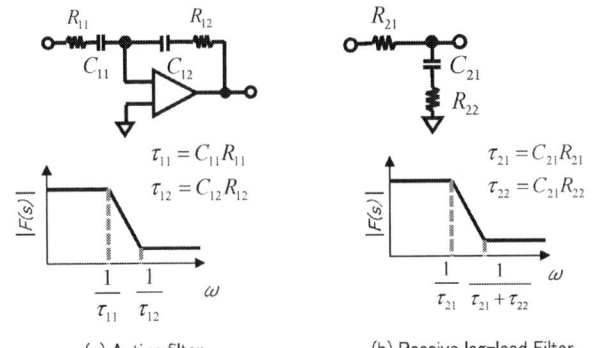

(a) Active filter   (b) Passive lag-lead Filter

Fig. 2. Conventional loop filter.

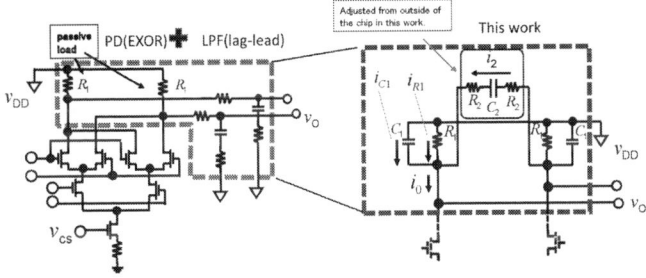

Fig. 3. Loop filter combined with the load resistance in PD.

these filters are expressed as

$$F_1(s) = K\frac{1+s\tau_{12}}{1+s\tau_{11}} \quad \text{(active)} \quad (1)$$

$$F_2(s) = \frac{1+s\tau_{22}}{1+s(\tau_{21}+\tau_{22})} \quad \text{(passive lag-lead)}. \quad (2)$$

On the basis of these relationships, the cutoff frequency of the transfer curve and the high frequency response can be designed. They can be changed independently in $F_1(s)$. However, an additional operation amplifier is used, which

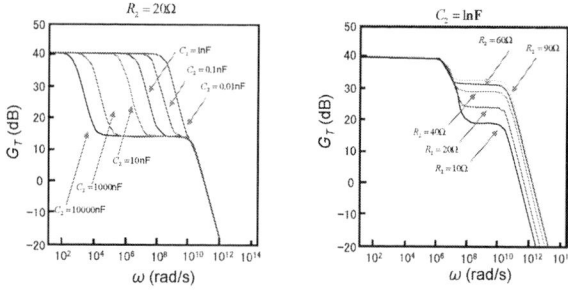

Fig. 4. R_2 and C_2 dependence of transconductance curve

leads to a larger chip area. The area for a lag-lead filter is not large. However, in $F_2(s)$, the cutoff frequency of the transfer curve and the high frequency response can not be changed independently.

B. Proposed filter

Figure 3 shows the configuration of the proposed passive filter in which the load resistance of the phase detector is combined into filter elements. Using a small-signal equivalent circuit analysis, we obtain

$$v_O = R_1 i_i = \frac{1}{j\omega C_1} \tag{3}$$

$$2v_O = i_2(2R_2 + \frac{1}{j\omega C_2}). \tag{4}$$

From these, the transconductance function with the filter is expressed as

$$|G_T| = \left|\frac{v_O}{i_0}\right| = \sqrt{\frac{\alpha}{\beta + \gamma}}, \tag{5}$$

where $\alpha = R_1^2 + 4R_1^2 R_2^2 \omega^2 C_2^2$, $\beta = (1 - 2R_1 R_2 \omega^2 C_1 C_2)^2$, and $\gamma = (2R_2 C_2 + R_1 C_1^2 + 2R_1 C_2)^2 \omega^2$. We assume that C_1 and R_2 are small enough compared with C_2 and R_1. On the basis of this, the ω dependence of G_T is described as follows.

When ω is small enough: $\alpha \approx R_1^2$, $\beta \approx 1$, and $\gamma \approx 4(R_1)C_2^2\omega^2$. The cutoff frequency in G_T depends on C_2.

When ω becomes larger: We obtain $\alpha = 4R_1^2 R_2^2 \omega^2 C_2^2$, $\beta = 1$, and $\gamma = 4(R_1)C_2^2\omega^2$. G_T depends on R_2.

When ω becomes much larger: $\alpha = 4R_1^2 R_2^2 \omega^2 C_2^2$, $\beta = 4R_1^2 R_2^2 C_1^2 C_2^2 \omega^4$, and $\gamma = 4(R_1)C_2^2\omega^2$. G_T gets smaller rapidly.

These indicate that the cutoff frequency and the high frequency response in the G_T curve can be controlled by setting C_2 and R_2, respectively. Figure 4 shows the R_2 and C_2 dependence of a transconductance curve calculated by (5). This shows that the cutoff frequency and high frequency response can be adjusted independently.

III. EXPERIMENTAL RESULTS

To verify our proposed method, we fabricated a CDR IC using the 65-nm MOSFET process. Figure 5 shows the chip layout and a photograph. C_2 and R_2 in Fig. 3 can be adjusted from the outside of the chip. Figure 5 also shows the output waveforms form when 36-Gb/s data was input. The extracted clock was 18GHz, which is the half rate of the input, and the recovered data (1:2 DEMUX) was 18Gb/s. Figure 6(a) shows the filter response (v_o in Fig. 3) when burst data was input.

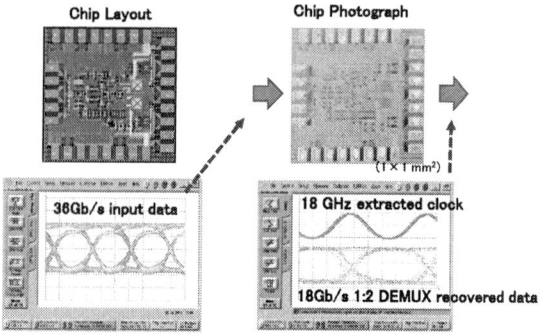

Fig. 5. Chip layout and output waveforms.

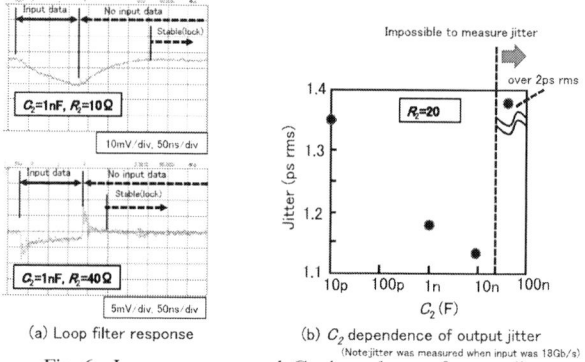

(a) Loop filter response   (b) C_2 dependence of output jitter.

(Note:jitter was measured when input was 18Gb/s)

Fig. 6. Loop responses and C_2 dependence of output jitter.

It shows the acquisition response changed in accordance with R_2. Figure 6(b) shows the C_2 dependence of the output jitter. It indicates that a too large or a too small cutoff frequency might lead to large jitter or falling out of lock. The minimum output jitter was lower than 1.15 ps rms. These results indicate that the adopted filter can be used to control the acquisition and jitter characteristics of the CDR. The total area of the chip including I/O buffer circuits is 1mm^2, and the power consumption is 290 mW.

IV. CONCLUSION

A 36-Gb/s clock and data recovery circuit with a passive loop filter, in which the load resistance of the phase detector and loop filter is combined, was presented. To confirm the validity of the proposed topology, we fabricated a 36-Gb/s CDR IC with the 65nm-MOSFET process. The acquisition characteristics and jitter can be controlled by setting R_2 and C_2. It also provides low jitter operation of 1.15ps rms. These results indicate that our proposed topology is applicable to the design of high-speed CDR ICs.

ACKNOWLEDGMENTS

A part of this research was supported by Grants-in-Aid for Scientific Research, The Japan Society for the Promotion of Science.

REFERENCES

[1] R. E. Best, 'Phase-Locked Loops' (MacGraw-Hill, 1999)

[2] F. M. Gardner, 'Phaselock Techniques' (Wiley, New York,2003)

[3] K. Kishine, H. Inaba, Ma. Nakamura, Mi. Nakamura, Y. Ohtomo and H. Onodera' Low-jitter design method based on ω_n-domain jitter analysis for 10-Gb/s clock and data recovery ICs, IET Electronics Letters, Vol. 45, No. 16, pp. 800-804, 2009.

[4] K. Kishine, et. al., 'PLL Design Technique by a Loop-trajectory Analysis Taking Decision-circuit Phase Margin into Account For Over-10-Gb/s Clock and Data Recovery Circuits', IEEE JSSCC, Vol. 39, Issue 5, pp. 740-750, 2004.

All-Synthesizable Transmitter Driver and Data Recovery Circuit for USB2.0 Interface

Kihwan Seong, Won-Cheol Lee, Byungsub Kim, Jae-Yoon Sim, and Hong-June Park

Department of Electronic and Electrical Engineering
Pohang University of Science and Technology (POSTECH)
Pohang, Korea
hjpark@postech.ac.kr

Abstract— **The transmitter driver and the data recovery circuit of receiver for 480 Mbps USB2.0 interface were designed with Verilog and synthesized to enhance design portability. The transmitter driver was implemented by using multiple tri-state inverter cells to generate a 0 ~ 400 mV swing for normal operation and a 0 ~ 800 mV swing for chirp operation. The data recovery circuit was implemented by using a blind over-sampling method with 5-phase 480 MHz clocks. The proposed transmitter driver and data recovery circuits were fabricated in a 65 nm CMOS process; the transmitter driver satisfies the USB2.0 eye-mask specification in the measured eye diagrams and the data recovery circuits gives a measured BER less than 1E-12 with PRBS 2E31-1 data and a USB2.0 5m cable. The transmitter driver and the data recovery circuit consume 24.3 mW and 1.6 mW, respectively at 1.2 V supply.**

Keywords; transmitter driver, data recovery circuit, USB2.0, all-synthesizable design

I. INTRODUCTION

Transceiver circuits such as USB2.0 high-speed transceiver are included as a small part in large digital chips such as CPU. Almost all the parts of the large digital chip are implemented by using digital synthesis. However, the transceiver circuits are implemented by using a full-custom design because of analog circuits such as PLL, CDR, transmitter driver, and receiver front-end circuits. If the analog circuits can be synthesized, the entire large digital chip including transceiver circuits can be implemented by using digital synthesis. Recently, the digital synthesis technique is applied to design PLL and CDR [1]-[4]. In this work, the transmitter driver and the data recovery circuit are implemented by using digital synthesis.

II. IMPLEMENTATION OF TRANSMITTER DRIVER

The transmitter driver of USB2.0 high-speed interface is required to generate a differential output signal; each node voltage swings between 0 and 400 mV for normal operation and between 0 and 800 mV for chirp operation during the initial high-speed handshaking step [5]. The tri-state inverter cells are used in this work to implement the transmitter driver

This work was supported by the National Research Foundation of the Ministry of Science, ICT and Future Planning (MSIP), Korea, under the contract numbers of NRF-2014R1A2A1A11052875 and IDEC.

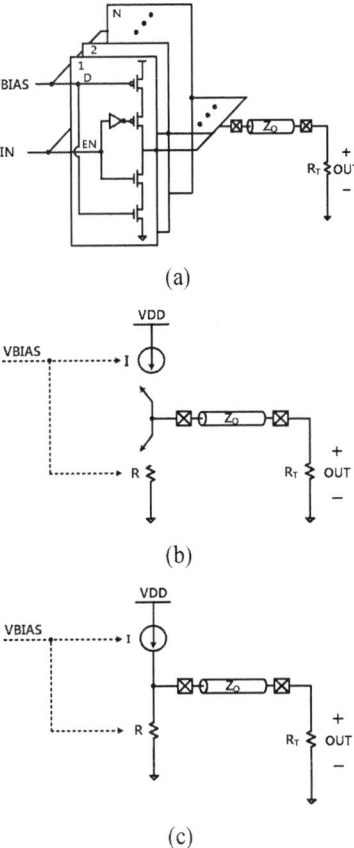

Fig. 1. Proposed transmitter driver (a) implementation (b) operation at IN = '1' (c) operation at IN = '0'

(Fig. 1(a)); the gate nodes of the top PMOS and the bottom NMOS are connected to VBIAS that is a DAC output [1] and the gate nodes of the center NMOS and PMOS are connected to the transmitter driver input (IN) and the inverted input, respectively. When IN = '1', both the center NMOS and PMOS are turned on and the top PMOS works as a current-source and the bottom NMOS works as a resistor because VBIAS is larger than 0.5 VDD (Fig. 1(b)). When IN = '0', both the center NMOS and PMOS are turned off (Fig. 1(c)).

978-1-5090-3220-4/16 $31.00 © 2016 IEEE   63   ISOCC 2016

III. IMPLEMENTATION OF DATA RECOVERY CIRCUIT

A blind over-sampling method is used in this work for digital synthesis of data recovery circuit because it can be implemented by using all digital circuits and its lock time is less than 1 data period [6]. A 5x sampler samples the received data by using 5-phase 480 MHz clocks.

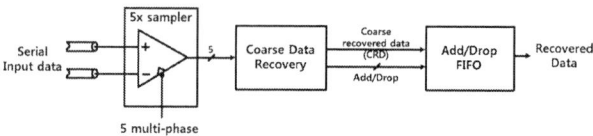

Fig. 2. All-synthesizable blind-oversampling data recovery circuit

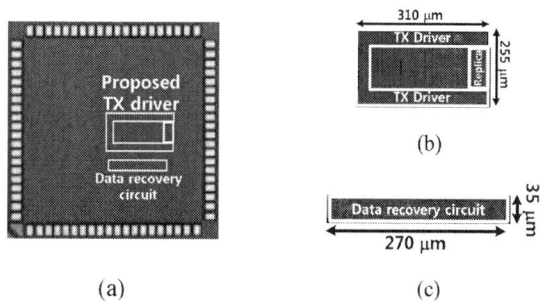

Fig. 3. (a) chip photo (b) layout of transmitter driver (c) layout of data recovery circuit

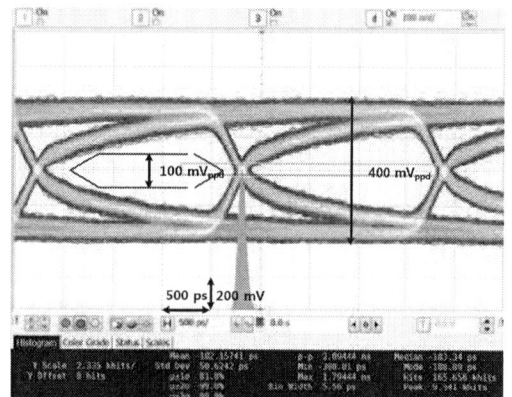

Fig. 4. Measured RX input eye-diagram and jitter with a USB2.0 5m cable

IV. MEASUREMENT RESULTS

The transmitter driver and data recovery circuit were designed in Verilog and fabricated in a 65nm CMOS process. The transmitter driver chip area including DAC is 0.02 mm^2 and the data recovery circuit occupies 0.01 mm^2 (Fig. 3). The power consumption is 24.3 mW and 1.6 mW for the transmitter driver and the data recovery circuit, respectively, with the

supply voltage of 1.2 V. The rms and peak-to-peak jitter of the RX eye were measured to be 50.6 ps and 210 ps with a 5-m USB2.0 cable (Fig. 4). The USB2.0 transceiver chip including the proposed transmitter driver and data recovery circuits was connected to a PC for test (Fig. 5). A measured BER of the data recovery circuit is less than 1E-12; a 480 Mbps 2E31-1 PRBS data is applied to input nodes of the 5x sampler shown in Fig. 2 by using a BER tester (Tektronix gigaBERT).

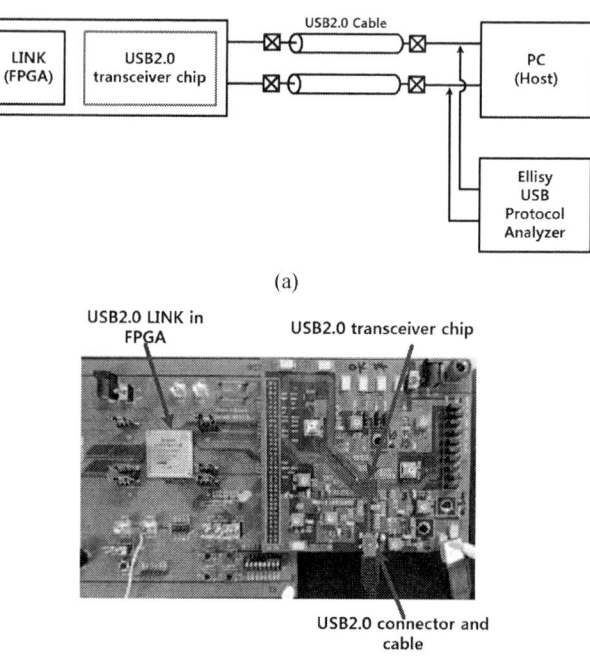

Fig. 5. (a) test setup (b) PCB photo

REFERENCES

[1] W. Deng, et al., "A 0.0066 mm^2 780 µW Fully Synthesizable PLL with a Current-Output DAC and an Interpolative Phase-Coupled Oscillator Using Edge-Injection Technique," *IEEE International Solid-State Circuits Conference (ISSCC)*, pp. 266-267, Feb. 2014

[2] W. Kim, et al., "A 0.032 mm^2 3.1 mW Synthesized Pixel Clock Generator with 30 psrms Integrated Jitter and 10-to-630MHz DCO Tuning Range," *IEEE International Solid-State Circuits Conference (ISSCC)*, pp. 250-251, Feb. 2013.

[3] A. T. Narayanan, et al, "A 0.011 mm^2 PVT-Robust Fully-Synthesizable CDR with a Data Rate of 10.05 Gb/s in 28nm FD SOI," *IEEE Asian Solid-State Circuits Conference (ASSCC)*, pp. 285-288, Nov. 2014

[4] K. Seong, et al, "All-Synthesizable 5-Phase Phase-Locked Loop for USB 2.0," *Journal of semiconductor technology and science (JSTS)*, pp 352-358, Jun. 2016

[5] "USB2.0 Transceiver Macrocell Interface Specification, Revision 1.05," Intel, Hillsboro, OR, Mar. 29, 2001

[6] S. Park, et al, "A Single-Data-Bit Blind Oversampling Data-Recovery Circuit With an Add-Drop FIFO for USB2.0 High-Speed Interface", *IEEE Transaction on circuits and system II (TCAS II)*, pp. 156-160, Feb. 2008

Power-Efficient Partially-Adaptive Routing in On-chip Mesh Networks

M. Jalili[1], J. Bourgeois[2], H. Sarbazi-Azad[1]

[1] Institute for Research in Fundamental Science (IPM), Tehran, Iran
[2] UFC/FEMTO-ST Institute, FC, France
majalili@ipm.ir, julien_bourgeois@pu-pm.univ-fcomte.fr, azad@ipm.ir

Abstract— The mesh network-on-chip (MNoC) is the most popular inter-processor communication infrastructures used in modern on-chip systems. Although many routing algorithms have been developed for MNoCs but almost all of them give better performance in cost of more complexity (more virtual channels) and hence extra power consumption. In this paper, we propose a partially adaptive routing algorithm for meshes that requires no virtual channels to ensure deadlock freedom. Simulation experiments show that the proposed method provides good performance while using less power consumption.

Keywords: *Mesh, Network-on-chip (NoC), Routing algorithm, Performance, Power consumption, Simulation.*

I. INTRODUCTION

One of the most important challenges in designing *network-on-chips (NoCs)* is routing algorithm. The routing algorithm determines the path for transferring a packet between the source and destination cores. Based on the freedom of choosing different paths toward the destination, routing algorithms are divided into two main categories: deterministic and adaptive. In deterministic routing, a packet always takes a predefined path from source to destination node. On the other hand, in adaptive routing, a packet has some options to take different paths to get closer to its destination.

Designers rely on Dimension Order Routing (DOR) to send the packet in the networks due to its simplicity [1]. In DOR algorithm, the data packet first routed to proper position in higher dimension before trying to transfer in next dimension. Although DOR is very simple and works well for uniform traffic load, it cannot satisfy the requirements of those with non-uniformity and hotspots in the traffic load. However, adopting fully adaptive routing algorithms requires high number of virtual channels and thus imposes more cost to the system in terms of system performance (more latency for the routing logic) and storage (the extra area imposed by the extra logic). So, partially adaptive routing algorithms have been proposed to make a balance between performance and storage.

Based on the insight that a partially-adaptive routing algorithm can offer better performance while imposing negligible storage and logic overheads [2], we introduce a new routing algorithm for mesh NoCs that requires no virtual channel to ensure deadlock freedom. Our idea is to traverse the

network fully adaptively in two different phases (first in positive and in negative direction) consecutively. This way, the packet almost enjoys the freedom of fully adaptive routing while not facing the complexities of using virtual channels. Our evaluation shows that the proposed routing algorithm can provide better performance and lower power consumption compared to the deterministic routing and fully-adaptive routing, respectively.

II. ROUTING ALGORITHMS

Routing algorithm plays a crucial role in NoC performance. Many on-chip networks employ deterministic routing due to its simplicity. The most popular deterministic routing algorithm used in mesh NoCs is Dimension Order Routing (DOR). According to DOR, the packet traverses the path between the source and destination in a predefined order of dimensions. According to DOR, a packet first is routed along dimension 1, then dimension 2, and so on, until it reaches to the destination node [1]. Due to the fixed dimension traversal scheme used in DOR, deadlock is avoided by nature.

A fully adaptive routing algorithm guarantees deadlock freedom while providing all possible paths to deliver a packet. This is usually achieved by using more virtual channels and more complex logic for routing. Duato's routing algorithm routes a packet by partitioning the virtual channels of each physical channel into two classes: adaptive and escape. This algorithm allows the packet to be routed fully adaptively through adaptive virtual channels of all physical channels that can bring the packet closer to its destination while guarantees the deadlock freedom using escape virtual channels [3]. The performance of fully-adaptive routing algorithms comes with the cost of more complexity and hence more static power consumption as a result of the added virtual channels and logic to insure deadlock freedom.

A partially adaptive routing algorithms enjoys a balance of the advantages of deterministic and fully adaptive routing methods, i.e. it provides good freedom to packets in traversing dimensions while using negligible or no extra hardware to avoid deadlocks [4]. We introduce a partially adaptive routing algorithm for mesh NoCs without using virtual channels. According to our algorithm, the dimensions that a packet must traverse to reach to its destination are partitioned in 2 sets: S+

and S-. The positive set, S+, includes those dimensions that the index of the source node at that dimension is less than the index of the destination. Similarly, S- includes those dimensions that the index of the source node is higher than that of the destination node. Then, we route the packet fully adaptively across the dimensions in S+; when all dimensions in S+ are traversed, we route the packet fully adaptively over the dimensions listed in S-.

III. EVALUATION RESULTS

We use Xmulator [5] for evaluating different routing algorithm. In order to show the effectiveness of the proposed scheme, we consider 3 sets of simulations: i) in the first scenario, we consider an 8x8x8 mesh NoC that uses no virtual channel (just a physical channel) with packet size of 8 and 16 flits. Fig.1 shows the average packet latency (APL) and power consumption of the proposed routing algorithm and DOR (both using no virtual channels). The proposed method postpones the networks saturation point by 2.01X and 1.96X the 2 packet sizes. For power consumption, before saturation point, the proposed scheme consumes almost the power DOR does, while after that point, considering the larger traffic load handled by the new routing algorithm, DOR consumes less power.

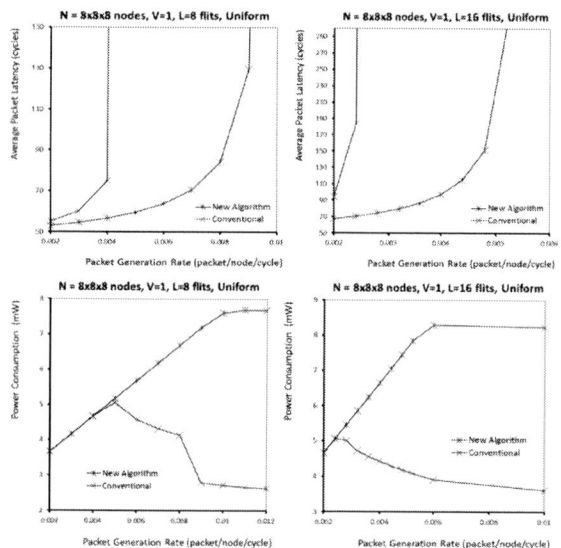

Fig.1 The avergae packet latency and poewr consumption of DOR and proposed routing schmes in an 8x8x8 NoC under unifom traffic

Lets consider 2 virtual channels per physical channel to include a fully-adaptive routing algorithm in the evaluation. Fig.2 shows the APL and power consumption in an 8x8x8 NoC under matrix-transpose traffic load. For packet size of 8 flits, although our scheme cannot outperform the fullyadaptive routing algorithm, it postpones the satuarion point compared to DOR by 10.5%. Moreover, our proposal consumes lower power than DOR (33.4% lower). For packet size of 16 flits, the performance of the proposed method increases noticeably compared to DOR and raches to 2X bridging the gap to the fully-adaptive algorithm. In the third senario, we evaluate an 8x8x8 mesh NoC under hotspot traffic load. The proposed routing algorithm exhibits close behaviour to the fully adaptive routing algorithm while working much better than DOR. A similar conclusion can be also drwan when considering power consumption.

IV. CONSLUSION

We introduced a partially routing algorithm for mesh NoCs that removes the need for extra virtual channels. Our method takes the advantages of adaptive routing and simplicity of deterministic routing schemes and provides a balanced solution in terms of power consumption and network latency. Future work in this line can consider evaluating the proposed scheme under real traffic loads and extending it for other network topologies.

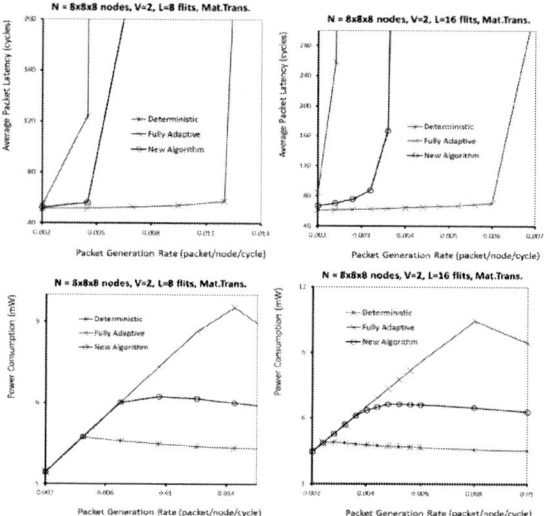

Fig.2 The avergae packet latency and power consumption of different routing algorithms in an 8x8x8 NoC under matrix-transpose traffic load.

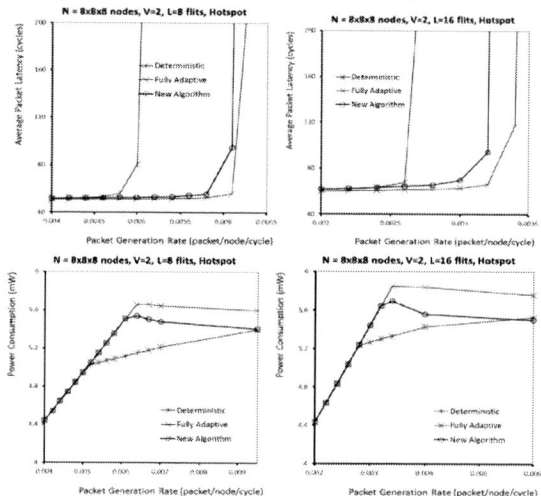

Fig.3 The avergae packet latency and power consumption of different routing algorithms in an 8x8x8 NoC under hotspot traffic pattern.

REFERENCES

[1] H. Sarbazi-azad, Performance analysis of wormhole routing in multicomputer interconnection networks, Ph.D. Thesis, University of Glasgow, U.K., 2002.

[2] L.S. Peh and W.J. Dally, "A delay model for router microarchitectures," IEEE Micro, vol. 21, no. 1, pp. 26-34, 2001.

[3] J. Duato, A new theory of deadlock-free adaptive routing in wormhole networks, IEEE transactions on Parallel and Distributed Systems, 1993.

[4] J. Duato, S. Yalamanchili, L. Ni, Interconnection Networks: An Engineering Approach, Morgan Kaufmann, 2003.

[5] A. Nayebi, S. Meraji, A. Shamaei, H. Sarbazi-Azad, XMulator: A Listener-Based Integrated Simulation Platform for Interconnection Networks, AMS, 2007.

[6] Fatemeh Aghaaliakbari,et al, "Efficient Processor Allocation in A Reconfigurable CMP Architecture for Dark Silicon Era," ICCD 2016.

Hash-Table and Balanced-Tree Based FIB Architecture for CCN Routers

Kenta Shimazaki[*†], Takashi Aoki[‡], Takahiro Hatano[‡], Takuya Otsuka[‡], Akihiko Miyazaki[‡],
Toshitaka Tsuda[§], and Nozomu Togawa[*]

[*]Dept. of Computer Science and Communications Engineering, Waseda University

[†]kenta.shimazaki@togawa.cs.waseda.ac.jp

[‡]NTT Device Innovation Center, NTT Corporation

[§]Global Information and Telecommunication Institute, Waseda University

Abstract—Recently, content centric networking (CCN) attracts attention as a next generation network on which every router forwards a packet to another router and also functions as a server. A CCN router has a forwarding table called FIB (Forwarding Information Base) but its table look-up can become a bottleneck. In this paper, we propose FIB data structure for CCN routers which can reduce the number of comparisons in its look-up table. Our proposed FIB is composed of a bloom filter and a hash table and each hash entry is connected to a balanced binary-search tree. By using our FIB, the number of comparisons cannot much increase even if hash collisions occur. Experimental results demonstrate the effectiveness of the proposed FIB over the several existing methods.

I. INTRODUCTION

Recently, traffic flow over the network is steadily increasing and content centric networking (CCN) attracts attention as a next generation network [3], on which every router forwards a packet to another router and also functions as a server. One of the major problems in CCN is that the CCN router requires many processing time. Speeding-up the CCN router much depends on how to reduce the FIB (Forwarding Information Base) processing time. FIB has a routing table and the most time-consuming process is table look-up. It is much required to reduce the number of comparisons in the look-up table in FIB.

In this paper, we propose hash-table and balanced-tree based FIB data structure for CCN routers reducing the number of comparisons in its routing table. Our proposed FIB is composed of a bloom filter and a hash table and each hash entry is connected to a balanced binary-search tree. By using our FIB, the number of comparisons cannot much increase even if hash collisions occur. Experimental results show that our proposed FIB can reduce the number of comparisons by 96.1%, 47.8%, and 14.1% compared to those implemented by using n-ary tree, hash table (linear list), and bloom filter + hash table (linear list), respectively.

II. FORWARDING INFORMATION BASE

A CCN router is composed of Content Store (CS), Pending Interest Table (PIT), and Forwarding Information Base (FIB). The memory size of the FIB can be an order of magnitude or more larger than that of PIT or CS. In order to speed up the processing time of the CCN router, it is necessary to speed up the FIB processing time.

FIB has a routing table, in which every entry has a pair of a content name and an output interface as shown in Fig. 1. Given a content name, FIB outputs its corresponding output interface. If FIB does not have a given content name, it performs the longest name match search. How to reduce the number of content name comparisons in the FIB table is the main concern. In this paper, we assume that we perform a single comparison at every entry in the FIB table.

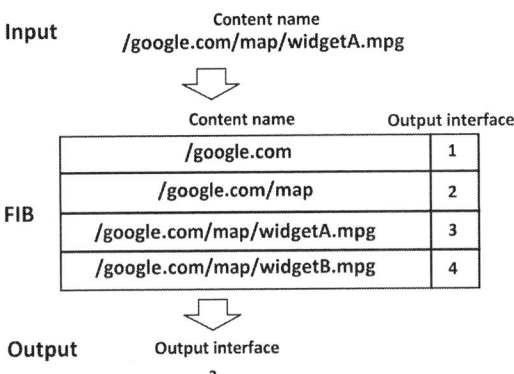

Fig. 1. Process in FIB.

III. HASH-TABLE AND BALANCED-TREE BASED FIB ARCHITECTURE

In this section, we propose an efficient data structure for FIB reducing the number of comparisons by effectively grouping and filtering the content names. We also propose its construction method and search method.

A. FIB data structure

There have been proposed several data structures for FIB; content addressable memory [4], n-ary tree [5], hash table (linear list) [3], and bloom filter + hash table (linear list) [2]. As in these methods, using a hash table is one of the good choices in terms of reducing the number of comparisons as well as hardware overheads.

One of the largest problems in using a hash table is how to deal with collisions. When we just use a linear list as in [2], [3], its search time must become longer in case a collision occurs in the hash table. Instead, we use a balanced binary search tree to construct a hash table as well as a bloom filter.

Fig. 2 shows our proposed FIB data structure, which is composed of a bloom filter and a hash table. Every hash entry in the hash table is connected to a balanced binary tree and hence we can reduce the number of comparisons even if a collision occurs in the hash table.

B. Construction method

When an empty FIB is prepared, every content name is inserted into our proposed FIB as follows:

1) Bloom filter: We first use a bloom filter [1], which shows whether a given content name is already inserted into the table in FIB or not.

The bloom filter is composed of B bits and m hash functions. Initially, all the B bits are set to be zero. When a

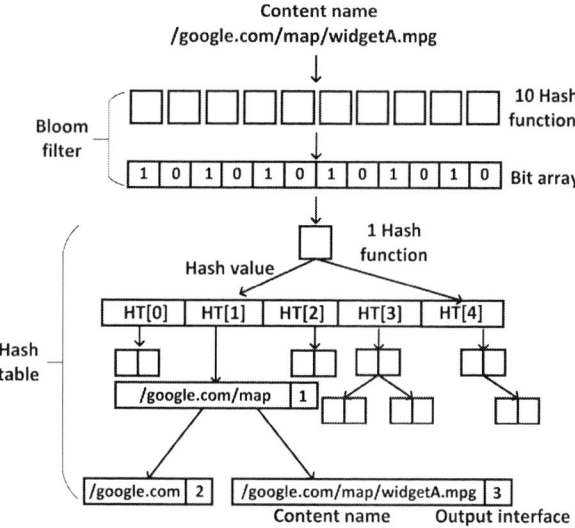

Fig. 2. Data structure of the proposed FIB.

TABLE I. THE NUMBER OF COMPARISONS ($S = 1000$).

FIB entries	Method	#searches	#comparisons
10	n-ary tree [5]	10	160 (1.00)
	Hash table (linear list) [3]		34 (0.21)
	Bloom filter + hash table(linear list) [2]		21 (0.13)
	Ours		18 (0.11)
100	n-ary tree [5]	100	10379 (1.00)
	Hash table (linear list) [3]		297 (0.03)
	Bloom filter + hash table(linear list) [2]		181 (0.02)
	Ours		175 (0.02)
1000	n-ary tree [5]	1000	234822 (1.00)
	Hash table (linear list) [3]		4580 (0.02)
	Bloom filter + hash table(linear list) [2]		2729 (0.01)
	Ours		2573 (0.01)
10000	n-ary tree [5]	10000	2571184 (1.00)
	Hash table (linear list) [3]		227644 (0.09)
	Bloom filter + hash table(linear list) [2]		138863 (0.05)
	Ours		69510 (0.03)

TABLE II. THE NUMBER OF COMPARISONS ($S = 10000$).

FIB entries	Method	#searches	#comparisons
10	n-ary tree [5]	10	160 (1.00)
	Hash table (linear list) [3]		34 (0.21)
	Bloom filter + hash table(linear list) [2]		21 (0.13)
	Ours		18 (0.11)
100	n-ary tree [5]	100	10379 (1.00)
	Hash table (linear list) [3]		282 (0.03)
	Bloom filter + hash table(linear list) [2]		172 (0.02)
	Ours		164 (0.02)
1000	n-ary tree [5]	1000	234822 (1.00)
	Hash table (linear list) [3]		3209 (0.01)
	Bloom filter + hash table(linear list) [2]		1912 (0.01)
	Ours		1806 (0.01)
10000	n-ary tree [5]	10000	2571184 (1.00)
	Hash table (linear list) [3]		61868 (0.02)
	Bloom filter + hash table(linear list) [2]		37739 (0.01)
	Ours		32185 (0.01)

content name n is given, every hash function H_i ($1 \leq i \leq m$) calculates its hash value $H_i(n)$ as follows:

$$H_i(n) = (hash_a(n) + hash_b(n) \times i) \% B \quad (1)$$

where $hash_a(n)$ and $hash_b(n)$ values are the upper half bits and the lower half bits of $H(n)$ calculated in Eqn. (2). Then we set the $H_i(n)$-th bit to one in the bloom filter when we insert the content name n into the FIB. When searching n, we can find out that n is already inserted into the FIB if all the bits calculated by $H_i(n)$ are already set to one.

2) Hash table: Let S be the size of the hash table, i.e., the hash table has S hash entries. Initially, every hash entry has an empty element. Now we assume that the content name n is just composed of several characters and every character has an 8-bit code. Let l be the length of n in character and let $str[i]$ be the i-th character in n. Then its hash value $H(n)$ is calculated by:

$$H(n) = \left(\sum_{i=0}^{l-1} str[i] \times 16^i \right) \% S \quad (2)$$

3) Balanced binary search tree: Even if a hash collision occurs, we keep all the FIB entries whose hash value is the same one using a balanced binary search tree. We construct a balanced binary search tree at the every hash entry in the lexicographical order of every content name. Note that, even when a content name includes "/", we just consider it to be one character.

C. Search method

Once the FIB data structure is constructed, we search the FIB table and find out the content name which matches a given content name. If we cannot find out a given content name, we perform a longest name match search in the same way.

IV. EXPERIMENT RESULTS AND CONCLUSIONS

The proposed method has been implemented in the C language and we have compared our proposed FIB to several existing methods. We set the hash size S to be 10 to 10000. Similarly, we set the number of stored FIB entries N to be 10 to 10000. In the bloom filter, we set $B = 14 \times N$ [bits]

and $m = 10$ [1]. We prepare 10–10000 content names to be searched ("#searches" in Tables I and II) in FIB whose length is 2 to 14 in every experiment. The miss rate, partial match rate, and complete match rate are set to be 40%, 35%, and 25%, respectively.

Experimental results demonstrate that our proposed FIB can reduce the number of comparisons by 96.1%, 47.8%, and 14.1% compared to those implemented by using n-ary tree, hash table (linear list), and bloom filter + hash table (linear list), respectively.

ACKNOWLEDGEMENT

The authors would like to thank Prof. Masao Yanagisawa of Waseda University for his insightful comments to this research.

REFERENCES

[1] A. Broder and M. Mitzenmacher, "Network applications of bloom filters: a survey," *Internet Mathematics*, vol. 1, no. 4, pp. 485–509, 2003.

[2] M. Fukushima, A. Tagami, and T. Hasegawa, "Efficient lookup scheme for non–aggregatable name prefixes and its evaluation," *IEICE Trans. on Communications*, vol. E96–B, no. 12, pp. 2953–2963, 2013.

[3] V. Jacobson, D. Smetters, J. Thornton, and M. Plass, "Networking named content," in *Proc. of ACM CoNEXT'09*, 2009, pp. 1–12.

[4] A. Ooka, S. Ata, K. Inoue, and M. Murata, "Design of a high–speed content–centric networking router using content addressable memory," in *Proc. of IEEE INFOCOM'14*, 2014, pp. 458–463.

[5] Y. Wang, K. He, H. Dai, W. Meng, J. Jiang, and B. Liu, "Scalable name lookup in NDN using effective name component encoding," in *Proc. of IEEE ICDCS'12*, 2012, pp. 688–697.

Low Latency IFFT Design for 3GPP LTE

Yeon-Jin Kim[1], Zheyan Piao[1] and Jin-Gyun Chung[1]
[1]Dept. of Electronic Engr., Chonbuk National University
Jeonju, South Korea
genie@jbnu.ac.kr

In-Gul Jang[2] and Kyung-Ju Cho[3]
[2]ETRI, Daejeon, South Korea
[3]Dept. of Electronic Engr., Wonkwang University
Iksan, Korea
sky4jjang@etri.re.kr, kjcho@wku.ac.kr

Abstract—**In this paper, a low latency IFFT architecture for 3rd Generation Partnership Project (3GPP) LTE is proposed. To reduce the latency, we reorder the IFFT input data. By using the reordered input data, both the latency and the memory in stage 1 are significantly reduced. Simulation results show that the latency for 2048-point IFFT is reduced about 42% compared with conventional architecture. The proposed architecture was verified using Altera Modelsim.**

Keywords; latency reduction, null signal, IFFT, memory size.

I. INTRODUCTION

To satisfy the large data processing and fast communication required by the increasing data traffics and new smart applications, the development of 5G communication is becoming more and more important.

One of the most crucial blocks in 3GPP LTE is FFT. Various efforts have been carried out for the efficient implementation of FFT blocks in terms of hardware area, power consumption and latency [1-3]. In conventional decimation in frequency (DIF) N-point IFFT, the memory size of stage i is $N/2^i$. Thus, it is important to reduce the memory size and consequently the latency in the first stage.

In this paper, we propose a low latency IFFT design method based on IFFT input data reordering.

II. PROPOSED LOW LATENCY IFFT DESIGN METHOD

Fig. 1 shows the radix-2^k based single-path delay feedback (SDF) IFFT architecture. The butterfly calculation in stage 1 can be expressed as

$$B(k_1, n_2) = x(n_2) + (-1)^{k_1}x(\frac{N}{2} + n_2), \quad (1)$$

where $k_1 = 0,1$ and $n_2 = 0,1,\cdots,N/2 - 1$.

In IFFT computations, for the first $N/2$ clock cycles, IFFT input signals are stored in the memory in stage 1. For the next $N/2$ clock cycles, the outputs corresponding to $k_1 = 0$ in (1) are sent to stage 2 while the outputs corresponding to $k_1 = 1$ are stored in the memory in stage 1.

In OFDM systems, IFFT length N can be expressed as

$$N = N_D + N_N, \quad (2)$$

where N_D and N_N represent the length of data and that of null signals, respectively. TABLE I shows N_D and N_N values for 3GPP LTE.

Fig. 2(a) shows the conventional arrangement of IFFT input signals. Fig. 2(b) shows the conventional IFFT input scheme expressed for butterfly computations. In Fig. 2(b), the length of Data 00 is the same as that of Null 1 and the length of Data 11 is the same as that of Null 0. The butterfly output $B(k_1, n_2)$ can be computed directly without addition or subtraction operations when either $x(n_2)$ or $x(N/2 + n_2)$ is a null signal. Thus, the butterfly operation in (1) is required only when n_2 satisfies the following condition:

$$\frac{N_N}{2} \leq n_2 \leq \frac{N_D}{2} - 1 \quad (3)$$

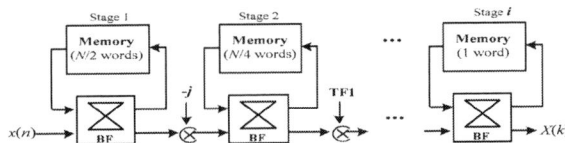

Figure 1. Radix-2^k based butterfly structure.

TABLE I. IFFT size (N), data length (N_D) and length of null signals (N_N) for 3GPP LTE

N	128	256	512	1024	2048
N_D	75	150	300	600	1200
N_Z	56	112	212	424	848

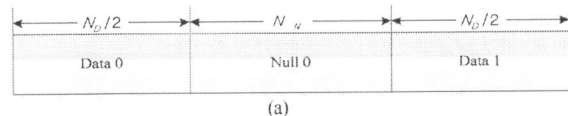

(a)

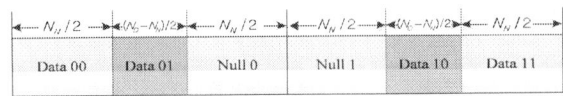

(b)

Figure 2. (a) Conventional IFFT input scheme, and (b) Conventional IFFT input scheme expressed for butterfly computations.

978-1-5090-3220-4/16 $31.00 © 2016 IEEE

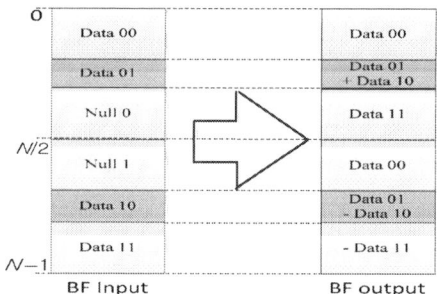

Figure 3. BF input and output signals in stage 1.

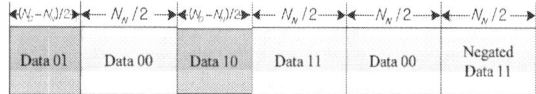

Figure 4. Reordered IFFT input.

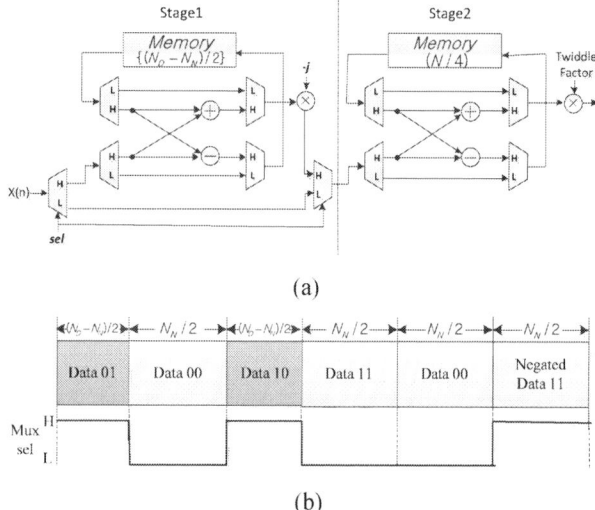

(a)

(b)

Figure 5. (a)Proposed butterfly architecture, and (b) Mux select signal for stage 1.

Fig. 3 shows the required BF input and output signals in stage 1. Addition or subtraction operations are performed only between Data 01 and Data 10. Based on this observation, latency of IFFT computation can be reduced by using the reordered IFFT input scheme in Fig. 4.

Fig. 5(a) shows the proposed butterfly architecture in stage 1. Fig. 5(b) shows the MUX selection signal for stage 1. As can be seen from Fig. 5(b), Data 00 can be sent to stage 2 after $(N_D - N_N)/2$ clock cycles while $N/2$ clock cycles are required in conventional architectures. In addition, the memory size in stage 1 is reduced from $N/2$ to $(N_D - N_N)/2$. Thus, the efficiency of the proposed method depends on the number of null signals in the IFFT input.

TABLE II. COMPARISION OF LATENCY AND MEMORY FOR 3GPP LTE

N	256		512		1,024		2,048	
Method	Conv	Prop	Conv	Prop	Conv	Prop	Conv	Prop
Latency	255 (1)	143 (0.56)	511 (1)	299 (0.58)	1,023 (1)	599 (0.58)	2,047 (1)	1199 (0.58)

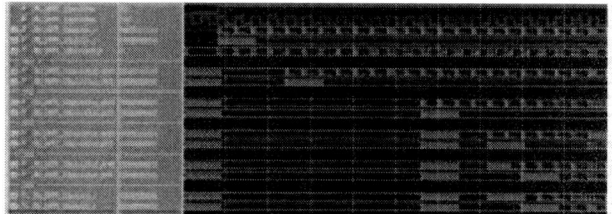

Figure 6. IFFT simulation results with $N = 32$.

By using the proposed architecture, the latency of IFFT can be derived as

$$T_{latency} = N - 1 - N_N. \qquad (4)$$

III. SIMULATION RESULTS

To verify the proposed IFFT architecture, we designed a 32-point IFFT using Verilog HDL. Fig. 6 shows the simulation results using Altera ModelSim. The simulation results show that we can achieve about 61% latency reduction in 32-point IFFT.

TABLE II compares the latencies for various IFFT computations for 3GPP LTE. As can be seen in TABLE II, when N is larger than 512 in 3GPP applications, latency reduction ratios are about 42%.

IV. CONCLUSIONS

In this paper, a low latency IFFT architecture was proposed. To reduce the latency, we reorder the IFFT input data based on the number of null signals in the IFFT input. By the proposed method, the latency is reduced about 42% in 3GPP applications.

REFERENCES

[1] B. Beheshti, "On Performance of LTE UE DFT and FFT Implementations in Flexible Software Based Baseband Processors", *Systems, Applications and Technology Conference, IEEE*, pp.1-4, 2009.

[2] In-Gul Jang, Kyung-Ju Cho, Yong-Eun Kim, Jin-Gyun Chung, "Memory Size Reduction Technique of SDF IFFT Architecture for OFDM-Based Applications", *IEICE TRANS. COMMUN.*, VOL. E95-B, No.6, pp.2059-2064, 2012.

[3] 3GPP LTE, "Evolved Universal Terrestrial Radio Access (E-UTRA); Base Station (BS) radio transmission and reception" 3GPP TS 36.104 v13.3.0, 2016-03

A 0.5V/22 μW Low Power Transceiver IC for Use in ESC Intra-body Communication System

Yuhwai Tseng, Tinyou Lin, Songwen Yau, Yingchieh Ho and Chauchin Su
Department of Electrical Engineering
National Chiao Tung University, Taiwan, R.O.C
yuhwaitseng@mail.nctu.edu.tw

Abstract—**This paper presents an integrated transciver chip for use in electro-static coupling intra body communication. A simplified circuit model was developed to analyze the channel characteristics. A Manchester code was used to increase signal energy. In front of the receiver, an inverter-based amplifier applies to amplify the received data. Then, a Clock and Data Recovery data develops to recover the transmitted data. The chip is fabricated using UMC 0.18um CMOS process with a chip area of 0.75 X 0.7 mm², power consumption of 22uW and a data transmission rate of 10M bit per second.**

Keywords— *Transceiver of Human Body Communication, Human Body Model, Intra-Body Communication*

I. INTRODUCTION

Recently, human body was functioned as a new wireless transmission scheme named as *Intra-body commutation* (IBC) [1]~[6]. IBC systems are divided into two categories - *electrostatic coupling* (ESC) and *electromagnetic waveguide* (EMW) systems. The ESC system is a ground free system in which the environment provides the signal return path. An EMW system produces electromagnetic waves using two electrodes, and treats the human body as a waveguide through which to transmit signals.

This work develops a low power transceiver integrated chip (IC) based on the ESC IBC system. A circuit model was proposed to analyze transmission characteristic of the ESC IBC system. The proposed IC is working in a supply voltage of 0.5V with a power consumption of 22 μW and a data transmission rate of 10M bit per second.

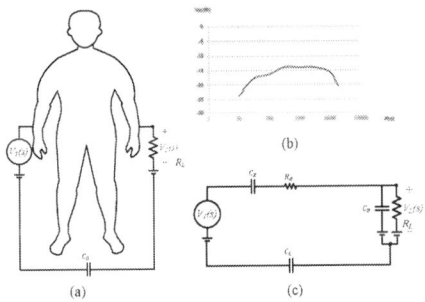

Figure 1. (a) simplified circuit model of an ESC IBC channel, (b) the frequency response of the channel, (c) equivalent circuit of the channel.

II. ESC IBC MODEL

Fig.1(a) shows a simplified circuit model of an ESC IBC channel. The transmitter and receiver are under different

ground potential. A signal return path between the transmitter and receiver is modeled as a capacitor C_G. The resistor R_L is the load resistor in front of the receiver. Figure 1(b) shows the frequency response of the channel. The result indicates that the channel is a band pass filter system with a gain of -14dB and a bandwidth of 12MHz. Figure 1(c) is an equivalent circuit of the channel in figure 1(a). C_X is a capacitor of body skin. C_B represents a capacitor between the body and the receiver ground. The evaluated result of the body parameters is shown in Table 1.

Table 1 The evaluated result of the body parameters.

R_B	0.8kΩ
C_X	35nF
C_B	65pF
C_G	16.5pF

III. ESC IBC TRANSCEIVER SYSTEM DESIGN

Figure 2 is the block diagram of the proposed transceiver system. The transmitter is composed of a counter, a preamble generator, a pseudo random data generator, a multiplexer, a Manchester encoder and an output buffer. For signal synchronization, a preamble code was transmitted before the data. The transmitted data is a pseudo random data made by a linear feedback shift register (LFSR) with a length of 2^7-1. Figure 3 is circuit diagram of the LFSR. The counter counts number of preamble code and controls the data passed through the channel. The preamble code and the data are encoded with a Manchester code. The output buffer sends the encoded data into the human body channel.

The receiver consists of a front end amplifier and a clock and data recovery circuit (CDR). This work devises a 0.5V low-voltage inverter-based amplifier. Figure 4 shows the architecture thereof, which is composed of a capacitor for isolating the operating voltage coupled into the human body, and two inverters, which include a gain stage for the amplification, and a biasing circuit to maximize the gain. All of the MOS transistors are designed to operate in the weak inversion region.

A signal coded with Manchester code is easy synchronized at receiver since the encoded signal contains clock information. Also, the Manchester encoded data has better noise immunity capability than the data without encoding. After the front end amplifier, a CDR with a four times over sampling rate was designed to recover the data. Figure 5 presents a diagram of the proposed algorithm of CDR. Where Φ_0, Φ_1, Φ_2, and Φ_3 represent four different sampling phases. If the received data is synchronized, the exclusive-or XOR value of the sampled data

---

The authors are grateful to United Microelectronics Corporation (UMC), Taiwan, for technology supports. This work is supported by Ministry of Science and Technology (MST) and Ministry of Science and Technology (MOST), R.O.C.

978-1-5090-3220-4/16 $31.00 © 2016 IEEE

at the phase Φ_1 and Φ_2 is true. If the phase of the received data is faster or slower than the local data, the XOR value of the sampled data at the phase Φ_1 and Φ_2 is false. In this situation, we check the XOR value of the phase of the sampled data at phase (Φ_0, Φ_2) and (Φ_1, Φ_3), respectively. If the phase of the received data is faster than the local data, the XOR value of the sampled data at phase (Φ_0, Φ_2) is false and the sampled data at (Φ_1, Φ_3) is true. On the contrary, If the phase of the received data is slower than the local data, the XOR value of the sampled data at phase (Φ_0, Φ_2) is true and the sampled data at (Φ_1, Φ_3) is false. A CDR is designed based on the above phase characteristic.

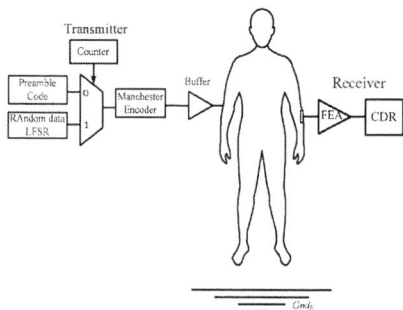

Figure 2. Block diagram of the proposed transceiver system.

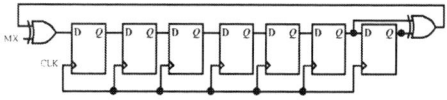

Figure 3. Circuit diagram of the LFSR.

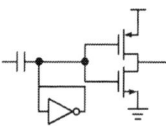

Figure 4. Architecture of the proposed inverter-based amplifier.

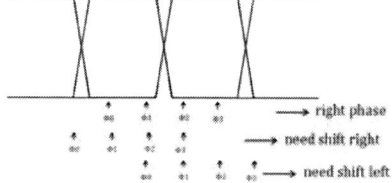

Figure 5. Diagram the proposed CDR algorithm.

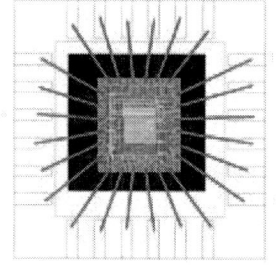

Figure 6. Chip layout of the proposed system.

IV. CHIP IMPLEMENTATION

The proposed transceiver is fabricated using UMC 0.18μm CMOS technology. Figure 6 displays a chip layout with the chip area of 0.75×0.7 mm^2. Figure 7 shows results of the post layout simulation. The results indicate the proposed chip is operated exactly under the worst case scenario. The proposed chip operates at 0.5V with a power consumption of 22 μW. Table 2 summarizes the performance of the proposed IC.

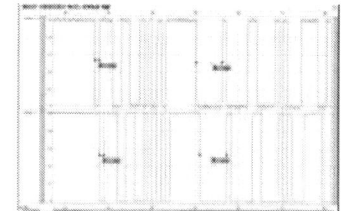

Figure 7. Results of the post layout simulation.

Table 2 Performance of the proposed IC

Item	Specification	Pre-Sim(TT)	Post-Sim(TT)
Technology	U18		
Supply Voltage	0.5V		
Power Consumption (AMP)	5uW	2.89uW	2.97uW
Power Consumption (CDR)	20uW	14.48 uW	22.12uW

V. CONCLUSIONS

This study develops an IC of ESC IBC system. A simplified circuit model of the ESC IBC system is used to analyze the transmission characteristic of the ESC IBC channel. A wide band transmission scheme using Manchester encoder is adopted in the developed band pass system. A 0.5V low operating voltage IC is developed to confirm the feasibility of a wideband ESC IBC system. The chip is implemented using UMC .18 CMOS process and its power consumption is only 22 μW, which supports a data transmission rate of more than 10Mbps.

REFERENCES

[1] T. G. Zimmerman, "Personal area network: Near-field intrabody communication," IBM System Journal, Vol. 35, No. 3&4, pp.609-617, 1996.

[2] S.-J. Song, N. Cho, H.-J. Yoo, "A 0.2-mW 2-Mb/s Digital Transceiver Based on Wideband Signaling for Human Body Communications," IEEE Journal of solid-state circuit, Vol. 42, Issue 9, Sept. 2007.

[3] Y. Tseng, C. Su, C.-N. Jimmy Liu, "Analysis and Design of Wide-Band Digital Transmission in an Electrostatic-Coupling Intra- Body Communication System," IEICE Trans. Comm, Vol. E92-B, No. 11, pp.3557-3563, Nov. 2009.

[4] Y. Lin, C. Chen, H. Chen, Y. Yang, S. Lu, "A 0.5-V biomedical system-on-a-chip for intrabody communication system," IEEE Trans. Ind. Electron 58: 690-698, FEB. 2011.

[5] N. Haga, K. Saito, M. Takahashi, K. Ito, "Equivalent circuit of intrabody communication channels inducing conduction currents inside the human body," IEEE Trans. Antennas Propag. 61: 2807-2816, May 2013.

[6] B. Kibret, M. Seyedi, TH. Daniel Lai, M. Faulkner, "Investigation of galvanic-coupled intrabody communication using the human body circuit model," IEEE J. Biomed. Health Inform. 18: 1196-1206, July 2014.

Area Efficient Neuromorphic Circuit Based on Stochastic Computation

Kiwon Yoon, Suhyeong Choi, and Youngsoo Shin

School of Electrical Engineering, KAIST

Daejeon 34141, Korea

Abstract— **Neuromorphic circuit can be simplified by applying stochastic computing, which uses a bit stream. A large number of stochastic number generators (SNGs) allows independent bit streams and hence secures accuracy, but outweighs the advantage of stochastic computing in circuit area. An area efficient SNG design method is proposed, in which a single linear feedback shift register (LFSR) is shared among a number of SNGs; independency of bit streams is made possible through shuffled wiring between LFSR and bit stream generators. Proposed design method is applied to a neuromorphic circuit that recognizes handwritten numbers; circuit area is reduced by 86% while prediction accuracy is sacrificed by 11% compared to a reference design in which LFSR is not shared.**

I. INTRODUCTION

A neuron in neuromorphic circuit consists of a few multipliers, an adder, and a threshold function. Since a large number of neurons are required in actual circuit, reducing the area of neuron is very important. Stochastic computing has been applied in neuron design for this purpose [1].

In stochastic computing, a real number $p \in [0, 1]$ is represented by the number of 1s (divided by the total number of bits) in a random bit stream S. Let S_x and S_y be two independent bit streams applied to AND gate and p_x and p_y be their corresponding real number respectively; the bit stream at the output of AND gate will represent $p_x p_y$ implying that AND gate can function as a multiplier.

A neuromorphic circuit with stochastic computing is shown in Fig. 1(a). At input layer, each real number p is converted to a bit stream through SNG, which is then applied to a hidden layer (or hidden layers); the output layer finally makes a decision. A typical structure of SNG is shown in Fig. 1(b), in which LFSR is employed to generate the final bit stream. Computation accuracy is mainly determined by how independent each bit stream is. This is intuitively shown in Fig. 2: correlated inputs yield inaccurate multiplication in (a), while accurate result is obtained with independent bit streams in (b). Independency of bit streams is achieved by a LFSR with its own unique seed. However, this causes large area occupied by LFSRs, e.g. 80% of area for LFSRs [2].

II. AREA EFFICIENT SNG DESIGN

Sharing a n-bit LFSR by all SNGs via rotated wiring was proposed [3], as shown in Fig. 3(a), where n is the bitwidth of SNG input. A random number is provided to every SNG, and is rotated by hardwiring in the middle so that each SNG

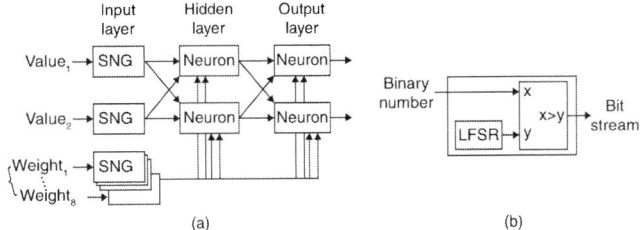

Fig. 1. (a) A neuromorphic circuit with stochastic computing, and (b) typical SNG structure.

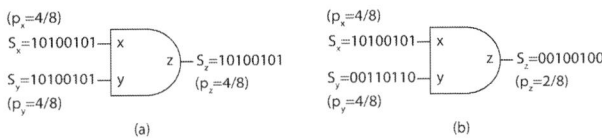

Fig. 2. Multiplication of two (a) correlated bit streams and (b) uncorrelated bit streams.

has different random numbers. Since no SNG contains LFSR anymore, area is drastically reduced. However, rotation gives only n different random numbers at a time, and the number of required SNGs ($>$1k; for a handwritten number recognition) is far more than n ($\leq$32; [3]). It means that there must be correlated bit streams, because some bit streams are born with same seed. Hence, prediction accuracy becomes low.

Proposed design let a m-bit ($m > n$) LFSR shared by all SNGs through shuffled wiring, as described in Fig. 3(b). Among m, n bits are randomly selected and shuffled before they are delivered to SNG. As a result, the maximum number of different random numbers at a time is $_mP_n$, which is much larger than n. Thus, bit streams are generated with less correlation, and prediction accuracy is ensured as unique seed design does. Also, area is decreased due to LFSR sharing.

III. EXPERIMENTS

We built an artificial neural network that consists of 196, 4 and 10 nodes at input, hidden, and output layer, respectively. Network was trained to predict handwritten numbers [4], and achieved 81% of prediction accuracy. Implementation was done with Verilog and circuit was synthesized with 28nm industrial library [5]. For prediction tests, l values from 10 to 19 were used, where 2^l is the length of bit stream.

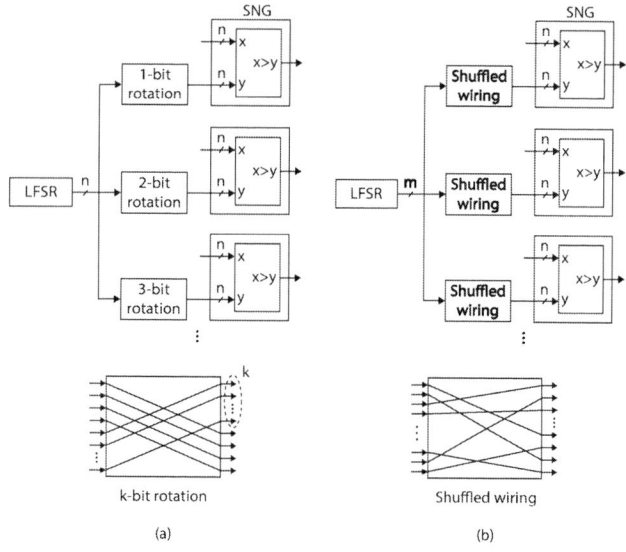

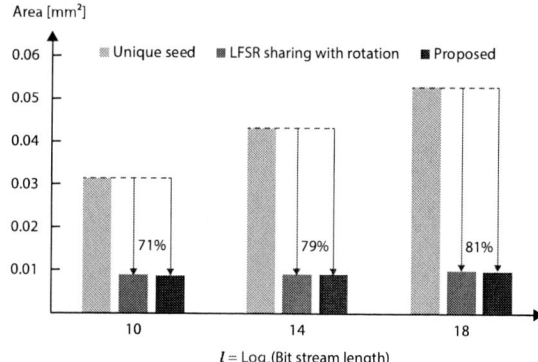

Fig. 4. Circuit area with three SNG implementation: unique seed, LFSR sharing with rotation, and proposed LFSR sharing with shuffled wiring; l is 10, 14, and 18.

Fig. 3. (a) LFSR sharing with k-bit rotation, and (b) proposed LFSR sharing with shuffled wiring.

We implemented the network by unique seed, LFSR sharing with rotation, and proposed design. In unique seed design, l-bit LFSRs were used. For the other designs, $(l-1)$-bit LFSR was shared by SNGs for pixel values, and l-bit LFSR and shared by SNGs for weights. In LFSR sharing with rotation design, n was equal to $l-1$ or l, but in proposed design, m was equal to $l-1$ or l, and n was fixed to 9.

We measure areas by three designs when l is 10, 14, and 18, as reported in Fig. 4. The number of LFSRs is 1,096 when unique seed design is applied, and it is reduced to 2 by LFSR sharing with rotation and proposed design. As a result, both designs reduce area by 71%, 79% and 81% when l is 10, 14, and 18, respectively. The number of registers in LFSR increases as l increases, thus impact of area reduction becomes bigger as l rises.

Prediction accuracies are measured through 200 test images while changing l from 10 to 19, as represented in Fig. 5. As stochastic computing is based on stochastic behavior, computation becomes more accurate as the length of bit stream increases, due to the law of large numbers. Since unique seed and proposed design provide uncorrelated random numbers for SNGs, prediction accuracies of both designs tend to increase as l increases. When l is extended to 19, degradations of prediction accuracy compared to original network are 6% and 11%, respectively. Prediction accuracy of proposed design is almost same as that of unique seed design, which implys that proposed design can be an alternative. Meanwhile, prediction accuracy of LFSR sharing with rotation design is below 30%, which is poor to use, even though area is reduced significantly. It is because of correlation between bit streams, which bring an inaccurate computation.

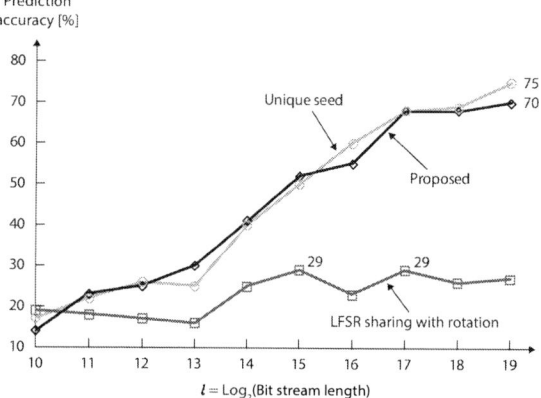

Fig. 5. Prediction accuracy from three SNG implementations while l is varied.

among a number of SNGs has been proposed; independency among bit streams is provided through shuffled wiring between LFSR and bit stream generators. The idea has been applied in neuromorphic circuit that recognizes handwritten numbers; circuit area is reduced by 86% while prediction accuracy is sacrificed only by 11%.

IV. CONCLUSION

Reducing the area of SNGs is a key in neuromorphic circuit design based on stochastic computation. Sharing an LFSR

ACKNOWLEDGEMENT

This work was supported by the National Research Foundation of Korea (NRF) grant funded by the Korea government (MSIP) (No. 2015R1A2A2A01008037).

REFERENCES

[1] V. Canals *et al.*, "A new stochastic computing methodology for efficient neural network implementation," *IEEE Trans. Neural Netw. Learn. Syst.*, vol. 27, no. 3, pp. 551–564, Mar. 2016.

[2] W. Qian *et al.*, "An architecture for fault-tolerant computation with stochastic logic," *IEEE Trans. Computers*, vol. 60, no. 1, pp. 93–105, Jan. 2011.

[3] H. Ichihara *et al.*, "Compact and accurate stochastic circuits with shared random number sources," in *Proc. Int. Conf. on Computer Design*, Oct. 2014, pp. 361–366.

[4] The MNIST database of handwritten digits. [Online]. Available: http://yann.lecun.com/exdb/mnist/

[5] *Design Compiler User Guide*, Synopsys, Mountain View, CA, June 2015.

A 4.1mA Adaptive Duty-Cycle Corrector Loop with Background Calibration in 45nm CMOS Process

Esther Kim, Deokgwan Jeong and Taehyoun Oh*

Department of Electronic Engineering
Kwangwoon University
Seoul, South Korea
E-mail: ohtaehyoun@kw.ac.kr

Abstract— **A mixed-mode adaptive duty-cycle corrector loop (DCC loop) architecture with background calibration method is proposed. The duty-cycle correction range is 20% - 80% with maximum 5.9ps duty error at 8GHz. The entire blocks of the DCC loop consume 4.1mA from 0.95 V supply voltage at 5GHz frequency. Successful operations of duty-cycle correction loop for various PVT conditions are verified via intensive corner simulations. The architecture is designed in 45nm CMOS process and occupies 0.0032mm².**

Keywords; Duty cycle correction loop (DCC); adaptive DCC loop; background calibration; clock

I. INTRODUCTION

In a circuit using both rising and falling edge of a clock signal for data retiming, a duty error results in imbalanced triggers and decreases timing margin. The achievable circuit speed is significantly reduced. A duty error results from mismatch between pull-up and pull-down time of PMOS and NMOS in a clock buffer at various corners. DCC schemes reduce the effect of such problems. In this paper, a mixed-mode adaptive duty-cycle corrector loop (DCC loop) with digital background calibration method is proposed. An analog RC duty cycle detector can sense duty error at the global clock driver (GCD) output regardless of PVT variations [1, 2]. Contrary to all analog DCC loops, digital loop architectures occupy small area and have high reconfigurability. A low speed digital loop consumes low power compared to DCC loops using SAR controller for fast convergence time [3, 4].

II. CIRCUIT IMPLEMENTATION

A. Overall circuit description

Fig. 1 (a) describes the proposed mixed-mode adaptive DCC loop architecture. The duty-cycle corrector (DCC) block adjusts either rising or falling timings. The output voltage level of the analog RC detector represents the duty of GCD output clock signal. VDD/2 voltage is interpreted as 50% duty. The other blocks in the DCC loop run at low speed. Depending on the sampler output that is determined by RC detector output, 5-bit counter output code is increased or decreased. This code is mapped to the number of enabled segments of DCC block.

The dotted blue line in Fig.1 (a) shows the DCC feedback loop to make the GCD output duty approach to 50% when the

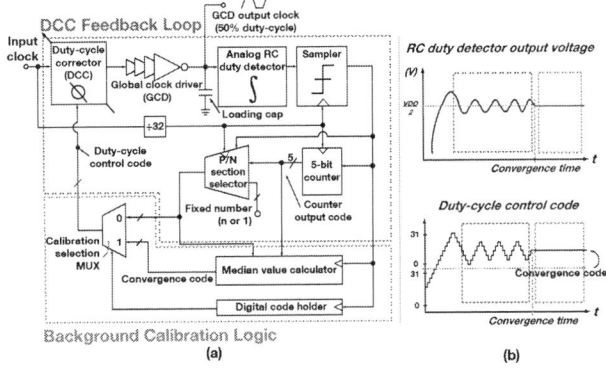

Figure 1. (a) Proposed adaptive DCC loop architecture with background calibration. (b) Illustrations of adpative convergence of the analog RC duty detector output (left-hand side) and duty-cycle control code (right-hand side).

input clock duty is not. However, the delay of the loop response causes fluctuation of RC duty detector output around VDD/2 as illustrated in the left-hand side graph of Fig. 1 (b). Then, the duty control code at DCC block input also fluctuates around the optimum code that makes GCD output duty 50% as depicted in the right-hand side graph of Fig. 1 (b). The fluctuation of RC detector node voltage and the code results in duty variation at GCD output and thus, lead to jitter on the clock signal.

The dotted yellow line in Fig. 1 (a) describes digital background calibration logic that reduces the jitter caused by fluctuation of codes, by fixing duty cycle control code at which the duty becomes 50%. The median value calculator averages the max/min values of the fluctuating code. The digital code holder counts the number of rising edges at the sampler output. If the occurrence number of rising edges exceeds a certain programmed number, the digital code holder makes "calibration selection MUX" choose the pre-calculated code from "median value calculator" and consequently the loop is locked. At this value after convergence, the background calibration blocks do not operate and consume zero power, and jitter caused by the loop fluctuation does not occur.

B. Duty-cycle corrector

Fig. 2 (a) shows the DCC block scheme that is composed of 32 PMOS and 32 NMOS segments. All segments have enable switches and resistors (R_P, R_N) at the drains of MOS switches

---

*Corresponding author

978-1-5090-3220-4/16 $31.00 © 2016 IEEE

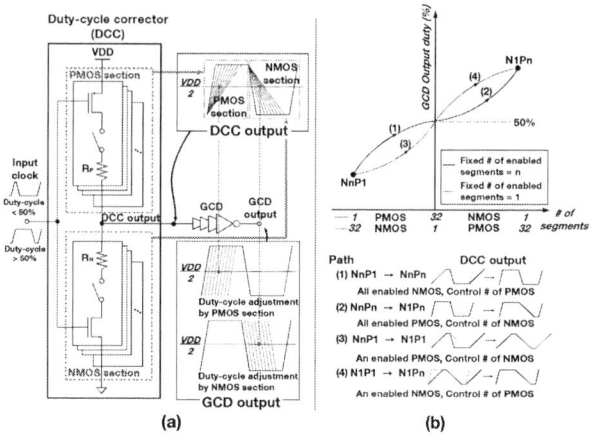

Figure 3. (a) Schematic of the proposed duty cycle corrector (DCC) block and GCD, variation of duty cycle at GCD output via DCC block. (b) Illustration of the duties at GCD output adjusted by number of enabled PMOS or NMOS segments of the DCC block.

and the input clock duty is corrected via adjustment of number of enabled segments. Fig. 2 (b) illustrates changes of the output duty cycle according to the number of the enabled segments. NnP1 stands for n (= 32) enabled NMOS segments and one enabled PMOS segment. The section of which the number of enabled segments is fixed is selected by "P/N section selector" in the DCC feedback loop. In this case, if fixed segments section is n segments of NMOS section, the duty variation at the GCD output depends on the number of PMOS enabled segments that changes rising time. For a high resolution dcc imposed on an input clock with a short period, the path (1) and (2) can be used programmably by setting the fixed number of segments with n. The values of R_P or R_N have a trade-off between wider input duty range (large R_P, R_N) and higher resolution (small R_P, R_N).

III. PERFORMANCE RESULT

Fig. 3 (a) presents the GCD output duties with the input clock duties swept for slow, typical, fast corners at 2GHz, 5GHz, and 8GHz frequencies. The results are obtained from post-layout simulations on proposed DCC loop. The achievable frequency range was 2 - 8GHz and the duty correction range was 20% to 80%. The duty error was up to 5.9ps at 8GHz clock. Fig. 3 (b) shows GCD output clock waveform and jitter at after the loop convergence. For an 8GHz 25%, 75% input clock at typical corner, the GCD output clock duties were all successfully compensated to 50% and jitter was 1.4ps after loop lock. A white noise SNR ratio was 30dB for all simulation cases. Fig. 3 (b) shows the layout of the proposed DCC loop and the entire blocks occupies 0.0032mm². The performance results are summerized in Table I and compared with other prior works.

TABLE I.   PERFORMANCE RESULT COMPARED WITH PRIOR WORKS.

	This work	**[1]**	**[3]**	**[4]**
Process	45 nm	45 nm	0.13 um	0.13 um
Supply	0.95V	1.8V+1.1V	1.2V	1.2V
Operating frequency	2 – 8GHz	2 – 1GHz	0.3125 – 1GHz	0.25 – 1GHz
Correction range	20%–80%	25%–75%	40%–60%	20%–80%
Duty error	Max.5.9ps @ 8GHz	Max. 20ps	Max.32ps	Max.36ps
Power per freq. (mW/GHz)	Max.0.78	1.4	3.2	Max.6.08
Jitter	1.4ps	N/A	15.5ps @1GHz	N/A

IV. CONCLUSION

We proposed an adaptive DCC loop architecture with a low power, wide duty correction and wide frequency range. The non-linear digital background calibration logics do not have influence on loop stability. The calibration method completely holds the optimal control code and does not generate the loop fluctuation after loop convergence. The performance of the DCC loop was simulated in various PVT conditions.

ACKNOWLEDGMENT

This research was supported by Basic Science Research Program through the National Research Foundation of Korea (NRF) funded by the Ministry of Education (NRF-2014R1A1A2056415).

REFERENCES

[1] Ravi Mehta, Sumantra Seth, Siddharth Shashidharan, et al, "A programmable, Multi-GHz, Wide-Range Duty Cycle Correction circuit in 45nm CMOS process," ESSCIRC, 2012, pp. 257–260.

[2] Sotirios Tambouris, Patent No. 7586349, Texas Instruments Deutschland Gmbh.:U. S., 2006.

[3] Young-Jae Min, Chan-Hui Jeong, Kyu-Young Kim, et al, "A 0.31-1 GHz Fast-Corrected Duty-Cycle Corrector with Successive Approximation Register for DDR DRAM Applications," IEEE VLSI system, 2011, pp., 20, 1524–1528.

[4] Chan-Hui Jeong, Ammar Abdullah, Young-Jae Min, et al, "All-Digital Duty-Cycle Corrector with a Wide Duty Correction Range for DRAM Applications.," IEEE VLSI system, 2015, pp., 24, 363–367.

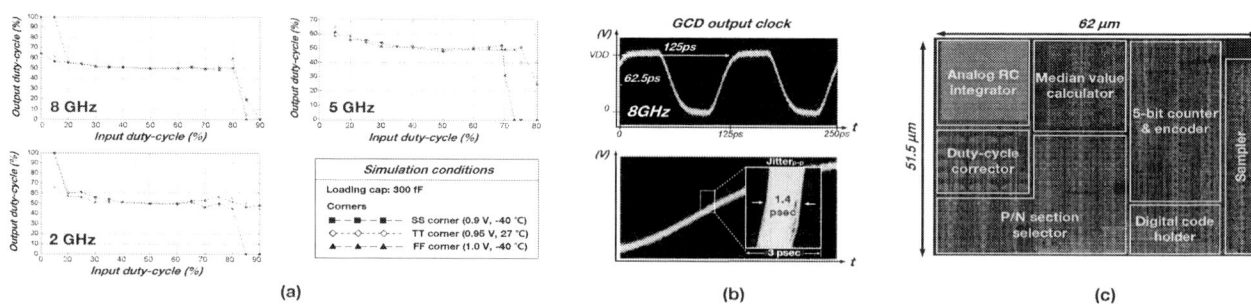

Figure 2. (a) Converged GCD output duties for various input duty and PVT conditions. (b) GCD output clock waveform and jitter at 8GHz after convergence. (c) Layout of whole architecture.

A New Approach to Binarizing Neural Networks

Jungwoo Seo[1], Joonsang Yu[1], Jongeun Lee[2], and Kiyoung Choi[1]

[1]Dept. of Electrical and Computer Engineering, Seoul National University, Seoul, Korea
{jungwoo.seo, joonsang.yu, kchoi}@dal.snu.ac.kr
[2]School of Electrical and Computer Engineering, UNIST, Ulsan, Korea
jlee@unist.ac.kr

Abstract— As deep neural networks grow larger, they suffer from a huge number of weights, and thus reducing the overhead of handling those weights becomes one of key challenges nowadays. This paper presents a new approach to binarizing neural networks, where the weights are pruned and forced to take degenerate binary values. Experimental results show that the proposed approach achieves significant reductions in computation and power consumption at the cost of a slight accuracy loss.

Keywords; image recognition; feedforward neural network; network pruning; weight compression

I. INTRODUCTION

Deep neural network research has made tremendous progress in the last several years. However, the use of deep neural network is sometimes restrictive due to large size and intensive computations. To address these issues, many different techniques have been suggested including vector quantization [1], weight pruning [2], and hashing trick [3].

Binarized neural network is proposed by Hwang et al. [4] and Courbariaux et al. [5]. Including a more recent incarnation [6], they all propose using only +1 and -1 for the degenerate values of weights of a deep neural networks, which may be considered to be an oversimplification. To alleviate the problem, we let different *neurons* have different degenerate weight values instead of +1 and -1. For this, we first prune near-zero weights and enforce the remaining weights to have two degenerate values, $+\alpha_i$ and $-\alpha_i$. The value of α_i can be different for ith neuron in a layer.

The pruning of near-zero weights reduces the number of multiplications. To further reduce the number of multiplications, we replace α_i to 1. To compensate for this replacement, the accumulation result of the products is multiplied by α_i. By doing so, all the multiplications of inputs and weights are replaced with simple sign changes and only one multiplication is left for each neuron.

In this paper, we present the detailed processes of binarizing weights in a feedforward neural network to significantly reduce the number of multiplications. We also show that high accuracy is maintained with much less computation and power consumption.

II. METHODOLOGY

A. Weight Compression

Before the weight compression, the neural network is trained in a normally way. It is a process of learning to see if a connection is important or not.

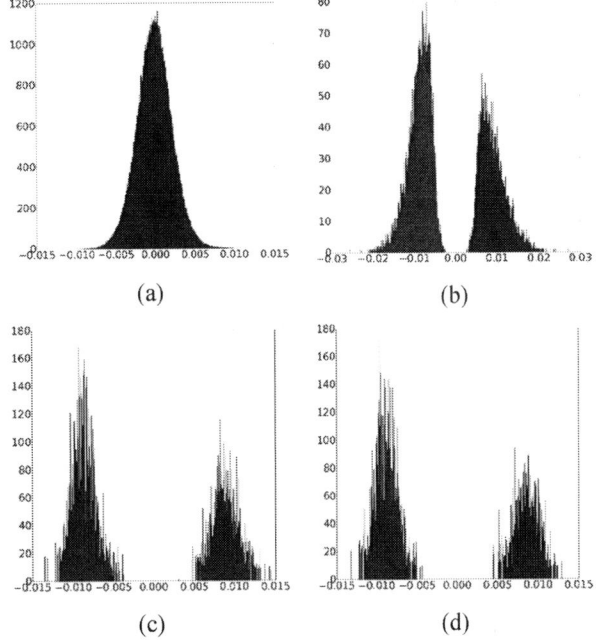

Figure 1. Change of weight distribution of third hidden layer. (a) Distribution of normally trained weights. (b) Distribution of weights after pruning near-zero weights. (c) Distribution of weights of all neurons after degeneration. (d) Distribution of weights after setting to $+\alpha_i$ and $-\alpha_i$.

The trained weights for each neuron typically form a Gaussian distribution and L2 regularization also leads weights closer to 0. Thus, most weights are gathered around 0 as shown in Fig. 1(a). So pruning near 0 weights reduces the number of weights effectively. The pruning threshold is determined by multiplying proper scaling factor to the standard deviation of weights.

Then the remaining weights after pruning form a bimodal distribution as shown in Fig. 1(b). In general, the pruned network shows accuracy as good as that of the original network. In this network, the remaining weights are converged to specific values, which are estimated by calculating mean values separately for the set of positive weights and that of negative weights.

After finding the mean values, all weights in each set are degenerated to their mean value as shown in Fig. 1(c). However, this simple binarization causes additional error. Thus, retraining is required to recover proper accuracy. This process of pruning and binarizing forms a cycle of compression. This cycle is iterated to get a minimal sized network and to find optimal mean values.

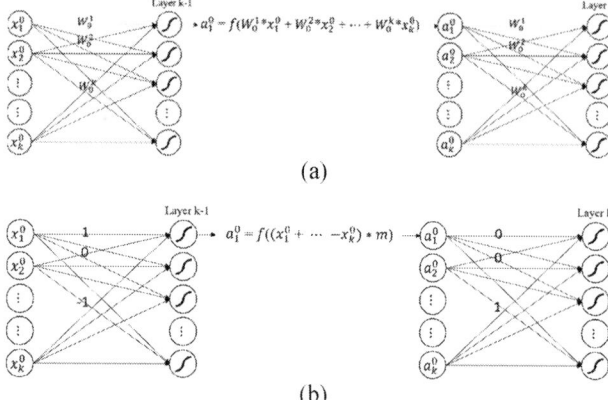

(a)

(b)

Figure 2. Multiplication can be reduced according to pruning and binarization (a) Conventional weight matrix multiplication. (b) Multiplication is delayed to activation function in pruned and binarized network

B. Multiplication in Activation Stage

Even after pruning near-zero weights and binarizing the remaining weights, lots of multiplications are still required to obtain the products of inputs and binarized weights. Fig. 2(a) shows the conventional case where the input vector is multiplied with the weight matrix. Each element of the resulting vector goes through an activation function. The total number of multiplications is the same as the number of elements in the weight matrix. The huge number of multiplications reduce operating speed and also consume lots of power.

These massive multiplications can be effectively reduced in our binarized network since all the weights for a neuron have the same absolute value $|\alpha_i|$ as shown in Fig.1(d). All weight values of $+\alpha_i$ or $-\alpha_i$ are changed to 1 or -1, respectively, and then the multiplication of α_i is placed after accumulation. In this way, we can reduce the number of multiplications.

III. EXPERIMENTAL RESULTS

A. Implementation details

We demonstrate the effectiveness of the proposed approach using a MLP-DNN and a CNN, both of which are designed for the MNIST benchmark dataset, running in Caffe on Nvidia TitanX GPUs.

The pruning threshold for each neuron is chosen as a quality parameter multiplied by standard deviation of the weights for the neuron. The quality parameter is empirically set to 0.8 in this work, but it can be adjusted to some other number. In the experiments, α_i is calculated by taking average of the absolute of the two degenerate values (positive and negative). Since their absolute values are already very similar, there is no significant accuracy loss when replacing them with the average.

B. CNN Model Result

The CNN model contains two convolution layers and one fully-connected layer with 1024 neurons. We apply our technique to the weights in the fully-connected layer. As shown in Table I, the baseline model of floating point implementation shows very high accuracy, but also has a large number of

weights and thus a large number of multiplications. After binarization about 50% of the weights are removed and the number of multiplications is reduced down to the number of neurons, while the accuracy increases slightly. It also achieves about 93.6% power reduction in the hidden layer.

C. MLP-DNN Model Result

For the MLP-DNN model, we compare the result with both of ternary FFDNN[4] and Binary connect[5] consisting of three fully-connected layers (see Table II). Note that they have different topologies—[4] has 500-500-2000 topology (TP1) and [5] has 1024-1024-1024 topology (TP2). Thus we implement our DNN on both topologies to obtain the comparison in Table II. Our work shows 0.41% of accuracy drop compare to [5], but achieves about 80% weights compression. And it also shows better result when comparing with [4], even though the baseline accuracy is different.

TABLE I.   CNN EXPERIMENTAL RESULTS

Method	Accuracy	Weights	Multiplications	Power
Baseline (FP)	99.09%	100%	1,058,816	100%
Proposed	99.1%	50.3%	1,034	6.4%

CNN Topology = {conv1: (5 * 5) * 20, conv2: (5 * 5) * 50, ip1: 1024, ip2. 10}

TABLE II.   COMPARISON WITH PREVIOUS WORK

Method	Accuracy			# of Weights
	Baseline (FP)	Binarized	Drop	
Ternary FFDNN[4]	99.03%	98.92%	-0.11pp	14.2%
Proposed (TP1)	98.70%	98.62%	-0.08pp	5.35%
Binary Connect[5]	98.70%	98.99%	+0.29pp	100%
Proposed (TP2)	98.70%	98.58%	-0.12pp	20.7%

DNN Topology = [4]: 500-500-2000-10, [5]: 1024-1024-1024-10

IV. CONCLUSIONS

We introduced a new binarized feedforward neural network, which has single-value weights and thus much less number of multiplications. As neural networks grow deeper and bigger, making the network smaller retaining high accuracy becomes more important. The sensitivity of the proposed approach to the size of neural networks can be a future research direction.

ACKNOWLEDGMENT

This work was supported by the KIST Institutional Program. (Project No. 2E26690-16-P023)

REFERENCES

[1] Yousha Gong, et al., "Compressing Deep Convoultional Networks using Vector Quantization," in arXiv preprint arXiv:1412.6115, 2014

[2] Song Han, et al., "Learning both Weights and Connections for Efficient Neural Networks," in NIPS, 2015.

[3] Wenlin Chen, et al., "Compressing Neural Networks with the Hashing Trick," in Proceedings of The 32nd ICML, 2015.

[4] Hwang Kyuyeon, and Wonyong Sung, "Fixed-Point Feedforward Deep Neural Network Design Using Weights +1, 0, -1", in SiPS, 2014

[5] Matthieu Courbariaux, Yoshua Bengio, Jean-Pierre David, "BinaryConnect: Training Deep Neural Networks with binary weights during propagations," in NIPS, 2015

[6] Syed Shakib Sarwar, et al., "Multiplier-less Artificial Neurons Exploiting Error Resiliency for Energy-Efficient Neural Computing", in DATE, 2016

978-1-5090-3220-4/16 $31.00 © 2016 IEEE

Customized SRAM design
for low power video code applications

Sangkyu Lee, Hoyoung Tang, Kyungrak Choi, and Jongsun Park

Korea University, Seongbuk-Gu, Seoul, 136-701, Republic of Korea, {sangkyu_lee, ho-2604, jongsun}@korea.ac.kr

Abstract— **In this paper, an embedded SRAM architecture of Video application is proposed to reduce the power consumption. By analyzing the general read and write access patterns, the embedded memory is customized to reduce power consumption while achieving general FIFO operations. Some of the signal activations and the Pseudo-read operations are removed in FIFO. According to the simulation results with 65nm CMOS process, the proposed embedded memory for line buffer achieves 17.62% power savings with 3.72% overhead compared to the conventional embedded SRAM approaches.**

Keywords-SRAM; Multimedia; Embedded memory; Low power operation; Line buffer

I. INTRODUCTION

Recently, portable devices such as smart-phones and video cameras are gaining popularity as well as making changes in every aspect of our daily lives. Multimedia data processing, including image/video applications is one of the key factors of the ever-increasing portable device market. However, image/video applications are very computationally intensive and require a large amount of embedded memory access, which results in significant power consumption and thus limits the battery lifetime of portable devices. In addition, random access based SRAM can result in significant power consumption due to unnecessary operation. Many previous research efforts have focused on reducing the power consumption of portable multimedia applications. In video application, the embedded memory access used for buffering is one of the primary sources of power consumption [1]. In order to tackle this problem, we propose an optimized architecture of SRAM to reduce power consumption by analyzing the access pattern of First-In-Fist-Out (FIFO) operation for video application.

II. VIDEO APPLICATION

In this section, we describe the embedded SRAM used as the line buffers [2] in video applications. Generally, the interface of the input data is sequential in video applications. Then, input data need to be lined up to be in the form of macro block using the line buffer as shown in Fig. 1. The line buffer generally consists of a number of FIFO memories using embedded SRAM banks and a number of registers. The bank number and address sizes of SRAM for line buffer are decided by the pixel size of macro block and the total image resolution. For example, in the case of the 4K UHD image where macro block and resolution is 8 x 8, 4096 x 2160, the line buffer is configured as 7 banks of SRAM of 4096 addresses (28 kB). Thus, video application requires a large amount of SRAM and the

embedded memory consumes a lot of power. Therefore, the embedded SRAM is necessary for reducing power consumption.

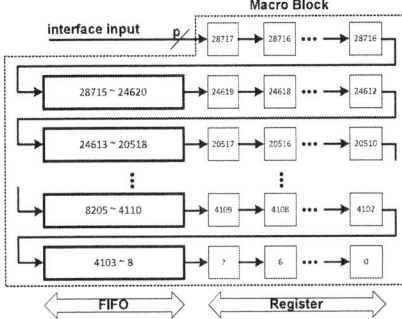

Fig. 1. Line buffer with FIFO memory

III. PROPOSED STRUCTURE

In this paper, we propose two methods that reduce power consumption for SRAM of FIFO operation. First, the unnecessary Pseudo-Read operations for Bit Line (BL), Bit Line Bar (BLB) are removed through the proposed SRAM structure. Second, the number of Word Line (WL) activation is decreased for reducing the power consumption while achieving the general FIFO operation.

A. Prevention of the unnecessary Pseudo-Read

The proposed SRAM structure is modified to have additional precharge_enable ports to conventional structure as shown in Fig. 2. This port can carry out the selection of Pseudo-Read operation. During the read operation, BLs and BLBs of the bit

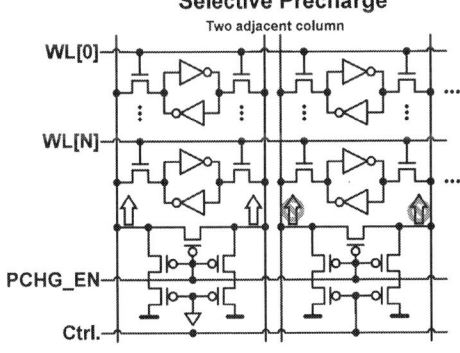

Fig. 2. Proposed SRAM structure with selective precharge

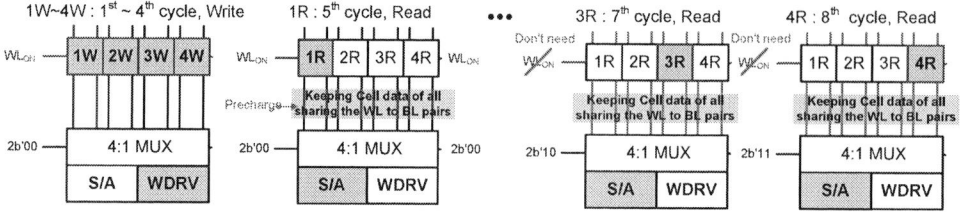

Fig. 4. Processing for unnecessary WL activation according to cycle

cells, which are connected with the same activated word line (WL), need to be precharged not to lose their stored data. This operation is called Pseudo-read operation. But the Pseudo-Read operation is not necessary for the data when it is previously accessed and will not be accessed again. In this work, using the selective precharge operation, the unnecessary Pseudo-read operations can be completely prevented. Fig. 3 shows the cases of the unnecessary Pseudo-Read elimination in case of 4 to 1 interleaving SRAM [3]. By using this selective precharge, the proposed SRAM structure can reduce power consumption without a large area overheads.

B. Removal of unnecessary WL activation

In order to operate read for sequentially increasing addresses, conventional SRAM has to activate the same WL N times. Because one WL is connecting N words, capacitances of the WL are significantly large. In addition, SRAM consumes power not only for WL but also for BL due to precharge for cell keeping data. Therefore, the unnecessary activations of WL, BL and BLB increase power consumption considerably [4]. In order to reduce these power consumptions, we propose the method that can keep data of BL, BLB after one WL activation through selective precharge control. As shown in Fig. 4, this method can reduce power consumption without area overheads when sequential read operations are processed.

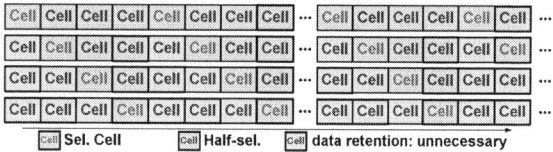

Fig. 3. Unnecessary cell for Pseudo-Read according to cycle

IV. IMPLEMENTATION RESULT

We present implementation results of the proposed SRAM structure for line buffer with the 8x8 pixels macro block and the 4096 x 2160 image resolution. The proposed line buffer is

Table I. Comparison of implementations results of conventional and proposed Line Block

	Power	Area
Conventional	8.40 mW	601223 um^2
Proposed.	6.92 mW	623588 um^2
Percent	-17.62 %	+3.72%

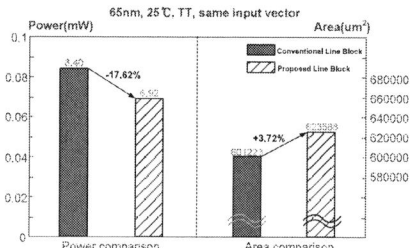

Fig. 5. Power and area comparison of Line block

designed with Samsung 65nm CMOS technology. The power and the area of SRAM for line buffer are compared in this Section. Table I shows the power and area comparisons between the proposed and conventional SRAM. Fig. 5 also shows a power consumption comparison of line buffers

V. CONCLUSION

In this paper, we propose a low power SRAM architecture that is optimized for the FIFO operations of line buffer. By analyzing FIFO memory access pattern for the video application, the numbers of BL, BLB precharge operations and WL activations are minimized during the FIFO operations. Therefore, the proposed embedded memory used for Video application achieves 17.62% of power savings with 3.72% of overhead compared to the conventional embedded SRAM approaches.

ACHKNALAGEMENT

This work was supported by the National Research Foundation of Korea (2011-0020128 and 2016R1A2B4015329). This work was also supported by information Technology Research and Development Program of Korea Evaluation Institute of Industrial Technology (KEIT) [10052716, Design technology development of ultralow voltage operating circuit and IP for smart sensor SoC].

REFERENCES

[1] C. P. Lin et al., "A 5 mW MPEG4 SP encoder with 2D bandwidthsharing motion estimation for mobile applications," in Proc. ISSCC Dig. Tech. Papers, Feb. 2006, pp. 1626–1635. motion estimation for mobile applications," in Proc. ISSCC

[2] F.-C. Huang, S.-Y. Huang, J.-W. Ker, and Y.-C. Chen, "HighPerformance SIFT Hardware Accelerator for Real-Time Image Feature Extraction," *IEEE Transactions on Circuits and Systems for Video Technology*, vol. 22, no. 3, pp. 340-351, Mar. 2012

[3] I. J. Chang, et. al, "A 32 kb 10T sub-threshold SRAM array with bit-interleaving and differential read scheme in 90 nm CMOS," *IEEE J. Solid State Circuits*, vol. 44, no. 2, pp. 650–658, Feb. 2009.

[4] K. Kim, et. al, "A low-power SRAM using bitline charge-recycling," *IEEE J. Solid-State Circuits*, vol. 43, no. 2, pp. 446–459, Feb. 2008.

ISFET with Built-in Calibration Registers through Segmented Eight-bit Binary Search in Three-Point Algorithm Using FPGA

Cyrel Ontimare Manlises [1], Febus Reidj G. Cruz [1,2], and Wen-Yaw Chung [2]

[1] School of Electrical Electronics and Computer Engineering, Mapua Institute of Technology, Manila, Philippines
[2] Department of Electronic Engineering, Chung Yuan Christian University, Chungli, Taiwan R.O.C.
ccontimare@mapua.edu.ph, frgcruz@mapua.edu.ph, eldanny@cycu.edu.tw

Abstract— **The calibration of chemical sensors is done manually or wherever the sensor is actually deployed. There are certain procedures that should be accurately followed, otherwise the pH measurement will produce an incorrect output. The main purpose of this paper is to create the circuit of an Ion-Sensitive Field-Effect Transistor (ISFET) with built-in calibration registers through segmented eight-bit binary search in three-point algorithm using FPGA. The circuit used the three-point calibration algorithm and the three standard buffers pH4, pH7, and pH10. The block diagram, schematic diagram, and the number of logic gates were derived after synthesizing the Verilog program in Xilinx/FPGA. An average of 0.30% error was computed to prove reliability of the created circuit in FPGA. Having an ISFET with built-in calibration registers will ease the work of the experts in performing calibrations and could be used as a pH level meter or a remote sensor node in several applications.**

Keywords; ISFET, segmented binary search, three-point algorithm, FPGA, calibration, pH level, ion sensitive field-effect transistor

I. INTRODUCTION

An electrochemical sensor like the ion-sensitive field-effect transistor (ISFET) plays an important role in the industry. Its broad application in environmental, agricultural, biomedical, pharmaceutical, food industries and many more makes it in demand and popular [1-5]. The ISFET is normally used for data acquisition and monitoring of pH levels [2-4]. Characteristics of pH sensors vary in time because of the changes in its electrodes. Changes in the performance of the electrodes might affect the accuracy of its measurements. Monitoring the pH level is necessary in order to avoid further chemical reactions in a substance [4]. It is in cases like this where a calibration takes place. The calibration is performed regularly, normally once a day or depending on the application, to maintain acceptable pH levels. Thru calibration, the sensitivity of the ISFET will be maintained, hence the measurements will be accurate and reliable. One of the techniques used was the two-point calibration. This technique was designed based on the binary search algorithm which utilized the standard buffer pH levels pH4 and pH7 [2-3].

The calibration of an ISFET is done where it is actually deployed. In such cases, an expert may be needed and a sequence of procedures should be followed. The procedure in calibrating an ISFET includes the measurement of a known solution and of an unknown solution. The pH's equivalent voltage of the measured known solution is saved into the memory or registers inside the system for reference during the calibration [1-2].

An ISFET with a built-in register can make the calibration easier because it would follow the plug and play standard especially when the sensor is due for replacement. It may also lessen the power consumption of the system because the registers are already included in the sensor itself [2]. By having built-in registers in a sensor, the logic gates and the design will be simplified, thus the area occupied by the circuit may also be reduced. In a segmented three-point calibration, the three standard buffers pH4, pH7, and pH10 were utilized.

II. METHODOLOGY

This section shows how the algorithm was implemented and how the circuit design was derived.

A. Three-point calibration algorithm

Figure 1 describes the segmented three-point calibration algorithm. In MEASURE, if RESET is *HI*, voltage unknown will be processed, otherwise, calibration will end. In CALIBRATION, if RESET is *LO*, mode parameters one to three will be tested and the process will end. In SEG3PT_ALGO, if STATE value is *IDLE* and START signal is set to *HI*, the voltages of the three standard buffers will be saved into the registers. If STATE value is *CHECK*, a segmented three-point calibration algorithm is performed.

The three standard buffers were assigned with 8-bit binary equivalents and is shown in Table I. The voltage for every change in pH unit is 58.16mV [6]. Equation (1) was used to compute for the interval, which is 0.02362pH/bit. This is how the look-up table library was derived.

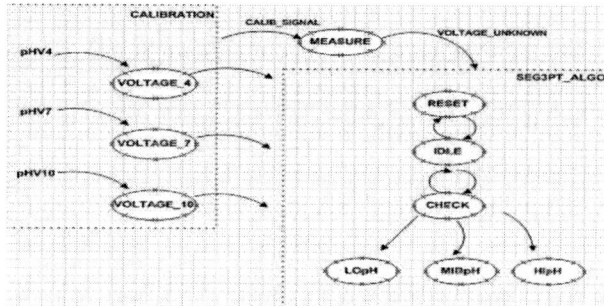

Fig 1. The process using the three-point algorithm

TABLE I. Eight-bit Assignment to pH

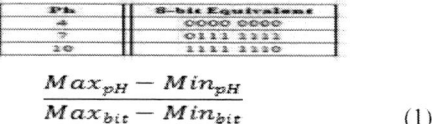

Ph	8-bit Equivalent
4	0000 0000
7	0111 1111
10	1111 1110

$$\frac{Max_{pH} - Min_{pH}}{Max_{bit} - Min_{bit}} \qquad (1)$$

B. Schematic Diagam

Figure 2 illustrates the schematic diagram for the ISFET with built-in registers. The number *Slices* is 142 (the area occupied by the circuit), the number of *Slice Flip-Flops* is just 84, the number of *4 input LUTs* is 263 (number of 4-input Boolean functions), the number of *bonded IOBS* is 43 (number of pins used). The clock period is 11.636ns while the maximum frequency is 85.937MHz.

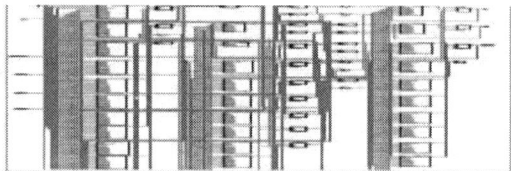

Fig 2. Schematic diagram of an ISFET with built-in registers

III. TESTS AND RESULTS

This section will prove that the design is working and is producing a correct output.

In Table II, the calculated pH was derived from the simulations thru test bench in Xilinx/FPGA. Equation (2) was used to compute for the percentage of error between the expected pH and the calculated pH. For pH4, pH7, and pH10 there was 0% error; for the pH5 there was 0.78% error; for the pH6 there was 0.52% error; for pH8 there was 0.489% error; and for pH9 there was 0.348% error. An average of 0.30% of error was concluded after the series of simulations.

TABLE II. Expected pH vs the Calculated pH

Expected pH	Voltage Unknown (Vrout) in mV	Calculated pH			% error
		In Decimal	In Binary	Output (pH)	
4	-173.4302745	254	11111110	4	0
5	-114.5320784	210	11010010	5.03937	0.78124845
6	-57.00360781	168	10101000	6.031496	0.52219217
7	0.52486277	127	01111111	7	0
8	59.42305884	83	01010011	8.03937	0.48971499
9	118.9515294	41	00101001	9.031496	0.34873514
10	174.48	0	00000000	10	0

$$\%error = \frac{MeasuredpH - ExpectedpH}{Measured} * 100 \qquad (2)$$

The measured pH 4.04, 5.99, 7.89 and 9.76 in a research ASIC for ISFET, which used a different algorithm, were considered as part of the data simulation to get the coefficient of determination (R^2) between the two different algorithms [3]. Figure 3 illustrates that the three-point algorithm produces a more linear output (R^2=0.9586) compared to that of the two-point algorithm (R^2=0.957).

Expected pH	pH in 3-pt	pH in 2-pt
4	4.04	4.04
5	4.188976	4.11811
5	4.377953	4.30709
5	4.661417	4.59055
5	4.944882	4.88764
6	6	5.99
8	7.566929	7.566929
8	7.92	7.89
8	8.03937	8.06299
8	8.464567	8.48819
9	9.102362	9.14961
9	9.338583	9.40945
10	9.78	9.76

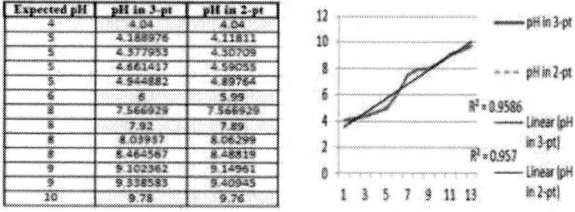

Fig 3. Data simulation using two different algorithms and their Linearity

IV. CONCLUSION

The ISFET with built-in calibration registers through segmented eight-bit binary search in three-point algorithm produced an average of 0.30% error. The calculated pH from simulations thru test bench in Xilinx/FPGA proves that the created circuit is reliable. Through synthesis, the schematic diagram was derived. The number of slices is 142. The number of slice flip-flops used is 84. The number of 4-input Boolean functions used is 263. The number of pins used in the circuit is 43. The clock period is 11.63ns and the maximum frequency of the whole circuit is 85.937MHz. The created circuit will ease the work of experts in performing calibrations and could be used as a pH level meter or a remote sensor node in several applications.

ACKNOWLEDGMENT

The researchers would like to express their sincerest gratitude to all Engineers at Mapua Institute of Technology who helped finish this study.

REFERENCES

[1] Jose Francisco Villapando Perez, Manuel Moises Miranda Velasco, Miguel Enrique Martinez Rosas, Horacio Luis Martinez Reyes, "ISFET Sensor Characterization", ScieVerse Science Direct *Procedia Engineering*, vol. 35, pp. 270-275. 2012

[2] Wen-Yaw Chung, Jian-Ping Chang, Febus Reidj G. Cruz, "Clock-Gated and Low-Power Standard Cell Library for ISFET Two-point Calibration Processor Chip", 978-1-4244-7456-1,IEEE, 2010

[3] Wen-Yaw Chung, Chung-Huang Yang, Yaw-Feng Wang, Wladyslaw Torbicz, Dorota G. Pijanowska, " A Signal Processing ASIC for ISFET-based chemical sensors",0026-2692, Elsevier Ltd, 2004.

[4] Miao Yuqing, Chen Jianrong, Fang Keming, "New Technology for the detection of pH ", *Journal of Biochemical and Biophysical Methods*, vol. 63, pp. 1-9. 2005.

[5] Wen-Yaw Chung, Yeong-Tsair Lin, Dorota G. Pijanowska, Chung-Huang Yang, Mng-Chia Wang, Alfred Krzyskow, Wladyslaw Torbicz, " New ISFET interface circuit design with temperature compensation",0026-2692, Elsevier Ltd, 2006.

[6] Erich K. Springer, "pH Measure Guide", HAMILTON, The Measure of Excellence,pp. 7-17.

Power-efficient Partitioning and Cluster Generation Design for Application-Specific Network-on-Chip

Jiayi Ma, Cong Hao, Wencan Zhang and Takeshi Yoshimura

Graduate School of IPS, Waseda University

Kitakyushu-shi, 808-0135 Japan

Abstract—Network-on-Chip (NoC) is a promising solution for System-on-Chip (SoC) challenges. In this work, we present a Decompose and Cluster generation Refinement (DCR) algorithm to find minimum power consumption simultaneously. A two-stage method is proposed for decompose and cluster generation step to generate solutions with lower power. Refinement step explores optimal positions and adjusts clusters for selected solutions to find balanced point between power consumption and CPU time. Experimental results show that the proposed method outperforms the existing work.

Keywords — Network-on-Chip; partitioning; clustering

I. Introduction

Power-efficient Application-Specific Network-on-Chip (ASNoC) design has become one of the key challenges for specific applications in nanoscale System-on-Chip (SoC) designs. The insertion and allocation of switches largely influence power consumption and a number of researches have been done to minimize power consumption by exploring optimal switch number.

W. Zhong [1, 2] integrated partitioning into floorplanning to find optimal switch number by two steps: used min-cut bi-partitioning algorithm in partition by tool hMetis [3] in the first step and proposed a floorplanning integrated with cluster generation (FCG) method to adjust clusters by inserting removed core into floorplan in the second step. The previous works consume CPU time for meaningless bi-partitioning and lack of flexibility with only one initial solution.

To conquer drawbacks, we proposed a Decompose and Cluster generation Refinement (DCR) algorithm to explore balanced point of minimum power consumption and running time. Main contributions are summarized as follows. (1) An efficient two-stage method for more optimal initial partitioning and cluster generation method is proposed. (2) Insertion after removed refinement is obtained to combine floorplan information with power reduction.

II. Proposed Algorithm

A. Overall Algorithm

Fig. 1 is the overall of algorithm with the input of cores and edges information. A fixed-outline floorplanning tool IARFP [4] is used to form initial floorplan. Section B introduces a two-stage initial partitioning and cluster generation method and Section C shows insertion after removed refinement method to reconstruct clusters by floorplanning. Minimized power consumption is calculated after refinement as output.

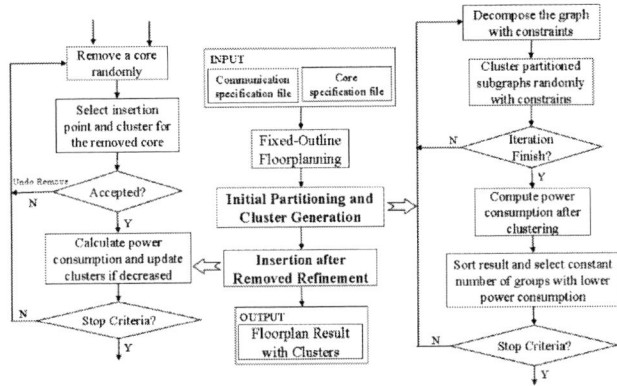

Figure 1. Overall of Algorithm Design

B. Partitioning and cluster generation

Initial partitioning and cluster generation algorithm has two parts. For partitioning parts, decompose the graph into subgraphs by removing edges after calculating weight of edges. For cluster generation parts, cluster subgraphs randomly to generate various solutions with different power.

At partitioning part, the weight $w_{i,j}$ of each edge $e_{i,j}$ connects core c_i and c_j can be calculated by (1), where $cr_{i,j}$ is the communication requirement between core c_i and core c_j, $dis\_c_{i,j}$ shows the distance between weighted cores. max_cr is overall maximum communication requirement. $mean\_dis$ is the average distance for connected cores, which is obtained from initial floorplanning. α, β, γ are used to adjust relative weight, where $\alpha + \beta + \gamma = 1$ and $rand()$ is a random number.

$$w_{i,j} = \alpha * \frac{cr_{i,j}}{max\_cr} + \beta * \frac{mean\_dis}{dis\_c_{i,j}} + \gamma * rand \qquad (1)$$

Edges are sorted in ascending order after computing. Decompose process begins from removing lower weighted edges and terminates when core number of each subgraph is less or equal to the set maximum cores number in one subgraph. The decompose part iterates for N times and iteration result is sent to cluster step for circulating M times.

For clustering part, clusters generate with randomly combination of decomposed subgraphs. One group of solution formed when all subgraphs clustered into clusters that core number of each cluster is less or equal to the set maximum cores number in one cluster. The cluster generation step iterates for M times and receive the next decomposed subgraphs solution. The result set has $N*M$ solutions with calculated power consumption and K (K < N * M) groups of

Figure 1. Comparison of the NoC synthesis results

Benchmark	V#	E#	Area (mm^2)		Total Power (mW)						Time (s)					
			DCR	TSF[2]	DCR Loop #					TSF[2]	DCR Loop #					TSF[2]
					1	3	5	10	20		1	3	5	10	20	
MPEG4	12	13	8.78	10.00	27.52	26.26	17.27	17.19	16.58	19.43	4.25	4.96	4.98	5.14	7.42	5.22
VOPD	12	14	16.53	19.58	21.18	19.53	16.80	16.80	16.80	21.06	2.80	3.79	3.85	4.27	4.50	4.78
263decmp3dec	14	15	27.70	30.37	102.86	98.60	86.50	77.03	76.95	113.66	2.80	6.55	6.67	6.78	6.88	7.97
263encmp3dec	12	12	21.70	24.32	1563.22	1392.78	992.50	982.67	978.36	1193.63	2.76	6.37	6.47	6.59	7.34	6.65
mp3encmp3dec	13	13	17.80	19.16	99.69	82.38	79.69	64.37	64.31	98.67	2.56	5.72	6.61	6.82	7.30	6.00
D_38_tvopd	38	47	33.27	37.98	135.00	132.87	99.53	97.92	94.70	104.25	20.73	29.75	30.85	31.25	33.85	35.31
D_50	50	57	41.56	47.60	241.17	211.68	177.79	163.04	160.75	165.45	30.02	49.97	52.64	54.24	63.92	70.98
Avg. IMP (%)			-11.36		19.99	10.03	-12.62	-17.71	-18.86		-48.53	-11.78	-10.20	-7.06	5.80	

solutions with lower power cost are selected for refinement process.

The power consumption computation considers both link power and switch power with their leakage and dynamic power, which is shown from (2) to (4). E_l^k and E_s^t represents bit energy of link k and switch t and lP_l^k and lP_s^t illustrates leakage power of link k and switch t separately. S and L is the total number of links and switches. The parameters of them are generated from Orion [5] under 70 nm technology.

$$P_{total}^n = P_{link}^k + P_{switch}^t \qquad (2)$$
$$P_{switch}^t = \sum_{t=1}^{S} cr_{i,j}^t * E_s^t + lP_s^t \qquad (3)$$
$$P_{link}^k = \sum_{k=1}^{L} cr_{i,j}^k * E_l^k + lP_l^k \qquad (4)$$

C. Insertion after removed refinement

Refinement procedure reconstructs clusters for selected solution by changing initial floorplan. The main process shows in the left part of Fig. 1, which is similar to [2]. A removed core is randomly selected and inserted into sequence pair to generate new floorplan with constraint of fixed-outline floorplan presented by (5) with width and height of floorplan outline. A is the total area calculated by initial floorplan, μ is a maximum whitespace fraction, δ is aspect ratio and generally $\delta = 1$. Power consumption is recalculated for candidate clusters when new floorplan generated and the lowest cost will be chosen as the output of final power consumption.

$$W = \sqrt{(1+\mu)A\delta} \ , \ H = \sqrt{(1+\mu)A/\delta} \qquad (5)$$

III. EXPERIMENTAL RESULT

We implemented our algorithm in C language and executed on a Windows with 2.50GHz CPU and 3 GB RAM. The clock frequency is set to be 750MHz and the corresponding maximum switch size is set to be 8. The data width of the NoC links is 32 bits. In experiment, the parameters in (1) are set as $\alpha = 0.5$, $\beta = 0.3$, $\gamma = 0.2$ and we set N = 20, M = 50 to generate 1000 solutions initially and K (K = 1, 3, 5, 10, 20) groups of solutions are used for insertion after removed refinement process. The results are average of 10 runs.

A two-step floorplanning (TSF) [2] is used for results comparison and benchmarks for experiment have the same source with [2]. In the table, the DCR Loop# shows selected groups numbers, which is K in the paper. V# and E# are core number and edge number for each benchmark. The proposed method shows the average improvement of 11.36% for floorplan area. Fig. 2 presents tendency for each benchmark with relationship of running time and power cost. Solid points in the curve presents the number of selected groups K (K = 1, 3,

5, 10, 20) from left to right. The third point (K=5) shows extreme descent and the curve changes smoothly from the fourth point (K=10) to the fifth point (K=20) for each benchmark. The optimal results presents when K ranges from 5 to 10 with reduced power consumption 12.62%, 17.71% and time cost 10.20%, 7.06%, respectively. This is because insertion points for each benchmark are constant for fixed-outline floorplanning even though solution groups are in a large number. Compared the last two figures with others, the proposed DCR algorithm has better result for small-sized applications benchmarks than large-sized. The reason is that benchmarks with large number of cores are more flexible and the balanced point is higher than small application benchmarks as the proposed algorithm using a random method to generate cluster and optimize solutions.

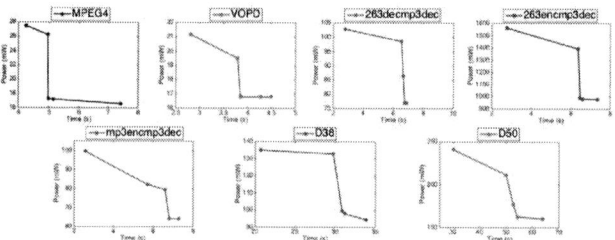

Figure 2. Tendency between running time and power cost for benchmarks

IV. CONCLUSIONS

In this paper, we proposed a DCR (Decompose and Clustering with Refinement) algorithm to minimize power consumption with considering of CPU time. The algorithm utilized characteristics of ASNoC for partitioning and clustering and also considered floorplan of NoC topology. Experimental results show that our algorithm is efficient to reduce power consumption, running time and floorplan area.

REFERENCES

[1] Zhong W, Yoshimura T, Bei Y U, et al. Cluster generation and network component insertion for topology synthesis of application-specific network-on-chips[J]. IEICE transactions on electronics, 2012, 95(4): 534-545.

[2] Zhong W, Song C, Huang B, et al. Floorplanning and topology synthesis for application-specific network-on-chips[J]. IEICE Transactions on FECCS, 2013, 96(6): 1174-1184.

[3] Karypis G, Aggarwal R, Kumar V, et al. Multilevel hypergraph partitioning: applications in VLSI domain[J]. Very Large Scale Integration (VLSI) Systems, IEEE Transactions on, 1999, 7(1): 69-79.

[4] Chen S, Yoshimura T. Fixed-outline floorplanning: Block-position enumeration and a new method for calculating area costs[J]. IEEE Transactions on CADICS, 2008, 27(5): 858-871.

[5] Wang H S, Zhu X, Peh L S, et al. Orion: a power-performance simulator for interconnection networks[C]. 35th Annual IEEE/ACM International Symposium on. IEEE, 2002: 294-30.

Computation of Modular Multiplicative Inverses Using Residue Signed-Digit Additions

Shugang Wei

Department of Mechanical Science and Technology, Gunma University

Ohta-shi,Gunma,Japan 373-0057   wei@gunma-u.ac.jp

Abstract—**This paper proposes a new algorithm of calculating modular multiplicative inverse numbers based on residue signed-digit(SD) number arithmetic. By introducing a p-digit radix-two SD number system into the residue arithmetic, a modular addition is implemented by using two SD adders for a modulus m, where $2^p - 1 \le m \le 2^{p+1} - 1$, and no carry propagations will arise during the additions. We give a new architecture with the residue SD adder to realize a faster modular multiplicative inverse computation. The design result shows that the proposed circuits using the SD arithmetic are faster than that based on the binary ones.**

I. INTRODUCTION

In a residue number system(RNS), each residue digit of addition/subtraction and multiplication is exclusively dependent on the digits of the operands[1]. Various methods of applications of RNS in digital signal processing have been presented[2]. When a residue addition and a modular multiplication in an RNS are performed by using the conventional binary system, the carry propagations arise and the speed of arithmetic operations will be limited[3], [4], [5], [6].

It is well-known that the carry propagation is limited to one position during additions in a signed-digit (SD) number system[7]. We have presented an approach method by introducing a radix-two SD number representation into residue arithmetic with a redundant residue number representation[8], [9], [10]. However, fast and compact circuits computing the multiplicative inverse numbers are required for some applications of RNS and encryption systems.

For this purpose, a sequential architecture for computing multiplicative inverse numbers is proposed by using a high speed SD adder, registers and a counter. Compared to the circuits using ordinary binary number system, the proposed circuits using the SD number system may work at a faster clock rate.

II. COMPUTATION OF MODULAR MULTIPLICATIVE INVERSES USING SD NUMBERS

Let m be a modulus in a residue number system, and x be a residue digit with respect to a modulus m. we consider that x is represented by a p-digit radix-two SD number representation as follows:

$$\begin{aligned} x &= x_{p-1}2^{p-1} + x_{p-2}2^{p-2} + \cdots + x_0, \\ & x_i \in \{-1, 0, 1\} \quad (i = 0, 1, \cdots, p-1), \end{aligned} \quad (1)$$

which can be denoted as $x = (x_{p-1}, x_{p-2}, \cdots, x_0)_{SD}$. When $2^p - 1 \le m \le 2^{p+1} - 1$, x has a value in the range of

$[-(2^p - 1), 2^p - 1]$ and is an element of l_m which is the set:

$$l_m = \{-(2^p - 1), \cdots, 0, \cdots, (2^p - 1)\}. \quad (2)$$

For an integer X, $x = \langle X \rangle_m$ is defined as an integer in l_m, where $2^p - 1 \le m \le 2^{p+1} - 1$. When $|X|_m \ne 0$, x has one of two possible values given by equations

$$x = \langle X \rangle_m = |X|_m, \quad (3)$$

and

$$x = \langle X \rangle_m = |X|_m - \text{sign}(|X|_m) \times m, \quad (4)$$

respectively, where

$$\text{sign}(x) = \left\{ \begin{array}{ll} -1 & x < 0 \\ 1 & x \ge 0 \end{array} \right. .$$

For the following relationship, an integer A is a modular multiplicative inverse number of x modulo m,

$$\langle A \times x \rangle_m = 1, \quad (5)$$

where x and A are relatively prime numbers.

We have an algorithm for the calculation of A from x:

Let $x(0) = 0$, then do the following addition modulo m repeatedly,

$$x(i+1) = \langle x(i) + x \rangle_m. \quad (6)$$

When $x(i+1) = 1$, then the multiplicative inverse number of x is $A = i + 1$.

III. IMPLEMENTATION OF RESIDUE MULTIPLICATIVE INVERSES

To implement the above algorithm, we can use an adder modulo m and a register for repeating the residue additions, and a counter for storing i.

The residue addition $\langle x + y \rangle_m$, where x and y are the SD numbers in the p-digit SD representation shown in (1), can be performed as follows: Let w_i and c_i be the intermediate sum and the carry of ith digit position, respectively. The values of them are determined by Table I with respect to the values of $x_i, y_i, x_{i-1}, y_{i-1}$. In Table I, $abs(x_i)$ is the absolute value of x_i. Thus the addition at each digit can be implemented in parallel. Therefore, we have the final sum with $(p+1)$-digit SD number representation.

$$S = x + y = (s_p, s_{p-1}, s_{n-2}, \cdots, s_0), \quad (7)$$

where $s_p = c_{p-1}$.

978-1-5090-3220-4/16 $31.00 © 2016 IEEE

TABLE I
RULES FOR ADDING SD NUMBERS

	$abs(x_i) = abs(y_i)$	$abs(x_i) \neq abs(y_i)$	
		$((x_i + y_i) \times (x_{i-1} + y_{i-1}) < 0)$ or $(x_{i-1} + y_{i-1} = 0)$ and $(x_i + y_i < 0)$	$((x_i + y_i) \times (x_{i-1} + y_{i-1}) > 0)$ or $(x_{i-1} + y_{i-1} = 0)$ and $(x_i + y_i > 0)$
w_i	0	$x_i + y_i$	$-(x_i + y_i)$
c_i	$(x_i + y_i)/2$	0	$x_i + y_i$

Let μ be a residue parameter and defined as $\mu = m - 2^p$. Then $\langle 2^p \rangle_m = -\mu$. Thus,

$$
\begin{aligned}
s &= \langle S \rangle_m \\
&= \langle \langle s_p 2^p \rangle_m + S_0 \rangle_m \\
&= \langle -s_p \times \mu + S_0 \rangle_m.
\end{aligned}
\tag{8}
$$

We select m as the modulus which is an integer in $[2^p - 1, 2^p + 2^{p-1} - 1]$. Then $\mu = m - 2^p$ is an integer in the range of $[-1, 2^{p-1} - 1]$. For $\mu > 1$, μ is expressed in a p-digit SD number representation as follows:

$$
\mu = (0, 1, \mu_{p-3}, \cdots, \mu_0)_{SD},
\tag{9}
$$

where $\mu_i \in \{-1, 0, 1\}$ for $i = 0, 1, \cdots, p - 3$. The largest value of μ is $2^{p-1} - 1$, which is $(0, 1, 1, \cdots, 1)_{SD}$. Therefore, the second addition for the modulo m SD addition can be performed as follows:

$$
s = \begin{cases}
S_0 & s_p = 0 \\
(-s_{p-1}, s_{p-2}, \cdots, s_0)_{SD} & s_p = -s_{p-1} \\
S_0 - \mu s_p & s_p \neq -s_{p-1}
\end{cases}.
$$

We use the above algorithm to compute the residue addition with μ repeatedly until the sum modulo m is 1. The 1 is not checked easly since the SD number system has redundancy. We subtract 1 from the sum and check if it is 0. The 0 has a unique represention.

The counter is designed by using the SD number representation. So that the clock rate of the multiplitive inverse system can be set optimally.

IV. PERFORMANCE EVALUATION

To evaluate the performance of the proposed modular arithmetic circuits, we suppose that the VLSI implementation is based on a gate array IC, because ASIC design on a gate array is a popular VLSI implementation method. We specify a binary representation $[x_i(1) x_i(0)]$ for a radix-two signed digit x_i, where $x_i(1)$ is the sign and $x_i(0)$ is the absolute value of x_i. Thus, a p-digit radix-two SD number a is represented by a vector with $2p$-bit length. For example, $(1, 0, -1, 0, -1)_{SD} = [0100110011]$. Then the modular SD arithmetic operations are described and simulated by VHDL, and using the VHDL description codes, the simulation and the logic circuit synthesis are performed by using VHDL synthesis software tool. In our experiments, the delay times are obtained by simulation under the condition of $0.18\mu m$ CMOS technology.

In Table II, the performance comparisons of the modular multiplicative inverse number circuits are illustrated. The

TABLE II
PERFORMANCE OF RESIDUE MULTIPLICATIVE INVERSE NUMBER CIRCUITS

Circuits	Length p	Area(μm^2)	Delay Time(ns)
Binary	8	3824.01	3.96
SD	8	5580.72	2.69
Binary	16	7409.23	7.15
SD	16	14323.94	2.90
Binary	32	15563.52	13.01
SD	32	30053.85	2.89

longest delay time is shown for each circuit with different word length p. Namely, the working clock rates of the circuits are dependent on the delay times. Thus, the modular multiplitive inverse circuits using the modulo m SD adders work at a fast clock rate which is much faster than that using binary adders.

V. CONCLUSIONS

In this paper, an architecture of modular multiplicative inverse opereations using the SD number aritimetic has been proposed. Since the delay time of the residue SD addition is independent on the word length, the proposed modular multiplicative inverse circuits with different moduli can work in a constant clock rate. The proposed circuit of modular multiplicative inverse operations is with compact structure and can be embedded in an encryption system for generating keys.

REFERENCES

[1] N.S.Szabo and R.I.Tanaka ,*Residue Arithmetic and Its Applications to Computer Technology*, New York: McGraw-Hill, 1967.
[2] M. A. Sonderstrand, W. K. Jendins, G. A. Junllien, and F. J. Taylor,*Residue Number System Arithmetic: Modern Applications in Digital Signal Processing*, IEEE Press, New York, 1986.
[3] D.P. Agrawal and T.R.N.Rao,"Modulo $(2^n + 1)$ arithmetic logic,"IEE J. Electronic Circuits and Systems, Vol.2, pp. 186-188, Nov. 1978.
[4] F.J.Taylor,"A VLSI residue arithmetic multiplier," IEEE Trans. Comput., Vol.C-31, pp.540-546,June 1982.
[5] A.Hiasat,"New memoryless, mod $(2^n \pm 1)$ residue multiplier," Electron. Lett., Vol.28, No.3, pp.314-315,Jan. 1992.
[6] C.Efstathiou, H.T. Vergos and D.Nikolos, " Modulo $2^n \pm 1$ adder design using select prefix blocks ", IEEE Trans. on comput. vol.52, no.11, pp.1399-1406, 2003.
[7] A.Avizienis,"Signed-digit number representations for fast parallel arithmetic,"IRE Trans. Elect. Comput., EC-10,pp.389-400, Sept. 1961.
[8] S.Wei and K.Shimizu,"Modulo $2^p - 1$ arithmetic hardware algorithm using signed-digit number representation," Trans. IEICE. INF. & SYST. Vol.E79-D, No.3, pp. 242-246, March 1996.
[9] S.Wei and K.Shimizu,"A novel residue arithmetic hardware algorithm using a signed-digit number representation", IEICE Trans.INF. & SYST., Vol.E83-D, No.12, pp. 2056-2064, Dec. 2000.
[10] S.Wei,"Residue checker using optimal signed-digit adder tree for error detection of arithmetic circuits", Proceedings of TENCON2014, PID. 00439, Oct., 2014.

Design Techniques for Ultra-Efficient Computing
(Review Paper)

Dongsuk Jeon

Graduate School of Convergence Science and Technology, Seoul National University
Seoul, Korea
djeon1@snu.ac.kr

Abstract— **This paper reviews various energy saving techniques at different design levels. Aside from conventional voltage scaling and low power circuit techniques, systematic approaches including energy-aware architecture design, system-level power optimization and application-specific low power circuit design are presented along with demonstration systems.**

Keywords; low power circuit, near-threshold computing, system co-optimization

I. INTRODUCTION

As the CMOS scaling has saturated recently, various power saving techniques are now aggressively adopted in energy-constrained systems such as smartphone and IoT devices. Traditional power reduction schemes including power and clock gating, voltage scaling and low power circuit topology effectively lower switching energy of the system. However, since the fabrication technology is moving towards more leaky process (especially for planar devices), the importance of leakage paths must be considered in the energy optimization process. In addition, recent complicated data processing algorithms for machine learning introduce more significant performance constraints, and hence we are no longer able to solely rely on the aforementioned methods which unavoidably result in large performance degradation. In this paper, systematic design approaches ranging across multiple design levels will be described. Through several design examples, techniques including energy-aware architecture optimization, novel low energy circuit topology based on algorithm characteristics and algorithmic optimization will be demonstrated in detail.

II. MULTI-LEVEL OPTIMIZATION METHODS

A. Energy-Aware Hardware Architecture

Advanced CMOS process provides excellent performance and switching energy improvements through smaller feature size, but only at the expense of enlarged leakage energy and variation. To maximize benefits of voltage scaling, one must find a tailored hardware architecture which can compensate for performance degradations and suppress leakage overhead. [1-2]. A design example on obtaining energy-efficient architecture is described in [3]. FFT (Fast Fourier Transform) is a widely used algorithm for signal processing and the authors in [3] propose energy-optimal FFT design based on a new parallelized architecture (Fig. 1).

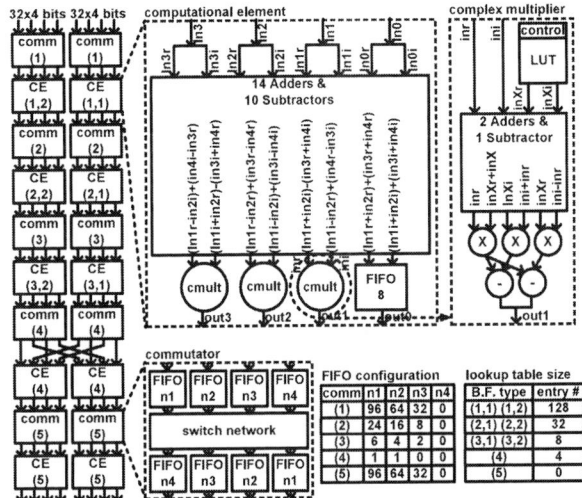

Figure 1. Low leakage parallel FFT architecture.

The proposed architecture enhances performance and suppresses leakage energy by incorporating multiple processing lanes. From detailed energy analysis, the 2-lane version was chosen as the final design, and the fabricated FFT core demonstrated >2× better efficiency than prior works.

B. System-Level Power Optimization

SoCs generally have multiple blocks tied to different power domains. Hence it is critical to consider all the blocks in the power analysis during initial system design. The authors in [4] propose a system-level power optimization methods and demonstrate in a syringe-implantable ECG monitoring device. Fig. 2 depicts a relationship between the noise level of the first stage low noise amplifier and the overall arrhythmia detection accuracy. As the input referred noise level of the amplifier is relaxed, the detection accuracy drops accordingly due to more noise in the amplified signal. Conventional designs generally target very low noise level on the order of 3μV, but it was observed that the proposed frequency domain detection algorithm achieves perfect accuracy with up to 15μV noise floor. Based on this observation, the authors designed an amplifier which has 9μV noise floor considering other noise sources and variations, which saved the power consumption of the amplifier by more than 6×. Along with other low power circuit techniques such as minimum energy operation, the design was fabricated in 40nm process and it achieved sub-100nW energy consumption during continuous monitoring.

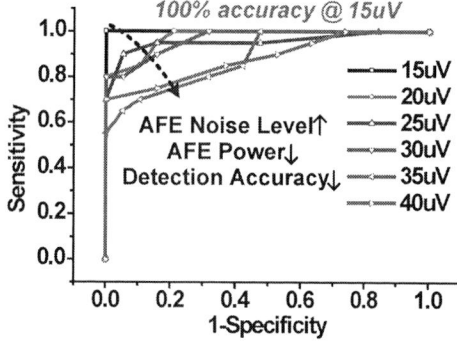

Figure 2. Trade-off between amplifier's noise level and detection accuracy.

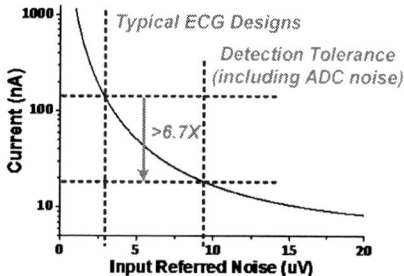

Figure 3. Energy-aware amplifier design.

C. Low Power Circuit Design Using Properties of Algorithm

Conventional low power circuit techniques target general-purpose systems. However, if we utilize the properties of the algorithm to be processed in the system then we may find opportunities for further energy savings. One example can be found in Reference [5].

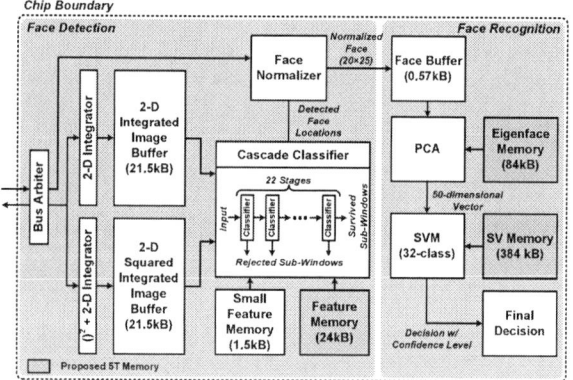

Figure 4. Face recognition hardware architecture.

Authors in [5] designed a face detection and recognition hardware accelerator aimed at battery-powered mobile systems. Due to power constraints, the accelerator must consume less than 100mW while continuously performing face recognition. The system requires more than 500kB on-chip memory, and hence memory access and leakage energy dictates overall system power. Once the detection and recognition algorithm is trained, all the algorithm coefficients are stored in the on-chip memory and remain unchanged. The authors utilize this property to design a mostly-read 5T memory design shown in Fig. 5.

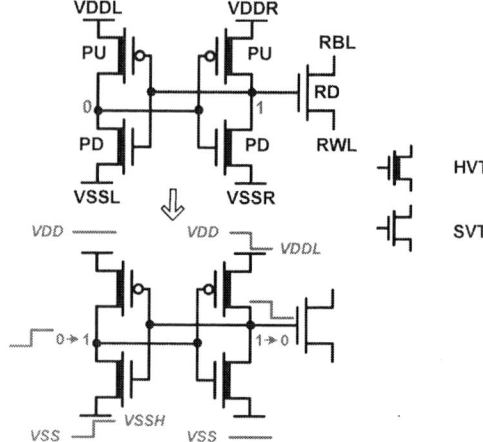

Figure 5. Proposed mostly-read 5T bit cell.

The proposed 5T bit cell is primarily optimized for read operation and provides similar read margin to conventional 7T or 8T designs. The write operation is accomplished through shared power rails. Hence the memory macro must be reset before writing new data, but its overhead is minimized since coefficient update is done only once at the beginning. The face recognition accelerator was fabricated in 40nm CMOS technology and it was able to operate down to 600mV power supply voltage. The accelerator processes HD input video at a throughput of 5fps while consuming only 23mW, making it a practical solution for mobile systems.

III. CONCLUSIONS

In this paper, low power design methodologies at multiple design levels are reviewed. To obtain maximum energy efficiency in mobile systems, one must consider those approaches altogether and also look into co-optimization opportunities across different levels.

REFERENCES

[1] S. R. Sridhara, M. DiRenzo, S. Lingam, S.-J. Lee, R. Blazquez, J. Maxey, S. Ghanem, Y.-H. Lee, R. Abdallah, P. Singh, and M. Goel, "Microwatt Embedded Processor Platform for Medical System-on-Chip Applications," *IEEE J. Solid-State Circuits*, vol. 46, no. 4, pp. 721-730, Apr. 2011.

[2] B. Zhai, D. Blaauw, D. Sylvester, and K. Flautner, "Theoretical and Practical Limits of Dynamic Voltage Scaling," in *Proc. Design Automation Conf.*, May 2005, pp. 868-873.

[3] D. Jeon, M. Seok, C. Chakrabarti, D. Blaauw, and D. Sylvester, "A Super-Pipelined Energy Efficient Subthreshold 240 MS/s FFT Core in 65 nm CMOS," *IEEE J. Solid-State Circuits*, vol. 47, no. 1, pp. 23-34, Jan. 2012.

[4] D. Jeon, Y.-P. Chen, Y. Lee, Y. Kim, Z. Foo, G. Kruger, H. Oral, O. Berenfeld, Z. Zhang, D. Blaauw, and D. Sylvester, "An Implantable 64nW ECG Monitoring Mixed-Signal SoC for Arrhythmia Diagnosis," in *IEEE Int. Solid-State Circuits Conf. Dig. Tech. Papers*, Feb. 2014, pp. 416-417.

[5] D. Jeon, Q. Dong, Y. Kim, X. Wang, S. Chen, H. Yu, D. Blaauw, and D. Sylvester, "A 23mW Face Recognition Accelerator in 40nm CMOS with Mostly-Read 5T Memory," in *Proc. IEEE Symposium on VLSI Circuits*, Jun. 2015, pp. C48-C49.

An Ultra-Low Power AES Encryption Core in 65nm SOTB CMOS Process

Van-Phuc Hoang[1] and Van-Lan Dao
Le Quy Don Technical University
236 Hoang Quoc Viet Str., Hanoi, Vietnam
Email: [1]phuchv@mta.edu.vn

Cong-Kha Pham
The University of Electro-Communications
1-5-1 Chofugaoka, Chofu-shi, Tokyo, 182-8585, Japan
Email: pham@ee.uec.ac.jp

Abstract— **This paper presents an efficient ASIC implementation of the low area and ultra-low power AES encryption core with an optimized S-box, Rcon and control blocks optimization, combined with a simple clock gating technique using an ultra-low power 65nm SOTB CMOS technology. The ASIC implementation results show that the proposed AES encryption core requires a small number of clock cycles with ultra-low power consumption and achieves higher resource usage efficiency compared with other designs.**

I. LOW-POWER, LOW-AREA AES CORE DESIGN

Advanced Encryption Standard (AES) is a well-known security standard for data encryption [1, 2] in many emerging wireless networks. Although the encryption is standardized, the efficient hardware architecture and implementation methods are the topics which many researchers are focusing on. Therefore, the objective of this paper is to design a low area and low power AES encryption core for emerging wireless networks.

In [2]-[5], authors have focused on optimizing AES encryption core for the low area implementation. However, they use an LUT-based (non-optimized) S-box that may result in a high area ASIC implementation. In some other papers [3]-[10], the authors focused on improving the S-box architecture, ultilizing FPGA embeded resources and some optimization techniques.

In this paper, AES encryption core processes data in 128-bit blocks with the key lengths of 128-bit. To reduce the AES encryption core area, we employ an 8-bit architecture with an optimized S-box so that the AES core encrypts an 8-bit data block in each clock cycle. The proposed AES encryption core architecture is shown in Fig. 1. This core includes a key expansion unit, a mixcolumn unit, a parallel to serial converter and a byte permutation unit. S-box 1 and S-box 2 blocks are the sub-blocks in the byte permutation unit and key expansion unit as described in [4].

For a low power consumption implementation, a simple clock gating technique is proposed by using *start_in* signal as shown in Fig. 1. The clock tree in the AES core is controlled by this signal. S-box is an important block in the AES core so that some papers on S-box optimization for the specific requirements have been published [2-5]. To reduce the complexity, we use the S-box architecture with the direct hardware implementation. In this paper, the S-box is transformed from $GF(2^8)$ to $GF(2^8)/GF(2^4)/GF(2^2)$ architecture [5].

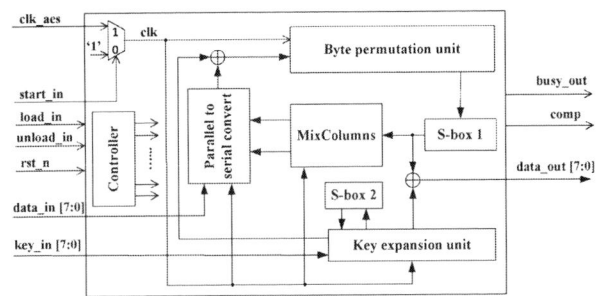

Fig. 1. The 8-bit AES encryption core architecture with iterative structure and simple clock gating technique

Moreover, according to [1], Rcon block takes the inputs from *r_in* signal which is the round index ranging from 0 to 9. Rcon also can be a multiplexer (MUX) circuit which uses *r_in* as the selection signal [4]. In our design, Rcon block is optimized by using the simple Karnaugh optimization.

For control part optimization, in the proposed AES encryption core, the control block in AES encryption core is a low power state machine. In this work, we propose a simple controller based on a counter as shown in Fig. 2. The control signal is generated from a simple 8-bit counter. The four maximal significant bits of counter output (counter[7:4]) are fed to key expansion block (*r_in* signal), and the 4-bit lower part (counter[3:0]) is used to select the operations in each AES encryption round. This simple control block is used to reduce the area and power consumption of the proposed AES encryption core.

Our main contribution in this paper is that an efficient AES encryption core for low area, ultra-low power systems is proposed based on the 8-bit iterative architecture with an optimized S-box, Rcon and control blocks optimization, combined with a simple clock gating technique.

II. IMPLEMENTATION RESULTS

The proposed 8-bit AES encryption core was modelled with VHDL, and then implemented with a 180 nm CMOS and 65nm SOTB CMOS [11] standard libraries by Synopsys Design Complier and Synopsys IC Complier tools. The ASIC implementation results of proposed and other AES encryption cores are shown in Table I in which the proposed AES encryption core area can be reduced to only 2.3kgates with 180nm CMOS library and requires the smallest number of cycles. Compared with other designs, the proposed AES encryption core power consumption can also be reduced significantly to only 7.1μW/MHz with 180nm CMOS technology and especially to 0.38μW/MHz with 65nm SOTB CMOS technology. The proposed 8-bit AES encryption core is the lowest power consumption AES encryption core presented in literature. Figure 3 is the layout of proposed 8-bit AES encryption core with 65nm SOTB CMOS technology.

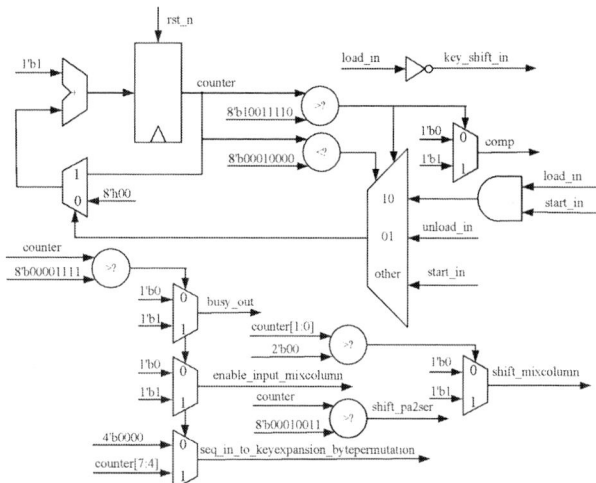

Fig. 2. The controller architecture for the 8-bit AES encryption core.

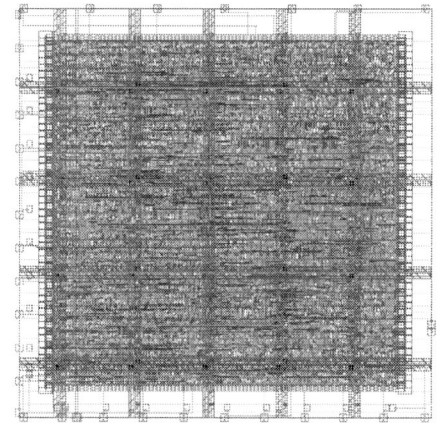

Fig. 3. Layout of the proposed 8-bit AES core using 65nm SOTB CMOS technology. The AES core layout dimension is 120μm×120μm.

TABLE I.   IMPLEMENTATION RESULTS OF PROPOSED 8-BIT AES ENCRYPTION CORE COMPARED WITH OTHER PAPERS.

Design	Technology	No. of cycles	Area (kgates)	Power consumption
Our work	180nm	160	2.3	7.1 μW/MHz
Our work	65nm SOTB	160	2.6	0.38 μW/MHz
[4]	130nm	160	3.1	37 μW/MHz
[6]	22nm	336	2.0	13 mW (*)
[7]	-	226	2.4	3.7 μA @ 100KHz
[8]	130nm	356	5.5	99 μW/MHz
[9]	65nm	200	0.012 mm²	4.6 μW @ 0.5V
[10] (**)	180nm	-	1.05×10³ μm²	39.1 μW/MHz

(*): @1.1GHz and 0.9V; (**): 8-bit architecture

III.   CONCLUSIONS

This paper has presented a low area, ultra-low power AES encryption core which can be used for emerging wireless networks. The implementation results in ASIC show that by using an optimized S-box, Rcon block and control blocks optimization, combined with a simple clock gating technique, the AES core area and power consumption can be reduced significantly. The proposed AES encryption core requires the smallest number of cycles and achieves lowest power consumption as well. Therefore, this AES encryption core is highly potential to be used in wireless network nodes which require long battery duration. In the future, we will further improve the AES encryption core and apply it for a wireless network application.

ACKNOWLEDGMENT

This research is funded by Vietnam National Foundation for Science and Technology Development (NAFOSTED) under grant number 102.02-2015.20.

This work is supported by VLSI Design and Education Center (VDEC), the University of Tokyo in collaboration with Synopsys, Inc. and Cadence Design Systems, Inc.

REFERENCES

[1] National Institute of Standards and Technology (NIST), "Advanced Encryption Standard (AES)," *FIPS Publication 197*, Nov. 2001.

[2] A. Satoh, S. Morioka, K. Takano, and S. Munetoh, "A compact Rijndael hardware architecture with S-box optimization," *Proc. ASIACRYPT 2001*, pp.239-254, Dec. 2001.

[3] D. Canright. "A very compact S-box for AES," *Proc. 7th Int. Workshop on Cryptographic Hardware and Embedded Systems (CHES2005)*, pp.441-455, Sep. 2005.

[4] P. Hamalainen, T. Alho, M. Hannikainen, T.D. Hamalainen, "Design and Implementation of Low-Area and Low-Power AES Encryption Hardware Core," *Proc. 9th EUROMICRO Conf. Digital System Design: Architectures, Methods and Tools (DSD2006)*, pp.577-583, 2006.

[5] T. Jarvinen, P. Salmela, P. Hamalainen, J. Takala, "Efficient byte permutation realizations for compact AES implementations," *Proc. 13th European on Signal Processing Conference*, pp.1-4, Sep. 2005.

[6] Sanu Mathew et al., "340mV-1.1V-289Gbps/W, 2090-gate NanoAES Hardware Accelerator with Area-optimized Encrypt/Decrypt GF(2⁴)² Polynomials in 22nm tri-gate CMOS," *Proc. Symposium on VLSI Circuits Digest of Technical Papers*, 2014.

[7] Amir Moradi et al., "Pushing the limits: a very compact and a threshold implementation of AES," *Advances in Cryptology - EUROCRYPT 2011*, *Lecture Notes in Computer Science*, vol. 6632, pp.69-88, 2011.

[8] T. Good and M. Benaissa, "692-nW advanced encryption standard (AES) on a 0.13-μm CMOS," *IEEE Trans. Very Large Scale Integr. (VLSI) Syst.*, vol.18, no.12, pp.1753-1757, Dec. 2010.

[9] Wenfeng Zhao, Yajun Ha, Massimo Alioto, "AES Architectures for Minimum-Energy Operation and Silicon Demonstration in 65nm with Lowest Energy per Encryption," *2015 IEEE International Symposium on Circuits and Systems (ISCAS)*, pp.1-4, May 2015.

[10] Liling Dong et al., "Low Power State Machine Design for AES Encryption Coprocessor," *Lecture Notes in Engineering and Computer Science*, vol.2216, no.1, pp.714-717, Mar. 2015.

[11] Koichiro Ishibashi, Nobuyuki sugii, Shiro Kamohara, Kimiyoshi Usami, Hideharu Amano, Kazutoshi Kobayashi, Cong-Kha Pham, "A Perpetuum Mobile 32bit CPU on 65nm SOTB CMOS Technology with the Reverse-Body-Bias Assisted Sleep Mode," *IEICE Trans. Electron.*, vol.E98-C, no.7, pp.536-543, Jul. 2015.

Cell-Based Delay Locked Loop Compiler

Pei-Ching Huang Shi-Yu Huang

Electrical Engineering Department, National Tsing Hua University, Taiwan

Abstract - *Digital Delay-Locked Loops (DLLs) have been widely used in today's ICs for all kinds of timing control. Even though a digital DLL circuit is much easier to design than its analog counterparts, our prior experience shows that weeks of efforts, if not months, could still be wasted in order to find a process resilient configuration for a specific DLL requirement. Thus, we propose in this work a cell-based DLL architecture and its compiler. According to a user's demand, our DLL compiler can generate a cell-based DLL circuit in just minutes, it can support easy process migration, and thereby saving a large amount of human efforts spent in tuning DLL designs for different manufacturing processes. Transistor-level simulation has been used to validate its ability in a 0.18 CMOS process and a 90nm CMOS process. It can support input clock frequency up to 1GHz in 0.18μm, and 1.25GHz in 90nm.*

Keywords: Delay-Locked Loop, Tunable Delay Line, Compiler, Process Migration, Cell-Based Timing Circuit

I. INTRODUCTION

It is a common perception that the process migration for a cell-based all-digital Delay-Locked Loop is effortless. But in our own experience in designing a Programmable Phase Shifter for a pulsed-RADAR SoC with a DLL inside [4], this may not be true.

In this work, we propose a cell-based DLL compiler to help the designer generate a DLL circuit rapidly. The DLL circuit this compiler generates can support a frequency range from 50MHz to 1250MHz (or a delay range from 0.8ns to 20ns) in a 90nm CMOS process. On the other hand, it can support a frequency range from 50MHz to 1000MHz (or a delay range from 1ns to 20ns) in a 0.18μm CMOS process.

II. PRELIMINARIES

A. Delay-Locked Loop (DLL)

A generic Delay Locked Loop (DLL) can be constructed based on an architecture as shown in **Fig. 1**

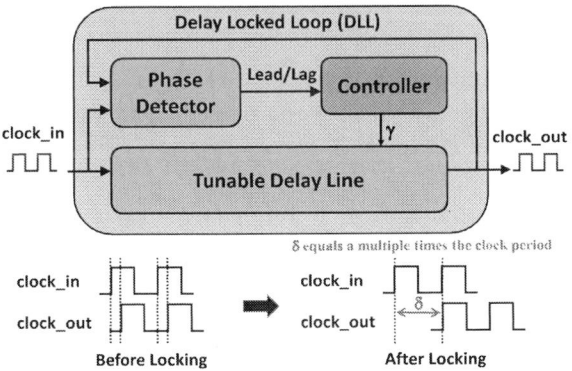

Fig. 1: Illustration of Cell-based Delay Locked Loop (DLL).

. It consists of three major blocks: (1) a Phase Detector, (2) a Tunable Delay Line (TDL), and (3) an overall controller.

This work was supported in part by Ministry of Science and Technology (MOST) of Taiwan under grant MOST-105-2221-E-007-121.

Initially, the output clock signal (namely *clock_out*) is not in-phase with the input clock signal (namely *clock_in*). But after the DLL is locked, their phase difference will disappear as shown in the figure. During the locking process, the output clock is successively compared with the input clock to decide the result of a one-bit signal, called *lead/lag* via a *Phased Detector*. Then this *lead/lag* signal will guide the controller to update the value of the tuning code of the Tunable Delay Line so that gradually the phase difference between the output clock and the input clock will be diminished. After locking, the delay across the Tunable Delay Line will be equal to a multiple times the clock period of the input clock signal.

III. PROPOSED PARAMETERIZED DLL

A. Parameterized Tunable Delay Line (TDL)

In this work, we adopted a parameterized TDL as shown in **Fig. 2** [2]. In order to increase its resilience to process variation, we inserted a **Shifting Stage**. The effect of the shifting stage is best illustrated by examining the so-called **Delay profiles** offered by the proposed parameterized TDL as shown in **Fig. 3**. In this example TDL, there are 6 Delay Stages, each having128 tri-state buffers (TBUFs). As a result, the total number of tuning bits required for the entire tri-state buffer array is 128× 6=768. The horizontal-axis of **Fig. 3** is the tuning code, namely $\gamma[767:0]$, while the vertical-axis is the delay across the overall Tunable Delay Line in pico-seconds. By adding the shifting stage, a family of delay profiles are offered due to various values of the **shifting selection code, $SS[5:1]$**. The *overall delay range* of the TDL after adding the shifting stage can cover a much wider delay range and thereby making the proposed TDL robust to process variation.

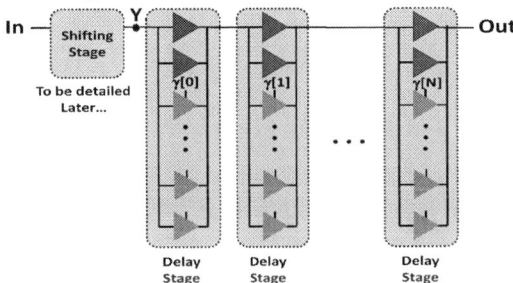

Fig. 2: Proposed parameterized Tunable Delay Line.

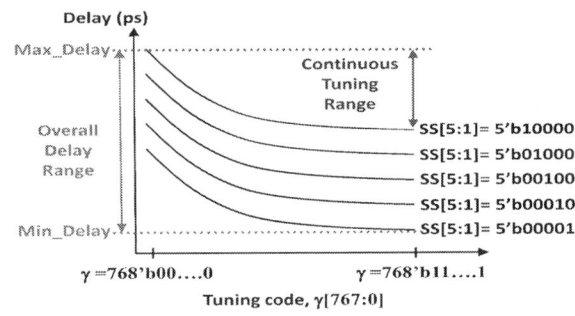

Fig. 3: Delay profiles versus various shifting selection codes, SS[5:1], and tuning codes, γ[768:0].

IV. DLL COMPILER

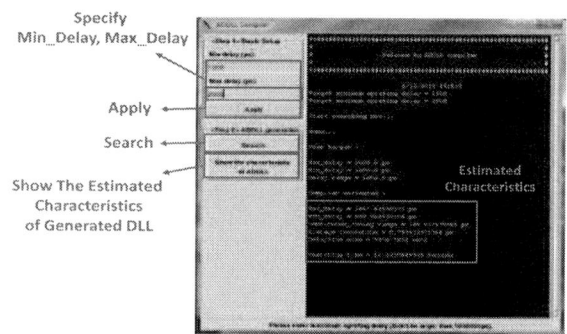

Fig. 4: Graphical User Interface (GUI) of our DLL compiler.

Our DLL compiler has a graphical user interface (GUI) as shown in **Fig. 4**. Its usage include the following steps: (1) It requires the user to provide the *Min_Delay* and *Max_Delay* of the overall delay range of a target Tunable Delay Line. (2) After pressing the APPLY button, the compiler will start to search for a few minutes, and then it will produces the DLL circuit in mixed format - with the Phase-Detector and the Tunable Delay Line in netlist files, and the controller in synthesizable RTL file. Along with it, some script files for running APR (Automatic Place and Route) tool and post-layout simulation will also be generated. The user can derive the accurate characteristics of the generated DLL by running the transistor-level simulation script. (3) Our tool can report or display on the screen a set of *estimated characteristics* through a **delay estimation method** similar to the one revealed in one of our previous works on cell-based Phase-Locked Loop (PLL) Compiler [3].

In the example shown in **Fig. 4**, the user has specified a *Max_Delay* of 2000ps and *Min_Delay* of 1000ps. The estimated characteristics of the produced DLL using a 90nm CMOS process are as follows:

- Estimated *Max_Delay* = 2847ps
- Estimated *Min_Delay* = 699ps
- Average tuning *resolution* = 0.75ps
- Estimated *Area* of the Tunable Delay Line = 5970 μm²

The left-hand side of **Fig. 5** is a *delay-range plot*, while the right-hand side is an *area plot* of a Delay Stage, under various configurations - from the uniform type (where all TBUFs are of the same size) towards a so-called **progressive type** (where the sizes of TBUFs follows an increasing pattern). It can be seen that these two plots are monotonically increasing to the right following a configuration order. **Our search** on these two plots starts from the **lower-left-most point** representing the uniform type and eventually stops at a point (during a trajectory towards the upper-right-most point representing the progressive type). At this final point, the delay range specified by the user is satisfied while the area overhead is the smallest.

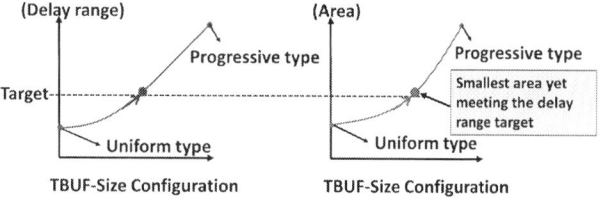

Fig. 5: Trajectory of our search for a small-area configuration of the Delay Stage while meeting the target delay range.

V. EXPERIMENTAL RESULTS

Table 1 shows 8 DLL instances produced by our compiler using a 90nm CMOS library. The clock frequencies for these DLLs range from 1.25GHz to 50MHz. It can be seen from the table that the area is larger for a DLL operating at a lower clock frequency due to the need of a larger TDL. For example, it is only (252 × 170) = 0.043μm² for a 1.25GHz DLL, while becomes larger at (640×120) = 0.077μm² for a 50MHz DLL. It is notable that for all DLLs, post-layout simulation has been conducted to validate that all target delay ranges are covered under 3 process corners (i.e., the target delay range is covered by the 3-corner overlapping range). The RMS jitters, Pk-to-Pk jitters of 1000 clock cycles after locking, and phase-errors are all estimated by post-layout transistor-level simulation. It is worth mentioning that the major contribution of this work is not to improve the performance of a DLL, but to minimize its design time, e.g., from weeks or even months [4] to just a few minutes.

Table 1. DLL instances in a 90nm CMOS cell library produced by our compiler, with the operating frequency ranging from 1.25GHz to 50MHz.

Process	A 90nm CMOS process @1V							
Clk frequency (MHz)	1250	1000	800	600	400	200	100	50
Clk period (ns)	0.8	1	1.25	1.67	2.5	5	10	20
# Shifting Stages	4	7	8	10	17	35	73	149
Area (μm²)	252*170	380*90	389*90	396*90	421*90	470*90	543*90	640*120
Locking Time (cycle)	93	104	176	313	656	1622	3612	6049
Target delay range (ps)	700~900	750~1250	1000~1500	1400~1900	2250~2750	4750~5250	9750~10250	19750~20250
3-corner overlapping range (ps)	664~1352	680~1732	863~2024	878~2374	924~3323	1035~6060	1237~11595	1635~22793
Power (mw)	18.5	13.5	11.2	8.8	7.4	5.6	4.7	4.3
RMS-jitter (ps)	1.1	1.4	2.1	1.4	1.7	4.1	2.6	3.2
Pk-Pk-jitter (ps)	13.0	11.0	12.0	14.0	11.0	14.0	8.0	9.0
phase-error (ps)	6.6	7.2	5.8	8.1	7.1	9.8	6.8	10.9

VI. CONCLUSION

In this work, we have proposed a parameterized cell-based All-Digital Delay Locked Loop architecture and its compiler which can generate a cell-based DLL circuits easily in minutes. With this new design technology, weeks or even months of tedious design efforts can be eliminated to speed up the design process. The delay range of the DLL circuit generated by our compiler can be tuned to have an ample margin according to a user's specification even under server process variation. Moreover, this DLL compiler supports easy process migration. To the best of our knowledge, this is the first push-button cell-based DLL compiler. For a 90nm CMOS process, it supports a delay range from 0.8ns to 20ns (or a frequency range from 50MHz to 1.25GHz).

REFERENCES

[1] Y.-J. Liao and S.-Y. Huang, "Temperature Tracking Scheme for Programmable Phase-Shifter in Pulsed Radar SoC", *Proc. of IEEE Symp. on VLSI Design, Automation, and Test* (VLSI-DAT), April 2016.

[2] T. Olsson and P. Nilsson, "A Digitally Controlled PLL for SoC Applications," *IEEE J. Solid-State Circuits*, vol. 39, no. 5, pp. 751-760, May 2004.

[3] C.-W. Tzeng, S.-Y. Huang, P.-Y. Chao, and R.-T. Ding, "Parameterized All-Digital PLL Architecture and Its Compiler to Support Easy Process Migration," *IEEE Trans. on VLSI Systems* (TVLSI), Vol. 22, No. 3, pp. 621-630, March 2014.

Hybrid GDI-NCL for Area/Power Reduction

Prashanthi Metku[1], Ramu Seva[1], Kyung Ki Kim[2], Yong-Bin Kim[3] and Minsu Choi[1]

[1]Dept of ECE, Missouri Univ of Science & Technology, Rolla, MO, USA, {pmcmc,rs2k6,choim}@mst.edu

[2]Dept of Electronic Eng., Daegu University, Gyeongsan, Korea, kkkim@daegu.ac.kr

[3]Dept of ECE, Northeastern University, Boston, MA, USA, ybk@ece.neu.edu

Abstract—**Null Convection Logic is a well-known paradigm for designing asynchronous logic circuits. The conventional CMOS-based NCL designs suffers larger area overhead and power consumption. A low power design technique called Gate Diffusion Input (GDI) has been adopted to overcome this limitation. In GDI technology, voltage swing exhibits significant voltage drop across the circuit. Therefore, not suitable for designing large combinational circuits. A novel HYBRID (CMOS+GDI) design is proposed in this work to efficiently address this issue. The HYBRID design utilizes both CMOS and GDI technology to reduce the number of transistor and power dissipation when compared to CMOS NCL circuits. The proposed approach is implemented in NCL Ripple Carry Adder (RCA) and simulated in Cadence Virtuoso for verification.**

I. INTRODUCTION

With the critical process variations and integration of millions of transistors in a single chip, complex synchronous integrated circuits (ICS) are facing the timing closure a complex task. Along with the timing issue, the increasing clock rate with decreasing size is causing a major problem of clock skew [1]. To achieve the tolerable skew, large part of the chip area is allotted for clock drivers. Hence, dissipating large amount of power prominently at the clock edge, where switching occurs. This problem has become the major limiting factor in the semiconductor industry where there is an increasing demand for low power devices. Thus, encouraging renewed interest towards asynchronous digital design [2].

One of the most commonly used methodology for asynchronous logic design is the Null Convection logic (NCL). NCL takes advantage of dual-rail (or quad-rail) encoding and local handshaking done by completion detection registers for clock-free operation. Normally, NCL chips are based on CMOS technology [3] which has the potential for high speed, low power and timing modularity. However, enhancement of one parameter has its effect on other parameters. One such example is the semi-static implementation of NCL circuits, reduces the area consumption but has a weak feedback loop [3]. To overcome the above limitations, a novel method hybrid approach leveraging GDI (Gate Diffusion Input) method is proposed in this work. GDI method was originally introduced to obtain low power synchronous designs [4] and has never been applied to NCL. The purpose of this study is to obtain low power and reduced area substitute to standard CMOS-based NCL designs. A GDI-NCL ripple carry adder has been designed to verify the proposed hybrid approach.

II. PRELIMINARIES AND REVIEW

Null Convection Logic

NCL is a novel methodology used for designing asynchronous circuits. It is a self-timed logic model where both data and control are integrated to a single signal, resulting to an asynchronous, clock-less and delay-insensitive circuits and systems. Communication between the circuits is carried out through local handshaking which also provides synchronization [5]. DATA and NULL states are used for synchronizing and I/O control.

Symbolic completeness and input completeness are the two criteria the NCL should possess to achieve delay-insensitive (DI) behavior. NCL also has the built-in hysteresis behavior to develop the DI circuits. NCL is implemented using the threshold gates. Th_{mn} gate is the primary type of threshold gate, where $1 \leq m \leq n.Th_{mn}$ gate contain n inputs and at least m out of n inputs should be asserted for the output to be asserted [5]. Because of the NCL's built-in hysteresis feature all the inputs must be de-asserted for the output to be de-asserted. Hysteresis ensures the completeness by not changing the previous output state till it receives the NULL wave front. NCL combinational circuitry consists of combinational logic between DI registers. Presently, NCL designs are implemented in CMOS technology [6]. However, CMOS implementation of these designs is known to consume larger area when compared to the synchronous designs. So, a novel hybrid GDI implementation of the NCL is proposed to address this issue in this work.

Gate Diffusion Input (GDI) Technique

Gate Diffusion Input (GDI) is a promising alternative to CMOS technology for considerable area/power reduction. GDI basic cell design [7] is depicted in Fig 1. It is a four-terminal device consisting of: 1) G: common gate input for nMOS and pMOS transistors; 2) P and N: pMOS and nMOS transistors outer diffusion node; 3) D: common drain for both the transistors. Bulks of the pMOS and nMOS are connected to P or N respectively. Depending on the design requirement P, N and D can be either used as input or output ports [8].

Different multiple-input gates can also be realized by combining several GDI basic cells. GDI technology can be successfully implemented in twin-well CMOS or silicon on insulator (SOI) technologies but only few functions can be implemented in standard p-well CMOS process [8]. In this paper, a hybrid GDI-NCL ripple carry adder (RCA) is designed based on OR, AND, NOT and MUX GDI functions.

III. PROPOSED HYBRID RCA

Implementation of NCL Full Adder [5] (consists of two $TH32$ and $TH34W2$ NCL gates) using CMOS technology is to dissipate more power and consume large area than

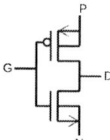

Fig. 1: Basic GDI cell. [7]

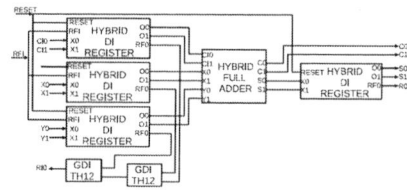

Fig. 5: HYBRID NCL Full Adder with DI Registers

GDI-NCL counterpart. Therefore, a new low-power design technique GDI can be used for designing the NCL Full Adder. However, GDI technique has voltage swing [9] issues leading to the significant voltage drop in the circuit. Hence, implementing large combination logic circuit solely based on GDI technology may be unreliable. To address this issue, a HYBRID (CMOS+GDI) technique is proposed. Designing NCL circuits using the HYBRID model results in reduced area and power dissipation when compared to the CMOS design model. The proposed HYBRID model is used in designing the four and eight bit NCL RCA for comparison.

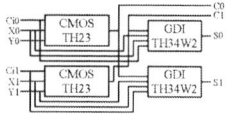

Fig. 2: Hybrid NCL Full Adder

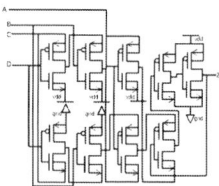

Fig. 3: GDI-TH34W2 NCL GATE

NCL RCA is the cascaded connection of Hybrid NCL full adders with HYBRID DI registers. Hybrid NCL full adder cell is a combination of CMOS and GDI based NCL gates as shown in Fig 2, where $TH23$ gate is based on conventional CMOS technology and $TH34W2$ gate is implemented in GDI as depicted in Fig 3.

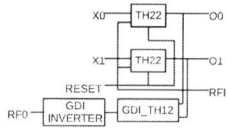

Fig. 4: HYBRID DI Register

Fig 4 illustrate HYBRID DI register where $TH12$ and inverter gate are designed in GDI technology. The complete model of single bit HYBRID NCL full adder with DI register is shown in Fig 5. Designing the RCA circuit using this HYBRID model not only reduces the number of transistors used but also reduces the power dissipation. The results of the proposed model are presented in the next section.

IV. SIMULATION RESULTS

Cadence $45nm$ technology is used for realization of the 4 and 8-bit NCL ripple carry adder using CMOS and HYBRID techniques. In HYBRID ripple carry adder the

$TH34W2$, $TH12$ and Inverter gates are designed by GDI technique. Performance comparison is done by simulating schematic using Cadence Virtuoso tool with $V_{DD} = 1V$ and temperature$= 27°C$. Performance comparisons are done by showing the power dissipation and the number of transistors used.

TABLE I: Simulation Results of NCL RCA implemented using CMOS and HYBRID technologies.

Design Methodology		Ripple carry Adder	
		4-bit	8-bit
CMOS	Power(μw)	5.21	10.42
	# of Transistor	1100	2212
HYBRID	Power(μw)	3.72	7.7
	# of Transistor	972	1956

As seen in Table I, the proposed HYBRID design has better performance when compared to CMOS design in terms of power dissipation and the number of transistor used. The HYBRID design results in 27% power reduction and # of transistors used are decreased by 11% in comparison to the conventional CMOS NCL design, while providing logically correct operation.

REFERENCES

[1] R. Mader, E. G. Friedman, A. Litman, and I. S. Kourtev, "Large scale clock skew scheduling techniques for improved reliability of digital synchronous vlsi circuits," in *Circuits and Systems, 2002. ISCAS 2002. IEEE International Symposium on*, vol. 1. IEEE, 2002, pp. I–357.

[2] A. Morgenshtein, M. Moreinis, and R. Ginosar, "Asynchronous gate-diffusion-input (gdi) circuits," vol. 12, no. 8, pp. 847–856, Aug. 2004.

[3] F. A. Parsan and S. C. Smith, "CMOS implementation comparison of ncl gates," in *Proc. IEEE 55th Int. Midwest Symp. Circuits and Systems (MWSCAS)*, Aug. 2012, pp. 394–397.

[4] A. Morgenshtein, A. Fish, and I. A. Wagner, "Gate-diffusion input (gdi): a power-efficient method for digital combinatorial circuits," vol. 10, no. 5, pp. 566–581, Oct. 2002.

[5] R. Bonam, S. Chaudhary, Y. Yellambalase, and M. Choi, "Clock-free nanowire crossbar architecture based on null convention logic (ncl)," in *Proc. 7th IEEE Conf. Nanotechnology (IEEE NANO)*, Aug. 2007, pp. 85–89.

[6] F. A. Parsan and S. C. Smith, "CMOS implementation of static threshold gates with hysteresis: A new approach," in *Proc. IEEE/IFIP 20th Int VLSI and System-on-Chip (VLSI-SoC) Conf*, Oct. 2012, pp. 41–45.

[7] A. Morgenshtein, A. Fish, and A. Wagner, "Gate-diffusion input (gdi)-a novel power efficient method for digital circuits: a design methodology," in *Proc. 14th Annual IEEE Int. ASIC/SOC Conf*, 2001, pp. 39–43.

[8] A. Morgenshtein, A. Fish, and I. A. Wagner, "Gate-diffusion input (gdi) - a technique for low power design of digital circuits: analysis and characterization," in *Proc. IEEE Int. Symp. Circuits and Systems ISCAS 2002*, vol. 1, 2002, pp. I–477–I–480 vol.1.

[9] A. Morgenshtein, V. Yuzhaninov, A. Kovshilovsky, and A. Fish, "Full-swing gate diffusion input logiccase-study of low-power cla adder design," *INTEGRATION, the VLSI journal*, vol. 47, no. 1, pp. 62–70, 2014.

A High-performance Circuit Design Algorithm using Data Dependent Approximation

Kazushi Kawamura[*†], Masao Yanagisawa[‡], and Nozomu Togawa[*]

[*]Dept. of Computer Science and Engineering, Waseda University
[†]kazushi.kawamura@togawa.cs.waseda.ac.jp
[‡]Dept. of Electronic and Physical Systems, Waseda University

Abstract—This paper proposes a high-performance circuit design algorithm using input data dependent approximation. In our algorithm, STEPCs (Suspicious Timing Error Prediction Circuits) are utilized for identifying the paths to be optimized inside a circuit efficiently. Experimental results targeting a set of basic adders show that our algorithm can achieve performance increase by up to 11.1% within the error rate of 2.1% compared to a conventional design technique.

I. INTRODUCTION

Performance improvement is becoming a challenging problem in LSI design due to PVT (Process-Voltage-Temperature) variations. In recent years, approximate computing [1], which enables us to improve LSI's performance and/or energy consumption with negligible error, has widely utilized for high-performance LSI design techniques [2], [4], [5]. These techniques are, in particular, expected to be applied in designing circuits for error-resilient applications like image processing. However, they cannot elicit sufficient performance because the input data set is not specified.

The data set input to an application specific circuit is not a non-biased (random) one but rather a biased one. Moreover, it has a large impact on propagation delays and signal transition probabilities inside a circuit. Based on these properties, each signal path in a circuit can be classified according to whether it is more likely to cause errors or not. For example, scaling the operating clock frequency or clock cycle, a signal path whose transition probability is very low rarely causes errors even if its propagation delay is large. Therefore, when designing a high-performance circuit within an acceptable error rate, it is important to identify, based on input data set, the paths to be optimized which are more likely to cause errors and then focus on optimizing those paths.

We propose, in this paper, a high-performance circuit design algorithm using input data dependent approximation. In applying our algorithm to a circuit, STEPCs (Suspicious Timing Error Prediction Circuits) [3] are preliminarily inserted into the circuit for efficient identification of the paths to be optimized. Experimental results targeting a set of basic adders show that our algorithm can achieve performance increase by up to 11.1% within the error rate of 2.1% compared to a conventional design technique.

II. A HIGH-PERFORMANCE CIRCUIT DESIGN ALGORITHM USING DATA DEPENDENT APPROXIMATION

In this section, we propose a circuit design algorithm using data dependent approximation. Given a circuit and an input

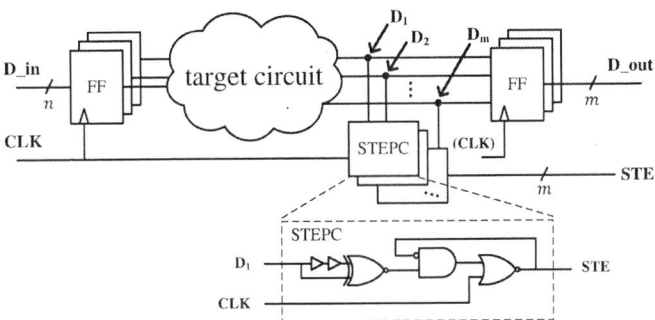

Fig. 1: The mechanism to identify the paths to be optimized.

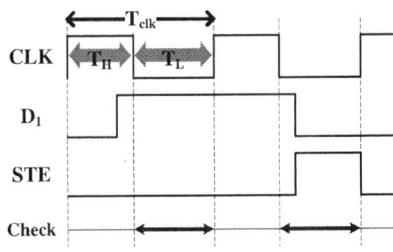

Fig. 2: Timing chart and check timing.

data set, our algorithm improves its performance based on the input data set under an error rate constraint by identifying the paths to be optimized and reconfiguring them. In order to identify those paths efficiently, STEPCs [3] are introduced to the target circuit as shown in Fig. 1. The target circuit is assumed, in this paper, to be a combinational circuit having n-bit input and m-bit output.

A STEPC, which has been originally developed for a timing error prediction scheme in [3], outputs a STE (Suspicious Timing Error) signal when some signal transitions are detected at its monitoring point while the clock signal is low. When our algorithm is applied to a circuit, each output point is monitored by a STEPC as shown in Fig. 1. A timing chart, when the mechanism of Fig. 1 works, is illustrated in Fig. 2. Given a clock signal whose period, rise time and fall time are T_{clk}, T_H and T_L, respectively, the target circuit operation for the operating clock period T_H is simulated by monitoring STE signals which are output from each STEPC. Specifically, by counting the number of STE signals for every signal path, the paths to be optimized can be identified without varying the clock period T_{clk}. Note that, in our algorithm, only the duty cycle (T_H, T_L) of the clock signal will be varied.

TABLE I: Adder configurations.

		0	1	2	3	4	5	6	7	8	9	10	11	12	13	14	15
Basic	RCA	HA	FA	FA	FA	FA	FA	FA	FA	FA	FA	FA	FA	FA	FA	FA	FA
	CLA based adder	4bit CLA				4bit CLA				4bit CLA				4bit CLA			
Optimized for random data set	Ours(1σ)	4bit CLA				4bit CLA				4bit CLA				4bit CLA			
	Ours(2σ)	4bit CLA				4bit CLA				4bit CLA				4bit CLA			
	Ours(3σ)	4bit CLA				4bit CLA				4bit CLA				4bit CLA			
Optimized for Gaussian distribution data set	Ours(1σ)	HA	FA	FA	FA	2bit CLA		4bit CLA				3bit CLA			FA	FA	FA
	Ours(2σ)	HA	FA	FA	FA	2bit CLA		4bit CLA				4bit CLA				FA	FA
	Ours(3σ)	HA	FA	3bit CLA			6bit CLA						3bit CLA			FA	FA

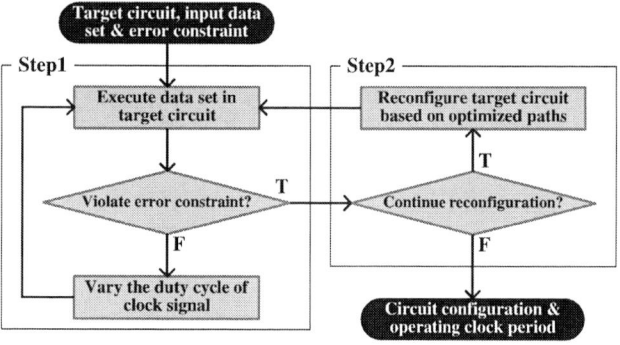

Fig. 3: The overall flow.

The mechanism described above enables us to identify the paths to be optimized in a target circuit quite efficiently because of the following reasons; Firstly, it is realized on an actual device and do not use any static analytical approaches. In addition, it is applicable to an actually operated circuit since the operating clock period does not vary and hence no errors occur in the target circuit during its operation. Therefore, its actual input data collection is achievable at the same time.

The overall flow of our algorithm is shown in Fig. 3. In **Step1** of our algorithm, gradually decreasing the rise time T_H of the clock signal, the input data set is executed in the target circuit into which STEPCs are preliminarily inserted. In this step, the paths to be optimized are efficiently identified as stated before. After that, in **Step2**, the target circuit is reconfigured to focus on optimizing those paths.

III. Experimental Results and Conclusion

In this section, the validity of our algorithm is verified by using it for 16 bit adder design. Given a 16 bit RCA (Ripple Carry Adder) which consists of 15 FAs (Full Adders) and 1 HA (Half Adder), our algorithm selects, based on an input data set and an error rate constraint, some carry chains in Step1, and then reconfigures each of them by using a CLA (Carry Look-ahead Adder) based multibit adder in Step2. Our algorithm has been realized on a Xilinx FPGA (xc7k325tffg900-2) by using Xilinx Vivado Design Suite 2014.1. In this experiment, we have finally obtained six adders by applying it to a 16 bit RCA based on the two different input data sets (random data set and Gaussian distribution data set) under the three different error rate constraints (3σ: 0.1%, 2σ: 2.1%, 1σ: 13.6%). Each data set has assumed to have 10,000 data. Let $P_{1,i}$ and $P_{2,i}$ be the switching probabilities of i-th bit in the first and second operands, respectively, where $0 \leq i \leq 15$. $P_{1,i}$ and $P_{2,i}$ in the

TABLE II: Implementation results of adder optimized for random data set.

	#LUT (Adder only)	#LUT (Total)	Clock period [ns]	Error frequency [%]
RCA	16	478	4.3	0.00
CLA based adder	20	482	3.0	0.00
Ours(1σ)	20	482	**2.7**	0.35
Ours(2σ)	20	482	**2.7**	0.35
Ours(3σ)	20	482	2.9	0.01

TABLE III: Implementation results of adder optimized for Gaussian distribution data set.

	#LUT (Adder only)	#LUT (Total)	Clock period [ns]	Error frequency [%]
RCA	16	478	4.3	0.00
CLA based adder	20	482	2.7	0.00
Ours(1σ)	17	479	**2.4**	0.05
Ours(2σ)	18	479	**2.4**	0.02
Ours(3σ)	19	481	2.7	0.00

Gaussian distribution data set are both equal to $\frac{1}{\sqrt{2\pi}}e^{-\frac{(i-7.5)^2}{2}}$, while those in the random data set are equal to 0.5.

Adder configurations, obtained by applying our algorithm to a 16 bit RCA, are shown in Table I, and the implementation result for each configuration is summarized in Tables II and III. "RCA" and "CLA based adder" in these tables show basic adders for comparison.

As shown in Table I, adder configurations have varied greatly depending on the input data set. The results in Tables II and III have shown that our algorithm can achieve performance increase by up to 11.1% within the error rate of 2.1% compared to a basic (CLA based) adder.

Acknowledgment

This research is supported in part by Grant-in-aid for JSPS Fellows (No. 15J07118).

References

[1] J. Han, "Approximate computing: An emerging paradigm for energy-efficient design," in *Proc. ETS*, 2013, pp. 1–6.

[2] A. Ranjan *et al.*, "Aslan: synthesis of approximate sequential circuits," in *Proc. DATE*, 2014.

[3] Y. Shi *et al.*, "Suspicious timing error prediction with in-cycle clock gating," in *Proc. ISQED*, 2013, pp. 335–340.

[4] D. Shin *et al.*, "A new circuit simplification method for error tolerant applications," in *Proc. DATE*, 2011.

[5] S. Venkataramani *et al.*, "Substitute-and-simplify: a unified design paradigm for approximate and quality configurable circuits," in *Proc. DATE*, 2013, pp. 1367–1372.

Discussion of Cost-effective Redundancy Architectures

Keewon Cho, Jooyoung Kim, Hayoung Lee and Sungho Kang
Dept. of Electrical and Electronic Engineering
Yonsei University
Seoul, Korea
{ckw1505, kimjy9850, yseehy214}@soc.yonsei.ac.kr, and shkang@yonsei.ac.kr

Abstract—To get a reasonable yield, memories incorporate redundancies to substitute for faulty cells. As the performance of repair algorithm reaches some saturation point, recent studies focus on various redundancy architectures for higher repair rate. In this paper, three kinds of spares, i.e., local, common, and global spares, are discussed to analyze the efficiency of redundancy architectures in respect of the repair cost. In order to estimate the impact of each spare, more than a hundred redundancy architectures are simulated with different faulty patterns. This paper performs a data analysis and suggests cost-effective redundancy architectures.

Keywords- built-in redundancy analysis (BIRA); repair cost; yield; redundancy architecture

I. INTRODUCTION

Memory yield and reliability problems become more sensitive because an advanced technology induces more defects. One powerful solution is a built-in redundancy analysis (BIRA) methodology which repairs faulty cells with prepared redundancies. However BIRA algorithms targeting a single block with local spares have some limitations in repair efficiency. Even though all repairable memories are fixed with the algorithms, the percentage of unrepairable memories is increased by multiple faults induced circumstances. Therefore, recent studies turn their attentions to change redundancy architectures [1, 2]. [1] introduces three different types of the redundancy constraints to achieve higher repair rate. [2] shows an optimal repair algorithm which utilizes a block-based redundancy architecture.

Because repair rate is directly connected with the manufacturing yield, a main purpose of previous studies is increasing repair rate. However, the memory repair cost is decided by various factors. Especially, area overhead is one of the major factor of the BIRA cost since the area of chip is limited. Furthermore, area overhead become more important in multi-memory blocks considering redundancy architectures.

This paper conducts a data analysis based on numerous experimental results from a simulation tool in c-language. In addition, cost-effective redundancy architectures targeting four memory blocks in parallel are suggested considering both repair rate and area overhead.

This research was supported by the MOTIE (Ministry of Trade, Industry & Energy) (10052875) and KSRC (Korea Semiconductor Research Consortium) support program for the development of the future semiconductor device.

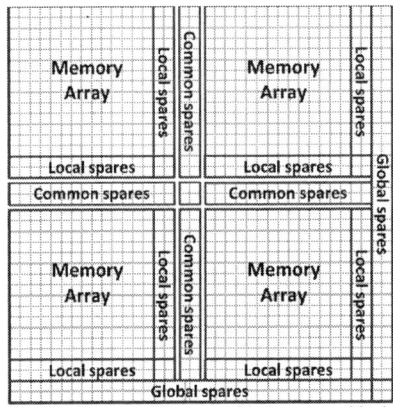

Figure 1. Basic redundancy architecture with 4-memory blocks.

II. PROPOSED ARCHITECTURES

This paper sets up redundancy architectures with 4-memory blocks utilizing three different kinds of spares. Local spares, which are commonly used in previous studies, can only be allocated in their located memory block. On the other hand, common and global spares can be allocated in multiple memory blocks. Common spares can be shared in their two adjacent memory blocks and global spares can be placed in multiple memory blocks. Figure 1 shows the basic redundancy architecture which utilizes all kinds of spares. Each memory block contains a local row and a local column spare. There are four common spares in Figure 1 and each of them can be used in its adjacent memory blocks. Global spare lines can repair two memory blocks simultaneously in this case. And also, these lines can be allocated wherever faults occurred.

Actual area overhead of various redundancy architectures can hardly be estimated without implementing all structures. Instead, previous study offers the storage requirement comparing method [3]. The area of storage requirement takes a most portion of the area of the whole BIRA and can easily be calculated with mathematical equations. To estimate area overhead of various redundancy architectures, storage requirements for each type of spare are evaluated. Since these values are varied depending on BIRA algorithms, this paper uses BRANCH algorithm [3] to observe the relationships among three kinds of spares.

978-1-5090-3220-4/16 $31.00 © 2016 IEEE

III. DATA ANALYSIS AND RESULTS

In order to acquire the data, we developed a simulation tool in c-language. When the number of each spare and fault counts is entered, the simulation tool generates faulty addresses at random. When the analysis is end, the tool shows each trial's repair rate and area overhead. Changing the number of faults and each spare, repair rate and area overhead are observed to accumulate the data. Since hundreds of redundancy architectures are simulated with different faulty patterns, it is hard to show all data. So this paper suggests new evaluation indexes to describe the performance of each type of spare.

Repair rate$_{RoI}$ denotes rate of increase in repair rate. It is the mean of expected values which are created from growth of the number of the specific type of spare. To get the *repair rate*$_{ROI}$ of each spare, 107 different redundancy architectures are used. And the basic redundancy architecture in Figure 1 is set as the reference set. Then, *repair rate*$_{RoI}$ can be expressed as

$$\frac{\sum_{i=1}^{m}(R\_1_i - R\_Ref) + \sum_{i=1}^{n}(R_2{}_i - R\_Ref) + \sum_{i=1}^{o}(R\_3_i - R\_Ref)}{m+n+o} \quad (1)$$

where R_Ref is repair rate of the reference set and R_1 is repair rate of the set which has one more additional spare than the reference set and m is the number of R_1 sets. R_2, R_3, n and o are decided as similar as R_1 and m.

Area overhead$_{ROI}$ denotes rate of increase in area overhead. It is derived from the difference between the number of storage elements of the reference set and that of R_1, R_2, and R_3 set. Then, it can be expressed as

$$area\ overhead_{-ROI} = \#\ of\ storage\ (R\_k - Ref). \quad (2)$$

This evaluation index helps to figure out the hardware impact of spare lines.

Figure 2 shows the performance of each spare's *repair rate*$_{ROI}$. Basically, injecting common spare lines shows great improvement in repair rate. This result is quite reasonable because common spare lines are more flexible than other spares. The fact that common spare lines can be shared in two memory blocks leads to a maximum 70% improvement in rate of increase in repair rate. On the other hand, global spare lines shows poor performance in *repair rate*$_{ROI}$. Since a global spare line is twice the size of a common (or local) spare line, we gives a penalty to the number of global spare lines. This penalty affects whole results in Figure 2.

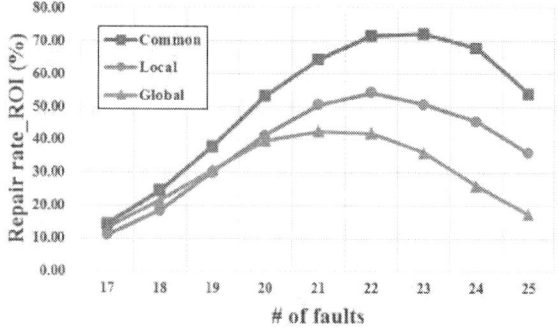

Figure 2. Repair rate$_{ROI}$ of each type of spare.

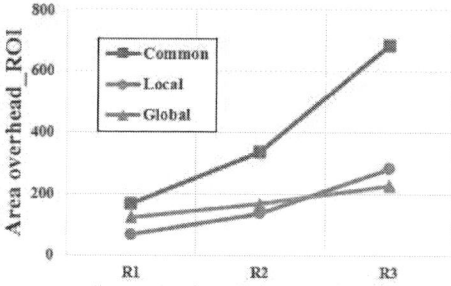

Figure 3. Area overhead$_{ROI}$ of each type of spare.

Figure 3 shows the performance of each spare's *area overhead*$_{ROI}$. The remarkable result in the graph is that the area increasing rate of common spare line is overwhelming. Also, we can observe that *area overhead*$_{ROI}$ of local spare lines is placed in the lowest position at first. However, as the number of adding spares is increased, its value moves up rapidly. It can be expected that the efficiency of global spare lines gets better in complicated faulty memories.

Because the faulty patterns can be varied in various memories, there is no perfect redundancy architecture. But cost-effective redundancy architectures can be recommended from the data analysis. For low area overhead, local and common spare lines should be injected to a minimum. Especially, the number of common spare lines should be carefully controlled. For high repair rate, enough number of global spare lines should be assigned. By doing so, a reasonable memory yield can be guaranteed with a proper area of BIRA module.

IV. CONCLUSION

A deep discussion of redundancy architectures in aspect of repair rate and area overhead is performed in this paper. Various redundancy architectures are used to simulate the performance of local, common and global spare lines. Two new evaluation indexes are suggested to describe the characteristic of each spare line. To improve both repair rate and area overhead, the number of each spare line should be treated properly. We expect that this data analysis helps to figure out cost-effective redundancy architecture.

ACKNOWLEDGMENT

This research was supported by the MOTIE (Ministry of Trade, Industry & Energy) (10052875) and KSRC (Korea Semiconductor Research Consortium) support program for the development of the future semiconductor device.

REFERENCES

[1] B.-Y. Lin, M. Lee, and C.-W. Wu, "Exploration Methodology for 3D Memory Redundancy Architectures under Redundancy Constraints," Test Symposium (ATS), 2013 22nd Asian., pp. 1-6, Nov. 2013.

[2] Štefan Krištofík, and Elena Gramatová, "Redundancy Algorithm for Embedded Memories with Block-based Architecture," IEEE 16th International Design and Diagnostics of Electronic Circuits & Systems (DDECS), pp. 271-274, Apr. 2013.

[3] W. Jeong, J. Lee, T. Han, K. Lee, and S. Kang, 2010. An advanced BIRA for memories with an optimal repair rate and fast analysis speed by using a branch analyzer. *IEEE Transactions on Computer-Aided Design of Integrated Circuits and Systems*, 29, 12, 2014-2026.

Equalization Scheme Analysis for High-Density Spin Transfer Torque Random Access Memory

Beomsang Yoo, Taehui Na, Byungkyu Song, and
Seong-Ook Jung

Department of Electrical and Electronic Engineering
Yonsei University
Seoul 03722, South Korea
Email: {rbs0412, lataehee, bksong, sjung}@yonsei.ac.kr

Jung Pill Kim and Seung H. Kang

Advanced Technology
Qualcomm Incorporated
San Diego, CA 92121, USA
Email: {jungpill and seungk}@qti.qualcomm.com

Abstract— **As the memory density increases for the big-data processing, the sensing speed is degraded because of the increased parasitic capacitive load. Thus, the equalization (EQ) scheme that is capable of improving the sensing speed has now become essential. This paper examines the effectiveness of EQ scheme on the sensing speed of offset-canceling dual-stage sensing circuit (OCDS-SC) in terms of cells per bit line (CpBL). The simulation results show that the OCDS-SC with EQ scheme achieves 3 times faster sensing time than that without EQ scheme in case of CpBL of 128. Additionally, the EQ scheme becomes more effective for reducing the sensing time according to the increase in the number of CpBL.**

Keywords; Equalization scheme; Nonvolatile memory (NVM); Sensing circuit; Spin transfer torque RAM (STT-RAM);

I. INTRODUCTION

A spin-transfer-torque random access memory (STT-RAM) is a promising candidate of high performance next generation memory for big-data processing by virtue of its great scalability characteristic [1]. However, as the memory density increases, the sensing speed is degraded due to the relatively larger capacitive load caused by a higher number of cells per bit line (CpBL). To overcome the poor sensing speed of high density memory, the equalization (EQ) scheme that is capable of improving the sensing time with only insertion of a single switch is introduced [2], [3]. In this paper, we analyze the effectiveness of EQ scheme by comparing the sensing time according to the presence of EQ scheme in the offset-canceling dual-stage sensing circuit (OCDS-SC), which is one of the powerful sensing circuits to achieve high read yield in deep sub-micrometer STT-RAM [4]. In addition, we analyze the EQ scheme in terms of the CpBL.

II. BACKGROUND

A. OCDS-SC and Equalization Effect

Fig. 1(a) represents the schematic of the OCDS-SC. The OCDS-SC operates in the two stages. During the first stage, SS1 is asserted and the data cell is connected to OUT_SC node. Then, the data voltage (V_{SA_data}) is developed and stored in C_{in_d}. During the second stage, on the other hand, SS2 is asserted and

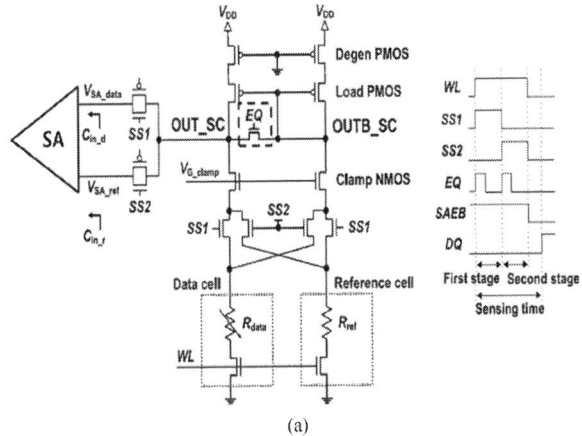

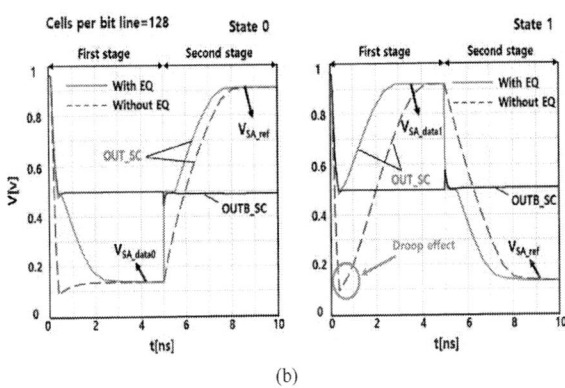

Fig. 1. (a) Schematic of the OCDS-SC and (b) transient response of OCDS-SC with and without EQ scheme at state 0 and 1, respectively.

the reference cell is connected to the OUT_SC node. Then, the reference voltage (V_{SA_ref}) is developed and stored in C_{in_r}. Then, the V_{SA_data} and V_{SA_ref} are compared and latched by sense amplifier (SA) to generate a digital signal (0 or 1). Because both V_{SA_data} and V_{SA_ref} are generated on the same branch at each stage, an offset caused by transistors mismatch between data and reference branches can be cancelled out. The EQ scheme, which connects the OUT_SC and OUTB_SC, operates at the beginning of each sensing stage to improve the sensing speed. Fig. 1(b) shows the transient responses of the OCDS-SC with and without the EQ scheme, respectively. When the word line (WL) is enabled initially, V_{SA_data} starts to discharge initially by the current flowing through the bit line (BL). In case without EQ scheme, V_{SA_data} initially almost drops to the ground level (droop effect) because OUT_SC node has smaller load capacitance compared to OUTB_SC node [3]. Thus, V_{SA_data} requires an additional voltage developing time to obtain a sufficient sensing margin, leading to the degradation of sensing speed. In case with EQ scheme, on the other hand, the parasitic capacitance mismatch between OUT_SC and OUTB_SC node is resolved, leading to the elimination of droop effect. As a result, the EQ scheme improves the entire sensing speed of the OCDS-SC.

B. Yield Estimation Methodology

When an SA senses the voltage difference (ΔV) between V_{SA_data} and V_{SA_ref}, the offset voltage of sense amplifier (V_{SA_OS}) can cause the read failure. Assuming the distributions of ΔV of OCDS-SC and V_{SA_OS} follow Gaussian distributions, then the read access pass yield for a single cell ($RAPY_{CELL}$) can be expressed in terms of the mean and standard deviation of ΔV and V_{SA_OS} as follows [4],

$$RAPY_{CELL} = \frac{\mu_{\Delta V} - \mu_{SA\_OS}}{\sqrt{\sigma_{\Delta V}^2 + \sigma_{SA\_OS}^2}}, \qquad (1)$$

where $\mu_{\Delta V}$ (μ_{SA_OS}) is the mean of ΔV and $\sigma_{\Delta V}$ (σ_{SA_OS}) is the standard deviation of ΔV.

III. SIMULATION RESULT

HSPICE Monte Carlo simulations are performed using the 22-nm PTM model parameters. A magnetic tunnel junction (MTJ) model for STT-RAM, which has a tunnel magnetoresistance ratio (TMR) of 100%, a low resistance (R_L) of 3kΩ and a high resistance (R_H) of 6kΩ is employed in this paper. Additionally, to consider the MTJ variation, a standard deviation of MTJ is adopted to be 6%. A supply and clamp voltages of 1.0 V and 0.6 V are used, respectively. Also, simulations are performed under the temperature ranged from −45 °C to 90 °C to consider the temperature corners. Fig. 2 shows the $RAPY_{CELL}$ according to the sensing time when the number of CpBL is 128. The OCDS with EQ scheme requires the sensing time of 1.43 ns to reach the target $RAPY_{CELL}$ of 3σ, whereas that without EQ scheme requires 4.25 ns due to the droop effect. It clearly shows that the EQ scheme improves the sensing time 3 times faster than without EQ scheme. Fig. 3 represents the sensing time satisfying the target $RAPY_{CELL}$ of 3σ ($t_{3\sigma}$) according to the number of CpBL. It shows that the enhancement of sensing time by the EQ scheme is continuously increasing according to the increase in the number of CpBL. Because the droop effect occurs more severely in the case of higher CpBL, the EQ scheme, which eliminates the droop effect, is absolutely required for fast sensing speed and high density STT-RAM.

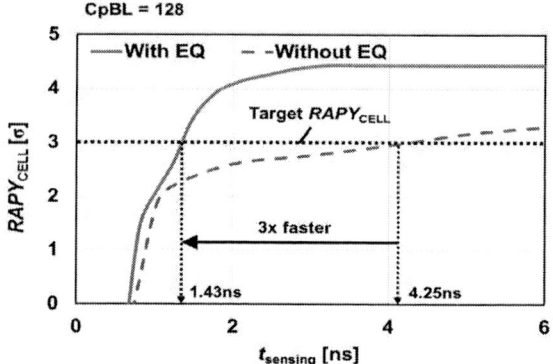

Fig. 2. $RAPY_{CELL}$ of OCDS-SC with and without EQ scheme according to the sensing time.

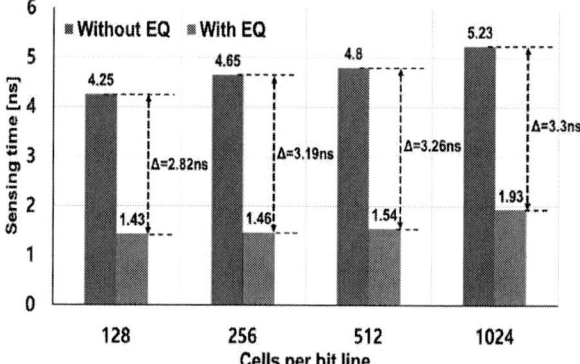

Fig. 3. Sensing time of OCDS-SC with and without EQ scheme according to CpBL.

IV. CONCLUSION

In this paper, the effectiveness of EQ scheme was analyzed by comparing the sensing time according to the presence of EQ scheme in the OCDS-SC. The simulation results proved that the EQ scheme reduces the sensing time significantly. Furthermore, the EQ scheme is more effective for reducing the sensing time in case of the higher number of CpBL. Therefore, we conclude that as the memory density increases, the EQ scheme is more essential to improve the sensing speed.

REFERENCES

[1] C. Kim et al., "A covalent-bonded cross-coupled current-mode sense amplifier for STT-MRAM with 1T1MTJ common source-line structure array," in *Proc. IEEE Int. Solid-State Circuits Conf. (ISSCC)*, Feb. 2015, pp. 1-3.

[2] J. P. Kim et al., "A 45nm 1Mb embedded STT-MRAM with design techniques to minimize read-disturbance," in *IEEE Symp. VLSI Circuits Dig. Tech. Papers*, 2011, pp. 296-297.

[3] B. Song, T. Na, and S. O. Jung, "Reference-circuit analysis for high-bandwidth spin transfer torque random access memory," in *Int. Symp. Low Power Electronics and Design (ISLPED)*, Jul. 22-24, 2015.

[4] T. Na et al., "A double-sensing-margin offset-canceling dual-stage sensing circuit for resistive nonvolatile memory," *IEEE Trans. Circuits Syst. II, Exp. Briefs*, vol. 62, no. 12, pp. 1109–1113, Dec. 2015.

[5] H. Nho et al., "Numerical estimation of yield in sub-100-nm SRAM design using Monte Carlo simulation," *IEEE Trans. Circuit Syst. II*, vol. 55, no. 9, pp. 907-911, Sep. 2008.

Variation-Tolerant and Low Power Look-Up Table (LUT) Using Spin-Torque Transfer Magnetic RAM for Non-volatile Field Programmable Gate Array (FPGA)

Kangwook Jo, Kyungseon Cho and Hongil Yoon
School of Electrical and Electronic Engineering
Yonsei University
Seoul, Korea

Abstract— **The non-volatile field programmable gate array (FPGA) is a promising candidate for the ultra-low-power computing due to its flexibility. However, the non-volatile devices have a critical drawback of reliability due to the process variations in read operation. We propose a novel look-up table scheme using the spin-torque transfer magnetic RAM with high variation tolerance and low power consumption. The proposed 8-input look-up table (LUT) has a 74.4% smaller read power consumption than that of the conventional 8-input SRAM-based LUT. The area of the proposed LUT is comparable to that of the conventional SRAM-based LUT.**

Non-volatile FPGA; look-up table; STT-MRAM

I. Introduction

With the dissemination of the internet-of-things (IoT) devices and the wearable devices, ultra-low-power application processors with high flexibility are required. The field programmable gate array (FPGA) with its basic element being the look-up table (LUT) usually based on SRAM suits the purpose. The conventional SRAM-based (SB) LUT, however, must be proceeded with a re-configuration procedure after the power is turned down, making its power management very difficult and inefficient. With the non-volatile FPGAs based on the spin-torque transfer magnetic RAM (STT-MRAM), re-programming of LUTs is not required for its power management with the instant-on capability to be exploited to the greatest extent. However, devices using STT-MRAM may encounter reliability problems in the read operation because the process to convert the cell resistance to the read-out voltage is very vulnerable to the process variations. A low power LUT scheme using STT-MRAM with variation tolerance is proposed in this paper.

II. Conventional Look-up Tables

One of the conventional LUTs using STT-MRAM is the non-volatile SRAM-based (NVSB) LUT [1]. The non-volatile SRAM (NVSRAM) cell is comprised of a standard SRAM cell, two magneto tunnel junctions (MTJs) and two transistors. If the LUT is not used for a long time, the power of LUT can be shut down. When the LUT needs to be re-used, the LUT is turned on and the NVSRAMs automatically read out the data stored in the MTJs. The latch-based STT-MRAM (LBS) LUT [2] uses the cross-coupled latch as the sense amplifier, but it must be synchronized with the clock signal, avoiding its use in the combinational logic with much limited functionality. In contrast, the voltage-divider-based (VDB) LUT [3] has better reliability and functionality in comparison with LBS LUT, but it has problems of large static current and area overhead.

III. Novel STT-MRAM Look-up Table

The novel STT-MRAM LUT is proposed to solve the aforementioned problems. The schematic of the proposed LUT is shown in Fig. 1. The proposed LUT is based on the VDB LUT. However, the sensing transistors, MP1 and MN1, are shared with all data cells to reduce the area. The select transistors, one NMOS and one PMOS, are added as shown in the dotted boxes of Fig. 1. This leads to the decrease of the parasitic capacitance and the improvement of the read speed in comparison to VDB LUT. In this work, the clock signal is used to reduce the static current. One NAND gate and one NOR gate are added and the outputs of two logic gate are connected to the gate of two sensing transistors as shown in Fig. 1. The sensing

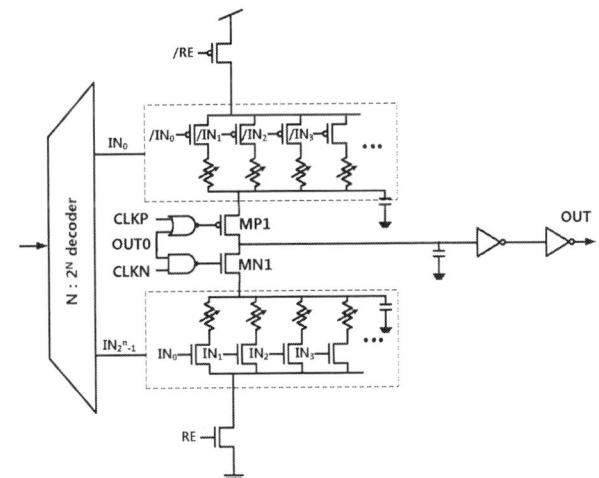

Fig. 1. Schematic of the proposed STT-MRAM LUT.

We acknowledge the CAD tool supports of the IC Design Education Center (IDEC) and the funding provided by SK Hynix.

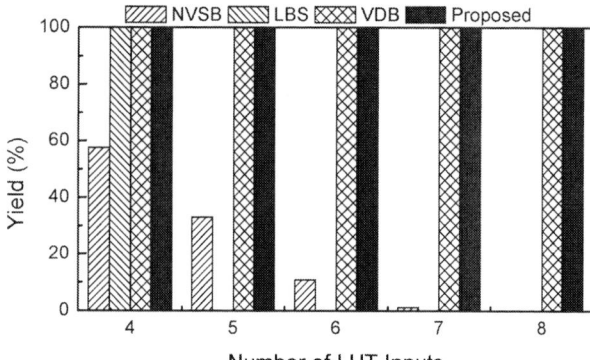

Fig. 2. LUT yield versus number of LUT inputs.

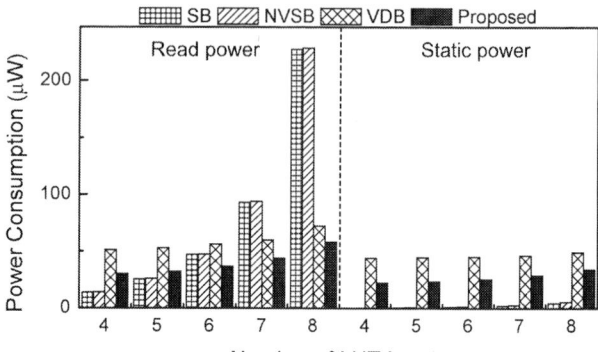

Fig. 3. Read and static power versus number of LUT inputs.

transistors are activated only in the half of the clock period. In the other half of the clock period, one of the two sensing transistors is turned off according to the output signal. For example, the pull-down NMOS transistor is turned off if the output is high. Therefore, nearly half of the static power decreases. This two-phase power reduction scheme can be adjusted to suit the architectural management in controlling the input signal changes only during the active period of the clock.

IV. COMPARISON OF PERFORMAMCE

For the proper comparison, all simulations reported herein are performed using the NCSU 45-nm process technology model parameters and the compact macro model of MTJ [4]. The supply voltage in this work is 1.2 V. It is assumed that the tunnel magnetoresistance ratio (TMR), defined by (R_{HIGH} − R_{LOW}) / R_{LOW}, is 100% where R_{HIGH} is the high state resistance and R_{LOW} is the low state resistance. Also, R_{LOW} of $3k\Omega$ is used for the NVSB, VDB and proposed LUTs, while R_{LOW} of $12k\Omega$ is used for the LBS LUT.

To assure proper reliability of LUT, the bit error rate (BER) should be nearly zero. In order to calculate the BER, the monte-carlo simulation of 1000 trials is performed with the variations in the threshold voltages and the MTJ resistances. The values of the matching factor, A_{Vth}, used for the NMOS and PMOS transistors are 2.82 mV·μm and 2.45 mV·μm, respectively. The standard deviations in the cell resistance are assumed to be 4%. Fig. 2 shows the LUT yields for a reliable read operation in various schemes as a function of the number of LUT inputs. The LUT yield is represented by $(1-BER)^N$ where N is the number of cells in the LUT. For example, BER of 3.4% is convertible to the LUT yield of 57.5% for 4-input LUT. The LUT yield of 8-input NVSB LUT is just 0.01%. BER of 4-input LBS LUT is zero. However, BERs of 5 to 8-input LBS LUTs are over 40% and the LUT yield is nearly zero. In contrast, VDB LUT and the proposed LUT show the perfect BER and the LUT yield for all the considered LUT inputs.

Fig. 3 represents the read and static powers. In this work, the read and static powers are defined by the power when the inputs and output change at every clock cycle of 100MHz and the power when the inputs do not change, respectively. NVSB LUT shows the similar result to SB LUT. The read powers of

SB LUT and NVSB LUT increase exponentially according to the number of the LUT inputs. The results of LBS LUT are omitted because LBS LUTs with larger than 5 inputs fail to operate correctly. The read powers of VDB LUT and the proposed LUT barely increase in accordance with the increased LUT inputs. As a result, the read power of the proposed 8-input LUT is smallest down to 25.6% that of SB LUT. The static power of the proposed LUT is reduced by more than 40% compared to VDB LUT. The total area is also comparable to that of SB LUT despite the additional circuit because the total cell area in the proposed LUT is under half of that of SB LUT.

V. CONCLUSION

A novel LUT using STT-MRAM for non-volatile FPGA is proposed to bear good variation-tolerance and small read and static powers. The power management on the architectural level is supported without any re-configuration process because of its non-volatility. Therefore, the proposed LUT can be promising for the applications needing both flexibility and low power performance.

REFERENCES

[1] Shuu'ichirou Yamamoto, Yusuke Shuto, and Satoshi Sugahara. "Nonvolatile power-gating field-programmable gate array using nonvolatile static random access memory and nonvolatile flip-flops based on pseudo-spin-transistor architecture with spin-transfer-torque magnetic tunnel junctions." Japanese Journal of Applied Physics, vol. 51, no. 11S, pp. 11PB02-1 - 11PB02-5, 2012.

[2] W. Zhao, E. Belhaire, C. Chappert, and P. Mazoyer, "Spin transfer torque (STT)-MRAM–based runtime reconfiguration FPGA circuit," ACM Trans. Embed. Comput. Syst., vol. 9, no. 2, pp. 14:1–14:16, Oct. 2009.

[3] Somnath Paul, Saibal Mukhopadhyay, and Swarup Bhunia. "Hybrid CMOS-STTRAM non-volatile FPGA: design challenges and optimization approaches." Proceedings of the 2008 IEEE/ACM International Conference on Computer-Aided Design. IEEE Press, 2008. pp. 589–592.

[4] Yue Zhang, Weisheng Zhao, Yahya Lakys, Jacques-Olivier Klein, Joo-Von Kim, Dafiné Ravelosona, and Claude Chappert, "Compact modeling of perpendicular-anisotropy CoFeB/MgO magnetic tunnel junctions," IEEE Trans. Electron Devices, vol. 59, no. 3, pp. 819–826, 2012.

Disturb-free 5T Loadless SRAM Cell Design with Multi-Vth Transistors Using 28 nm CMOS Process

Chua-Chin Wang [†], Senior Member, IEEE, and Chia-Lung Hsieh
Department of Electrical Engineering National Sun Yat-Sen University
Kaohsiung, Taiwan 80424
Email: ccwang@ee.nsysu.edu.tw

Abstract—**A single-ended load SRAM cell composed of multi-Vth transistors is proposed in this study. Particularly, the PDP (power-delay product) performance of the loadless SRAM cell is enhanced by a write assistant loop and an isolated wordline-controlled transistor (WLC). Additionally, a shared bitline inverter is added on the column-wise bitline to boost the read access speed at the minimal expense of area cost. The energy dissipation per write/read operation is found to be 96.624/8.104 fJ provided that the SRAM cells is driven a 0.8 V VDD power supply using a typical 28 nm CMOS technology.**

Keywords—**single-ended SRAM cell, loadless, power-delay product (PDP), multi-Vth transistor, disturb-free**

I. INTRODUCTION

As well noted, memory devices have long been the core of digital systems next to CPUs. Particularly, the cache of CPUs is usually composed of SRAMs to ensure high throughput. Thus, they have undoubtfully consumed a great portion of overall power consumption. 4-T loadless SRAM has been recognized as a possible solution for the lower power demanding SRAM [1], where low-V_{th} transistors are used as bit line drivers and high-V_{th} transistors are the data latch components. However, the read/write disturbance is queationable in such a cell due to lack of bitline isloation mechanism. The degration of the SNM (static noise margin) pointed out by [2] has verified such a potential hazard. In this study, a loadless SRAM with a write-assist loop to decouple the noise from bitlines during write operation to achieve disturbance free is analyzed. Furthermore, a shared read inverter is used to reject the potential bitline voltage variation so as to enhance the disturb-free feature. Notably, the SRAM cell itself is a 5-T design with a pair of ultra-high-V_{th} PMOS transistors to serve as the latch-like storage.

II. DISTURB-FREE 5T SRAM CELL

We propose a novel loadless SRAM cell in this study to resolve the R/W disturbance predicament in loadless SRAMs. Referring to Fig. 1, to reject the potential disturbance from the bitlines, we propose to insert one WL-controlled transistor ($\approx$WLC), namely N3, between BLB and the cell. Besides, the sources of N1 and N2 are coupled to become the "write-assist loop", where the common mode noise coupled from GND will likely be rejected. Notably, N1 and N2 are driven by WA

[†]Prof C. —C. Wang is with Department of Electrical Engineering, National Sun Yat — Sen University, Kaohsiung, Taiwan 80424. (e — mail : ccwang@ee. nsysu. edu. tw)

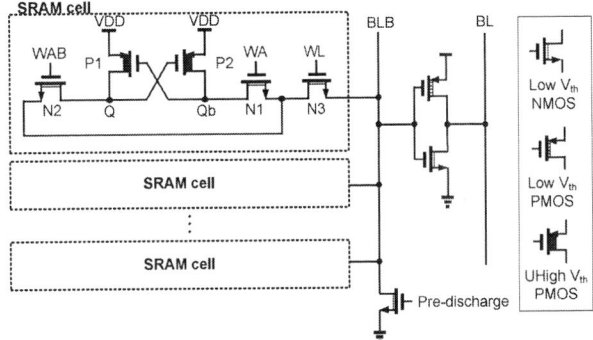

Fig. 1. Proposed disturb-free loadless SRAM cell

(word access) and WAB (word access bar), respectively.

· **Read access** : The read operation timing diagrams are very much similar to those of 4T loadless SRAM [1] As soon as the address lines are valid, WL is asserted to turn on N3. Meanwhile, WA is high and WAB is low. The state at Qb will be passed to BLB via N1 and N3 to generate valid Data_out. Notably, a shared inverter is inserted between BLB and BL to reject the nose coupled in BL. Besides, the shutoff N2 will ensure Q is free from the disturbance of BLB.

· **Write access** : By contrast, the write operation timing diagrams are shown in Fig. 2. When "1" is to be written, WL and WA are asserted to turn on N1, and N3, respectively. BLB is pre-discharged to ground such that a "0" is stored at Qb, which in turn switch on P1 to charge Q to high. By contrast, When "0" is writen, WL and WAB are pulled on to turn on N2 and N3, resptively, and BLB is also pre-discharged at the same time such that Q is pulled low.

Therefore, the proposed loadless 5T SRAM cell is a single-ended disturb-free design. Besides the R/W noise margins can be enhanced, the area cost is also drastically reduced compared with the known disturb-free design approaches. Notably, the inverter bewteen BLB and BL is shared by many SRAM cells, since only one cell in a single column is allowed to access the bitline pair in any single R/W cycle. The area overhead is obscure.

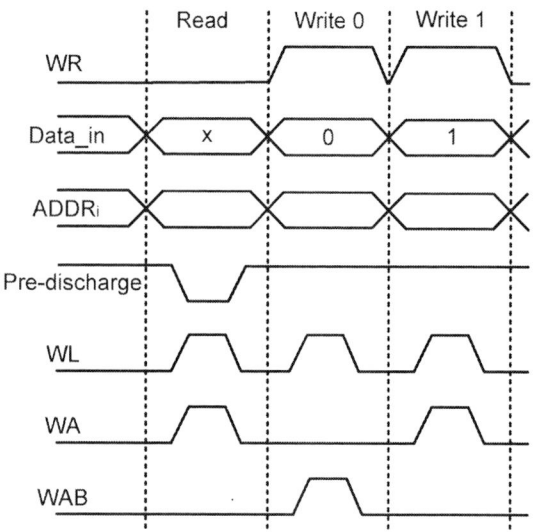

Fig. 2. Write timing diagram of the proposed loadless SRAM cell

III. IMPLEMENTATION AND SIMULATION

Referring to Fig. 3, the area of the proposed cell is $0.52 \times 1.63 \ um^2$ using a typical 28 nm CMOS process (TSMC).

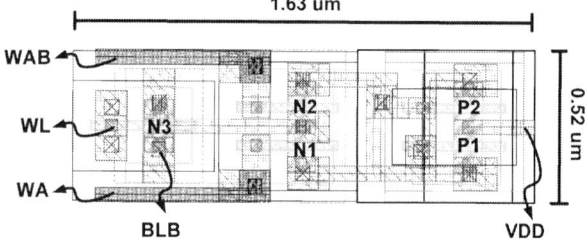

Fig. 3. Layout of the proposed SRAM cell

The worst-case read SNM of the proposed cell are shown in Fig. 4. Since one of the WAB and WA will be low to shut off one of the access NMOS, either Q or Qb will remain the same during the read operation, which is different from the conventional 6T-based SRAM cells. The worst-case SNM is found to be 363.60 mV (read), 800.31 mV (write), and 445.21 mV (hold), by all-PVT corner simulations, which is far better than the other single-ended SRAM update date. Most important of all, the write operation of the proposed cell is disturb-free in any case. The comparison with prior works based on all-PVT-corner post-layout simulations with VDD=0.8 V is tabulated in Table I. Notably, FOM is estimated by the following equation,

$$FOM = \frac{\frac{SNM}{VDD}}{\frac{Write\,PDP}{VDD} \times \frac{Read\,PDP}{VDD} \times normalized\,CA} \quad (1)$$

where CA denotes cell area. The proposed design outperforms the others with respect to the SNM, cell area, and write disturb-free. Besides, our design also attains the edge of quite balanced PDP in read/write operations.

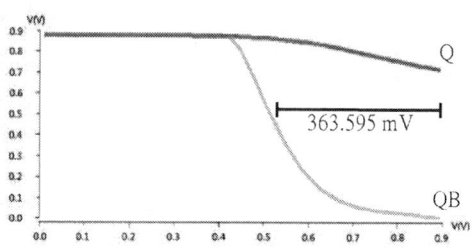

Fig. 4. Worst-case read SNM of the proposed SRAM cell

TABLE I

COMPARISON WITH PRIOR WORKS

	[3]	[4]	[6]	This work
Transistors	8T	9T	5T	5T
SNM (mV)	283.54	267.85	326.90	363.60
WM (mV)	281.90	N/A	N/A	N/A
Write PDP (fJ)	1.08	4.37	0.187	0.124
Read PDP (fJ)	0.248	0.039	0.130	0.0638
Cell Area (um^2)	3.96 ×0.8	3.625 ×1.18	2.05 ×1.12	1.63 ×0.52
VDD (V)	0.6	0.35	0.6	0.8
FOM	1.62	1.84	28.46	34.19
Year	2010	2011	2014	2016

ACKNOWLEDGMENT

This investigation is partially supported by Ministry of Science and Technology under grant MOST 104-2622-E-006-040-CC2. The authors would like to express their deepest gratefulness to Chip Implementation Center of National Applied Research Laboratories, Taiwan, for their thoughtful chip fabrication service and EDA tool support.

REFERENCES

[1] C.-C. Wang, Y.-L. Tseng, H.-Y. Leo, and R. Hu, "A 4-Kb 500-MHz 4-T CMOS SRAM using low- V_{THN} bitline drivers and high- V_{THP} latches," *IEEE Trans. on Very Large Scale Integration (VLSI) Systems*, vol. 12, no. 9, pp. 901-909, Sep. 2004.

[2] C.-C. Wang, C.-L. Lee, and W.-J. Lin, "A 4-Kb Low Power SRAM Design with Negative Word-Line Scheme", *IEEE Trans. on Circuits & Systems - I : Regular Papers*, vol. 54, no. 5, pp. 1069-1076, May 2007.

[3] M.-H. Tu, J.-Y. Lin, M.-C. Tsai, S.-J. Jou, and C.-T. Chuang, "Single-ended subthreshold SRAM with asymmetrical write/read-assis," *IEEE Trans. on Circuits & Systems - I : Regular Papers*, vol. 57, no. 12, pp. 3039-3047, Dec. 2010..

[4] M.-H. Tu, J.-Y. Lin, M.-C. Tsai, C.-Y. Lu, Y.-J. Lin, M-H. Wang, H.-S. Huang, K.-D. Lee, W. Shih, S.-J. Jou, and C.-T. Chuang, "A 72Kb single-ended distrurb-free subthreshold SRAM with cross-point data-aware write wold-line, negative bitline," *Subthreshold Microelectronics Conference*, pp. 31, Sep. 2011.

[5] E. Seevinck, F. J. List, J. Lohstroh, "Static-noise margin analysis of MOS SRAM cells," *IEEE Journal of Solid-State Circuits*, vol. SC-22, no. 5, pp. 748-754, Oct. 1987.

[6] C.-C. Wang, C.-H. Liao, and S.-Y. Chen, "A single-ended disturb-free 5T loadless SRAM with leakage sensor and read delay compensation using 40 nm CMOS process," *2014 IEEE Inter. Symp. on Circuits and Systems*, pp. 1126-1129, June 2014.

Single-Flux-Quantum Cache Memory Architecture

Koki Ishida[1], Masamitsu Tanaka[2], Takatsugu Ono[3], and Koji Inoue[3]

[1]Graduate School of Information Science and Electrical Engineering, Kyushu University, Fukuoka, Japan
[2]Department of Quantum Engineering, Nagoya University, Nagoya, Japan
[3]Faculty of Information Science and Electrical Engineering, Kyushu University, Fukuoka, Japan
[1]Email: koki.ishida@cpc.ait.kyushu-u.ac.jp

Abstract— **Single-flux-quantum (SFQ) logic is promising technology to realize an incredible microprocessor which operates over 100 GHz due to its ultra-fast-speed and ultra-low-power natures. Although previous work has demonstrated prototype of an SFQ microprocessor, the SFQ based L1 cache memory has not well optimized: a large access latency and strictly limited scalability. This paper proposes a novel SFQ cache architecture to support fast accesses. The sub-arrayed structure applied to the cache produces better scalability in terms of capacity. Evaluation results show that the proposed cache achieves 1.8X fast access speed.**

Keywords; single flux quantum; cache memory; shift register

I. SUPERCONDUCTIVE COMPUTING AND ITS PROBLEM

CMOS microprocessors have been faced with a limitation of clock speeds because of increasing computing power, i.e., known as "power-wall problem". Single-flux-quantum (SFQ) devices and circuits are promising to solve the problem due to its ultra-fast-speed and ultra-low-power natures. SFQ circuits use superconducting devices, namely Josephson junctions (JJs) to process digital signals [1]. In SFQ logic, information is stored in the form of magnetic flux quantum and transferred in the form of picoseconds-duration SFQ voltage pulse. Unlike CMOS designs, it operates in pulse logic fashion, and two types of signals are used: *"sync pulse"* and *"data pulse"*, and SFQ logic gates recognize input signal level as '0' or '1' by means of examining the existence of a data pulse between two consecutive sync pules. Fig. 1(a) shows an operation example of an SFQ AND gate. Here, *data pulse A* at time *T2* and *data pulse B* at time *T1* are stored inside of the SFQ gate as logical input level of '1'. If no pulse appears as *data pulse A* at *time T1*, it is stored as '0'. This means that each logic gate supports latch function that is a unique feature of SFQ logics compared to conventional CMOS gates. *HoldTime* and *SetupTime* are conditions that have to be satisfied to ensure correct operations.

Some SFQ microprocessors have so far been demonstrated and one of them, called CORE1β (Fig. 1(b)), successfully operated over 25 GHz [2][3][4]. In the design, an SFQ L1 cache has been prototyped in order to realize high-speed memory [3]. The cache uses an SFQ shift register [5] and operates in bit-serial fashion, as does the microprocessor core, to reduce hardware complexity. However, such bit-by-bit fine-grained operations make the cache access time much longer, resulting in poor microprocessor performance. In addition, the large scale of the selector logic used to pick up referenced data strictly limits the scalability of the cache's capacity. Since several shift register based SFQ memories have so far been proposed [3][6][7], two serious problems exist in the traditional

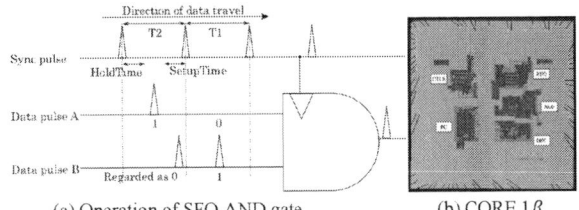

(a) Operation of SFQ AND gate.　　(b) CORE 1β.

Figure 1. An SFQ AND gate and microphotograph of CORE 1β.

cache design. Fig. 2(a) illustrates a high level model of existing SFQ cache memory. First, its bit-serial access, i.e., reading bit-by-bit makes extremely cache access time longer, e.g., 64 times bit reads are required to perform 64-bit word load. Second, its scalability in terms of cache capacity is quite low. This is because the scale of multiplexer to read a data prohibits increasing the number of cache entries.

II. BIT-PARALLEL SFQ CACHE ARCHITECTURE

To solve the problems explained in Section I, we propose a novel SFQ cache memory architecture that supports bit-parallel access. Our cache employs a shift register that consists of SFQ circuits. We use a circular buffer in the sub-arrayed structure to realize low-latency non-destructive accesses. In addition, sub-arrayed structure mitigates the negative effects of multiplexer logics in terms of access latency. Purpose of designing architecture is realizing low latency and large capacity SFQ cache memory.

Fig. 2(b) illustrates a high level model of our proposed architecture. The architecture uses loop-shaped, and sub-arrayed shift registers to realize non-destructive-access, and low-latency-access. The shift register is composed by cascading flip-flops and the data are shifted to next flip-flop by an input of clock pulse. A cache entry corresponds to one set of flip-flops in the shift register, and the length of the shift register (the number of cache entries) is related to the access time. Thus, to reduce the access time, a long shift register is divided into sub-arrayed shift registers. High-order bits of the index correspond to the addresses of the entries, while low-order bits of the index are associated with the sub array.

When read/write accesses are occurred, the index address is decoded and the number of *shift pulses* is calculated from the index address and shift register's position. *Shift pulses* are clock pulses, which are used to move data inside of shift register. For example, if shift register consists of 4 entries, the maximum number of *shift pulses* is 3 (the minimum number of *shift pulse* is 0). In read access, data is selected by multiplexer. In write access, data and tags are stored in shift register.

978-1-5090-3220-4/16 $31.00 © 2016 IEEE

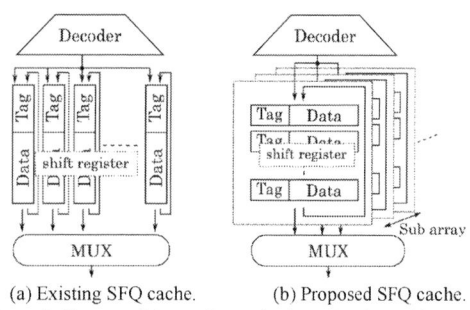

(a) Existing SFQ cache.　　(b) Proposed SFQ cache.

Figure 2. Two candidates of organization of cache architectures.

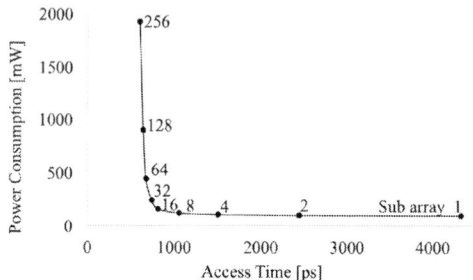

Figure 3. Trade-off between power consumption and access time in changing the number of sub arrays.

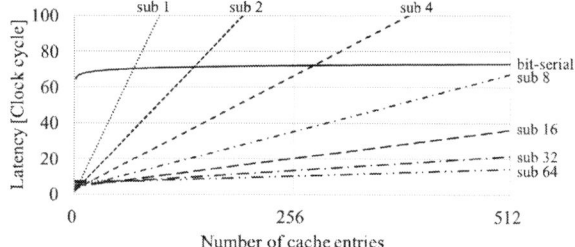

Figure 4. Estimated latency of the multiplexer.

III. EVALUATION

In this section, we evaluate the access time, power consumption, and area of proposed architecture. We compare the results with existing SFQ shift register memory to unveil the effectiveness of our approach. We assume that the cache size of evaluation target is 16 Kbits. In this case, the index is 8 bits and the tag is 22 bits. The number of entries is 256. Fig. 3 shows the trade-off between access time and power consumption in changing the number of sub arrays. There is a sweet spot considering both of access time and power consumption. According to Fig. 3, we evaluate 32 sub-arrayed shift register cache. In our evaluation, the following models are introduced. Access time is given by Equation 1,

$$T = T_{diff} + T_{rot} + T_{sel}, \quad (1)$$

where T_{diff} is the required time for calculating the number of shift pulses, T_{rot} is the required time for shifting data, and T_{sel} is the required time for selecting data. They are given by the following equations.

$$T_{diff} = CCT_{diff} \times N_{gate-diff} \quad (2)$$

$$T_{rot} = CCT_{rot} \times N_{gate-rot} \quad (3)$$

$$T_{sel} = CCT_{sel} \times N_{gate-sel} \quad (4)$$

Here, CCT_{diff}, CCT_{rot}, CCT_{sel} are clock cycle time of logic gates which includes delay of logic gates, and $N_{gate-diff}$, $N_{gate-rot}$, $N_{gate-sel}$ are the number of logic gates included data path of each functional unit, respectively. We evaluate the area based on the number of JJs and past design results [6][7], because we did not layout the design. The number of JJs considering wiring costs are obtained based on the past design results [8]. Power consumption is given by the following.

$$P = (\alpha \Phi_0 I_c f + V I_{bias}) \times N_{JJ} \quad (3)$$

In this equation, α is switching probability, Φ_0 is flux quantum, I_c is the critical current of JJs, f is clock frequency, V is voltage of bias current, I_{bias} is bias current, and N_{JJ} is the number of JJs.

The estimated access time is 736.6 ps that outperforms the existing SFQ shift register memory by 1.8X. Fig. 4 shows relationship between access time and the number of cache entries (sub2 means bit-parallel shift register which uses 2 sub arrays). According to Fig. 4, sub-arrayed shift registers reduce the latency of the multiplexer compared with an existing SFQ shift register by 7X with 256 cache entries. Moreover, even if

cache entries increase more, multiplexer's impact on the access time of sub-arrayed shift register is smaller than that of the traditional shift register. The number of JJs to be required is 624,406 JJs, and its estimated area is 543.6 mm². Compared to the existing SFQ memory, the proposed architecture is larger by 2.1X. The power consumption is estimated 240.9 mW.

ACKNOWLEDGMENT

This work was supported a part by JSPS KAKENHI Grant Number 16H02796 and by ALCA-JST

REFERENCES

[1] K. K. Likharev, et al. "RSFQ logic/memory family: A new Josephson-junction technology for sub-terahertz-clock-frequency digital systems," *IEEE Transactions on Applied Superconductivity*, Vol. 1, No. 1, pp. 3–28, Mar. 1991.

[2] M Tanaka, et al. "Design of a pipelined 8-bit-serial single-flux-quantum microprocessor with multiple ALUs," *Superconductor Science and Technology*, Vol. 19, No. 5, p. S344, Mar. 2006.

[3] M Tanaka, et al. "Design and implementation of a pipelined 8 bit-serial single-flux-quantum microprocessor with cache memories," *Superconductor Science and Technology*, Vol. 20, No. 11, p. S305, Oct. 2007.

[4] Y Yamanashi, et al. "Design and implementation of a pipelined bit-serial SFQ microprocessor, CORE 1β," *IEEE Transactions on Applied Superconductivity*, Vol. 17, No. 2, pp. 474–477, Jun. 2007.

[5] P. Yuh, et al. "Design and testing of rapid single flux quantum shift registers with magnetically coupled readout gates," *IEEE Transactions on Applied Superconductivity*, Vol. 2, no. 4, pp. 214–221, Dec. 1992.

[6] K Fujiwara, et al. "Design and high-speed test of (4 × 8)-bit single-flux-quantum shift register files," *Superconductor Science and Technology*, Vol. 16, No. 12, p. 1456, Nov. 2003.

[7] K. Fujiwara, et al. "Design and component test of SFQ shift register memories," *IEEE Transactions on Applied Superconductivity*, Vol. 13, No. 2, pp. 555–558, Jun. 2003.

[8] M. Tanaka et al. "High-density shift-register-based rapid single-flux-quantum memory system for bit-serial microprocessors," *IEEE Transactions on Applied Superconductivity*, Vol. 26, No. 5, p. 1301005, Aug. 2016.

Parallel Decoding for Multi-Stage BCH decoder

Prashanthi Metku[1], Ramu Seva[1], Kyung Ki Kim[2], Yong-Bin Kim[3] and Minsu Choi[1]

[1]Dept of ECE, Missouri Univ of Science & Technology, Rolla, MO, USA, {pmcmc,rs2k6,choim}@mst.edu
[2]Dept of Electronic Eng., Daegu University, Gyeongsan, Korea, kkkim@daegu.ac.kr
[3]Dept of ECE, Northeastern University, Boston, MA, USA, ybk@ece.neu.edu

Abstract—**3D heterogeneous processor (commonly termed as 3DHP) integrating multiple processor (such as CPU/GPU) and DRAM dies vertically interconnected by a massive number of Through-Silicon Vias (TSVs) is expected to address the limited bandwidth, high latency and energy consumption of off-chip DRAM. However, spatial and temporal variability due to hotspots in on-chip thermal gradient may result in wide bit error rate variation in DRAM dies. A multi-path BCH decoder has been recently proposed to efficiently address this issue. In this paper, a novel parallel decoding approach for the Multi-Stage BCH decoder is proposed and validated. The proposed approach efficiently leverages the multiple decoding paths to decode multiple words and minimizes the overall decoding latency.**

I. Introduction

Processors are evolving toward a 3D heterogeneous integration (3DIC) of CPU, GPU and DRAM dies vertically interconnected by TSVs (Through-Silicon Vias) to alleviate power, bandwidth and latency bottlenecks. The four heterogeneous dies (i.e., CPU, GPU, analog and DRAM) are stacked and interconnected by TSVs. TSVs enable a massive number of vertical channels among CPU, GPU and DRAM dies while providing much shorter distance of data travel. Therefore, 3DHP (3D Heterogeneous Processor) technology is anticipated to inherently provide much higher bandwidth, low latency and power consumption [1].

The conventional 2D integration/packaging technology is mature enough to assume a near-constant bit error probability (BEP) over time in DRAM. Hence, EDAC (Error Detection and Correction) engine does not need to be designed to adapt to a varying BEP over time. However, the same EDAC strategy cannot be directly applied to 3DHP, since it is anticipated to have a varying BEP caused by hotspots (i.e., spatial/temporal variation in temperature). Various design-time solutions are available to tackle hotspots in 3DIC designs but the transient nature of thermal hotspots cause the design time solutions less effective. Multi-Stage BCH decoder [2] has been designed to anticipate the varying BEP and provide just-enough DRAM error protection while minimizing the decode latency. In this paper, a novel parallel decoding approach for the Multi-stage BCH decoder [2] is proposed to decode multiple words and further reduce the decoding latency.

II. Preliminaries and Review

In 3DHP, the increase in power density of 3D stacking causes much elevated temperature, which normally results in exponential rise in charge leakage of DRAM cells and requires significant increase in refresh frequency [3] to retain data at the expense of additional power and performance overhead. Also, the spatial and temporal variability in temperature (i.e., hotspots) further complicates the DRAM reliability issues [4]. Due to hotspots, leakage current increases which discharges the charge stored by the capacitor, increasing the probability of the memory errors [5]. ECC (Error Correction Code) [6] are used to address this issue in 3DIC. They are different ways the ECC can be employed depending on the number of errors detected and corrected.

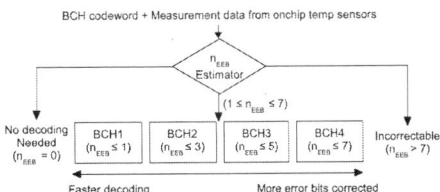

Fig. 1: The proposed multi-path BCH Decoder. The estimated number of error bits for the incoming BCH codeword is calculated from the measurement data from onchip temperature sensors and is denoted as n_{EEB} [2].

Bose-Chaudhuri-Hocquenghem (BCH) codes are one of the strong eficient error-correcting codes [7] used to detect and correct multibit errors occurred in the memory. The Multi-Stage BCH decoder design consists of four decoding path with variable target t and decoding latency as shown in Fig 1. In 3DHP, hotspots are to show spatial/temporal localities as they are mainly caused by aggressive switching activities in CPU and GPU processor dies. To ensure thermal integrity among 3D-stacked dies, onchip temperature sensors are placed to detect hotspots. When the Multi-Stage BCH decoder reads a word to decode, temperature measurement data from the distributed onchip temperature sensor network is also read [2] and used to calculate n_{EEB}, which is the Estimated number of Error Bits for the incoming word. Then, the fastest decoding path which can be used to correct n_{EEB} number of error bits gets adaptively selected to decode the incoming codeword with the minimum decoding latency.

III. Parallel Decoding for Multi-Stage BCH Decoding

The main advantage of the Multi-Stage BCH decoder to single stage BCH decoder is the reduced decoding latency with area overhead. In addition to the serial decoding (i.e., decoding one codeword at a time), the Multi-Stage BCH decoder can be utilized to correct multiple words in parallel provided incoming codewords have different n_{EEB}. Even though one or more decoding paths are occupied, if there is an unoccupied path with minimum $t \geq n_{EEB}$ of the incoming codeword, that path can start decoding it in parallel for further reduction in decoding latency. If there is no idle decoding path that can be used to decode the incoming codeword, it is temporarily stored in the storage buffer until an appropriate vacant decoding path becomes available.

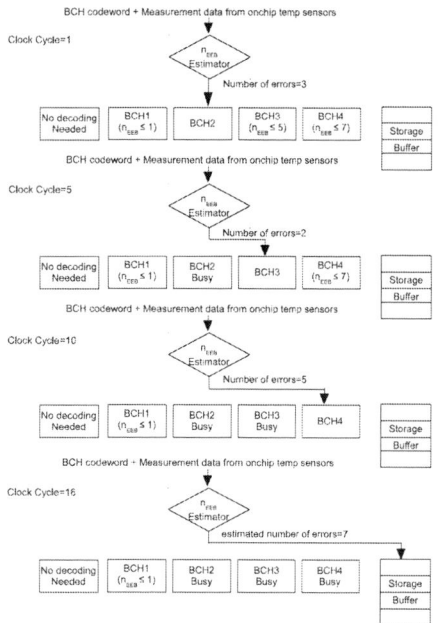

Fig. 2: An example of the proposed parallel decoding approach leveraging multiple decoding paths.

Fig. 2 illustrates an example of the proposed parallel decoding process. As seen in the first clock cycle ($\phi = 1$), n_{EEB} of the incoming codeword is 3 the decoding path with the minimum $t \geq n_{EEB}$ is BCH2 (i.e., $t = 3$). So, BCH2's availability is checked and it is currently idle. Therefore, the word is given to BCH2 to be decoded. In $\phi = 5$, another word is read from the memory, which is estimated to contain two error bits. It is given to BCH3 to be decoded, because BCH2 is currently decoding the first word and the next higher error correcting decoder path available is BCH3. Similarly, the next incoming codeword estimated to have five error bits is read from the memory in $\phi = 10$ is given to BCH4, since BCH3 is busy decoding the second word. As seen from the figure for $\phi = 16$ when the codeword with $n_{EEB} = 7$ is coming for decoding, it is stored in the storage buffer instead, since BCH4 is currently busy decoding another word. When the respective decoder becomes available the word stored in storage buffer is fetched to be decoded.

IV. SIMULATION RESULTS

The adaptive 4-path BCH decoder design has been used to evaluate the proposed parallel decoding technique. A cycle-accurate simulator has been implemented in Matlab to generate simulation results for the proposed parallel decoding technique. Bit error variation caused by spatial/temporal hotspots is simulated by introducing the following user-provided simulation parameters: p_{BE_H}: the increased bit error probability due to hotspots. p_{BE_C}: the baseline bit error probability unaffected by hotspots. f_{hot}: the relative frequency of codewords subject to p_{BE_H}. f_{cold}: the

relative frequency of codewords subject to p_{BE_C}, where $f_{hot} + f_{cold} = 100\%$. $size_{buf}$: the storage buffer size.

TABLE I: Parallel decoding simulation results showing the average decoding latency by varying p_{BE_H}/p_{BE_C}, f_{hot}/f_{cold} and $size_{buf}$.

p_{BE_H}/p_{BE_C}	$size_{buf}$	$\dfrac{f_{hot}/f_{cold}}{40/60}$	$\dfrac{f_{hot}/f_{cold}}{60/40}$	$\dfrac{f_{hot}/f_{cold}}{80/20}$
0.003/0.002	4	15.7 ns	15.8 ns	16.8 ns
	8	15.6 ns	15.69 ns	16.4 ns
	16	15.7 ns	15.8 ns	16.2 ns
0.009/0.002	4	17.1 ns	20 ns	23.3 ns
	8	16.8 ns	19.7 ns	23.1 ns
	16	16.6 ns	19.63 ns	22.8 ns
0.011/0.005	4	23.2 ns	26.9 ns	30.2 ns
	8	22.9 ns	26.7 ns	29.8 ns
	16	22.5 ns	26.5 ns	29.6 ns

In 3DHP, DRAM dies are subject to p_{BE} variation due to hotspots showing spatial/temporal localities. Depending on the user-provided simulation parameters summarized above, n_{EEB} is calculated for each incoming codeword for decoding. Table I shows the average decoding latency for various p_{BE_H}/p_{BE_C}, f_{hot}/f_{cold} and $size_{buf}$ values chosen arbitrarily. The obtained results also indicate that the average decoding latency for the proposed adaptive multi-path BCH decoder leveraging the parallel decoding technique has less decoding latency when compared to the static BCH. Thus, it can be concluded that the proposed parallel decoding of the adaptive 4-path BCH decoder can achieve significantly lower decoding latency ranging from 15.7 ns to 29.6 ns for p_{BE_H}/p_{BE_C}, f_{hot}/f_{cold} and $size_{buf}$ values chosen with area overhead of 47.97%.

ACKNOWLEDGMENT

This material is based upon work supported by the National Science Foundation under Grant No. CCF-1337167 and CCF-1539840, in part.

REFERENCES

[1] W. Yun, K. Kang, and C. M. Kyung, "Thermal-aware energy minimization of 3D-stacked L3 cache with error rate limitation," in *Proc. IEEE Int. Symp. of Circuits and Systems (ISCAS)*, May 2011, pp. 1672–1675.

[2] P. Metku, R. Seva, K. K. Kim, and M. Choi, "Multi-stage BCH decoder to mitigate hotspot-induced bit error variation," in *Proc. Int. SoC Design Conf. (ISOCC)*, Nov. 2015, pp. 47–48.

[3] J. H. Ahn, B. H. Jeong, S. H. Kim, S. H. Chu, S. K. Cho, H. J. Lee, M. H. Kim, S. I. Park, S. W. Shin, J. H. Lee, B. S. Han, J. K. Hong, P. B. Moran, and Y. T. Kim, "Adaptive Self Refresh Scheme for Battery Operated High-Density Mobile DRAM Applications," in *Proc. IEEE Asian Solid-State Circuits Conf. ASSCC 2006*, Nov. 2006, pp. 319–322.

[4] T. Hamamoto, S. Sugiura, and S. Sawada, "On the retention time distribution of dynamic random access memory (DRAM)," *IEEE Trans. Electron Devices*, vol. 45, no. 6, pp. 1300–1309, Jun. 1998.

[5] M. H. Cho, Y. I. Kim, D. S. Woo, S. W. Kim, M. S. Shim, Y. J. Park, W. S. Lee, and B. I. Ryu, "Analysis of Thermal Variation of DRAM Retention Time," in *Proc. IEEE Int Reliability Physics Symp*, Mar. 2006, pp. 433–436.

[6] R. W. Hamming, "Error detecting and error correcting codes," *The Bell System Technical Journal*, vol. 29, no. 2, pp. 147–160, Apr. 1950.

[7] S. Y. Wong, C. Chen, and Q. M. J. Wu, "Low Power Chien Search for BCH Decoder Using RT-Level Power Management," *IEEE Trans. VLSI Syst.*, vol. 19, no. 2, pp. 338–341, Feb. 2011.

A RAM cache approach using Host Memory Buffer of the NVMe interface

JuHyung Hong, SangWoo Han and Eui-Young Chung
Department of Electrical and Electronic Engineering
Yonsei University
Seoul, Korea
{jh.hong, swhan0330}@dtl.yonsei.ac.kr, eychung@yonsei.ac.kr

Abstract— **This paper proposes new methods with Host Memory Buffer to improve IO performance in NVMe interface. Although Host Memory Buffer is a versatile memory architecture, it has been considered limitedly as metadata cache such as Logical-to-Physical table. The proposed architecture uses Host Memory Buffer as data cache with modification of an NVMe command process and the additional DMA path between system memory and Host Memory Buffer. The proposed architecture improves the performance of IO request by 23% in case of sequential writes compared to a device buffer architecture.**

Keywords; NVMe, Host Memory Buffer (HMB), RAM cache

I. INTRODUCTION

NAND Flash Memory (NFM)-based storage has become the mainstream with many innovation techniques despite of some drawbacks such as high cost per bit, limited lifecycle and reliability. Especially, the appearance of host interface suitable to NFM stimulates the growth of NFM-based storage. The interface of Solid State Drive (SSD) for client or database computing is changing from SATA based on AHCI or SAS to NVMe over PCIe [1]. In addition, Universal Flash Storage (UFS) is emerging for consumer devices. NVMe provides parallel operation by supporting each up to 64K commands within 64K queues and reduces latency with efficient command protocols. The maximum performance of NVMe is limited by physical IO bandwidth of the PCIe. Nowadays, Intel Skylake has PCIe 3.0 4 lanes in Platform Controller Hub (PCH) chipset and its unidirectional bandwidth is 3.91GB/s.

RAM Disk is a software which takes exclusively a portion of system memory and shows it to user as a drive. Therefore, the performance is an order of magnitude faster than a storage device. RAPID mode of SAMSUNG SSD eliminates IO bottlenecks by using main memory as a write/read cache [2]. This mode is used primarily to accelerate write performance or to decrease queue depth 1 latency. Though the cache-based acceleration improves the completion time for IO transactions, RAPID mode is available by setup of software and a kernel driver of a host and has to manage the large buffer. In addition, this mode only takes effects in one SSD installed OS.

II. HOST MEMORY BUFFER

In specification of NVMe 1.2, Host Memory Buffer (HMB) enables using main memory as the device working buffer memory. The device can use HMB exclusively like belonging to the device logically whatever an application is. However, it has been considered merely as L2P table [3]. This is because of a Remote DMA (RDMA) function for the NVMe device and high bandwidth of PCIe. The host submits command into a Submission Queue (SQ) in Host memory. The physical memory address used for data transfers is located in Physical Region Page (PRP) entries of SQ. As the NVMe device fetches SQ, PCIe DMA transfers data between the host and the device through Transaction Layer Packets (TLPs). It enables a zero-copy operation by handing over only the address of data. If we exploit HMB to apply RAM cache, the additional copy operation is required. The bandwidth of PCIe offers higher bandwidth and lower latency compared to a SATA interface. However, the bandwidth of 2.4GHz 20 lanes QPI system bus is 19.2GB/s, which is twice or more higher than PCIe 3.0 4 lanes. Therefore, it is plausible to apply HMB as RAM cache.

UFS extents its specification to use Host memory as if it would be in the device [4]. Unified Memory (UM) Extention adds new commands to transfer data between UM and the device. New commands can copy either from Host memory to UM or within UM. However, NVMe does not support schemes like UM commands. This paper proposes new methods which exploit HMB as RAM cache to improve performance in NVMe.

III. THE PROPOSED METHODS

A. The Proposed NVMe command process

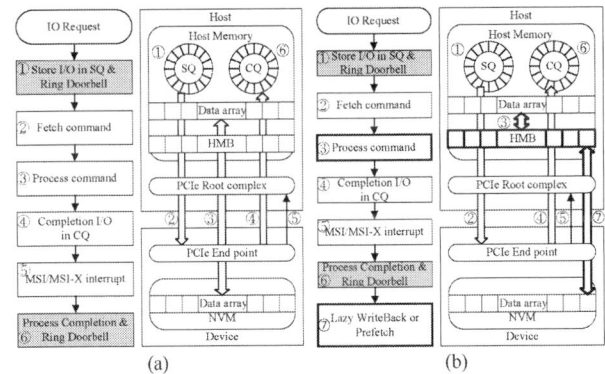

Figure 1. NVMe command process. (a) the conventional command process, (b) the proposed command process

Fig. 1 (a) shows the NVMe command process. First, the host stores command entries in SQ and writes the updated SQ tail pointer to doorbell ①. The device fetches commands and processes read/write commands ②-③. And the device writes completion to Completion Queue (CQ) and sends interrupt to the host ④-⑤. Lastly, the host processes completion ⑥. Fig. 1 (b) represents the modified command process to apply HMB as RAM cache. When the device processes read/write commands. The device does not transfer data between Host memory and the device but between Host memory and HMB ③. When HMB is full or the device needs to store data into permanent storage, the device tries to do lazy write-back from HMB ⑦.

B. The proposed architecture for RAM cache

We leverage a DMA controller of Host to exploit the new path between system memory and HMB as depicted in Fig. 2. Once the host notifies the device of the specific memory mapped addresses through a register level interface of NVMe, the device can access those addresses. When the host sends Set Features command to enable HMB feature, the host let the device know the specific function register of DMA controller additionally. Finally, the device can run DMA and check the status of DMA controller inside the host.

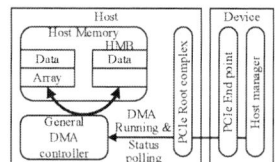

Figure 2. The proposed architecture for RAM cache using HMB

IV. EXPERIMENTS

A. Experimental setup

We used QEMU x86_64 virtual machine to implement the modified NVMe interface because we cannot recognize the memory mapped address of a real host system. We modified the VSSIM/eVSSIM simulator to evaluate the performance of the proposed architecture with timing parameters of NFM and the access latency of Host memory through NVMe Interface [5]. The simulator used an 8 channel 4 way NFM array of which parameters are specified by values such as Table 1.

TABLE I.   SPECIFICATIONS OF SIMULATOR

Parameters	Values	Parameters	Values
Page size	8192B	NFM Read latency	40us
Pages per a block	256	NFM Write latency	900us
Number of Blocks	256	NVMe W/R latency	2.8us
Number of Planes	2	NAND CH bandwidth	400MB/s
HMB Write buffer	64MB	PCIe bandwidth	3.91GB/s
HMB Read buffer	512KB	Host DMA bandwidth	4.8GB/s

B. Experimental results

We run FIO to generate synthetic IO events into Linux 3.13.0 kernel on Ubuntu 12.04. FIO is one of the most popular benchmark tools for IO performance measurement. We compare the HMB buffer architecture to both the device embedded buffer and the device without buffer. HMB improves the bandwidth up to 6.0x and 23% respectively compared to Bufferless and the device buffer in sequential writes as shown in Fig.3 (a). There are the performance

improvement up to 6.1x and 15% corresponding in random writes as shown in Fig. 3 (b). Write performance shows remarkable enhancement than read cases because of eliminating long write latency of NFM and the low hit ratio within buffer in a read operation.

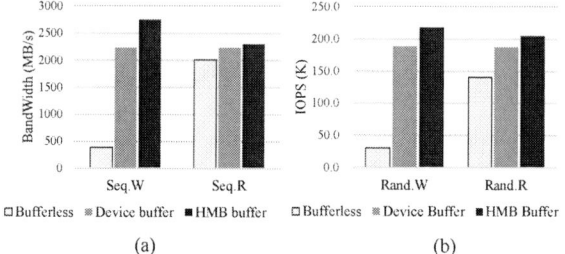

(a)   (b)

Figure 3.   Performance comparision. (a) Average Bandwidth in Seq.W/R, (b) Average IOPS in Rand W/R

The limitation of HMB buffer is that the bottleneck is changed to bandwidth of NFM after the buffer overflow as shown in Fig. 4. Therefore, the NVMe device should expand the parallelism of NFM or the host manager of the device should handle HMB buffer entries efficiently.

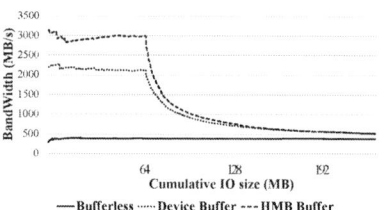

Figure 4.   Buffer overflow impact on cumulative IO size

V. CONCLUSION

We exploit HMB as RAM cache to overcome the bandwidth limitation of NVMe over PCIe. We leverage the Host DMA controller which is controlled by the device. We implement the proposed architecture into QEMU and modify the released SSD simulator to evaluate the improvement. Experiments show that the average bandwidth with HMB buffer is increased up to 12.8% compared to device buffer.

ACKNOWLEDGMENT

This work was supported by the ICT R&D program of MSIP/IITP 2016(R7177-16-0233) and by Samsung Electronics.

REFERENCES

[1] "NVMe 1.2 Specifications." http://www.nvmexpress.org/specifications/.

[2] "RAPID mode white paper," http://www.samsung.com/semiconductor /minisite/ssd/download/consumer.html.

[3] J. G. Hahn, E. Erez, S. A. Jean, G. B. Desai and V. K. Nadh, "Methods systems and computer readable media for proving flexible host memory buffer," U.S. Patent US20160026406, Jul. 30, 2015.

[4] "UFS UNIFIED MEMORY EXTENTION,Version 1.0," https:// www.jedec.org/standards-documents/results /jesd220-1/.

[5] J. Yoo, Y. Won, J. Hwang, S. Kang, J. Choi, S. Yoon, and J. Cha, "VSSIM: Virtual machine based SSD simulator," in IEEE Symposium on Mass Storage Systems and Technologies (MSST), 2013, pp. 1–14.

A Passband Lock Loop Circuit System for Band Pass Filter

Hung-Wen Lin
Department of Electrical Engineering
Yuanze University
Jhong-Li City, Taoyuan County ,Taiwan
hwlin@saturn.yzu.edu.tw

Jin-Yi Lin
Design Service Department
Wpisil Technologies Inc.
Hsin-Chu City, HsinChu County ,Taiwan
archin_lin@episil.com

Abstract— **This paper proposes a passband lock loop for tunable band-pass-filter (BPF). AC-coupling, swing amplification, peak conversion and low-noise S/H circuit are designed to differentiate several hundred uV of swing difference. All-digital peak-tracking -controller and FSM are also realized for easy verification and low area. The proposed loop was integrated with a BPF cell with 7-bits controls and 30M~50MHz of tuning range. In 0.18um technology, the proposed loop and BPF cell respectively occupied active area of 0.0046mm² and 0.027mm², and consume currents of 0.51mA and 1.44mA. With a 625kHz reference clock, the maximum period for passband tracking is 192 clock cycles.**

Keywords; Band Pass Filter, Passband, Peak Detector

I. INTRODUCTION

For the reported automatic tuning systems for *band-pass-filter* (BPF), usually their hardware overhead are about equal to or even higher than BPF. [1] uses a *phase-lock-loop* (PLL), different gain compensators to find and tune the gain at different frequency. [2] designs a resonant *voltage-controlled-oscillator* (VCO)-based BPF and uses a PLL system to adjust the both central frequency of VCO and BPF. [3] utilizes amplifiers, analog-to-digital converter and processor-based controller to obtain and adjust the parameters of BPF.

This work proposes a passband lock loop which features a lower hardware overhead. Fig.1(b) shows the concepts of the proposed passband lock loop, in which the passband of BPF could be adjusted via external digital controls to compensate the process variation. A sin wave signal at target frequency f_{ref} is applied to BPF with random initial passband controls. As the adjustment of passband becomes close to f_{ref}, a higher output swing could be detected at the outputs of BPF. As the output swing becomes its maximum value, the passband of BPF is about equal to target frequency f_{ref}. The passband lock loop, as shown in Fig.1(a), includes an input buffer, a *peak-detector* (PD), a *sample-and-hold* circuit (S/H), a difference comparator, a *confidence-counter* (CC) [4], a *peak-tracking-controller* (PTC) and a *passband-control-finite-state-machine* (PC FSM).

As the passband of BPF is far different from the reference clock, the output swing of BPF would be highly decayed and hardly detected by PD. So, the input buffer was added to amplify the differential swing but isolate the common mode level. PD detects the signal swing and generates present peak voltage. The present peak voltage was compared to the previous peak voltage stored in S/H circuit by the difference

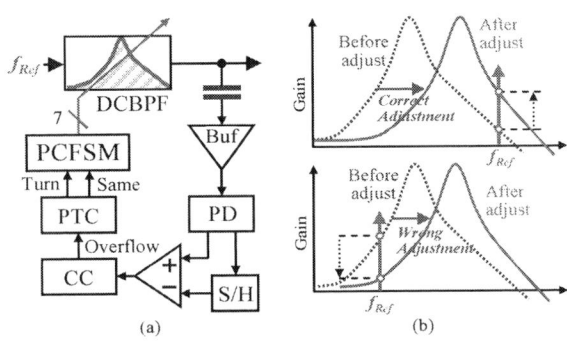

Fig.1 Passband lock loop (a) System architecture (b) Operations.

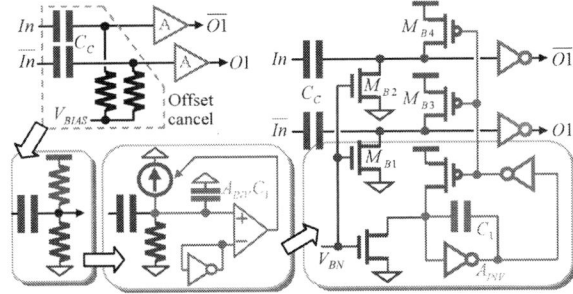

Fig.2 Schematic of input buffer before PD.

comparator. The compared results was accumulated by CC to filter the short-term noise in the transient response of BPF and PD. CC outputs reliable compared results, denoted as 'same' and turn', to show that the present peak voltage is higher and lower than the previous peak voltage, respectively. The following PTC maintains the same adjustment direction as same=1, or turn around as turn=1. PCFSM receives the outputs of PTC to add or reduce 7-btis passband control word by 1, and the passband of BPF is thus changed. Before PCFSM submits new control word, S/H circuit will store the present peak voltage for the next-time peak voltage comparison.

Fig.2 illustrates the schematic of input buffer before PD. To reduce the loading effects on BPF, input buffer utilizes simple inverters biased at threshold level to provide high gain, high swing. The gain of inverter is highly varied with input common mode level, so two AC-coupling capacitors (C_C) are added to block input common mode level. A common-mode feedback loop made of an inverter comparator (A_{INV}), an active miller-capacitor (C_1) and transistor-based voltage-dividers (M_{B1}

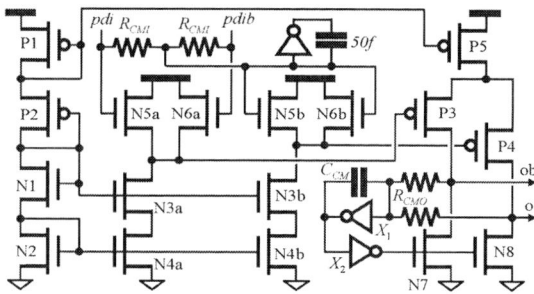

Fig.3 Schematic of PD.

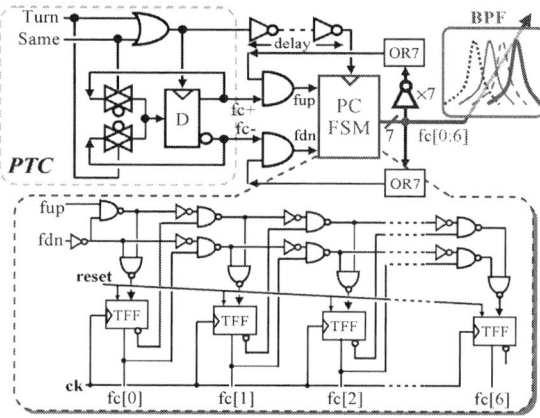

Fig.4 Schematic of PTC and PCFSM.

~M_{B4}) are inserted to set the common-mode level be equal the threshold level of the inverter.

Fig.3 is the schematic of proposed PD, in which the main detector (N3a~N6a) and the duplicated detector (N3b~N6b) generate the differential mode and the common mode levels, respectively. To isolate the loading effects of S/H, a buffer consisted of p-type differential-pair (P3~P5, N7~8) and output common-mode feedback loop (R_{CMO}, X_1, X_2, C_{CM}) are placed.

As the passband lock loop is locked to target frequency f_{ref}, the difference voltage between S/H and PD might be less than several mV. A 3-stage difference comparator with a total gain of 60dB enlarges the difference voltage to several hundred mV. The circuit noise of S/H is lowered by using differential controls, complementary signal switches and by increasing the sampling capacitance to several hundred fF. Both passive MOS capacitors and active Miller-capacitors are used to reduce the area of the sampling capacitors.

Fig.4 is the schematic of PTC and PCFSM. In PTC, the value stored in the internal DFF decides the direction of adjustment of passband control word fc[0:6]. As same=1, DFF keeps its value, and PCFSM would keep the adjustment of fc[0:6] (+1→+1, -1→-1). As turn=1, DFF changes its value, and PCFSM would inverse the adjustment of fc[0:6] (+1→-1, -1→+1). Fig.5 shows the test circuit layout realized in 0.18μm CMOS technology, in which the proposed passband lock loop is integrated with 1-stage BPF cell. With the same controls another 6-stage 7-bits 30MHz~50MHz programmable BPF [5] was with the same adjustment. The passband lock loop, replica BPF cell and the 6-stage BPF core occupied active area of 0.0046 mm², 0.027 mm² and 0.1134mm², respectively.

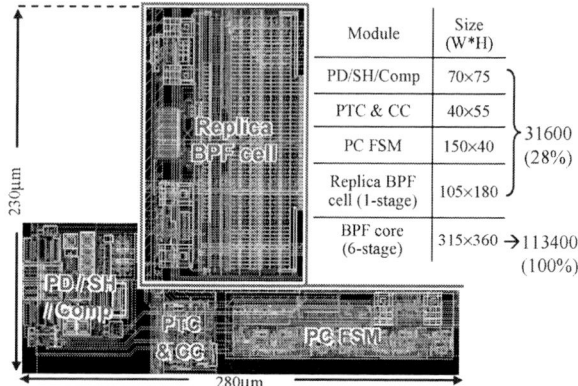

Module	Size (W*H)	
PD/SH/Comp	70×75	
PTC & CC	40×55	31600 (28%)
PC FSM	150×40	
Replica BPF cell (1-stage)	105×180	
BPF core (6-stage)	315×360 →113400 (100%)	

Fig.5 Circuit layout of the passband lock loop system.

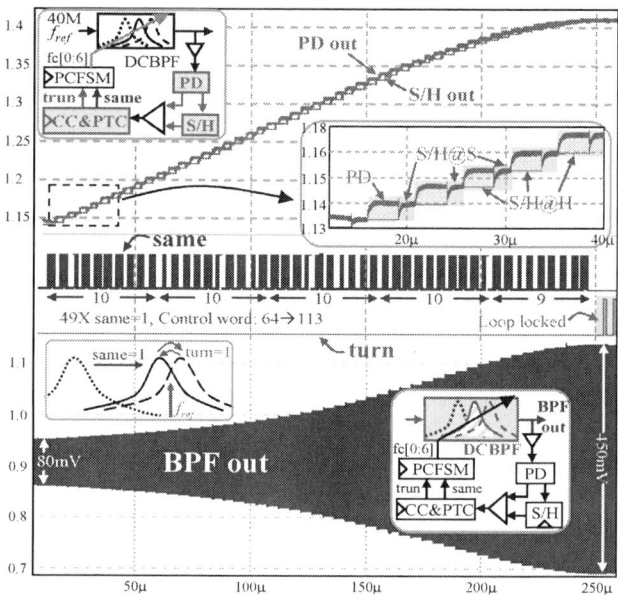

Fig.6 Simulation results of overall passband lock loop system.

Fig.6 is the simulation result of the passband lock loop. The target passband f_{ref} is 40MHz, the system clock is 625kHz and the overflow number of CC is 3. At initial the control word fc[0:6] of BPF is reset to 1000000 (64), and the output swing of BPF is 80mV. After 49 times of control word adjustments, equal to 147 clock cycles, the swing is tracking to a maximum of 460mV and then keep itself. Under 1.8V of supply voltage, the proposed loop system consumes a total current of 1.95mA.

REFERENCES

[1] H. Liu et al., "An accurate automatic tuning scheme for high-Q continuous-time bandpass filters based on amplitude comparison," *IEEE T-CASII*, Aug. 2003.

[2] O. Omeni et al., "A micropower CMOS continuous-time filter with on-chip automatic tuning," *IEEE T-CASI*, Apr. 2005.

[3] Y.-H. Kim et al., "Automatic tuning circuit for Gm-C filters," *IEEE ICECS*, Dec. 2005.

[4] H.-W. Lu et al.," All digital 625Mbps & 2.5Gbps deskew buffer design," *IEEE VLSI-DAT*, Apr. 2005.

[5] H.-W. Lin et al.," A Low-Area Digitalized Channel Selection Filter for DSRC System," *IEEE VLSI-DAT*, Apr. 2014.

A $11mV$ Single Stage Thermal Energy Harvesting Regulator with Effective Control Scheme for Extended Peak Load

Priya.V, Murali. K. Rajendran, Shourya Kansal and Ashudeb Dutta
Dept. of Electrical Engineering
Indian Institute of Technology, Hyderabad
ee14resch11012@iith.ac.in

Abstract—A thermo-electric energy harvesting based regulator system with output power maximization for high conversion ratio and very low input voltages, suitable for low power biomedical applications is presented. An optimal control topology for an inductor based regulator is implemented. Zero Current Switching is achieved by Pulse Width Modulation. Feedback mode control regulates output voltage to 1V with maximal load support. The system delivers peak power of 1.5mW and 19uW at 100mV and 11 mV input, respectively. The post-layout simulations done in UMC 180nm CMOS, show peak efficiency of 68% at 11mV input to boost converter.

Index Terms—thermal energy harvesting, single stage regulator, zero current switching.

I. INTRODUCTION

Advanced self-powered body sensor nodes come with in-built real time wireless data transmission and on-board data processing which are power hungry operations [1]. In such scenarios, there is a need to improve the peak load that could be supported by energy harvesting systems. This need can be met if maximum available power is extracted from the source when the load demands. In scenarios where the available power is adequate to supply a regulated load, single stage architecture can be used instead of general charger-regulator multi-stage topology. In this context, there is a need to maximize the output load that could be supported with single stage. Single stage converter system as proposed in [2], can support limited peak load since the system is not structured to extract maximum available power. To address these issues, a single stage, high conversion ratio inductor based DC-DC converter in boost topology with a mixed-signal controller is proposed, which works as regulator for an input range of 11mV-100mV. By controlling high side (HS) switch instead of low side (LS) switch for achieving Zero Current Switching (ZCS), LS switch is made available to acheive maximum power extraction from source when load demands, thus improving the maximum load that system could support. The paper is organized as : proposed architecture and implementation are discussed in Section II. Section III gives analysis and results.Section IV concludes the paper.

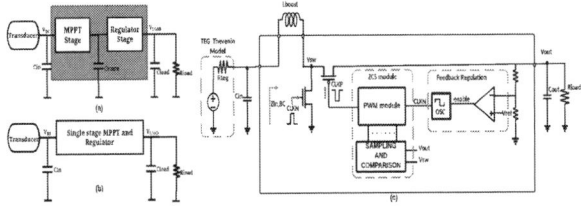

Fig. 1: (a) Conventional regulator system for energy harvesting (b) Proposed architecture (c) Overall architecture of the proposed single stage thermal energy harvesting system

II. PROPOSED ARCHITECTURE AND IMPLEMENTATION

The motive of the proposed regulator topology is to extend the supportable peak load by extracting maximum power from the source when load demands. The topology is demonstrated with the system comprising of an inductor based DC-DC converter circuit, digital control circuitry for ZCS and feedback control for output voltage regulation. The energy source considered is a thermoelectric generator (TEG). Over all architecture is given in Fig. 1(c). The fundamental switching frequency of the converter f_{sw} is fixed according to (1) [4] to ensure maximum power extraction at peak load.

$$f_{sw} = \frac{R_{TEG}}{8L} \qquad (1)$$

where, R_{TEG} is the internal impedance of transducer and L is the value of inductor. In this work, L=22uH and R_{TEG}=4 Ohms(single transducer) are chosen. Thus f_{sw}= 22 KHz. CLKN signal is given from fixed frequency(f_{sw}) oscillator masked by enable from output voltage feedback control. [2] uses the LS switch to acheive ZCS and fails to extract the maximum power available and end up in supporting narrow range of load. In this work, LS switch is dedicated for extraction of the maximum available power when load demands and we control the off time of HS switch to achieve ZCS. Also, since the conversion ratio targeted is very high, the HS switch on-time is very less compared to LS-switch on time, so it becomes easier on power and circuitry to realize required delay control for HS switch than that required for LS switch. The HS switch on time is decided by the digital ZCS controller. To determine whether the on time of HS switch has to be increased or decreased for the next cycle, switch node voltage, V_{sw} and V_{out} are compared just after the HS switch is off. The HS switch on-time is varied from 190 ns to 2 us,with 28ns per step precision, as calculated from the peak inductor current I_{Lmax} obtained for the fixed inductor charging time τ_N as given by (2).

$$I_{Lmax} = \frac{V_{in}}{L}\tau_N = \frac{V_{out} - V_{in}}{L}\tau_P \Rightarrow \tau_P = \frac{\tau_N}{\frac{V_{out}}{V_{in}} - 1} \quad (2)$$

where, τ_P is on time of HS switch and V_{in} is the input voltage. A discrete comparator compares the voltages at both sides of the HS switch just after the switch is off for every cycle. The pulse width is modulated using an up/down counter and delay select module with programmable capacitor bank as shown in Fig.2.

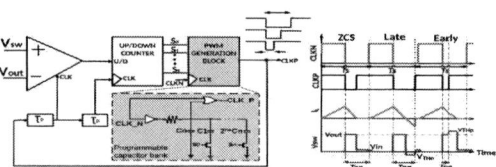

Fig. 2: Circuit implementation of PWM module for ZCS control and associated timing waveforms

III. ANALYSIS AND RESULTS

By controlling the input voltage as proposed, system is able to support a maximum load of 1.5mW at 1 V output voltage. In case of light loads,the switching frequency settles as per the load requirement, due to the feedback. And as the load increases, system pushes itself towards maximum power extrcation. The system performance is characterized for different loads at various input voltages Fig.3(a). The efficiency is dropping at higher input voltages Fig.3(a). This is due to the constant inductor charging time; which causes higher conduction loss at higher input voltage due to larger peak current. The power lost by conduction is plotted along with peak efficiency of HS switch controlled ZCS system in Fig.3(b). Even though the system exhibits a low efficiency curve, output power it can deliver is high. The simulation results obtained for full load to minimum-load dynamic variation is shown in Fig.4(a). The performance comparison of proposed architecture with existing literature is given in Table 1. As compared to other works, the system exhibits lesser efficiency at higher input voltages, owing to conduction loss, inspite of which it is able to deliver a higher output power. Layout of the full system in UMC 180nm is shown in Fig.4(b). The total area consumed in the chip is $0.047 mm^2$.

IV. CONCLUSION

A thermo electric energy harvesting system for input voltages from 11mV with peak load extension is designed in UMC 180nm CMOS technology. The system exploits the advantage of maximum power point tracking in a regulator system to extend the load range. Maximum output power of 1.5mW and 19uW is delivered at 100mV and 11mV input voltage

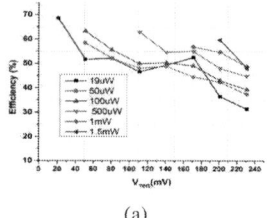
(a)

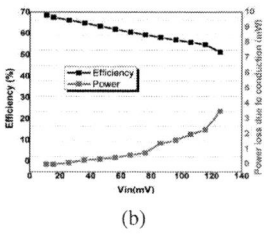
(b)

Fig. 3: Efficiency Analysis: (a) System efficiency versus open circuit input voltage V_{TEG} for different loads for V_{out}=1 V (b) Efficiency degradation due to conduction loss versus boost converter input voltage, V_{in}

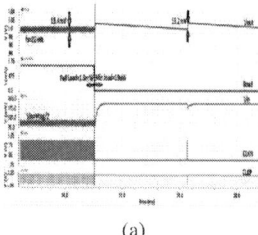
(a)

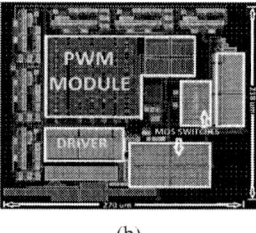
(b)

Fig. 4: (a) Simulation results of V_{out} regulation for dynamic load change along with corresponding timing waveforms of CLKN, CLKP, input voltage V_{in} and load current I_{load} (b) Chip layout in UMC 180nm CMOS

respectively, with output voltage regulated at 1V. System peak efficiency of 68% is obtained at 11mV input voltage. Post-layout simulation results validate the proposed scheme of controller for DC-DC converters in low power energy harvesting applications.

TABLE I: Performance Comparison

	This work	[2]	[3]	[4]
	Regulator	Regulator	Regulator	Charger+ Regulator
Vin in mV	11-100	20-250	50	25
Vout	1 V	1 V	1.2V	1.8V
η	68%@22mV	46%@20mV	45%@35mV	-
Peak load	1.5mW	175uW	282uW	300uW
@available Pin	@4mW	@3mW	@4088uW	
Stages	Single	Single	Three	Two
Process	180nm	130nm	65nm	350nm

REFERENCES

[1] C. C. Wu et al., "A pliable and batteryless real-time ECG monitoring system-in-a-patch," VLSI Design, Automation and Test (VLSI-DAT), 2015 Int. Symposium on, Hsinchu, 2015, pp. 1-4

[2] Carlson, E.J.; Strunz, K.; Otis, B.P., "A 20 mV Input Boost Converter With Efficient Digital Control for Thermoelectric Energy Harvesting," in Solid-State Circuits, IEEE Journal of , vol.45, no.4, pp.741-750, April 2010

[3] Weng, P.-S.; Tang, H.-Y.; Ku, P.-C.; Lu, L.-H., "50 mV-Input Batteryless Boost Converter for Thermal Energy Harvesting," in Solid-State Circuits, IEEE Journal of , vol.48, no.4, pp.1031-1041, April 2013

[4] Ramadass, Y.K.; Chandrakasan, A.P., "A Battery-Less Thermoelectric Energy Harvesting Interface Circuit With 35 mV Startup Voltage," in Solid-State Circuits, IEEE Journal of , vol.46, no.1, pp.333-341, Jan. 2011

Buffer With Neuron MOSFETs for Class-G Headphone Driver

Yuki Matsuda, Akio Shimizu, and Yohei Ishikawa
National Institute of Technology, Ariake College
Fukuoka, Japan
Email: e49231@ga.ariake-nct.ac.jp

Sumio Fukai
Saga University
Saga, Japan
Email: fukais@cc.saga-u.ac.jp

Abstract— **A buffer with Neuron MOSFETs is proposed to simplify a Class-G amplifier. For high efficiency and high fidelity, general Class-G amplifier switches high and low supply voltages depending on an input-signal. Therefore, general Class-G amplifier consists of Class-AB and Class-B buffers. This paper proposed that Class-G amplifier could be high efficiency and high fidelity without Class-B buffer by using Neuron MOSFETs. The performance of the proposed Class-G amplifier with Neuron MOSFETs is evaluated by Spectre using ROHM 0.18µm CMOS device parameters.**

Keywords; Class-G amplifier; Neuron MOSFET; buffer

I. INTRODUCTION

Recently, audio amplifiers are widely used in electronic devices having communication function such as mobile phones and computers [1], [2]. To increase battery lifetime, high efficiency amplifier is needed. Therefore, a Class-D amplifier has been developed [3]. However, Class-D amplifier provides a level of switching noise that interferes with RF functions such as electric devices by Electro-Magnetic Interference (EMI). On the other hand, Class-AB amplifier does not occur any switching noise. However, power efficiency of Class-AB amplifier is lower. A Class-G amplifier has been developed to realize high efficiency and high fidelity sound without EMI problems. Therefore, Class-G amplifier can have advantage of both Class-AB and Class-D amplifiers [4]. General Class-G amplifier switches high and low supply voltages depending on input-signal [5].

In this paper, the Class-G amplifier with the Neuron MOSFETs is evaluated by Spectre using ROHM 0.18µm CMOS device parameters. Proposed Class-G amplifier could be high efficiency and high fidelity without Class-B buffer by using Neuron MOSFETs. The Neuron MOSFETs can adjust gate potential when supply voltage switches from high supply voltage to low supply voltage [6]. This paper is provided to simplify circuit configuration of amplifier.

II. PROPOSED CLASS-G AMPLIFIRE

A. Architecture of Class-G Amplifiers

Fig. 1 shows the proposed Class-G amplifier architecture.

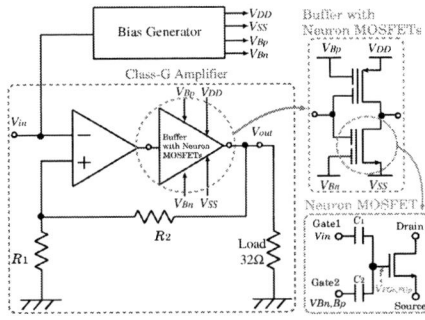

Fig. 1. Schematic of Class-G amplifier

Proposed Class-G amplifier consists of the buffer with the Neuron MOSFETs and an op-amp. The buffer with the Neuron MOSFETs needs both high and low supply voltages to adjust gate potential in each of Neuron MOSFET. Depending on input-signal, bias generator provides optimum supply voltage to buffer with Neuron MOSFETs.

B. Buffer With Neuron MOSFETs

The buffer consists of N and P type Neuron MOSFETs with two gate terminals. The Neuron MOSFETs adjust gate potential by applying voltage from DC-voltages ($V_{Bn,Bp}$). Schematic model of the Neuron MOSFET is shown in Fig.1.

Floating-gate potentials, $V_{FGn,FGp}$, are given by

$$V_{FGn,FGp} = \frac{C_1 V_{in} + C_2 V_{Bn,Bp}}{C_1 + C_2} \qquad (1)$$

where V_{in} is an input-voltage of the Neuron MOSFET, C_1 and C_2 are input-capacitance, and $V_{Bn,Bp}$ are DC-voltages. Gate potentials, $V_{FGn,FGp}$, are controlled by $V_{Bn,Bp}$. Drain current is determined by $V_{FGn,FGp}$. Therefore, the buffer with the Neuron MOSFETs can operate as both Class-AB and Class-B buffers. Floating-gate of the Neuron MOSFET is charged with an electric charge while the Neuron MOSFETs is produced. Therefore, initial charge method of elimination is devised [7].

C. Bias Generator

To switch high and low supply voltages, buffer with Neuron MOSFETs uses bias generator.

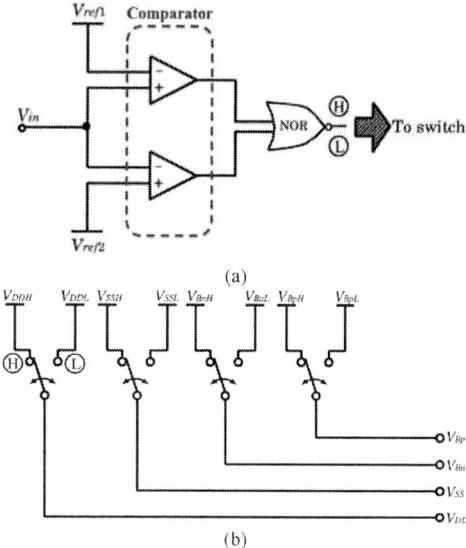

(a)

(b)

Fig. 2. Bias Generator (a) Input-signal monitoring circuit (b) Bias switching circuit

Schematic of bias generator is shown in Fig.2. Bias generator monitors V_{in} in the monitoring circuit. Fig. 2 (a) shows monitoring circuit consisting of NOR gate and comparator. When V_{in} exceeds $V_{ref1,ref2}$, supply voltages switch from low supply voltage to high supply voltage in the switching circuit. Therefore, bias generator provides both high and low supply voltages depending on the input-signal.

III. SIMULATION RESULTS

When the audio signal frequency reaches 1kHz, the transient responses of some supply voltages and the output of amplifier are shown in Fig .3. When the input-signal amplitude exceeds $V_{ref1,ref2}$ (±450mV), V_{DD} switches from 0.9V to 1.77V and V_{SS} switches from -0.8V to -1.79V. The output-signal of amplifier is amplified about 1.5 times as much as the input-voltage amplitude. In addition, proposed Class-G amplifier can drive load resistance of 32Ω. Therefore, proposed Class-G amplifier can amplify without distorting when supply voltages switched by bias generator. Also, a quiescent power is 57.3mW at a supply voltage of 3.6V.

The THD performance of the overall Class-G amplifier is shown in Fig. 4. When the audio signal frequency reaches 20kHz, the THD is in the range of 0.19% to 0.71%. When the audio signal frequency reaches 1kHz, the THD is in the range of 0.10% to 0.6%. This has confirmed that the proposed Class-G amplifier is adequate for the typical audio applications.

To confirm performance of buffer with Neuron MOSFET, comparator and part of amplifier use basic operating amplifier consisting of common source circuit and differential amplifier.

IV. CONCLUSION

The Class-G buffer with the Neuron MOSFETs is presented to be high efficiency and high fidelity without Class-B buffer.

The Class-G buffer is incorporated in Class-G amplifier. The proposed Class-G amplifier was simulated using ROHM 0.18μm CMOS device parameters.

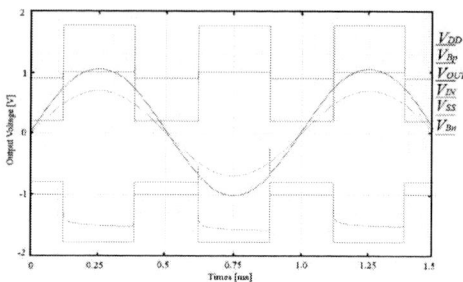

Fig. 3. Transient analysis of the amplifier with the Neuron MOSFETs (frequency is 1kHz)

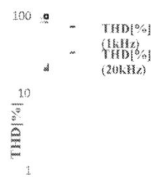

Fig. 4 THD performance of the Class G amplifier

Proposed Class-G amplifier uses only one buffer for both high and low supply voltages by using Neuron MOSFETs that can operate as both Class-AB and Class-B buffers. This paper proposes that Class-G buffer can simplify circuit of amplifier.

ACKNOWLEDGMENT

This work is supported by VLSI Design and Education Center (VDEC), the University of Tokyo in collaboration with Synopsys and Cadence Design Systems, Inc..

REFERENCES

[1] Jianlong Chen, Sasi Kumar Arunachalam, Todd L. Brooks, Iuri Mehr, Felix Cheung, and Hariprasath Venkatram, "A 62mW Stereo Class-G Headphone Driver with 108dB Dynamic Range and 600μA/Channel Quiescent Current," ISSCC Dig. Tech. Papers, pp. 182-183, 2013.

[2] Sherif Galal, Hui Zheng, Khaled Abdelfattah, Vinay Chandrasekhar, Iuri Mehr, Alex Jianzhong Chen, John Platenak, Nir Matalon, and Todd L.Brooks, "A 60 mW Class-G Stereo Headphone Driver for Portable Battery-Powered Devices," IEEE J. Solid-State Circuits, vol. 47, No. 8, 2012.

[3] Alex Lollio, Giacomino Bollati, and Rinaldo Castello, "A Class-G Headphone Amplifier in 65 nm CMOS Technology," IEEE J. Solid-State Circuits, vol. 45, No. 12, 2010.

[4] Alex Lollio, Giacomino Bollati, Rinaldo Castello, "Class-G Headphone Driver in 65nm CMOS Technology," ISSCC Dig. Tech. Papers, pp. 84-85, 2010.

[5] Huiyuan Zhang, M. T. Tan, and P. K. Chan, "A Single-Supply Inductorless CMOS Class G Speaker Amplifier," IEEE J. Solid-State Circuits, pp. 1-4, 2011.

[6] Tadashi Shibata and Tadahiro Ohmi, "Neuron MOS Binary-Logic Integrated Circuits- Part I: Design Fundamentals and Soft-Hardware-Logic Circuit Implementation," IEEE J. Solid-State Circuits, vol. 40, No. 3, 1993.

[7] E.Rodrigues-Villegas and H.Brnes, "On Dealing With the Charge Trapping in Floating-Gates MOS (FGMOS) Transistors," IEEE Trans. Circuits and Syst. II, Exp. Briefs, vol. 54, No. 2 ,2007.

978-1-5090-3220-4/16 $31.00 © 2016 IEEE

Energy-efficient spread second capacitor capacitive-DAC for SAR ADCs

Sung-min Lee, Ju Eon Kim, Dong-Hyun Yoon and Kwang-Hyun Baek
Department of Electrical and Electronics Engineering
Chung-Ang University
Seoul, Korea
kbaek@cau.ac.kr

Abstract— **An energy-efficient capacitive-digital-to-analog converter (C-DAC) switching with low common-mode voltage variation switching method is proposed for successive approximation register analog-to-digital converters (SAR ADCs). In the proposed spread second capacitor capacitive-DAC (SSC C-DAC), a role of capacitor for second conversion step is spread to all capacitors except the most significant bit (MSB) capacitor. The proposed SSC C-DAC achieves 98.1% more efficient switching energy and can be comprised of the number of quarter unit capacitor compared with conventional scheme. Furthermore, in order to achieve low variation of common-mode voltage, the proposed switching method adopts only one side of tri-state switching method in differential-type SAR ADC after second conversion.**

Keywords- Energy efficient DAC, Low power ADC, SAR ADC, Common-mode variation

I. INTRODUCTION

In these days, SAR ADCs have been preferred for low-power applications and have advantage in area. Due to the progress of process and no preamplifier in general SAR ADC, the C-DAC is considered as the most power-hungry block in SAR ADC. Thus, the researches for reducing the switching energy of the C-DAC have been issued recently and that efforts have led to the development of various switching schemes [1-3]. In the proposed SSC C-DAC scheme, the sampling state and first conversion follow the switching scheme of previous scheme for reducing switching energy of C-DAC [2].

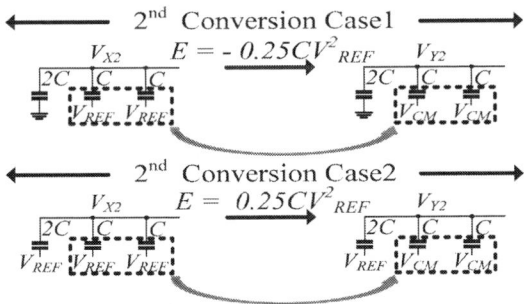

Fig. 1 Examples of second conversion in SSC-CDAC

Fig. 1 shows the examples of second conversion in the proposed SSC- CDAC. As shown in fig. 1, role of capacitors for second conversion step is spread to all capacitors except the

MSB capacitor. Due to that switching procedure, the proposed switching scheme does not need to connect MSB capacitor to V_{CM} unlike already proposed switching scheme [2]. Therefore, the burden of buffer for driving MSB capacitor can be reduced. And also, this switching makes reduction of common-mode voltage variation more than switching scheme [2].

II. PROPOSED SSC C-DAC SCHEME

Fig. 2 shows examples of switching procedure of 4-bit SSC C-DAC.

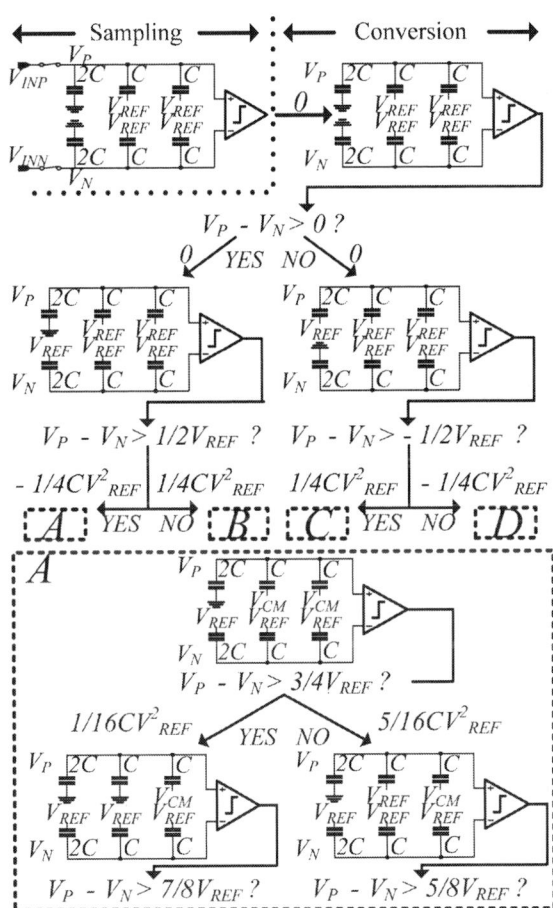

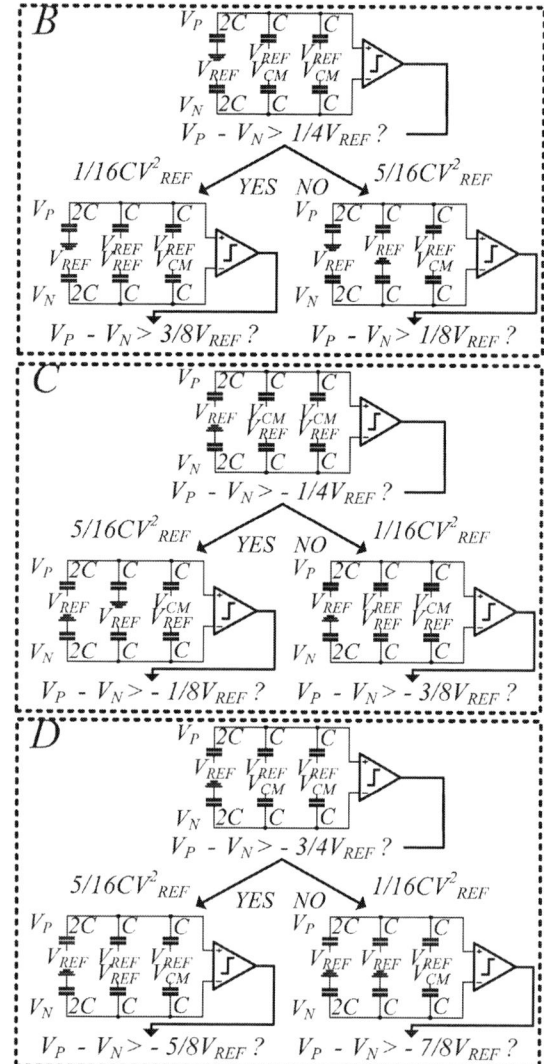

Fig. 2 Examples of switching procedure of 4-bit SSC C-DAC

Fig. 3 shows the waveform of C-DAC of [2] and proposed switching procedure.

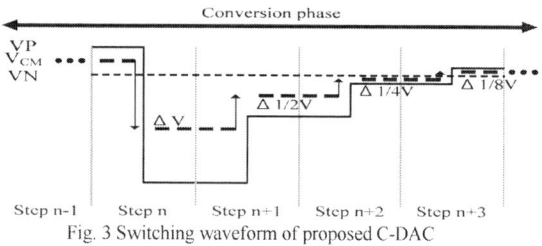

Fig. 3 Switching waveform of proposed C-DAC

Because one-side of tri-state switching is followed after second conversion, variation of common-mode voltage can be mitigated. The switching energy of proposed SSC C-DAC is verified in MATLAB. Fig. 4 shows the comparison of switching energy of the proposed SSC C-DAC with previously developed C-DACs.

III. SIMULATED RESULTS AND COMPARISON

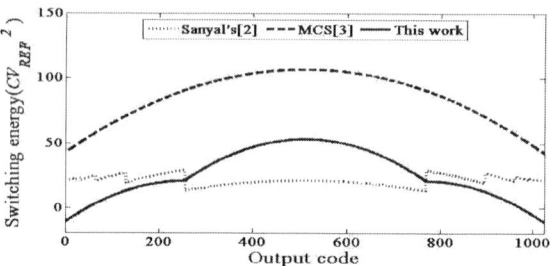

Fig. 4 Switching energy versus output code

Table I shows the comparison table of switching scheme for 10-bit SAR ADC. The number of total unit capacitor is also reduced by 75% compared with conventional C-DAC.

Table I. Comparison of switching scheme for 10-bit SAR ADC

Switching scheme	Average Energy (CV_{REF}^2)	Energy saving	Total Unit Capacitor
Conventional	1363.3	Reference	2048
MCS[3]	82.1	93.7%	1024
Sanyal[2]	21.3	98.4%	512
This work	26.4	98.1%	512

IV. CONCLUSION

Energy-efficient SSC C-DAC is proposed. In the proposed C-DAC, variation of common-mode voltage can be more mitigated. The switching energy and the number of total unit capacitor are reduced by 98.1% and 75%, respectively, compared with conventional C-DAC.

ACKNOWLEDGMENT

This work was supported by Institute for Information & communications Technology Promotion(IITP) grant funded by the Korea government(MSIP) (No.B0101-16-1348,PON Stick Development for 1Gbps Symmetric Internet Access). And also was supported by the Korea Institute for Advancement of Technology(KIAT) grant funded by the Korean government (Motie:Ministry of Trade, Industry & Energy, HRD Program for Software-SoC convergence)(No.N0001883). And also This research was supported by Basic Science Research Program through the National Research Foundation of Korea(NRF) funded by the Ministry of Education (No. 2015R1D1A1A 01060031)

REFERENCES

[1] B. P. Ginsburg and A. P. Chandrakasan, "An energy-efficient charge recycling approach for a SAR converter with capacitive DAC", IEEE ISCAS, May. 2005, pp. 184 – 187

[2] A. Sanyal, N. Sun, "SAR ADC architecture with 98% reduction in switching energy over conventional scheme", Electronics Letters, Vol.49 Issue4 P248-250, Feb.14.2013

[3] V. Hariprasath, J. Guerber, "Merged capacitor switching based SAR ADC with highest switching energy-efficiency", Electronics Letters, Vol.46 Issue 9 pp. 620-621, Apr.29.2010.

A design of new voltage to current converter with high linearity and wide tuning

Yui-Hwan Sa, Pyo-Hoon Son, Ki-Hong Kim, Hi-Seok Kim, and Hyeong-Woo Cha,

Department of Electronics Eng., Cheongju University
298 Deaseong-ro Sangdang-gu, Chongju-city, Chungbuk, 360-764, Republic of Korea
jmh35795@naver.com, khs8391@cju.ac.kr, hwcha@cju.ac.kr

Abstract— A novel voltage-to-current (V-I) converter with high linearity and wide tuning range was designed. The V-I converter consist of differential amplifier, voltage attenuator, source follower, and adaptive current mirror. The simulation result shows that the V-I converter has tuning range from 0.2V to supply voltage of 3.0V. The nonlinearity error of the V-I converter was less than 0.5% for the same range. Power dissipation of the CMOS and BJT V-I converter has 1mW.

Keywords- Analog; Integrated; Circuits; Highly linearity; Voltage-to-curtent(V-I) converter.

I. INTRODUCTION

Voltage-to-current(V-I) converters are basic building blocks in many analogue and mixed signal designs, such as multipliers, continuous-time filters, data converters, high-performance sensor interfaces, or variable gain amplifiers [1]-[2]. In these applications, the overall system performance depends largely on the performance of the V-I subcircuits. This leads to the need for a time-, temperature- and supply voltage-independent transconductance with a highly linear range and an appropriate bandwidth.

Recently, V-I converter with high linearity and wide control range until supply voltage was reported [3]-[4]. The converter has good characteristics for applying VFC and VCO at single supply voltage. However, the converter has complex circuit configuration because two rail-to-rail operation transconduct-ance amplifier(OTA) and voltage attenuator using two resistors. Therefore, the converter has also large power dissipation.

In this paper, we design a new and simple V-I converter without OTAs(or OP-AMPs). The proposed V-I converters realized circuit configuration of CMOS transistor and BJT, respectively.

II. CIRCUIT DESCRIPTION

Proposed CMOS linear V-I converter with wide tuning to supply voltage shown in Figure 1. The circuit consists of differential amplifier($M_1 \sim M_4$), voltage attenuator(R_1, R_2), source follower(M_5), adaptive current mirror($M_6 \sim M_8$), and MOS capacitor(M_{10}). MOS capacitor M_{CAP} stabilizes basis voltage V_{BIAS}.

Because the differential amplifier operates as a voltage follower due to connecting input with output, $V_A \approx V_{CON}$ and then the gate voltage of M_5 was as follow,

$$V_{G5} = \alpha V_{CON}, \quad \alpha = R_2/(R_1 + R_2) \quad (1)$$

The voltage V_{SG5} for variation of M_5 is almost same as V_{GS6} because of adaptive Wilson current mirror composed with $M_6 \sim M_8$ [5]. Therefore, $V_B = V_{G5}$ and then output current I_O is given by

$$I_O = \frac{V_B}{R} = \alpha \frac{V_{CON}}{R} = \frac{R_2}{R_1 + R_2} \frac{V_{CON}}{R} \quad (2)$$

If we use current mirror configured with M_8 and other PMOS transistor of gain K, a highly linear V-I converter with wide tuning from 0V to supply voltage can be realized. To stabilize current output of the V-I converter, the MOS capacitor M_{CAP} was used.

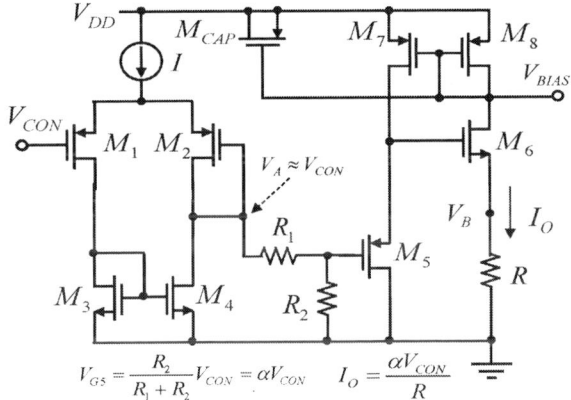

Figure 1. Circuit diagram of a new CMOS linear V-I converter with power supply range

Proposed BJT linear V-I converter with wide tuning to supply voltage shown in Figure 2. The circuit consists of differential amplifier ($Q_1 \sim Q_4$), voltage attenuator (R_1, R_2), emitter follower (Q_5), adaptive current mirror(Q_6~Q_8), and a capacitor(C). The capacitor C stabilizes basis voltage V_{BIAS}.

Because this converter has same configuration like CMOS V-I converter, the operation principle and output current are same.

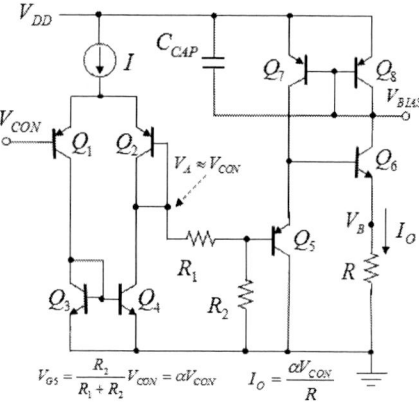

Figure 2. Circuit diagram of a new BJT linear V-I converter with power supply range.

III. SIMULATIONS AND MEASUREMENTS RESULTS

The proposed V-I converters shown in Figure 1 and 2 simulated with Cadence PSpice Tool. The model parameters was the full parameters of $0.35\ \mu m$ CMOS process by TSMC and Q2n2222(npn) and Q2n2907(pnp). The value of the device are set by $V_{DD} = 3.0V$, $R_1 = 100k\Omega$, $R_2 = 50k\Omega$, and $M_{CAP}(W/L) = (200/200)\mu m$, and $C_{CAP} = 1pF$. The all bias current I was set by $I = 100\mu A$ in the V-I converters.

Figure 3 shows linearity and control range characteristics of the V-I converters shown in Figure 1 and 2 for resistor of R=20kΩ. The solid line was theoretical result for equation (2). The dot and dash-dot line were simulation result of CMOS and BJT V-I converter, respectively.

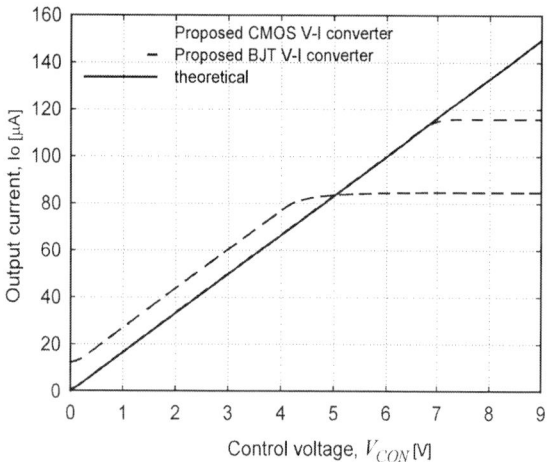

Figure 3. Linearity and control range characteristics of the V-I converters.

The results shows that the CMOS V-I converter has control range from 0.2V to $3V_{DD}$-($V_{M5(sat)}$+$V_{M7(sat)}$). We think CMOS V-I converter has offset current because of finite gain of differential amplifier and mismatching between M5 and M6. However, the proposed BJT V-I converter has control range from 0V to $3V_{DD}$-($V_{M5(sat)}$+$V_{M7(sat)}$) and identify theoretical equation. The linearity error was 0.5% control voltage range from 0.2V to 2.8V at supply voltage V_{DD}=3.0V.

Table 1 show the performance results of the V-I converters shown in Figure 1 and 2, respectively.

TABLE 1.
PERFORMANCE RESULTS OF THE V-I CONVERTERS

Contents	Proposed		Conventional
	CMOS	BJT	Paper[3]
Supply Voltage[V]	3.0	3.0	3.0
Linear range[V]	0.2~3.0	0.2~0.3.0	0.2~2.9V
Power dissioation[mV]	1mW	1mW	3mW
Linearity error[%]	0.5	0.5	0.5
Number of transistor	8	7	26

IV. CONCLUSION AND FUTURE SUBJECT

Novel CMOS and BJT V-I converters with high linearity and wide tuning range was designed and it applied CMOS VCO. The V-I converters have a high linearity and wide tuning range from 0.2V to power supply. In the future, we will optimize the CMOS V-I converter using the fabrication model parameters of $0.35\ \mu m$ TSMC CMOS process and fabricate chip.

ACKNOWLEDGMENT

This work was supported by the Industrial Core Technology Development Program (10049192, Development of a smart automotive ADAS SW-SoC for a self-driving car) funded By the Ministry of Trade, industry & Energy.

REFERENCES

[1] A. Demosthenous and M. Panovic, "Low-voltage MOS linear transconductor/squarer and four-quadrant multiplier for analog VLSI," *IEEE Trans. Circuits Syst. I, Reg. Papers*, vol. 52, no. 9, pp.1721–1731, Sep. 2005.

[2] J. M. Algueta-Miguel, A. J. Lopez-Martin, L. Acosta, J. Ramírez-Angulo,and R.Gonzalez-Carvajal, "Using floating gate and quasi-floating gate techniques for rail-to-rail tunable CMOS transconductor design," *IEEE Trans. Circuits Syst. I, Reg. Papers*, vol. 58, no. 7, pp. 1604–1614, Jul. 2011.

[3] C. Azcona, B. Calvo, S. Celma and N. Medrano.: 'Highly-linear rail-to-rail 1.2 V–0.18 mm CMOS V-I converter', *ELECTRONICS Letters*, vol. 47 no. 18. Sep. 2011.

[4] C. Azcona, B. Calvo, S. Celma, N. Medrano, and Pedro A. Martinez, " Low-Voltage Low-Power CMOS Rail-to-Rail Voltage-to-Current Converters," *IEEE Trans. Circuits Syst. I, Reg. Papers*, vol. 60, no. 9, pp. 2333–2342, Sep. 2013.

[5] H.-W. Cha and K. Watanabe, "Wideband CMOS Current Conveyor, "*IEE Electronic Letters*, vol. 32, no. 14, pp. 1245-1246, 1996.

A Fully Integrated High-efficiency Step-up DC-DC Converter for Energy Harvesting Applications

Seyed Mohammad Noghabaei, and Mohamad Sawan

Polystim Neurotech Laboratory, Department of Electrical Engineering
Polytechnique Montreal
Montreal, Canada
mohammad.noghabaei@polymtl.ca

Abstract— In this paper, a novel low-voltage and fully integrated Step-up DC-DC converter for energy harvesting applications, designed in 130 nm CMOS technology is presented. Simulation results proved that the proposed step-up converter consisting of a differential cross-coupled architecture along with a latched charge pump enable us to boost up very low input voltages beyond 50 mV. The input voltage of 80 mV is converted to 1 V at a load resistance of 1 MΩ with a conversion efficiency of about 24%. The proposed converter consumes only 4.2-µW.

Keywords- DC-DC converter; energy harvesting; differential cross-coupled oscillator; charge pump

I. INTRODUCTION

Recently, energy harvesting has been identified as a practical option for self-power microsystems such as wireless sensors, medical devices, and remote weather stations [1]. In energy harvesting, a DC/DC converter is utilized to boost very low output voltage of the harvesters. There are different topologies and techniques for DC/DC converters. An LC oscillator with minimum startup voltage of 73 mV is utilized [2]; nevertheless, due to high power consumption of LC oscillator a very low efficiency (around 1 %) is obtained. A step-up system includes an enhanced swing ring oscillator and a Dickson charge pump with Schottky diodes to convert the DC voltages below 100 mV (with minimum start-up voltage of 10 mV) to 1 V [3]. However, as a partially integrated system, the circuit has significant power consumption, loss, and extremely low efficiency.

In the current study, we present a fully integrated step-up converter with a novel structure consisting of a differential cross-coupled oscillator, signal conditioning building block and a modified latched charge pump (LCP) in order to convert 80 mV DC to 1 V for energy harvesting applications.

II. PROPOSED STEP-UP CONVERTER STRUCTURE

A. Oscillator Architecture

The proposed step-up converter is designed around a low-voltage oscillator with minimum power consumption and optimum swing at its output with sinusoidal waveforms (Figure 1a). As a significant advantage, the proposed oscillator is fully on-chip and self-startup topology (ambient noise can start the oscillation). Besides, the output swing of this oscillator is two times greater than other structures. Thus, by increasing the amplitude of oscillator clocks ($\varphi 1$ and $\varphi 2$, Figure 1a), the generated clocks are able to drive MOS switches in latched charge pump (LCP) in the subthreshold region.

This work is supported by NSERC, Canada Research Chair on Smart Medical Devices and CMC Microsystems.

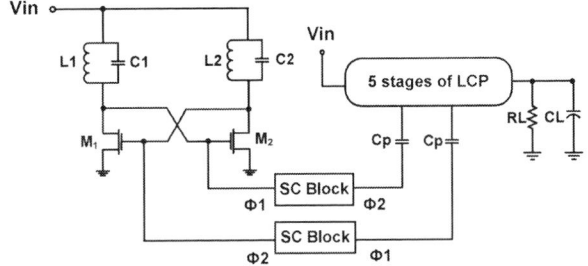

(a)

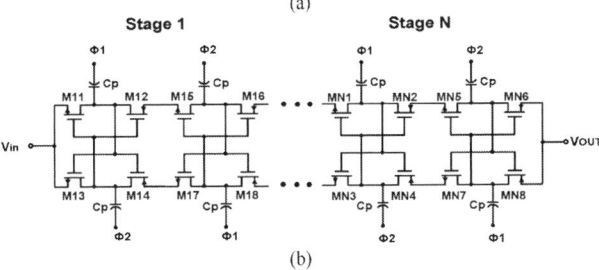

(b)

Figure 1. (a) The proposed step-up converter, and (b) The proposed latched charge pump (LCP)

Two zero threshold voltage transistors (ZVT) are used in the proposed oscillator (Figure 1a) to occur the oscillation with low-input voltage (below the threshold voltage). Besides, two passive LC resonant circuits with on-chip components are designed to generate the desired oscillation frequency.

The Signal Conditioning block (SC block, Figure 1b) is designed to isolate the oscillator from the load variations of the charge pump circuit. Since increasing the number of stages in the charge pump, change the negative resistance of the proposed oscillator as well as cancelling the oscillation. Hence, in order to have stability in oscillation, instead of increasing the width (W) of M1 and M2, the SC block is designed to eliminate the loading effect and damping effect of Rp (a parallel resistor with an inductor in practice). Thus, the power consumption decreases and efficiency significantly increases.

B. Charge Pump

Oscillator generates clocks, and these clocks start the boosting process. Five stages LCPs (Figure 1b) is employed instead of conventional charge pumps (CPs). The advantages of the LCPs to the conventional CPs (diode connected) is eliminating the threshold drop voltage and unlike the bootstrap CPs two clocks instead of four clocks are needed [4].

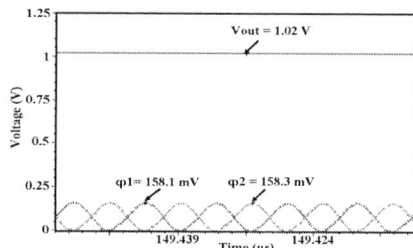

Figure 2. Simulated results of the proposed step-up, Vout versus output clocks of the oscillator (φ1 and φ2) at RL= 1 MΩ and CL= 10 pF

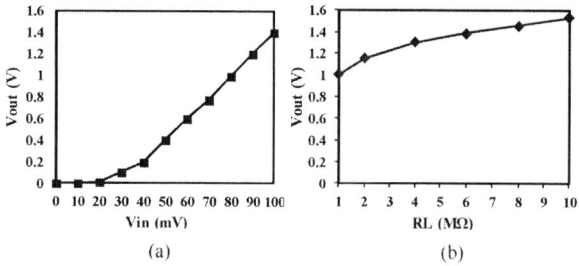

(a)   (b)

Figure 3. (a) Vout vs Vin at load resistor RL= 1 MΩ, and (b) Vout vs load resistor variation at Vin = 80 mV

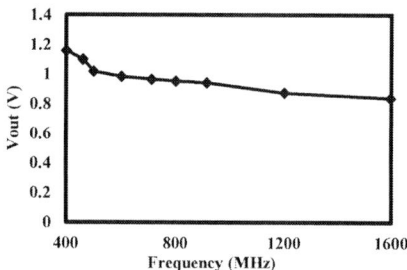

Figure 4. Vout vs oscillator frequency variation

The bulk of NMOS and PMOS switches are connected to the source in order to decrease the threshold voltage of the switches and modify to work with very low voltage. For this, the ZVT and LVT (low threshold voltage transistor) are employed for the NMOS and PMOS switches, respectively (Figure 1b). Therefore, less power loss across the transistors are expected.

III. SIMULATION RESULTS

The simulation results of the proposed DC-DC (Figure 2) show two differential outputs of the oscillator (φ1 and φ2, Figure 1a) with the amplitude of around 158 mV and the output voltage of 1 V at the output of the converter with Vin = 80 mV.

To find the optimum performance point of the proposed step-up, we carried out sensitivity analyses (Figures 3 and 4). The parameter analysis on Vin reveals that the system works when the DC input voltage is greater than 50 mV, and the performance of the system increases as the input voltage increases (Figure 3a). The output voltage versus the load (variation between 1MΩ and 10 MΩ) is presented, Figure 3b. By increasing the load resistor, the Vout is increased from 1 to 1.5 V for loads of 1 MΩ and 10 MΩ respectively. The frequency variation is performed between 400 MHz to 1.6 GHz (Figure 4) to find the optimum oscillation frequency for 1 V at the output of the proposed step-up converter. Therefore, at the resonance frequency of 500 MHz,

the output voltage of 1 V is obtained (Figure 4) with optimum on-chip values L1 = 100 nH and C1 = 1 pF. The oscillation frequency of the proposed oscillator (Figure 1a) is:

$$f = \frac{1}{2\pi\sqrt{L1.C1}}$$

Besides, the computed power consumption and power conversion efficiency at the resistor load of 1 MΩ are 4.2 μW and 24 %, respectively.

Table I summarizes the performance and comparison of the proposed step-up converter with previously published works.

TABLE I.   PERFORMANCE COMPARISON OF THE PROPOSED STEP-UP

Metric	[2]	[5]	This work
Arch. & Tech	LC oscillator + SI converter & (130 nm)	LC-Transformer oscillator + SC converter & (65 nm)	Fully differential cross-coupled oscillator + LCP & (130 nm)
On-chip/Off-chip	Fully Integrated	Fully Integrated	Fully Integrated
Start-up voltage	73 mV	85 mV	80 mV
Vout (V) @Load	1 @ NA	1.2 @ NA	1 @ 1 MΩ
Efficiency @Load	1% @ NA	1.5%@ NA	24 % @ 1 MΩ

IV. CONCLUSION

In this paper, a new step-up DC-DC converter is presented. A low voltage fully integrated topology is adopted to boost the low DC voltage from 80 mV to 1 V at a resistor load of 1 MΩ. The proposed step-up operates in beyond the voltage of 50 mV. However, due to constraints such as fully on-chip design and power consumption, the system is optimized for input voltage 80 mV with optimum sizes of the transistors aspect ratio, inductor, and capacitors.

ACKNOWLEDGMENT

Authors acknowledge support from NSERC, Canada Research Chair on Smart Medical Devices and CMC Microsystems.

REFERENCES

[1] Visser, Hubregt J., and Ruud JM Vullers. "RF energy harvesting and transport for wireless sensor network applications: Principles and requirements."Proceedings of the IEEE 101.6 (2013): 1410-1423.

[2] M.B. Machado, M.C. Schneider, M. Sawan, and C.G. Montoro, "Fully-Integrated 86 mV–1V Step-up Converter for Energy Harvesting Applications," in Proc. NEWCAS, 2014, pp. 452–455.

[3] M. Bender Machado, M. Sawan, M. Cherem Schneider and C. Galup-Montoro, "10 mV – 1V step-up converter for energy harvesting applications," Proc. Symp. on Integrated Circuits and Systems Design (SBCCI), pp.1-5, Sept. 2014.

[4] Palumbo, Gaetano, & Pappalardo, Domenico. (2010). Charge pump circuits: An overview on design strategies and topologies. IEEE Circuits and Systems Magazine, First Quarter, 2010, 31–45.

[5] H. Fuketa, Y. Momiyama, A. Okamoto, T. Sakata, M. Takamiya and T. Sakurai, "An 85-mV input, 50-μs startup fully integrated voltage multiplier with passive clock boost using on-chip transformers for energy harvesting," Proc. European Solid State Circuits Conf. (ESSCIRC), pp.263-266, Sep. 2014.

An Integrated Optical Parallel Adder as a First Step Towards Light Speed Data Processing

Tohru Ishihara†, Akihiko Shinya‡, Koji Inoue§, Kengo Nozaki‡ and Masaya Notomi‡

†Graduate School of Informatics, Kyoto University, Yoshida-honmachi, Sakyo-ku, Kyoto 606-8501, JAPAN

‡NTT Nanophotonics Center / NTT Basic Research Laboratories, 3-1 Morinosato Wakamiya, Atsugi 243-0198, JAPAN

§Graduate School of Inf. Sci. and Elec. Eng., Kyushu University, 744 Motooka, Nishi-ku, Fukuoka 819-0395, JAPAN

Abstract—**Integrated optical circuits with nanophotonic devices have attracted significant attention due to its low power dissipation and light-speed operation. With light interference and resonance phenomena, the nanophotonic device works as a voltage-controlled optical pass-gate like a pass-transistor. This paper first introduces a concept of the optical pass-gate logic, and then proposes a parallel adder circuit based on the optical pass-gate logic. Experimental results obtained with an optoelectronic circuit simulator show advantages of our optical parallel adder circuit over a traditional CMOS-based parallel adder circuit.**

I. Introduction

Today's highly information-oriented society, with its easy access to multimedia and the Internet, would not be realizable without CMOS LSI technologies. With the CMOS downscaling over the past five decades, gate delays in the LSI circuits have drastically decreased. Historically, the delays of local level wires also decreased with transistor downscaling. At ultra-scaled dimensions, however, the effective resistivities, hence delays, of local level wires increase rapidly due to size effects [1]. Post-layout analysis using predictive technology models [2] shows that interconnect performance degradation may dominate over the device speed improvement in a 22 nm technology node and below [1]. This means that technology scaling itself cannot resolve the latency issue of CMOS LSI circuits in advanced technology nodes such as 7 nm and below.

Optical communication technologies, in contrast, have been rapidly growing over the past several decades. With much of the backbone of the Internet deployed using optical fiber, it is no surprise that fiber optics are the fastest form for long-distance broadband communication. Although optical technologies dominate the long-distance communications, electronics still remain as major players that transfer and process information at the local level. Recent advances in nanophotonics, however, make it possible to migrate power-efficient light-based communication into ever-shorter distances and move onto silicon chips as on-chip optical interconnects.

In this paper, we introduce a concept of optical pass-gate (OPG in short hereafter) logic which enables the optical interconnects to employ a logic functionality. Based on the OPG logic, this paper proposes an integrated optical parallel adder as a first step towards light speed data processing. The rest of the paper is organized as follows. Section II introduces a concept of OPG logic. A circuit architecture of integrated optical parallel adder is proposed in Section

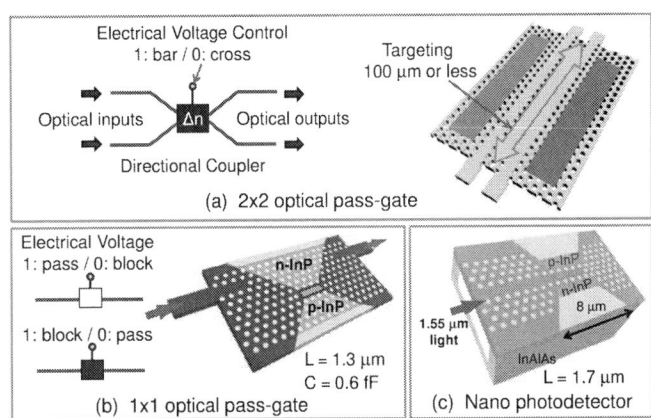

Fig. 1. Examples of optical pass gate with photonic crystal

III. Several previous work on the optical logic circuits are summarized in Scetion IV. Section V concludes this paper.

II. Optical Pass Gate Logic

For constructing logic circuits, one is usually interested in logic gates containing nonlinear elements. In this work, as the logic gates, we use electric voltage-controlled optical switches as shown in Fig. 1 (a) and (b), which we refer to as 2×2 OPG and 1×1 OPG, respectively. The structure of the OPG is similar to that of pass-transistors used in MOS LSI circuits. The pass-transistors are used as switches to pass logic signals between nodes of a circuit, instead of as switches connected directly to supply voltages. Similar to the pass-transistors, our 2×2 OPG shown in Fig. 1 (a) passes two optical input signals through the OPG. At a zero-voltage bias on the control input, the optical input signals pass to the outputs after crossing each other while they are passing straight to the outputs when a voltage bias is given to the control input. The 2×2 OPG can be easily functioned as XOR and multiplexer (MUX) gates. Similarly, the 1×1 OPG shown in Fig. 1 (b) works as an optical switch. Depending on the voltage on the control input, it selects pass or block mode. In the pass mode, the optical input signal is passing through the OPG to the output while it is blocked in the block mode. Although the OPG is similar to the pass-transistor, the delay characteristics are different from each other. The path delay in the OPG physically depends on the speed of light passing through the gate while that in the pass-transistor depends on the parasitic RC on the circuit. With the OPG, therefore, the light-speed data processing is possible.

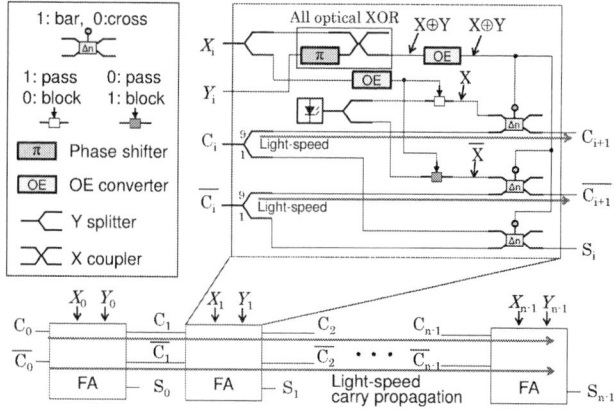

Fig. 2. Design example of full adder based on optical pass-gate

The nano-photodetector shown in Fig. 1 (c) is an essential element for optoelectronic circuits. We use a functional model of the photonic-crystal nano-photodetector presented in [3] as an OE converter in our parallel adder.

III. PARALLEL ADDER CIRCUIT BASED ON OPTICAL PASS GATE

A design example of a full adder circuit using the OPGs is depicted in Fig. 2. Three $2{\times}2$ OPGs and two $1{\times}1$ OPGs are used in this full adder. Since the OPG is not good at making negation of optical logic signal, the carry bar signal is also propagated from the lowest digit to higher digits along with the carry propagation. The $2{\times}2$ OPGs used for the carry and carry bar work as a MUX. When a voltage bias is given to the control input, which means that XOR of X_i and Y_i is 1, the carry and carry bar signals are passing through the OPGs. This means that light-speed carry propagation is possible in this full adder architecture. Another important feature of the architecture is that the carry propagation path is composed of a single $2{\times}2$ OPG and a Y splitter only. Therefore, the delay of the carry propagation, which determines the performance of the adder circuit, is dominated by the delay of the $2{\times}2$ OPG. If we reduce the delay of the OPG, the performance of the adder is directly improved. We are aiming at developing a $2{\times}2$ OPG with a path delay of less than 1 ps by miniaturizing the OPG using photonic crystal. Once the sub-picosecond OPG is realized, per digit delay of the adder circuit can be also less than picosecond which is more than one order of magnitude faster than that of CMOS-based full adder circuits.

By simply connecting the full adder cells in series, a ripple carry adder (RCA in short hereafter) which is one of the most popular parallel adder implementations can be constructed as shown in the bottom of Fig. 2. We designed an 8-bit RCA based on the OPG and simulated it using an optoelectronic circuit simulator. We assume that the gate length, input capacitance of the control port, and the switching delay of the $2{\times}2$ OPG are 100 μm, 2 fF, and 5 ps, respectively. Under this assumption, per digit delay of the circuit estimated is around 1 ps while the initial delay including OE conversion of X_i and Y_i signals and switching delay of the $2{\times}2$ OPG is more than 10 ps. As a result, the critical path delay of the 8-bit optical RCA is about 20 ps in total. We also designed

and evaluated a CMOS-based 8-bit RCA using a 16 nm high-performance CMOS transistor model. The critical path delay of the CMOS RCA is about 180 ps. This means that the OPG-based RCA is 9 times faster than the CMOS-based one.

IV. RELATED WORK

A concept of optical directed logic (DL in the following) is proposed in [4]. Similar to our approach, DL uses $2{\times}2$ OPG as a basic building block and reduces the latency in calculating a complicated logic function by taking advantage of fast and low-loss propagation of light. However, in DL design, the number of switches needed scales super-linearly with the total number of logic operations in the function. In [5], a 2-stage DL architecture is proposed, which drastically reduces the number of pass-gates needed in a logic function. However, it still needs a large number of pass-gates for realizing a given function since the logic operation available for the second stage is limited to a logical OR only. Although the architecture of our OPG-based RCA is similar to the 2-stage DL architecture, our architecture has more flexibility in the second stage, which reduces the critical path delay, the number of fanout branches, and the number of pass-gates used in the circuit. A similar approach based on the binary decision diagram (BDD in short) is proposed in [6]. The basic logic structure of a BDD-based circuit is the same from that of DL-based circuits. Therefore, our RCA design is superior to the BDD-based RCA presented in [6] as well in terms of the critical path delay, the number of fanout branches and the number of pass-gates.

V. CONCLUSION

In this paper, we introduced a concept of optical pass-gate logic, which makes it possible to integrate logic functionalities on the optical interconnects. If we apply the OPG logic to parallel adders, a light-speed carry propagation can be realized. Moreover, thanks to the nanophotonics technologies, the length of state-of-the-art OPG reaches to a few tens of micrometer scale. This results in a sub-picosecond delay in the OPG, which is more than one order of magnitude less than that of the CMOS pass-transistors. Our future work will be devoted to apply our idea to more complicated logic functions.

ACKNOWLEDGMENT

This work is partly supported by JST CREST project.

REFERENCES

[1] A. Ceyhan, et al., "Impact of Size Effects in Local Interconnects for Future Technology Nodes: A Study Based on Full-Chip Layouts," in *Proc. of IEEE Interconnect Technology Conference / Advanced Metallization Conference*, pp. 345–348, May 2014.

[2] PTM, Predictive Technology Model. Available: http://ptm.asu.edu

[3] K. Nozaki, et al., "Photonic-Crystal Nano-Photodetector with Ultrasmall Capacitance for On-Chip Light-to-Voltage Conversion without an Amplifier," in *Optica*, vol. 3, pp. 483–492, April 2016.

[4] J. Hardy, and J. Shamir, "Optics Inspired Logic Architecture," in *Optics Express*, vol. 15, no. 1, pp. 150–165, January 2007.

[5] Q. Xu, and R. Sorei, "Reconfigurable Optical Directed-Logic Circuits Using Microresonator-Based Optical Switches," in *Optics Express*, vol. 19, no. 6, pp. 5244–5259, March 2011.

[6] T. Asai, Y. Amemiya, and M. Kosiba,"A Photonic-Crystal Logic Circuit Based on the Binary Decision Diagram," in *Proc. of Int'l Workshop on Photonic and Electromagnetic Crystal Structures*, T4-14, March 2000.

Integrated Circuits Design Using Carbon Nanotube Field Effect Transistor

Yong-Bin Kim, Senior Member, IEEE

Abstract— Complementary metal-oxide-semiconductor (CMOS) technology scaling has been a main key for continuous progress in silicon-based semiconductor industry over the past three decades. However, the bulk CMOS technology has approached the scaling limit due to the increased short-channel effects as technology scales down to 90 nm and below. Last about a decade witnessed a dramatic increase in nanotechnology research, especially the nano-electronics. These technologies vary in their maturity. Carbon nanotubes (CNTFETs) are at the forefront of these emerging technologies because of the unique mechanical and electronic properties. This paper discusses and reviews the feasibility of the CNTFET's application at this point of time in integrated circuits design by investigating different types of circuit blocks considering the advantages that the CNTFETs offer.

Keywords; Carbon Nano Tube FET(CNTFET), integrated circuyits design, low power circuits, emerging technology

I. INTRODUCTION

As complementary metal-oxide semiconductor (CMOS) technology scaling has been enabling higher integration capacity in very large scale integration designs. Over the last few years, devices at 16 nm have been manufactured and the deep sub- micron/nano range of below 10 nm is foreseen to be reached in the near future as technology continues to scale down. In the sub-10nm technology era, the leakage current has substantially increased, and the sensitivity to process variations in manufacturing is considerably increasing. Due to the lower power supply and the smaller node capacitance, the amount of charge stored on a circuit node is becoming increasingly smaller, thus making circuits more susceptible to spurious voltage variations caused by externally induced phenomena such as cosmic ray neutrons and alpha particles. Therefore, new materials and devices have been investigated to replace silicon in nanoscaled transistors from the year 2015 and beyond. As one of the promising new devices, Carbon Nano Tube FETs (CNTFETs) avoid most of the fundamental limitations for traditional silicon devices, due to their unique one-dimensional band-structure that suppresses backscattering and makes near- ballistic operation a realistic possibility [1][2].

In this paper, the basic device structure of CNTFET is reviewed, device performance is assessed, and advantages over the conventional CMOS are summarized. Based on the assessment and comparison with CMOS, combinational logic gates design and sequential logic gates design methodologies are proposed to maximize the benefits of using the new technology.

"The detail data will be presented in the oral session."

II. CNTFET BASED LOGIC GATE PERFORMANCE ANALYSIS

To measure the actual performance of a CNTFET compared to a MOSFET, it is necessary to compare performance at a circuit level under PVT variations. In this paper, logic gates and benchmark circuits are designed at 32 nm for both CNTFET and CMOS technologies; delay, power, PDP, leakage current and frequency response are simulated and compared [3].

For comparing the performance of CNTFETs with MOSFETs at circuit level, the inverter as a fundamental logic gate is considered first; the inverter is designed with minimal width and a number of tubes in 32 nm technology. For Si CMOS, a PMOS/NMOS ratio between 2 and 3 is used for compensating the difference in mobility between PMOS and NMOS. In this paper, 3 to 1 (PMOS:NMOS) ratio is used when designing the inverter because the voltage transfer characteristic (VTC) of the MOSFET inverter shows a more symmetrical shape in the center of the logic threshold voltage (VDD/2) for a ratio of 3 in 32 nm. Simulation results show that the PDP of the 32 nm MOSFET is about 100 times higher than that of the 32 nm CNTFET.

The maximum and minimum leakage power for 32 nm MOSFET and CNTFET based logic gates. The maximum leakage power of the MOSFET-based gates is 75 times larger than for CNTFET gates. The minimum leakage power of the MOSFET is about three times larger than for CNTFET.

In terms of frequency response, CNTFET based inverter shows nearly 3dB more voltage gain and three time higher 3dB frequency than the MOISFET based inverter, thus confirming the superiority of CNTFET. On the other hand, the sensitivity of CNTFET based gate to the process variation is mush less that the CNTFET based gates: The On-current change of MOSFET device is about 30%(13%) for 10% change of length(width) while the On-current change of CNTFET based design is 0.5% for the same change. The reduction in power consumption due to voltage scaling is also confronted with the increased sensitivity to voltage variations. The overall PDP of the CNTFET based gates is significantly lower than for the MOSFET based gates. As the circuit speed increases, a larger power consumption is often encountered, thus resulting in more heat at chip level. Circuits with an excessive power dissipation are more susceptible to run-time failures and account for serious reliability problems [4]. The study of this research shows that the PDP of the MOSFET gates increases with temperature; however, the PDP of the CNTFET logic gates is constant. Moreover, the maximum leakage power of the MOSFET gates increases linearly with temperature, while for the CNTFET based gates this increase is exponential.

III. TERNARY LOGIC GATE DESIGN USING CNTFET

Digital computation has been performed on two-valued logic, i.e., there are only two possible values (0 or 1, true or false) in the Boolean space. Multiple-valued logic (MVL) replaces the classical Boolean characterization of variables with either finitely or infinitely many values such as ternary logic [5]) or fuzzy logic. Ternary logic (or three-valued logic) has attracted considerable interest due to its potential advantages over binary logic for designing digital systems. For example, it is possible for ternary logic to achieve simplicity and energy efficiency in digital design since the logic reduces the complexity of interconnects and chip area [6]. In CNTFET, The threshold voltage of the transistor is determined by the diameter of the CNT. Therefore, a multi-threshold design can be accomplished by employing CNTs with different diameters (and, therefore, chirality) in the CNTFETs. A resistive-load CNTFET based ternary design has been proposed in [7]. However, in this configuration, large off-chip resistors (of at least 100 MOhms values) are needed due to the current requirement of the CNTFETs. The design technique proposed in [8] eliminates the large resistor by employing active load with P type CNTFET. In this research, the multi valued logic design based on multi-threshold CNTFETs is assessed and compared with existing multivalued logic designs based on CNTFETs by designing primitive gates and benchmark circuits such as adders. It turns out that the ternary logic using CNTFET achieves significant savings in PDP comparing with those proposed before. By replacing ternary gates with binary gates in the internal logic, the proposed design also achieves power and delay savings due to the reduced number of transistors in the binary gates. The simulation results confirm that the ternary logic based half adder's PDP is reduced by 90% compared with the conventional multi-valued logic family.

IV. MEMORY DESIGN

Design of fast and power efficient memory structures continues to be of the highest priority, and ballistic transport operation and low off current make the CNTFET a suitable device for high performance and increased integration density of SRAM design. Moreover, the MOSFET-like model of the CNTFET is likely to be scalable down to 10 nm channel length, thus providing a substantial performance and power improvement compared to the MOSFET model. With today's aggressive scaling, substantial problems such as power consumption and stability have already been encountered when the 6T SRAM cell configuration is utilized in CMOS at nanoscale ranges. In this paper, the 6T SRAM cell of is designed using CNTFETs and its performance is assessed comprehensively with a newly proposed figure of merit denoted as "SPR" to compare stability, power dissipation, and write time with other existing SRAM cell designs.

The metric SPR provides design metrics for a SRAM cell in terms of delay, stability, and power. Simulation confirms that the SPR of the CNTFET of 6T SRAM cell is four times higher than its CMOS counterpart, hence attaining low power, high stability, and low delay within the comprehensive metric provided by the SPR under write conditions. Monte Carlo simulation by HSPICE has been performed ton investigate the impact of the random variations on the delay and power of the CNTFET and CMOS SRAM cells at 0.9 V power supply and room temperature. Simulations results shows that the write time of the proposed dual-diameter CNTFET 6T SRAM cell has much better tolerance to process variations compared to its CMOS counterpart. Due to the significantly low standby power consumption and fewer parameters causing changes in power consumption, the CNTFET SRAM cell shows a significantly better tolerance to process variations.

V. CONCLUSION

In this paper, the most promising technology (CNTFET) based design's feasibility has been tested by assessing the technology performance and design metrics on combinational and memory circuits along with the ternary logic design to take advantage of the merits that CNTFETs offer. Based on the review and observation of the CNTFET based digital logic circuits, it is fair to say that CNTFET based integrated circuit design is a viable solution to replace the conventional bulk CMOS technology and it turns out to be an effective choice of future technology.

REFERENCES

[1] A.Rahman and et. al, IEEE Trans. Electron Devices 50, pp.1853 (2003)

[2] A. Akturk, G. Pennington, N. Goldsman , and A. Wickenden , IEEETrans. Nanotechnology 6, pp. 469 (2007).

[3] K. Kim and Y. Kim, IEEE Trans Very Large Scale Integrated Systems 17, pp.517 (2009)

[4] K. K. Kim and Y. B. Kim , IEEE International Symposium on CircuitS and Systems , New Orleans, LA 2007 May, pp. 1161.

[5] M. Mukaidono, IEEE Transactions on Computer C.-35, pp. 179 (1986).

[6] P.C. Balla and A. Antoniou, IEEE Journal Solid-State Circuits SC-19, 739 (1984)

[7] A. Raychowdhury and K. Roy, IEEE Transactions on Nanotechnology Vol. No. 4, pp. 168 (2005).

[8] S. Li n , Y. B. Kim , and E Lombardi, IEEE Transactions on Nanotechnology, Vol. 10, pp. 217 (2011)

Memory Efficient Hardware Accelerator for Kernel Support Vector Machine Based Pedestrian Detection

Asim Khan and Chong-Min Kyung

Department of Electrical Engineering
Korea Advanced Institute of Science and Technology (KAIST),
Daejeon, South Korea
Email: {asimkhan, kyung}@kaist.ac.kr

Abstract—**Pedestrian detection being a vital as well as complex problem poses a unique challenge from accuracy and complexity point of view. On-chip memory requirement is one of the key issues for sliding window based detectors. In this paper a memory efficient hardware architecture is proposed which estimates the weights from a partially stored model at runtime. It uses a simple and robust feature with histogram intersection classifier. The implementation results show 80% reduction in logic resources and 46% reduction in memory without sacrificing accuracy as compared to the state of the art hardware implementations.**

Keywords—Pedestrian Detection; FPGA; Memory Efficient;

I. INTRODUCTION

Pedestrian Detection (PD) is a key component in many vision applications ranging from Advanced Driver Assistance Systems (ADAS) and Unmanned Ground Vehicles (UGV) to robotics and surveillance. It is however, considered to be one of the most complex problems due to large variations in human postures, multiple textures and different illumination conditions. A wide range of sophisticated and complex pedestrian detectors have been proposed in the past decade. Histogram of Oriented Gradients (HOG) [1] is still the most discriminative feature proposed till date and is used as a stand-alone detector as well as in conjunction with multiple features and cascaded detectors. Linear Support Vector Machine (linSVM) is generally used as a classifier with HOG due to its simplicity in implementation. It has been shown that Histogram Intersection Kernel (HIK) based SVM gives better results [3] as compared to linSVM. The complexity of HIK-SVM is however a major bottleneck for real time and low power systems. To this end we proposed a simpler yet accurate feature named Histogram of Significant Gradients in conjunction with a fast Look-Up Table (LUT) based HIK-SVM to replace HOG [3]. Our proposed framework requires much less resource usage still producing better results than conventional HOG on all standard datasets. Moreover, it is highly suitable for dedicated real-time hardware accelerator as it does not require any floating point operations. On-chip memory requirement is however the major bottleneck in achieving this goal as memory required to store the trained model is much higher in comparison to HOG. In this paper we propose a unified memory architecture for all the pipelined modules. In addition, we utilize the statistics of trained model to approximate the weights at runtime from a partially stored model in memory reducing the memory requirement significantly. The rest of the paper is organized as follows. We briefly explain the HSG-HIK framework in Section II followed by the memory efficient hardware architecture in Section III. Section IV presents the

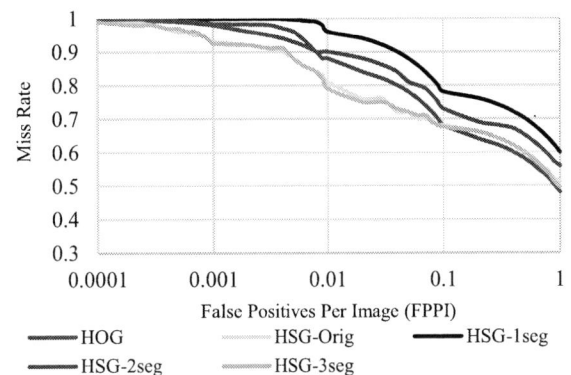

Fig. 1. ROC curve depicting miss rate of HOG and memory efficient HSG

complexity and accuracy results and finally Section V concludes the paper.

II. MEMORY EFFICIENT HSG-HIK FRAMEWORK

HOG feature captures the shape of an object using gradient histograms in localized cells within a detection window. We proposed a simpler feature named Histogram of Significant Gradients (HSG) [3] which uses average gradient magnitude as a threshold to cast a binary vote in the orientation histogram. This is shown in Fig. The advantages of using HSG over HOG are twofold: 1) It is algorithmically simpler and does not involve any floating point operations. 2) The resultant feature vector can be classified using LUT based HIK SVM because of integer only operands. The inner summation of HIK SVM can be computed beforehand in case of HSG as it takes only integer values with a limited dynamic range so and can be written as,

$$T(i,k) = \left(\sum_{j=1}^{m} \propto_j min(k, SV_j(i)) \right) \quad (1)$$

$$h_{HIK}(x) = \sum_{i=1}^{n}(T(i, x(i)) + b \quad (2)$$

Here, '*T*' stores the values of inner summation in a 2D LUT resulting in just 'n' additions per classification. *SV* is the support vector, '*α*' is the learned coefficient and '*b*' is the learned bias. For a histogram cell of width $'w_c'$ and height $'w_h'$, size of 2D LUT is $n \times w_c \times w_h$ which is $'w_c \times w_h'$ times the memory required for linear SVM. We analyzed the statistics of learned model and realized that instead of storing weights for all combinations, weights can be approximated using piecewise linear segments. The slope and intercept for each segment needs to be stored in memory. The size of memory is therefore, $n \times segments$. This approximation results in significant reduction in memory without sacrificing the accuracy. Fig. 1 shows the

Fig. 2. Accuracy Results on Caltech Dataset

ROC curve on Caltech Dataset [2] for different number of segments and Fig. 2 shows some qualitative results. We can see using three segments produces the same results as original HSG.

III. HARDWARE ARCHITECUTRE

Our pedestrian detection framework is designed to use minimal resources when implemented on FPGA. Fig. 3 shows the proposed pipelined architecture with a conventional CMOS images sensor generating pixel data in raster scan order. Major blocks in pipeline includes gradient magnitude and orientation filter, block based HSG feature generator and HIK classifier.

The proposed memory efficient hardware architecture utilizes two unified memory modules as shown in Fig. 3 instead of using two line buffers for gradient calculation and five line buffers for feature calculation. In addition, an extra segment selector block is added with the HIK classifier which comprises of a simple comparator to select the segment for each input feature value. The second modification is calculating the trained model at runtime. This is done by shifting the input feature value according to the slope of selected segment and adding the intercept to the resultant value. In a nutshell we have used an extra comparator, shifter and adder as compared to original implementation. This just adds a negligible overhead to logic resources in comparison to a noteworthy reduction in memory requirement.

IV. PERFORMANCE AND COMPLEXITY

HSG-HIK core is implemented on an Altera Cyclone IV EP4C115 FPGA. Each frame is grabbed from the CMOS image sensor using an *Avalon Streaming Protocol* interface as shown in Fig. 4. HSG-HIK IP core computes the detection results for overlapped windows in the frame and store them in the DRAM using *Avalon DMA IP Core*. *Nios-II* processor finally applies non-maximum suppression and outputs the results to video display. Table I compares the resource utilization of proposed memory efficient HSG-HIK framework with state of the art HOG based pedestrian detection systems. Our proposed

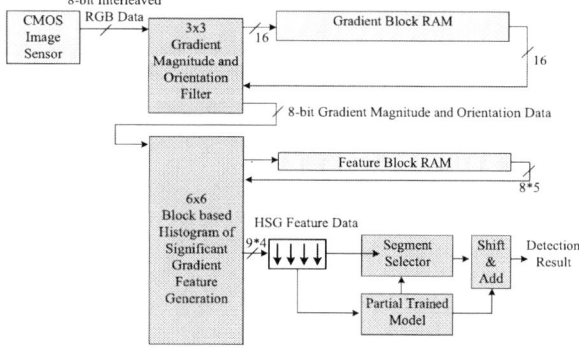

Fig. 3. Proposed Memory Efficient pipeline

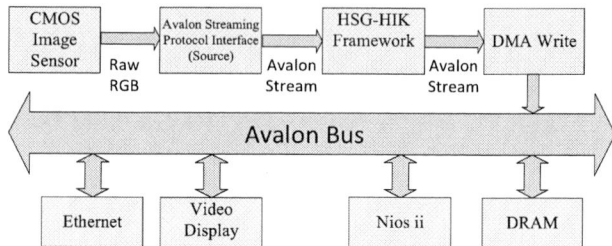

Fig. 4. HW-SW co-design of custom IP core

framework does not require any DSP blocks and has 6x less logic resource utilization. Moreover, memory requirement is reduced to 0.34 Mbits only.

V. CONCLUSION

A memory efficient pedestrian detection accelerator is proposed by approximating the classification weights at runtime. FPGA implementation shows 6x reduction in logic and 2x reduction in memory while having the same accuracy on standard datasets.

ACKNOWLEDGEMENT

This work was supported by the Center of Integrated Smart Sensors funded by Ministry of Science, ICT & Future Planning as Global Frontier Project (CISS-2013M3A6A6073718).

REFERENCES

[1] N. Dalal and B. Triggs, "Histograms of Oriented Gradients for Human Detection," presented at the Proceedings of the 2005 IEEE Computer Society Conference on Computer Vision and Pattern Recognition (CVPR'05) - Volume 1 - Volume 01, 2005.

[2] *Caltech Pedestrian Dataset.* Available: http://www.vision.caltech.edu/Image_Datasets/CaltechPedestrians/.

[3] M. Bilal; A. Khan; M. U. K. Khan; C. M. Kyung, "A Low Complexity Pedestrian Detection Framework for Smart Video Surveillance Systems," in *IEEE Transactions on Circuits and Systems for Video Technology*, vol.PP, no.99, pp.1-1

[4] K. Mizuno, Y. Terachi, K. Takagi, S. Izumi, H. Kawaguchi, and M. Yoshimoto, "An FPGA Implementation of a HOG-based Object Detection Processor," *IPSJ Transactions on System LSI Design Methodology,* vol. 6, pp. 42-51, 2013.

[5] P. Y. Chen, C. C. Huang, C. Y. Lien and Y. H. Tsai, "An Efficient Hardware Implementation of HOG Feature Extraction for Human Detection," in *IEEE Transactions on Intelligent Transportation Systems*, vol. 15, no. 2, pp. 656-662, April 2014.

[6] M. Hemmati, M. Biglari-Abhari, S. Berber and S. Niar, "HOG Feature Extractor Hardware Accelerator for Real-Time Pedestrian Detection," *Digital System Design (DSD), 2014 17th Euromicro Conference on,* Verona, 2014, pp. 543-550.

TABLE I
COMPARISON OF HARDWARE RESOURCES

Design	[4]	[5]	[6]	Proposed
Device	Cyclone IV	-	Virtex 5	Cyclone IV
LUT Resources	34,403	7226	5188	843
Registers	23,247	12462	5176	535
Block Ram (Mbits)	0.63	-	1.2	0.34
DSP Blocks	68	-	49	0
Operating Frequency (MHz)	76.2	125	270	190
Video Resolution	1920x1080	1920x1080	1920x1080	1920x1080
Frame Rate (fps)	30	60	64	75

A TSV Test Structure for Simultaneously Detecting Resistive Open and Bridge Defects in 3D-ICs

Young-woo Lee, Junghwan Kim, Inhyuk Choi and Sungho Kang

Dept. of Electrical and Electronic Engineering
Yonsei University
Seoul, Korea
{roberto, kjhcz, ihchoi}@soc.yonsei.ac.kr and shkang@yonsei.ac.kr

Abstract— **After the 3D stacking process, TSV-based 3D-ICs are required to perform the post-bond testing in order to detect TSV faults or device functional defects. To detect the resistive open and bridge defects, various effective TSV testing techniques have been studied. At an early stage of TSV manufacturing, it is important to consider that the TSV testing is required not only determining whether each TSV is defective or non-defective, but also digitizing the fault degree into the TSV resistance value during the silicon debugging. In this paper, we propose a new TSV test structure for simultaneously detecting the resistive open and bridge defects with supporting the debug mode to analysis the characteristic of the specific TSV. It can highly reduce the test time by detecting TSV defects at the same time without compromising test quality.**

Keywords; TSV test; 3D-ICs; resistive open and bridge defects; characteristic of TSVs

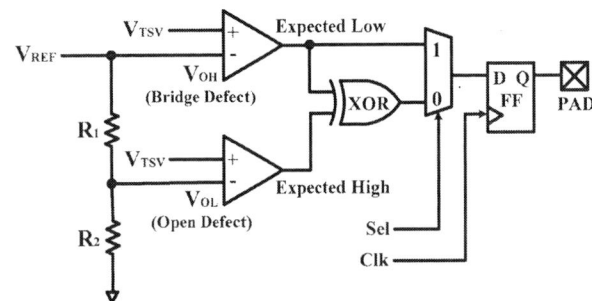

Figure 1. Proposed test structure.

II. PROPOSED IDEA

A. Proposed Test Structure

The test structure of the previous work [3] consists of a shared comparator and a flip-flop to measure the voltage across a TSV (V_{TSV}), depending on the resistance of the TSV (R_{TSV}). Figure 1 represents the proposed test structure which adds a shared comparator, a XOR gate and a multiplexer in order to reduce the total test time in half, and it can be divided into three main parts. The first main parts consist of two shared comparators. The shared comparator compares two input voltage signals V_{TSV} and V_{REF}. According to the experimental results of [3], the voltage level of the TSV-to-TSV bridge defects is greater than the standard voltage level of fault-free TSVs. In contrast, the voltage level of the resistive open defects is smaller. For this reason, the reference voltage (V_{REF}) of each shared comparator has the different value, and each upper and lower comparator is only capable of detecting the TSV-to-TSV bridge defects and the resistive open defects, respectively. The different V_{REF} can be supplied by a single external pin, according to the voltage division rule. The second part is to determine whether the TSV has any defects or not. The outputs of each upper and lower comparator are supposed to be a logic 0 (low) and logic 1 (high) if V_{TSV} is a fault-free value, as described in TABLE I. The final result is classified as a pass (1: High) or a fail (0: Low) by the XOR gate. Lastly, the third part is to select the test/debug mode by controlling the multiplexer. The default mode is the test mode (0: Low), but it can change to the debug mode (1: High) when it needs the fault diagnosis in detail whether the TSV defect is the resistive open defect or the TSV-to-TSV bridge defect. If the test/debug mode is changed from logic 0 to 1 right after the TSV is classified as a fail, the type of TSV defects can be diagnosed by the output

I. INTRODUCTION

The device performance and functionality in 3D-ICs are highly impacted by the fidelity of signals through TSVs [1]. For this reason, the post-bond TSV testing is important to guarantee the quality and yield for mass production. The previous works [2, 3] are based on the voltage divider structure, which inspect the resistance-related delays by measuring the voltage across a TSV. These test methods can support to characterize the resistance of the defective or fault-free TSV for yield improvement during the silicon debugging and also to perform as a part of at-speed test to qualify timing specifications of 3D-ICs. However, the previous work [2] is only capable of testing the resistive open defects, which has inability to detect TSV-to-TSV bridge defects. In contrast to [2], the previous work [3] can detect both defects. It also provides lower hardware overhead and peak current consumption than [2] by using a shared comparator and a sequential pulse-transfer. However, the drawback of [3] is that each test for detecting the resistive open and bridge defects cannot be tested together at the same time. Each TSV testing needs the one test clock period per a single TSV, respectively. The total test time of [3] is 2 × (*the number of TSVs × the test clock period*. The proposed test structure is designed based on the previous work [3]; the voltage divider structure including a shared comparator and a sequential pulse-transfer. The proposed test structure can effectively reduce the total test time by using DFT (Design for Testability) techniques that provide the parallel test method without compromising test quality.

This work was supported by the National Research Foundation of Korea(NRF) grant funded by the Korea government(MISP) (No. 2015R1A2A1A13001751).

TABLE I. Expected output results of the proposed test structure.

	Resistive open defects	TSV-to-TSV bridge defects	Fault-free
Upper Comparator	0 (Low)	1 (High)	0 (Low)
Lower Comparator	0 (Low)	1 (High)	1 (High)
XOR	0 (Low/Fail)	0 (Low/Fail)	1 (High/Pass)

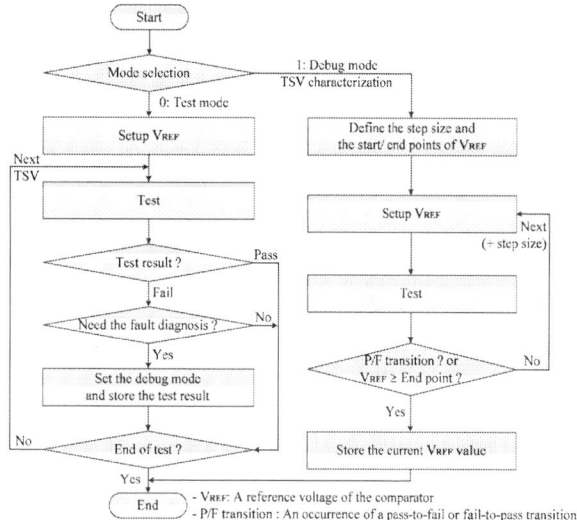

Figure 2. Test flow diagram of the proposed test structure.

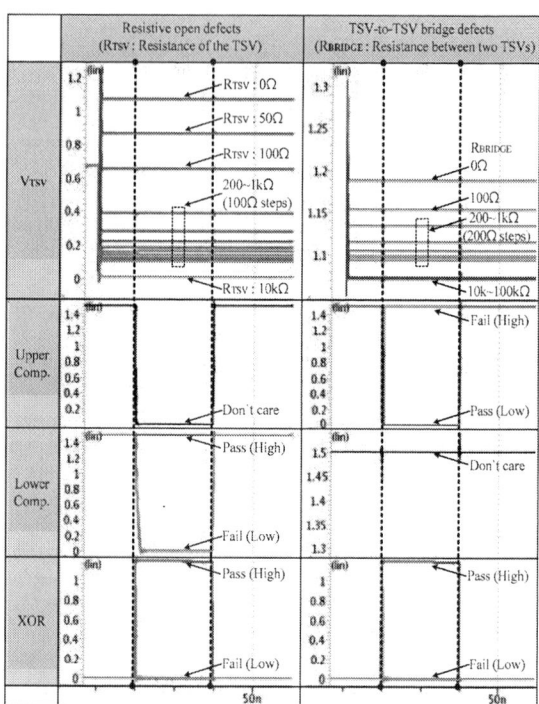

Figure 3. Voltage profiles as a function of different R_{tsv} and R_{bridge}.

result of the next test cycle; the TSV-to-TSV bridge defect (1: High) and the resistive open defect (0: Low), respectively.

B. Debug mode for analyzing the characteristic of TSVs

Figure 2 shows the test flow diagram of the proposed test structure. The test flow of the normal mode is the same procedure as described in the previous subsection until it reaches the last TSV or the end of test. In addition, the proposed test structure can support the debug mode for analyzing the characteristic of a specified TSV by detecting the voltage across the TSV. To digitize the fault degree into the actual TSV resistance value, the voltage across a TSV can be measured by detecting the pass-to-fail or fail-to-pass transition point with sequentially increasing or decreasing the V_{REF} value of the upper comparator. Consequently, the V_{TSV} is the current V_{REF} value at the transition point of the test result and it returns the end point value if there is no transition point. The V_{TSV} value can be converted to the resistance of the TSV by *HSPICE* simulation results.

III. EXPERIMENTAL RESULTS

The proposed test structure is evaluated by using *HSPICE (Nangate 45 nm Library)*. The representative value of TSV and FET specifications in the *HSPICE* simulation was extracted from published data [3]. Figure 3 shows the voltage profiles as a function of different R_{TSV} and R_{BRIDGE}; each blue/red line represents the pass/fail condition, respectively. The V_{REF} was determined by below test conditions. The TSV with the resistive open defect is presumed if the R_{TSV} is greater than 500Ω. In this case, the upper comparator output is staying the low regardless of the R_{TSV}, but the lower comparator changes from high (Pass) to low (Fail) when the R_{TSV} exceeds 500Ω.

The TSV is considered to be the bridge defect if the resistance between two TSVs is less than 10kΩ. In such a case, the upper comparator output switches from high (Fail) to low (Pass) when the R_{BRIDGE} is greater than 10kΩ, but the lower comparator is staying the high regardless of the R_{BRIDGE}. The outputs of both comparators transfer to the XOR gate which returns to the low if the TSV has any defects. In summary, experimental results guaranteed the test quality of the proposed test structure which can detect both the resistive open and bridge defects in parallel. Based on experimental results, the proposed test structure can reduce the total test time in half, compared to the previous work [3].

IV. CONCLUSION

In this paper, a new TSV test structure for detecting the resistive open and bridge defects and for supporting the debug mode is proposed. It can reduce the total test time in half with slightly increasing the hardware overhead, compared to the previous work [3]. Consequently, the test cost will dramatically decrease without compromising test quality.

REFERENCES

[1] M. Cho, C. Liu, D. H. Kim, S. K. Lim, and S. Mukhopadhyay, "Pre-bond and post-bond test and signal recovery structure to characterize and repair TSV defect induced signal degradation in 3-D system," IEEE Transactions on Components, Packaging and Manufacturing Technology, vol. 1, no. 11, pp. 1718–1727, 2011.

[2] F. Ye and K. Chakrabarty, "TSV open defects in 3-D integrated circuits: Characterization, test, and optimal spare allocation," in Proceedings of IEEE Design Automation Conference, pp. 1024–1030, 2012.

[3] H. Sung, K. Cho, K. Yoon, and S. Kang, "A Delay Test Architecture for TSV with Resistive Open Defects in 3D-Stacked Memories," IEEE Transactions on Very Large Scale Integration Systems, vol. 22, no.11, pp.2380-2387, 2014.

Novel Pixel Calibration Circuit for Bolometer-Type Uncooled Infrared Image Sensor

Sang-Hwan Kim[1], Byoung-Soo Choi[1], Jang-Kyoo Shin[1*], Jae-Hyoun Park[2], and Kyoung-Il Lee[2]

[1]School of Electronics Engineering, Kyungpook National University
[2]Korea Electronics Technology Institute
*jhshin@ee.knu.ac.kr

Abstract—**Lately, research on bolometer type uncooled infrared image sensor has been increasing for industrial applications. But it is hard to remove a Fixed Pattern Noise (FPN) of bolometer sensor. In this paper, a novel average-current calibration algorithm is presented for reducing bolometer resistance offset. A resistor which is produced standard CMOS process, on the average, has a deviation, respectively. We compensate for deviation of each active resistor using average-current calibration algorithm. The algorithm consists of as follows: bolometer pixel array, average current generator, current-to-voltage converters (IVCs), digital-to-analog converter (DAC), and analog-to-digital converters (ADCs). These bolometer-resistor array and readout circuit were designed and manufactured by 0.35μm standard CMOS process.**

Keywords: **average-current calibration, bolometer, resistance offset, readout circuit**

I. INTRODUCTION

Recently, uncooled infrared image sensors have enormous potential for apply to various civilian applications, such as security systems, automotive night vision system, and medical applications [1]. This is because uncooled infrared image sensors have many outstanding advantages, such as low cost, small volume, and low power consumption [1]. Among various uncooled infrared image sensors, bolometer sensors based on micromachining technology have a higher temperature coefficient of resistance (TCR) than others. Moreover, these infrared sensors absorb the thermal radiation emitted by a target rather than the incident photons. Thus they have a wider spectral response from X-rays to milli-meter waves than the infrared photon detectors [2]. There are many research activities about integration uncooled infrared sensors with readout circuits and signal processors. A monolithic infrared image sensor which uses the bolometer sensor based on micromachining technology compatible with CMOS technology, is one of the main issue for achieving low-cost, low-noise, and good-uniformity sensor [1, 2].

Undesirably, the resistance offsets due to the process non-uniformity are much more than 100 times the resistance variations by infrared radiation among pixels for a bolometer sensor array [3]. This causes a relatively high and serious fixed pattern noise (FPN) which can be occurred from the characteristics difference between each bolometer-based pixel in a 2-dimensional sensor array. For CMOS readout circuits and signal processors, it is necessary to compensate these excessive offsets for useful infrared images. In order to solve the offset

problems, various approaches have been reported. However, these compensation methods suffer from many difficulties for low cost implementation [3-4].

In this paper, we propose a new average-current calibration method to compensate the resistance offsets for readout circuits of bolometer type uncooled infrared sensors. This calibration algorithm makes a feature of calibration the resistance offset using difference between an average dark current of the bolometer arrays and a dark current of each bolometer-based pixel. In order to verify this algorithm, bolometers and its controllable offsets are modeled by poly-Si resistors and MOS switches. The proposed readout circuit uses an active pixel array, a reference pixel array, an average current generator, current-to-voltage converters (IVCs), calibration current generators with a digital-to-analog converter (DAC), and analog-to-digital converters (ADCs). This calibration algorithm is implemented using a 0.35μm mixed signal CMOS process.

II. CIRCUIT DESIGN

Fig. 1 shows a block diagram of the proposed calibration algorithm, which is composed of a bolometer pixel array with a calibration circuit using DAC, IVC, and ADC.

The principle algorithm of the proposed readout circuit is as follows: An average dark current (I_{REF}) is generated by a reference pixel array is subtracted by a dark current (I_{ACT}) of each active pixel. After that, the current difference ($I_{REF} - I_{ACT}$) with information of pixel deviation is converted to the voltage signal and the digital value through an IVC and an ADC, respectively. To obtain the output current of approximately zero or nano-scale, a proper calibration current (I_{CAL}) proportional to a DAC output should be added to I_{ACT}. The adequate DAC input is acquired by the ADC output for the initial current difference and proper calculation. Through this calibration algorithm, nano-scale dark currents with small deviations can be obtained for the all active bolometer pixels with different resistance offsets.

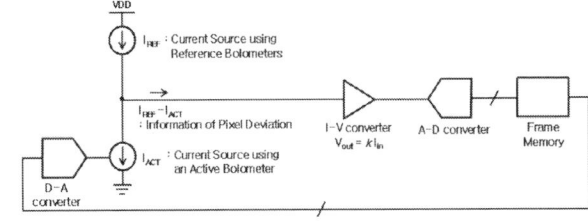

Figure 1. Block diagram of algorithm circuit array

This work was supported by the Industrial Core Technology Development Program, (Project No. 10052933) funded by the Ministry of Trade, Industry & Energy (MOTIE, Republic of Korea).

The readout circuit with the proposed calibration algorithm is designed and fabricated by a 0.35μm mixed signal CMOS process. Fig. 2 shows the layout of the designed readout circuit. Each active bolometer is implemented by a poly-Si resistor with different resistance, such as 90kΩ, 95kΩ, 100kΩ, and 105kΩ. Reference bolometer array is designed by a poly-Si resistor of 110kΩ with its process variations.

III. RESULTS AND DISCUSSION

In a readout circuit for a bolometer type uncooled infrared image sensor, it is important that each pixel has same integration interval for an integrator-type IVC at the same infrared radiation. Thus, the proposed average-current calibration algorithm can be verified by comparing integration time between no-calibrated pixel with different offset and calibrated pixel.

Fig. 3 represents a no-calibrated output voltage waveform of the integrator-type IVC with its reset pulse at 90kΩ pixel. Table 1 shows the integration time as a function of the pixel resistance before calibration. The integration time is defined as a time when the output voltage is discharged to 300mV after reset. We confirm that the integration time is increased as the pixel resistance is increased. This means that a pixel with more resistance offset has shorter integration time. The calibrated pixel, therefore, should have longer integration time than the no-calibrated pixel.

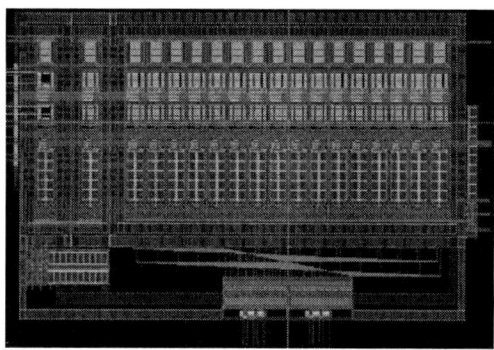

Figure 2. Layout of the designed readout circuit

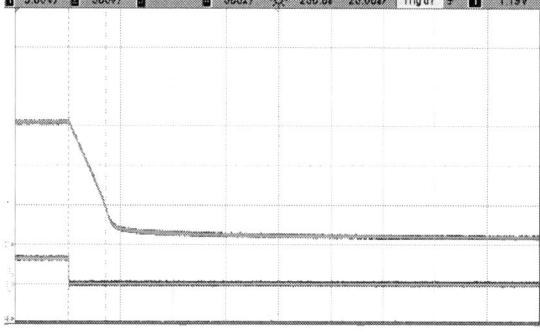

Figure 3. No-calibrated output voltage waveform of the IVC at 90kΩ pixel

Table 1. Integration time as a function of pixel resistance before calibration

Pixel Resistance [kΩ]	Integration Time [μs]
90	12.4
95	16.8
100	26.8
105	37.2

For verifying proposed algorithm, we measure the output voltage waveforms of the integrator-type IVC as the calibration current increased. The calibration current is controlled by the digital input value of DAC with 10-bit resolution. Fig. 4 shows the calibrated output voltage waveforms of the IVC. As shown in the waveforms, the integration time is increased when the calibration current or DAC output is increased. This means that the calibration current can reduce the resistance offset. This measurement results accord well with Table 1.

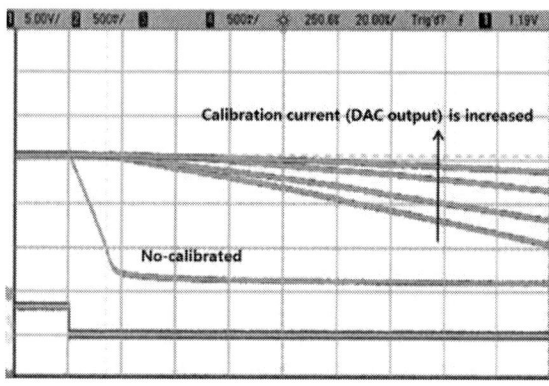

Figure 4. . Calibrated output voltage waveform of the IVC at 90kΩ pixel

IV. CONCLUSIONS

A novel average-current calibration algorithm for a bolometer type uncooled infrared image sensor is successfully implemented. The characteristic of algorithm was shown by experiment. It is concluded that this calibration algorithm can be effectively applied to the bolometer type uncooled infrared image sensor with excessive resistance offsets.

REFERENCES

[1] S. Eminoglu, M. Y. Tanrikulu, and T. Akin, "A Low-Cost 128 X 128 Uncooled Infrared Detector Array in CMOS Process," IEEE J. Microelectromechanical Systems, vol. 17, no. 1, pp. 20-30, Feb. 2008

[2] T. H. Yu, C. Y. Wu, Y. C. Chin, P. Y. Chen, F. W. Chi, J. J. Luo, C. D. Chiang, and Y. T. Cherng, "A New CMOS Readout Circuit for Uncooled Bolometric Infrared Focal Plane Arrays," Proc. IEEE Int'l Symp. on Circuits and Systems, pp. II493-II496, May 2000

[3] S. J. Hwang, A. Shin, H. H. Shin, and M. Y. Sung, "A CMOS Readout IC Design for Uncooled Infrared Bolometer Image Sensor Application," Proc. IEEE Int'l Symp. on Industrial Electronics , pp. 2788-2791, Jul. 2006

[4] S. J. Hwang, H. H. Shin and M. Y. Sung, "A New CMOS Read-out IC for Uncooled Microbolometer Infrared Image Sensor", Int. J. Infrared Millimeter Waves, pp. 953-965, Jul. 2006

A Novel Frequency-shift Readout System for CEA Concentration Detection Application

Deng-Shian Wang, Yun-Shen Liu, and Chua-Chin Wang, *Senior Member, IEEE*
Department of Electrical Engineering, National Sun Yat-Sen University
Kaohsiung, Taiwan 80424
Email: ccwang@ee.nsysu.edu.tw

Abstract—**This paper demonstrates a novel frequency-shift readout system for CEA (carcinoembryonic antigen) concentration detection. The proposed system comprises a scanning signal generator, a power detector, and a control circuit. Since the frequency-shift range of FPW (flexural plate wave) sensors is around 1 ~ 20 MHz, the proposed power detector contains an amplitude to voltage convertor to carry out the required sensing function. The proposed system doesn't need any high bandwidth OPAs. The measurement results based on FPGA verification show that the linearity of the proposed system is 0.983, where the maximum output error is 0.15 MHz.**

Keywords—Analog processing circuit, resonant frequency, power detector, peak detector, frequency-shift readout circuit

I. INTRODUCTION

According to the statistical data published by the WHO in 2014, there are approximately 14 million new cases and 82 million deaths related to cancer in 2012. The 5 most common slides of cancer diagnosed in 2012 were lung, livers, stomach, colorectal, and breast. In order to cure and discover cancers at very early stage, the CEA blood test has recently been widely used. Notably, CEA is mainly used as a tumor marker for cancers of lung, livers, stomach, colorectal, and breast, etc [1] – [3]. Most of the healthy people, even the smoking ones, the CEA concentration in blood should be lower than 5 ng/mL. If a person whose CEA concentration in blood is higher than 10 ng/mL, he or she may have high stake of various cancers.

Currently, the ELISA (enzyme-linked immunosorbent assay) and QCM (quartz crystal microbalance) have long been used to measure the CEA concentration [4]. However, the ELISA or QCM is not suitable for the clinics in the remote and suburban areas because of long operating time and high cost. To resolve the problems, based on the fact that the resonant frequency-shift of the FPW sensor is proportional to the CEA concentration, Liao *et al.* reported a CEA concentration measurement system in 2013 [5]. However, Liao's readout system has two problems.

1. OPA with high bandwidth input range: Since the resonant frequency-shift range of the FPW sensor is about 1 ~ 20 MHz, Liao's system needs two high bandwidth OPAs, where one follows with the sensor and the other is used to be an unit gain buffer in the peak detector.

2. Too many OPAs in the peak detector: In Liao's peak detector, the filter stage is required to filter out high-frequency components of the sensor output, where at least two OPAs are needed.

This paper presents a novel frequency-shift readout system. The high input bandwidth requirement has been relaxed by an amplitude to voltage convertor (AVC). Since the AVC will generate a DC voltage proportional to the input amplitude, the filter stage is no longer needed. The proposed readout system will be introduced in the next section and the measurement result is given in Section IV.

II. CIRCUIT DESIGN AND ANALYSIS

Fig. 1 shows the architecture of the proposed frequency-shift readout system. The proposed frequency-shift readout system contains three parts: scanning signal generator, power detector, and control circuit. The frequency of the scanning signal, which is generated by the scanning signal generator, increases with the time (namely, counter). Since the sensor has different frequency response at different input frequency, the amplitude of the $V_{in}(t)$, sensor's output, will vary along with the scanning signal frequency. The AVC in the power detector filters out the AC components and generates a DC voltage. After the gain stage enlarges the AVC output, the peak detector detects the maximum input voltage and sends a flag signal to the control circuit. Finally, the control circuit will calculate the CEA concentration by the resonant frequency difference between before and after the standard test solutions which is titrated into the FPW sensor. Because the scanning signal generator is our previous work reported in 2011 [6], there is no need to rephrase hereby. The details of the power detector will be introduced in following text.

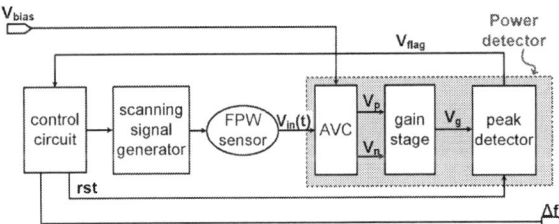

Fig. 1. The architecture of the proposed frequency-shift readout system.

III. POWER DETECTOR IN THE PROPOSED SYSTEM

The proposed power detector is composed of an AVC, a gain stage and a peak detector. Fig. 2 shows the proposed AVC, which is used to convert input amplitude into a proportional DC voltage. Since the M_{101} is driven into saturation region by

978-1-5090-3220-4/16 $31.00 © 2016 IEEE

V_{bias}, and C_{101} blocks the input DC term, the current of M_{101} is shown as follows.

$$i_{M_{101}}(t) = \frac{1}{2} \cdot \beta_n \cdot (V_{in}(t) + V_{bias} - V_{TN})^2 \qquad (1)$$

where $\beta_n = \mu_n C_{ox}(W_{101}/L_{101})$ is the MOSFET transconductance parameter. The C_{102} and C_{103} are utilized to ground out the AC terms of the output voltages. Therefore, the differential output voltages is shown as follows.

$$V_{diff} = V_p - V_n = \frac{\beta_n}{2} \cdot R \cdot V_{in}(t)^2 \qquad (2)$$

To increase the sensitivity of the proposed frequency-shift detector, the gain stage is required to amplify the AVC output. When the sensor has a maximum output amplitude, which indicates that the frequency of the scanning signal is equal to the resonant frequency of the FPW sensor, the peak detector will send a flag signal to trigger the register in the control circuit to store the scanning signal information. After the standard test solutions is titrated into the FPW sensor, the scanning process will be repeated to calculate the CEA concentration of the solution.

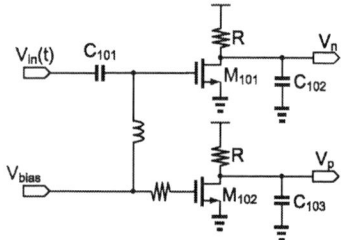

Fig. 2. The schematic of the proposed AVC.

IV. IMPLEMENTATION AND MEASUREMENT

The proposed frequency-shift readout system is realized using discretes and an FPGA board (including the scanning signal generator and control circuit), as shown in Fig. 3. Fig. 4 shows the measurement results, where the linearity is $R^2 = 0.983$ and the maximum error is 0.15 MHz. The comparison with our previous work is tabulated in Table I. The proposed frequency-shift readout system only needs two OPAs (one is in gain stage, and the other is in the proposed peak detector) such that it effectively reduces the complexity. More importantly, there is no need of any high bandwidth OPAs.

V. CONCLUSION

This paper presents a novel frequency-shift readout system for CEA concentration detection. The proposed circuit is able to detect the resonant frequency of the FPW sensor with only maximal 0.15 MHz error. By the measurement results, the linearity of the proposed frequency-shift detector is proves to be $R^2 = 0.983$.

ACKNOWLEDGMENT

This investigation is partially supported by Ministry of Science and Technology under grant NSC 104-2622-E-006-040-CC2, NSC 102-2221-E-110-081-MY3, and NSC 102-2221-E-110-083-MY3. The authors would like to express their deepest gratefulness to Chip Implementation Center of

National Applied Research Laboratories, Taiwan, for their thoughtful chip fabrication service and EDA tool support.

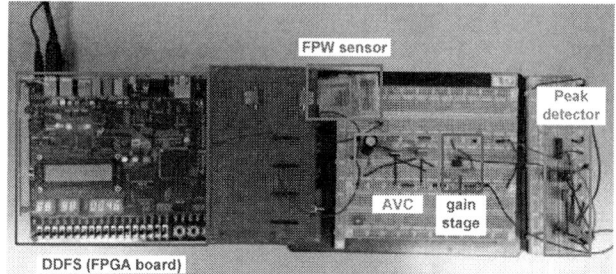

Fig. 3. Photo of the experimental environment.

TABLE I. COMPARISON WITH PRIOR WORKS

	[5]	[7]	This work
year	2013	2014	2016
technology	FPGA & chip	FPGA & chip	FPGA & discretes
linearity	None	0.9772	0.983
maximal error	None	0.12 MHz	0.15 MHz
number of OPA	4 (at least)	4 (at least)	2

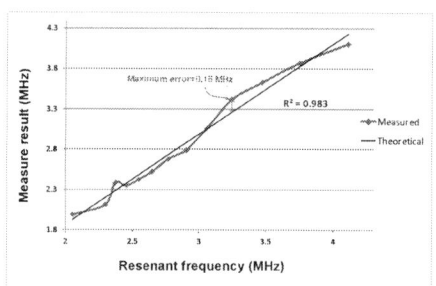

Fig. 4. The measurement results of the proposed system.

REFERENCES

[1] M. Grunnet and J. B. Sorensen, "Carcinoembryonic antigen (CEA) as tumor marker in lung cancer," *World Journal of Surgery*, vol. 76, no. 2, pp. 138, May 2012.

[2] M. J. Duffy, "Carcinoembryonic antigen as a marker for colorectal cancer," *Clinical Chemistry*, vol. 47, no. 4, pp. 624-630, Apr. 2001.

[3] D. Pectasides, A. Mylonakis, M. Kostopoulou, M. Papadopoulou, D. Triantafillis, J. Varthalitis, M. Dimitriades, and A. Athanassiou, "CEA, CA 19-9, and CA-50 in monitoring gastric carcinoma," *World Journal of Surgery*, vol. 20, no. 4, pp. 348, Aug. 1997.

[4] I.-Y. Huang, M.-C. Lee, C.-H. Hsu, and C.-C. Wang, "Development of a FPW allergy biosensor for human IgE detection by MEMS and cystamine-based SAM technologies," *Sensors & Actuators B: Chemical*, vol. 132, no. 1, pp. 340-348, May 2008.

[5] C. -C. Wang, T. -C. Sung, C. -H. Liao, C. -M. Chang, J. -W. Lan, and I. -Y. Huang, "A CEA concentration measurement system using FPW biosensors and frequency-shift readout IC," *IEEE Inter. Conf. on Electronics, Circuits and Systems* (ICECS), pp. 27-30, Nov. 2013.

[6] C. -H. Hsu, Y. -C. Chen, and C. -C. Wang, "ROM-less DDFS using non-equal division parabolic polynomial interpolation method," *IEEE Inter. Symposium on Integrated Circuits*, pp. 59,62, Dec. 2011.

[7] C. -C. Wang, C. -H. Liao, C. -M. Chang, J. -W. Lan, and I. -Y. Huang, "A fast CEA analyzer prototype for point of care testing," *IEEE Inter. Conf. on Electron Devices and Solid-State Circuits* (EDSSC), pp. 1-2, Jun. 2014.

Gap in pagination due to withheld papers.

Pages 135-138

Design of a Configurable Bit-Resolution CMOS Image Sensor for the Image Depth Extraction

Seongjoo Lee and Minkyu Song

Dept. of Semiconductor Science, Dongguk University, Seoul, KOREA

E-mail: mksong@dongguk.edu

Abstract— **Design of a configurable bit-resolution CMOS image sensor(CIS) is described. Recently, CIS pixel matrixes composed of both RGB and Infrared(IR) color filters are used for the implementation of an image depth extraction. However, in order to compensate the light density between RGB and IR, the single-slope ADC inside of CIS must have a configurable bit-resolution. For example, RGB signal has a 8-bit resolution, while IR signal has an 12-bit resolution. The proposed CIS has 4 different bit resolutions for RGB pixel, such as 8-bit, 6-bit, 4-bit and 2-bit. The proposed ADC has a maximum resolution of 12-bit for IR pixels with the architecture of two-step single-slope(TS SS) type. The proposed CIS has a 100MHz clock, and it has been designed with 0.18μm CIS technology.**

Keywords; CMOS image sensor; configurable bit-resolution; image depth extraction; RGB pixels and IR pixel; single-slope ADC

I. INTRODUCTION

Recently, Virtual Reality(VR) system is now widely used in our life with new technology developments. It is expected that VR system is applied to game, entertainment, medical system, and so on [1]. Especially, the demands on 3-dimensional (3-D) recording equipment are rapidly increasing, because entertainment and medical systems with the VR are based on 3-D scene. In order to implement 3-D scene, VR does not need only 2-D imaging, but also 3-D depth map for vertical dimensions at the same time [2]. Thus a CMOS image sensor (CIS) is needed to satisfy the image depth requirements. Fig. 1 shows a scheme of CIS pixel matrix with RGB and IR color filter to implement the image depth extraction [3]. A conventional green filter of the A group is replaced to an IR filter for 3-D scene. As a result of imaging program, when brightness of a RGB imaging is 1/16 lower than brightness of an IR scene, contrast is almost same between RGB imaging and IR scene. That is, a requirement of ADC resolution for IR scene has a factor of 4-bit higher than ADC resolution for RGB imaging. If this method is applied to column-parallel CIS with a conventional ADC, there would be a problem at the fixed bit resolution [4-6]. When a column ADC converts the analog signal of RGB and IR pixels into digital bits, it is impossible for a conventional ADC to process a different resolution on each pixel to make a balanced scene. Therefore, as a first step of the image depth extractable CIS, a variable bit-resolution ADC which is able to change the bit resolution

is proposed. For example, the proposed ADC converts a signal of RGB filter with an 8-bit resolution, while the proposed ADC converts a signal of IR filter with a 12-bit resolution. The proposed ADC has 4 different resolutions for RGB signals which are 8-bit mode, 6-bit mode, 4-bit mode and 2-bit mode. A two-step single-slope ADC (TS-SS ADC) for the column-parallel CIS has an advantage of simple architecture composed of a counter, a comparator and a ramp generator. When the resolution of TS-SS ADC is increased, the conversion time of this ADC is also increased exponentially. To avoid this problem, the proposed ADC is based on the TS-SS architecture composed of a coarse ADC and a fine ADC [7]. The circuit design and implementation are discussed in section II. Measurement results and conclusions are summarized in Section III and IV, respectively.

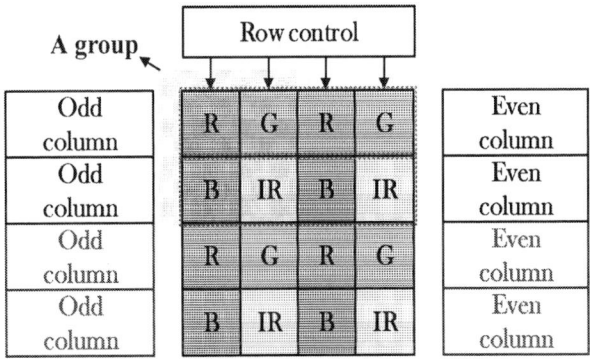

Fig.1 : CIS Pixels with RGB+IR Filters for Image Depth Extraction

II. CIS SCHEME AND CIRCUIT DESIGN

Fig. 2 shows the CIS block diagram that was designed in this paper. The CIS structure is composed of pixels, column ADCs, and digital control blocks. The pixel converts the amount of light into the corresponding voltage, which then becomes the input for the ADC. The pixel output voltage is transformed into a digital code in the block of ADC. The digital control block controls the pixel, ADC, and output interfaces, respectively. Further, a configurable bit-resolution CIS for the image depth extraction is also included. The scheme of ADC is based on a two-step single-slope ADC, which composed of coarse block and fine block.

978-1-5090-3220-4/16 $31.00 © 2016 IEEE   139   ISOCC 2016

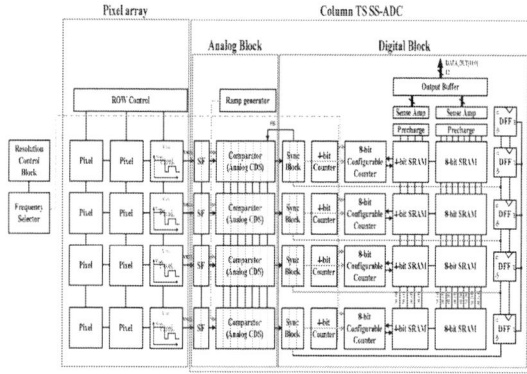

Fig. 2 Block Diagram of the Proposed CIS

III. EXPERIMENTAL RESULTS

Fig. 3 shows the chip layout and microphotograph of the designed CIS, and the core area of image sensor is about 4.14mm² (2.3 mm × 1.8mm) with a total chip area of 6.25mm² (2.5 mm×2.5 mm). The chip in this study has been fabricated with a Towerjazz 0.18μm CIS process, which uses an active pixel sensor (APS) with a 4-Tr structure. The odd column array is placed on the left side of the pixel array, and the even column array is placed on the right side of the pixel array. Further, to reduce fixed pattern noise, all the columns are designed to have the same repetitive pattern. The CIS has the pixel array of QCIF resolution which has 220×176 pixels.

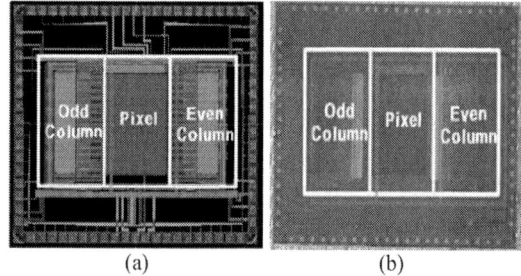

(a)   (b)

Fig.3 (a) Chip Layout (b) Chip Microphotograph for CIS

Fig.4 shows the measured QCIF sample images with the condition of the 4 different bit resolutions. It achieves the frame rates of 30frame/s at a main clock speed of 100MHz. Fig.4(a) shows the image of 8-bit resolution mode for RGB pixels, while the bit resolution of IR pixel is 12-bit. Fig.4(d) shows the image of 2-bit resolution mode for RGB pixels, while that of IR is 6-bit. Tab. 1 shows the performance comparison among the bit resolution modes.

IV. CONCLUSIONS

An image depth extractable CIS was discussed. To implement the technique, for example, RGB signals had 8-bit resolution, while IR signal had 12-bit resolution. Based on this technique, a high quality image depth map was obtained. Further, a two-step single-slope ADC was also discussed.

ACKNOWLEDGEMENTS

This research was supported by Basic Science Research Program through the National Research Foundation of Korea (NRF) funded by the Ministry of Education, Science and Technology(2015R1A1A2001455), and by the MSIP(Ministry of Science, ICT and Future Planning), Korea, under the ITRC(Information Technology Research Center) support program (IITP-2016-H8501-16-1010) supervised by the IITP(Institute for Information & communications Technology Promotion), SIDRC.

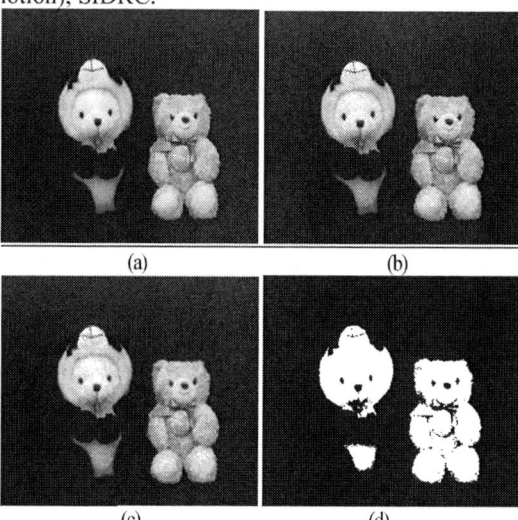

Fig. 4. Measured Sample Images with the variation of RGB bit
(a) 8-bit   (b) 6-bit   (c) 4-bit   (d) 2-bit

Tab.1 Performance Summary of the Prototype CIS.

	PFPN	CFPN	RFPN	RN	DR[dB]	Power[uW]
8-bit	0.48	0.3	0.06	0.88	60.5	72.5
6-bit	1.76	0.29	0.16	5.03	55.6	45.4
4-bit	3.67	0.30	0.21	12.65	36.6	38.6
2-bit	5.49	0.34	0.32	19.04	8.12	36.3

REFERENCES

[1] A. Rizzo, A. Hartholt, M. Grimani, A. Leeds and M. Liewer "Virtual Reality Exposure Therapy for Combat-Related Posttraumatic Stress Disorder," *IEEE Computer*, Vol. 47, no. 7, pp.31-37, Jul., 2014.

[2] E.R. Fossum, "CMOS Image Sensors: Electronic Camera-On-a-Chip," *IEEE Transactions on Electron Devices*, Vol. 44, no. 10, pp. 1689-1698, Oct., 1997.

[3] K. Fife, A. E. Gamal and H.-S. P. Wong. "A 3D Multi-Aperture Image Sensor Architecture," *IEEE Custom Integrated Circuits Conference*, 2006, pp. 281-284.

[4] T. Sugiki et al., "A 60 mW 10 b CMOS image sensor with column-to-column FPN reduction," *in Proc. IEEE ISSCC Dig. Tech. Papers*, pp. 108-109, 2000.

[5] S. Lim, J. Cheon, S. Ham, and G. Han, "A new correlated double sampling and single slope ADC circuit for CMOS image sensors," *in Proc. Int. SoC Des. Conf.*, pp. 129–131, Oct. 2004.

[6] M. F. Snoeij et al., "Multiple-ramp column-parallel ADC architectures for CMOS image sensors," *IEEE J. Solid-State Circuits*, vol. 42, no. 12, pp. 2968–2967, Dec. 2007.

[7] J. Lee, H. Park, B. Song, K. Kim, J. Eom, K. Kim and J. Burm, "High Frame-Rate VGA CMOS Image Sensor Using Non-Memory Capacitor Two-Step Single-Slope ADCs," *IEEE Trans. on Circuits and Systems-1: Regular papers*, Vol. 62, no. 9, pp. 2147–2155, Sep. 2015.

P-Backtracking: A New Scan Chain Diagnosis Method with Probability

Tae Hyun Kim, Hyun Yul Lim, Sungho Kang
Yonsei University
Electrical and Electronic Engineering
Seoul, Korea, republic of
{incendio9, lim8801}@soc.yonsei.ac.kr, shkang@yonsei.ac.kr

Abstract— Scan chain architecture is a common and major DFT for SoCs. Scan chains must be flawless for reliable SoC test. If there is a fault in a scan chain, eliminating the fault and its cause is important for high yield of SoC. In this paper, a new scan chain diagnosis method, "p-backtracking", is proposed. This method uses only software to backtrack the logic circuit and calculates the probability of fault in scan chain. The experimental results using ISCAS'89 benchmark circuits show that p-backtracking can find single fault location with higher diagnosis accuracy and smaller diagnosis resolution compared to the conventional diagnosis methods.

Keywords; Scan chain diagnosis; probability; backtraking

I. INTRODUCTION

Scan chain architecture is the most widely used DFT technique for SoC test. When there is a fault in the scan chains, locating the scan chain faults helps IC manufacturers to remove the cause of the faults and improve the yield of SoC. The procedure to find the locations of scan chain faults is called scan chain diagnosis. Many scan chain diagnosis techniques have been researched. [1][2][3][4] The diagnosis techniques can be classified into two groups. One is hardware-based technique, and the other is simulation-based technique.

Hardware-based diagnosis techniques [1][2] focus on modifying the scan cells, while simulation-based diagnosis techniques [3][4] utilize software to find fault locations. Since the simulation-based techniques require no additional hardware and complexed control plan, the simulation-based techniques are flexible to apply for many types of SoCs and advantageous when chip area is limited than hardware-based methods. However, simulation-based methods often fail to pinpoint the location of fault and provide a list of scan cells with large diagnosis resolution.

In this paper, a new simulation-based diagnosis method called "p-backtracking" is proposed. Through backtracking and probability calculation, the probability that the input scan cells contained logic 1 and logic 0 when captured to the logic circuit will be found. Comparing the probability to the input value of the test pattern, the probability if the value in the scan cell is corrupted or not can be found. The probability of corruption can be used as an additional indicator of a fault, which helps reducing diagnosis resolution.

II. PROPOSED IDEA

The proposed idea, called "p-backtracking" is based on backtracking and probability calculation. Logic circuit backtracking is useful to find the input cone of a certain output of the circuit. Especially in the circuits using scan chain architecture, the input-ends and the output-ends of logic circuit are connected to the scan cells. Thus, the input scan cells which affected the failed scan cell can be found by backtracking the failed scan cell.

A. Probability for Scan Chain Diagnosis

For more precise scan chain diagnosis, "p-backtracking" uses probability. The probability of corruption is another indicator of fault. For an input scan cell, which is found by backtracking the failed scan cell, if the probability that the input scan cell is high, the probability that the fault lies upstream of the input scan cell is also high.

B. Probability Calculation for Single Logic Gate

The probability that the input is corrupted or not will be calculated while backtracking. To calculate the probability of corruption, the probability for logic gate should be modeled first. When the output of a logic gate is known, the probability of logic 1 and logic 0 on its each input port can be calculated. For example, if the output of a two input OR gate is 1, the sets of its possible input are {0, 1}, {1, 0}, and {1, 1}. Thus, the probability that logic 1 is assigned to its input port is 2/3 and logic 0 is 1/3, respectively.

The probability for a logic gate can be represented in a *probability matrix* form in (1).

$$P_{GATE} = \begin{pmatrix} p_{00} & p_{01} \\ p_{10} & p_{11} \end{pmatrix} \tag{1}$$

$$P_{OR} = \begin{pmatrix} 1 & 0 \\ 1/3 & 2/3 \end{pmatrix}, P_{AND} = \begin{pmatrix} 2/3 & 1/3 \\ 0 & 1 \end{pmatrix}$$

$$P_{NOR} = \begin{pmatrix} 1/3 & 2/3 \\ 1 & 0 \end{pmatrix}, P_{NAND} = \begin{pmatrix} 0 & 1 \\ 2/3 & 1/3 \end{pmatrix} \tag{2}$$

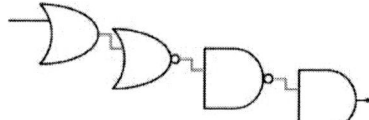

Figure 1. Backtracked Logic Path

In (1), p_{ij} represents the probability that the input value is logic j when the output of the gate is logic i. For instance, the probability matrix of OR gate is P_{OR} in (2). The probability matrices of other gates are also described in (2).

C. Probability Calculation for Logic Path

To calculate the probability of input values for a logic circuit, the probability matrix of each logic gate in the logic path should be multiplied.

For example, a logic path found by backtracking is described in Figure 1. In this example, the order of the gates in the path from the output to the input is AND-NAND-NOR-OR. Thus, if we multiply the probability matrices P_{AND}, P_{NAND}, P_{NOR}, P_{OR} in sequence, the probability matrix for the logic path input and output can be calculated. The result is described in (3).

$$P_{AND} P_{NAND} P_{NOR} P_{OR} = \begin{pmatrix} 73/81 & 8/81 \\ 19/27 & 8/27 \end{pmatrix} \quad (3)$$

D. Scan Chain Diagnosis with P-Backtracking

For scan chain diagnosis, "p-backtracking" requires three Scan chain diagnosis can be performed by following diagnose steps.

Step 1. Use fail log of flush test to find faulty scan chain, *f_chain*.

Step 2. Use fail log to make the list (*f_list*) of the failing scan cells (*f_cell*).

Step 3. Calculate upper bound *UB* and lower bound *LB* of fault candidate. [2]

Step 3. Backtrack each *f_cell* in the *f_list*. Find the input scan cells *input_cell* of *f_cell*. If *input_cell* is in *f_chain* and between *UB* and *LB*, make the list of the input scan cells and their logic path in *input_list*.

Step 4. Follow the logic path in *input_list* and multiply P_{GATE} sequentially. The probability that the input scan cell contains logic 0 is *pi_0* and logic 1 is *pi_1*.

Step 5. Compare input value of test pattern to *pi_0* and *pi_1*. Find the probability *p_c*, which implies that the input is corrupted. If the input value that was expected to be stored in a scan cell is 1, *p_c* is *pi_0*. If the expected input value is 0, *p_c* is *pi_1*.

Step 6. If *p_c* for an *input_cell* is higher than any other cells in *input_list*, mark the *input_cell* to *marked_input_cell*. If the same *input_cell* is registered in *input_list* more than

once, find highest *p_c* for the *input_cell*. The location of fault will be at the upstream of the *marked_input_cell*.

Step 7. Considering *UB* and *LB* of fault candidate and *marked_input_cell*, find the location of fault. If the result does not provide a precise location of fault, a list of possible fault location should be provided.

III. EXPERIMENTAL RESULTS

The experiment was performed with ISCAS'89 benchmark circuits and Synopsys 32nm educational library. S5378 and S9234 have five scan chains. S15850 and S38584 have ten scan chains. The average results are shown in Table 1. Diagnosis accuracy (DA) is 1 if the actual fault is in the result list, or 0 if it is not. Diagnosis resolution (DR) is the size of result list. Smaller DR means the better result. The results show that DR gets larger as the length of scan chain grows. Adjusting the number of scan chains can solve this problem. If the number of scan chain is increased, the length of the scan chain is reduced, resulting in smaller DR.

TABLE I.   EXPERIMENTAL RESULTS

Benchmark Circuit	[3]	P-Backtracking	
	DR	DA	DR
S5378	1.92	1.00	1.90
S9234	6.72	1.00	6.08
S15850	8.77	1.00	8.07
S38584	19.27	1.00	16.41

IV. CONCLUSION

P-backtracking, a new method for scan chain diagnosis is proposed in this paper. P-backtracking uses only software to backtrack the logic circuit and calculate the probability of fault in the scan chain. The probability helps reducing the list of fault candidates. Experimental result shows that "p-backtracking" finds fault location with high DA and small DR compared to the conventional diagnosis method.

ACKNOWLEDGMENT

This work was supported by the National Research Foundation of Korea (NRF) grant funded by the Korea government(MSIP) (No. 2015R1A2A1A13001751).

REFERENCES

[1] Samantha Edirisooriya and Geetani Edirisooriya, Diagnosis of Scan Path Failures, VTS, 1995

[2] Fei Wang, et al., A Design-for-Diagnosis Technique for Diagnosing both Scan Chain Faults and Combinational Circuit Faults, ASPDAC, 2008

[3] Ruifeng Guo and Srikanth Venkataraman, A Technique for Fault Diagnosis of Defects in Scan Chains, ITC, 2001

[4] Yu-Long Kao, et al., Jump Simulation: A Technique for Fast and Precise Scan Chain Fault Diagnosis, ITC, 2006

Design-Time Energy Optimization for Asymmetric Multiprocessor System-on-Chip

Yonghee Yun and Young Hwan Kim
Department of Electrical Engineering
POSTECH
Pohang, Republic of Korea
{ynpwh, youngk}@postech.ac.kr

Abstract—This paper proposes a static task mapping algorithm for energy-efficient embedded system design. The proposed algorithm is based on a genetic algorithm, which generates a set of solution candidates and evolves the candidates using genetic operations. During the evolvement, the proposed approach evaluates the time slack of each candidate, and then updates the candidates using a novel adaptive generation method. Experimental results show that the proposed method reduced the energy consumption by up to 16.1% and 34.1% compared to the existing methods and outperformed them in most cases.

Keywords—design space exploration; static task mapping; energy optimization; design-time task mapping

I. Introduction

As the performance requirements of embedded applications are increasing, multiprocessor system-on-chip (MPSoC) has become widespread in modern embedded system design. In such MPSoC systems, the performance and energy consumption largely depend on the utilization of resources. Thus, mapping application tasks onto a given architecture, called task mapping, is a major step in the embedded system design [1]. The task mapping problem, which is an NP-complete problem [1], has generally been done in a static fashion at design-time. To solve the design-time task mapping problem, many heuristic methods have been used. Among the heuristic methods, the genetic algorithms (GA) are known to generally find the best mapping for most cases [2]. Existing GA-based task mapping algorithms use crossover and mutation operators to generate the next candidates. However, they apply the same crossover and mutation operators to all the candidates without considering the characteristics of the candidates. Thus, the problem is not effectively optimized. In this paper, we present a GA-based static task mapping algorithm considering the characteristics of the candidates to reduce the energy consumption of the system. For the generation of next candidates, the proposed algorithm considers the time slack of each candidate and generates the next candidates based on the time slacks adaptively. The remainder of this paper is organized as follows. Section II defines a problem and presents the system model. Section III presents the proposed task mapping algorithm. Section IV presents the experimental results. Finally, Section V concludes the paper.

II. System Model and Problem Statement

This research was supported by Samsung Electronics Company, Ltd., and the MSIP (Ministry of Science, ICT and Future Planning), Korea, under the "ICT Consilience Creative Program" (IITP-R0346-16-1007) supervised by the IITP (Institute for Information & communications Technology Promotion)

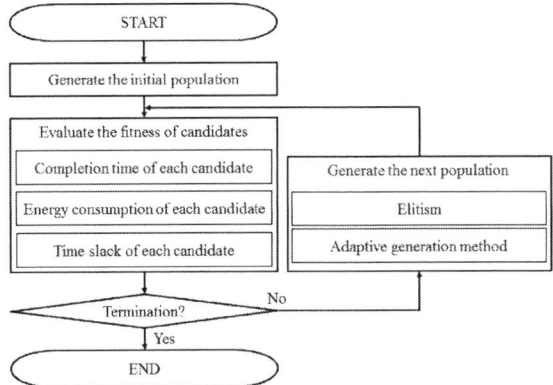

Fig. 1. Overview of the proposed method.

The system model consist of an application model and a hardware architecture model. For the application model, we use a directed acyclic graph (DAG) to represent a parallel application. In the DAG, each node represents an application task and each directed edge represents a dependency between a pair of tasks. For the hardware architecture model, we employ an asymmetric MPSoC which contains different types of cores. The cores in the system have the same instruction set architecture but their performance, size, and power characteristics are different. The cores are connected using a mesh-type network on chip (NoC) interconnect.

Given the system model, our goal is to find the best mapping between the application tasks and hardware cores to minimize the energy consumption while satisfying the given timing constraint. Unlike the existing task mapping algorithms, the proposed algorithm considers the time slack of each candidate to adaptively apply a generation method to each candidate.

III. Proposed Task Mapping Algorithm

Fig. 1 shows a flowchart of the proposed static task mapping algorithm. In this section, each process of the proposed algorithm is explained in detail.

A. Initial Population Generation

The proposed algorithm randomly generates an initial group of candidates, called population. Each candidate represents a mapping of the tasks into the cores.

B. Evaluation

The proposed algorithm evaluates the candidates in the population; the completion time (T), total energy consumption, and the number of tasks (N_T), which is needed to be modified based on the time slack, are calculated for each candidate. T can be obtained by calculating a finishing time of all tasks and the total energy consumption can be obtained by summing the energy consumed by all cores and the communication network. Using the evaluated values of T and total energy consumption, the proposed algorithm determines the ranking of the candidates. Finally, N_T which is based on the time slack is calculated by

$$N_T = \begin{cases} \left\lceil M \cdot \dfrac{T-T_C}{T_C} \right\rceil, & \text{if } T > T_C \\ \left\lfloor M \cdot \dfrac{T-T_C}{T_C} \right\rfloor, & \text{otherwise} \end{cases} \tag{1}$$

where M is the number of total tasks and T_C is the given timing constraint to finish the given all tasks.

C. Generation

The proposed algorithm generates the next candidates using two strategies: elitism and adaptive generation method. Firstly, the proposed algorithm directly copies the top one percent of candidates to the next generation using the elitism strategy. Then, the adaptive generation method is applied to each candidate except for the bottom one percent of candidates. Adaptive generation method moves randomly selected N_T tasks from one type of cores to the other type of cores, generating a new candidate. For example, if N_T of a candidate is a positive value, the completion time is higher than the timing constraint. This means that timing violation occurs. In this case, the proposed algorithm randomly selects N_T tasks assigned to slower cores, and then re-assigns the selected tasks to the randomly selected faster cores. This would effectively reduce the completion time of the candidate. In contrast, if N_T of a candidate is a negative value, the completion time is smaller than the timing constraint, which means that timing violation does not occur. In this case, the proposed algorithm randomly selects N_T tasks assigned to faster cores, and then re-assigns the selected tasks to the randomly selected slower cores, expecting to reduce the energy consumption of the system while satisfying the timing constraint.

D. Termination

The iteration process is terminated if the number of generations reaches 1000, or the best solution is not updated for one percent of population times consecutively.

IV. EXPERIMENTAL RESULTS

We compared the proposed algorithm with GeneS [3], which is based on the conventional genetic algorithm, and the mapping avoidance (MA) approach [4]. We used the Embedded System Synthesis Benchmarks Suite (E3S) [5] as an application

TABLE I. ENERGY CONSUMPTION OF THREE MAPPING ALGORITHMS

Applications	Timing Constraint (ms)	Algorithms		
		GeneS (mJ)	MA (mJ)	Proposed (mJ)
Auto-4	0.350	42.5	37.4	38.1
	0.525	22.5	24.5	19.9
	0.700	20.5	24.5	16.4
Consummer-4	4.110	921.0	-	846.8
	6.165	730.7	792.4	612.8
	8.220	356.5	469.2	327.5
Networking-4	2.070	596.2	-	540.8
	3.105	450.4	455.3	448.1
	4.140	356.7	-	358.6
Office-4	0.800	139.0	131.5	129.3
	0.900	103.9	131.5	100.1
	1.000	93.4	131.5	86.6
Telecom-4	0.270	79.2	73.4	72.9
	0.405	62.5	57.3	60.5
	0.540	52.8	-	49.1

model. We performed the experiments using the real data obtained from PowerPC [5]. The benchmark MPSoC we used has the 8-core big.LITTLE architecture, consisting of four cortex-A15 and four cortex-A7. As a means to reflect the core differences, we adjusted the voltage and frequency of the real data. The NoC energy model proposed in [4] was used for a network. As shown in Table I, the proposed method outperformed the benchmark methods in most cases. In the best cases, the proposed method reduced energy consumption by 16.1% compared to GeneS and by 34.1% compared to MA.

V. CONCLUSION

This paper presented a new approach to static task mapping, which can effectively reduce the energy consumption of the system while satisfying the timing constraint. The proposed algorithm generates the next candidates using an adaptive generation method, based on the time slack of each candidate. The experimental results showed that the proposed method reduced the energy consumption compared to the existing methods in most cases. We expect that the proposed algorithm would be effectively used for the design of an energy-efficient embedded systems.

REFERENCES

[1] W. Quan, "Scenario-based run-time adaptive Multi-Processor System-on-Chip," *Ph.D Thesis*, 2015.

[2] T.D. Braun et al., "A Comparison of Eleven Static Heuristics for Mapping a Class of Independent Tasks onto Heterogeneous Distributed Computing Systems," *J. Parallel and Distributed Computing*, vol. 61, no. 6, pp. 810-837, June 2001.

[3] Y. Wang, H. Liu, D. Liu, Z. Qin, Z. Shao, and E. Sha, "Overhead aware energy optimization for real-time streaming applications on multiprocessor system-on-chip," *ACM Trans. Design Autom. Electron. Syst.*, vol. 16, no. 2, pp. 14:1–14:32, Apr. 2011.

[4] C. Lee, S. Kim, and S. Ha, "A Systematic Design Space Exploration of MPSoC Based on Synchronous Data Flow Specification," *J. Signal Process. Syst.*, vol. 58, no. 2, pp.193-213, Feb. 2010.

[5] The Embedded System Synthesis Benchmarks Suite (*E3S*). [Online]. Available: http://ziyang.eecs.umich.edu/~dickrp/e3s/

eFuse based IC Authentication Architecture

Seung-Yeob Lee and Joon-Sung Yang
Department of Semiconductor and Display Engineering
Sungkyunkwan University
Suwon, Korea
{lee.sy, js.yang}@skku.edu

Abstract— **In the era of IoT (Internet of Things), more devices now frequently carry confidential data. At the same time, the risk of invasive attacks also increases. Hence, the security against these attacks has become more critical. In this paper, we propose eFuse-based authentication logic. This logic generates a unique value of each IC using eFuse trimming information. In the authentication process, the unique value is used to distinguish authentic ICs from copied ones. Since the eFuse trimming information resides in ICs, the proposed method can be implemented with lightweight architectures. Experimental results in 0.13-um technology show that the proposed method can reliably authenticate ICs in various operating conditions.**

Keywords; eFuse, Security, IC, Authentication

I. INTRODUCTION

As IoT technology has emerged as a future of electronic devices, the IoT devices would carry confidential information. Since the conventional authentication methods are vulnerable to invasive attacks, PUF (physical unclonable function) [1-4] has been introduced as a promising technique. Manufacturing flow causes a process variation and this can be used to identify each IC (integrated circuit). In PUF, instead of secret keys stored in IC devices, CRPs (challenge-response pairs) [1] are used as the secret keys.

In this paper, we propose an eFuse-based authentication architecture. The eFuses are a type of on-chip fuses that can be electrically programmed [5]. The proposed method utilizes the eFuse information used to trim the reference voltages. The reference voltages in IC are significantly influenced by process variations and they are trimmed using eFuses. This paper exploits the fact that a number of eFuses in IC hold the voltage trimming information, hence, this feature can be used to represent a unique characteristic of the ICs.

II. EFUSE BASED AUTHENTICATION LOGIC ARCHITECTURE

Due to process variations, a voltage supplied to each block in IC would be different from what is initially intended. To avoid instable operation, the reference voltages need to be adjusted by eFuses (i.e., voltage trimming). Let's assume that there are two chips A and B generated from the same wafer and each has two reference voltages (V_i and V_j), V_{A_i}, V_{A_j} and V_{B_i}, V_{B_j}. V_{A_i} and V_{B_i} are supposed to be the same level, however, they would be different owing to the process variations. For each voltage, trimming is performed and adjusted voltages can be represented as V_{TA_i}, V_{TA_j}, V_{TB_i} and

V_{TB_j}. After trimming, $V_{TA\_i} = V_{TB\_i}$ and $V_{TA\_j} = V_{TB\_j}$ is achieved, and these voltages are supplied to each block. The voltage difference before and after trimming can be expressed as $\Delta V_{A\_i}(= V_{TA\_i} - V_{A\_i})$, ΔV_{A_j}, ΔV_{B_i} and ΔV_{B_j}.

The eFuse trimming is a post-manufacturing process which is differently performed for each IC. Hence, if attackers duplicate the chip A, new trimming operation is necessary for the copied one. The reference voltages before and after trimming from the copied chip A can be denoted as $V_{copiedA\_i}$, $V_{copiedA\_j}$, $V_{copiedTA\_i}$ and $V_{copiedTA\_j}$. Since the original and counterfeit IC are manufactured through different fabrication processes, the process variations on the both chips would significantly differ. Due to these different process characteristics, $V_{copiedA\_i}$ is not the same as V_{A_i}. The level of $V_{copiedA\_i}$ needs to be trimmed for a correct operation, hence, $V_{copiedTA\_i}$ becomes very close to V_{TA_i}. In the same manner, the other reference voltages in the copied IC are trimmed. As the reference voltage after trimming is the same from the original and counterfeit ICs, the voltage difference before and after trimming, $\Delta V_i(= V_{T\_i} - V_i)$ and $\Delta V_{copied\_i}(= V_{copiedT\_i} - V_{copied\_i})$ would show a big difference. The proposed method defines an authentication region to distinguish the original and counterfeit IC using this distinction (Fig. 1). However, a tail

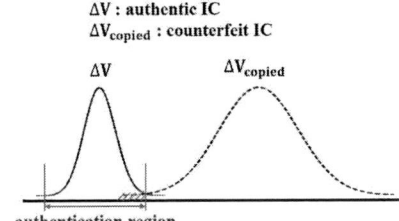

Figure 1. Reference voltage delta distribution (before and after trimming) from the authentic and counterfeit ICs

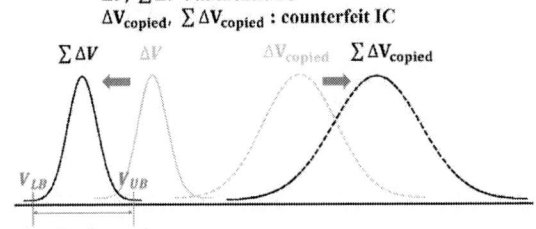

Figure 2. Multiple ΔV's to improve authentication accuracy

This research was supported by Basic Science Research Program through the National Research Foundation of Korea(NRF) funded by the Ministry of Education (NRF-2015R1D1A1A01058856) and Samsung Research Fund.

from a counterfeit IC delta distribution may overlap with an authentication region and it causes a wrong decision. To resolve this problem, the proposed method uses multiple ΔV's. As in Fig. 2, the distribution of $\sum \Delta V$ $(=\Delta V_i + \Delta V_j + \ldots + \Delta V_q)$ would make a larger separation between the authentic and counterfeit ICs, and improve an authentication accuracy by removing overlapping tails.

Fig. 3 depicts an architecture for the proposed logic that consists of three components – voltage subtractor, adder and comparator. The voltage subtractor generates a reference voltage difference (ΔV). The voltage adder receives inputs (-ΔV_i, -ΔV_j,..., -ΔV_r). It generates a summed voltage of inputs, $\sum \Delta V$. For each IC, an authentication region can be set through measurements. A lower and upper bound of the authentication region (V_{LB} and V_{UB}) are fed into a comparator. The comparator checks whether $\sum \Delta V$ belongs to the authentication region, and generates an authentication result output.

III. SIMULATION RESULTS

The proposed logic is simulated with 0.13-um technology. 5 different simulations using 3, 5, 10, 15 and 25 reference voltages are run to generate $\sum \Delta V$ (each block has one reference voltage). We assume that the target $\sum \Delta V$ is -0.45V for all cases and the reference voltages before trimming are 0.85V, 0.79V, 0.745V, 0.73V and 0.718V for 3 to 25 blocks respectively. The reference voltages are trimmed to 0.7V. It should be noted that $\sum \Delta V$ can be set to any values.

Fig. 4 gives experimental results with Monte Carlo simulations for 3 and 15 blocks cases at 25°C. Other test cases also show a Gaussian distribution. Using the $\sum \Delta V$ distribution results, V_{LB} and V_{UB} can be defined for authentication. Because the copied IC is fabricated by a untrusted manufacturer, its trimming information does not reflect original IC's process characteristics. Hence, $\sum \Delta V_{copied}$ distribution is expected to be far different from the $\sum \Delta V$ distribution in authentic ICs.

Fig. 5 describes median voltages and distribution lengths of a $\sum \Delta V$ distribution with respect to temperature. A larger median voltage difference is found as a temperature increases with more number of blocks. This happens since more blocks suffer from more process variations. The distribution length between the largest and the smallest $\sum \Delta V$ also decreases as temperature increases. Using the median voltage and

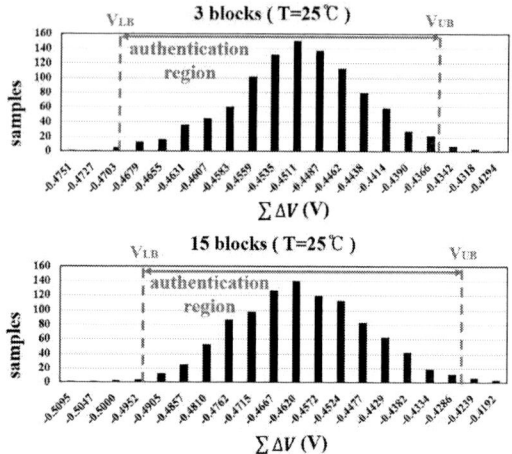

Figure 4. Monte Carlo simulation results for 3 and 15 blocks at 25°C

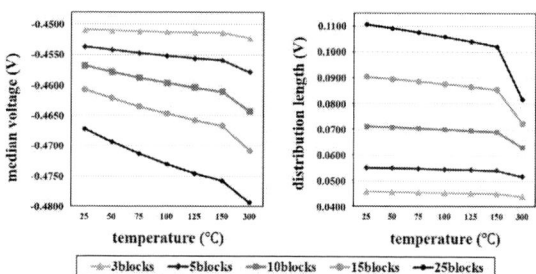

Figure 5. Median voltage and distribution length variations with respect to temperatures

distribution length with respect to temperature, the authentication region is set and this can correctly authenticate ICs regardless of operating conditions. Based on the result in Fig 5, the authentication regions for 3 and 15 blocks are illustrated in Fig. 4

IV. CONCLUSION

The eFuses are used to configure manufactured ICs, hence, they can uniquely characterize the ICs. The proposed method exploits this for authentication. Since the counterfeit ICs has new trimming information, the proposed method can be used to grant authentication.

REFERENCES

[1] G. Edward Suh, Srinivas Devadas, "Physical unclonable functions for device authentication and secret key generation," ACM/IEEE Design Automation Conference (DAC), pp.9-14, 2007

[2] 2 Herder, Charles, et al. "Physical unclonable functions and applications: A tutorial." Proceedings of the IEEE 102.8 (2014): 1126-1141

[3] Rostami, Masoud, et al. "Quo vadis, PUF?: trends and challenges of emerging physical-disorder based security." Proceedings of the conference on Design, Automation & Test in Europe. European Design and Automation Association, 2014

[4] 4 Delvaux, Jeroen, and Ingrid Verbauwhede. "Attacking PUF-based pattern matching key generators via helper data manipulation." Topics in Cryptology–CT-RSA 2014. Springer International Publishing, 2014. 106-131

[5] Robson, Norman, et al. "Electrically programmable fuse (eFUSE): from memory redundancy to autonomic chips." 2007 IEEE Custom Integrated Circuits Conference. 2007

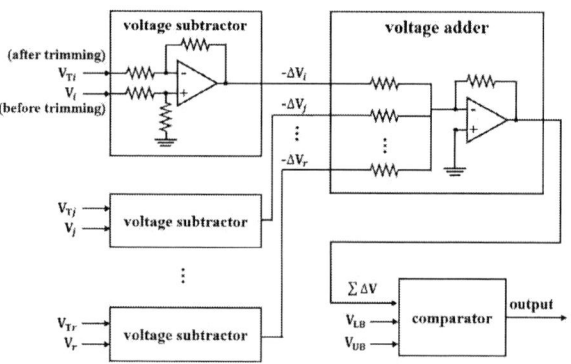

Figure 3. Proposed eFuse based authentication logic architecture

978-1-5090-3220-4/16 $31.00 © 2016 IEEE

Process Variation-aware Bridge Fault Analysis

Heetae Kim, Inhyuk Choi, Jaeil Lim, Hyunggoy Oh and Sungho Kang
Electrical & Electronic Engineering
Yonsei University
Seoul, Korea
{kht2161, ihchoi, limji, kyob508}@soc.yonsei.ac.kr and shkang@yonsei.ac.kr

Abstract— **Bridge faults are important that cause a reliability concern. Since process variation affects the bridge faults, it should be considered for bridge fault analysis. This paper proposes a new analysis method for resistive bridge faults considering process variation. The proposed method analyzes defect coverage for resistive bridge faults by using circuit level modeling. The proposed method uses the lower level analysis and it reduces redundant test patterns for bridge test.**

Keywords; Bridge fault; bridge resistance; process variation;

I. INTRODUCTION

Bridge faults are unexpected resistive connections between two or more wires and they decrease the reliability as various faulty behaviors. Moreover, process variation makes new types of faulty behavior caused by one bridge fault [1]. The new types of bridge behaviors generated by the process variation can make test escapes by traditional method [2, 3]. Therefore, a new method for the bridge fault test under process variation is needed.

The work in [2] presented a circuit level bridge fault modeling and analyzed the logic behaviors of bridge faults as the bridge resistance. However, it didn't consider the process variation, so it couldn't solve the test escape problem by the process variation. In [3], defect coverage was defined which considered the relationship between the logic behaviors for one bridge fault. One bridge fault can be observed as various logic behaviors and [3] analyzed the logic behaviors including the new types of logic behaviors by process variation. However, since it analyzed the bridge faults by using logic behavior level, only the types which have the same victim nodes can be considered.

This paper proposes a bridge fault analysis method using circuit level analysis. The proposed method can consider the types of bridge behaviors when the victim nodes or the value of two nodes are exchanged. Since the proposed method extends the analysis level from behavior level to circuit level, it allows more reduction of bridge test patterns.

II. BRIDGE FAULT BEHAVIOR

A bridge fault is observed as various logic behaviors. To make the defected circuit as fault-free, the bridge resistance(R_{br}) should be large enough and the minimum value of the bridge resistance for normal operations is called critical resistance(CR). Fig. 1 shows an example of a bridge fault

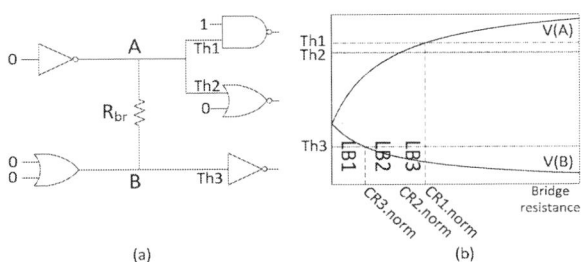

Figure 1. Example of a bridge fault (a) Circuit level modeling (b) Voltages of node A and B without process variation

between node A and B. Th1, Th2 and Th3 are logic threshold of driven gates, NAND, NOR and NOT gates, respectively, and CR1, CR2 and CR3 are critical resistances corresponding to the logic thresholds.

Fig. 1(b) shows three types of logic behaviors(LBs) without process variation, LB1, LB2 and LB3. The value which doesn't satisfied the logic threshold has faulty value at the corresponding gate inputs, therefore the intervals of the LBs are decided by CRs. For instance, since the interval of LB1 is lower than all CRs, all gate inputs for Th1, Th2 and Th3 have faulty values in LB1.

Process variation can generate new types of LBs which can't observed when nominal case. Fig 2(a) shows an example when Th2 is increased by process variation. In this example, CR2 becomes larger than CR1 and it generates the new types of behaviors, LB4, LB5 which intervals are [CR3.norm, CR1.norm] and [CR1.norm, CR2.PV], respectively. Therefore, process variation may cause test escapes and decreases the reliability of the test.

III. BRIDGE FAULT ANALYSIS

This paper proposes a new method for bridge fault test by analyzing the defect coverage which is proposed in [3]. This paper defines the defect coverage as bridge resistance based defect coverage(BRDC) which is defined as covered bridge resistance interval(CBRI) over global bridge resistance interval(GBRI). The CBRI and GBRI are union intervals of the covered and detectable LBs, respectively.

Table 1 shows the detectable LBs from Fig. 1 and Fig. 2. Assume a test pattern that excites LB1 – LB5 and observes only Th2. Since the effect of LB3 can't be observable by Th2, the test pattern can check only LB1 – LB5 except LB3. For this

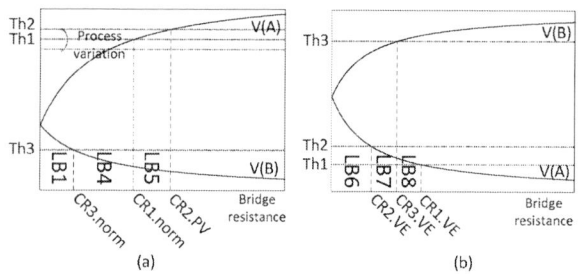

Figure 3. Various logic behaviors. (a) under process variation (b) the values of node A and B are exchanged with each other

TABLE I. LOGIC BEHAVIORS FOR NOMINAL, PROCESS VARIATION(PV) AND VALUE EXCHANGED(VE) CASES.

	Nominal			PV			VE		
	LB1	LB2	LB3	LB1	LB4	LB5	LB6	LB7	LB8
Th1	0 x	0 x	0 x	0 x	0 x	1 √	1 x	1 x	1 x
Th2	0 x	0 x	1 √	0 x	0 x	0 x	1 x	0 √	0 √
Th3	1 x	0 √	0 √	1 x	0 √	0 √	0 x	0 x	1 √

TABLE II. NUMBER OF TEST PATTERNS FOR TABLE 1

	[2]	[3]	Proposed
# of test patterns	7	3	1

test pattern, CBRI is calculated as [0, CR2.PV] by union of LB1, LB2, LB4 and LB5. Since GBRI for LB1 – LB8 is calculated as [0, max(CRs)], BRDC for the test pattern is CR2.PV over max(CRs).

The example test pattern can't cover LB3, so another test pattern is needed for LB3 in logic behavior level analysis. However, in circuit level analysis, the range of LB3 is covered by that of LB4. Testing the bridge fault means checking whether the bridge resistance is larger than the critical resistance. For instance, a test pattern for LB4 checks that the bridge resistance is larger than the range of LB4. Since the range of LB3 is included by LB4, the test pattern for LB4 also can test LB3. This relationship also can be considered between types when the value of bridged two nodes are exchanged. Table 1 also shows LB6 – LB8, which are logic behaviors when the value of node A and B are exchanged with each other. If CR1.VE is smaller than CR2.PV, LB6 – LB8 are covered by LB1, LB4 and LB5. Therefore, in the circuit level analysis, the test pattern that can detect LB5 covers all types of LBs on Table 1.

Table 2 shows that the number of test patterns for LBs in Table 1. The work in [2] doesn't consider the process variation. Therefore [2] needs 7 patterns for LB1 – LB8, because LB2 and LB4 are distinct only in the proposed method. The work [3] considers the process variation and behavior level analysis. Therefore [3] needs three patterns which cover LB1 – LB4, LB5 and LB6 – LB8 respectively. However, the proposed method needs only one pattern for LB1 – 8 as shown above.

Since the process variation is not always the same amount, another LB that is different with LBs from Table 1 can be

```
Input: A test pattern tp, Set of process variation P
Output: BRDC(tp, P) for a bridge fault
1   for all bridged node i
2     for all driven gate threshold th
3       GBRI = GBRI+1
4       Calculate the left and right end of CR
        LCR_{i,th} RCR_{i,th} under process variation P
5   for all bridged node i
6     for all driven gate threshold th
7       if tp can detected th
8         CBRI = CBRI+1
9         DLCR = max(DLCR, LCR)
10      else
11        CBRI = max(0, (RCR-LCR)/DLCR)
12  return CBRI/GBRI
```

Figure 2. Pseudo code for calculating BRDC

generated. For instance, LB9 can be generated by increasing Th1 under the process variation. But we can't assure that LB9 is covered by LB5 or LB5 is covered by LB9. In this case, test patterns for both LB5 and LB9 are needed.

The pseudo code for calculating the proposed BRDC which considers circuit level analysis is presented in Fig. 3. In CBRI calculation, the proposed analysis method under circuit level is considered. If a test pattern can detect the LB totally, 1 is added to CBRI. However when the pattern can't detect the LB, the amount of interval that detectable LB covers is added to CBRI.

IV. CONCLUSION

Bridge faults are important manufacturing faults and the process variation brings test escapes for bridge fault testing. This paper proposed the analysis method for the bridge faults under the process variation. Since the proposed method analyzes the bridge faults on the circuit level, it allows reduction of test patterns for bridge faults.

ACKNOWLEDGMENT

This work was supported by the IT R&D program of MOTIE/KEIT. [10052716, Design technology development of ultra-low voltage operating circuit and IP for smart sensor SoC].

REFERENCES

[1] H. Villacorta, J. G. Gervacio, J. Segura and V. Champac, "Low VDD and body bias conditions for testing bridge defects in the presence of process variations," Microelectronics Jounal, Vol. 46, pp.398-403, 2015.

[2] Z. Li, X. Lu, W. Qui, W. Shi and D. M. H. Walker, "A circuit level fault model for resistive bridges," ACM Trans. Design Automation of Electronic Systems, Vol. 8, pp.546-559, 2003.

[3] U. Ingelsson, B. M. Al-Hashimi, S. Khursheed, S. M. Reddy and P. Harrod, "Process variation-aware test for resistive bridges," IEEE Trans. Computer-Aided Design Integration Circuits and Systems, Vol. 28, pp.1269-1274, 2009.

A Test Methodology to Screen Scan-Path Failures

Junghwan Kim, Young-woo Lee, Minho Cheong, Sungyoul Seo and Sungho Kang
Department of Electrical & Electronic Engineering
Yonsei University
Seoul, Korea
{kjhcz, roberto, cmh9292, sungyoul}@soc.yonsei.ac.kr and shkang@yonsei.ac.kr

Abstract— **It is important to screen scan-path failures because scan-path failures affect product yield even though these are not related to the device functional operations. However, the additional efforts such as diagnosis are required to screen scan-path failures. In this paper, we propose a new test methodology to screen scan-path failures under the Automatic-Test-Pattern-Generation (ATPG) constraints and the multi-capture-clock condition without diagnosis. Experimental results show that the proposed methodology efficiently screens scan-path failures with high test coverage.**

Keywords; scan-path failure; diagnosis

I. INTRODUCTION

Nowadays, a scan-based testing is a widely used methodology for the higher test coverage and the faster test time. The number of scan chains in Very-Large-Scale-Integrated (VLSI) circuits increases proportionally to gate counts and it is reported that 10% to 30% of all defects cause the scan chains to fail, the scan chain failures account for almost 50% of chip failures [1]. However, if there are scan chain failures, these are not always related to the device functional operations. It depends on the fault location such as the internal of flip-flop or the scan-path; scan-path failures are defined as the scan chain failures but not the functional failures, as shown in Figure 1. In order to screen scan-path failures, the detection of the fault location is required. Many diagnosis techniques have been researched to find the fault locations efficiently. Commonly, the diagnosis techniques are classified into the hardware-assisted, the software-based and the signal profiling [2]. However, these have the following drawbacks. The hardware-assisted diagnosis requires the additional hardware overhead, and the software-based diagnosis needs the additional simulation time. Lastly, the signal profiling diagnosis can find the most accurate the fault location, but it requires the high cost equipment and the long analysis time. Therefore, the objective of this paper is to screen scan-path failures without diagnosis.

II. PROPOSED METHODOLOGY

Normally, a scan-capture vector is not executed, if the scan chain has a failure, because a capture operation is affected by the shift operations in any scan chains. In the proposal, in order to screen scan-path failures, a specific capture vector is executed even though the scan chain has a failure. Furthermore, a test flow with the specific capture vector is proposed.

"This work was supported by the National Research Foundation of Korea (NRF) grant funded by the Korea government(MSIP) (No. 2015R1A2A1A13001751)."

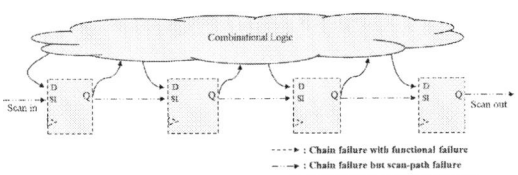

Figure 1. Scan failure according to the fault location

A. Concept of the proposed vector

For the proposed vector, the first step is to find the failed scan chain. It is reasonable to assume that there is only a failed scan chain at a time. There is a simple way to find it with the chain-check vector. Most of ATPG tools support to generate the chain-check vector without the scan-capture operation. Each scan chain has its own scan-in/out pins, respectively. So it is possible to detect which scan chain is failed by analyzing the test result from the scan-out pin using the Automatic-Test-Equipment (ATE). After finding the failed scan chain, it is stored in the ATE memory. Next, it is necessary to generate the specific capture vector called the screening vector using the ATPG tool in order to distinguish scan-path failures from the scan chain failures; the state of flip-flops of the failed scan chain can be observed even though the scan-path is not tested by the screening vector. The scan-in data cannot be loaded into all the flip-flops of the failed scan chain under the SI-Mask constraints. The definition of SI-Mask constraints is that all values of the Q-to-SI path replace with the "X" value. D-ports and Q-ports should not have any constraints to make capture operation free. Normally, a capture vector is generated by the ATPG tool under the single-capture-clock condition. However, the state of flip-flops of the failed scan chain cannot be observed because of the SI-Mask constraints under the single-capture-clock condition. In contrast, the multi-capture-clock vector can solve this problem. Figure 2 represents the comparison of the data route between two vectors when the chain-2 is failed, as an example. The value of the 2-C (flip-flop) is captured from the 1-B at a capture-clock. After the capture operation, the value of the 2-C is shifted to the 2-D and it is changed to the "X" value because of the SI-Mask constraints. So it is not possible to observe the state of 2-C under the single-capture-clock condition. Under the multi-capture-clock condition, the value of 2-B which is captured from the 1-A sequentially transfers to the 3-C at the second-capture-clock. Therefore, it is possible to observe the state of 2-B through the chain-3 although the chain-2 has the SI-Mask constraints.

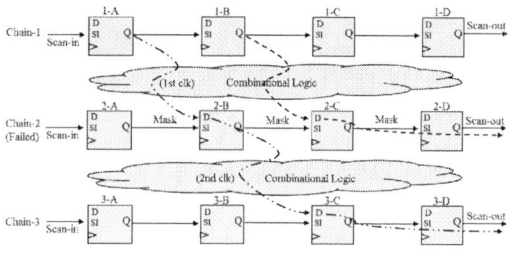

Figure 2. Data route of the two vectors

- - - - - → : Data route of the single-capture-clock vector
- · - · - · → : Data route of the multi-capture-clock vector

B. Concept of the proposed test flow

The proposed test flow is shown in Figure 3. At first, the screening vector sets which correspond to all of the scan chains are required before the test. Initially, the chain-check vector set is executed. If the chain-check vector set is passed, the normal capture vector set is executed. If the chain-check vector set is failed, the failed scan chain is detected through the failed scan-out pin using the ATE. After that, the screening vector set which corresponds to the failed scan chain is executed. If the screening vector set is passed, it means the failure location is in the scan-path and it does not affect the functional operations. Conversely, if the screening vector set is failed, the fault location is in the internal of flip-flops and it affect the functional operations.

III. EXPERIMENTAL RESULTS

In the proposed vector, long scan chain needs more the number of SI-Mask constraints. Normally, test coverage decreases if the vector has too many constraints. Therefore, the test coverage of the proposed vector according to the scan chain length is confirmed in this experiment. Table 1 shows the experimental results using the benchmark circuit B19 by *Synopsys TetraMAX*. The columns of each constraint show the test coverage. When the scan chain length decreases, the test coverage increases because the number of flip-flops of the failed scan chain decreases. When the number of flip-flops under these constraints is less than 1.4% of total flip-flops counts, and the test coverage is reached to 98%.

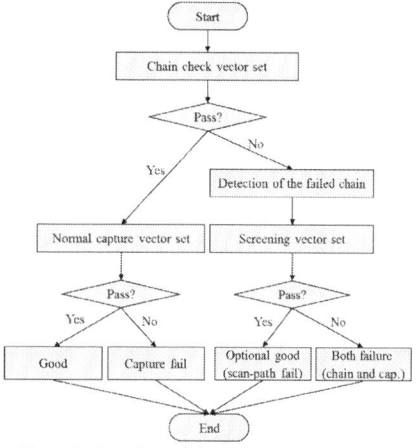

Figure 3. Test flow of the proposed methodology

TABLE I. THE COMPARISON OF TEST COVERAGE

The number of chains / Chain length	Normal vector	Proposed vector (multi-capture-clock)				
	No constraint	Chain-1 SI constraint	Chain-2 SI constraint	Chain-10 SI constraint	Chain-20 SI constraint	Average
4/1649	99.40%	74.11%	70.50%	-	-	72.31%
10/660		87.62%	87.19%	80.51%	-	85.11%
20/330		93.44%	87.94%	90.35%	87.79%	89.88%
30/220		93.73%	92.38%	94.59%	94.96%	93.92%
40/165		97.10%	94.03%	93.93%	94.05%	94.78%
50/132		97.47%	94.13%	93.50%	96.26%	95.34%
60/110		97.68%	94.19%	95.31%	98.46%	96.41%
70/95		97.83%	97.81%	98.91%	98.57%	98.28%

TABLE II. THE COMPARISON OF TEST CYCLES

The number of chains / Chain length	Normal vector	The number of test cycles				
	No constraint	Chain-1 SI constraint	Chain-2 SI constraint	Chain-10 SI constraint	Chain-20 SI constraint	Average
4/1649	2,020,391	1,889,905	1,808,970	-	-	1,849,438
10/660	794,271	837,375	775,084	1,037,633	-	883,364
20/330	397,932	397,282	404,631	427,582	478,525	427,005
30/220	271,164	274,526	269,196	268,060	278,989	272,693
40/165	203,277	209,682	204,133	208,166	199,759	205,435
50/132	163,750	165,791	165,929	161,349	176,462	167,383
60/110	137,743	140,471	139,453	140,132	143,077	140,783
70/95	118,576	122,325	127,023	125,751	128,809	125,977

In order to confirm the number of cycles, additional experiment is performed. The proposed vector set needs more test cycles than normal vector set in general. There are two reasons. First the ATPG tool makes more vectors than the normal vector set to satisfy the higher test coverage. Secondly, the addition of multi-capture cycles into a test vector causes a slight increase in test cycles [3]. Table 2 shows the comparison of test cycles between two kinds of vectors.

The proposed vector needs 2.4% more test cycles than normal vector. Also, all of test cycles can be wasted because the proposed methodology cannot salvage all of the scan chain failures. Although the proposed requires more test cycles as shown in Table 2, the main advantage is that some failed chips can turn into optional good chips without the extra diagnosis time. For the case of 70 scan chains and 95 scan chain length, 9.5 % of total area is accounted for by the scan-path. Assuming that there are the uniformly defects, 9.5% of failed chips can be usually turned into optional good chips. For example, if there are 100 failed chips, 9.5 chips can be salvaged and if there are 1000 failed chips, 95 chips can be salvaged. In other words, the more scan chain failures increase, the more the proposed methodology can be effective.

IV. CONCLUSION

In this paper, a new test methodology that can screen scan-path failures without diagnosis is proposed. The experimental results show that as the scan chain length decreases and the number of scan chain failures increases, the proposed methodology becomes more efficient. It can be a practical solution if the diagnosis time is taken into consideration.

REFERENCES

[1] Y. Huang, R. Gui, W-T. Cheng and J. C-M. Li, "Survery of Scan Chain Diagnosis," *IEEE Trans. Design and Test of Computers*, vol.25, no.3, pp. 240–248, 2008.

[2] Helen-Maria Dounavi and Yiorgos Tsiatouhas, "Stuck-at Fault Diagnosis in Scan Chains," in *Proc. Design & Technology of Integrated Systems In Nanoscale Era (DTIS)*, pp.1-6, 2014.

[3] Gaurav Bhargava, Dale Meehl and Jamse Sage, "Acheving serendiptous N-detect mark-offs in Multi-Capture-Clock scan patterns," in *Proc. International Test Conference*, pp. 1-7, 2007.

CMOS Energy Efficient Integrated Radios for Emerging Low Power Standards

Mustafijur Rahman and Ramesh Harjani

Department of Electrical & Computer Engineering

University of Minnesota, USA. Email: harjani@ece.umn.edu

Abstract— **This paper describes IEEE 802.15.6 standard compliant 2.36-2.484GHz transmitter and receiver frontends suitable for emerging low power standards. The transmitter digitally multiplexes the appropriate phases from an 800MHz poly-phase filter output to generate π/4 DQPSK signals at 2.4GHz. Modulation at 1/3rd the RF frequency reduces the total power consumption and enables channel selection using a PLL running at 800MHz. The prototype transmitter has an energy efficiency of 2.5nJ/bit at 1.2Mbps while delivering -10dBm RF power. The receiver uses a passive coupling based frequency translated mutual noise cancellation technique which allows for low supply operation thereby saving power. The FOM is 10dB higher than the state-of-the-art and the power consumption is 194μW which is 0.5X lower than the state-of-the-art.**

WBAN; 802.15.6; low power; injection locking; transmitter; receiver; noise cancellation

I. INTRODUCTION

With the proliferation of the Internet of Things the demand for low power radios continues to increase. This paper presents 802.15.6 compliant transmitter and receiver frontends. On the transmit side we propose a robust energy efficient architecture using a passive phase generation technique that utilizes subharmonic injection locking (IL). On the receive side we propose a frequency translated passive mutual noise cancellation technique that solves the power and noise figure issues normally found in sub 1V mixers.

II. ARCHITECTURE

A. Transmitter architecture

The block diagram for the proposed transmitter is shown in Fig. 1 [1]. A MUX-based architecture is proposed where modulation occurs at 800MHz (i.e. 1/3rd the RF frequency). A passive polyphase filter centered at 800MHz generates all the 8 phases necessary for π/4 DQPSK modulation at 2.4GHz. These phases are appropriately selected by the phase multiplexer (MUX) based on the digital baseband data. Modulation at 1/3rd the RF frequency reduces the power consumption and also enables us to employ simpler low power circuits which are difficult to operate at the higher RF frequencies. In addition channel tuning can be achieved by a PLL running at 1/3rd the RF frequency which further reduces power. A pulse slimmer enhances the third harmonic content at 2.4 GHz of the phase that is selected by the MUX. The ILO, tuned to 2.4GHz, locks on to this third harmonic and functions as both a high-Q bandpass filter and a frequency multiplier. Fig. 1 also shows the constellation at 800 MHz and 2.4GHz that is mapped by

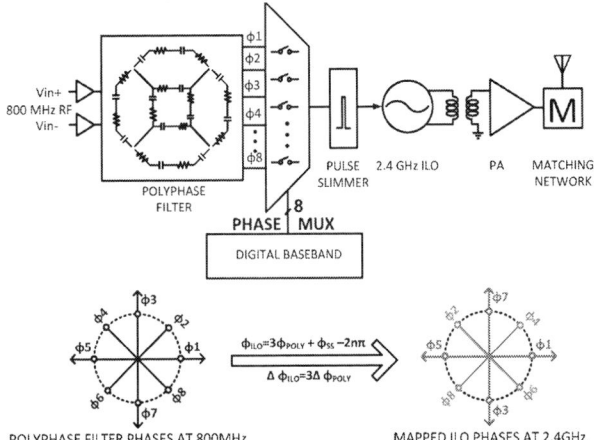

Fig. 1. Block diagram for the proposed transmitter.

this 3X frequency/phase multiplication. This design does not require capacitor bank calibration, can support a large number of channels even at high GHz frequencies and has no nonlinear phase mapping issues unlike existing IL techniques [2].

B. Receiver architecture

As shown in Fig. 2, we propose an active transconductance mixer requiring low LO power to achieve low Vdd operation [3]. The output SNR of a system can be expressed as $SNR_{out}=\epsilon VDD^2/N_{out}$, where $0<\epsilon<1$ and N_{out} is the output noise power. Low Vdd operation decreases the SNR_{out} and hence increases the effective noise figure. Additionally, lower power operation with devices operating in weak or moderate inversion increases the noise figure. Therefore, for low power operation noise cancellation techniques become essential.

In this paper, a frequency translated mutual noise cancelling (FTMNC) mixer is proposed. Unlike traditional techniques [4], we perform symmetrical noise cancellation of a fully differential structure where each path cancels the noise of the other side after downconversion to IF. The RF and LO is combined and applied to the mixer in differential form. A symmetric center tapped differential inductor couples the noise current of one side of the differential topology to the other side thereby making it common mode but retains the signal voltage in differential form. Each side has a transconductance mixer which downconverts the RF to baseband by generating in-phase IF signals which are added at the baseband. On the other hand, noise from each half, is in common-mode and gets downconverted out of phase (by +LO and −LO) and hence gets cancelled due to the addition after frequency translation.

This work is supported by Semiconductor Research Corporation (SRC) through Texas Analog Center of Excellence at the University of Texas at Dallas (Task ID:1836.98).

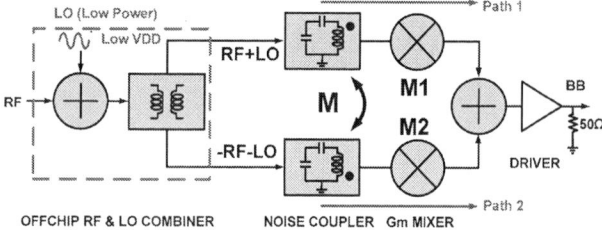

PROPOSED DESIGN: Gm MIXER USING FTMNC → LOW LO POWER & LOW NF

Fig. 2 Block diagram for the receiver frontend.

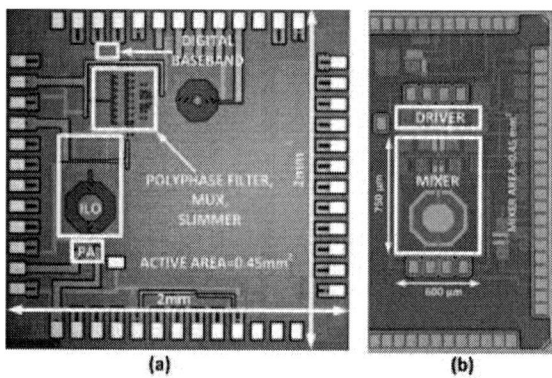

Fig. 3. Die-micrograph of transmitter and receiver frontends.

III. MEASUREMENTS

The transmitter was implemented in IBM's 130nm CMOS technology. Fig. 3 (a) shows the die micrograph. The measured RMS EVM for π/4 DQPSK modulation is 3.21% as shown in Fig. 4. The complete transmitter including estimated PLL power can be implemented with a power consumption of roughly 3mW resulting in a 2.5nJ/bit efficiency which is 1.5X lower than the state-of-the-art [5]. The receiver frontend was implemented in TSMC's 65nm CMOS technology. Fig. 3 (b) shows the die micrograph. The proposed circuit has a noise figure of 6.55dB while consuming the lowest power of 194μW from a 0.7V Vdd in comparison to state-of-the-art. The proposed design has the highest FOM of 31dB which is 10dB higher than the state-of-the-art [6]. Fig. 5 shows measured and simulated noise figures for the receiver with and without noise cancellation.

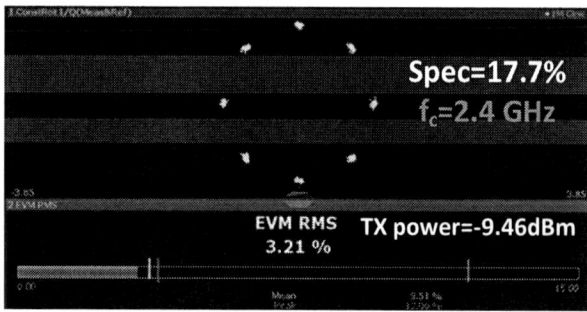

Fig. 4. Measured EVM of transmitter output at 2.4GHz.

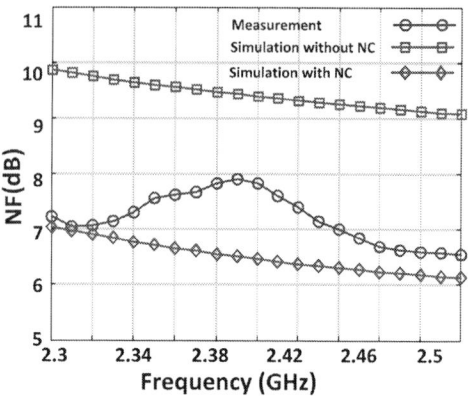

Fig. 5. Measured and simulated NF with and without NC.

IV. CONCLUSIONS

Modulation at 800MHz i.e. 1/3rd the RF frequency using a sub-harmonic injection locking technique results in a simple and low power design. Additionally, operating the PLL at 800MHz for frequency synthesis results in improved energy efficiency. The design is flexible and can incorporate other harmonics provided the phase mapping is appropriately thought through. Though the techniques developed here have focused on the 802.15.6 standard they are easily portable to other emerging low power standards with potentially tighter specifications.

On the receive side, the frequency translated mutual noise cancellation technique proposed here improves the noise figure and power consumption of sub 1V mixers. This allows for the development of extremely low power receivers. The proposed noise cancellation technique is mutual where one path cancels the noise of the other and vice-versa. The mixer has the lowest noise figure and power consumption in comparison to the state-of-the-art and its FOM is 10dB higher than the state-of-the-art.

REFERENCES

[1] M. Rahman, M. Elbadry and R. Harjani, "An IEEE 802.15.6 Standard Compliant 2.5 nJ/Bit Multiband WBAN Transmitter Using Phase Multiplexing and Injection Locking," IEEE JSSC, vol. 50, no. 5, pp. 1126-1136, May 2015.

[2] S. Diao et al., "A 50-Mb/s CMOS QPSK/O-QPSK Transmitter Employing Injection Locking for Direct Modulation," IEEE TMTT, vol. 60, no. 1, pp. 120-130, Jan. 2012.

[3] M. Rahman, R. Harjani, "A Sub-1-V 194-μW 31-dB FOM 2.3–2.5-GHz Mixer-First Receiver Frontend for WBAN With Mutual Noise Cancellation," IEEE TMTT, vol. 64, no. 4, pp.1102-1109, April 2016.

[4] S. C. Blaakmeer, E. A. M. Klumperink, D. M. W. Leenaerts, and B. Nauta, "Wideband Balun-LNA with Simultaneous Output Balancing, Noise-Canceling and Distortion-Canceling," IEEE JSSC, vol. 43, no. 6, pp. 1341–1350, June 2008.

[5] Liu, Y-H, et al., "A 1.9nJ/b 2.4GHz multistandard (Bluetooth low energy/Zigbee/IEEE802.15.6) transceiver for personal/body-area networks," IEEE ISSCC, Feb. 2013.

[6] J. Deguchi, D. Miyashita, and M. Hamada, "A 0.6V 380μW -14dBm LO-Input 2.4GHz Double-Balanced Current-Reusing Single-Gate CMOS Mixer with Cyclic Passive Combiner," in IEEE ISSCC, Feb 2009, pp. 224–225.

A 400MHz 3-10Mbps Transceiver IC with ~0.3 nJ/bit TX/RX Energy Efficiency for Body Area Applications

Zhaoyang Weng[1], Jingjing Dong[1], Hanjun Jiang[1], and Zhihua Wang[1,2]

[1]Tsinghua National Laboratory for Info. Sci. & Tech., Inst. of Microelectronics, Tsinghua Univ., Beijing, China
[2]Shenzhen Engr. Lab. on Wireless Healthcare IC Tech., Research Inst. of Tsinghua Univ. in Shenzhen, Guangdong, China
Email: jianghanjun@tsinghua.edu.cn

Abstract— **This paper presents a 400MHz energy-efficient transceiver for body area applications. The transmitter employs a FIR-embedded phase modulator to shape signal and suppress unwanted side-lobe energy. It achieves 5.9% EVM value at 10Mb/s data rate. The receiver adopts zero-IF architecture and achieves a sensitivity of -86dBm at 10Mb/s HS-OQPSK modulation. Fabricated in 65nm CMOS process, the transmitter and the receiver consumes 2.98mW and 3.1mW from 1V supply, respectively. The TX and RX energy efficiency of the transceiver is 0.3 nJ/bit and 0.31nJ/bit, respectively.**

Keywords — *energy efficiency, phase modulator, wireless transceiver, Wireless Body Area Network (WBAN)*

I. INTRODUCTION

Wireless communication plays an important role in wireless body area network (WBAN) applications. Among recent mobile healthcare applications, some demands not only fairly high data rate, but also extremely low power consumption.

In the design of transceivers for such applications, ASK modulation with incoherent demodulation is usually adopted for its simplicity in implementation, but its poor immunity against interference limits its applications [1]. Besides, FSK and PSK modulations are also popular in low power design. PLL-based transceiver is a favorable topology adopted in such designs. However, the data rate of the PLL-based transceiver is limited by the loop bandwidth. Although two-point modulation breaks the constraint in theory, gain and delay mismatch still limits the performance.

In this paper, a 400MHz transceiver is proposed to achieve high data rate and low power simultaneously. The transmitter is based on the phase-switching topology and the receiver employs zero-IF architecture.

II. TRANSCEIVER DESIGN

Fig. 1 shows the block diagram of the proposed transceiver. It consists of a phase switching transmitter, a zero-IF receiver and a ring VCO based PLL which provides the quadrature local frequency. The frequency of the ring VCO ranges from 350MHz to 500MHz to cover the concerned 400~457MHz frequency. The half-sine shaped offset-QPSK (HS-OQPSK) modulation is adopted to achieve high spectral

This work was supported, in part, by NSFC under contract #61474070, #61431166002, Guangdong International Collaboration Program # 2014A050503017, and Beijing Engr. Research Center #BG0149

efficiency and reduce spectral regrowth. The data rate varies from 3Mbps to 10Mbps.

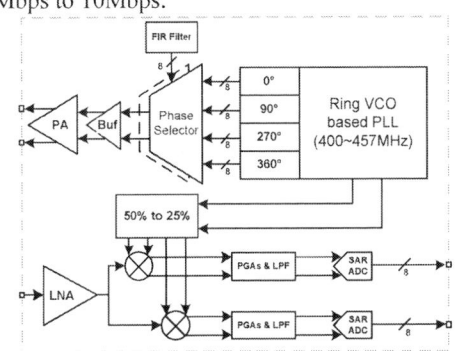

Fig. 1. Transceiver block diagram

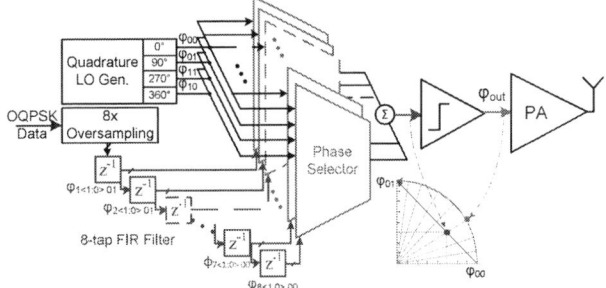

Fig. 2. Transmitter architecture

A. Transmitter

The phase-switching transmitter composes of a 8-tap FIR filter, phase selectors, a limiter and a power amplifier as shown in Fig. 2. The OQPSK data controls the phase of RF signals directly by phase selectors. The phase selection logic ensures only one phase of the quadrature LO signals will be sent to the output of the phase selector. To avoid phase hopping at the switching moments and to realize half-sine shaping, a 8-tap FIR filter is employed. The data are oversampled by 8 times firstly and then fed into the 8-tap FIR filter. The filter is implemented with a few cascade D-type flip-flops. Each tap with 2-bit control words selects the output phase of corresponding phase selector and all the output signals are added in the current domain by V-to-I convertor. When input data change, the ratio of quadrature signals of 8 phase selectors output changes gradually, and so does the phase of the summed signal, which means half-sine shaping for OQPSK modulation is achieved approximately. As the amplitude of the summed signal is not constant, a limiter is added subsequently and the final stage is a power amplifier.

B. Receiver

The receiver architecture is shown in Fig. 3. It adopts a zero-IF architecture for its low power consumption. A single-ended cascode LNA is employed to save power and improve reverse isolation. An additional bypass branch shunt with the load resistor reduces the DC voltage drop on the load to improve the output range while barely affecting the gain. The passive mixer converted the single-ended signal to differential form. With 25% duty cycle LO signal driven, the gain loss has been reduced to -3.1dB and the power consumption of the mixer is less than 0.1mW in simulation.

Each branch of the IF circuits consists of three programmable amplifiers (PGAs) and a 3rd-order active LPF. Every PGA provides 0-20dB voltage gain with 1dB step. The GBW of the opamps in the PGAs are configurable to meet different bandwidth requirements, which saves power at low data rate. The AGC circuits detect the amplitude of the ADC output and adjust the PGAs gain. To remove the DC offset which degrades the performance of the zero-IF receiver, a novel PGA-gain-uncorrelated calibration method is adopted. Since the offset is independent of PGAs gain in this method, the calibration process operates only one time and the calibration words will be fixed then. The successive approximation (SAR) ADC converted the IF output into 8 bit digital data.

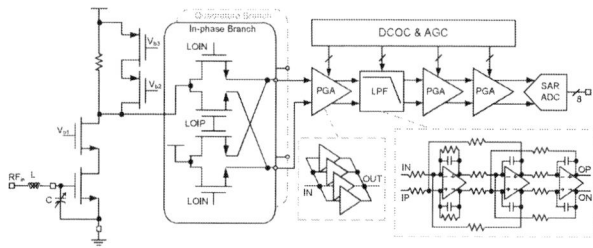

Fig. 3. Receiver architecture

III. EXPREMENTAL RESULTS

The proposed transceiver is implemented in 65nm CMOS process. The transmitter and the receiver consume 2.98mW and 3.1mW from 1V supply voltage, respectively.

Fig. 4 shows the measured 8-tap OQPSK modulation spectrum at 10Mbps data rate. With the 8-tap FIR pulse shaping, the side lobe is 20dB lower than the main lobe. The transceiver has a measured EVM of 5.9%, which is equivalent to 24.5dB SNR. The measured S11 parameter of the receiver is below -12dB across the 400-457MHz range, which shows a good matching performance of the RF port. The sensitivity of the receiver is -86dBm at 10Mbps data rate.

Table I summarize the performance of the proposed transceiver and compares it with other recent works. The transmitter achieves an energy efficiency of 0.298nJ/bit and the receiver achieves 0.31nJ/bit. Both perform better than other works.

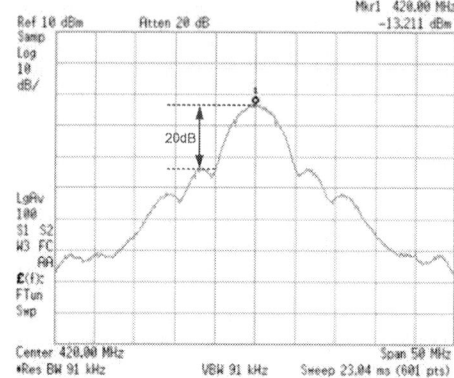

Fig. 4. Measured 8-tap OQPSK modulation spectrum

TABLE I. PERFORMANCE SUMMARY AND COMPARISON

	This work	TCAS-I 2015 [2]	ISSCC 2014 [3]	TCAS-I 2013 [4]
Frequency (MHz)	400-457	402-405	402-457	402-457
Data Rate (Mbps)	3-10	0.2/0.446	0.01-4.5	4
RX sensitivity (dBm)	-86 @10Mbps	-96.5 @0.446 Mbps	-93.5 @0.5625 Mbps	-
TX Power (mW)	2.98 @-15dBm	-	2.28 @-17dBm	2.6 @-8dBm
TX Energy/bit (nJ/bit)	0.298	-	0.4	0.65
RX Power (mW)	3.1	1.3	2.19	-
RX Energy/bit (nJ/bit)	0.31	2.91	0.33	-

IV. CONCLUSION

In this work, we propose a high-efficiency transceiver for wireless body area network that compose of a phase-switching transmitter and a zero-IF receiver. The transceiver works at a maximum data rate of 10Mbps but both the transmitter and the receiver consume about 3mW power, resulting in an energy efficiency of 0.3nJ/bit.

REFERENCES

[1] Ma C, Hu C, Cheng J, et al. A Near-Threshold, 0.16 nJ/b OOK-Transmitter With 0.18 nJ/b Noise-Cancelling Super-Regenerative Receiver for the Medical Implant Communications Service, IEEE Transactions on Biomedical Circuits and Systems, 2013, 7(6): 841-850.

[2] Cruz H, Huang H Y, Lee S Y, et al. A 1.3 mW low-IF, current-reuse, and current-bleeding RF front-end for the MICS band with sensitivity of -97 dBm, IEEE Transactions on Circuits and Systems I: Regular Papers, 2015, 62(6): 1627-1636.

[3] Vidojkovic M, Huang X, Wang X, et al. A 0.33nJ/b IEEE802.15.6/Proprietary-MICS/ISM Band Transceiver with Scalable Data-Rate from 11kb/s to 4.5Mb/s for Medical Applications, 2014 IEEE International Solid-State Circuits Conference, ISSCC, 2014: 170-171.

[4] Liu Y H, Chen L G, Lin C Y, et al. A 650-pJ/bit MedRadio Transmitter With an FIR-Embedded Phase Modulator for Medical Micro-Power Networks (MMNs), IEEE Transactions on Circuits and Systems I: Regular Papers, 2013,60(12): 3279-3288.

Time-Varying Circuit Approaches for Software Defined and Cognitive Radio Applications

S. Pamarti, N. Sinha, S. Hameed

Electrical Engineering Department
University of California, Los Angeles
Los Angeles, CA, U.S.A.
{spamarti,nsinha,sameed}@ucla.edu

M. Rachid

Silvus Technologies
Los Angeles, CA, U.S.A.
mansour.k.rachid@gmail.com

Abstract— **A brief overview of the recently developed "Filtering by Aliasing" (FA) technique is presented. FA utilizes a linear, periodically time varying (LPTV) circuit followed by a sampler to achieve sharp apparent filters from a continuous-time (CT) input to a discrete-time (DT) output. Applications to RF front-ends and spectrum scanners along with the effect of circuit parasitics on achievable performance is also described.**

Keywords: SDR, Cognitive Radio, LPTV, Spectrum Sensing, Recievers

I. INTRODUCTION

As communications technology migrates towards wideband software-defined radios (SDR) and cognitive radios (CR), receivers are expected to dynamically acquire or monitor one or more signals of interest whose frequency locations, bandwidths, and interference scenarios may not be known *apriori*. Direct digitization at the antenna itself has long been a cherished goal but analog-to-digital converter (ADC) performance is improving far too slowly for this to be feasible in the foreseeable future. So, sharp programmable analog filters are essential to the operation of such receivers. Off-chip filters (e.g., SAW filters), which are employed in traditional receivers, provide the requisite sharp filtering but are too bulky, costly, and inflexible. Conventional integrated filters e.g., gm-C or active-RC are programmable but exhibit strict trade-offs between power, area, noise, linearity and filter sharpness and are difficult to implement in fine CMOS processes. Techniques such as DT analog signal processing [1] and N-path filters [2], allow for sharp programmable filters at baseband and RF, respectively, but suffer from non-linearity.

The authors recently developed a new linear periodically time varying (LPTV) circuit technique [3] and demonstrated agile RF front-end [4] and a passive spectrum scanner [5] with best reported linearity (and in the latter case, power consumption) of comparable prior art. This paper presents (a) an overview of the LPTV technique and its application to RF front-ends and spectrum scanners, and (b) examines the role of transistor parasitics in their performance. Section II describes the basic idea, referred to as "Filtering by Aliasing (FA)" while Section III summarizes its application to RF front-ends and spectrum scanners. Section IV presents a discussion on the effects of transistor parasitics on LPTV circuit performance.

This work was supported, in part, by National Science Foundation Award ECCS-1408647.

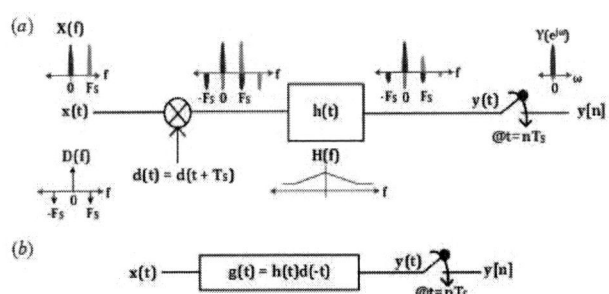

Figure 1: (a) Block diagram of Filtering by Aliasing (FA) (b) Equivalent LTI system for samples of y[n].

II. FILTERING BY ALIASING

Fig. 1(a) shows a simplified block diagram of Filtering by Aliasing [3]. The input, $x(t)$, is multiplied (or "spread") by a periodic signal, $d(t) = d(t+T_s)$, and filtered by a simple analog filter e.g., an integrator or 1st order RC; the result is sampled uniformly at a rate, $F_S = 1/T_S$. The system is equivalent to the input being passed through an equivalent LTI filter with impulse response, $g(t) = h(t)d(-t)$, followed by sampling (see Fig. 1(b)). Essentially, multiplication with a periodic $d(t)$ causes weighted and frequency translated copies of the input spectrum. Subsequent sampling causes these weighted spectral copies to alias on top of each other. By choosing $d(t)$ appropriately for the given $h(t)$, the low-frequency in-band signals, $[0, F_S/2)$, will remain unaltered whereas high-frequency out-of-band signals, $[F_S/2, \infty)$, will be significantly suppressed owing to the aliasing spectral copies canceling out each other, as illustrated. Effectively, desired sharp equivalent filters, $G(f)$, can be realized by designing $d(t)$ for given, simple filters, $h(t)$.

The authors have demonstrated that multiplication by $d(t)$ can be effectively achieved with high linearity and low power consumption by employing an LPTV resistor, $R(t) = R(t+T_S)$. The following section briefly describes the application of this LPTV technique to (a) an agile RF front end [4], and (b) a passive spectrum scanner [5].

978-1-5090-3220-4/16 $31.00 © 2016 IEEE

III. EXAMPLE PRACTICAL LPTV CIRCUITS

A. Agile RF Front End

Fig. 2(a) shows a simplified schematic of an agile RF front end: the LPTV resistor, R(t), connects the input to the virtual ground of an inverter-based integrator, through a passive square wave mixer. Momentarily ignoring the F_O mixer, it is relatively straight forward to observe that the remaining circuit implements a sharp FA-based LPF with $d(t) = \{R_S + R(t)\}^{-1}$ and h(t) being an integrate-and-dump block. The F_O mixer simply translates the LPF to a BPF centered at F_O.

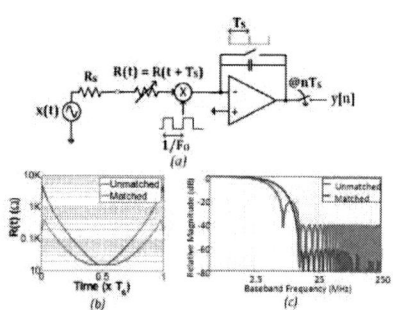

Figure 2: (a) Simplified schematic of BW and center frequency programmable RF front end with best reported close-in blocker tolerance, demonstrated in [4], (b) example R(t) choices and (c) equivalent filter responses without and with the impedance matching constraint.

As described in Section II, sharp equivalent filters can be realized by choosing R(t) appropriately. However, for use directly at the antenna, input impedance match is required. It was shown that $S_{11} = $ mean($[R(t)-R_S]/[R(t)+R_S]$). Desired maximum S_{11} places a constraint on the optimal choice of R(t). Sharp filters with >40dB stop-band attenuation can be achieved even for $S_{11} < -20$dB (see Fig. 2(b),(c)). Measurements of a 65nm CMOS prototype [4] demonstrated a receiver with programmable BW and center frequency while achieving >17dBm IIP3 even when the blocker is just 1.25x BW away.

B. Passive Spectrum Scanner

A spectrum scanner can be readily realized using a sharp programmable BPF by sweeping the filter center frequency. To achieve the required high linearity and low power consumption, a purely passive LPTV circuit, shown in Fig. 3(a), was employed [5]. Compared to Fig. 2(a), in the absence of a virtual ground, proper integration is not achieved here.

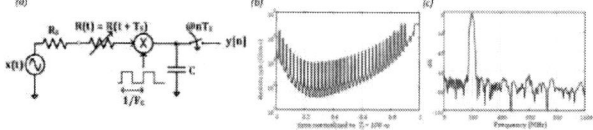

Figure 4: (a) Simplified block diagram of passive spectrum scanner demonstrated in [5] with lowest power consumption and best linearity reported. Example R(t) choice and equivalent filter response ((b) and (c)) are also shown.

However, sharp filtering is still achieved by FA principles since R(t) modulates the current and hence the charge integrated on the capacitor, thereby realizing d(t). This LPTV circuit offers many benefits. Foremost, a passive implementation without amplifiers results in very low power consumption and excellent linearity. In a 65nm prototype

where R(t) is realized as a 2GS/s, 9-bit resistor DAC, <8mW and >31dBm IIP3 were demonstrated. Secondly, R(t) is chosen such that the effective impulse response, g(t), includes the sinusoidal carrier, $g(t) = \cos(\omega_c t)g_0(t)$ where $g_0(t)$ is the impulse response of a sharp LPF. Consequently, images at multiples of Fc, in the filter response are suppressed.

IV. IMPLEMENTATION OF LPTV ELEMENTS

As expected, the performance of the LPTV filters depends on errors and parasitic effects in the realization of R(t). It was demonstrated that 10-bit DAC is sufficient for up to 60dB of filter stop band attenuation.

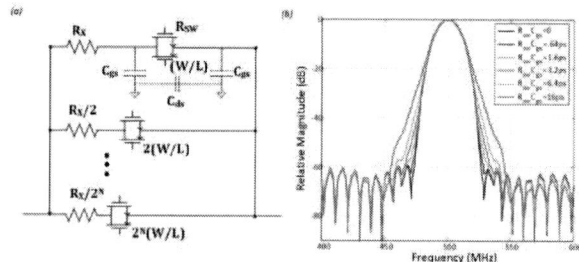

Figure 3: (a) Binary weighted resistor DAC schematic showing example transmission gate switch parasitics. (b) Effect of switch non-idealities (R_{SW}, C_{gs}) on the performance of the filter in Fig. 2(a).

Consider the binary weighted single-ended resistor DAC running at F_{CK} shown in Fig. 4(a). Resistor errors are caused by sizing errors and unequal variation (with PVT) of the value of the poly resistors and the transmission gates R_{ON}. The former errors dominate and can be easily calibrated using DC measurements. The latter can be minimized by setting $R_{sw}/R_X = 0.2$ or less. However, using large transmission gates increases capacitive parasitics, particularly C_{gs}. This limits the speed of the resistor DAC. Note that even though R(t) runs at F_{CK}, actual R(t) is a lower bandwidth signal. However, C_{gs} degrades filter sharpness and stop band attenuation. Example simulated filter responses for the RF front end described in Section II.A are plotted for a range of $R_{sw}C_{gs}$ products to quantify the effect of switch parasitics. Noticeable degradation is observed for $R_{sw}C_{gs} > 3$ps. Note that $R_{sw}C_{gs}$ gets progressively smaller with CMOS technology scaling.

ACKNOWLEDGMENT

The authors thank Dr. M.C. Chang, TSMC for fabrication.

REFERENCES

[1] M. Tohidian, et al, "A 2mW 800MS/s 7th-order discrete-time IIR filter with 400kHz-to-30MHz BW and 100dB stop-band rejection in 65nm CMOS," ISSCC Dig. Tech. Papers, pp.174-175, 2013.

[2] M. Darvishi, et al, "A 0.1-to-1.2GHz tunable 6th-order N-path channel-select filter with 0.6dB passband ripple and +7dBm blocker tolerance," ISSCC Dig. Tech. Papers, pp. 172-173, 2013.

[3] M. Rachid, S. Pamarti, and B. Daneshrad, "Filtering by Aliasing," IEEE Trans. Signal Processing, vol.61, no.9, pp.2319-2327, 2013

[4] S. Hameed, et al, "A programmable receiver front-end achieving >17dBm IIP3 at <1.25xBW frequency offset," ISSCC, 2016, pp. 446-447.

[5] N. Sinha; et al., "An 8mW, 1GHz Span, Passive Spectrum Scanner with > +31dBm Out-of-Band IIP3", RFIC, 2016 IEEE ,pp.278-281

A 0.5-V Sub-mW Energy-Efficient Receiver in 0.18-μm CMOS for IoT Applications

Tse-Wei Wang, Yi-Lin Tsai, Chong-Rong Lee, Fu-Lian Hung, and Tsung-Hsien Lin

Graduate Institute of Electronics Engineering, National Taiwan University, Taipei, Taiwan 10617

Abstract - **A 0.5-V differential BPSK (D-BPSK) receiver (RX) realized in 0.18-μm CMOS is presented in this paper. This RX adopts the injection-locking technique to demodulate the received signal. The core of this RX is an injection-locked oscillator which converts the input phase transition to envelope variation for demodulation. This work is fabricated in TSMC 0.18-μm CMOS technology. The proposed RX consumes 0.97 mW from a 0.5-V supply. The sensitivity is -45 dBm. At 10-Mbps data rate, the energy efficiency is 97 pJ/b.**

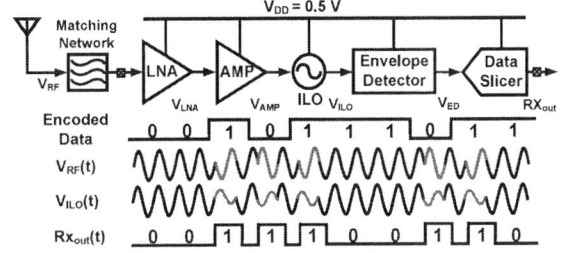

Fig. 1. The proposed D-BPSK RX.

I. INTRODUCTION

Short-distance energy-efficient telemetry is an important enabler for various IoT and wearable applications. In these applications, limited energy sources impose great challenges on circuit designs. Several prior works, for example [1, 2], have demonstrated low-power receiver (RX) designs. However, their supply voltages are still high, which limits the power consumption and energy efficiency. To further facilitate low-power consumption, low-voltage operation is preferred.

This work presents an energy-efficient RX working from a 0.5-V supply. It adopts a phase-to-amplitude conversion scheme, which is based on the injection-locking operation reported in [3], for demodulation. The previous design in [3] can only work with a supply higher than 0.9 V. In this work, design techniques to enable 0.5-V operation are presented.

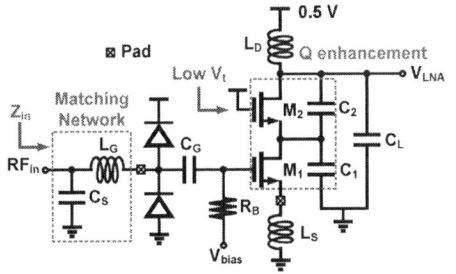

Fig. 2. LNA with Q-enhancement circuit.

II. PROPOSED ARECHITECTURE

Fig. 1 shows the block diagram of the proposed RX, which includes an LNA, a multi-stage amplifier (AMP), an injection-locked oscillator (ILO), an envelope detector, and a data slicer. LNA and the following AMP amplify the received D-BPSK signal and then injects the signal to the ILO. The ILO performs dynamic phase-to-amplitude conversion which translates the phase variation into oscillation amplitude variation. The oscillation envelope is then captured by the envelope detector. Finally, the data slicer completes the demodulation process.

The LNA implementation is depicted in Fig. 2. The Q-enhancement technique, which is based on the Colpitts topology [4], is adopted to form a positive feedback path for canceling the effect of parasitic loss of inductor L_D. The cascode common-source LNA combined with the Q-enhancement technique provides sufficient gain in ultra-low-voltage design without consuming excess power.

The ILO, composed of a digitally-controlled oscillator (DCO) and a G_m stage (M_7/M_8), is shown in Fig. 3. The ILO adopts the dual-conduction PN-complimentary LC-tank topology [5]. One main pair is used for sustaining class-C operation, while the other is an auxiliary pair where an additional resistor R_S is inserted for minimizing the total current consumption after oscillation start-up.

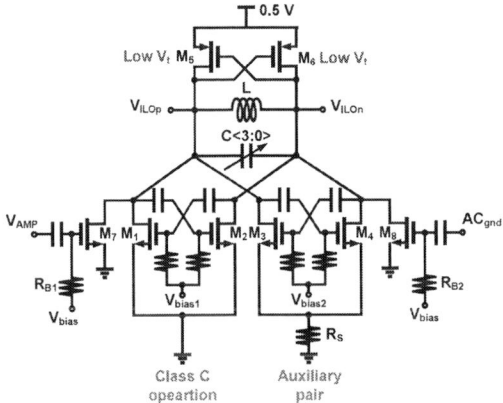

Fig. 3 Class-C dual-conduction ILO.

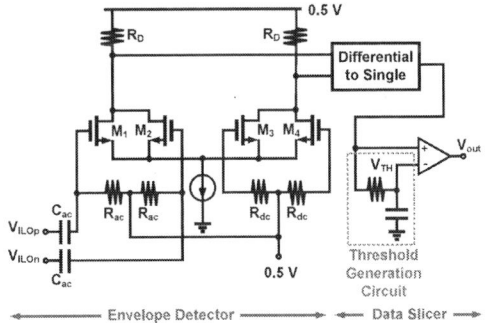

Fig. 4 Baseband Circuit.

This work is supported in part by MOST and MTK (MOST 104-2622-8-002-002). The authors thank National Chip Implementation Center (CIC), Taiwan, for the chip fabrication.

For a proper operation of the proposed phase-to-amplitude conversion, the ILO frequency must be near the frequency of the incoming RF signal. This can be achieved by adjusting the DCO via a control loop, which senses the oscillation envelope.

Fig. 4 displays the baseband circuits including an envelope detector and a data slicer. The envelope detector is realized as an active rectifier which feeds the data slicer to generate the raw data [6]. The data slicer is composed of comparators with threshold generation circuit.

III. EXPERIMENTAL RESULTS

The proposed RX is fabricated in TSMC 0.18-μm CMOS process. Operated from a 0.5-V supply, this RX consumes 0.97 mW. Fig. 5 shows the D-BPSK encoder and decoder, and illustrates the recorded timing waveforms which demonstrates the operation of the proposed RX. Fig. 6(a) displays the measured bit error rate (BER) versus input power. For 10-Mbps data rate with a BER of 10^{-3}, the RX sensitivity is about -45 dBm. At 10 Mbps, the RX energy efficiency is 97 pJ/bit. Fig. 6(b) plots the measured sensitivity versus frequency offset (between RX and TX) of this D-BPSK RX. For a frequency offset of +/- 0.2 MHz, the sensitivity degrades by about 1 dB. This sets the DCO frequency resolution.

Fig. 7(a) shows the chip micrograph. Fig. 7(b) presents the receiver benchmark in terms of energy efficiency and sensitivity. Table I. summarizes the performance of the proposed RX and compares with other similar works. The proposed RX can operate at 0.5-Vsupply and results in a good energy efficiency.

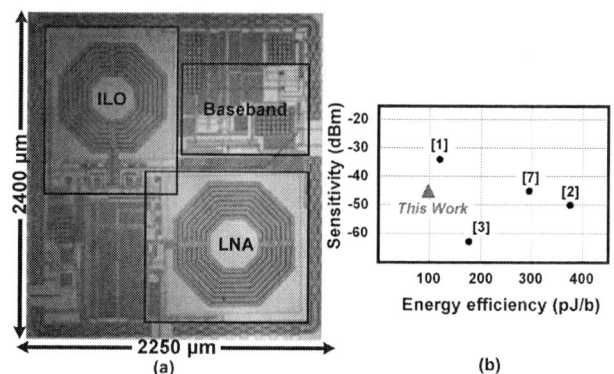

Fig. 7. (a) Chip micrograph; (b) receiver benchmark.

TABLE I. PERFORMANCE SUMMARY AND COMPARISON

Reference	[1]	[2]	[3]	[7]	This Work
Technology (nm)	90	110	180	180	**180**
Supply (V)	1.2	1.2	0.9	1.8	**0.5**
Modulation	FSK	OOK	DPSK	OOK	**DPSK**
Freq. Band (MHz)	300	433	430	400	**417**
Data Rate (Mbps)	1	2	10	2	**10**
Power (μW)	120	750	1770	590	**970**
Sensitivity (dBm)	-34	-50	-63	-45	**-45**
Efficiency (pJ/bit)	120	375	177	295	**97**

REFERENCES

[1] H. Yan et al., *IEEE T- MTT*, vol. 59, no. 5, pp. 1339-1349, May 2011.

[2] J.-H Park et al., *IEEE ISOCC*, pp. 206-207, 2014.

[3] Y.-L. Tsai et al., *IEEE Symp. VLSI Circuits*, pp. 70-71, 2014.

[4] J. Bae et al., *IEEE Symp. VLSI Circuits*, pp. 36-37, 2009.

[5] K. Okada et al., *IEEE Symp. VLSI Circuits*, pp. 228-229, 2009.

[6] J. Lee, *IEEE JSSC*, pp. 2058-2066, Sept. 2006.

[7] L.-C Liu et al., *IEEE BioCas*, pp.153-156, 2011.

[8] B. Razavi, *IEEE JSSC*, pp. 1415-1424, Sept. 2004

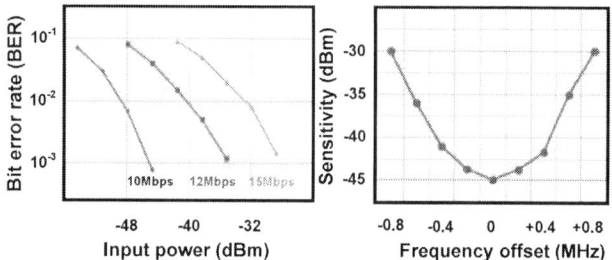

Fig. 6. (a) BER versus input power; (b) sensitivity versus frequency offset.

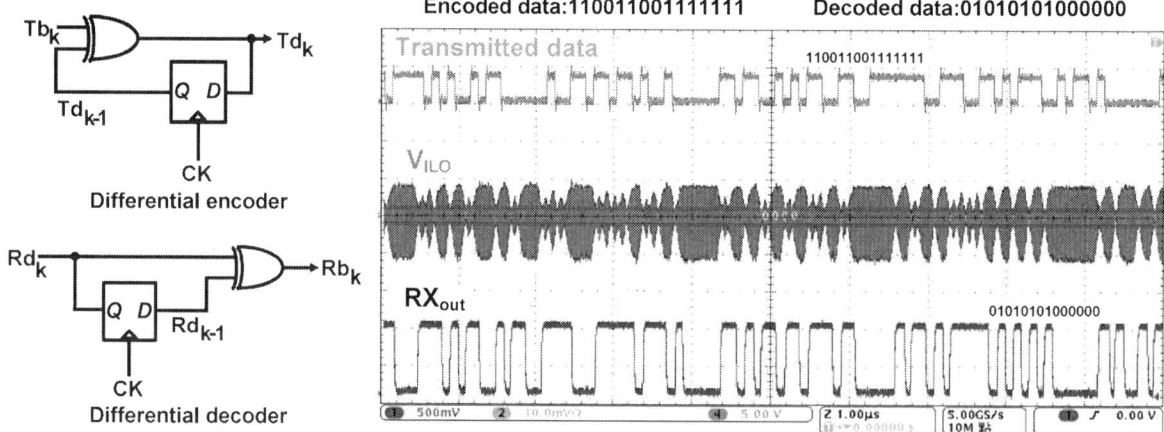

Fig. 5. Measured RX waveforms, (Top: transmitted encoded data; middle: ILO envelope; bottom: output of Data Slicer)

Automatic Image Deviation Detection for AVM Auto-Calibration

Jiwon Bang, Junghwan Pyo
Electronics and Communications Engineering
Kwangwoon Univ.
Seoul, Korea
hecarrie@naver.com, higre_pyo@hanmail.net

Yongjin Jeong
Electronics and Communications Engineering
Kwangwoon Univ.
Seoul, Korea
yjjeong@kw.ac.kr

Abstract— **Around View Monitoring(AVM) images are widely used in Advanced Driver Assistance Systems(ADAS). In order to generate an AVM image, we need to obtain four coordinates from the front, left, right, and rear camera images installed to a vehicle to perform perspective transform. Two coordinates can be extracted from the car body and the other two can be extracted from lane end points of the images. However, due to external factors such as collision, drift, etc., the physical position or angle of the installed cameras may change. This leads to the corruption of the initially obtained coordinate's actual location inside the image. In this paper, we propose an AVM auto-calibration algorithm which uses automatic image deviation detection. We compare the current car body coordinates to the initially obtained car body coordinates for deviation detection. For four 640x480 input images, it takes 2400ms for deviation detection and 3875ms for the whole AVM auto-calibration algorithm at Intel i5 3.5GHz processor environment.**

Keywords; AVM; Auto-Calibration; Lane Edge Detection; ADAS;

I. INTRODUCTION

Around View Monitoring(AVM) images are crucial for ADAS, thus most of the vehicles manufactured nowadays possess AVM systems. AVM images are generated from fish-eye lens cameras installed to the front, left, right, and rear of a vehicle. We perform radial distortion correction which occurs due to the characteristics of fish-eye lens cameras, and manually extract four points from the corrected image in order to perform perspective transform. The perspective transform generates bird's eye view images of the four input images, and we stitch the four bird's eye view images to generate the final AVM image. However, due to external factors such as collision, drift, etc., the position or angle of the cameras installed to the vehicle may change. This leads to the corruption of the initially obtained four coordinate's actual location inside the input image. AVM auto-calibration enables on-road coordinate modification to solve this problem.

Only few researches related to AVM auto-calibration methods are introduced to the public since this issue is manufacturer-friendly. Nevertheless, some researches attempted to perform this algorithm due to the inconvenience of the manual coordinate extracting procedure of AVM image generation. [1] tried using specific calibration ground patterns and a lookup table for comparison for auto-calibration for deviation rates up to ± 2 degrees in X-axis, Y-axis, and angle shift. [2] attempted to invent a semi-auto calibration algorithm also by using known ground patterns and corner detection. [3] proposed an auto-calibration algorithm for the front camera image of the AVM system for driving states using the vanishing point of the image provided from the lanes and the license plate region of the car body. In this paper, we propose an AVM auto-calibration algorithm which uses automatic image deviation detection in order to provide accurate AVM images in on-road situations, which means we don't have to go to a specific shop or area to fix the deviation. We detect image deviation comparing the car body points of the current images and the initial state images. When image deviation is detected, we use these two coordinates and two more coordinates extracted from lane edge points of the deviated image region.

II. AVM AUTO-CALIBRATION ALGORITHM

A. Algorithm Flow

The brief algorithm flow for the entire AVM auto-calibration is shown in Fig. 1.

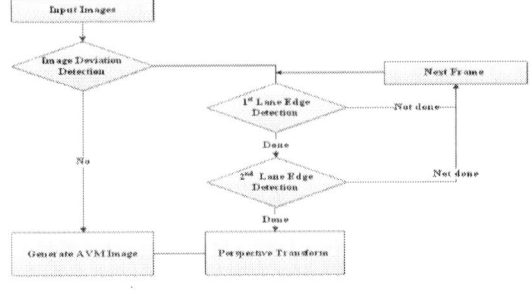

Figure 1. Algorithm flow

Input image deviation detection is performed by extracting the two current car body coordinates and comparing them to the initially obtained car body coordinates. When a deviated image is detected, we detect two lane edge coordinates in order to correct the corrupted images. Using the obtained two car body points and two lane edge points, we perform a perspective transform to generate corrected bird's eye view images.

B. Image Deviation Detection

Since AVM auto-calibration isn't a system which is required for every frame, we developed an algorithm that indicates the system to perform auto-calibration only when deviation is detected. By comparing the initial state car body point coordinates and the current, respectively, we detect which image is deviated. We use the two license plate edge points for the front image, the front bumper region edge and the front door handle for the left and right images, and the two screws which tightens the license plate for the rear image. Fig. 2 shows the detected car body points.

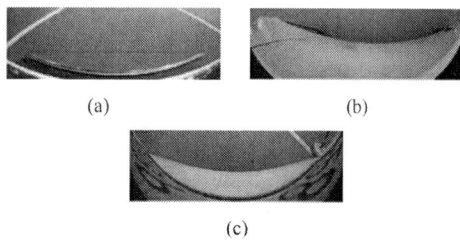

(a)   (b)

(c)

Figure 2. Car body points. (a) Front, (b) Right, (c) Rear

C. Lane Edge Detection

Normally, we can only find one lane end point from one frame since conventional road lanes appear in turn. So, we detect one end point, save the coordinate, and wait until we find another end point. Fig. 3 shows an example how two lane end points are extracted.

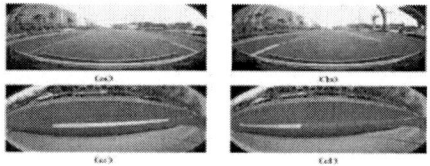

Figure 3. Lane end detection with multiple images, (a) – (b) Front image case, (c) – (d) Left image case

D. Perspective Transform

After performing car body point extraction and lane end detection, we obtain four coordinates which makes the system capable of performing a perspective transform that corrects the corrupted image.

III. RESULTS AND DISCUSSION

To check the results of the proposed algorithm, we recorded ±3 degrees deviated images for x-axis, y-axis, and angle shift. The driving state images were recorded using ImageNEXT 2D VGA fish-eye lens cameras installed to the four sides of a vehicle. Since there is no quantitative measuring method of AVM auto-calibration, we used PSNR to check the proposed algorithm results. Fig. 4 shows the auto-calibration algorithm results and Table 1 shows the PSNR results, respectively. We achieved an average 10.71 increase of PSNR for on-road condition AVM images, a significant increase considering that PSNR is a log scale

measuring index. The execution time, measured with an Intel i5-4690 3.5GHz processor environment, of the proposed image deviation detection algorithm was 2400ms, and the entire auto-calibration algorithm 3875ms, respectively. Since we don't have to perform auto-calibration for every frame, the execution time result shows promising on-vehicle implementation potential compared to [2] which takes 60 seconds in a specific patterned environment.

TABLE I.   PSNR OF DEVIATED AND CORRECTED IMAGES

Input	Deviated image	Corrected image
Front	22.10	35.02
Left	15.74	29.12
Right	23.07	28.55
Rear	28.78	39.83

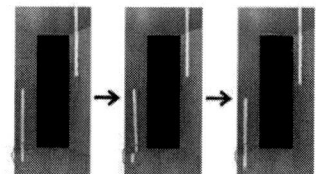

Figure 4. Auto-calibration result, (a) Original, (b) Deviated, (c) Corrected image

IV. CONCLUSION

In this paper, we developed a full AVM auto-calibration for all directions using an additional perspective transform as a preprocessing procedure to correct deviated images. We detect which image is deviated by using car body point coordinate extraction. We obtained the other two coordinates required for the perspective transform by using lane end detection with multiple images. For experimental results, we used ±3 degree complex deviated case input images and achieved a 10.71 PSNR improvement for on-road environment AVM images. The PSNR result around or over 30 for on-road environment also indicates that the proposed system fulfills its practical purpose for users. The execution time of the entire algorithm was 3875ms, which shows promising on-vehicle implementation potential since the algorithm is performed only when deviation is detected.

ACKNOWLEDGMENT

This work was supported by the IT R & D program of Ministry of Trade, Industry and Energy (10049192, Development of a Smart Automotive ADAS SW-Soc for a Self-Driving Car).

REFERENCES

[1] Yu-Lung Chang, Li-You Hsu, and Oscal T.-C. Chen, "Auto-Calibration Around-View Monitoring System", IEEE Conf. Vehicular Technology, Sep., 2013, p. 1-5

[2] S.M. Santhanam, V. Balisavira, and S. H. Roh, "Lens distortion correction and geometrical alignment for Around View Monitoring system",18th IEEE Int. Symposium on Consumer Electronics, June. 2014, p. 22-25

[3] Jiwon Bang, Junghwan Pyo, and Yongjin Jeong, "AVM Auto-Calibration for Driving Vehicle Images" Information Science and Applications(ICISA), 15-18 Feb. 2016, p.471-480

Hardware implementation of fast traffic sign recognition for intelligent vehicle system

Eunchong Lee, Sang-Seol Lee, Youngbae Hwang and Sung-Joon Jang

Korea Electronics Technology Institute
Daewangpangyo-ro 712beon-gil, Bundang-gu, Seongnam-si
Gyeonggi-do, South Korea
{ elee, sslee81, ybhwang, sjjang0626}@keti.re.kr

Abstract— In intelligent vehicle systems, the traffic sign recognition (TSR) has become very important because of the rapid increase in demand of intelligent transportation system (ITS). However, most of them were implemented by expensive systems due to the high complex computation for processing the real-time segmentation and recognition process. In this paper, the preprocessing with the red region extraction in the RGB color space has been applied to improve TSR performance. To implement proposed architecture, we adopt the Xilinx Virtex-7 V2000T Field Programmable Gate Array (FPGA) with a fully pipelined structure. Experimental results show that the implemented preprocessing block use 1,212 lookup table (LUT) and 11.14 KB sized internal memory. The proposed architecture enables a real-time TSR processing with pipelining for a full-HD (FHD) video (2 Mega pixels) at 60 frames per second.

Keywords; Traffic Sign Recognition (TSR); Color Extraction; Intelligent vehicle systems; Segmentation; FPGA;

I. INTRODUCTION

In recent years, the advanced driver assistance system (ADAS) and self-driving technologies are very important. Especially, the TSR function is very important application for informing the drivers about current road conditions and reducing the traffic accidents [1]. Most of the TSR algorithms are composed of pre-processing (color extraction, labeling, region-of-interest (ROI)), feature extraction, recognition and classification. The most of the state-of-the-art feature descriptors of object recognition for TSR use the histogram of oriented gradients (HoG) with support vector machines (SVM) for classification [2-4]. But, applying HoG and SVM to entire image consumes too much computational power. So, the ROI extraction is the foundation technique of TSR to reduce complexity. For identifying the ROI, a binary image is used with color segmentation [5]. Especially red region detection to create binary image is very important since most of the traffic signs such as speed limit signs and warning signs are red. Moreover, the low-complexity labeling process is very important to segmentation of red region without additional iteration process and storing process of the interim results. Thus, to put the label together into one which has been separated are a significant challenge of hardware implementation for real-time processing. Our goal is to design a fast pre-processing hardware with low latency and reasonable logic size.

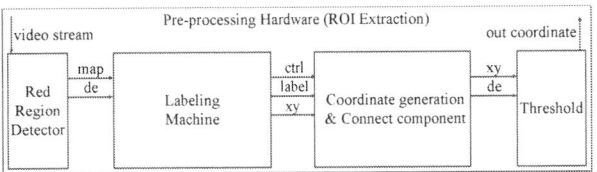

Figure 1. Pre-processing Hardware Architecture

II. PROPOSED ARCHITECTURE

The proposed hardware is divided into 4 sub-block modules (red region detector, labeling machine, coordinate generation & connect-component and threshold) shown as Fig 1. The role of red region detector is finding red pixels and creating binary image map. The labeling machine assign a label to distinguish the binary image component. Also, it put together labels that has been separated, if needed. The coordinate generation makes x,y-coordinate for unique component's with pipeline connect-component technique. Finally, the threshold module reduce ROI candidates to minimize computational power for hog processing.

A. Labeling Machine

It was applied labeling algorithm using the extra index buffer to specify the candidate region for the traffic sign recognition. Labeling machine assign a label to distinguish if there is a valid pixel in the binary image. Also, if the label has been assigned to the adjacent pixel already, current pixel assign the same label, and finally to generate a candidate region for the traffic sign recognition.

In Fig. 2 (a), is showing a process of generating the label map by using the index buffer. First of all, if the effective pixel is present at the location of the scan, the pixel is assigned a new label. If the effective pixel is present in the next scan position is assigned the same value as the label previously assigned to pixel. In some cases, more than one effective pixel regions are met at the same pixel location. In this case, the smaller value of the adjacent label is assigned to current pixel. However, when connecting the adjacent label on the current location, the area that has been marked as connect to current pixel cannot be combined into one. To solve this problem, the label that is changed is stored to the index buffer that is referred to generate the coordinates of the candidate region.

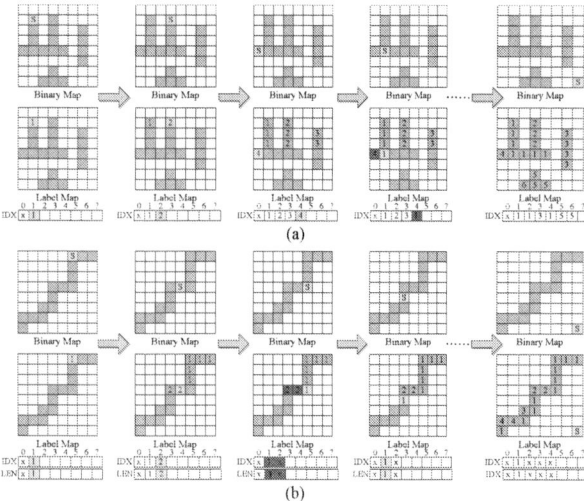

(a)

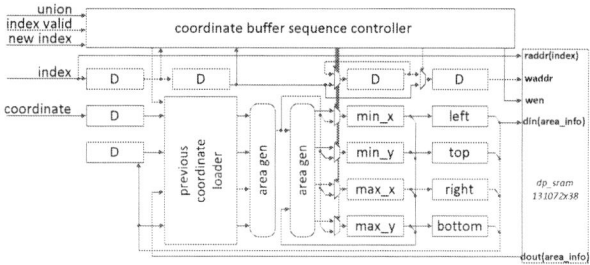

(b)

Figure 2. Labeling sequence

In Fig. 2 (b), there is presented a number of different areas connected to a single label. For the hardware processing, the variable latency is occurred to determine the root label from the multiple read process at the index buffer. In this case, hardware processing is struggle to manage scheduling. If two or more regions are met, it is necessary to change the value of the label of entire pixel which has been labelled same value in a row. However, it is not possible to change all the label that is stored to different sram address at the same time. To solve this problem, the only first position of label in a row but also the length of label are stored together. For the next row, the connect-component process is performed by valid label and length that were stored to first pixel of the same label location. After applying it, we do not need to store label for entire frame. We need to store only the label and the length of the previous row, so it is possible to reduce the size of index buffer as 1 / image-height.

Figure 3. Coordinate generator

B. Coordinate generator

The size of the final candidate region within the image is impossible to estimate. Therefore, it is common to generate the coordinates of the candidate region after complete label map is generated. Our proposed hardware architecture reduces unnecessary waste cycle through the real-time update coordinate and connect-component. In Fig. 3, the information of coordinate and label are stored to the dual-port sram by coordinate generator. If the same label comes in,

coordinate update process is performed between org from the coordinate buffer and new input coordinate data. Also, if the connect-component label comes in, same update process is performed and unnecessary label is marked as invalid. The proposed hardware is required 5cycle to store coordinate data from the input but, data hazard problem has arisen when the current coordinate is read before completely finishing coordinate update sequence. Thus, we solve the data hazard problem by receiving feedback data from intermediate results of the coordinate update sequence. Furthermore, appropriate pipeline technique has been applied pixel-by-pixel processing structure which was designed to make faster working frequency and easier IP application for various video sizes.

III. IMPLEMENTATION AND RESULTS

The proposed hardware was implemented in Verilog-HDL and was simulated with Cadence ncverilog and Synopsys Verdi-3 tool. A dedicated hardware IP is synthesized by the place and route tool using Xilinx Vivado for Virtex-7 V2000T FPGA. The proposed hardware is working on maximum frequency at 200 MHz. It is enough to generate the real-time FHD image 60 frames per second on operating frequency at 148.5 MHz.

TABLE I. IMPLEMENTATION RESULTS

	Used		Utilization
Logic utilization			
Number of Occupied Slices	256		0.08%
Number of FIFO/BRAMs	3.5		0.27%
	LUT	FF	Memory
Complexity distribution			
Red Region Detector	29	35	-
Labeling Machine	147	219	1.64KB
Coordinate generator	350	310	9.5KB
Threshold	44	78	

ACKNOWLEDGMENT

This work is supported by the Industrial Core Technology Development Program of MOTIE/KEIT, KOREA. [10060387, Development of realtime processing ADAS algorithm IPs based on HD Camera image for Autonomous driving car].

REFERENCES

[1] Yunxiang Ma and Linlin Huang, "Hierarchical traffic sign recognition based on multi-feature and multi-classifier fusion", International conference on information science and electronics technology, 2015.

[2] Sheldon Waite and Erdal Oruklu, "FPGA-Based Traffic Sign Recognition for Advanced Driver Assistance Systems", Journal of Transportation Technologies, 2012.

[3] Greenhalgh, Jack, and Majid Mirmehdi, "Real-time detection and recognition of road traffic signs", IEEE T. ITS, 2012.

[4] Zaklouta F. and Stanciulescu B., "Real-time traffic-sign recognition using tree classifiers", IEEE T. ITS, 2012.

[5] D. Soendoro, and I. Supriana, "Traffic Sign Recognition with Color-based Method Shape-arc Estimation and SVM", Proc. of 2011 International Conference on Electrical Engineering and Informatics, Bandung, Indonesia, pp. 1-6, July 2011.

Front Collision Warning based on Vehicle Detection using CNN

Junghwan Pyo, Jiwon Bang
Dpt. of Electronics and Communications Engineering
Kwangwoon Univ., Seoul, Korea
higre_pyo@hanmail.net, hecarrie@naver.com

Yongjin Jeong
Dpt. of Electronics and Communications Engineering
Kwangwoon Univ., Seoul, Korea
yjjeong@kw.ac.kr

Abstract— **Front Collision Warning(FCW) is a critical safety function of Advanced Driver Assistance System(ADAS). Recently, many researches related to FCW systems which use monocular camera image processing have been introduced. In this paper, we propose an FCW system for highway environment based on vehicle detection using Convolutional Neural Network(CNN) as a classifier. Adaptive Region-of-Interest(ROI) is set using lane detection to enhance speed and detection performance of the system. We measure the distance between our vehicle and the detected vehicle in front by calculating the ratio between the lane width of the position of the detected vehicle and our vehicle, respectively. Time-to-Collision(TTC) is used as a collision warning index. For FHD(1920x1080) black-box camera images taken in highway environment, the detection rate of the proposed CNN is 99.1%, and the execution time of the system is 19.8ms per frame.**

Keywords; FCW; Vehicle Detection; CNN; Lane Detection;ADAS;

I. INTRODUCTION

As Advanced Driver Assistance System(ADAS) has emerged as a major field of interest to engineers, many researches and applications related to the area have been introduced to the public. Moreover, ADAS has become a requirement, not a recommendation to vehicle manufacturers. Euro New Car Assessment Program(NCAP) is a representative program which requires ADAS functions as a safety indicator for manufactured vehicles. Front Collision Warning(FCW) is one of the safety functions required for Euro NCAP. FCW provides the user a visual or voice warning when collision with the vehicle in front is expected. For most FCW systems, Time-to-Collision(TTC) is used as a collision warning index[1].

The FCW system is usually composed of two stages; the vehicle detection stage and distance measuring stage. In the vehicle detection stage, Hypothesis Generation(HG) is performed to find candidates for cars, and Hypothesis Verification(HV) to verify the candidate, respectively[2]. To calculate TTC, we measure the distance between the detected vehicle in front of the user's current lane[3]. Warning signals are provided to the user based on TTC result.

Many previous researches have been made related to on-road vehicle detection. Most of the early works used a combination of feature extraction and machine learning. The various combinations of the two are well organized in [4]. For recent works, Convolutional Neural Network(CNN)s are

being applied to vehicle detection systems[5]. Compared to its predecessors, CNN-based vehicle detection doesn't require a feature extracting phase since the network itself learns its own features.

In this paper, we propose an FCW system with FHD black-box images taken in highway environment. CNN is used for vehicle detection, and CalTech rear-viewed vehicle database (1999) is used to evaluate the detection rate of the CNN algorithm. An adaptive Region-of-Interest(ROI) setting stage using lane detection is added before the vehicle detection stage to enhance speed and detection performance of the system. TTC is calculated using the distance measured by the method proposed in [6] to provide warnings.

II. FCW ALGORITHM

A. Algorithm Flow

The entire algorithm flow for the FCW is shown in Fig. 1.

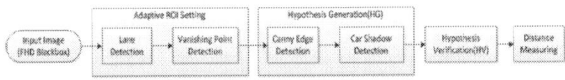

Figure 1. Algorithm flow

Adaptive ROI setting is performed by detecting the lane the driver is currently in, and the Vanishing Point(VP) of the image is determined using the two lanes. After the ROI is set, we use Canny edge detection and car shadow detection for HG. HV is performed using a pre-trained CNN. Distance measuring and collision warning is performed using the verification result and the detected lanes.

B. Adaptive ROI Setting

This stage is performed for three reasons; to reduce computation time, to eliminate hypothesizes outside the current lane, and to detect VP using lanes. Lane detection is performed using line Hough transform, and the VP of the image is set as the intersection point of the two lanes. Fig. 2 (b) shows the result of this stage. Since FCW systems only require cars in front of the current driving lane, we can reduce the number of candidates

C. Hypothesis Generation / Hypothesis Verification

We find car-like candidates by using edge figures and car shadows. Since we restrict the scanning area in the previous stage, the number of HGs is fewer than 5 in most highway cases. We verify the extracted candidates using a CNN

978-1-5090-3220-4/16 $31.00 © 2016 IEEE   163   ISOCC 2016

similar to Lenet[7]. Fig. 2(c) shows the verification result of the provided input image.

D. Distance Measuring / Warning

In order to measure the distance between the user's vehicle and the vehicle in front, we use the "width-ratio" model proposed in [3]. With the measured result, TTC is calculated every 0.5 seconds to provide the user visual warning signals when collision is expected 3 seconds later[1].

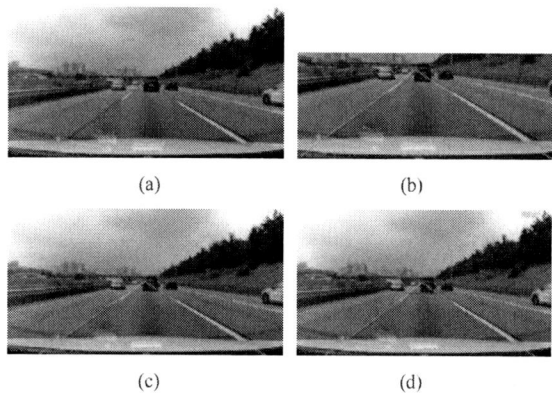

(a)   (b)

(c)   (d)

Figure 2. FCWS. (a) Input image, (b) Adaptive ROI result, (c) Verification result, (d) Full system result

III. RESULTS AND DISCUSSION

The proposed system provides the user the position of the vehicle in front, the distance between the vehicles, and collision warning based on TTC. We checked the vehicle detection rate of the CNN in order to verify the system. 2,128 vehicle rear samples were collected from Korean highways and [8], and 5,130 non-vehicle samples were collected from [9] and slices of Korean highway roads, respectively. The training time of the CNN was 1200 seconds. For general vehicle detection cases, we used the CalTech 99 rear-viewed vehicle dataset which is composed of 126 samples, and achieved 100% detection rate. For actual FCW environment, we counted the number of true positive and false positive results for 3140 frames. We achieved a 99.36%(3120 true positive results) precision rate for daytime highway environment images. Table 1 shows the detection rate values for our CNN and other vehicle detection methods compared in [2].

Due to the lack of additional sensors such as LIDAR or RADAR, we couldn't check the distance error of our system's result and the ground truth distance. Instead, we assumed that the result of [6] is accurate up to 30 meters. For collision warning, the TTC of people in general researched in [1], which varies from 1.1 seconds to 1.8 seconds, is used. Our system provides warning for cases when TTC is less than 3 seconds. The 1 second redundancy is given since the proposed system isn't an Automatic Emergency Braking(AEB) system, but a user-aiding warning system. The execution time for the proposed system was 19.8ms per

frame in Intel i5 3.4GHZ CPU environment. The basic outline of the entire system is shown in Fig. 2(d).

TABLE I.   TRUE POSITIVE RATE FOR CALTECH DATASET

Methods	True Positive Rate
PCA + SVM	86.59%
Gabor + SVM	90.24%
Wavelet + SVM	90.24%
Wavelet + Gabor + SVM	91.06%
Haar-like + Cascaded AdaBoost	92.38%
Haar-like + AdaBoost	92.89%
Haar-like + AdaBoost + SVM	94.41%
Proposed System	100%

IV. CONCLUSION

In this paper, we developed an FCW system using CNN for vehicle detection. We added an adaptive ROI setting stage to reduce execution time and eliminate redundant vehicle-like candidates. A Lenet-like CNN is used for vehicle detection. We achieved a 100% detection rate for the CalTech 99 dataset and 99.36% for actual daytime highway environment test images, respectively. The system can measure distance up to 30 meters from the user vehicle, and sends out warning signals when TTC is below 3 seconds. The execution time of the system, which is 19.8ms, indicates that the system can be executed in real-time in PC environment.

ACKNOWLEDGMENT

This work was supported by the IT R & D program of Ministry of Trade, Industry and Energy (10049192, Development of a Smart Automotive ADAS SW-SoC for a Self-Driving Car).

REFERENCES

[1] Kusano K. and Gabler H., "Method for estimating time to collision at braking in real-world, lead vehicle stopped rear-end crashes for use in pre-crash system design," *SAE Int. J. Passeng. Cars – Mech. Syst.*4(1):435-443, 2011

[2] Xuezhi Wen, "Efficient feature selection and classification for vehicle detection", IEEE Trans. Circuits and Systems for Video Technology, Vol. 25, pp. 508-517, March 2015

[3] John C. Hayward, "Near-miss determination through use of a scale of danger", Highway Research Record, vol. 384, pp. 24-34, 1972

[4] S. Sivaraman, M. Trivedi, " Looking at vehicles on the road: a survey of vision-based vehicle detection, tracking, and behavior analysis", IEEE Trans. ITS, vol. 14, No. 4, pp.1773-1795, December 2013

[5] Y. Gao, H. Lee, "Vehicle make recognition based on convolutional neural network", ICISS 2015, pp.1-4, 14-16 December 2015

[6] J. Kim, "Effective road distance estimation using a vehicle-attatched balck box camera", Journal of KIICE, vol.19, No.3, pp.651-658, 2015

[7] Y. LeCun et al., "Comparison of learning algorithms for handwritten digit recognition", International converence on artificial neural networks, pp. 53-60, 1995

[8] Linjie Yang, Ping Luo, Chen Change Loy, Xiaoou Tang, "A large-scale car dataset for fine-grained categorization and Verification", IEEE Conference of CVPR, pp.3973-3981, 7-12 June 2015

[9] J. Arróspide, L. Salgado, M. Nieto, "Video analysis based vehicle detection and tracking using an MCMC sampling framework", EURASIP Journal on Advances in Signal Processing, vol. 2012, Article ID 2012:2, Jan. 2012

Development of Burst Error Effect Reduction Algorithm for CAN using Interleaver Method

Ronnie O. Serfa Juan
Electronic Engineering Department
Cheongju University and TUP
South Korea and Philippines
ronnie71@naver.com

Min-Woo Jeong
Electronic Engineering Department
Cheongju University
Cheongju City, South Korea
jeongminwoo@cju.ac.kr

Hi-Seok Kim
Electronic Engineering Department
Cheongju University
Cheongju City, South Korea
khs8391@cju.ac.kr

Abstract— **A controller area network (CAN) controller is a critical component for immediate response in correcting certain flaws in any electronic control unit (ECU) especially in advanced driver assistance system (ADAS) application. This paper uses interleaving scheme to minimize the effect of Electro-Magnetic Interference (EMI) and aims to improve the system efficiency by improving the system's frame rate. This proposed method interleaved the original CAN data frame in order to minimize or totally eliminate the needed stuff bits that causes by the effect of EMI. The probability of error is reduced, thereby reducing also the probability of retransmission. The proposed scheme is synthesized on the Xilinx Virtex-5 FPGA.**

Keywords: Interleaver; EMI; CAN; burst error

I. INTRODUCTION

Controller Area Network (CAN) is a serial communication protocol that has evolved in automotive and industrial applications with a high level of fidelity and security [1]. A CAN bus is the backbone network for communication between networking controllers for engine timing, transmission, and brakes, etc. Its complexity has been increasing due to recent trends in the automotive industry, which is also proportional to the increasing numbers of nodes connected to the CAN bus that requires a high data rate.

An environment like the automotive and industrial applications are exposed in Electro-Magnetic Interference (EMI) and Electrostatic Discharge (ESD). CAN's physical layer must able to survive high energy transients produced by a number of disturbances like ignition system noise. The effects of EMI on the CAN bus can reduce the efficiency by introducing bit errors into the frame. These bit errors can be single bit or burst errors depending on the severity of the EMI. As the number of inductive loads increases, the probability of a frame getting corrupted by EMI, also increases. CAN itself has its own self-correcting method that is used for error checking in each frame's contents which is called as the Cyclic Redundancy Checking (CRC) code. The corrupted frames are checked for errors at the receiver and if errors are detected, the receiver node raises an error flag and request for retransmission [2]. An increase in the number of retransmissions has a bad impact on the throughput of the system. This paper proposes a technique to reduce the effect of EMI that causes the usage of stuffed bits and its impact to CAN bus performance.

This work was supported by the IT and R&D Program of Ministry of Trade, Industry and Energy (No. 10049192, Development of Smart Automotive ADAS-SW SoC for Self-Driving Car) and partially supported by Cheongju University Research Scholarship Grants from 2015 to 2017.

II. ERROR MANAGEMENT TECHNIQUE OF CONTROLLER AREA NETWORK

Error management in CAN protocol is one of the important aspects in evaluating the CAN bus performance. The error handling schemes aims to detect errors in the transmitted message frames over the bus and request for retransmission. The node that detected an error raises an error flag, halting the transmission on the bus [3]. The five different error detection mechanisms used by CAN protocol: *bit monitoring* that sends a bit on the bus and monitors it continuously; *bit stuffing* that inserts a complementary bit at the sixth bit position of the five consecutive identical bits; *frame check/form error* raises a from error flag upon detection of a wrong bit level in one of the fixed field frames; *acknowledgement check* monitors a dominant bit in acknowledge slot; and *cyclic redundancy check (CRC)* is a 15-bit CRC sequence used for error checking of frames between start of frame (SOF) to data field [4] of CAN's frame.

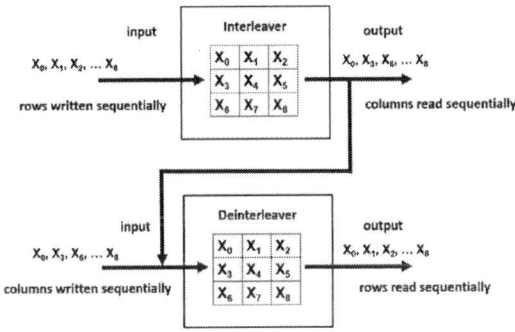

Figure 1. Interleaver and De-interleaver mechanism

III. PROPOSED INTERLEAVER METHOD

CAN application like in automotive environment is exposed to EMI's effect that resulted a burst error in the transmitted data. Interleaving technique is one type of error correcting codes (ECCs) that can correct burst error especially for random error occurrences. Fig. 1 shows the interleaver and de-interleaver mechanism. It rearranges the input data such that the consecutive data are spaced apart. Then, at the receiving

---

978-1-5090-3220-4/16 $31.00 © 2016 IEEE   165   ISOCC 2016

end, the interleaved data is arranged back into the original sequence by the de-interleaver [5]. As a result, burst errors that was introduced during transmission appears to be independent at the receiver and thus allows a better error correction. The proposed algorithm is shown in Fig. 2, first, the data bit length needs to be determine, then the input data sequence from start of frame (SOF) to data frame (K). The suggested row (m) and column (b) sizes of the rectangular matrix is shown in Table I for this interleaver scheme.

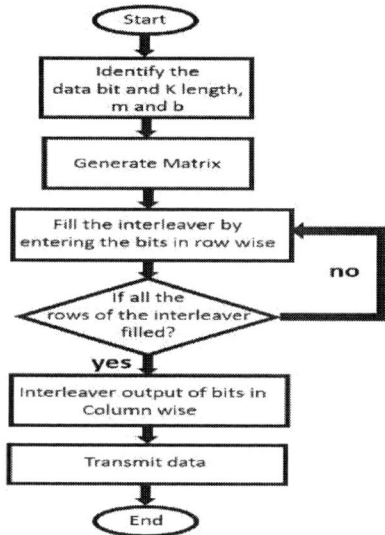

Figure 2.   Proposed Interleaver Algorithm

TABLE I.   RECTANGULAR MATRIX SIZE IDENTIFICATION

Number of data bits	K (bits)	m (bits)	b (bits)
8	42	7	6
16	50	10	5
24	58	29	2
32	66	11	6
40	74	37	2
48	82	41	2
56	90	15	6
64	98	14	7

Figure 3.   Stuffed bits usage between Standard Serial Transmission to the proposed Interleaver method

IV.   EXPERIMENTAL RESULTS

For simulation and verification purposes of this proposed implementation, 2.0-A Frame of CAN was tested and synthesized to Xilinx Virtex 5 FPGA by using Xilinx ISE at 125 kbps of baud rate and tested in an 8-bytes data of 2.0A Frame. Fig. 3 shows that the stuffed bits in the interleaver method was reduced compared to the standard serial transmission. While in Table II shows a decreased in frame length variations between the original serial transmission and to the proposed interleaver method. And Table III shows the frame rate results of the network.

TABLE II.   FRAME LENGTH VARIATION BETWEEN SERIAL TRANSMISSION AND INTERLEAVER METHOD

	Serial Transmission	Interleaver Method
number of frame bits (without stuffed bits)	111	111
number of stuffed bits used	19	14
total number of frame bits	130	125

TABLE III.   FRAME RATE COMPARISON BETWEEN SERIAL TRANSMISSION TO THE PROPOSED INTERLEAVER METHOD

	Frame Rate (frame per second)
Serial Transmission	961
Interleaver Method	1,000

V.   CONCLUSION

In this paper, the proposed architecture can minimize the effect of EMI, which can result to the usage of stuffed bits in CAN system especially in automotive application where the sources of burst errors is severe. The simulation results aimed the target of minimizing the needed stuffed bits as shown in Table II. Table III shows an increased in Frame Rate that improves the system throughput and minimizes the required system's payload.

ACKNOWLEDGMENT

This work was supported by the IT and R&D Program of Ministry of Trade, Industry and Energy (10049192, Development of Smart Automotive DAS SW-SoC for Self-Driving Car) and partially supported by Cheongju University Research Scholarship Grants from 2015 to 2017.

REFERENCES

[1] "CAN specification version 2.0," 1991. The Bosch IC Design Center, Reutlingen, Germany. Available

[2] CiA, "Controller Area Network (CAN) - protocol," Available Online: http://www.cancia.org/can/protocol/.

[3] KVASER, "The CAN Protocol Tutorial,"

[4] "CAN Specification Version 2.0," 1991. The Bosch IC Design Center, Reutlingen, Germany. Available Online: esd.cs.ucr.edu/webres/can

[5] LogiCORE IP Interleaver/De-Interleaver v7.1 Product Guide

Hardware Implementation of aggregated channel features for ADAS

Hohyon Song,   Bosun Jeong,   Hyunkyu Choi,   Taeho Cho,   Hweihn Chung

Nextchip Co. Ltd.
Seongnam, Korea

hodal@nextchip.com   jsun@nextchip.com   chk0905@nextchip.com   thcho@nextchip.com   hweihn@nextchip.com

Abstract— **In this paper, we propose the hardware detector architecture implemented in the semiconductor level to achieve the higher speed and performance efficiently as pre-processor for ADAS vision system compared to the existing solution which is done by ECU side only or S/W implemented intently. Herein the architecture represents the higher speed as real time that we implement a hardware multi-scale pedestrian detector operating in real time (30fps on 640x480 images, full-search) and performance as ACF based for detection algorithm in a highly integrated manner. Its advanced ADAS algorithms deliver highly improved detection rate eventually. For the efficient method, we construct the image pyramid directly rather than using the approximate features at nearby scale for providing greater accuracy. To actualize it in an effective way, we design the detector separately as two parts – H/W part and S/W part. In other words, H/W part generates pyramid images and extracts features then does classification. S/W part does clustering from the H/W classification result using NMS. As a simulation result, the performance is 18%@10^{-1}FPPI in the INRIA DB. According to well-defined system partitioning, it offers faster calculation and securing higher detection rate.**

Keywords; pedestrian detection, ACF, ADAS, hardware implementation

I.   INTRODUCTION

Pedestrian safety in the road is an important issue. So many automotive car makers are developing ADAS (Advanced Driver Assistance Systems) using various sensors to reduce traffic accident. The trend of ADAS is sensor fusion. As each sensor has unique pros and cons, so using only one sensor can't cover all situations. Nevertheless, vision sensor should be used by default in sensor fusion systems, because the vision sensor is similar to human visual system and video camera technology is matured and has cost effectiveness.[1] Various approaches have been tried to detect pedestrians from visible light images, up to now. The shape and appearance of the pedestrians are used to separate them from the background. Some of the features used for pedestrian detection are Haar wavelet, Gabor filter outputs, symmetry, intensity gradients and their histogram, etc. One article surveyed pedestrian detectors with focusing on sliding window approaches.[2] In ADAS, it's very important to make high recognition performance and real-time processing speed. So, basically we used size, aspect ratio and symmetry of human body and the Haar or HOG features of body model. But, as it is seen from [2], detectors which uses above features have some performance limitation. Actually we implemented two-

stage hardware detector using Haar feature [3] and HOG feature [4] but its performance was not good for pedestrian detection. And nowadays the detector's research trend is CNN (Convolutional Neural Network). But although the CNN has high performance, it's not easy to make the hardware architecture. So currently that's not suitable to use in ADAS systems. In this paper, we show the performance and speed of hardware implemented detector which is based on ACF (Aggregated Channel Feature).[5]

II.   HARDWARE IMPLEMENTATION

We design the hardware detector, which is based on ACF algorithm. Basically, the ACF algorithm is suitable for hardware implementation and has relatively high performance. Furthermore some research showed that we can add more feature channels to ACF original 10 feature channels to get more improved results.[6]

A.   High-speed Processing Architecture

The detector gets the image data from system bus interface. Then the image scaler makes 24-step down-scaled images (6 scales per one octave) from original image. In order to use this detector to ADAS application (long range detection) and to reduce hardware size, detector's detection size is 32x64. And the image buffer does not store all down-scaled image entirely, but stores only the required area image. Then, feature extraction and classification task can be done with operating as a pipelined form. Figure 1 shows hardware architecture for that.

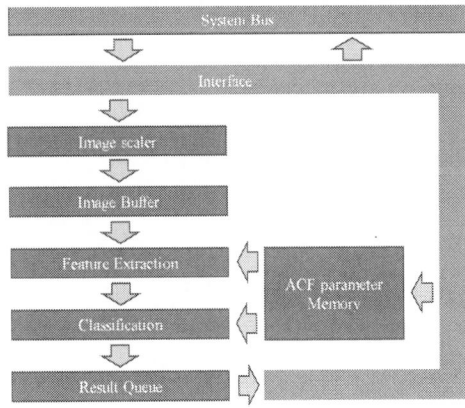

Figure 1.   hardware architecture

This work is supported by the World Class 300 Project R&D program of SMBA/KIAT. [S2367453, Development of UHD ISP for Surveillance and ADAS Integrated SoC for Automotive]

As the detector's ACF parameters can be uploaded from system bus interface, so when training results are changed we can update the ACF parameters. And the result queue stores down-scale number, x and y position when the classification result is true. After then the software part can run the NMS with this result queue data.

B. Optimization of Feature Extraction and Classification

The ACF generates 10 channel features (1-gradient magnitude, 6-gradient histograms, 3-color information), so about the 32x64 image, all features have 5120 (16x32x10) feature vectors. After classification for one 32x64 image is finished, slide-window shifts 4 pixels and update only 32x4 next image. So, feature extraction part only needs to update 320 (16x2x10) feature vectors.

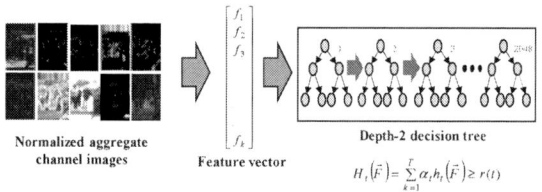

Normalized aggregate channel images

Feature vector

Depth-2 decision tree

$$H_t(\vec{F}) = \sum_{k=1}^{T} \alpha_t h_t(\vec{F}) \ge r(t)$$

Figure 2. classification flow

The classification uses soft cascade way. The default step is 2048, but we check the weighted sum after set a step count. At that point if the weighted sum is lower than rejection threshold, the classification do not go to the next step. So we can reduce the processing time. Figure 2 shows the classification flow.

C. Hardware and Software Partition

We separate the tasks with repetitive and complex tasks for the entire image and with intermittent and simple tasks for the selective area. So, the hardware part generates the pyramid images, extracts the features and does classification. By using the NMS, the software part clusters the results that's classified by hardware.

III. RESULTS

The hardware detector is implemented with Verilog HDL (Hardware Description Language). It works at 166MHz clock speed, and it has 30fps performance at VGA size full-search. To evaluate the detection performance, we use the INRIA DB. It results 18% miss rate at 10^{-1} FPPI (False Positive Per Image).

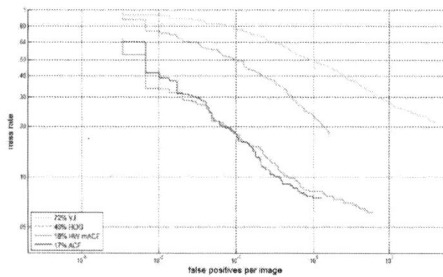

Figure 3. Quantitative Evaluation (INRIA DB)

We modify the original ACF algorithm to be suitable for hardware implementation. First, the original ACF algorithm uses the LUV 4:4:4 color space but our ACF algorithm uses the YCbCr 4:2:2 color space. Second, we approximate the down scale scheme to reduce the running time, from 8 scale per one octave to 6 scale per one octave. Finally, our detector's detection size is 32x64, so our result is a little worse than original's result. On the other hand, we get a real-time operation speed.

IV. CONCLUSIONS

Nowadays the safety of the VRUs (vulnerable road users), consisting of pedestrians, bicyclists becomes important factors of ADAS. For reliability of ADAS, the operation speed and computing power are as important as its performance essentially.

In this paper, we propose the core technology, hardware detector in ADAS to provide real time operation and higher detection rate. It is achieved these targets by well-defined H/W and S/W system partitioning supporting simultaneous ADAS processing. In addition, it provides the effective way onto the system by minimizing the issues of the system such as chronic/serious software burden issues and the computing power issues for pedestrian detection due to numerous operations, either as a pre-processor.

In the future with advanced ISP (Image Signal Processing) technology, we will research the improvement of the detection rate for reliable image quality enhancement and tracker using the motion information for occlusion problem solution for the safety.

REFERENCES

[1] T. Gandhi, and M. M. Trivedi, "Pedestrian Protection Systems: Issues, Survey, and Challenges," IEEE. Trans. Intelligent Transportation Systems, vol. 8, no.3, pp. 413–430, September 2007.

[2] P. Dollár, C. Wojek, B. Schiele, and P. Perona, "Pedestrian Detection: An Evaluation of the State of the Art," IEEE. Trans. Pattern Analysis and Machine Intelligence, vol.34, no.4, pp. 743–761, April 2012.

[3] P. Viola, and M. Jones, "Rapid object detection using a boosted cascade of simple features," IEEE. Proc. Computer Vision and Pattern Recognition, vol.1, pp. 511–518, December 2001.

[4] N. Dalal, and B. Triggs, "Histograms of Oriented Gradients for Human Detection," IEEE. Proc. Computer Vision and Pattern Recognition, vol.1, pp. 886–893, June 2005.

[5] P. Dollár, R. Appel, S. Belongie, and P. Perona, "Fast Feature Pyramids for Object Detection," IEEE. Trans. Pattern Analysis and Machine Intelligence, vol.36, pp. 1532–1545, August 2014..

[6] C. Toca, C. Pătrașcu, and M. Ciuc, "Performance testing and functional limitations of Normalized Autobinomial Markov Channels," IEEE. Proc. Intelligent Computer Communication and Processing, pp. 401–405, September 2015.

Improvements in Parallel SIMD Implementation of Single Image Defogging

Kristofor B. Gibson and Truong Q. Nguyen
Electrical and Computer Engineering
University of California, San Diego
La Jolla, CA, USA
Email: kboydgibson@gmail.com, tqn001@eng.ucsd.edu

Hannoh Yoon
R&D Center
MtekVision Co., Ltd.
Seoul, Korea
yunhr@mtekvision.com

Abstract— **In our previous work, we showed how to employ a single image defogging method by using a parallel <u>SIMD</u> architecture on a System on Chips (<u>SoC</u>). In this work, owing to improvements of hardware, we provide an update to this implementation with more optimization steps. In this article we also present a fog detection capability by using a simple fog metric.**

Keywords---SIMD architecture; defogging; contrast enhancement

I. INTRODUCTION

Many will experience that using a camera outside there are many artifacts that need to be addressed. One of these artifacts is low contrast due to fog or haze. In order to account this, a spatially varying contrast enhancement would be required [1]. At the time of writing this article, almost a decade ago enhancing imagery that has fog or haze was a difficult problem and required multiple images and/or require concrete prior information. This problem subsided when researchers found ways to remove the fog and hence enhance an image using a single foggy image thus the name "Singe Image Defogging".

We now repeat the introduction of the fog model as in [2]; we show that in [1], a dichromatic model was employed that used the findings from Duntley [3] and Koschmieder [4]. A common model used for foggy images is

$$\widehat{x}_i = t_i x_i + (1 - t_i)a, \qquad (1)$$

with $\widehat{x}_i \in {}^3$ as the foggy color image at pixel location *I*, the clear-day image x_i, a transmission image $t \in$, and an arlight color $a \in {}^3$. One can see in (1) that the foggy image (low contrast image) is a blending of both the desired clear-day image **x** and the airlight **a** color with the transmission *t* as the blending weight. On an overcast day, the airlight is the color of an object significantly far away. The transmission image (or map) is based on the idea that as photons enter an area of thin slice of scattering medium the light is scattered by a certain ratio β where it is modeled with an exponential decay function

$$t_i \approx e^{-\beta r_i} \qquad (2)$$

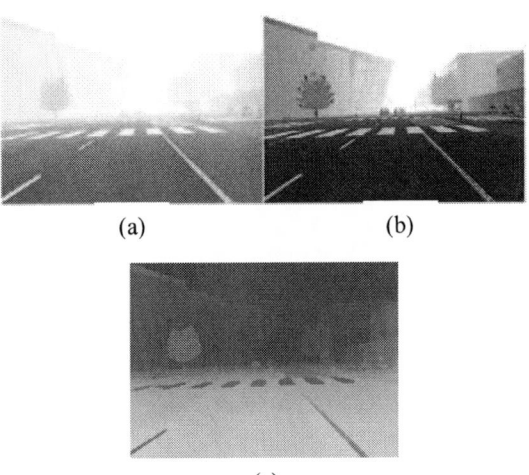

(a)   (b)

(c)

Figure 1. Example of single image defogging. (a) Foggy image $\tilde{x}$ from Tarel et al. [5] (b) Defogged image **x**. (c) The estimated transmission image *t*.

where *r* is the range of the object at spatial location *i*. We do need to mention that (2) is an approximation; this is because many assumptions are made. The scattering of light is wavelength dependent [3], [6]. In this work we make the same assumption as in [1] that scattering in foggy scenes—light is scattered equally—thus the reason we see white colored fog. With this assumption we can drop the dependence on wavelength and use the same transmission for each color channel for RGB color formats.

For single image defogging, the main problem is to estimate *t* given $\widehat{x}_i$. An example of single image defogging is in Fig. 1.

In [7] a key finding was that many single image defogging methods use a prior θ to estimate the transmission

$$t = 1 - w\theta \qquad (3)$$

where $w = 0.95$ is chosen for strong enhancement and $w = 0.75$ for less aggressive enhancement. The many ways to estimate the prior is discussed in [7]. Most methods do not account for the limitations on embedded systems. In [2] the limitations and ways to counter them with a parallel SIMD implementation were discussed. In this work, we show an even more advanced

---

This work is supported by the Technology Development Program for Commercializing System Semiconductor funded by the Ministry Of Trade, Industry and Energy (MOTIE, Korea). (No 10041126, Title: International Collaborative R&BD Project for System Semiconductor.)

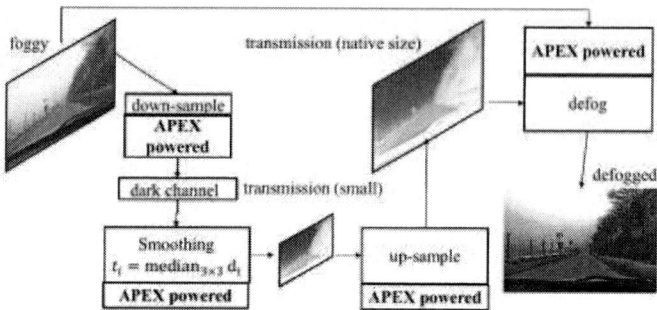

Figure 2. Schematic of SIMD implementation of single image defogging. Each algorithm implemented in the APEX-2 processor is labeled with "APEX powered"

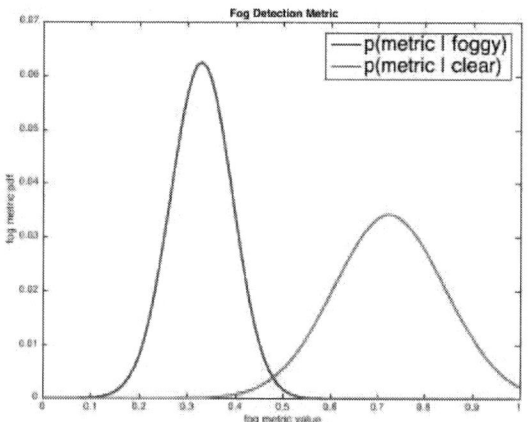

Figure 3. Marginal PDFs of the fog detection metric

method owing to the improvements of hardware and our choices of calculations in the image processing pipeline. One significant difference between our new method and our previous method in [2] is that we are using the RGB color format instead of the YUV color format which allows for more rich color enhancement capabilities.

II. APPROACH

We wish to develop a defogging capability in a hardware platform so that it can be used in near real-time applications. In our approach, we use the S32V234 SoC from NXP™. The two essential components for implementing the defogging algorithm reside on the Quad ARM-9 processors and the dual single instruction multiple data (SIMD) based technology, APEX-2. APEX-2 provides a programmable massively parallel array processing engine capable of handling computation-intensive video, imaging and graphics applications at higher performance but with lower power than traditional DSPs.

For defogging the image sequence, the defogged image **x** is computed by estimating the transmission. In order to efficiently estimate the transmission, the RGB image is down-sampled to a smaller size frame buffer. This is a more efficient way to estimate the dark prior [8] because less pixels are evaluated. Most of our approach is similar to the method in [2] with only a few changes. Our approach is illustrated in Fig. 2. A significant change is we now have the recovering of the image (multiple division operations) powered by APEX-2 which is significantly faster.

We employ the fog detection metric discussed in [9] and measure the marginal probability distribution functions (given clear image or foggy image) using the database provided in [10]. The results are in Fig. 3. Note that we use a threshold value of 0.5 as in [9].

III. CONCLUSION

In our work we show a more optimized single image defogging capability by using the S32V234 SoC and powered by APEX-2. We can achieve 30 frames-per-second on 1280×720 resolution color images. We also show how to discriminate between clear and foggy images using the metric proposed in [9].

REFERENCES

[1] S.G. Narasimhan and S. K. Nayar, "Vision and the atmosphere," *Internation Journal of Computer Vision*, vol. 48, no. 3, pp. 233-254, 2002

[2] K. B. Gibson, T. Q. Nguyen, and H. Yoon, "A parallel SIMD implementation of single image defogging," in *2015 International SoC Design Conference (ISOCC)*. IEEE, 2015, pp. 167-168.

[3] S. Q. Duntley, A. R. Boileau, and R. W. Preisendorfer, "Image transmission by the troposphere I," JOSA, vol. 47, no. 6, pp. 499-506, 1957.

[4] H. Koschmieder, "Luftlicht und sichtweite," Naturwissenschaften, vol. 26, no. 32, pp. 521-528, 1938.

[5] JJj.P. Tarel, N. Hautiere, A. Cord, D. Gruyer, and H. Halmaoui, "Improved visibility of road scene images under heterogeneous fog," in *Intelligent Vehicles Symposium (IV), 2010 IEEE*, IEEE, 2010, pp. 478-485.

[6] W. E. K. Middleton, *Vision Through the Atmosphere*. University of Toronto Press., 1952.

[7] K. B. Gibson and T. Q. Nguyen, "An analysis of single image defogging methods using a color ellipsoid framework," *EURASIP Journal on Image and Video Processing*, May 2013.

[8] K. He, J. Sun, and X. Tang, "Single image haze removal using dark channel prior," in *Computer Vision and Pattern Recognition, 2009. CVPR 2009. IEEE Conference on*. IEEE, 2009, pp. 1956-1963.

[9] K. B. Gibson and T. Q. Nguyen, "A perceptual based contrast enhancement metric using adaboost," in *Circuits and Systems (ISCAS). 2012 IEEE International Symposium on*. IEEE, 2012, pp. 1875-1878

[10] J.-P Tarel, N. Hautière, L. Caraffa, A. Cord, H. Halmaoui, and D. Gruyer, "Vision enhancement in homogeneous and heterogeneous fog," *IEEE Trans. Intell. Transp. Syst.*, vol. 4, no. 2, pp. 6-20, 2012

Dehazing in Color Filter Array Domain

Yeejin Lee, Truog Q. Nguyen

ECE Dept. University of California, San Diego
9500 Gilman Dr. La Jolla, CA, USA
yel031@eng.ucsd.edu, tqn001@ucsd.edu

Changyoung Han
Core Logic, Korea
changyoung.han@gmail.com

Abstract— **This paper introduces an alternative image processing pipeline for efficiently implementing dehazing algorithms. In the proposed pipeline framework, demosaicking is performed after dehazing process and transmission is estimated in a color filter array plane itself. The proposed method reduces implementation complexity and suppresses demosaicking artifacts amplification. Experimental results verify that the proposed pipeline framework requires less memory resources and reduces computational complexity while preserving visual quality.**

Image restoration; de-weathering; demosaicking; image processing pipeline.

I. INTRODUCTION

In the presence of haze or fog, the radiance $\mathbf{x}_p \in \mathbb{R}^3$ from an object is attenuated by atmospheric scattering corresponding to distance from the camera to the object, at a spatial location p. When the scattered light $\mathbf{a} \in \mathbb{R}^3$ along an atmospheric path combines with the attenuated object radiance, the captured scene radiance $\mathbf{y}_p \in \mathbb{R}^3$ is degraded as [1]

$$\mathbf{y}_p = t_p \mathbf{x}_p + (1 - t_p)\mathbf{a} , \qquad (1)$$

where $t_p \in \mathbb{R}$ is transmission, which is exponentially decayed with distance and invariant to wavelengths.

Dehazing is a post-processing step to enhance visibility in a digital camera's image processing pipeline. The degraded visibility is classically improved by statistically estimating scene depth in (1) and by adjusting scene radiance to it [2, 3, 4]. In addition, it is well known that statistical features in natural scenes are spatially varying slowly (i.e. scale-invariant) [5]. In this paper, we propose an alternative image processing pipeline framework based on performing dehazing in a color filter array (CFA) pattern image, which is spatially undersampled. The proposed method requires less memory resources and reduces computational complexity. Furthermore, applying a dehazing algorithm before demosaicking suppresses demosaicking artifact amplification. Any artifacts caused by other steps in a pipeline can be exacerbated by the subsequent defogging process.

In the next section, we present the proposed pipeline framework. Then we demonstrate that the proposed pipeline reduces computational complexity without degrading visual quality of the restored images.

This work is supported by the Technology Development Program for Commercializing System Semiconductor funded by the Ministry Of Trade, Industry and Energy (MOTIE, Korea). (No 10041126, Title: International Collaborative R&BD Project for System Semiconductor)

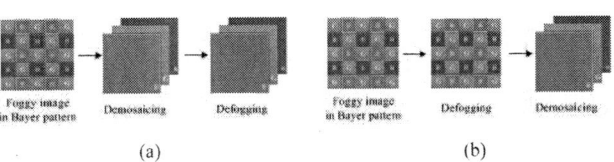

Figure 1. (a) Typical digital camera processing pipeline. (b) Proposed pipeline

II. DEHAZING IN COLOR FILTER ARRAY

In a typical digital camera's image processing pipeline, demosaicking step is placed in the early stage as shown in Figure 1(a). On the other hand, in the proposed image processing pipeline framework, an input image is first dehazed and then demosaicked as shown in Figure 1(b). The proposed pipeline framework requires less memory resources and reduces the required number of operations. Note that the memory requirements (compared to the original image size in a CFA pattern) as well as the implementation complexity are increased by a factor of 3 for RGB images. Though the proposed framework can be generalized to any CFA pattern, the Bayer pattern will be assumed for investigation throughout this paper. The Bayer pattern is one of the common realizations of CFA in which the filter pattern is 50% green, 25% red, and 25% blue.

Transmission is commonly measured from the dark channels in a single foggy image [2, 3, 4]. As described in the locally piece-wise constant image model [5], a new color channel region is defined as a 2 x 2 block (connected 4-pixel to upper left location) of the CFA pattern so that all color values are able to be measured. Likewise, dark channels d_p^B in a CFA image can be measured within the same spatial region Ω_p^B of a RGB image

$$
\begin{aligned}
d_p^B &= \min_{p \in \Omega_p^B} y_p^B \\
y_p^B &= \sum_{c \in \{r,g,b\}} y_p^S(c)
\end{aligned}, \qquad (2)
$$

where y_p^S is the raw image sampled by the image sensor according to the CFA pattern. Similarly, any statistical values are measured in the same local regions of a RGB image. The transmission t_d^B using (2) in the CFA image is computed as

$$t_d^B = 1 - \gamma \min_{p\Omega_p^B} \frac{y_p^B}{a^B}, \qquad (3)$$

where y_p^S is divided by the corresponding atmospheric light color a^B measured in the CFA pattern. The c-th color of atmospheric light is individually estimated from the most haze-opaque regions using bright channels or dark channels [2, 3]. The weighting coefficient controls the natural appearance of the dehazed images.

III. Experimental Results

This paper considered a patch-based DCP [2] method and a pixel-based adaptive Wiener filter method [4] for comparing the quality of dehazed images processed in a CFA pattern against that of RGB plane. In DCP method, guided image filter [6] was used to refine transmission instead of soft matting to reduce time complexity. The different sample patch sizes were selected to be 15 x 15 and 32 x 32 pixels following the suggestions in [2] and [4], respectively. The weighting coefficient was set to 0.95 and 0.9 for DCP and Wiener filter methods, according to the corresponding papers. RGB images were reconstructed from Bayer images using the gradient-corrected bilinear interpolated filter [7], which is computationally efficient and fast. The performance of the proposed image processing pipeline framework was evaluated by using Naturalness Image Quality Evaluator (NIQE) [8] and the blind contrast measurements [9]. NIQE measures image quality based on natural scene statistics (NSS) without reference images. The metric e represents the restored visible edges that were invisible in hazy images. The value of $\bar{r}$ is the ratio of the gradients at visible edges before and after restoration. The examples of dehazed images are shown using test algorithms in Figure 2 and objective results are tabulated in Table I. The experimental results demonstrate that the proposed image pipeline can be used instead of the typical pipeline with less computational complexity and minimal impact on image quality.

IV. Conclusion

We proposed an efficient digital camera's image processing pipeline framework based on applying a dehazing algorithm in color filter array domain. The proposed method reduces implementation complexity while preserving image visual quality.

References

[1] McCartney, E. J, "Optics of the Atmosphere: Scattering by Molecules and Particles," New York John Wiley & Sons Inc., 1976.

[2] He, K., Sun, J. and Tang, X , "Single Image Haze Removal Using Dark Channel Prior, " in Proc. IEEE Conf. Comput. Vis. Pattern Recognit. (CVPR), p.1956-1963.

[3] Philippe, T. J-. and Hautière, N., "Fast Visibility Restoration from a Single Color or Gray Level Image," in " in Proc. IEEE Conf. Comput. Vis.(ICCV), 2009, p.2201-2208.

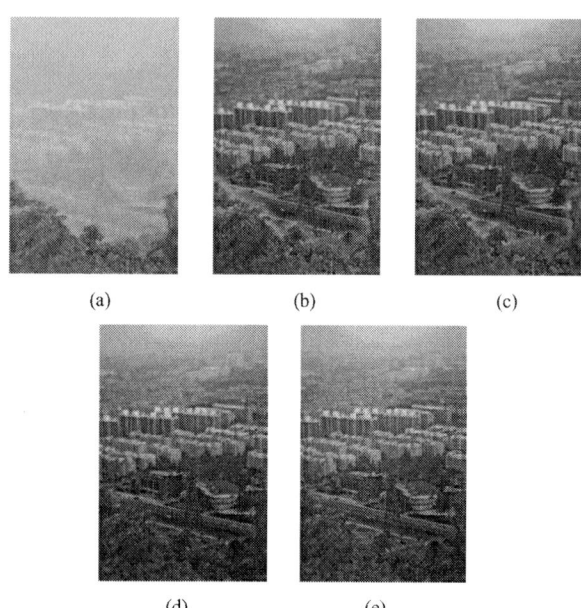

Figure 2. Examples of defogged images (the first image in Table 1). (a) Hazy image. (b) Dehazed image of RGB image using the DCP method. (c) Dehazed image of the Bayer image using the DCP method. (d) Dehazed image of RGB image using the Wiener filter method. (e) Dehazed image of the Bayer image using the Wiener filter method.

TABLE I. COMPARISON OF DEHAZING ALGORITHMS PERFORMANCE ON RGB IMAGES AND CFA IMAGES. EVALUATED USING THE NIQE AND THE BLIND CONTRAST MEASUREMENTS $(e, \bar{r})$. THE SIZE OF TEST IMAGES ARE SPECIFIED BELOW THE IMAGE NUMBERS.

		1 (400 x 600)			2 (1024 x 768)		
		NIQE	e	$\bar{r}$	NIQE	e	$\bar{r}$
DCP	RGB	26.900	2.927	3.947	20.405	0.504	1.714
	CFA	27.227	2.803	4.219	21.790	0.508	1.917
Wiener filter	RGB	23.671	2.698	3.557	21.065	0.501	2.288
	CFA	22.898	2.443	3.744	21.572	0.465	2.437

[4] Gbson, K. B. and Nguyen, T. Q., "Fast Single Image Fog Removal Using the Adaptive Wiener Filter," Proc. IEEE Int. Conf. Image Process. (ICIP), 2013, p. 714-718, April 1955.

[5] Ruderman, D. L. and Bialek, W., "Statistics of natural images: Scaling in the woods," in Neural Info. Process. Sys., 1993, p. 551-558.

[6] He, K., Sun. J. and Tang, X., "Guided Image Filtering," in IEEE Trans. Pattern Anal. Mach. Intell., 2012, 35, p. 1397-1409.

[7] Malvar, H. S., He, L.W., and Cutler, R., "High-quality linear interpolation for demosaicing of Bayer-patterned color images," in Proc. IEEE Int. Conf. Acoust., Speech, Signal Process. (ICASSP), 2004, 3, iii-485-8.

[8] Mittal, A., Soundararajan, R. and Bovik, A. C., "Making a 'Completely Blind' Image Quality Analyzer." in IEEE Signal Process. Lett., 2012, 20, p.209-212.

[9] Hautière, N., Tarel, J. P., Aubert, D., and Dumont, E., "Blind Contrast Enhancement Assessment by Gradient Ratioing at Visible Edgese," in Image Analysis & Stereology Journal, 2008, 27, p. 87-95

Nighttime Image Enhancement Applying Dark Channel Prior to Raw Data From Camera

Yan Gong, Yeejin Lee, and Truong Q. Nguyen

ECE Dept. University of California, San Diego

9500 Gilman Dr. La Jolla, CA, USA

yag018@ucsd.edu, yel031@eng.ucsd.edu, tqn001@eng.ucsd.edu

Abstract— **This paper presents an alternative approach to enhance visibility of images captured at night. In order to improve visual quality of nighttime image, we perform image enhancement in linear sensor space, which best represents image formation models. The analyses show that the proposed approach is helpful to reduce error amplification and computational complexity. In addition, experimental results verify that the proposed method enhances visibility for real-world dataset.**

Night vision; image enhancement; contrast enhancement; digital image processing pipeline; advanced driver assistance system

I. INTRODUCTION

Image quality and contrast become worse in poor visibility condition such as low-light and bad weather. Since this contrast degradation negatively impacts image understanding, it is necessary to improve the image quality before the captured images are used in computer vision applications of driver assistance system. Various visibility restoration algorithms have been proposed in order to capture as many features as possible in poor conditions [1, 2]. One approach is to use an image formation model and to improve image quality by learning parameters of the model, for example, image degradation factor in low-light condition or light transmission factor in the presence of fog [3, 4]. Recently, Dong *et al.* found connectivity of two degradation models in low-light and fog by observing statistics on the negative image of a nighttime image [5].

Image enhancement step takes place as a post-processing in a classical image processing pipeline. In the pipeline, scene radiance may not be properly restored by using the image degradation model [6]. The model is best described in terms of light intensity captured by image sensors, and light intensity is non-linearly scaled after gamma correction. To this end, we proposed an alternative image processing pipeline based on performing enhancement step in a linear sensor space (see Figure 1). In addition, performing this enhancement step in the early stage of a pipeline brings advantages to reduce possible artifacts amplification occurred by reversing the model and computational complexity.

The rest of this paper is organized as follows. The proposed approach is addressed in Section II. Then, experimental results

---

This work is supported by the Technology Development Program for Commercializing System Semiconductor funded by the Ministry Of Trade, Industry and Energy (MOTIE, Korea). (No 10041126, Title: International Collaborative R&BD Project for System Semiconductor)

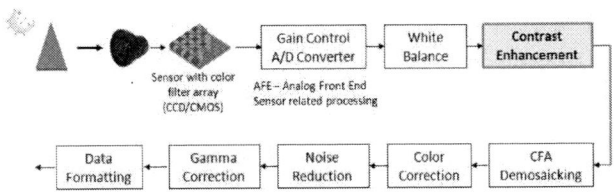

Figure 1. Proposed image processing pipeline

are presented in Section III. Finally, we conclude this paper in Section IV.

II. PROPOSED APPROACH

A. Image Degradation Model

The scene radiance of the negative image (i.e. $J_N^c(x) = 1 - J^c(x)$) is degraded as

$$J_N^c(x) = d(x)I^c(x) + (1 - d(x))A^c, \qquad (1)$$

where $I^c(x)$ is the scene radiance of a color channel ($c \in \{R, G, B\}$) without degradation at the pixel location x, and $J^c(x)$ is an input image captured at night. Here, $d(x)$ is the degradation factor, and A^c is atmospheric light from a global source. The atmospheric light is typically measured from the brightest part of $J_N^c(x)$. The degradation factor $d(x)$ is measured according to the dark channel prior (DCP), which assumes that one or more of the color components of natural scene is very dark

$$d(x) = 1 - \omega \min_{x \in R} \left(\min_{c \in \{R,G,B\}} \left(\frac{J_N^c(x)}{A^c} \right) \right), \qquad (2)$$

where ω is a weighting coefficient, and R is a local region at the center pixel location x.

Then, the radiance is restored by reversing (1) as follow

$$I^c(x) = \frac{J_N^c(x) - A^c(x)}{d(x)} + A^c(x). \qquad (3)$$

In the proposed pipeline, though all color components are inaccessible at a pixel location, the degradation factor in (2) is measurable in the same spatial region R of a CFA image, due to image nature of piecewise constant.

B. Advantages of the proposed pipeline

First, we can reflect light intensity more precisely by performing enhancement step in linear sensor domain. The image degradation model in (1) is mismatched with the already-processed images through a image processing pipeline [6]. The scene radiance is no longer scaled linearly after gamma correction. This makes it more difficult to recover the natural scene radiance in the processed images.

Second, the required operations and memory resources are less in the proposed approach. Raw sensor data typically come in the form of a color filter array pattern. Since each pixel is filtered to record only one of color channels, the amount of data in CFA images is 3 times less than that in RGB images. This means that the implementation cost of the enhancement step in the proposed pipeline is approximately decreased by a factor of 3, comparing to methods using RGB images.

Finally, error amplification effect by enhancement step could be relaxed. The recovery process in (3) could amplify any artifacts and errors generated by other steps in the subsequent enhancement step, because the degradation factor d is in $(0,1]$. This possible overall artifacts amplification is relieved by performing enhancement in the early stage of the pipeline.

III. EXPERIMENT RESULTS

We validated the proposed approach for the test data captured using Point Grey Grasshopper 3 in a raw format and RGB format. The acquired raw sensor data was processed following the proposed image processing pipeline in Figure 1. Note that camera's image processing pipeline varies from camera models. The proposed pipeline considers tasks that are shared by most manufacturers. The white balance was implemented using gray world assumption [7]. The gamma correction was moderate ($\gamma = 1.25$). The raw sensor images were demosaicked using the gradient-corrected bilinear interpolated filter [8] and denoised using block-matching and 3D filtering [9]. The sample patch size was selected to be 15 x 15 pixels, and the weighting coefficient was set to 0.98 in (2).

The results in Figure 2 demonstrate that the proposed approach produces more natural color with better perceptual quality in the enhanced images.

IV. CONCLUSION

We proposed an alternative approach to enhance night vision. The proposed method more accurately reflects light intensity in the recovered images. Furthermore, the proposed

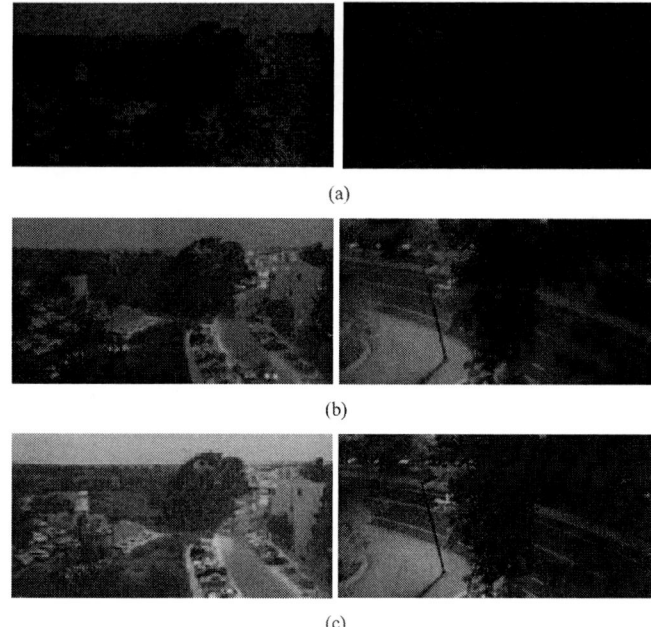

Figure 2. Examples of enhanced images. (a) Input images captured at night. (b) Enhanced images of input RGB images. (c) Enhanced images following the proposed pipeline in Figure 1.

method reduces computational operations and artifacts amplification.

REFERENCES

[1] Y. Rao and L. Chen, "A survey of video enhancement techniques," Journal of Information Hiding and Multimedia Signal Processing, vol. 13, no. 1, p. 71–99, 2012.

[2] S. Lee, S. Yun, J.-H. Nam, C. Won, and S.-W. Jung, "A review on dark channel prior based image dehazing algorithms, " EURASIP Journal on Image and Video Processing, 2016.

[3] Y. Yoo, J. Im, and J. Paik, "Low-light image enhancement using adaptive digital pixel binning," Sensors, vol. 15, p.14917-14931, 2015.

[4] He, K., Sun, J. and Tang, X , "Single Image Haze Removal Using Dark Channel Prior, " in Proc. IEEE Conf. Comput. Vis. Pattern Recognit. (CVPR), 2009, p.1956-1963.

[5] X. Dong, G. Wang, Y. Pang, W. Li, J. Wen, W. Meng, and Y. Lu, "Fast efficient algorithm for enhancement of low lighting video," in Proceedings of International Conference on Multimedia and Expo. IEEE, 2011, p. 1-6.

[6] Y. Lee and T. Q. Nguyen, "Analysis of color rendering transformation effects on dehazing performance," in Proc. IEEE Int. SoC Design Conf. (ISOCC), 2016.

[7] E. H. Land, "Recent advances in retinex theory and some implications for cortical computations: color vision and the natural image," Proc. Natl. Acad. Sci. USA, vol. 80, no. 16, p. 5163–5169, Aug. 1983.

[8] Malvar, H. S., He, L.W., and Cutler, R., "High-quality linear interpolation for demosaicing of Bayer-patterned color images," in Proc. IEEE Int. Conf. Acoust., Speech, Signal Process. (ICASSP), 2004, 3, iii-485-8.

[9] K. Dabov, A. Foi, V. Katkovnik, and K. Egiazarian, "Image denoising by sparse 3-D transform-domain collaborative filtering," IEEE Trans. Image Process., vol. 16, no. 8, p. 2080–2095, 2007.

Pedestrian Detection Aided by Temporal Prior

Zhaowei Cai, Matthew Jacobsen and Nuno Vasconcelos
University of California, San Diego
9500 Gilman Drive
La Jolla, California, USA
zwcai@ucsd.edu, mdjacobs@cs.ucsd.edu, nuno@ucsd.edu

Abstract— **In this paper, we developed an algorithm to construct temporal priors to improve the stability of pedestrian detections. The proposed temporal prior has no additional computation cost and could be applied to any image based pedestrian detector. We also present a FPGA implementation that can process VGA video at a frame rate of 30~40 frames per second in real time. The design is implemented on a Xilinx Zynq FPGA. It performs all steps of the algorithm in the FPGA fabric including: color space conversion, frame rescaling, HOG feature extraction, and candidate evaluation in a sliding window. The design uses parallel pipelines to evaluate multiple scales per frame.**

Pedestrain detection; temporal prior; hardware design; FPGA

I. INTRODUCTION

Pedestrian detection is a very important research area, especially for driver assistance systems and autonomous vehicles. However, it is still very challenging since it has high requirements for speed and accuracy. Currently, the typical pedestrian detection process is based on individual images, even though the applications are usually on videos. There is a couple of consequent problems: 1) the detections are not stable, e.g. the bounding boxes are flickering; 2) the same object will be missed at current frame even though it is detected in previous frames. These problems happen because of no temporal constraint. In this paper, we introduce temporal priors to make the detections over continuous frames stable. The experiments have also shown substantial improvements by temporal priors. Another advantage of the proposed temporal prior is it has no additional computation cost and can be applied to any other pedestrian detectors. We have also implemented the pedestrian detector using a FPGA. Our implementation maintains the same level of detection accuracy as the C++ software version. It achieves a 2× speed up over the software version to provide detections between 30~40 frames per second (FPS) on VGA video.

II. ALGORITHMS

The pedestrian detector of [1] is used in this paper, which uses cascaded framework, runs in real time and is based on individual images.

A. Temporal Priors

A detection at time t is defined by a bounding box (x_t, y_t, h_t, w_t). The estimated detection at time $t+1$ is

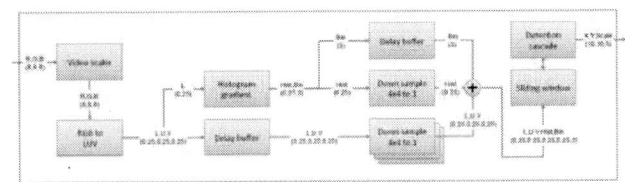

Fig. 1. The detection pipeline of hardware design.

$$x_{t+1} = x_t + \Delta x_t \qquad (1)$$

where Δx_t is the displacement at time t. The displacement depends on the speed of the object, which is $\Delta x_t = x_t - x_{t-1}$.

The priors of this object present around x_{t+1} is subjected to a Gaussian distribution,

$$p(x' \mid x_{t+1}) \sim N(x_{t+1}, \sigma_x^2).$$

The same model holds for the other coordinates. The final score at location (x, y, h, w) is

$$s(x, y, w, h) = p(x, y, w, h) * K + s_d(x, y, w, h)$$

where p is the prior, K is a constant and s_d is the detection score of the detector of [1]. With temporal prior, the locations where objects were present will have higher scores.

B. Nearest Neighbor Tracking

Since the object speed is need in (1), a tracking algorithm is required to obtain Δx_t. For an object detected in previous frame, there may be multiple detections which are candidate matches to the object. The tracking algorithm will help to find the best matched detection. The traditional single or multiple objects tracking algorithms could be used here, but most of them require expensive computational costs, which is inconsistent to our goal, real-time pedestrian detection. To save computation, we resort to nearest neighbor tracking algorithm. Given a detected object bounding box at time t, the tracked bounding box is

$$bb_{t+1} = \arg\max_i (IOU(bb_t, bb^i))$$

where bb^i is one candidate bounding box, and IOU is the "intersection over union" between two bounding boxes. The speed is known once two bounding boxes are matched.

This work is supported by the Technology Development Program for Commercializing System Semiconductor funded by the Ministry Of Trade, Industry and Energy (MOTIE, Korea). (No 10041126, Title: International Collaborative R&BD Project for System Semiconductor)

978-1-5090-3220-4/16 $31.00 © 2016 IEEE

Method	detector (no prior)	detector (with prior)
MR	30.92%	26.47%
FPS	6	6

TABLE I. DETECTION EVALUATION

III. HARDWARE DESIGN

The hardware design implements our algorithm in single pass stream processing fashion, which is based on [2]. Fig. 1 is a block diagram of the detection data flow.

A. Video Scaler

The video scaler uses bi-linear interpolation to scale input video frames by an arbitrary factor. Input pixels are line buffered into on-chip RAM blocks and rescaled as they are read out. The interpolation is accomplished by calculating a set of coefficients for each pixel and using the surrounding pixel values to create an output pixel. The scaling parameters are also used to identify the number of pixels per line and total lines to output.

B. RBG to LUV Converter

The conversion between RGB and LUV requires linear combinations of the RGB components, multiplication, division, and cubed root operations. The cubed root operation was represented using a pre-calculated lookup table (LUT).

C. Histogram Gradient

The HOG features for each pixel are a measure of the magnitude and direction of the dominate gradient in a pixel. A pixel's value is determined by the value of the four surrounding pixels in the cardinal directions. We mapped the magnitude value directly to one of the 6 directions using a high precision LUT. We then found the 6 thresholds for entries in the LUT and encoded them directly into the module. Only square root operations were needed at runtime after simplification.

D. Sliding Window

Our implementation uses a set of line buffers to hold n pixel lines, the frame width. The value of n is determined by the height of the window (in our case 16). New pixels fill the buffers from the bottom up, pushing the old pixels out the top. This technique is similar to that used by Cho in his Viola-Jones FPGA implementation [3]. The key difference is that pixels are accessed from the line buffers directly instead of from a synchronously maintained register array.

E. Detection Cascade

Once features are extracted, they must be compared to thresholds. A weak learner contributes a value to a running score. The detection cascade handles reading feature data from the sliding window and cascade parameters (thresholds, scores, etc.) from a pre-loaded on-chip RAM. Comparisons are made, feature scores are calculated, and the running score is updated. Afterwards, it is compared to the current stage threshold to see if an early rejection is possible. This process will continue until

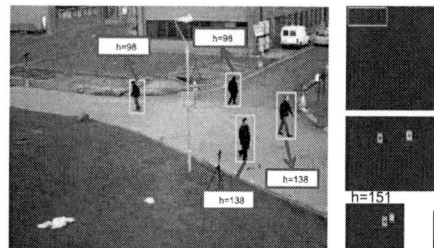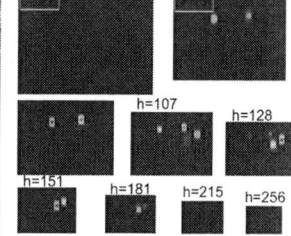

Fig. 2. The temporal prior maps over different detection scales.

all the stages have passed or the window is early rejected. If passed, the window location and scale is outputted. If rejected, the process resets and the next window begins evaluation. The process is fully pipelined to extract a feature value each cycle.

IV. EXPERIMENTS

The proposed algorithm is evaluated on the Caltech pedestrian dataset [4]. The typical evaluation on Caltech pedestrian dataset is image based. To show the improvement of temporal prior, the new algorithm is tested on continuous video frames. The prior maps over different scales are illustrated in Fig. 2. It is reasonable to assign high priors around the locations where objects are present in the previous frame. The detection performance comparison with/without temporal prior is shown in TABLE I. It can be found that the temporal priors improve temporal stability of detections, enabling better and more stable detection results. The speed comparison also shows that the introduction of temporal prior has no additional computation costs. Although the detector of [1] is used in this paper, the proposed algorithm has no limitation to be applied to any other detectors. It is a useful general tool.

We also implemented our hardware design in Verilog using Xilinx Vivado 2014.3 targeting 200 MHz on a Zynq XC7Z045 FPGA. We used a ZC706 development board. Depending on the number of pedestrians in the scene, the detector was able to run at 30~40 FPS.

V. CONCLUSION

We proposed an efficient temporal prior to obtain stable pedestrian detections in real-time videos. We also presented a hardware based implementation on an FPGA which runs at a rate of 30~40 FPS on VGA video.

REFERENCES

[1] Z. Cai and N. Vasconcelos. "A real-time cascade pedestrian detection based on heterogeneous features," International SoC Design Conference, 2015.

[2] M. Jacobsen, Z. Cai and N. Vasconcelos. "FPGA implementation of HOG based pedestrian detector," International SoC Design Conference, 2015.

[3] J. Cho, S. Mirzaei, J. Oberg, and R. Kastner, "Fpga-based face detection system using haar classifiers," in FPGA. ACM, 2009.

[4] P. Dollar, C. Wojek, B. Schiele, and P. Perona. "Pedestrian detection: An evaluation of the state of the art". IEEE Transactions on Pattern Analysis and Machine Intelligence, 34(4):743–761, 2014.

Moving Objects Detection using Classifying Object Proposals for Driver Assistance System

Kunyao Chen*, Subarna Tripathi*, Youngbae Hwang†, Truong Nguyen*
*University of California San Diego, USA
†Korea Electronics Technology Institute, Korea

Abstract—We present a new framework for driver assistance system, detecting moving objects in the street scene. Our algorithm supports a wide range of objects including vehicles, cyclists, pedestrian etc. Based on candidate bounding boxes detected by object proposals, our classifier only responds to the objects truly moving, which is more practical for real applications. Using unified features of color, structure and motion information, our system runs in real time with 66% detection rate in CamVid dataset. In addition, our method can be implemented efficiently with pipelined function blocks.

Index Terms—Moving Objects detection, Driver assistance system, Optical flow

I. INTRODUCTION

Moving object detection is one of important features that gives drivers better awareness of the surroundings. It can be used to alert the drivers when some object moving towards and avoid the collisions when the drivers are distracted.

Many recent works focus on specific objects, like pedestrian detection [1]. Their applications are limited as, in real world application, we have to consider different moving objects in different driving conditions. Pedestrian detection may be not useful when we are driving on the highway. There have been numerous researches on semantic segmentation, which enable us to point out those potentially moving objects in one frame, i.e. vehicles, cyclists, pedestrian etc. However, those semantic segmentation methods[2][3] are too complicated to work in real time cases, while real time approaches[4] contain much noise and error. In addition, segmentation methods only show us the objects. Without motion information, we cannot even tell if the object is moving, if we should take action for it.

The most relevant work comes from Zhun Zhong etc [5]. They re-rank object proposals those include moving vehicles on KITTI dataset[6]. However, in their framework, they use many complex features such as semantic segmentation results, CNN features, stereo information etc. This work is not amenable for hardware-implementation. Additionally, it is not clear how the above method can perform in sparsely annotated dataset such as CamVid [7].

Therefore, in this paper, we present a hardware friendly framework for moving objects detection. Instead of using complex features, we focus on using three types of hand crafted features to achieves a quite satisfying detection rate. We will explain dataset and tools we used in next section, as well as the most important part, feature engineering. And we will briefly demonstrate the results of our experiment.

II. METHOD

A. Dataset and Tools

We use CamVid dataset [7] for training and evaluation of our methodology. Camvid dataset provides semantic segmentation ground truths on four street view videos with about 700 frames. For each of these frames, we generate bounding box around moving objects region, i.e. *car*, *truck*, *cyclist*, *pedestrian*. These boxes are taken as the ground truth annotations of our classification problem.

First, we generate object proposals with EdgeBoxes [8] and leverage motion information from optical flow between consecutive frames. Among many of the hardware friendly optical flow algorithms, we choose T.Brox[9] method for its best accuracy.

CamVid is sparsely annotated i.e. one frame per second has segmentation ground truth. In order to get meaningful motion features we compute forward optical flow of every ground truth video frame with its immediate next frame from the raw video. The later one does not have an annotation.

B. Features

As we pursue less computational cost implementation and also training in a relatively small dataset, we choose three types of hand-crafted features for this particular application.

(1) Box properties

First, we extract the features purely related to the box itself, which include bottom y and center x coordinate, normalized height, width and box area, as well as aspect ratio. We assume that the objects too far away need no attention, therefore with these box properties, we aim to ignore the objects running in lanes close to the boundaries of the frame, also those running far ahead.

(2) Color and structure

We also consider color and structure information inside the boxes. For the color feature, we create LAB histogram i.e. 20 bins representation for each L,A,B component. Besides, we choose HOG features for the structure information. After PCA, we keep only 50 components without losing much accuracy. One thing special here is we also extract LAB histogram for the bottom patches(with same size of the proposal box). The idea is the objects we interest are always on the road.

(3) Motion

After applying the real-time optical flow, we can obtain magnitude and direction of the motion for each pixel. With the intuition that moving objects should have different motion pattern with its sounding, we consider four neighboring patches of the proposal box. First calculate the mean magnitude difference with the four neighbors, then combine the direction histogram (20 bins each) of all five patches as the final motion features.

C. Classification

In the training set, features extracted from those ground truth boxes are taken as positive samples. For the negative sample, we first apply hard negative mining. We generate candidate windows with decreasing scores using Edgeboxes[8]. Only the windows which have less than 30% IOU with any ground truth are considered as negative samples. As with R-CNN [10], we also set negative to positive sample ratio around 3 : 1. Then, we learn an SVM classifier with RBF kernel.

D. Evaluation and Result

We repeat the same box and positive and negative sample generation process in the test set. As we cannot control the number of negative sample in this step, the negative to positive sample ratio can reach 7 : 1. With the features we design, the overall classification accuracy is 81.4%.

As Edgeboxes still generate many overlapped boxes, non maximum suppression is applied to remove those overlapped boxes and only keep the boxes with largest area in one region. After non maximum suppression, remaining boxes with more than 50% IOU are taken as true detections. In this criterion, we can achieve 66.2% detection rate. Two frame results are shown in Fig. 1. The top image demonstrate that we successfully detects different objects in the single framework. As we merely take the results of semantic segmentation as ground truths, some missing detections are actually what we try to ignore on purpose. As shown in the bottom image, the pedestrian walking on the left boundary, the bus driving on the other lane and the truck running far ahead of us do not have any influence on our driving. In this sense, our framework is more intelligence on practice and the detection rate of our algorithm is actually even higher.

III. CONCLUSION

Experimental results show that we can achieve a satisfying detection rate even with simple SVM classifier and hand crafted features. We believe our work can enable a broad range of application for driver assistance system, such as general object alert, general collision avoidance. We cannot deny that many problems still remain. In the future, we will work on cleaning up the ground truth, and test the framework in other datasets such as KITTI. Moreover, we'd like to explore other hardware friendly features such as integral channel features, and try new evaluate the methods.

ACKNOWLEDGMENT

This work is supported by the Technology Development Program for Commercializing System Semiconductor funded

Fig. 1. Sample results of Moving Object Detection. Green boxes are ground truths, blue boxes are true detections while red ones means missing detections.

by the Ministry Of Trade, Industry and Energy (MOTIE, Korea). (No 10041126, Title: International Collaborative R&BD Project for System Semiconductor)

REFERENCES

[1] R. Benenson, M. Omran, J. Hosang, and B. Schiele, "Ten years of pedestrian detection, what have we learned?" in *European Conference on Computer Vision*. Springer, 2014, pp. 613–627. 1

[2] L.-C. Chen, G. Papandreou, I. Kokkinos, K. Murphy, and A. L. Yuille, "Semantic image segmentation with deep convolutional nets and fully connected crfs," *arXiv preprint arXiv:1412.7062*, 2014. 1

[3] J. Long, E. Shelhamer, and T. Darrell, "Fully convolutional networks for semantic segmentation," in *Proceedings of the IEEE Conference on Computer Vision and Pattern Recognition*, 2015, pp. 3431–3440. 1

[4] J. Shotton, M. Johnson, and R. Cipolla, "Semantic texton forests for image categorization and segmentation," in *Computer vision and pattern recognition, 2008. CVPR 2008. IEEE Conference on*. IEEE, 2008, pp. 1–8. 1

[5] Z. Zhong, M. Lei, S. Li, and J. Fan, "Re-ranking object proposals for object detection in automatic driving," *CoRR*, vol. abs/1605.05904, 2016. [Online]. Available: http://arxiv.org/abs/1605.05904 1

[6] A. Geiger, P. Lenz, C. Stiller, and R. Urtasun, "Vision meets robotics: The kitti dataset," *International Journal of Robotics Research (IJRR)*, 2013. 1

[7] G. J. Brostow, J. Fauqueur, and R. Cipolla, "Semantic object classes in video: A high-definition ground truth database," *Pattern Recognition Letters*, vol. xx, no. x, pp. xx–xx, 2008. 1

[8] P. Dollar and C. L. Zitnick, "Edge boxes: Locating object proposals from edges," *ECCV*, 2014. 1, 2

[9] T. Brox and J. Malik, "Large displacement optical flow: Descriptor matching in variational motion estimation," *PAMI*, 2011. 1

[10] R. Girshick, J. Donahue, T. Darrell, and J. Malik, "Rich feature hierarchies for accurate object detection and semantic segmentation," *CVPR*, 2014. 2

Dense Stereo-based Real-time ROI Generation for On-road Obstacle Detection

Soon Kwon[1,2] and Hyuk-Jae Lee[1]

[1]Department of Electrical and Computer Engineering, Seoul National University, Seoul, Korea
[2]Convergence Research Center for Future Automotive Technology, DGIST, Daegu, Korea
{soonyk, hjlee}@capp.snu.ac.kr

Abstract—The use of 3D visual information has become widespread as an essential cue for detecting on-road obstacles in ADAS. In this paper, we propose an accurate dense-stereo-based system for the generation of on-road obstacle ROIs. To balance the concerns of computation overhead and algorithm accuracy, this paper presents an efficient depth map generation that combines global stereo matching with depth up-sampling. The entire system has been implemented with a hardware and software partitioning method running on an FPGA and embedded CPU for real-time processing. The implementation results verify that the proposed stereo vision system efficiently outputs accurate ROI candidates for on-road obstacle detection.

Keywords: stereo matching, depth map genration, region of interest, obstacle detection, advanced driver assistance system

I. INTRODUCTION

In the field of advanced driver assistance systems (ADAS) and autonomous driving applications, 3D computer vision methods are widely used to recognize free space and on-road obstacles [1]. Using depth information, the region of interest (ROI) search can be simplified and is more robust than with conventional 2D search methods [1, 2]. For example, a road surface profile used for free space detection can be easily analyzed by applying the V-disparity method with dense depth map information.

However, the computational overhead in a dense stereo matching process, which finds corresponding positions for left and right image pixels, presents challenging issues to be overcome. A number of studies have shown algorithmic or architectural improvements for speeding up the stereo matching process. Compared with algorithmic approaches [3] that deal with computation efficiencies, architectural approaches [4] are more suitable for real-time computation purposes. By considering hardware-friendly algorithm architectures, they can achieve remarkable reductions in computation time.

We illustrate here a real-time stereo-vision-based ROI generation system for the detection of on-road obstacles. The main contribution is in designing the system to use hardware and software partitioning for efficient implementation as an embedded vision system. The paper is organized as follows. Section II introduces the proposed algorithm. In Section III, the result of the implementation of the proposed hardware and software partitioning is described in detail. Finally, Section IV concludes the paper.

II. OVERVIEW OF THE PROPOSED ALGORITHM

Figure 1 shows the proposed dense-stereo-based ROI generation for the detection of on-road obstacles. In the depth map generation step, the rectified stereo image pair is firstly downscaled to reduce the increase in computation cost, which is proportional to the image size. For the downscaled stereo image, a global stereo matching process based on a 2D Markov random field (MRF) generates a smooth and accurate depth map. Finally, the resolution of the depth map result is reconstructed to an original size by using a depth map up-sampling method [5]. Note that we intend to minimize the quality difference between the original depth map and the up-sampled one.

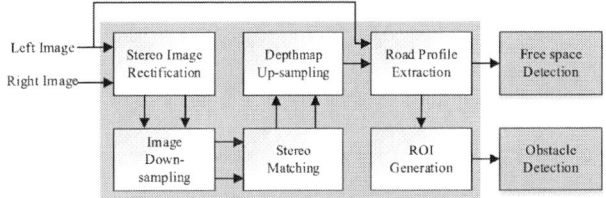

Figure 1. Flow of proposed algorithm

In designing a robust stereo matching algorithm for on-road scenes, we should consider the variation in illumination or the possibility of large textureless regions, which frequently appear in the road or on the vehicle's planar surface. As shown in Figure 2, we use the census transform as a local cost function and iteratively optimize the global matching cost by using a four-level hierarchical belief propagation (HBP) algorithm [3]. Note that the iterative cost update scheme in HBP is effective for the estimation of a good depth value for ambiguous regions such as those mentioned above.

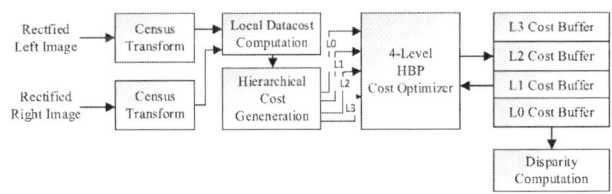

Figure 2. Stereo matching process

In the ROI generation step as shown in Figure 3, the generated depth map and the color image are used to extract the road surface profile and ROI candidates. Using the U-V-

978-1-5090-3220-4/16 $31.00 © 2016 IEEE   179   ISOCC 2016

disparity method, the initial ROI estimation results are further refined and merged by using scale information and the class-specific features of the obstacles such as the aspect ratio and size.

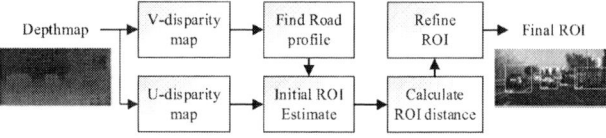

Figure 3. ROI generation

III. IMPLEMENTATION RESULTS AND EVALUATION

In order to overcome the enormous computational overhead of the global stereo matching process, a dedicated hardware accelerator has been designed [4]. Our parallel processing architecture based on hierarchical structures in HBP and using hundreds of processing elements (PEs) is hundreds of times faster than the original HBP algorithm. As shown in Figure 4, we designed an integrated hardware architecture that includes image rectification, cost computation, and HBP-based global optimization logic. The hardware architecture was implemented on the Xilinx Kintex-7 field-programmable gate array (FPGA) chip, and the implementation results are summarized in Table I.

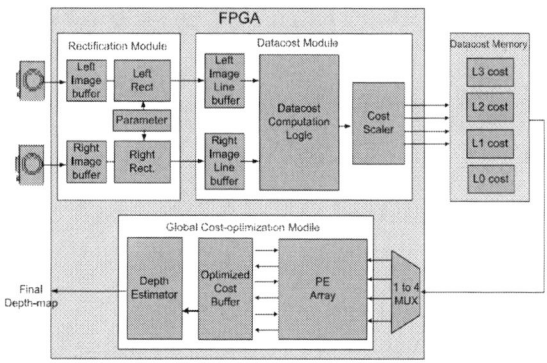

Figure 4. Hardware architecture of stereo matching IP core

TABLE I.   RESULTS OF IMPLEMENTATION ON A KINTEX-7

		Implementation Results
clock frequency		53 Mhz
on-chip memory requirement		646KB
block-ram usage	Total	68%
	Globalcost	33.4%
	Rectification /localcost	16.3%
	Memory I/F	18.3%
LUT usage	Total	62%
	Globalcost	42.1%
	Rectification /localcost	5.7%
	Memory I/F	14.2%

The remaining processes, including image down-sampling, depth up-sampling, and ROI generation, have been implemented in the NVIDIA TX1 embedded processor. Among these processes, the depth scaling processes have been implemented on the TX1 graphics processing unit (GPU) core using the CUDA parallel programming language.

TABLE II.   COMPUTATION TIMES FOR REAL-TIME PROCESSING

Process	Execution Time (ms)		
	CPU*	FPGA	GPU
IMAGE DOWN-SAMPLING	3	-	1.5
STEREO MATCHING	2,576	27.0	-
DEPTH UP-SAMPLING	307	-	9.9
Total	2,886	38.4 (26 FPS)	

* Intel i-7 2.8Ghz CPU

Table II presents the total computation time of the proposed depth map generation system using the FPGA-based hardware accelerator and GPU-based parallel programming. The total time is markedly less than that using only CPU-based execution. Note that the proposed stereo matching process achieves a real-time processing rate of 26 frames per second (FPS) for an HD-sized stereo image sequence.

Finally, the total execution time as executed by hardware and software partitioning achieved a real-time processing rate of 15 FPS. The ROI generation portion consumed only about 5 ms on TX1's CPU core.

IV. CONCLUSION

We have proposed an efficient stereo-vision-based on-road obstacle ROI generation system for use in ADAS. In order to reduce computational costs, an efficient depth map generation process has been implemented. In addition, in order to accelerate the execution, FPGA-based hardware and GPU-based parallel programming are utilized. The implementation results show that the proposed system outputs an accurate dense depth map and multiple ROI candidates in real time.

ACKNOWLEDGMENT

This work was supported by the DGIST R&D Program of the Ministry of Education, Science and Technology of Korea (16-IT-02). It was also supported by the MSIP(Ministry of Science, ICT and Future Planning), Korea, under the ITRC(Information Technology Research Center) support program (IITP-2016-H8601-16-1008) supervised by the IITP(Institute for Information & communications Technology Promotion).

REFERENCES

[1] H. Zhencheng, and K. chimura. "UV-disparity: an efficient algorithm for stereovision based scene analysis." IEEE Intelligent Vehicles Symposium(IVS), 2005.

[2] J. M. Alvarez and M. L. Antonio, "Road detection based on illuminant invariance." IEEE Transactions on Intelligent Transportation Systems 12.1: 184-193, 2011.

[3] P. F. Felzenszwalb and D. P. Huttenlocher, "Efficient belief propagation for early vision", IJCV, 70(1): 41-54, 2006.

[4] S. Kwon, C-H. Lee, Y-C. Lim and J-H. Lee, "A sliced synchronous iteration architecture for real-time global stereo matching." IS&T/SPIE Electronic Imaging. International Society for Optics and Photonics, 2010.

[5] S. B. Lee, S. Kwon and Y. S. Ho, "Discontinuity adaptive depth upsampling for 3D video acquisition", Electronics Letters, 49.25: 1612-1613, 2013.

High Bandwidth Memory(HBM) with TSV Technique

Jong Chern Lee, Jihwan Kim, Kyung Whan Kim, Young Jun Ku, Dae Suk Kim, Chunseok Jeong, Tae Sik Yun, Hongjung Kim,
Ho Sung Cho, Sangmuk Oh, Hyun Sung Lee, Ki Hun Kwon, Dong Beom Lee, Young Jae Choi, Jaejin Lee, Hyeon Gon Kim, Jun
Hyun Chun, Jonghoon Oh, and Seok Hee Lee
HBM Design Team, SK Hynix Inc.
Icheon, Gyeong-gi, South Korea
Jclee2@sk.com

Abstract— **In this paper, HBM DRAM with TSV technique is introduced. This paper covers the general TSV feature and techniques such as TSV architecture, TSV reliability, TSV open / short test, and TSV repair. And HBM DRAM, representative DRAM product using TSV, is widely presented, especially the use and features.**

I. INTRODUCTION

These days, demand for higher bandwidth and speed increases in high-end graphics memory and high-performance computing (HPC) market. TSVs are the most proper solution because of lower capacitance than wire bonding or PCB connection, smaller area than pads, and low power consumption. In this paper, TSV architecture, TSV reliability and TSV repair techniques are explained, and then HBM DRAM is also introduced [1].

II. TSV TECHNIQUE

A. TSV Architecture

General TSV cross section view is shown in Figure.1. There are 5 chips which connected by TSV. TSV is composed by the parallel capacitors and serial resistors [2]. The TSV connection path is shorter and thinner than wire bonding or PCB connection. Therefore, its stacking architecture achieves high speed and low power consumption.

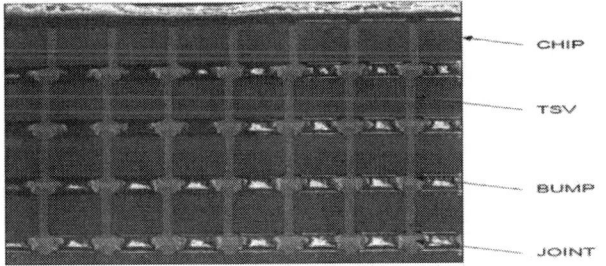

Figure 1. TSV cross section view

B. TSV Reliability

TSV and micro-bump process is difficult technique for manufacturing. And this process handles a large portion of the known good stacked die (KGSD) yield. Therefore, the more the number of TSVs increase, the more important TSV reliability became. Figure 2 shows the TSV open/short test circuit. The condition of TSV connection is checked with

current which flows from top slice to the bottom slice. First, current sources on the top slice are turned on by the test-mode signal. And the current from the current source flows to monitoring pads on the bottom die. Test equipment measure and store the values of the current and it judges the pass/fail condition [3].

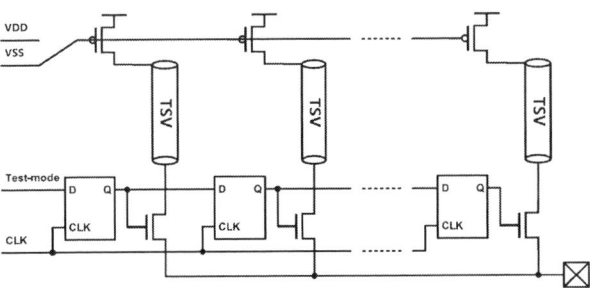

Figure 2. TSV Open/Short Test circuit

C. TSV Repair

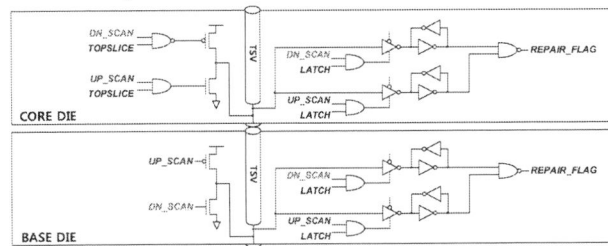

Figure 3. TSV self-repair scheme

Typically, failed TSVs have been repaired by the fuse cutting. TSV self-repair technique which is shown in figure 3 does not have to execute TSV open/short test and to repare fuses. During the boot-up time, TSV self-repair sequence starts operating automatically TSV open/short test and repairing fail TSV using redundant TSV. Each TSV has a pair of PMOS and NMOS. PMOS is a strong current source and NMOS is a weak leaker. Basic concept of the technique is a fighting of these PMOS and NMOS. If TSVs in TSV set are connected themselves well, voltage level of the TSV set is high. Repair flag which informs each TSV to repair itself, is generated by two sequence, upward scan and downward scan. The TSV repair technique has advantages of not only a testability, but a post packaging failure. With this technique, fuse cutting does not needed, and even we do not have to know whether the TSVs are opened or shorted.

978-1-5090-3220-4/16 $31.00 © 2016 IEEE   181   ISOCC 2016

III. HBM DRAM

A. System using HBM DRAM

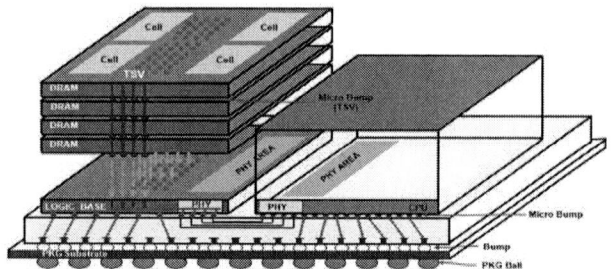

Figure 4. HBM system overview

Figure4 shows general SiP (System in Package) using HBM DRAM. The SIP contains stacked HBM DRAM, Controller (CPU or GPU), Interposer and PKG substrate. The HBM DRAM and the controller have micro-bumps for connection to the interposer. PHY area of HBM DRAM and the controller are connected through the interposer and they communicate each other directly. Some micro-bumps of HBM DRAM and the controller are connected PKG Balls through the interposer and PKG Bumps. They are used for direct access to the HBM DRAM or the controller.

B. HBM Architecture

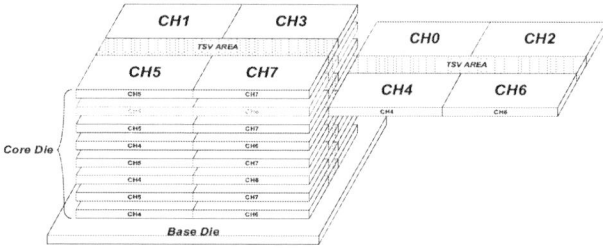

Figure 5. HBM architecture

Figure5 shows 8-hi stack HBM DRAM architecture. The HBM DRAM consist of a base die and 8 core dies. The base die controls the core dies and interfaces to the controller. And the core dies have memory cells. Each core die has 4 channels with 8 banks. Therefore, only 2-hi stack can make full 8-channel bandwidth.

C. HBM Speed, Bandwidth, and Power Efficiency

Figure 6 shows various DRAM's per pin speed, bandwidth, and I/O power efficiency. The per pin speed is growing twice faster than previous generation. However, latest HBM2's per pin speed is 2Gbps, same as DDR3. Low per pin speed achieves a good signal quality and lower power consumption. Furthermore, HBM1 and HBM2 bandwidth are 4.6 times and 9.1 times higher than GDDR5. Because it has the large number of I/O pins (1024 I/O in 8-channels). And HBM2 I/O power is less than 42% GDDR5 x32. Because micro-bump's parasitic RLC is less than PKG ball. And low I/O speed without a pre-emphasis or DFE circuits takes advantage to reduce I/O current.

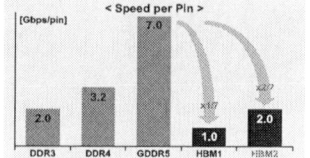
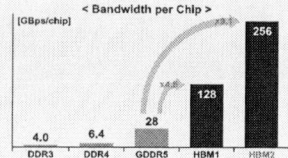
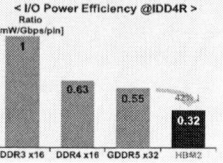

Figure 6. Per pin speed, bandwidth, and power efficiency of DRAMs

IV. DIE PHOTO AND CONCLUSION

A. HBM2 Die Photo

The HBM2 die photos are shown in Figure 7. HBM2 chip is fabricated with 21nm CMOS DRAM process. HBM2 Base die size is 71.64mm^2.

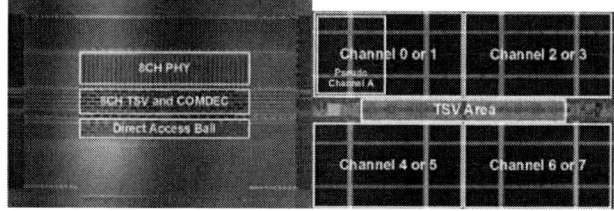

Figure 7. Base and core die photos of HBM2

B. Conclusion

This paper presented TSV techniques that will be the most important techniques for future semiconductor devices. The TSV connection path is consist of TSVs, micro-bumps, and joints. It achieves high speed and low power. TSV reliability can be checked with current which flows through TSV path. The measured values indicate the TSV connection. Presented TSV self-repair technique executes both TSV open/short test and repairing without test equipment and fuses. And the HBM DRAM using thousands of TSVs is introduced. Which has low per pin speed, highest bandwidth, and most efficient I/O power consumption in every DRAM products.

REFERENCES

[1] J. C. Lee, et al., "A 1.2V 64Gb 8-channel 256GB/s HBM DRAM with peripheral-base-die architecture and small-swing technique on heavy load interface," in IEEE ISSCC Dig. Tech. Papers, pp. 318-319, Feb. 2016.

[2] H. Kim, et al., "Measurement and Analysis of a High-Speed TSV Channel," IEEE Trans. Comp., Packag. Manuf. Technol., vol. 2, no. 10, pp. 1672-1685, Oct. 2012

[3] D. U. Lee, et al., "An Exact Measurement and Repair Circuit of TSV Connections for 128GB/s High-Bandwidth Memory(HBM) Stacked DRAM," in Symposium on VLSI Circuits, pp. 1-2, Jun. 2014.

Emulation of Processing in Memory Architecture for Application Development

Jin-San Kwon, Tae-ho Hwang, and Dong-Sun Kim
Korea Electronics Technology Institute
Seongnam, Rep. of Korea
jinsan.kwon@keti.re.kr

Abstract— Since the new technologies like big data and cloud computing require tremendous amount of transactions between processors and memory, researches on a new memory system called Processing in Memory (PIM) architecture has been suggested as a solution for those memory intensive applications. To make software utilize the new architecture, a development environment with tool chain and debug infrastructures supplementing extended instruction sets and target emulation is needed. This paper introduces a way to emulate a target platform having PIM architecture memory device. The emulated platform may use the PIM device as a system memory or simple memory mapped bus device. The actual PIM architecture is implemented in both cycle-based memory simulator and FPGA based hardware platform. The emulator has a bridge for these implementations so the emulated target platform may be freely used with early stage application development for the PIM architecture.

Keywords: Processing in Memory; QEMU; memory simulator

I. INTRODUCTION

In the point of memory architecture, concurrent and distributed processing of huge amount of data like cloud computing and big data is not a suitable usage for traditional processor-memory architecture. Because of memory latency, performance improvement is largely rely on cache which buffers the latency and maximizes overall performance. However, those applications utilizing distributed resources require tremendous amount of transactions between processors and memory, resulting inefficient resource utilization. To alleviate this problem, researches on a new memory system called Processing in Memory (PIM) architecture has been suggested as a solution for those memory intensive applications [1]. It contains a small portion of arithmetic unit on every memory array, therefore, relatively simple computation can be done in the memory system without carrying the data on system bus which may result the aforementioned latency.

Besides the main processor, using PIM arithmetic units requires additional instructions and runtime environment. The new instructions target the PIM elements and results of the instructions are immediately available on the memory without any interventions from the main processor. To support software development utilizing those added instruction sets, we implemented the PIM architecture in both simulator and FPGA hardware form, and a platform emulator which uses PIM

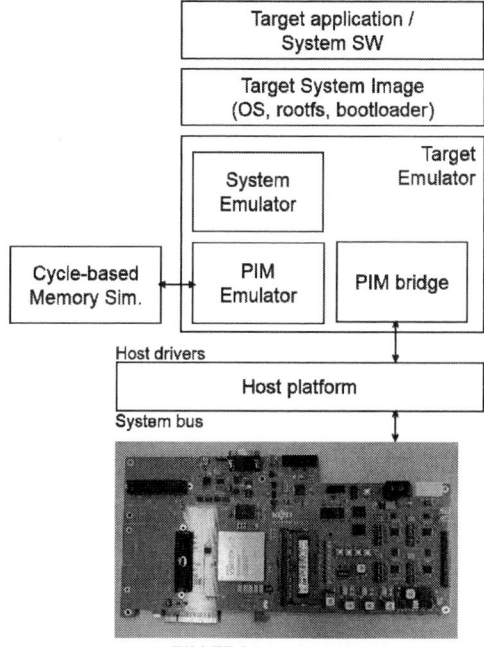

Figure 1. Processing in Memory emulation

architecture as either system or device memory as shown in the Fig. 1. The actual memory operations are done by either a cycle-based memory simulator or an external FPGA based implementation. Those memory components are bridged to the platform emulator, and since the bridged memory device is in general form, it can be attached on any processor architectures supported by the platform emulator.

In this paper, we introduce the memory emulator, FPGA based implementation, and the platform emulator. As an example, we use x86-64 for the main processor architecture of the emulated target system, and FPGA based PIM memory device is connected into the platform. The emulated platform boots Linux kernel and accesses the PIM memory as memory device.

II. PLATFORM EMULATION

In our implementation, as described earlier, two types of PIM memory devices are available. To accommodate both, we

978-1-5090-3220-4/16 $31.00 © 2016 IEEE

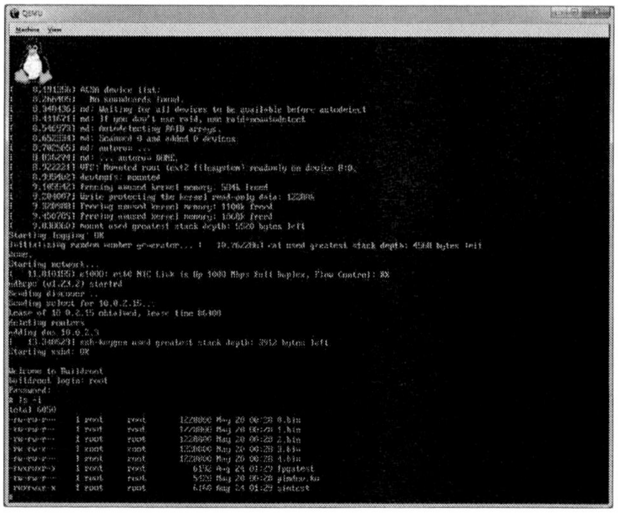

Figure 2. Target running on the emulator

implement general parts of the target platform including main processors, system buses, and peripherals using QEMU [2]. Although it is possible to use the PIM architecture for the system's main memory, we put it onto the system bus and treat it as memory device so it can be manipulated, inspected, and debugged within the device region freely.

On the QEMU's perspective, the PIM device is a simple memory device attached on the system bus. It is not important for the emulator whether the actual device is a memory simulator or a FPGA implementation. Therefore, we create a bridge device in the QEMU holding specific size, and put the bridge on a platform's system bus with designated address. In our example, we reserved 0x80000000 ~ 0x8FFFFFFF for the bridge on an i386 platform implementation. This will hold 256MiB of address space for the PIM bridge. The emulated target machine runs a Linux kernel holding the reserved bridge area as shown in the Fig. 2.

Since the PIM is a generic memory device, the only permitted actions from the emulator are read and write operations. These are defined in QEMU's MemoryRegionOps structure and registered on the emulator during device initialization sequence. After initialization, the registered read and write functions are called back from the emulator in case of reading or writing attempts are made on the device's address space. This event can be made by programs running on a target requesting such operations on the device memory region. Since we put the bridge device on the region, when a program requests memory input or output, the bridge device context receives the requests and has to act on it by returning data or caching it on local registers.

The bridge device, now on behalf of the Processing in Memory architecture device, relays the requested action to the actual implementation. The emulator currently has two PIM implementations by default which are memory simulator and hardware implementation. The simulator is based on

DRAMsim [3] and MARSSx86 [4] which are cycle accurate simulators, thus every aspect for the memory are simulated accurately without actual hardware implementation. On the other hand, FPGA based PIM hardware implementation can be attached on the bridge and every memory transactions can be processed in the hardware implementation. Currently, the FPGA platform is connected to the host machine via PCIe bus and memory packet can be transferred by either programmed I/O or DMA transfer operations for better processor utilization efficiency.

For debugging the applications using the PIM address region, GNU debugger suit can be used in both local and remote debugging mode. Bare metal applications without operating systems on the target machine can be remotely debugged by using GDB server stub shipped with QEMU [5]. More debugger options are available if the program is running on Linux kernel. GDB server is ported in the target machine with the operating system by default, and connection to the server using cross GDB on the host machine can be made as usual. If the device driver should be debugged, the debug server stub in the emulator can be used as the driver and kernel is a huge bare metal application. To simplify this process by skipping bootloader stage, a port of GDB stub is also included in the kernel space, acting like a GDB server with full debug capability on kernel objects. In case of local debugging solely on the target machine, the target debugger and supporting library and sources can be ported with the operating system and debugger can be attached locally.

III. CONCLUSION

This paper proposed a platform emulator with a Processing in Memory device. The PIM device can be either simulated or hardware implemented memory device, and those are bridged into the emulator. The emulated target machine detects the memory as conventional memory device reside on the system bus, therefore it can be treated as system memory or simple memory region. Detailed debug perspectives were also discussed for bare metal, native, and kernel space applications.

REFERENCES

[1] J. Ahn, S. Hong, S. Yoo, O. Mutlu, and K. Choi, "A scalable processing-in-memory accelerator for parallel graph processing," In Proceedings of the 42nd Annual International Symposium on Computer Architecture, 2015.

[2] F. Bellard, "QEMU, a Fast and Portable Dynamic Translator," FREENIX Track: 2005 USENIX Annual Technical Conference, 2005.

[3] D. Wang, B. Ganesh, N. Tuaycharoen, K. Baynes, A. Jaleel, and B. Jacob, "DRAMsim: A memory-system simulator," SIGARCH Computer Architecture News, vol. 33, no. 4, pp. 100-107. September 2005.

[4] A. Patel, F. Afram, S. Chen, and K. Ghose, "MARSSx86: A Full System Simulator for x86 CPUs," Design Automation Conference 2011 (DAC'11), 2011.

[5] P. Dovgalyuk, D. Dmitriev, and V. Makarov, "Don't panic: reverse debugging of kernel drivers," In Proceedings of the 2015 10th Joint Meeting on Foundations of Software Engineering (ESEC/FSE 2015), 2015.

[6] M. Petullo, and S. Jon, "The lazy kernel hacker and application programmer," Presentation at the 3rd ACM workshop on Runtime Environments, Systems, Layering and Virtualized Environments, 2013.

This work was supported by the IT R&D program of MOTIE/KEIT. [10052653, Development of processing in memory architecture and parallel processing for data bounding applications]

Implementation of a Low-Overhead Processing-in-Memory Architecture

Young-Jong Jang, Byung-Soo Kim, Dong-Sun Kim, Tae-ho Hwang
SoC Platform Research Center
Korea Electronics Technology Institute (KETI)
Sungnam-Si, KyungGi-Do, Korea
youngjong.jang@keti.re.kr, bskim4k@keti.re.kr, dskim@keti.re.kr, taeo@keti.re.kr

Abstract— Since the new technologies like big data and cloud computing require tremendous amount of transactions between CPU and memory, researches on a new memory system called PIM (processing in memory) architecture has been suggested as a solution for those memory intensive applications. This paper introduces a low-head PIM architecture. And we introduce the processing techniques of the PIM's instructions on the PIM architecture. And we have designed the PIM architecture using Verilog HDL, and the operation of the PIM architecture verified through RTL simulation.

Keywords-Processing in memory; PIM; 3D-stacking PIM;

I. INTRODUCTION

As big data and cloud computing are become relatively commonplace, the amount of data transfer between the CPU (central processing unit) and the memory has skyrocketed over the past few years. But since memory wall problem, the data processing speed of the memory is significantly slower than the processing speed of the CPU. So it was difficult to improve the overall performance of the computer system [1]. To solve these problems, various PIM (processing in memory) architecture were discussed [2-8]. The PIM provides high bandwidth, massive parallelism, and high energy efficiency by implementing the processing unit in the main memory, so it was eliminating the overhead of data movement between the CPU and the memory. In this paper, we introduce low-overhead PIM architecture. And we introduce the processing techniques of the PIM's instructions on the PIM architecture.

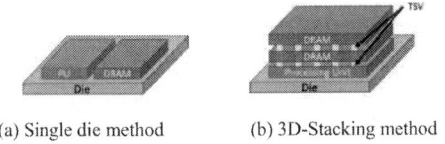

(a) Single die method	(b) 3D-Stacking method

Figure 1. The Implementation methods of the PIM

II. PIM(PROCESSING IN MEMORY) ARCHITECTURE

There are two way of implement methods of the PIM. One is to implement by combing all of the processing unit and the DRAM cells in a single die, as shown Fig. 1(a) [2-4]. And the other is a method for 3D-stacking the processing unit and the

DRAM cells, as shown Fig. 1(b) [5-8]. Our PIM architecture was also used 3D-stacking technique. The 3D-stacking technique is placing directly the processing unit layer on the die, and DRAM cell layer is placing above it by using TSV (through silicon vias).

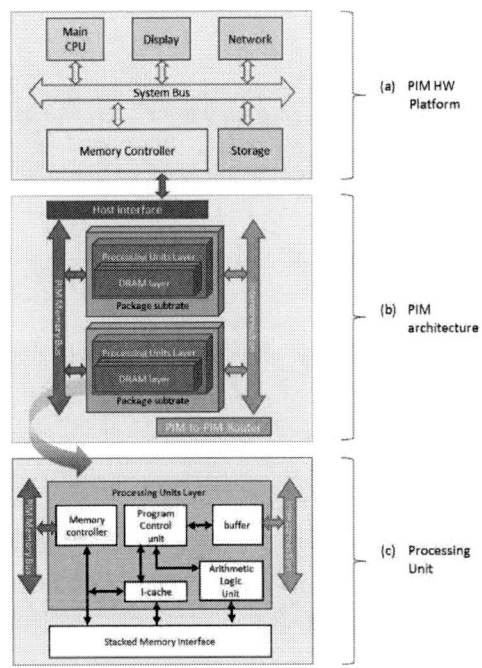

Figure 2. The PIM architecture: (a) The PIM hardware platform, (b) PIM memory architecture, (c) PIM processing unit

PIM hardware platform is the same as the conventional computer system, as shown in Fig. 2(a), and replace only the memory controller for the PIM. As shown in Fig. 2(b), the PIM architecture is connected by two interconnector. One is the PIM memory bus connector that is connected the memory controller in PIM hardware platform, and the other is a PIM-to-PIM interconnector for connecting between each processing units. A transfer sequence of PIM command is as follows. The main CPU sends a data read/write command or other operation commands to the memory controller of PIM hardware platform. This memory controller interprets the transfer command, and

---

This work was supported by the IT R&D program of MOTIE/KEIT. [10052653, Development of processing in memory architecture and parallel processing for data bounding applications]

978-1-5090-3220-4/16 $31.00 © 2016 IEEE   185   ISOCC 2016

determine to transmit to any PIM. And then the memory controller of PIM hardware platform sends a command to each PIM through the host interface. As shown in Fig. 2(c), the memory controller of PIM is interprets the transmitted command, and then it read a data from DRAM cells, or it write a data to DRAM cells, or it perform arithmetic operations in arithmetic logic unit of PIM.

III. PIM INSTRUCTION

The PIM command is divided into a header part and a tail part, as shown in Fig. 3. The PIM header consist of CUB (cube id), ADRS (address), tag, LNG (packet length), CMD (command). And the PIM tail composed of CRC (cyclic redundancy check), RTC (return token count), SLID (source link ID), Pb (poison bit), SEQ (sequence number), FRP (forward retry pointer), RRP (return retry pointer). Table 1 shows the Boolean atomic command that is processed in the PIM architecture.

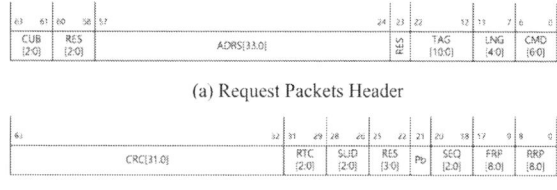

(a) Request Packets Header

(b) Request packets Tail

Figure 3. The PIM architecture request: (a) header, (b) tail

TABLE I.   BOOLEAN ATOMIC COMMANDS

Command Description	command [6:0]	Request Packet length	ATOMIC FLAG
16-bytes XOR	7'b1000000		
16-bytes OR	7'b1000001		
16-bytes NOR	7'b1000010	2	Yes
16-bytes AND	7'b1000011		
16-bytes NAND	7'b1000100		

16-bytes XOR request processing procedure is as follows. Processing unit confirm the 16-bytes XOR command by decoding the header of the request. And then processing unit is receiving the next 16-bytes data, and stores it in the buffer. And

simultaneously it reads the stored data in the DRAM cell of the request address. And the processing unit processes the XOR operation into the read data and the stored data in buffer, and stores XOR data into DRAM cell of request address. We have designed a PIM structure using Verilog HDL, the operation of the PIM structure were verified through the RTL simulation, as shown in Fig. 4.

IV. CONCLUSION

This paper proposed a PIM architecture that used 3D-stacking technique, and we were explained a request operation on a PIM architecture by a XOR request processing procedure. It was verified using Verilog HDL for RTL simulation. Other operations (for example, large burst data read/write, add, subtract, etc.) are expected to be added later. It is our future work.

REFERENCES

[1] B. Rogers, A. Krishna, G. Bell, K. Vu, X. Jiang, and Y. Solihin, "Scaling the bandwidth wall: challenges in and avenues for CMP scaling, " In ACM SIGARCH Computer Architecture News, vol. 37, no. 3, pp. 371-382. June 2009.

[2] H. S. Stone, "A Logic-in-Memory Computer, " IEEE Transactions on Computers, vol. 19, no. 1, pp. 73-78, January 1970.

[3] M. Gokhale, B. Holmes, and K. Iobst, "Processing In Memory: the Terasys Massively Parallel PIM Array," IEEE Computer, vol. 28, no. 4, pp. 23-31, April 1995.

[4] J. Draper, J. Chame, M. Hall, C. Steele, T. Barrett, J. LaCoss, J. Granacki, J. Shin, C. Chen, C. W. Kang, I. Kim, and G. Daglikoca, "The architecture of the DIVA processing-in-memory chip, " In Proceedings of the 16th international conference on Supercomputing, ACM, pp. 14-25, June 2002.

[5] J. Thomas Pawlowski, "Hybrid memory cube (HMC), " Hot Chips. vol. 23. 2011.

[6] J. Ahn, S. Hong, S. Yoo, O. Mutlu, and K. Choi, "A scalable processing-in-memory accelerator for parallel graph processing, " In ACM SIGARCH Computer Architecture News, ACM, vol. 43, no. 3, pp. 105-117, June 2015.

[7] J. Ahn, S. Yoo, O. Mutlu, and K. Choi, "PIM-enabled instructions: a low-overhead, locality-aware processing-in-memory architecture, " In 2015 ACM/IEEE 42nd Annual International Symposium on Computer Architecture (ISCA), IEEE, pp. 336-348, June 2015.

[8] E. Azarkhish, D. Rossi, I. Loi, and L. Benini, "Design and Evaluation of a Processing-in-Memory Architecture for the Smart Memory Cube, " In International Conference on Architecture of Computing Systems, Springer International Publishing, pp. 19-31, April 2016.

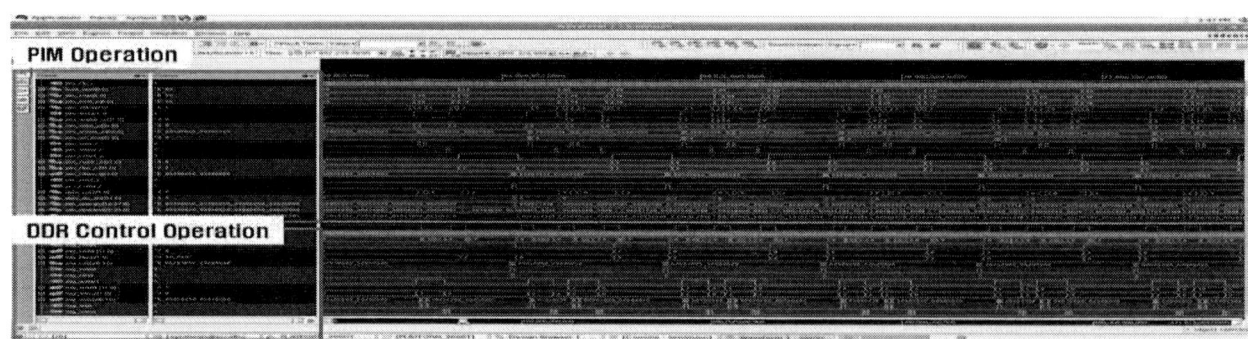

Figure 4. A RTL simulation of The PIM architecture

High Density PCM(Phase Change Memory) Technology

Hongsik Jeong

E. E. Department and Center for Brain Inspired
Computing Research of Tsinghua University
Beijing China, China
hongsikjeong@tsinghua.edu.cn

Abstract— Nowadays emerging memory devices have been much paid attention due to needs for the performance improvement and energy efficiency of computing in big data era for better computing hierarchy. Among them, Phase Change Memory (PCM) is being persistently spotlighted due to its fantastic characteristics such as high scalability, high endurance, long retention and so on. Moreover the technologies for high density PCM beyond the density of 128Gb have been successfully developed for commercialization, which can meet the requirement of new applications for more efficient computing hierarchy like storage class memory and neuromorphic computing. In this paper key technologies for high density PCM will be discussed along with applications.

Keywords; PCM(phase change memory); MLC; 3D integration; storage class memory, neuromorphic computing

I. Introduction

Legacy computer architectures consist of CPU, DRAM as a main memory, and NAND or HDD as storages. Unfortunately there has been a big latency gap between memory (<100ns) and storage (~100us), which masks the system be complicated and inefficient. Phase Change Memory (PCM) is considered to be most proper for a hybrid memory subsystem so called storage class memory (SCM) to reduce the performance gap between DRAM and Flash in memory hierarchy. This innovative system technology can make computing system more cost-effective, more energy-saving, and much faster. However to meet the requirement of SCM, high density and high-performance technologies of PCM have to be developed. To fulfill these needs, a cost-effective and high-speed phase change memory cell scheme will be introduced at 19nm technology node, which is directly scalable down to 10nm node and can be extendable to stacked array for high density [1]. Moreover MLC and stacking technology will be discussed for higher density. In this paper I will suggest the future technology roadmap of high density PCM in big data era based on these technology prospect and new applications.

II. Key Technologies

Now, Technical pathways for high density PCM by 3D stacking will discussed. To satisfy this goal, two technical pathways should be considered. One is simple process integration for high density PCM. We should design a process integration scheme well in order to achieve low cost as well as to be highly scalable and easily stackable. The other is to have proper phase change materials which can determine cell performances such as write speed and reliability.

Similar to other memories, a cell in PCM consists of one switch and one storage (1T+1R). Here, we should consider how process integration scheme is designed in cell array to have good cost-competitiveness, how can the on-current as high as possible, the off-current as small as possible in a cell switch be achieved, and how to meet required cell performances using proper phase change materials.

The other approach to achieve a higher density PCM is studying multi-level cell (MLC) technology. There are two key barriers to be overcome for MLC technology. One is large resistance distribution due to the process variations provoked from contacts, transistors, interconnections and so on. Fortunately a lot of design technologies to overcome this problem for NAND Flash already have been developed well. Therefore these technologies like Incremental step pulse programming (ISPP) will be also very useful for PCM.

The effect of ISPP technology for PCM is shown in Fig. 1. The cell resistance distribution before adopting ISSP is dotted by red line in Fig.1(a) which is very large and is not applicable for MLC. After adopting ISPP dotted blue line in Fig.(a), we can see the drastic improvement of cell distribution which makes the MLC possible as shown in Fig.1 (a). Also, the write disturbance was checked in Fig.1(b) and we can see that the weakest programming state made by ISPP is very reliable.

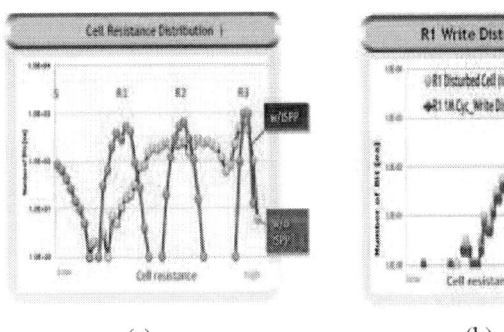

(a)   (b)

Fig.1. (a) Cell resistance distributions before and after adopting ISPP. (b) The result of write disturbance test for R1 state (weakest program state made by ISPP).

The other barrier for MLC is the resistance drift of PCM. Recently many technologies to overcome this problem have been developed successfully by design [2] and process method [3]. Combining ISSP and technologies for suppress the resistance drift, we can achieve MLC operation of PCM successfully.

III. Technology Roadmap of High Density PCM

Already the successful development of 3D X-point memory which can be presumed as PCM has been announced, where density is 128Gb with stacking technology [4]. Moreover the possibility of stackable PCM cell integration technology beyond 19nm technology node and MLC are shown in this paper. Considering this developing trend of PCM technology, I can suggest future high density PCM technology roadmap as shown in Fig. 2.

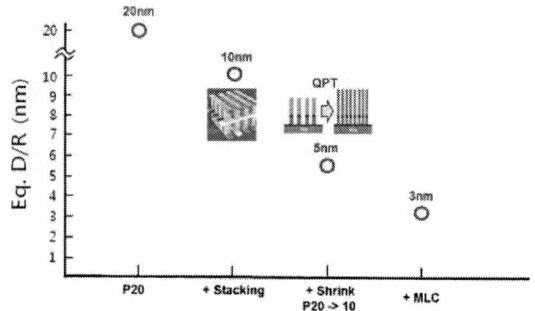

Fig.2. Technical Roadmap of High Density PCM. Equivalent D/R means technology node of PCM storing actual bits which corresponds to same bits by SLC NAND technology node (D/R).

The PCM has same cell feature size($4F^2$) of NAND Flash. Moreover PCM has better intrinsic scalability. Therefore if we succeed in developing stacking and MLC technology, we can make PCM have better cost effectiveness and performance comparing NAND Flash in near future, corresponding to beyond 5nm equivalent design rule(D/R) of SLC NAND Flash. This roadmap shows the bright future of PCM for new applications which makes computing efficiency improve drastically and memory hierarchy change innovatively for big data era

Memory(DRAM) Latency <100ns	M-type ← SCM → S-type			Storage(NAND) Latency ~100us

Requirement*			PCM status	Remark
Storage Type	M-type	S-type		
Cost	3~5x Cost of HDD		Possible	3D X-Point
Speed	<200nsec	<1usec	150ns	-
Random IOPS	> 100k IOPS		O.K.	Strongest Point
Sequential Write	>1GB/sec	> 100MB/sec	130MB	-
Endurance	10E12	10E9	> 10E11	
Power	< 1/10 Enterprise HDD		Possible	-

*Source from IBM

Fig.3. Current PCM technology status comparing requirement of SCM application. M-type means memory like application and S-type means storage like application.

The SCM application which was suggested in IBM [5] is very close to commercialization, already current PCM technology shows the possibility in this application as shown in Fig.3. Even though some technical barriers remain to cover all range of SCM application, for example, cheaper cost for storage type SCM and better sequential performance for memory type SCM are required, however it is expected that those will be overcome in near future, being considered the speed of PCM development.

Another attractive application is neuromorphic computing. Nowadays PCM has taken center stage in neuromorphic devices. A nucleation-dominant phase change material such as $Ge_2Sb_2Te_5$ undergoes the nucleation and growth of crystallites due to Joule heating after the applied current or voltage is higher than threshold value and its conductance is rapidly increased when those crystallites meet together and make a percolating path in amorphous matrix, which is quite analogous to the integrate-and-fire model (spiking) in biological neuron [6]. If we utilize this characteristics, we can emulate many kinds of neuron functions such as synaptic plasticity, non-linear function of neuron, massive parallelism and learning procedure of neural networks.

Recent big progress of high density PCM technology have made the opportunity to open new applications for more efficient computing in big data era, which can provide innovative ways to overcome the deterioration of Moore's law trend.

References

[1] Dae-Hwan Kang, et.al., "Considerations on Highly Scalable and Easily Stackable Phase Change Memory Cell Array for Low-Cost and High-Performance Applications," Non-Volatile Memory Technology Symposium (NVMTS), 2014 14th Annual, pp.1–5.

[2] N. Papandreoul, et.al., "Drift-Resilient Cell-State Metric for Multilevel Phase-Change Memory", Electron Devices Meeting (IEDM), 2011, pp. 3.5.1–3.5.4.

[3] Wabe W. Koelmans1, Abu Sebastian1, Vara Prasad Jonnalagadda1, Daniel Krebs, Laurent Dellmann1and Evangelos Eleftheriou, "Projected phase-change memory devices," Nature Comm., 6, 2015

[4] Press Released from Micron 2015.

[5] R. F. Freitas W. W. Wilcke, "Storage Class Memory: The Next Storagy System Technology", IBM J. Res. & Dev., 52, 2008

[6] Tomas Tuma, et.al., "Stochastic phase-change neurons", Nature Nanotechnology, 11, 2016

Robust Optical Fingerprint Sensor to Moisture Fingerprints

Young-Hyun Baek
Research Institute
UnionCommunity Co., Ltd.
Seoul, Korea
neural76@unioncomm.co.kr

Abstract— **We proposes a robust optical fingerprint sensor to moisture fingerprints. This sensor a new lens and prim design and optical-path change structure is proposed, which shows improvement of difference between a ridge and valley sensing image than typical fingerprint sensor. Simulations show that the optical fingerprint sensor an effective performance in the moisture ten-fingerprints and the result of NIST quality map.**

Keywords; fingerprint; optical sensor; moisture finger; NIST quality map

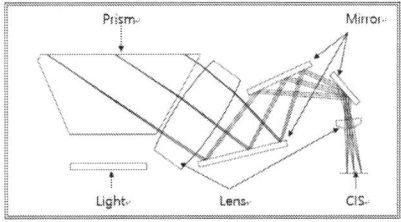

Figure 1. Feature of the device optical design

I. INTRODUCTION

Fingerprint sensing is one of the most widely deployed of all biometric technologies. There are a number of different techniques for capturing a fingerprint image including optical, capacitive, radio frequency, ultrasound, and thermal methods. One common shortcoming of many conventional fingerprint sensing technologies are the frequent occurrence of poor-quality images under a variety of common operational circumstances. Though each particular imaging method has different sensitivities, in general poor images may result from conditions such as dry skin, worn surface features of the finger, poor contact between the finger and sensor, bright ambient light, and moisture on the sensor [1, 2].

In this paper, we propose a robust optical fingerprint sensor with a moisture fingerprint based an optical-path change structure and new prim and lens deign. In this paper is organized as follows. In Section 2, we proposed the optical-path structure and design. Section 3 presents the experimental results and conclusions.

II. PROPOSED OPTICAL FINGERPRINT SENSOR

The finger touches a glass prism and the prism is illuminated with diffused light. The light is reflected at the valleys and absorbed at the ridges. The reflected light is focused onto a CCD or CMOS sensor. Optical fingerprint sensors provide good image quality and large sensing area but they cannot be miniaturized because as the distance between the prism and the image sensor is reduced, more optical distortion is introduced in the acquired image. Finger 1 show feature of the device optical design [3~5].

A. Proposal Lens Design

Figure 2 is the optical design output applying the spherical lens to satisfy the standard FBI PIV Spec. In order to satisfy the PIV must be able to decompose the line of 0.05mm prism fingerprint surface. 0.05mm line of the prism surface is reduced through the image formation lens system to the image sensor there is in early value to be dismantled should enter the one line to a pixel of the image sensor. But consider that the line of machining tolerances and tolerances enclosures, an assembler 1line differential is 1 pixel is put in danger. It is most desirable to enter the 1 line to 2 pixel. If 1 line is designed to fall into the 3 pixel image. This is because the larger the angle of incidence to the micro-lens over each pixel in the image CMOS sensor.

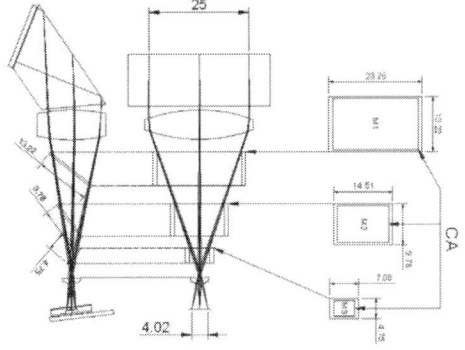

Figure 2. New spherical lens design

The prism surface is at least 0.05mm to be suitable for the more than 2.8um, 5.6um or less, when incident on the image sensor through the lens system. At this time, the ratio of the line is between 0.112 times 0.056 times. To enter the 1 line 21 pixel image differs, so the vertical direction is 25mm cis shall be imaged by 2.8mm. Figure 2 is applying a spherical lens design.

B. Robust ladder-type distortion is corrected in the design and moisture fingerprints.

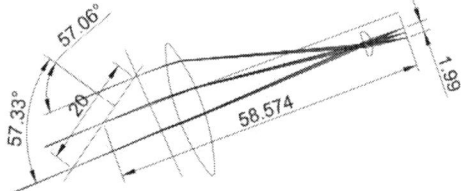

Figure 3.   Robust lens design in moisture fingerprints

C. The prism design is optimized for moisture fingerprints.

The prism is set to start because it should be separated so that they are at the water. In addition, the surface of the fingerprint image without being affected by the external light source should be uniform. The following is a structure in which external light incident to the prism does not enter the lens g2. If the external light is incident to the prism, Figure 4 (a) not enter more than a spectacle of light, such as. Caught by a first angle to a present angle of the prism Figure 4 (b) to set the material of the prism to *sf4*.

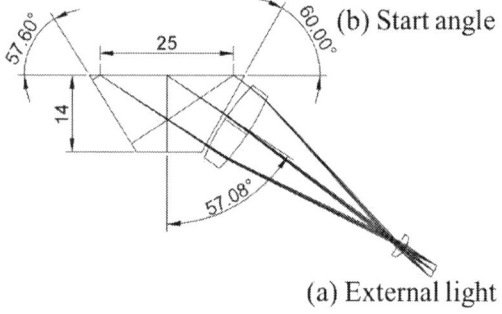

(b) Start angle

(a) External light

Figure 4.   The prism design in moisture fingerprints

III.   EXPERIMENTS AND CONCLUSION

This section focuses on evaluating the proposed sensor. We start by describing the test image data. Then, the experiment results of the proposed robust sensor is presented for moisture fingerprints image. Finally, the effectiveness of the proposed sensor is discussed through comparison with other fingerprint sensor.

Figure 5 (a) is proposed sensor images to captured a moisture fingerprints. Figure 5 (b) is traditional optical sensor images to captured a moisture fingerprints.

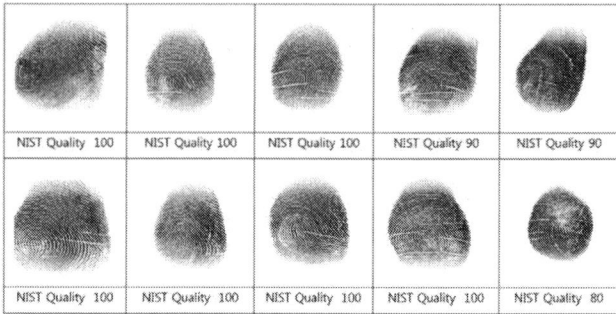

(a) The proposed sensor result images (include NIST Quality score)

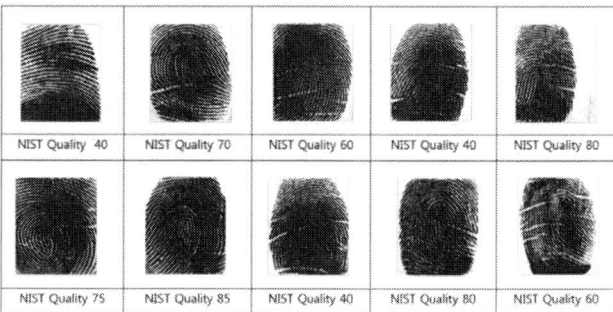

(b) The other sensor "Model: HFDU06S_Nitgen" result images (include NIST Quality score)

Figure 5.   The comparison result images

Finally, the results show that the proposed sensor performance has been improved more than image quality of conventional fingerprint sensor.

ACKNOWLEDGMENT

This research was supported Advanced Technology Center (ATC) grant funded by the Korean government (Ministry of Trade, Industry and Energy) (No. 10051484).

REFERENCES

[1] Sarat C. Dass, Anil K. Jain, "Fingerprint Classification Using Orientation Field Flow Curves", Michigan State University.

[2] Maltoni, D., Maio, D., Jain, A.K., Prabhakar, S.: "Handbook of Fingerprint Recognition", Springer-Verlag, New York, 2003.

[3] S. -J. Kim, K. -H. Lee, S. -W. Han and E. Yoon, "A CMOS Fingerprint System-on-a-Chip With Adaptable Pixel Networks and Column-Parallel Processors for Image Enhancement and Recognition", Journal of Solid-state Circuit, vol. 43, no. 11, (2008), pp. 2558-2567, 2008.

[4] J. W. Lee, D. J. Min, J. Y. Kim and W. C. Kim, "A 600-dpi Capacitive Fingerprint Sensor Chip and Image-Synthesis Technique", IEEE J. of Solid-state circuits, vol. 34, no. 4, (1999) April, pp. 469-475.

[5] S. -M. Jung, J. -M. Nam, D. -H. Yang and M. K. Lee, "A CMOS Integrated Capacitive Fingerprint Sensor with 32-bit RISC Microcontroller", IEEE Journal of Solid-state Circuit, vol. 40, no. 8, (2005), pp. 1745-1750.

Deep Learning Application Trial to Lung Cancer Diagnosis for Medical Sensor Systems

Ryota Shimizu, Shusuke Yanagawa, Yasutaka Monde, Hiroki Yamagishi,
Mototsugu Hamada, Toru Shimizu, and Tadahiro Kuroda
Faculty of Science and Technology
Keio University
Yokohama, Japan
r_shimizu@kuroda.elec.keio.ac.jp

Abstract— **Personal and easy-to-use health checking system is an attractive application of sensor systems. Sensing data analysis for diagnosis is important as well as preparing small and mobile sensor nodes because sensing data include variations and noises reflecting individual difference of people and sensing conditions. Deep Neural Network, or Deep Learning, is a well-known method of machine learning and it is effective for feature extraction from pictures. Then, we thought Deep Learning also can extract features from sensing data. In this paper, we tried to build a diagnosis system of lung cancer based on Deep Learning. Input data of the system was generated from human urine by Gas Chromatography Mass Spectrometer (GC-MS) and our system achieved 90% accuracy in judging whether the patient had lung cancer or not. This system will be useful for pre- and personal diagnosis because collecting urine is very easy and not harmful to human body. We are targeting installation of this system not only to gas chromatography systems but also to some combination of multiple sensors for detecting gases of low concentration.**

Keywords; Deep Learning, Deep Neural Network, Stacked Autoencoder, Gas Chromatography Mass Spectrometer(GC-MS)

I. INTRODUCTION

Health diagnosis is very difficult because it needs knowledge or experience of medical science but pre-diagnosis via sensing data maybe easy to do. For example, body temperature is a sensing data generated by thermometer, simple and harmless sensing system, and we have already known that high temperature is a sign of fever or some diseases. Like thermometer, pre-diagnosis should be done without any difficulties and impacts to the human body. Therefore, our major goal is to build an easy health checking system which can detect some diseases by analyzing sensing data, such as breath, saliva, and urine, which can be collected without hurting human body.

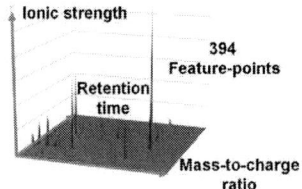

Figure 1. Human Urine Data via a GC-MS

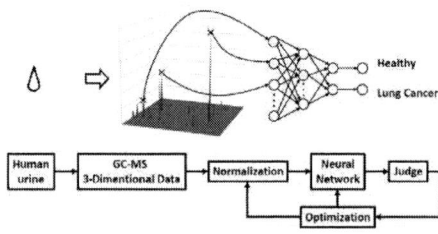

Figure 2. Schematic of the System

It is known that urine of lung cancer patients contains many specific chemical substances, so-called biomarkers. Human urine can be numerical data by Gas Chromatography Mass Spectrometer (GC-MS). GC-MS data of human urine have 3-dimensional feature, retention time, mass-to-charge ratio, and ionic strength (Figure 1). Deep Neural Network can classify pictures into some classes by extracting features from pixel data of the picture and we had assumed that Deep Neural Network is also effective to classify lung cancer patients from GC-MS data of human urine.

As far as we know, there are few examples of applying Deep Learning to human vital data in spite of many examples of applying it to image recognition or sound recognition. Our another goal is to prove that Deep Learning is also effective for health diagnosis by detecting lung cancer without any special knowledge or skills.

II. METHODOLOGY

Figure 2 shows the schematic of the system. First, human urine is converted into 3-dimensional data by GC-MS, which are the input of Neural Network. Before calculation, we normalize input data in two ways described later. Then, normalized input is used in the first layer of Neural Network and calculated in multi hidden layer. After the calculation, the output of the Neural Network is expressed in two values: the patient has lung cancer, or not. The Neural Network is trained again and again by backpropagation method. We evaluated the results and optimized the way of normalization, learning parameters, and network structure.

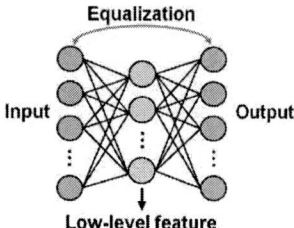

Figure 3. Autoencoder

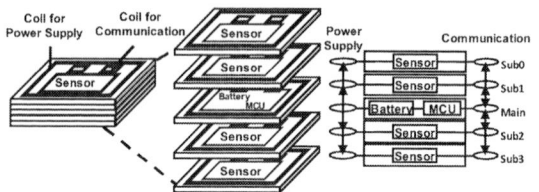

Figure 4. Flexible Multi-Sensor Node

A. Autoencoder

In this work, we used Stacked Autoencoder for the hidden layer of the Neural Network. Stacked Autoencoder means multi-layer Autoencoder, so we describe Autoencoder at first. Autoencoder (Figure 3) is one of the Neural Network structure whose input value and output value must be same. By utilizing hidden layer, it can learn how to express the input data by smaller number of feature values, i.e. low-level feature expression. This procedure looks like encoding, so that is why this structure is called Autoencoder. By stacking Auto-encoder and reducing the nodes in hidden layer, the Neural Network can learn less number of feature values than former layer. It is known that Stacked Auto-encoder is effective for pre-training of Neural Network because it can reduce dimensions of the data.

B. Normalization

As shown in Figure 1, GC-MS can distinguish substances of human urine into 364 feature points by retention time and mass-to-charge ratio. Each feature point has a feature value of ionic strength. We used Stacked Autoencoder and its activation function is a sigmoid function, whose output value must be from 0 to 1. So we normalized input data to be from 0 to 1. We had two ways of normalization: (1) divide each feature value of substance by maximum feature value of that substance in all patients, (2) divide all feature value of each patient by maximum feature value of that patient. By the experiment, we tried to decide which normalization works better.

Table 1. Condition and Results of the Experiment

Structure	394(Input)-200(Hidden)-100(Hidden)-2(Output)	
Pre-training	300 times with learning rate 0.1	
Supervised learning	300 times with learning rate 0.5	
Training data	57 patients' urine data	
Test data	10 patients' urine data	
Normalization	(1)	(2)
Sensitivity	40 %	80 %
Specificity	40 %	100 %
Accuracy	40 %	90 %

III. RESULTS

We evaluated our approach by comparing sensitivity, specificity, and accuracy. Sensitivity is the ratio of estimated cancer patients to real cancer patients, specificity is the ratio of estimated healthy patients to real healthy patients, and accuracy is the ratio of correctly estimated patients to all patients. Table 1 shows the experiment condition and results.

This table shows that proposed method worked well and normalization (2) works better, then achieved 90 % accuracy. That accuracy is enough for pre-diagnosis and it means that we showed possibility to detect lung cancer without any medical knowledge or experience but only Deep Neural Network.

IV. APPLICATION SYSTEM CONCEPT

As noted on last section, our system successfully detected lung cancer from human urine GC-MS data. However, GC-MS system instrument is too large to use in daily life. Our next goal is to make our system to be mounted on a small device without GC-MS. Our team is developing a flexible sensor node which can combine several small sensors onsite and process sensing data (Figure 5). Our challenge should be how to find out an effective and smaller set of chemical sensors and a mapping of the Deep Neural Network function. We are analyzing the Deep Neural Network we generated to make it more compact for this purpose. In addition, we are also trying to apply the multi-sensor node to detect other features of human breath.

V. CONCLUSIONS

In this work, we tried to apply Deep Learning to human urine data and achieved 90 % accuracy in the determination of lung cancer. Our work proved that Deep Learning is also effective for human vital data analyzation and we can do pre-diagnosis without any special medical knowledge.

ACKNOWLEDGMENT

This work is supported by CREST, Japan Science and Technology Agency.

REFERENCES

[1] Muxuan Liang, Zhizhong Li, Ting Chen, and Jianyang ZengIntegrative "Data Analysis of Multi-platform Cancer Data with a Multimodal Deep Learning Approach," IEEE/ACM Transactions on Computational Biology and Bioinformatics, pp. 928-937, Feb. 2011.

Normally-off Power Management for Sensor Nodes of Global Navigation Satellite System

Takashi Nakada and
Hiroshi Nakamura
The University of Tokyo
Tokyo, Japan
{nakada, nakamura}
@hal.ipc.i.u-tokyo.ac.jp

Toshifumi Nakamoto
Core Corporation
Osaka, Japan
t-nakamoto@core.co.jp

Toru Shimizu
Keio University
Kanagawa, Japan
toru.shimizu.jp@ieee.org

Abstract - **Global Navigation Satellite System (GNSS) is a well known method to detect the location. A GNSS sensor is integrated with wireless sensor nodes to know its location distributed in geographically wide area, along with some energy harvesting device such as a solar cell. However, the GNSS sensor consumes relatively larger energy and requires longer sensing time for tracing satellites everytime, compared with other sensors. Although some power-down modes are used to reduce the power consumption, such as a "sleep-mode" and a "standby-mode", accumulation of running time and waiting time power is not small enough for a smaller capacity solar cell. "Normally-off" is a way of computing control which aggressively powers off components of computer systems. It uses power gating mechanisms and non-volatile memories (NVMs) to turn off the power supply component by component. We applied the Normally-off control to a wireless sensor node to achieve extra low power for a smaller solar cell. This paper describes our Normally-off Computing application to the GNSS aware sensor nodes and evaluation of the performance under solar cell power supply.**

Keywords: Power Management, Non-volatile Memory, Global Navigation Satellite System, Normally-off Computing

I. NORMALLY-OFF GNSS SENSOR NODE

A. A GNSS sensor system

Location information is important for sensor network system. Our solar powered Normally-off GNSS sensor system is shown in Fig.1

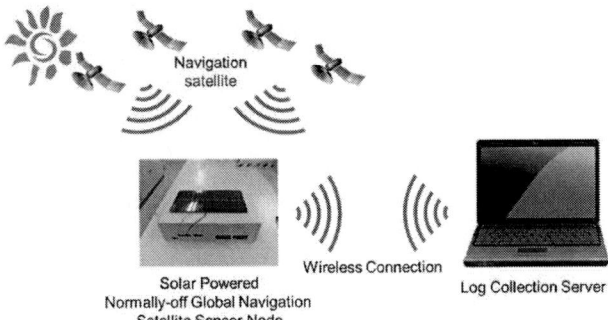

Figure 1 Normally-off GNSS system

The sensor node collects location information periodically and transfers location information to the log collection server. The transfer is less frequent than sensing. Multiple data are transferred at once to reduce energy consumption of wireless connection.

In addition to hardware optimization, aggressive power management is certainly required in order to minimize total energy consumption.

B. Normally-off GNSS sensor node

1) System configuration and its behavior

A block diagram of our Normally-off sensor node of Fig.1 is shown in Fig.2. Yellow arrows specify flows of power supply and gray arrows specify power management signals. Blue arrows are for communication between the components

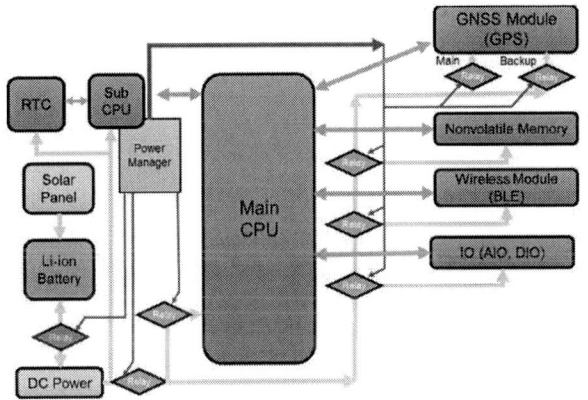

Figure 2 Normally-off GNSS sensor node configuration

A solar panel is connetcted via a Li-ion battely. Sub CPU and Realtime-Clock (RTC) are always turned on, i.e. power supplied. Power manager, a software running on the Sub CPU, controls power supply of other component.

For the GNSS module, a low power GPS module is used and a dual power supply mechanism is employed, in which a backup power supply is added with a main power supply. Even when the main power is turned-off, the GPS module wakes up quickly by "Hot start", if the backup power is supplied. If both the main and the backup powers are off, power consumption gets lower, but the wake up of GPS module takes longer time by "Cold start". A Bluetooth Low Energy (BLE) is used for the wireless module.

2) Power model

We constructed a power model of the Normally-off GNSS sensor node. The system power mode is defined as a combination of power mode of each component. Different from other sensors, GPS module requires longer and variable time for sensing locations, as show in Fig. 3, because the required time depends on locations of satellites in the sky.

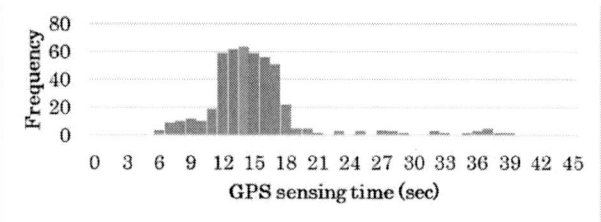

Figure 3 GPS sensing time (Hot start)

3) Normally-off power management for GNSS sensor node

We propose a novel scheme of the Nomally-off power management to solve the sensing time variation issue. Power-on of CPU is postponed until the GPS sensing completes.

We introduce a log-normal distribution function f(x) to express distribution of GPS sensing time as follows. The "x" is GPS sensing time, and μ and σ are obtined by fitting.

$$f(x) = \frac{1}{\sqrt{2\pi}\sigma x} e^{-\frac{(\ln x - \mu)^2}{2\sigma^2}}, 0 < x < \infty$$

In order to find the optimal delay time for the CPU power-on, we made a model of average energy consumption with f(x), as following:

$$D_{OPT} = \arg\min_D \int_0^D P_{GPS} x f(x) dx$$
$$+ \int_D^\infty (P_{GPS} x + P_{CPU}(x - D)) f(x) dx$$

P_{GPS} and P_{CPU} indicate power consumption by GPS and CPU, and D is power-on delay. As a result, D that minimizes this equation becomes the optimal delay D_{OPT}.

II. EVALUATION

Table 1 Current Consumption

Component	Current
Main CPU	57.58mA (on) / 8.63mA (stand-by) / 0.10mA (off)
GPS module (main)	20.45mA (on) / 2.0mA(off)
GPS module (backup)	0.052mA (on) / 0.019mA(off)
BLE module	4.44mA (on) / 0.004mA(off)

We evaluated energy consumption of a system based on the Fig. 1. Intervals of sensing and data transfer are fixed to 10 and 60 minutes respectively. Since the intervals are long enough, every component except for Sub CPU and RTC are stand-by or turned-off between sensing and data transfer. The results are shown in Table 1.

The average current for GPS and CPU is shown in Fig. 4. In this evaluation, the delay time of CPU power-on is changed. This figure clearly shows that the tradeoff certainly exists between the GPS and the CPU current consumption depending on the CPU power-on delay. As seen from this result, the optimal delay time of CPU power is 14 seconds.

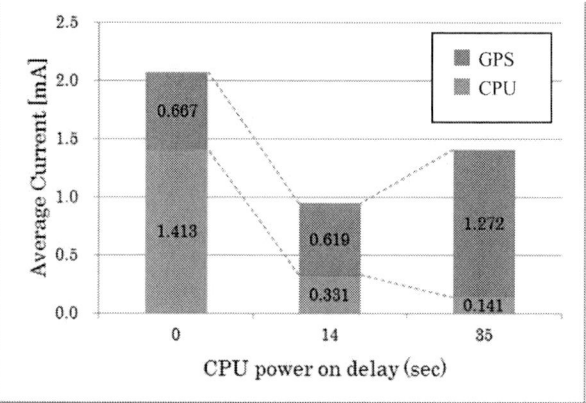

Figure 4 Power on delay vs. Average current

Average currents under different power management modes are shown in Table 2. (The power for Sub CPU and RTC is placed outside.) If no agressive power management is applied, average current is 11.59mA. When basic Normally-off management is applied, in which components not used are aggressively power-off, average current becomes 2.137mA. When we apply the optimal CPU power-on delay time, the average current becomes only 0.974mA, which is 8.4% of that of no aggressive power management.

Table 2 Average Current Consumption

Power management	Avg. cur.
Normally-on (with CPU stand-by)	11.59 mA
Ordinary Normally-off without CPU power-on delay	2.137 mA
Proposed Normally-off with optimal CPU power-on delay	0.974 mA

III. CONCLUSIONS

We have concluded that our Normally-off power management can reduce the average energy consumption by up to 91.6%. Based on this result, the size of solar panels for GNSS sensor nodes can be reduced to fit to wider applications.

ACKNOWLEDGMENT

This work is supported by Normally-Off Computing Project of NEDO in Japan and JSPS KAKENHI Grant Number JP16K12405.

Low-Power Multi-Sensor System with Normally-off Sensing Technology for IoT Applications

Masanori Hayashikoshi
Renesas Electronics Corporation
Kodaira-shi 187-8588 Japan.
Kanazawa University
Kanazawa-shi 920-1192 Japan

Hideyuki Noda
Renesas Electronics Corporation
Kodaira-shi 187-8588 Japan
Kanazawa University
Kanazawa-shi 920-1192 Japan

Hiroyuki Kawai
Tokushima Bunri University
Sanuki-shi 769-2193 Japan

Hiroyuki Kondo
Renesas Electronics Corporation
Kodaira-shi 187-8588 Japan

Abstract— **We propose the low-power multi-sensor system with normally-off sensing technology for IoT applications, which achieves almost zero standby power at the no-operation modes. A hierarchical power management scheme with activity localization can be reduced the number of transitions between power-on and power-off modes with re-scheduling and bundling task procedures. We also propose autonomously standby mode transition control by selecting optimum standby mode of microcontrollers, reducing total power consumption in the multi-sensor network. The prototyping evaluation boards with sensors are developed and demonstrated, observing 91% power reductions by adopting the proposed power gating and autonomously standby mode transition control.**

Keywords-component; Low-Power; Normally-off; Nonvolatile Memory; Sensor Network

I. INTRODUCTION

The cyber-physical system [1], which combines the large-scale data processing by the server and cloud on the Internet and the sensor nodes controlling sensors and actuators by the network, is a figure of next generation embedded systems aim. Future, information equipment, human and things are connected in the network, and the large amount of data is handled for the realization of advanced services. Thus, sensor nodes are used extensively in order to gather real-time information in the social environment and natural environment, and these volumes will be increased with the development of cyber-physical systems. Therefore, it becomes important how to reduce the power consumption of huge sensor nodes without reducing the processing performance.

In this paper, we show the normally-off sensing technology of low-power multi-sensor system to realize low-power consumption of huge sensor nodes without reducing the processing performance.

II. NORMALLY-OFF ARCHITECTURE FOR LOW-POWER MULTI-SENSOR SYSTEM

Normally-off architecture for low-power multi-sensor system is consists of normally-off power manager (a), multi-sensors, sensor controller (b), sensor data buffers (c), and microcontroller. Sensor module and microcontroller are normally-off and/or intermittent, and normally-off power manager and real time clock (RTC) are normally-on (Figure 1).

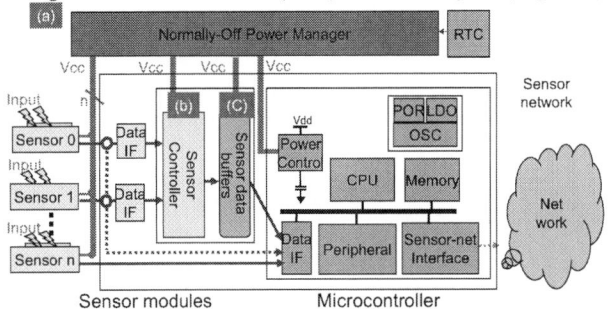

Figure 1. Normally-Off Architecture

III. NORMALLY-OFF POWER MANEGEMENT METHOD

The target of normally-off power management in this work is to achieve optimal power control in consideration of the breakeven time (BET) in sensor nodes and new ideas of two are proposed. One is a task scheduling technology and another is an autonomous standby mode transition technology. By these ideas, it becomes possible of energy optimization in a multi-sensor network system and of usability improvement by ease of application software development.

A. Task Scheduling Technology

In the conventional power gating (PG) control, the microcontroller is activated when every data sampling process of sensors is performed. Therefore, total of power consumption energy is large with increase of number of power-on/off cycles of microcontroller.

In this work, hierarchical power gating (PG) control for task

---

Identify applicable sponsor/s here. If no sponsors, delete this text box.
(sponsors)

978-1-5090-3220-4/16 $31.00 © 2016 IEEE

scheduling technology is proposed, in which sensing data is buffered in sensor data buffer once, and after then microcontroller is activated and performed the process together (Figure 2). Therefore, it is possible to optimize the number of power on/off cycles and much decrease total of power consumption energy.

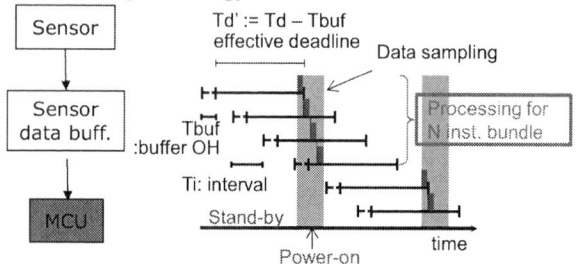

Figure 2.   Hierarchical power-gating (PG) control

B.   Autonomous Standby Mode transition technology

In some applications, it may not be possible to power-off because of shorter sampling period. In order to maximize low-power effects, the autonomous standby mode transition technology is proposed, which is to make select and transition the optimum standby mode of MCU autonomously to minimize power consumption energy by quantifying the constraints of hardware and software.

Understanding of the constraint condition of hardware side and software side is needed in order to select the optimal standby mode. The constraint of the hardware side is the breakeven time (BET), which is derived from standby power consumption specific to each device and the energy overhead of the state transition of the standby and active. The constraint of the software side is the waiting time depending on the application.

Industrial available microcontrollers for sensor network applications are supported for several standby modes. Each standby mode has different BET (Figure 3). It should be dynamically selected optimum standby mode according the task workloads in use cases to minimize the total power consumption.

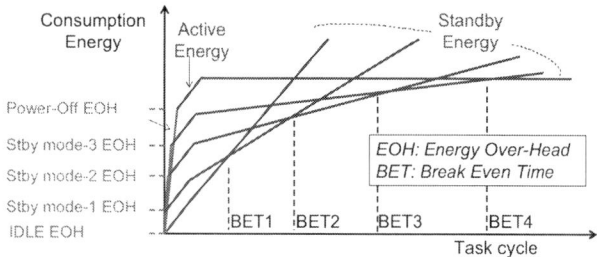

Figure 3.   Break-even time of each standby mode

IV.   EVALUATED RESULT

We demonstrate in case of the intruder detection monitor system as a use case of IoT applications (Figure 4). The prototyping evaluation boards with multi-sensors are developed and demonstrated, observing around 91% power reductions by adopting the proposed power gating and autonomously standby mode transition control (Figure 5), where the evaluation period is 1,000s, and the detection intruder time of 3s per evaluation period of 1,000s.

In this use case, the sampling period of the infrared sensor and standby mode of microcontroller are determined with the distance of approaching object by using the autonomous standby mode transition technology, and the judgement of intruder detection is relaxed with the distance of approaching object by using the task scheduling technology, though the sampling period of motion sensor is still kept to maintain the detection accuracy.

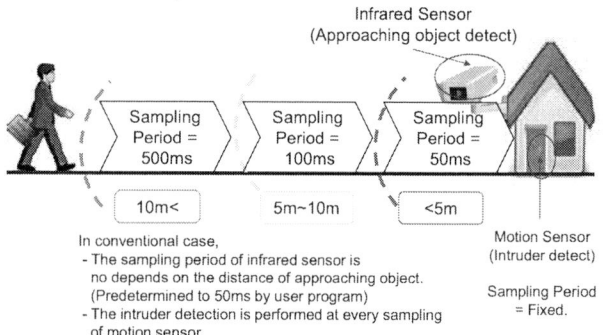

Figure 4.   Intruder detection system

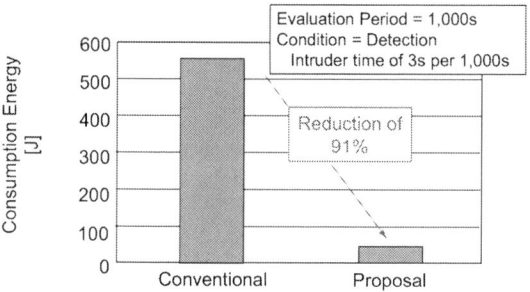

Figure 5.   Evaluated Result.

V.   SUMMARY AND CONCLUSIONS

We propose the low-power multi-sensor system with normally-off sensing technology, for IoT applications, in which the task scheduling technology and the autonomous standby mode transition technology are applied. And the possibility of power reduction of 91% are demonstrated.

ACKNOWLEDGMENT

This work is supported by Normally-Off Computing Project of the New Energy and Industrial Technology Development Organization (NEDO) in Japan.

REFERENCES

[1]   Cyber-Physical Systems Executive Summary, Prepared by the CPS Steering Group, March 6, 2008.

[2]   Normally-Off Computing Project: Challenges and Opportunities, The 19th Asia and South Pacific Design Automation Conference, 2014

A large scale access-control list for IoT security comprising embedded IP-core and DDR DRAM

Kazunari Inoue [1] and Yuji Yano [2]

[1] National Institute of Technology, Akashi College, Japan,
[2] Osaka-city University, Department of Engineering Graduate School of Engineering, Japan,

Abstract— **A large scale access-control list (ACL) comprising of 2K-entry times 512bit-key is reported. That ACL consists of embedded IP-core on SoC/FPGA and external DDR DRAM for having further entry size. Those inexpensive components are dedicated to low-cost industrial sensor network and IoT security. Unlike conventional TCAM-base in routers, plural hash codes with DRAM take place of search in this paper's ACL, which is effective for power saving as well. Verified prototype design on Xilinx Kintex-7 platform indicates over 10Gbps performance.**

Keywords; IoT; Security; ACL; Embedded IP, DDR DRAM

I. INTRODUCTION

According to the Internet Usage Statistics in June 2016, the world internet users is 3.6 billion that is 49.2% of the world wide population. Although the penetration in north America and Europe are relatively high, the number of internet user is still growing especially in Asia and Africa. [1] On the other hand, the number of internet device is predicted to indicate tremendous growth, where 50 billion IoT sensor nodes appear in 2025. [2] Those packet traffics departed from sensor nodes are aggregated by routers namely IoT gateway. The features of IoT gateway is similar, but is different from conventional routers. Since the sensor node is rather cost-sensitive than today's PC, tablet and smartphone, the network security should be operated by IoT gateway instead of each node. Robust and cost effective security is serious matter for IoT sensor network.

II. IoT GATEWAY AGGREGATES PACKET TRAFFIC

A. Packet traffic in IoT sensor network

Today's TCP/IP protocol is not perfectly fitted for IoT sensor network. While IP-packet length is defined by min. 64-byte and max. 1.5K-byte, typical packet length in sensor network is much smaller, e.g. 14-byte in CAN (Controller Area Network in vehicle), where most of the traffics by sensor nodes are short packets. In fact, a short packet processing is much more difficult task for routers than a long packet. For example, in 1G-bps performance network environment, the number of packet arriving at the router is 8.9M packet-per-sec.(-pps) and 0.81M-pps. with the packet length is 14-byte and 1.5K-byte, respectively. Typical IP packet length is not always 1.5K-byte, however the number of processing in 14-byte traffic is almost 10x as busy as that of IP packets. Although IoT sensor network does not require high-performance such that 1G-bps seems to be good enough, the number of packet arriving IoT gateway is greater than that of IP routers. Therefore, the desired performance for IoT gateway should be over 10Gbps.

B. Access-control at IoT gateway

Unlike today's PC, tablet and smartphone, industrial IoT sensor devices are lower-priced components such as $1 and/or disposable. It is too difficult to implement any security tool like virus scan in all sensor nodes. Therefore, the cost effective security is desired for IoT gateway instead of each node. The network security is well researched area, and an access-control list (ACL) is to monitor the flow between the source (src.) and the destination (dst.) with packet by packet manner is an effective security tool for IoT gateway. In this paper, we propose a large scale ACL integration in IoT gateway to monitor the flow of all sensor devices. The ACL consists of numbers of entries, where src., dst., for L2/L3/L4 and protocols are described in each key. Thus, typical key size in ACL is much longer than simple forwarding information base, e.g. 512bit ACL vs. 32bit FIB. Also, each ACL entry should involve the rule, whether the packet is permitted or denied. In conventional network routers, having large scale ACL is difficult since TCAM is tremendously expensive and power hungry device. [3-4] This paper's ACL aims to integrate large scale table as well as achieving the cost and the power reduction.

III. HASH BASED ACCESS-CONTROL

We propose hash-base ACL instead of TCAM-base. The advantage in hash-base is to save the cost, because it does not need any expensive memory. Also, regardless the key size, which is 512bit key in this case, total number of bit is given by the entry size. For example, when total entry size is 2K-entry, 11bit address can indicate each entry. Thus, the hash code in this paper should convert external 512bit into 11bit.

A. Congestion probability

Despite benefits carried out by hash-base, the congestion is the serious concern. Having plural hash codes alleviates and reduces the congestion. Table 1 shows that the number of entry registered in memory is increased by having plural hash codes.

Table 1   #Entry registered (T/L 2K-entry)

	ξ	σ	#Entry regsitered	#Entry remained
Hash #1	1.000	0.6321	1294	754
Hash #2	0.368	0.8366	630	124
Hash #3	0.061	0.9703	120	4
Hash #4	0.002	0.9990	3	1
Hash #5	0.000	0.9998	0	1
Hash #6	0.000	0.9998	0	1

Here, σ means the expected value, where the entry is successfully registered in the memory with corresponding Hash code, and is given by following formula. [5-6]

$$\sigma = \frac{1 - e^{-\xi}}{\xi} \qquad (1)$$

Regarding ξ, k means the entry size in each memory, where 2,048-entry is used in this case. Also, p means the number of bit used in each hash code, where 11bit is used in this case.

$$\xi = \frac{k}{2^p} \qquad (2)$$

$$p = \log_2 N \qquad (3)$$

When single hash code is used (Hash #1 only), 63.2% out of 2,048-entry can be registered. The expected value σ, where the entry is successfully registered in memory can be improved 83.6%, 97.0% and 99.9% by implementing more than four hash codes.

B. Badwidth demanded by plural hash codes

Having numbers of hash codes reduces the congestion probability and improves expected value to be registered. However, additional concern is the memory bandwidth. Since each hash code needs to read data from the different bank is external DDR DRAM, demanded bandwidth is getting worse 2x, 3x, when the number of hash code is increased 2x, 3x, respectively. Since 512bit key in ACL is relatively long and also plural hash codes require multiple reads from DRAM, the bandwidth is the considerable performance bottleneck. Figure 1 shows the relational trade-off between the improvement according to the expected value σ and the performance drop caused by memory bandwidth demanded by plural Hash codes.

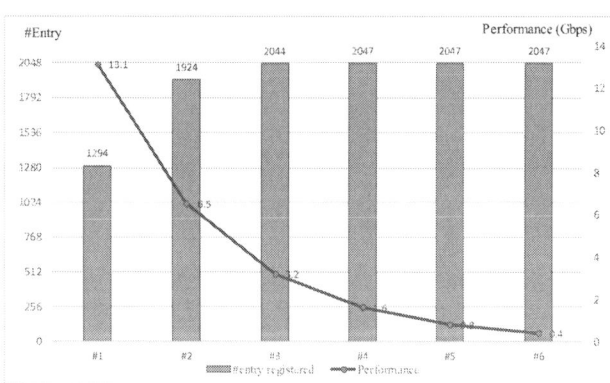

Figure 1 Plural hash codes and Bandwidth

IV.   EMBEEDED IP AND PRE-COMPARE

A. functional atchitecture

Figure 2 illustrates the functional architecture according to this paper's embedded IP-core performed on FPGA and external DDR DRAM.

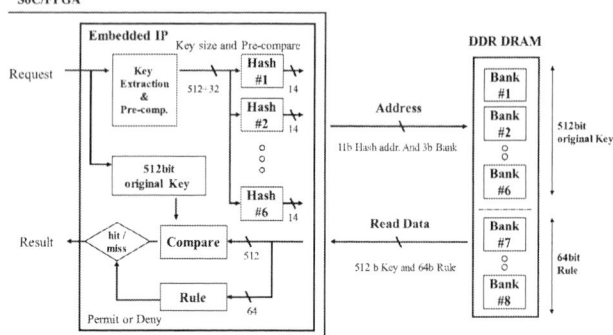

Figure 2 embedded IP and external DDR DRAM

There are six hash codes in embedded IP-core, and hash code #1, #2 through #6 independently communicates Bank#1, #2 through #6 in external DDR DRAM, respectively. As described previously, having multiple hash codes reduces the congestion risk and the number of entry registered in memory can be improved. On the contrary, it causes the serious performance bottleneck due to the memory bandwidth.

Figure 3 shows this paper's pre-compare in IP-core to save the memory bandwidth and keeps the performance high.

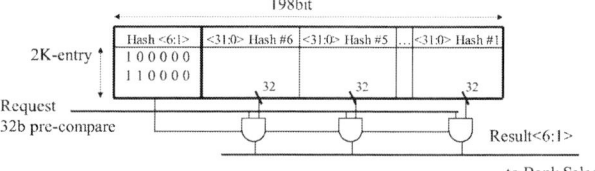

Figure 3   Pre-compare function in embedded IP

Even though hash codes #6 and #5 state multiple hits in the second row, either <31:0>Hash #6 or #5 results "hit" in most cases, because of 32bit pre-comparing out of 512bit. Thus, the corresponding Bank address should be read from external DDR DRAM. The pre-compare function in embedded-IP significantly saves the bandwidth. The performance verified by the prototype design on Xilinx Kintex-7 platform indicates 13.1G-bps by the effect of pre-compare function even if the multiple hits occur. Although IoT sensor network typically states low-performance such as 1G-bps, verified 13.1G-bps is fitted to short-packet traffic in the IoT sensor networks.

REFERENCES

[1] http://www.internetworldstats.com/stats.htm

[2] Dave Evans, "The Internet of Things" cisco white paper, April 2011.

[3] P. Gupta, S. Lin , and N. McKeown, "Routing lookups in hardware at memory access speeds," Proc. INFOCOM, IEEE Press, Piscataway, N.J., 1998, pp. 1240-1247.

[4] H.Iwamoto, Y.Yano, Y.Kuroda, K.Yamamoto, S.Ata and K.Inoue, "250Msps 0.5W eDRAM-Based Search Engine Dedicated Low Power FIB Application", IEICE TRANS. Electronics, Vol. E96-C, No. 8, pp. 1076-1082, Aug. 2013.

[5] T. Sasao, "A Design method of address generators using hash memories," IWLS-2006, pp. 102-109, Vail, Colorado, U.S.A, June 7-9, 2006.

[6] T.Sasao, M.Matsuura and H.Nakahara, "A Realization of Index Generation Functions Using Modules of Uniform Sizes", 19th International Workshop on Logic and Synthesis, pp.201-208.

$3D^2$ Processing Architecture

- High Reliability and Low Power Computing for Novel Nano Tactile Sensor Array -

Kiyotaka Komoku*, Kazutami Arimoto, Tomoyuki
Yokogawa, Hitoshi Yamauchi, and Yoichiro Sato
Faculty of Computer Science and Systems Engineering
Okayama Prefectural University
Soja, Okayama, JAPAN
komoku@c.oka-pu.ac.jp

Hidekuni Takao
Faculty of Engineering
Kagawa University
Kagawa, JAPAN

Abstract—We have proposed new instrumental techniques to quantify human touch feelings by nano tactile sensor array system with $3D^2$ processing architecture. In the computation of quantification of human touch feelings, high-level semantics features are calculated from low-level features which are the result of FFT, wavelet translation, etc. For some application, especially medical application, high reliability is required for the recognition results. Furthermore, real-time processing and low power computing are also required. $3D^2$ Processing architecture provides high reliable and low power computing for the accelerator. The architecture has three features: Parallel Computation of Multiple Recognition Algorithms for High Reliability, Spatial-Parallel Temporal-Pipeline Streaming Processing for High Energy Efficiency Processing, and Current Reuse Energy Pipeline for Low Power Processing.

Keywords; tactile sensor array, human touch feelings, low power computing, high reliability computing

I. INTRODUCTION

Many tactile sensors have been developed in recent years. However, it is not achieved to realize quantification of human touch feeling using these sensors. If it is possible to quantify human touch feelings of various targets, like human skin, fabric, and interior materials, etc., the technique will apply to materials development and medical applications.

We have proposed new instrumental techniques to quantify human touch feelings by "nano tactile sensor array system with $3D^2$ processing architecture". In this paper, the $3D^2$ processing architecture which executes signal processing of sensor array data is described.

II. A SMART NANO TACTILE SENSING ARRAY SYSTEM WITH $3D^2$ PROCESSING ARCHITECTURE

The configuration of the novel tactile sensor used in the proposed system is shown in Fig.1 [1]. The tactile sensor is fabricated by MEMS, and consists of a contactor, suspensions, and a chip frame. Two-axis movements of the contactor-tip are independently detected by the two independent suspensions. Therefore, it can detect the surface shape and local frictional force of the measuring target simultaneously.

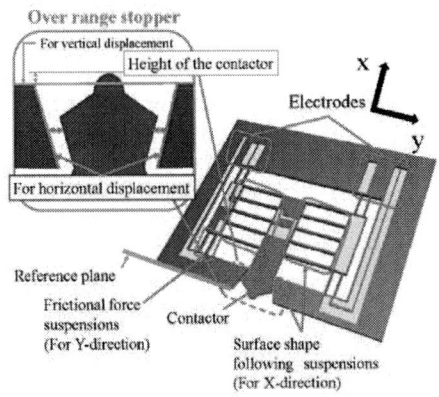

Figure 1. Configuration of novel tactile MEMS sensor [1]

The concept of proposed system for quantifying human touch feelings is shown in Fig.2. The system is consist of a MEMS tactile sensor array and the data processing accelerator. The tactile sensor can detect micro surface roughness at a high resolution. By using a lot of the nano tactile MEMS sensors as an array, it may collect the plane information of the measuring target in detail. A significant amount of sensing data from the sensor array is processed with the data processing accelerator. "$3D^2$ processing architecture" provides the high reliability and low power computing of the accelerator.

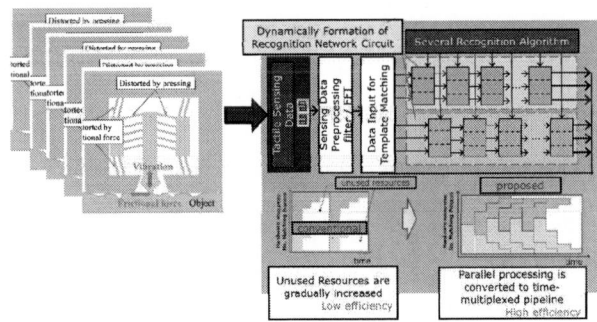

Figure 2. nano tactile sensor array system with $3D^2$ processing architecture

III. 3D² PROCESSING ARCHITECTURE

In the computation of quantification of human touch feelings (recognition), high-level semantics features are calculated from low-level features which are the result of FFT, wavelet translation, etc. For some application, especially medical application, high reliability is required for the recognition results. To guarantee the reliability, concurrent calculation using multiple algorithms is efficient. On the other hand, real-time processing and low power computing are also required.

3D² Processing architecture provides high reliable and low power computing for the accelerator. This architecture consists of following three key features. 3D² Processing is a combination of Spatial-Parallel Temporal-Pipeline Streaming Processing (3D processing) and Current Reuse Energy Pipeline (virtual 3D implementation).

A. Parallel Computation of Multiple Recognition Algorithms for High Reliability

The first feature is the parallel computation of multiple recognition algorithms for high reliability. To achieve high-reliability of the recognition result, various recognition algorithms are executed concurrently.

B. Spatial-Parallel Temporal-Pipeline Streaming Processing for High Energy Efficiency Processing

The second is spatial-parallel temporal-pipeline streaming processing for high energy efficiency processing. Enormous computing power is required for the execution of multiple recognition algorithms. In conventional parallel computing implementation, there are unused hardware resources and the energy efficiency is low. By converting parallel processing to time-multiplexed pipeline processing, the unused hardware resources decrease and the energy efficiency is improved (Fig.2). The hardware resources are configurable so that various algorithms can be implemented. As the algorithm implementation framework, SVIRAL's TruStream Virtual machine[2] is considered. This framework is a kind of multicore/multiprocessor parallel programming model based on stream processing. The Virtual machine consists of Core and Network node (Fig.3) and is mapped programs (tasks) efficiently.

C. Current Reuse Energy Pipeline for Low Power Processing

The last is Current Reuse Energy Pipeline for low power processing (Fig.4). For the enormous computation of recognition processing of large tactile sensor net, many stream processing systems (processing boards) are required. By Stacking the streaming processing systems connecting their power supply in serial (virtual 3D implementation), upper stage current is reused, and low power processing is realized.

IV. CONCLUSIONS

3D² Processing architecture for novel tactile sensor array system is described. The architecture has three features:

Parallel Computation of Multiple Recognition Algorithms for High Reliability, Spatial-Parallel Temporal-Pipeline Streaming Processing for High Energy Efficiency Processing, and Current Reuse Energy Pipeline for Low Power Processing. We will substantiate the proposed architecture using FPGA.

ACKNOWLEDGMENT

This work was partly supported by CREST, Japan Science and Technology Agency.

REFERENCES

[1] R. Kozai, K. Terao, T. Suzuki, F. Shimokawa, and H. Takao et.al.., "A Novel Configuration of Tactile Sensor to Acquire the Correlation Between Surface Roughness and Frictional Force," Tech. Dig. IEEE Micro Electro. Mech. Syst., vol28., pp.245-248, 2015.

[2] http://www.sviral.net/

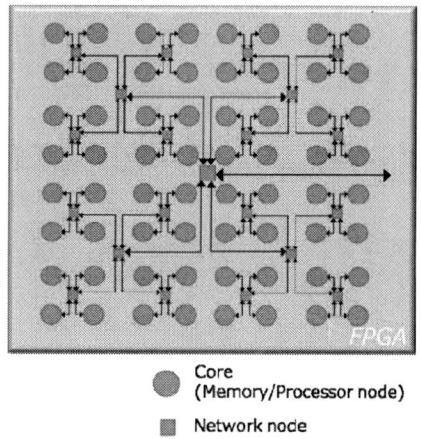

Figure 3. SVIRAL's TruStream Virtual machine [2]

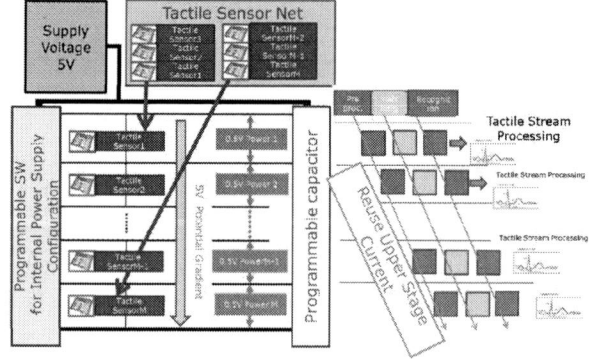

Figure 4. Current Reuse Energy Pipeline

Attack Sensing against EM Leakage and Injection

Noriyuki Miura
Graduate School of System Informatics
Kobe University
Kobe, Japan
miura@cs.kobe-u.ac.jp

Shivam Bhasin
Physical Analysis and Cryptographic Engineering
Temasek Labs, Nanyang Technological University
Singapore
sbhasin@nut.edu.sg

Abstract— **This paper introduces two circuit-level reactive sensor-based countermeasures against malicious implementation attacks exploiting EM-radio leakage from and injection to cryptographic processors. An LC-oscillator-based sensor detects a tiny micro-EM probe approaching proximity to the processor for a passive EM analysis attack. A PLL-based sensor detects intentional high-level EM power injection for an active EM fault injection attack. Hardware-level security enhancement of the crypto processor has been demonstrated with successful sensor operations silicon-proven by ASIC and FPGA test-chip measurements.**

Keywords; hardware security; sensor; EM information leakage

I. INTRODUCTION

Information security is one of the most critical issues in today's advanced information society and coming IoT/IoE eras, where trillion intelligent things with sensors will be distributed and collect precious private information and autonomously communicate over the internet. Malicious attackers struggle to steal the precious information by exploiting not only in cyber (software) but also physical (hardware) domain security holes. One serious threat is an EM-based implementation attack onto a cryptographic processor. Cryptography is a powerful tool to protect information in the software domain. However, by either statistically analyzing EM radiation from the processor hardware during crypto operation [1] or intentionally injecting EM power into the hardware to induce fault operations [2], the precious information can be stolen in the hardware domain. EM-based attacks are considered a serious threat as it can be conducted from a distance, without establishing direct measurement connections to the target or without any need of decapsulation and other reverse engineering techniques. A hardware-domain security enhancement technique is in strong demand. This paper introduces two sensor-based circuit-level countermeasures against the EM attacks.

II. PASSIVE EM-PROBE SENSOR

The first countermeasure is an EM-probe sensor against a passive EM Analysis (EMA) attack [3-5]. Figure 1 illustrates the conceptual sketch of the sensor. For the EMA attack, a malicious attacker utilizes a tiny micro EM-probe to capture EM radiation from the cryptographic processor core. By statistically analyzing the correlation between the EM waveforms and crypto core operation, a secret information such as secret key can be disclosed [1]. More seriously, the

This work is supported by Japan Society for the Promotion of Science (JSPS) Grants-in-Aid for Scientific Research (KAKENHI) Grant Number 26240005.

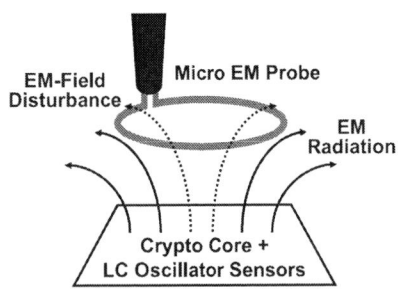

Fig.1 Conceptual sketch of EM-probe sensor.

attacker brings the probe close proximity to the chip surface to precisely capture the local EM radiation from the core [6]. An on-board level EM leakage equalization technique [7,8] can be bypassed since this local EMA can directly access the EM radiation from the on-chip crypto core surface. However, it is impossible to observe EM radiation without disturbing itself. Dual LC oscillators are integrated together with the crypto core. EM field disturbance due to the probe approach can be detected as the LC oscillation frequency shift. No reference clock which may be exploited by the attacker is needed since one of the oscillators acts as the reference to the other for measuring relative frequency difference. The sensor circuits can be implemented in fully digital CMOS as the oscillator core and the detection logic consist of a cross-coupled inverter and two counters respectively, resulting in a small hardware overhead [3-5]. This digital implementation also makes it possible to integrate this EM-probe sensor in an FPGA platform which is widely used for embedded security applications.

III. ACTIVE EM-INJECTION SENSOR

The second countermeasure is an EM-injection sensor against active EM Fault Injection (EMFI) attack [2]. The attacker injects high-level EM power through an active EM probe to the crypto chip to intentionally induce fault operation and compares the faulty cipher and correct output to steal the secret key. Again, the countermeasure is to detect this abnormal EM-power injection by using on-chip sensor. Figure 2 depicts the proposed EM-injection sensor [9]. A PLL macro block is used as the sensor core, which is widely available almost on all the FPGA platforms. A ring oscillator is formed with a single inverter and routing buffers encapsulating the crypto core to be protected. The oscillation clock is fed back to the reference clock input of PLL as *Rclk*. A generated core

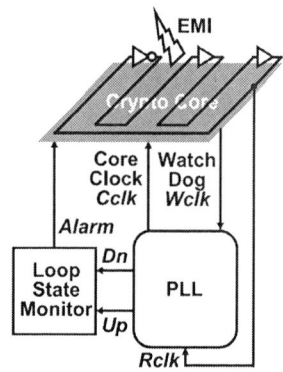

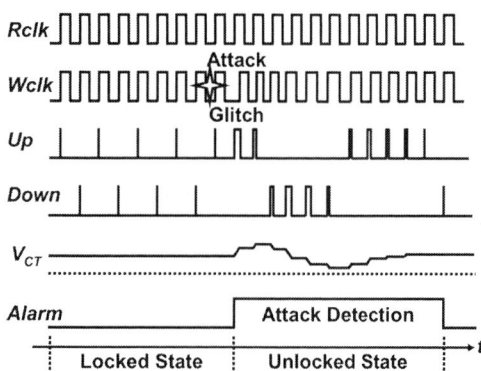

Fig.2 Conceptual sketch of EM-injection sensor.

oscillator and PLL with the loop-state monitor were place and route automatically by using a Xilinx standard design platform with a small supporting custom program [9]. A 400MHz 300W (55dB) broadband class-A amplifier was used to inject an EM pulse with 1.5ns width for the EMFI attack. A remotely-controllable XYZ micro-positioner scans an EM injection spot across an entire chip surface in XY directions to obtain a 30x30-matrix attack-detection-rate map. A 100% detection coverage was confirmed with the proposed PLL-based EM-injection sensor. The detailed measured results will be presented at the conference site.

V. CONCLUSION

In this paper, two IC-level sensor-based countermeasures are introduced against malicious hardware implementation attack onto cryptographic processors. An EM-probe sensor detects a passive EM analysis attack by observing EM-field disturbance due to a micro EM probe approach. Fully digital LC-oscillator-based sensor circuits measure relative oscillation frequency shift without a reference clock for small hardware overhead implementation. An EM-injection sensor detects an active EM fault injection attack by reacting against intentional EM-power injection. An existing macro block is utilized in the FPGA to amplify the EM injection event as a digital alarm signal to protect the cryptographic processor. A silicon prototype demonstrations in 180nm ASIC and 45nm Xilinx FPGA implementations exhibit successful sensor operations to enhance hardware security for future intelligent things.

clock generated in PLL, *Cclk*, is fed and distributed in the crypto core and again feedback to PLL as *Wclk*. All three clock signals behave as watch dogs to monitor abnormal EM injection. The intentional EM injection as the EMFI attack causes clock glitches in the watch-dog clock paths, resulting in unlock of the PLL control loop. A digital loop state monitor finds explicit pulses in up/down signals of the loop and raises alarm as attack detection. Both with EM-probe and EM-injection sensors, the cryptographic processors turns into lock or dummy-key operation mode to protect the secret key.

IV. TEST-CHIP MEASUREMENT

An ASIC test chip was designed and fabricated in 180nm standard CMOS as the EM-probe sensor demonstration prototype. A 128-bit AES cryptographic processor with composite field S-box was integrated together with the proposed EM-probe sensor. Successful attack detection was demonstrated with maximum 5% LC oscillation frequency shift due to 0.5mm-ϕ micro EM probe approach within 0.1mm close proximity. Thanks to the fully digital sensor circuit implementation with the high-frequency LC oscillator operation, the sensor hardware overhead is suppressed within only 1.6%, 0.2%, and 7.6% in area, performance, and power penalty, respectively.

Xilinx Spartan-6 FPGA was used for demonstrating the EM-injection sensor operation. An existing PLL macro block is used as the sensor core. All the circuits including the ring

REFERENCES

[1] K. Gandolfi, C. Mourtel, and F. Olivier, "Electromagnetic Analysis: Concrete Results," *CHES Lecture Notes in Computer Science*, vol.2162, pp.251–261, May 2001.

[2] D. Boneh, R. A. DeMillo, and R. J. Lipton, "On the Importance of Checking Cryptographic Protocols for Fault," *EUROCRYPTO Lecture Notes in Computer Science*, vol. 1233, pp.37-51, May 1997.

[3] N. Miura, D. Fujimoto, D. Tanaka, Y. Hayashi, N. Homma, T. Aoki, and M. Nagata, "A Local EM-Analysis Attack Resistant Cryptographic Engine with Fully-Digital Oscillator-Based Tamper-Access Sensor," *IEEE Symposium on VLSI Circuits Digest of Technical Papers*, pp. 172-173, June 2014.

[4] N. Homma, Y. Hayashi, N. Miura, D. Fujimoto, D. Tanaka, M. Nagata, and T. Aoki, "EM Attack is Non-Invasive? - Design Methodology and Validity Verification of EM Attack Sensor," *CHES Lecture Notes in Computer Science*, Vol. 8731, No. 1, pp. 1-16, Sept. 2014..

[5] N. Miura, D. Fujimoto, M. Nagata, N. Homma, Y. Hayashi, and T. Aoki. "EM Attack Sensor: Concept, Circuit, and Design-Automation Methodology," *Proceedings of DAC*, pp. 1-6, June 2015.

[6] T. Sugawara, D. Suzuki, M. Saeki, M. Shiozaki, and T. Fujino, "On Measurable Side-Channel Leaks Inside ASIC Design Primitives," *CHES Lecture Notes in Computer Science*, vol.8086, pp.159-178, Aug. 2013.

[7] C. Tokunaga and D. Blaauw, "Secure AES Engine with a Local Switched-Capacitor Current Equalizer," *ISSCC Digest of Technical Papers*, pp.64-65, Feb. 2009.

[8] N. Miura, D. Fujimoto, R. Korenaga, K. Matsuda, and M. Nagata, "An Intermittent-Driven Supply-Current Equalizer for 11x and 4x Power-Overhead Savings in CPA-Resistant 128bit AES Cryptographic Processor," *A-SSCC Digest of Technical Papers*, pp.225-228, Nov. 2014.

[9] N. Miura, Z. Najm, W. He, S. Bhasim, X. Ngo, M. Nagata, and J. Danger, "PLL to the Rescue: A Novel EM Fault Countermeasure," *Proceedings of DAC*, pp. 1-6, 2016.

How to Design Hardware Prime Field Multipliers for Bilinear Pairing

Daisuke Fujimoto, Yusuke Nagahama, Tsutomu Matsumoto
Yokohama National University,
Kanagawa, Japan
fujimoto-daisuke-ht@ynu.ac.jp

Abstract— **Advanced cryptographic functionalities such as searchable encryption, aggregate signatures, proxy re-encryption, and attribute-based encryption can be realized by adopting bilinear pairings based on elliptic curves over a prime finite field GF(p). To achieve enough security the size of p should be at least a-few-hundred bit. Thus the "good" GF(p) hardware multiplier is necessary to construct the "good" bilinear pairing computation. We introduce a design methodology of such hardware multipliers with the Montgomery multiplication algorithm.**

Keywords: Finite Field Multiplier, Montgomery Multiplication, Pairing Cryptography Introduction

I. INTRODUCTION

Online service such as cloud almost become common. The data on online service should be saved by security technique like encryption. It is difficult to achieve function such as search on server avoiding decryption. Decrypted data increase security risks. To achieve advanced function without decryption on cloud service, advanced cryptographic functionalities such as searchable encryption, aggregate signatures, proxy re-encryption, and attribute-based encryption is proposed. The major tool for realizing these cryptography is bilinear pairing based on elliptic curves over a finite field. The small characteristic size of finite field should be avoid for achieving enough security. Therefore the prime finite field should be used. The size of prime p should be at least a-few-hundred bit to achieve enough security. Therefore the computation cost of bilinear pairing becomes high. To achieve flexible use of cloud data it is necessary to accelerate bilinear pairing computation.

Fast software implementation on Intel CPU is proposed in [1][2]. However the architecture unit of Intel CPU is not suitable high-radix computation such as prime field arithmetic. Hardware aiming high-radix arithmetic is expected faster and energy efficient. In this paper we introduce design of hardware prime field modular multiplier.

II. BASIS OF PAIRING HARDWARE

Bilinear pairing is map defined by equation (1). G_1 is group over finite field GF(p). G_2, G_3 are groups over extension of finite field GF(p).

$$G_1 \times G_2 \rightarrow G_3 \qquad (1)$$

The algorithm of bilinear pairing consist of miller loop, final addition, and final exponentiation. These parts are realized by

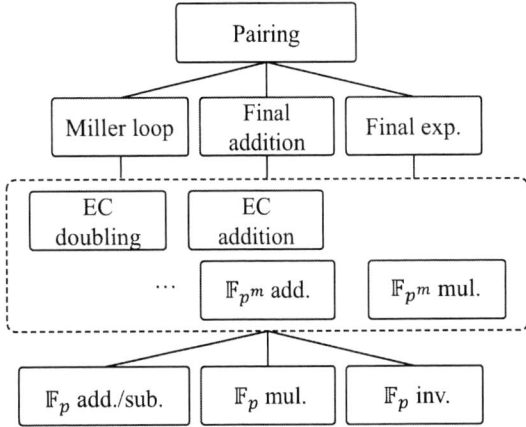

Figure 1.   Computation overview of bilinear pairing.

Algorithm 1: k bit Montgomery multiplication

input : Mult. size:k; block num.:n; multiplicand:A; multiplier:B;
moduler:M; $M > 2$, $\gcd(M,2) = 1$, $(-MM')\mathrm{mod}2^k = 1$,
$4M < 2^{kn} = R$, $(2^{kn}R^{-1})\mathrm{mod}M = 1$, $0 \leq A,B \leq 2M$,
$B = \sum_{i=0}^{n-1}(2^k)^i b_i$, $b_i \in \{0,1,\ldots,2^k - 1\}$.
output: $S_n \equiv ABR^{-1}(\mathrm{mod}M), 0 \leq S_n \leq 2M$.

1 $S_0 := 0$;
2 **for** $i := 0$ **to** $n - 1$ **do**
3 $\quad L : q_i \quad := (((S_i + b_i A)\mathrm{mod}2^k)M')\mathrm{mod}2^k$;
4 $\quad S_{i+1} := (S_i + q_i M + b_i A)\mathrm{div}2^k$;
5 **end**
6 **return** S_n

Figure 2.   Algorithm of k-bit Montgomery multiplication.

Algorithm 2: k bit quotient pipelining Montgomery multiplication

input : mult. size:k; delay param.:d; block num.:n; multiplicand:A;
multiplier:B; moduler:M; $M > 2$, $\gcd(M,2) = 1$,
$(-MM')\mathrm{mod}2^k = 1$, $\tilde{M} = (M'\mathrm{mod}2^{k(d+1)})M$, $4\tilde{M} < 2^{kn} = R$,
$M'' = (\tilde{M} + 1)/2^{k(d+1)}$, $0 \leq A,B \leq 2\tilde{M}$, $B = \sum_{i=0}^{n+d}(2^k)^i b_i$,
$b_i \in \{0, 1,\ldots,2^k - 1\}$, and $b_i = 0$ for $i \geq n$.
output: $S_{n+d+2} \equiv ABR^{-1}(\mathrm{mod}M), 0 \leq S_{n+d+2} \leq 2\tilde{M}$.

1 $S_0 := 0$; $q_{-d} := 0;\ldots; q_{-1} := 0$;
2 **for** $i := 0$ **to** $n + d$ **do**
3 $\quad L_1 : q_i \quad := S_i \mathrm{mod}2^k$;
4 $\quad L_2 : S_{i+1} := S_i/2^k + q_{i-d}M'' + b_i A$;
5 **end**
6 $S_{n+d+2} := 2^{kd}S_{n+d+1} + \sum_{j=0}^{d-1} q_{n+j+1}2^{kj}$;
7 **return** S_{n+d+2}

Figure 3.   Algorithm of k-bit quotient pipelining Montgomery multiplication.

combination arithmetic of extension field operations shown in figure 1. All arithmetic computations are based on addition, subtraction, multiplication, and inversion over GF(p). In bilinear pairing algorithm inversion appears only few times. Therefore prime field multiplications are bottleneck of bilinear pairing computation. To accelerate pairing computation it is important to design faster hardware prime field multiplier.

III. DESIGN OF GF(P) MULTIPLIER

Multiplication over GF(p) is realized by modular multiplication with modulo p. A well-known fast algorithm of modular multiplication is Montgomery multiplication. Montgomery multiplication can be obtained by modular multiplication result without division. Normally Montgomery multiplication is implemented by loop structure for reusing small multiplier among rounds (algorithm 1,k-bit MM). This algorithm is suitable for software implementation. However there are some data dependency daring round and critical path becomes long (Figure 4(a)). To avoid data dependency, quotient pipelining Montgomery multiplication (QPMM) algorithm is proposed [3] (algorithm 2). Quotient q_i is evaluated at next cycle. Therefore critical path become only two multiplier in parallel and one adder chain. Number of computation cycle and output data size increase but that's effect is smaller than critical path reduction.

Main parameter of QPMM is the size of multiplier. The implement results on KC705 FPGA board are shown in table 1. We choose 254 bit as the size of p for optimal ate pairing on BN curve satisfying 126 bit security level. QPMM implement achieves smaller critical path delay than that of k-bit Montgomery multiplication at same size of multiplier. Among QPMM implementation smaller size of multiplier increases the number of operation cycles. Critical path of QPMM mainly depend on the critical path of one multiplier stage. Therefore larger size of multiplier increase critical path of QPMM. On the other hand, latency of larger multiplier is shorter than that of small multiplier because effect of multiplier size is smaller than that of cycle number. The number of DSPs (Digital Signal Processor in FPGA) and LUTs (Lookup Table in FPGA) is almost linear to the size of multiplier. Designer of cryptographic system should consider these trade-off of circuit area and latency depending on the size of multiplier. Clock frequency is decided by critical path delay and designer should consider matching with other system element such as memory, prime field adder/subtractor, and prime field inverter.

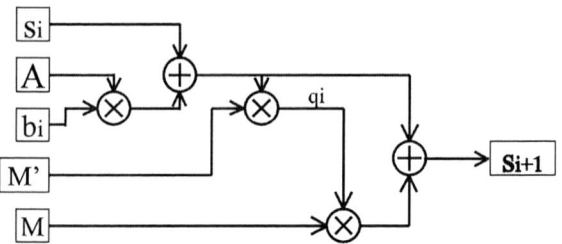

(a) k-bit Montgomery multiplication

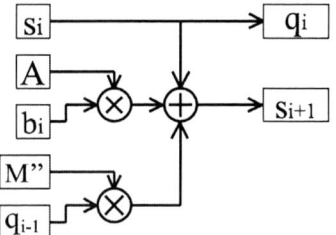

(b) k-bit quotient pipelining Montgomery multiplication

Figure 4. Data path of montgomery multiplication.

IV. CONCLUSION

In this paper we introduced the bottleneck of bilinear pairing computation and how to design fast hardware multiplier and demonstrate QPMM design on KC705 FPGA board. To optimize system performance the designer should choose the parameter of QPMM considering circuit area and latency.

ACKNOWLEDGMENT

A part of this paper is based on results obtained from a project commissioned by the New Energy and Industrial Technology Development Organization (NEDO).

REFERENCES

[1] D. F. Aranha, P. S. L. M. Barreto, P. Longa, and J. E. Ricardini,"The Realm of the Pairings," in Proc. of Selected Areas in Cryptography (SAC 2013), pp.3 – 25, 2013.

[2] E. Zavattoni, L. J. D. Perez, S. Mitsunari, A. H. S`anchez-Ram`ırez, T. Teruya, and F.Rodr`ıguez-Henr`ıquez,"Software implementation of an Attribute-Based Encryption scheme", IEEE Trans on Computers, vol.64, issue.5, pp.1429 –1441, 2015.

[3] H. Orup, "Simplifying quotient determination in high-radix modular multiplication, " Proc.12th IEEE Symposium on Computer Arithmetic, pp.70-77, 1999.

Algorithm	Mult. size	Data size	Critical path delay [ns]	Num. of cycles	Latency[ns]	LUTs	DSPs
k bit MM	k=32	255	18.38	8	147.04	1531	67
k bit QPMM	k=32	318	10.15	11	111.65	2628	60
k bit QPMM	k=16	286	8.51	19	178.71	1973	31
k bit QPMM	k=64	388	12.12	7	84.84	3431	112

TABLE I.   IMPLEMENTATION COMPARISON OF DESIGN PARAMETERS

A Lightweight Metric for The Evaluation of Network Congestion in NoC-based MPSoC

Yang Huang, Letian Huang
University of Electronic Science and Technology of China
Chengdu, China
Corresponding Author: huanglt@uestc.edu.cn

Xiaohang Wang
South China University of Technology
Guangzhou, China

Abstract— **An effective run-time application mapping algorithm needs to reduce network congestion due to its significant impact on network performance. In this paper, we propose a lightweight congestion evaluation metric for application mapping algorithms in NoC-based MPSoC. We obtain information from task graphs and the mapping results, and use it to calculate a congestion metric to evaluate the link interference for a given communication flow. The congestion metric is the number of flows which share the same link with the given flow, it reflects both internal and external congestion. The congestion metric is accurate and lightweight, it can be used in runtime application mapping algorithms and other real-time tasks. Simulation result shows a strong correlation between the metric value and the average latency. Compared with other similar metrics, our congestion metric shows much greater accuracy.**

Keywords— many-core system; Network-on-Chip; mapping algorithm; congestion

I. INTRODUCTION

Network on Chip (NoC), due to its efficient communication and high scalability, has become mainstream many-core interconnect facility. Network congestion is the main cause for performance degradation of NoC, many works try to reduce or eliminate it from different aspects.

Many existing task mapping algorithms aim to reduce the network congestion, these algorithms try to map the tasks belong to the same application onto the adjacent positions. In these algorithms, network congestion is usually divided into internal and external congestion. ``Internal" and ``external" means the congestion occurs between the communication flows belong to the same application or different applications.

Mapping algorithms usually break the mapping process into two stages: region selection and task allocation. Region selection tries to find the best region to reduce the external congestion. Task allocation tries to optimize the locations of tasks to reduce internal congestion. Several metrics are proposed to evaluate the mapping algorithm performance, such as Average Packet Latency (APL), Average Weighted Manhattan Distance (AWMD), Mapped Region Dispersion (MRD)[1], etc. These metrics assist in evaluating the performance of mapping algorithms. Some of them can be adopted to guide mapping process, such as ICR[1], PL[2], and ICEB[3].

The congestion-related metrics used in many mapping algorithms only consider the internal congestion, they cannot reflect network congestion especially with high system utilization. When system utilization becomes higher, the mapped region of an application will become more fragmented and the congestion tends to occur within different applications. In this paper, a metric for the evaluation of network congestion is proposed which is able to reflect both internal and external congestion, we call it Global Congestion Metric (GCM). GCM is lightweight and accurate, it can be used to evaluate the congestion of a given communication flow and the performance of a given mapping algorithm. It can also be used in runtime application mapping algorithms and other real-time tasks.

II. RELATED WORK

Chou et al. in [4] analyze the impact of different network contention on the packet latency, and they show that the path-based network contention has a great impact on packet latency. They proposed an ILP-based mapping algorithm to minimize the inter-tile network contention. In [5], they break the mapping process into two stages: region selection and task mapping. They use MRD to evaluate the continuity of the selected region and try to minimize the AWMD.

Authors in [1], [3], [6] follow the two stages mapping process, Fattah et al. in [1] try to keep the mapped region contiguous and place the communicating tasks in a close neighborhood. They use MRD and Internal Congestion Ratio (ICR) as the indicators of congestion probability, MRD factor is defined as the mean value of all possible node pairs Manhattan Distance (MD) in the mapped region, ICR is the number of edges of an application using the same communication channel (according to the XY algorithm) with respect to its total number of edges. In [3] they proposed ICE (Internal Congestion and Energy) and ICEB (ICE per Bit) metric to guide their mapping algorithm. ICE is calculated by adding up all the communication value along the path of a communication flows, ICEB is the sum of all ICE value with respect to the sum of all communication flows. The two metrics only consider the internal congestion.

Carvalho et al. in [2] proposed MMC, MAC, and PL heuristic targeting reduction of network congestion, PL metric is best among them. PL is the sum of all channel loads along the path of a communication flow. The PL metric is used to assist in choosing a better mapping node instead of the default method in NN (Nearest Neighbor) mapping algorithm.

III. THE DETAIL OF GLOBAL CONGESTION METRIC

The data dependency of an application can be presented by directed acyclic graph (DAG) which is the primary graph model used for program representation [7]. When deterministic routing is used, the route of packets in each communication flow is calculable using the information of DAG and mapping result of an application. For each communication flow, the Global Congestion Metric (GCM) is defined by equation 1:

$$GCM_i = \Sigma_{\forall e_j \in E} S_{i,j} \qquad (1)$$

$S_{i,j} = 1$ if edge e_i and edge e_j share at least one link when they are mapped, else $S_{i,j} = 0$. The accuracy of GCM is based on three aspects:

a) Non-continuous mapping will inevitably cause fragmentation which leads to more external congestion. Internal and external congestion are both considered in GCM.

b) Path-congestion has a significant impact on packets latency and it occurs with high probability [4].

c) The communication flows and their routes are known in advance under a deterministic routing policy, and the quantity of flows is limited. This brings feasibility to predict congestion with low complexity.

IV. EVALUATION

In this section, we conduct a detail simulation to evaluate the accuracy of GCM. An 8*8 homogeneous many-cores system is simulated. The XY routing algorithm is applied. PEs inject packets (5 flits/packet) into the network at the maximum packet injection rate of 0.2 packet/cycle/router. Applications are generated by TGFF [8]. The number of tasks is 4~20. The communication volumes between two tasks are 10~30 packets (application set 1), 20~60 packets (application set 2). The period is between 200 and 400 cycles. Applications enter the platform at a rate of 0.002 application/cycle and are scheduled sequentially. An application set contains 2000 applications. Simulation time is long enough to run one application set.

We use CoNA [1] mapping algorithm in the first simulation. Average packet latency (APL), GCM value, Manhattan Distance for all communication flows are recorded. Figure 1(left) shows the GCM distribution, the congestion probability, figure 1(right) shows average latency for the flows with the same GCM value. The larger the GCM value, the greater the probability of congestion, and it leads a larger average latency. GCM shows a strong correlation with network congestion.

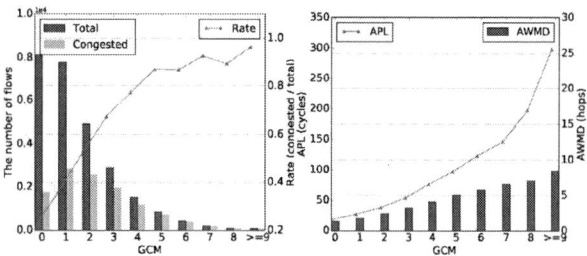

Figure 1 GCM with congestion probability (left), with average latency (right)

To compare with other similar metrics (ICE [3], ICR without external congestion [1], and PL [2]), we use different mapping algorithms [2], [6] to study the relationship between these metrics and the performance of mapping algorithm. The average metrics value and the average latency for all communication flows are recorded. Figure 2 shows the different metrics and the average latency, all metrics and latency are normalized to random mapping algorithm. As figure 2 shows, GCM exhibits greater accuracy comparing with other metrics.

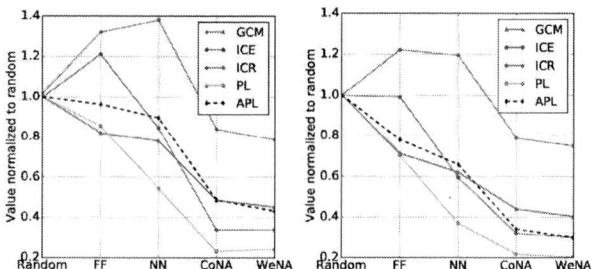

Figure 2 GCM with other similar metrics, application set 1 (left), set 2 (right)

V. CONCLUSION AND FUTURE WORK

In this paper, we propose a lightweight and accurate metric (GCM) for network congestion. It calculates the number of flows which share the same link with the given flow. It predicts the congestion for a communication flow, and it can be used to evaluate the performance of mapping algorithms. Due to its low complexity, it is practical in many real-time tasks, such as mapping algorithm. We will explore them in the future.

REFERENCES

[1] M. Fattah, M. Ramirez, M. Daneshtalab, P. Liljeberg, and J. Plosila, "CoNA: Dynamic application mapping for congestion reduction in many-core systems," in Computer Design (ICCD), 2012 IEEE 30th International Conference on, 2012, pp. 364–370.

[2] E. L. de Souza Carvalho, N. L. V. Calazans, and F. G. Moraes, "Dynamic task mapping for MPSoCs," Design & Test of Computers, IEEE, vol. 27, no. 5, pp. 26–35, 2010.

[3] M. Fattah, P. Liljeberg, J. Plosila, and H. Tenhunen, "Adjustable contiguity of run-time task allocation in networked many-core systems," in Design Automation Conference (ASP-DAC), 2014 19th Asia and South Pacific, 2014, pp. 349–354.

[4] C.-L. Chou and R. Marculescu, "Contention-aware application mapping for network-on-chip communication architectures," in Computer Design, 2008. ICCD 2008. IEEE International Conference on, 2008, pp. 164–169.

[5] C.-L. Chou and R. Marculescu, "Incremental run-time application mapping for homogeneous NoCs with multiple voltage levels," in Proceedings of the 5th IEEE/ACM international conference on Hardware/software codesign and system synthesis, 2007, pp. 161–166.

[6] L. Huang, H. Dong, J. Wang, M. Daneshtalab, and G. Li, "WeNA: Deterministic run-time Task Mapping for Performance Improvement in Many-core Embedded Systems," IEEE Embedded Systems Letters, vol. 7, no. 4, pp. 93–96, 2015.

[7] O. Sinnen, Task scheduling for parallel systems, vol. 60. John Wiley & Sons, 2007.

[8] R. P. Dick, D. L. Rhodes, and W. Wolf, "TGFF: task graphs for free," in Proceedings of the 6th international workshop on Hardware/software codesign, 1998, pp. 97–101.

An Efficient FPGA Implementation for odd-even sort based KNN algorithm using OpenCL

Hai Peng, Letian Huang
School of Communication and Information
University of Electronic Science and Technology of China
Chengdu, China
Corresponding Author: huanglt@uestc.edu.cn

John Chen
PSG University Program
Intel
Chengdu, China

Abstract—Owing to the rising demands for resources integration, a lot of research efforts has been devoted to accelerating the data classification techniques. K-Nearest Neighbor(KNN)algorithm, as one of the most important data classification algorithms, is widely used in text categorization, predictive analysis etc. Two most vital issues of improving KNN are accelerating the speed of convergence and efficiently optimizing the parallel implementation. In this paper, we propose an efficient implementation on FPGA based heterogeneous computing system using OpenCL. An odd-even sort based KNN is designed to make full use of the parallel pipeline structure of FPGA. The results has shown that the performance of the proposed algorithm is much better than traditional GPU based KNN .

Index Terms—OpenCL; KNN; Odd-even Sort; FPGA; Heterogeneous Computing;

I. INTRODUCTION

With the increasingly growing demands for data classification in recent decades, Field Programmable Gate Arrays(FPGAS), which are highly suitable for parallel implementation of a lot of classification algorithms ,have combined with CPU and formed a new kind of heterogeneous computing architecture. Open Computing Language (OpenCL) [1]is a framework which relies on advanced programming language and at the same time provides supports to FPGA targets. KNN [2] has been widely used in the filed of data classification , but the cost of KNN is extremely intensive.

In this paper, an efficient odd-even sort based KNN is presented. By using OpenCL, it makes full use of properties of odd-even sort algorithm and redesign the KNN algorithm in parallel. Thus, the efficiency of distance calculation and distance rank has been improved.

Traditional KNN algorithms firstly need to sort all of N distances and then choose the k minimum distances from N. However, due to the fact that in KNN only k minimum distances are necessary, we redesign the KNN in a parallel way using odd-even sort algorithm. For each query object, only k work-items are used instead of N. As a consequence, the efficiency of KNN has been obviously improved.

The rest of this paper is organized as follows. Section II describes the KNN algorithms and the OpenCL program framework. In Section III we propose an efficient FPGA implementation for odd-even based KNN algorithm using OpenCL. Section IV details the performance results and comparisons.

II. BACKGROUND

A. K Nearest Neighbor Algorithm

KNN is used to classify objects by a majority vote of k nearest reference objects [3].Distance calculation and distance rank are the two most time-consuming processes in KNN. Firstly, N distances are calculated between query objects and reference objects. The distance rank process is the most important part in KNN. Based on the distance matrix obtained by distance calculation part, distance rank part sorts each row of the matrix and chooses the k smallest distances for each row.

B. OpenCL Architecture

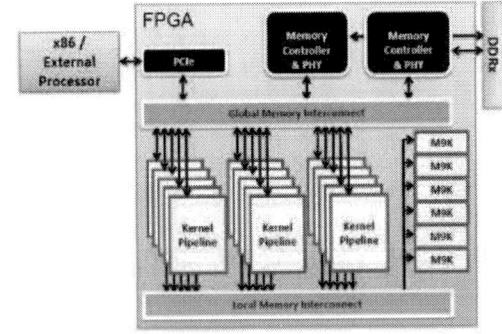

Fig. 1. OpenCL platform

OpenCL is a framework for parallel programming on heterogeneous platforms.it is based on a runtime host library and C99 extensions for device programming adapted to support vectorized data types. As shown in Fig 1, the framework consists of two main parts. The first part is the host which handles the data-flow generated by applications and transfers it to the OpenCL devices through PCIe. The second part is the FPGA which works as the OpenCL device. The work-items on a device are organized into work-groups which share a common memory region and synchronization points. The memory of the device consists of three levels: global memory, local memory and private memory.

III. ODD-EVEN SORT BASED KNN ALGORITHM USING OPENCL

A. Distance Calculation Kernel

Distances between query objects and reference objects are calculated in this kernel. As shown in Fig 2,for the reason that distance calculations are independent, this kernel can be fully parallel mapped on FPGA [4].Due to the low efficiency of access to global memory, we load the reference dataset into local memory from global memory in advance. The calculation results are stored in distances matrix.

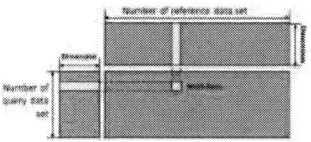

Fig. 2. Distance Calculation Kernel

B. Distance Rank Kernel

Once the distances matrix is generated from distance cal culation kernel, distance rank kernel is launched to find the k smallest distances in each row of the matrix. As shown in Fig 3, the entire sorting process is divided into three stages. In the first stage, k work-items sort the first 2k distances in a row using odd-even sort. Each work-item compare 2 elements at one time. For example, in the even turn, the first work- item compare the 1st and 2nd distances in each row. While in the odd turn, the first work-item compare the 2nd and the 3rd distances. The k smallest distances out of the 2k distances are gained after a comparison up to 2k+1 times. In the second stage, parallel bubble sort [4] is used to enable k work-items to carry the k smallest distances from the head to the tail of the row. The third stage is the same with the first stage. Odd-even sort is applied to ensure the k smallest distances of the row are chosen out. During the whole process, only k work-items are used.

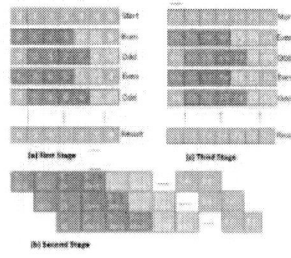

Fig. 3. Distance Rank Kernel

IV. EVALUATION

A. Test Environment

Three target devices are considered: a CPU, a GPU and an FPGA. The CPU is an Intel Core i7-3770K running at 3.5GHZ with the reference software written in matlab. The GPU is an AMD Radeon HD7950 running for comparison with FPGA. The FPGA board is a Terasic DE5 with a Stratix V 5SGX.

B. Hardware Resources Management and Utilization

KDD-CUP 2004 quantum physics dataset is used to test our KNN algorithm. The number of dimensions of each record is 40.Since K is usually not large compared to the number of reference objects, we set it 64. TABLE 1 shows resources usage condition of the KNN system when implemented on the DE5 FPGA.

TABLE I
RESOURCE USAGE

Stratix V	
Logic utilization	70%
Dedicated logic registers	34%
Memory bits	45%
DSP block(18-bit)	6%
Clock Frequency	131.42MHZ

C. Comparison

TABLE 2 shows the performances for each kernel on GPU and FPGA, along with the software reference results. Through comparison between GPU and FPGA, we have verified that the new KNN based on FPGA has better performances than traditional GPU devices. By comparing the energy efficiency ratio(EER),FPGA based KNN is more than 3 times higher than traditional GPU based KNN.

TABLE II
PERFORMANCES

Platform	CPU	GPU	FPGA
Feature size/nm	22	28	40
Runtime/ms	11020.42	29.28	77.59
Objects/s	1.85	699.45	263.96
Speedup	/	378	142
Power/w	130	200	24
Objects/J	0.014	3.497	10.998
EER	/	249	786

V. CONCLUSION

This paper has presented an efficient FPGA implementation for an old-even sort based KNN algorithm using OpenCL.The EER of FPGA based KNN is more than 3 times higher than traditional GPU based KNN.

REFERENCES

[1] M. Aaftab, "The opencl specification," *Khronos OpenCL Working Group*, vol. 1, 2011.

[2] G. Nolan, "Improving the k-nearest neighbour algorithm with cuda," *Honours Programme, The University of Western Australia*, 2009.

[3] S. Liang, C. Wang, Y. Liu, and L. Jian, "Cuknn: A parallel implementation of k-nearest neighbor on cuda-enabled gpu," in *Information, Computing and Telecommunication, 2009. YC-ICT'09. IEEE Youth Conference on.* IEEE, 2009, pp. 415–418.

[4] Y. Pu, J. Peng, L. Huang, and J. Chen, "An efficient knn algorithm implemented on fpga based heterogeneous computing system using opencl," in *Field-Programmable Custom Computing Machines (FCCM), 2015 IEEE 23rd Annual International Symposium on.* IEEE, 2015, pp. 167–170.

An Address Remapping Algorithm to Reduce Power Consumption in NoC-based Chip-Multiprocessors

Shuyu Chen, Letian Huang
School of Communication and Information
University of Electronic Science and Technology of China
Cehngdu, China
Corresponding Author: huanglt@uestc.edu.cn

Song Li
Big Date Department
Inspur Group
Jinan, China

Abstract—NoC-based Chip-Multiprocessors(CMPs) are promising mechanisms which are essential to satisfy a growing need for easily scalable and high-performance in recent decades. However, in the process of running the programs, there are some data that misses from L1 cache to L2 cache. These cache misses have some effects on network congestion and access time. In this paper, we use the average number of hops to describe the fairness of the thread resources and explain each thread's access to each bank is fair. After then, we described an address remapping algorithm for cache miss in NoC-based CMPs. We evaluate our mechanism using Simplescalar and Cacti power simulation tool. Experimental results show that address remapping achieves significant improvement in the system performance and power consumption.

Keywords—CMPs; Network-on-Chip; remapping algorithm; cache miss

I. INTRODUCTION

At one time single-core processor is the mainstream market, but with the advances in computer architecture and semiconductor innovation, the development of the single-core processor is seriously limited by three major hindrances, which are power consumption, frequency and system reliability of processor.

At this point, NoC-based CHIP MULTIPROCESSORS (CMPs) attract a lot of attention and they become the dominant trend in parallel processing systems. The main design idea of CMP architectures is many Processing Elements (PEs) integrated on one model [1]. Power consumption of Networks on Chip (NoC) is becoming an important design concern. In the context of parallel processing systems, the reduction of power consumption can extend battery life and reduce system temperature.

In CMP, the chip area is split into many tiles and each tile contains networking elements and PEs, the interactions among tiles is accomplished by sending messages using routers. Although L1 caches are likely to remain private to each processor, the L2 caches may be shared [2] by all processors. There exist some cache misses which have some effects on system performance.

This paper investigates a new mechanism for increasing the performance of the system and reducing power consumption under a cache miss. The rest of this paper is organized as follows. Section 2 reviews background information on Typical NoC-based CMP and fairness analysis of thread resource

allocation. Section 3 discusses address remapping algorithm. Section 4 presents the experimental results, and Section 5 presents the conclusions.

II. BACKGROUND

As shown in Figure 1, the CMP of 16 tiles [3] uses a mesh network to provide communication channels for the distributed shared cache. Large-capacity L2 cache are divided into many banks for parallel access, and the cache banks belonging to different tiles are connected together via the mesh network, thus any processor can flexibly access the banks of L2 cache in local or remote tiles.

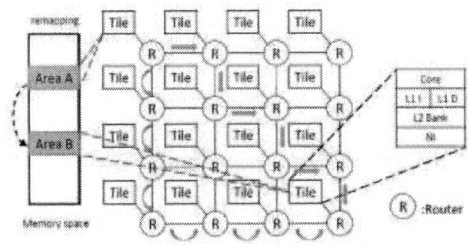

Fig. 1. the CMP of 16 tiles

In NOC-based CMPs, each tile corresponds to a node, which includes a single processor core, a private L1 cache, a shared L2 cache bank and a network interface NI. Each thread corresponds to a processor core and each core corresponds to a cache bank. The cache bank access is fair that the first thread will access to their nearest local bank, if the data are required for the thread is not in this bank, thread will access to the next cache according to the access distance. We need to measure the standard access distance and we use hop number to measure the access time. Thus, the average number of hops can measure the performance of the program access memory.

For example, in the figure1, If core 0 want to access the core 15 as indicated in the graph, no matter how the path are selected, the least number of routers experienced by thread and the router hop number is 6. This is the most extreme example and it takes much longer time than others. If core 0 and core 15 visit the shared L2 cache bank by mapping the core 0 physical address space to core 15, it can reduce access time and improve the performance of the system.

978-1-5090-3220-4/16 $31.00 © 2016 IEEE
209
ISOCC 2016

III. The detail of Address Remapping Algorithm

Consider a system with a miss cache [4]. When a miss occurs, data is loaded into L2 cache bank. This kind of situation will lead to the long runtime of the program and the network congestion, and this paper proposes an address remapping algorithm of private data address space. The specific address remapping algorithm is shown in Figure 2.

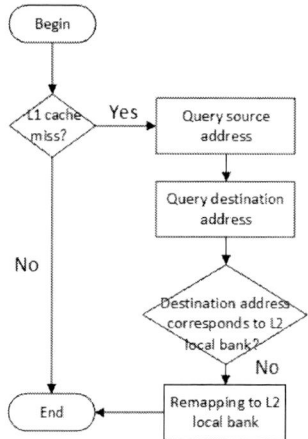

Fig. 2. Block diagram of the proposed methodology

As shown in Figure 2, first of all, we track the miss data from L1 cache that scatter to the cach L2 bank, then search the user space address and record the user ID of the corresponding L1 nuclear, and query the corresponding network address. If the network address does not correspond to the local L2 bank, the destination address is remapped to its corresponding local bank L2.

For example, the data in the first core L1 bank is lost to the last core L2 bank, then you can calculate the access distance from the 0 jump to the 6 which is the farthest access distance in our network structure. This is an extreme assumption, but we still can get a conclusion that using such a remapping algorithm will reduce thread memory latency and shorten the access distance without cross the adjacent or distant router to access other core cache bank, thus reduce the network congestion.

IV. Evaluation

A. Simulation tool and configuration

To evaluate the effectiveness and efficiency of our approach, we extend the simplescalar and Cacti power simulation tool to implement the address remapping algorithm. SimpleScalar is a set of tools that model a virtual computer system with CPU, Cache and Memory Hierarchy.Cacti power simulation tool is developed by the WRL company's HP laboratory and put into use an extremely powerful power consumption assessment model.

In this experiment we have to use a number of test program set for our modified simulator to carry out a complete test. The test program set configuration is shown in Table I.

TABLE I
The test program set configuration

test program set	Problem size	test program set	Problem size
Blackscholes	96000	Canneal	200000.nets
Cholesky	8096	Fft	256K complex points
Fmm	20000000	Fluidanimate	In_35K.fluid
Lu	1024*1024 matrix	Ocean	R 2000
Radiosity	50000	Radix	R 1024
Raytrace	1370000000	Shallow	S 2048
Streamcluster	4096*1000	Water	512 molecules

B. Result analysis

Figure 3 shows the result of network power consumption after the address remapping. The horizontal coordinate is the 14 test program set to test the simulator and the vertical coordinate is the ratio of the power consumption of network system compared to the normal operational system. We can draw the conclusion that the network power consumption of the system can be reduced by an average of 3%, by using such a thread private data address space remapping algorithm.

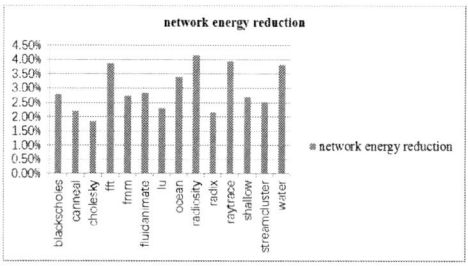

Fig. 3. The result of network power consumption after the address remapping

V. Conclusion and Future Work

In this paper, we explored the system architecture of NoC-based Chip-Multiprocessors. We described fairness analysis of thread resource allocation, including the specific measure of fairness and an example with farthest distance. The main contribution of the paper is a new thread private data address space remapping algorithm was used to get around address space miss in cache and to optimize the performance of the system and the power consumption. We will test on the actual hardware in the future and improve optimization algorithm.

References

[1] D. Zydek, H. Selvaraj, L. Koszałka, and I. Poźniak-Koszałka, "Evaluation scheme for noc-based cmp with integrated processor management system," *International Journal of Electronics and Telecommunications*, vol. 56, no. 2, pp. 157–168, 2010.

[2] J. Huh, C. Kim, H. Shafi, L. Zhang, D. Burger, and S. W. Keckler, "A nuca substrate for flexible cmp cache sharing," *IEEE transactions on parallel and distributed systems*, vol. 18, no. 8, pp. 1028–1040, 2007.

[3] Y. Wang, L. Zhang, Y. Han, H. Li, and X. Li, "Address remapping for static nuca in noc-based degradable chip-multiprocessors," in *Dependable Computing (PRDC), 2010 IEEE 16th Pacific Rim International Symposium on.* IEEE, 2010, pp. 70–76.

[4] N. P. Jouppi, "Improving direct-mapped cache performance by the addition of a small fully-associative cache and prefetch buffers," in *Computer Architecture, 1990. Proceedings., 17th Annual International Symposium on.* IEEE, 1990, pp. 364–373.

Neural Network based Seizure Detection System using Raw EEG Data

Tianchan Guan , Xiaoyang Zeng
School of Micro-Electronic
Fudan University
Shanghai, China

Letian Huang
School of Micro-Electronic
University of Electronic Science and
Technology of China
Chengdu, China
huanglt@uestc.edu.cn

Tianchan Guan, Mingoo Seok
School of Engineering
Columbia University
New York, U.S.A

Abstract—**In this paper we present a new seizure detection system based on neural networks. The system takes raw EEG data without any explicit feature extraction. The removal of explicit feature extraction steps can improve flexibility and also mitigate hardware overhead and high clock rate often associated with feature extraction steps. We also propose a scheme to construct training data from raw EEG data which naturally has a significant length imbalance between seizure and non-seizure state waveform. The proposed detection system is validated with the CHB-MIT database. The results show that the proposed system can achieve detection performance compared to existing systems having explicit feature extractors.**

Keywords; Seizure detection, Neural Network, Electroencephalogram (EEG)

I. INTRODUCTION

Epilepsy is one of the most common chronic neurological diseases, which affects around 1\% of the world's population. To provide more timely treatment, EEG seizure detection SoCs are developed recently. By continuously acquiring and analyzing EEG data, the SoC can detect the seizure and then use integrated stimulator to suppress it [1] or record and transmit abnormal data for further treatment [2],[3]. The typical architecture of existing seizure detection SoCs consists of a front end for data acquiring, a feature extractor and a classifier. Among those, a feature extractor imposes several challenges. Firstly, it is among the most hardware-intensive components [2]. Furthermore, as it is often implemented hard-wired logic for speed and energy-efficiency, it reduces system flexibility. Moreover, a number of features of EEG for seizure detection have been proposed and the efficacy of them varies from a patient to a patient [4]. Obviously, it is a less attractive option to integrate multiple feature extractors due to hardware overhead.

To solve this problem, in this paper, we propose a new seizure detection system using neural networks (NN). To detect seizure, it directly takes raw EEG waveform without an explicit feature extraction step. NN is a non-linear classifier which is widely used in many applications like speech and image pattern recognition. Recently it is applied to the seizure detection problem and achieves excellent performance outperforming all other classifiers [5]. But it is still relying on a hand-picked set of features that may not be as effective across different patients and over time. In addition, we devise a scheme to construct training dataset for our NN classifier. The raw data acquired from epileptic patients are mostly in non-

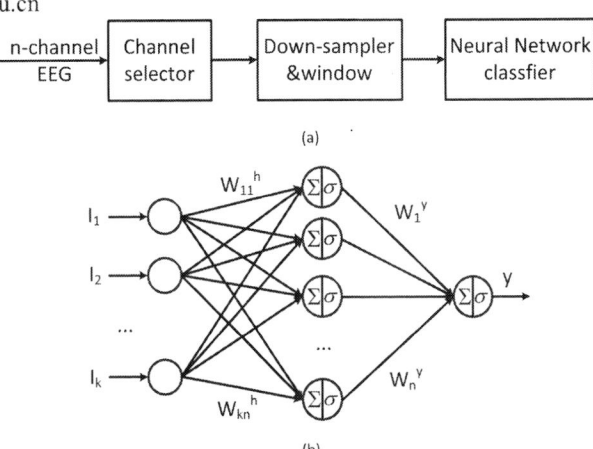

Fig.1. Structure of proposed seizure detection system (a), and neural network classifier (b)

seizure state. Therefore, the raw data contains only a small amount of seizure state waveforms. We find this large length imbalance between seizure and non-seizure state waveforms makes it difficult to train the NN classifier. Therefore, from the original raw waveform, we construct a training dataset that has multiple copies of seizure state waveforms but only a subset of non-seizure state waveforms. With this training dataset, we can effectively train the NN classifier.

We perform the detection sensitivity and specificity simulations for the seizure detection system with the CHB-MIT database [6],[7]. The results show that the proposed system achieves the sensitivity and specificity that is comparable to the state-of-the-art systems. The removal of explicit feature extractors makes the proposed systems significantly more flexible than existing systems.

II. SEIZURE DETECTION SYSTEM

A. System Architecture

Fig. 1(a) shows the proposed seizure detection system. The system selects one of n EEG channels for seizure detection. Then, the system downsamples incoming EEG waveform since the signal in the 0-20Hz band is of interest to detect seizure. The original EEG is often sampled at 256 Hz while the data fed to the NN classifier can be as low as 40Hz. The architecture of NN classifier is shown in Fig. 1(b). It is based on a multi-layer perceptron (MLP) NN. The classifier produces an output (y)

Table I

Design Parameter List

	specificity	sensitivity	overhead	value
downsample rate	N	N	N	4
Window size	P	N	P	1s
Train window overlap	N	P	P	0.875s
Hidden layer number(n)	N.A	N.A	P	1
Hidden layer order	N.A	N.A	P	600
Detect window overlap	N.A	N.A	P	0
Detect window number(x)	P	N	P	3

that determines if the input x-s window waveform contains seizure-like signals. The system checks if certain number of consecutive windows are classified as seizure and finally makes its decision. We use float-point for each synaptic weight but we expect further optimization can be possible. The downsample rate and window size x which determine number of input units (k), number of hidden units (n), and other design parameters will be analyzed in last part of this section.

B. Training

We train the classifier based on the well-known back-propagation algorithm. One of obstacles of training the classifier is the amount imbalance between seizure and non-seizure state waveforms. Epileptic patients are mostly in non-seizure (normal) state. Specifically, the duration of the normal waveform can be as long as tens of hours while seizure-state data lasts only a few minutes. Such large imbalance can make it impossible to train the NN classifier. To address this issue, we resample data by sampled seizure data in highly overhead time window. To be specific, for a seizure lasts for 5 minutes, if the time window is 2-second and overlap is as much as 1.875s, the number of seizure data samples is 2385 which is 16 times the 0-overlap.

C. Design Consideration

Table I shows the system-level parameters of the proposed system used in this work. We also include the correlation between system parameters and system performances and overhead. "P" means that a parameter and a performance (or overhead) are positively-correlated. "N" means negatively-correlated. Some correlations are not available (N.A) for some of the parameters as they show non-linear relationship, requiring further study and optimization.

III. PERFORMANCE

The proposed seizure detection system is tested with CHB-MIT database [6],[7].We train the NN classifier with "leave-one-out" cross-validation scheme. Fig. 2 summarizes detection sensitivity and specificity over 24 patients. Performance of other patients can hopefully be improved by adjusting design

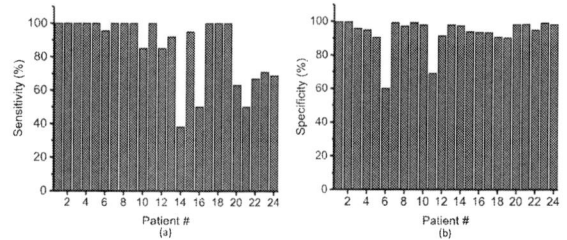

Fig. 2. Seizure detection sensitivity (a) and specificity (b) with the CHB-MIT database

parameters according to previous section. The system can detect all the seizures of 11 patients in the database with specificity higher than 90%. The average detection sensitivity and specificity is 82% and 95.5% respectively. The performance is comparable with 86% sensitivity and 96% specificity demonstrated in [2], while the work in [2] uses filter bank running at data-sampling rate as feature extractor.

IV. CONCLUSION

In this paper we propose a neural-network based seizure detection system. By introducing a NN as a seizure detector, we remove the explicit feature extraction steps, which can be intensive in area and power dissipation and also can degrade the flexibility of the system. We also propose a scheme to prepare training data from raw measurements, which can effectively train the proposed NN based classifier. The simulation shows that the detection performance on the CHB-MIT dataset is comparable to existing systems employing an explicit inflexible and power-hungry feature extractor.

REFERENCES

[1] W. M. Chen, H. Chiueh, T. J. Chen, C. L. Ho, C. Jeng, M. D. Ker, C. Y. Lin, Y. C. Huang, C. W. Chou, T. Y. Fan, M. S. Cheng, Y. L. Hsin, S. F. Liang, Y. L. Wang, F. Z. Shaw, Y. H. Huang, C. H. Yang, and C. Y. Wu, "A fully integrated 8-channel closed-loop neural-prosthetic cmos soc for real-time epileptic seizure control," IEEE Journal of Solid-State Circuits, vol. 49, no. 1, pp. 232–247, Jan 2014

[2] J. Yoo, L. Yan, D. El-Damak, M. A. B. Altaf, A. H. Shoeb, and A. P. Chandrakasan, "An 8-channel scalable eeg acquisition soc with patient-specific seizure classification and recording processor," IEEE Journal of Solid-State Circuits, vol. 48, no. 1, pp. 214–228, Jan 2013

[3] M. A. B. Altaf and J. Yoo, "A 1.83 μ j/classification, 8-channel, patient-specific epileptic seizure classification soc using a non-linear support vector machine," IEEE Transactions on Biomedical Circuits and Systems, vol. 10, no. 1, pp. 49–60, Feb 2016.

[4] L. Logesparan, A. J. Casson, and E. Rodriguez-Villegas, "Optimal features for online seizure detection," Medical & biological engineering & computing, vol. 50, no. 7, pp. 659–669, 2012.

[5] S. S. Alam and M. I. H. Bhuiyan, "Detection of seizure and epilepsy using higher order statistics in the emd domain," IEEE journal of biomedical and health informatics, vol. 17, no. 2, pp. 312–318, 2013

[6] A. L. Goldberger, L. A. Amaral, L. Glass, J. M. Hausdorff, P. C. Ivanov, R. G. Mark, J. E. Mietus, G. B. Moody, C.-K. Peng, and H. E. Stanley, " Physiobank, physiotoolkit, and physionet components of a new research resource for complex physiologic signals," Circulation, vol. 101, no. 23, pp. e215–e220, 2000.

[7] A. H. Shoeb, "Application of machine learning to epileptic seizure onset detection and treatment," Ph.D. dissertation, Massachusetts Institute of Technology, 2009.

Low-Cost Concurrent Error Detection Schemes for Logarithmic Converters

Tso-Bing Juang, Ying-Ren Lee*
Department of Computer Science and Information Engineering
National Pingtung University (NPTU)
Pingtung City, Taiwan, R. O. C
tsobing@mail.nptu.edu.tw
*boy020319@gmail.com

Chin-Chieh Chiu
Department of Computer Science and Information Engineering
National Chao Tung University (NCTU)
Hsinchu City, Taiwan, R. O. C
ufo.keb@gmail.com

Abstract—In this paper, low-cost concurrent error detection (CED) schemes for logarithmic converters are proposed. By adopting our previously proposed logarithmic converters with developed converters with the same functions, the proposed schemes can perform logarithmic conversion with CED ability. Simulation results show that our proposed CED schemes for logarithmic converters can achieve at most 61.65% area and 32% delay reductions. Our proposed low-cost CED schemes can be applied to real-time computation-intensive computations for achieving high-speed logarithmic conversions with CED ability.

Keywords: Concurrent error detection (CED), logarithmic number system (LNS), computer arithmetic, VLSI design.

I. INTRODUCTION

Since the chip sizes shrink as the progress of VLSI technology, the problems of soft errors and single faults arise and attracted by IC designers again [1]. One of the remedies for solving such kind of errors or faults is to apply concurrent error check (CED) technology which can detect if the functions for circuits under protection are correct or not. The simple implementation way the for CED is to use dual module redundancy (DMR), wherein two same copies of circuits under protection are used, and one comparator to initiate the error flag if inconsistent results for the circuits occur. However, the area costs will be getting higher applying DMR technique. Therefore, in previous years, there are some works proposed trying to reduce the area cost while achieving the CED ability at the same time [1-5]. For example, in [1] the authors proposed some novel area-efficient CED schemes for complex multiplications.

In the meantime, for many computation-intensive applications, one of the methods for simplifying tremendous computation overheads is to apply logarithmic number system (LNS) [6] which can greatly boost the performance with high computation throughputs. In LNS-based computation environments, it always consists of three stages, including logarithmic conversions, simple computations and anti-logarithmic conversions, while the corresponding architecture

is shown in Fig. 1. In previous work, there are many work about logarithmic/anti-logarithmic converters were been proposed since they play the key parts in deciding the overall performances of LNS-based computation units.

In this paper, our focus is to propose some low-cost concurrent error detection schemes for logarithmic conversions. This paper is organized as follows. In section II we introduce the basic concept of concurrent error detection scheme for logarithmic converters. Our proposed four low-cost logarithmic converters with CED ability are given in Section III. Comparisons for proposed schemes are listed in Section IV, and Section V concludes the paper.

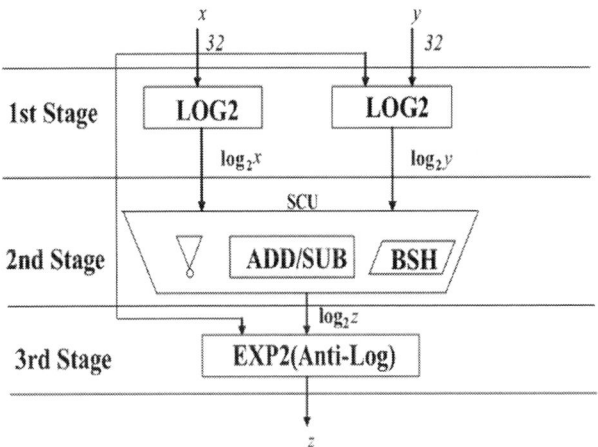

Fig.1: The architecture for LNS-based computations. [6]

II. BASIC CONCEPT OF LOGARITHMIC CONVERTERS AND ITS CED IMPLEMENTATIONS

In our previous work [6], we have proposed a low-error logarithmic conversion method given in follows names as Type I, where x is the input, where $x_{4MSBits}$ is denoted as the four most significant bits (MSB) after the point of x.

Type-I

$$\log_2(1+x) \approx \begin{cases} x+(2^{-2}\text{-}2^{-4})x_{4\text{MSBits}}, & \text{if } 0 \leq x < 0.5 \\ x+\left(2^{-3}\text{-}2^{-5}\right)x_{4\text{MSBits}} + \left(2^{-3}\text{-}2^{-5}\right), & \text{if } 0.5 \leq x < 1 \end{cases}$$

To increase CED ability on our logarithmic converter in [6], we can apply the scheme as shown in Fig. 2, while two logarithmic converters with same functions performing conversions at the same time, and output the error flag if two outputs from converters are not the same.

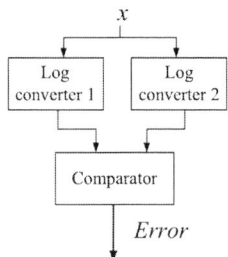

Fig. 2: CED scheme for logarithmic converters

III. PROPOSED CED SCHEMES FOR LOGARITHMIC CONVERSIONS

Based on Type I described in Section II, we have derived three alternative logarithmic converters with same functions and accuracies listed as follows. These four different types of converters can be used in Fig. 2 to become alternative CED schemes for logarithmic conversion.

Type II

$$\log_2(1+x) \approx \begin{cases} x+(2^{-2}\text{-}2^{-4})x_{4\text{MSBits}}, & \text{if } 0 \leq x < 0.5 \\ x+\left(2^{-2}\text{-}2^{-4}\text{-}2^{-5}\right)x_{4\text{MSBits}} + \left(2^{-2}\text{-}2^{-4}\text{-}2^{-5}\right), & \text{if } 0.5 \leq x < 1 \end{cases}$$

Type III

$$\log_2(1+x) \approx \begin{cases} x+\left(2^{-1}\text{-}2^{-2} + 2^{-5}\right)x_{4\text{MSBits}}, & \text{if } 0 \leq x < 0.5 \\ x+\left(2^{-2}\text{-}2^{-3}+2^{-5}\right)x_{4\text{MSBits}} + \left(2^{-2}\text{-}2^{-3}+2^{-5}\right), & \text{if } 0.5 \leq x < 1 \end{cases}$$

Type IV

$$\log_2(1+x) \approx \begin{cases} x+(2^{-1}\text{-}2^{-2}\text{-}2^{-3}+2^{-4})x_{4\text{MSBits}}, & \text{if } 0 \leq x < 0.5 \\ x+\left(2^{-1}\text{-}2^{-2}\text{-}2^{-4}\text{-}2^{-5}\right)x_{4\text{MSBits}} + \left(2^{-1}\text{-}2^{-2}\text{-}2^{-4}\text{-}2^{-5}\right), & \text{if } 0.5 \leq x < 1 \end{cases}$$

Since there are six different schemes for Type I to IV if we apply alternative schemes in Fig. 2 to perform logarithmic conversion, we can make a good choice between these possible alternative implementations for CED schemes. Section IV demonstrates our simulation results.

IV. SIMULATION RESULTS AND COMPARISONS FOR ALTERNATIVE CED ARCHITECTURS OF LOGAIRHIMTC CONVERSIONS

We have coded our proposed alternative logarithmic converters for Type I to IV using Verilog HDL; all are synthesized by TSMC 0.18μm standard cell library. Synthesis results and comparisons are given in Table I, where Type (A, B) is denoted as using Type A and Type B as implementations in

Log converter 1 and converter 2 in Fig. 2, respectively. As for area (delay) savings are all obtained compared with the maximum area (delay) costs among alternative implementations. According to Table I, we can observe that the implementation for Type (3,4) which is the most area and delay consumptions among all alternative implementations. Type (1,2), Type (1,4) and Type (2,3) consume the lowest area costs while Type (1,4) save the most delay cost (. Although Type (2,4) could achieve the fastest among them, the area cost is still higher than that of Type (1,2), Type (1,4) and Type (2,3).

Table I: Simulation results and comparisons for alternative CED implementations

Alternative implementations	Delay (ns)	Area (μm²)	Area Savings	Delay Savings
Type (1,2)	0.9	93.13	-61.65%	-28.00%
Type (1,3)	1.25	239.50	-1.37%	0.00%
Type (1,4)	0.88	93.13	-61.65%	-29.60%
Type (2,3)	1.09	93.13	-61.65%	-12.80%
Type (2,4)	0.85	182.95	-24.66%	-32.00%
Type (3,4)	1.25	242.82	0.00%	0.00%

V. CONCLUSIONS

In this paper, low-cost CED schemes for logarithmic converters are proposed. By comparing our proposed six alternative schemes, we can choose the candidate scheme with most area and delay consumptions. The proposed schemes can be applied to real-time computation-intensive computations for achieving high-speed logarithmic conversions with CED ability.

ACKNOWLEDGEMENTS

This work was supported partly by the Ministry of Science and Technology (MOST) in Taiwan under contract number MOST 104-2221-E-153 -001 – and MOST105-2221-E-153 -007 -.

REFERENCES

[1] R. Baumman, "Soft Errors in Advanced Computer Systems," IEEE Design and Test of Computers, vol. 22, no. 3, pp. 258-266, May/June 2005.

[2] M. Nicolaidis, "Design for Soft Error Mitigation," IEEE Trans. Device and Materials Reliability," vol. 5, no. 3, pp. 405-418, Sept. 2005.

[3] M. Hunger and D. Marienfeld, "New Self-Checking Booth Multipliers," Int'l J. Applied Math. and Computer Science, vol. 18, no. 3, pp. 319-328, 2008.

[4] B. Shim and N. Shanbhag, "Energy-Efficient Soft Error-Tolerant Digital Signal Processing," IEEE Trans. Very Large Scale Integration Systems, vol. 14, no. 4, pp. 336-348, Apr. 2006.

[5] Salvatore Pontarelli, Pedro Reviriego, Mbr., Chris J. Bleakley, and Juan Antonio Maestro, Mbr., "Low Complexity Concurrent Error Detection for Complex Multiplication," IEEE Transactions on Computers, vol. 62, no, 9, pp. 1899-1903, September 2013.

[6] Tso-Bing Juang, Sheng-Hung Chen and Huang-Jia Cheng, "A Lower-Error and ROM-Free Logarithmic Converter for Digital Signal Processing Applications," IEEE Transactions on Circuits and Systems II, vol. 56, no. 12, pp. 931-935, December 2009.

Digital Image Preprocessing and Hair Artifact Removal by using Gabor Wavelet

Uzma Jamil

Department of Computer
Engineering, Bahria University,
Islamabad, Pakistan.
uzma_gcuf@yahoo.com

Shehzad Khalid

Department of Computer
Engineering, Bahria University,
Islambad, Pakistan.
shehzad_khalid@hotmail.com

M.Usman Akram

National University of Sciences &
Technology, Islamabad, Pakistan.
usmakram@gmail.com

Abstract—**Diagnosis of skin cancer needs specialized equipment, doctors and continuous monitoring to treat it well. Patients living in remote areas normally cannot access such facilities. To overcome these barriers of access, Computer Aided Diagnostics, an emerging field in computer science, often called telemedicine, is being considered a promising approach. In this research, preprocessing and hair artifact removal experiment was performed on dermatoscope images by using Morphological and Gabor wavelet-based techniques. It has been found that, in some cases, wavelet transformations provide better results as compared to other techniques like gel, water bubbles and dark hair around the surface affected by cancer, i.e. these artifacts are removed with less effort.**

Index Terms— **Illumination, Melanoma, dermoscopy, image enhancement, hair Inpainting, hair segmentation, skin cancer.**

INTRODUCTION

Computer added diagnostic (CAD) system has three important steps i.e. image segmentation, feature extraction and classification [1]. Better segmentation leads better features extraction, classification and diagnosis [3]. In order to perform skin lesion segmentation effectively some predefined post and pre-processing steps are required and these steps are discusses in detail by Abbas [2]. As dermoscopic image contain some artifact like hairs, gel, skin lines, dark spots and water deposits which can affect the segmentation results. In order to handle these artifact some per-processing steps of segmentation are applied and then perform the segmentation.

RELATED WORK

To perform skin lesion segmentation effectively some pre defined post and pre-processing steps are required and these steps are discusses in detail by Abbas [2]. Abbas discussed these steps in detail. Illumination correction is a process to adjust image color so that segmentation process could be performed effectively , as skin lesion dermoscopic image could have low tones and smooth transition between healthy and effected skin [3, 4] which cause the segmentation task difficult. Illumination estimation is a first step to be performed and this concept is widely researched. The earlier algorithm used this model is called retinex algorithm. It use Gaussian filters on image to estimate the illumination component. In skin lesion images segmentation, one of the most important task is to remove the irrelevant artifacts from the image as these artifacts could leads to wrong segmentation, so different approached has been developed and tested. Skin lesion may contain different artifacts which can cause the bad segmentation results. These artifacts could be Hairs, Blood Vassals, Water Bubbles or dark spots. However hairs are the most challenging task in image segmentation in order to remove them different approaches are purposed [2,4].

METHODOLOGY

Proposed Methodology for dermoscopic image for this paper is shown as in figure 1.

A. Input Image to the System

We have used European and PH2 data-sets as input images to the proposed system. Images are well diversified in nature and contain number of artifacts which can cause the segmentation a hard task.

B. Image Pre-Processing

The first step in image segmentation is to prepare the image for segmentation by pre-processing step. Details of pre-processing steps performed are given below.

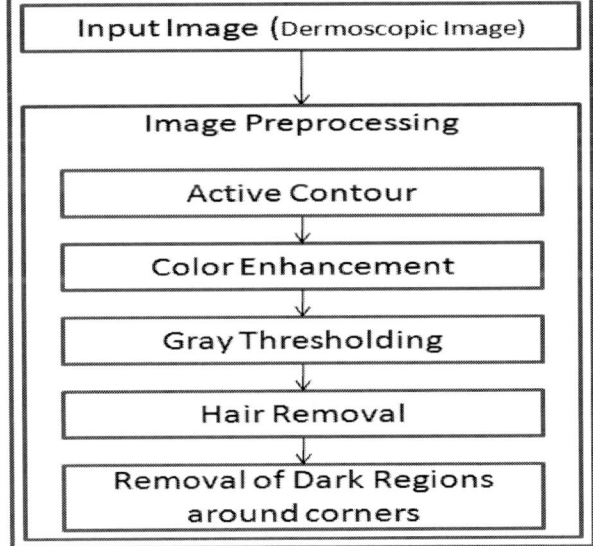

Figure 1. Proposed Methodology

- Active Contour

  An active contour or a simple elastic snake can be represented by the energy function defined by Eq.1.

  Esnake = $\int 1$ o Esnake(V (s)) = $\int 1$ o (Einternal(v(s)) + Eimage(V (s)) + Econ(v(s)))ds (1)

  Energy function of snake is sum of its external and internal energy. As internal energy Einternal is composed of continuity in contour and smoothness of contour and Esnake is the sum of all the forces due to the image itself E image and the constant force introduce by user i.e. Econ .

- Color Enhancement

  Color Enhancement is next step where RGB values are linearly combined into the single value called Luminance on the basis of following formula presented in Eq.2.

  Luminance = R * 2989 + G * 5870 + B * 114 (2)

- RGB with Highest Entropy

  In this method the entropy of each color is computed on the basis of following formula as given in Eq.3.

  $E(c) = \sum_{i=0}^{L-1} p(i) . \log_2 p(i)$ (3)

  after the RGB color component with highest entropy is selected as in Eq. 4.

  $i = \arg \max S(i)$ (4)

- Selection of L*A*B Color

  L*a*b color space by breaking it into L, a and b component has also been experimented as L*a*b is representation of CIE L*, u*, v* color space where L represent Lightness and a, b are color opponent dimensions.

- Blue Color Selection

  During this experiment it is found that the selection of blue color gives batter result as compare to other techniques as blue component gives clear color segmentation between the lesion and normal skin.

- Gray Thresholding

  We used Otsu's method to find the intensity which exhaustively finds for the threshold to minimize the intra-class variance as below in Eq. 5.

  $\sigma^2_x (t) = w1 (t) \sigma^2_1 (t) + w2 (t) \sigma^2_2 (t)$ (5)

C. Hair Artifact Removal

The next step is to remove the hairs from the image as hairs can affect the overall performance of skin lesion segmentation adversely. The hair removal process involves three steps i.e. hair improvement, hair segmentation and hair in-painting presented in Figure 2.

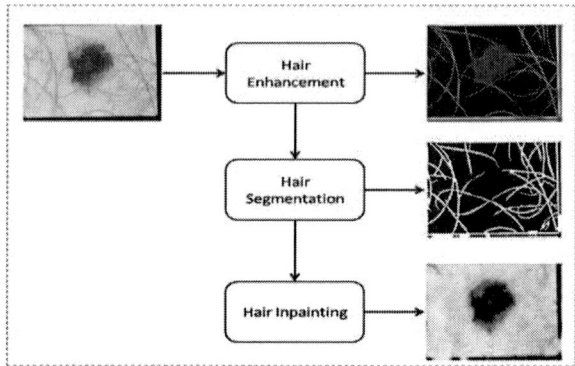

Figure 2. Flow diagram for hair artifacts removal algorithm

D. Detection of four Dark Corners

Otsu method is a thresholding technique to separate the image and background by apply the threshold. The binary mask created by this method is used to mask the dark corners, and is also used to determine dark corners in the image.

CONCLUSION

In this paper, the proposed approach handles the problems of unwanted artifacts such as hairs and tiny vessels by employing directional wavelet filters, enhancing and detecting pixels representing these artifacts. The detected hairs and vessel pixels and filtered and the missing pixels are inpainted using neighborhood information. The problem of uneven luminance problem is also addressed by estimating non-uniform illumination and performing equalization in luminance information. This article presents the information that will help investigators to critic the importance of high-level concepts which need to be more of an effort to make the correct diagnosis of skin cancer.

ACKNOWLEDGMENT

We are really thankful to Higher Education Commission of Pakistan to give the indigenous Phd scholarship to Ms. Uzma Jamil to complete her studies that is the part of this research article.

REFERENCES

[1] M. E. Celebi Q. Abbas and I. F. Garcia. Hair removal methods: A comparative study for dermoscopy images. Biomedical Signal Processing and Control, 6(4):395404, 2011.

[2] M Kruk1, B Świderski1, S Osowski, J Kurek1, M Słowińska and I Walecka, Melanoma recognition using extended set of descriptors and classifiers, Journal on Image and Video Processing, DOI 10.1186/s13640-015-0099-9, 2015.

[3] U Jamil, S Khalid, Analysis of Valuable Techniques and Algorithms Used in Automated Skin Lesion Recognition Systems, International Journal of Privacy and Health Information Management, 3(2), 95-111, July-December 2015.

[4] O Abuzaghleh, B. D. Barkana, M. Faezipour, Noninvasive Real-Time Automated Skin Lesion Analysis System for Melanoma Early Detection and Prevention, IEEE Journal of Translational Engineering in Health and Medicine, Vol.3, 2015.

A Flexible Software Defined Radio-based UHF RFID Reader Based on the USRP and LabView

Wang Yuechun, Ka Lok Man
Department of Computer Science
and Software Engineering
Xi'an Jiaotong-Liverpool University
Suzhou, China
yuechun.wang@xjtlu.edu.cn,
ka.man@xjtlu.edu.cn

Robert G. Maunder
School of Electronics and Computer
Science
University of Southampton
Southampton, United Kingdom
rm@ecs.soton.ac.uk

Jin Kyung Lee, Kyung Ki Kim
Department of Electronic
Engineering,
Daegu University, Gyeongsan,
South Korea
jklee@live.daegu.ac.kr,
kkkim@daegu.ac.kr

Abstract— **This paper presents a UHF RFID Reader designed for recognition and tracking in IoT domain. It is built by NI USRP software radio platform and NI LabVIEW with flexible physical/MAC layer parameters, which can be modified easily and monitored clearly from front panel of this Reader compared to commercial RFID Reader. Queried random number sequence from a commercial Tag can be detected within half meter using this UHF Reader. All designs of this Reader are based on EPC Gen-2 RFID protocol, any further research based on this Reader can be easily connected and tested with commercial Tags.**

Keywords; UHF RFID Reader, EPC Gen-2 RFID protocol, NI USRP, LabVIEW, Internet of Things

I. INTRODUCTION

Radio frequency identification (RFID) is one of the techniques that widely applied in recognition and item-level tracking tasks in Internet of Things domain [1]. UHF RFID supports significant wider read range, higher inventory rates, and rewritable product IDs [2]. Protocols used by RFID community in recent years are standardized for Electronic Product Code™ (EPC) to support the use of RFID [3]. Some researches aimed to provide complete control of physical and MAC layer based on these protocols have been done with NI USRP and Python-programmed GNU Radio [4,5], which will be challenge for some researchers. As a tool for teaching, LabVIEW is much easier with the advantages of using visual programming language.

This paper presents a UHF RFID Reader using USRP with LabVIEW software. An attempt to build flexible RFID Reader for less C++ experienced researcher is presented in this paper. A UHF RFID Reader that can achieve several basic functions such as generating and sending commands, receiving and analyzing responses is presented. This Reader has flexible controlled parameters, which fits for studying principle of RFID Reader's operation on physical and MAC layer.

II. FLEXIBLE RFID READER

In this section, the theory and implementation of designed RFID Reader based on block diagram of Reader-Tag system (Figure 1) is presented. This system contains 6 parts and the Reader parts are core of this system.

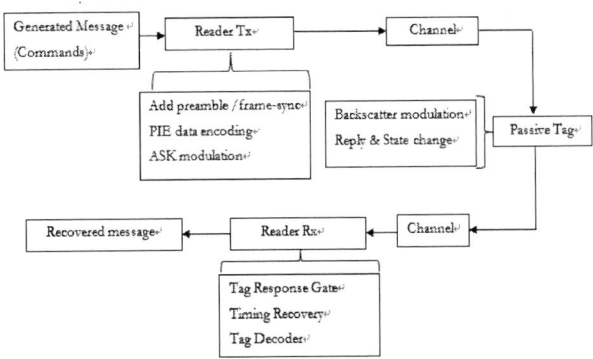

Figure 1. Block diagram of Reader-Tag system

A. Overall Transceiver

Structure of hardware part is built with two NI USRP, two RFID Circular antennas (PN6-868LCP/RCP), and one RFID passive tag (EURUHFT4928). Besides, Gigabit Ethernet cable is used to connect USRP with host PC. Compatible MIMO cable is used to link two USRPs and share the Ethernet. Antennas are PN6-868LCP/RCP which are chosen based on requirements of Gen-2 protocol. Avonwood EURUHFT4928 Tag is used for all experiments in this paper.

B. Reader transmitter

There are two functions supported by Reader transmitter: transmitting commands and providing energy to power up Tag.

As one significant character of passive Tag is that it acquires energy from transmitted signal of Reader, which denotes that Reader is required to transmit continuous waveform (CW) between commands in order to keep Tag powered up. The simplest method is to keep USRP transmitting sequence of ones when the Transmit button is released. The size of transmitted continuous sequence is 100 bits in this paper. The choice of this value considered the system latency as well as the computing speed of LabVIEW. If this value is too high, it may cause the system latency because the receiver should receive all these ones before it stopped even when the expected signal has already been detected. If this value is too small, the USRP may stop working because the transmitter buffer becomes empty before the new command transmitted.

C. Reader receiver

Block diagram of Reader Receiver chain is illustrated in Figure 2. Reader receiver catches both signals transmitted by Reader transmitter and backscattered by passive Tags.

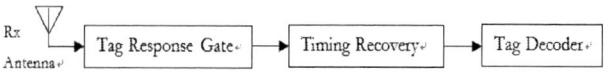

Figure 2.   Block diagram of Reader Receiver

Tag Response Gate is used to detect and pass the useful information sequence to the downstream block. The detection technique is setting threshold in this paper. This block can reduce the computation of receiver because timing recovery and tag decoding only deal with expected information sequence rather than entire received signal which may include quantity of latency time. Synchronization of tag response is done in Timing Recovery block by calculating the cross-correlation of known preamble sequence and received tag response. Mission of Tag Decoder is to decode received RN16 sample sequence. Supposing backscatter from tag uses FM0 encoding format, key point of decoding FM0 encoded sequence is to identify the level reversal within one symbol. By observing the sequence of neighbor samples' difference within every data bit, original bit stream can be decoded. To match the transmitter and receiver, IQ rate and Carrier frequency in Tx parameters and Rx parameters should be the same. In this paper, IQ sampling rate is fixed as 400 kSps and Carrier frequency is fixed as 867.6 MHz.

III.   OUTCOMES AND EVALUATION

Parameter setting of Tx and Rx is listed in Table I, and entire signal sequence caught by receiver is illustrated in Figure 3, including Query and ACK command, Tag backscattered RN16, and the extended waveform of Query and ACK respectively, which shows no response to ACK command from Tag.

TABLE I.   PARAMETERS SETTING OF TX AND RX

Parameters	Tx	Rx
Session Handle	10.0.0.3	10.0.0.2
IQ sampling rate (Sps)	400 k	
Carrier Frequency (Hz)	867.6 M	
Gain (dB)	25	0
Tari (µs)	25	-
# of samples	-	300

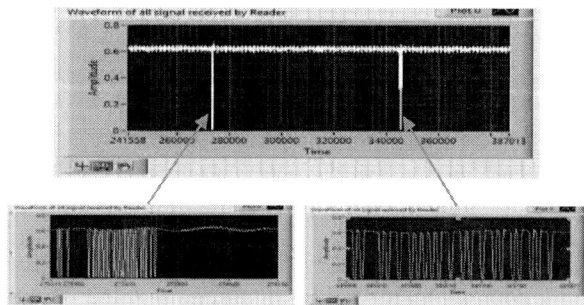

Figure 3.   Entire signal sequence received by Reader

Reason of failed receiving Tag ID sequence is the delay between transmitted ACK command and RN16 sequence. Based on Gen-2 protocol, the maximum link timing after the last bit of RN16 and before the first bit of ACK command is 20TPri, where TPri denotes duration of one FM0 symbol or a single subcarrier cycle. In this case, delay between RN16 and ACK was over 500 µs/ 200 samples, Tag will ignore ACK and no response will be backscattered.

To solve this problem, two attempts have been done. One attempt is to reduce the number of samples in each received frame. Another is to decline the size of array that transmitted to provide CW to Tag. Both are tried to decline that length of transmitted CW did not carry any information but just provide energy to Tag. However, when collecting the overall signal information using a feedback node in LabVIEW, both of these two attempts slowed done PC computing speed which finally resulted in LabVIEW stopped working because the transmit buffer became empty before the new data was provided. Besides, there are still several potential solving methods: one method is to use the 'Abort' function in LabVIEW which will stop receiver USRP keep receiving when expected information had been detected. This can allow transmitter sending ACK command as quick as possible after RN16 being detected rather than waiting all the CW sequences in transmit buffer have been received. Another method is to program LabVIEW in a lower level such as writing some C codes to allow the USRP sending CW rather than keep silence. If possible, using faster PC may solve this problem to a certain extent as well.

IV.   CONCLUSION AND FUTURE WORK

Implemented system achieved function of sending correct Reader-to-Tag commands under any adjustable parameter based on Gen-2 and decoding received Tag response. Due to computing speed of PC, response to ACK command from Tag has not finished till writing this paper. The potential research directions are studied and would be (1) to test actual environmental performance of Reader under variety communication channel by adding simulated noise to received signal in LabVIEW, (2) to build a passive Tag using USRP to simulate channel and two more USRPs may be used to act as transmitter and receiver of Tag.

REFERENCES

[1] Kubo, The Research of IoT Based on RFID Technology, in 2014 7th International Conference on Intelligent Computation Technology and Automation (ICICTA). 2014, 2014 7th IEEE International Conference: Changsha. p. 832-835.

[2] Chiu, S., et al., A 900 MHz UHF RFID Reader Transceiver IC. Solid-State Circuits, IEEE Journal of, 2007. 42(12): p. 2822-2833.

[3] Buettner, M. and D. Wetherall. A Flexible Software Radio Transceiver for UHF RFID Experimentation: UW TR: UW -CSE- 09- 10- 02.

[4] Buettner, M. and D. Wetherall. A software radio-based UHF RFID reader for PHY/MAC experimentation. in RFID (RFID), 2011 IEEE International Conference on. 2011.

[5] GS1, EPC Radio-Frequency Identity Protocols Generation-2 UHF RFID Specification for RFID Air Interface Protocol for Communication at 860 MHz - 960 MHz Version 2.0.1 Ratified. 2015.

Radio Frequency Energy Harvesting Technology

Lanxiang Wang[1], Menglong He[1], Zhao Wang[1], Mark Leach[1], Jingchen Wang[1], Kalok Man[2], Eng Gee Lim[1]

[1]Dept. of Electrical and Electronic Engineering,
[2]Dept. of Computer Science and Software Engineering,
Xi'an Jiaotong-Liverpool Univeristy
Suzhou, P.R.China
Enggee.lim@xjtlu.edu.cn

Abstract— **Radio frequency (RF) energy harvesting is a promising technique to energize low power electronic devices due to the sustainability it could offer resulting from the surge in ambient wireless signals it could utilize. In this paper, a concise literature survey on this technique is presented. Firstly, the architecture of a RF energy harvesting network is briefly introduced. Secondly, background relating to the antenna and rectifier designs is provided. Finally, some state-of-the-art designs from recent years are presented.**

Keywords: RF energy harvesting; Rectenna; Antenna; Rectifier.

I. INTRODUCTION

With the appearance of an increasing number of wearable electronic devices, the advent of Internet of Things (IoT), and the maturity of RF identification (RFID), the demand for suitable power supplies for such electronic devices has surged. Harvesting energy from the environment to satisfy this demand is considered to be an eco-friendly and self-sustainable method. A variety of energy sources have been utilized, such as solar energy, thermal energy, piezoelectric energy, and radio frequency (RF) energy. Among these sources, the RF energy harvesting technique has recently received increasing interest due to the plentiful and unutilised ambient wireless signals in the environment, produced by sources such as TV towers, cellular base stations, and wireless routers [1]. The amount of energy available for reception could be used to energize low power electronic devices; this is of particular convenience when it is difficult or even dangerous to replace batteries. The crucial component in the RF energy harvester is the rectenna, which is composed of an antenna and a rectifying circuit. A rectenna can collect electromagnetic energy and convert it into DC power.

This paper briefly reviews the development of the rectenna based RF energy harvesting techniques and is organized as follows. Section II demonstrates the architecture of a rectenna RF energy harvesting network. Section III focuses on several typical antenna and circuit designs. Finally, a brief conclusion is drawn in Section IV.

II. RF ENERGY HARVESTING NETWORK

A typical architecture for an RF energy harvesting network is shown in figure 1, which consists of three main components including an antenna, a matching network, and a rectifier. The receiving antenna can harvest ambient RF energy from various transmitting sources. The RF energy can be roughly described by both the determinate Friis equation model and the

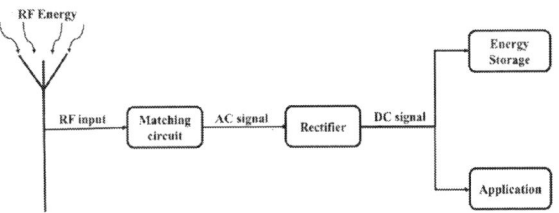

Fig. 1: RF Energy Harvesting Network

probabilistic Rayleigh model. According to [2], the latter is considered to be more realistic and practical. A matching network is used to ensure maximum power is delivered from the antenna into the rectifier section and avoiding power reflection [3]. The function of a RF-DC converter is to convert the AC type RF energy received by an antenna into a DC voltage, which can be used to supply electronic devices directly or to be stored for future use.

III. RECTENNA DESIGN

A. Antenna

The antenna, a vital element of an RF energy harvesting system, is used to capture radiated RF energy from sources transmitting in the antennas design bandwidth. The designs of antennas vary for use in different situations including dipole antennas, microstrip patch antennas, etc. [4]. The designs are optimized to minimise size and improve antenna gain. However, these two main requirements cannot always be fulfilled simultaneously. Antenna such as the Yagi-Uda or multiband antenna arrays could provide the relatively high gains desired for a rectenna design, but are generally large in size. The overall dimensions need to be miniaturized for deployment and application in mini-type devices including wearable devices and unmanned aerial vehicles (UAV) [4][5]. Furthermore, the bandwidth and cost requirements should also been considered in design.

The sources of signals used for harvesting cover a variety of frequency bands and include more recent broadcasts such as: GSM, UMTS, LTE, ISM and wifi, as well as more traditional broadcasts like FM/AM radio and television [5]. Focussing on the arears of the RF spectrum with relatively higher spectral density, maximises harvesting potential. Early receiving antenna designs of mainly focused on the reception of a single frequency band [6], while recently dual band and tri band antennas are also obtaining popularity in some works [5][7].

Research into antenna designs for RF energy harvesting systems has primarily focussed on microstrip based antennas due to their lightweight, robustness, flexibility and low cost [4]. Patch antennas are widely applied in the design of wearable devices. The author in [8] designed a microstrip patch antenna operating over the frequency range 4-9.5 GHz which exhibited an average gain above 7.4 dBi and efficiency of more than 85%. The microstrip patch antenna in [9] with wearable substrate was designed to operate in the ISM band (2.4 GHz). Moreover, the fractal Koch antenna is considered to be an efficient methodology with smaller size when compared to the more common Euclidean geometry based antennas, and a typical fractal geometry is shown in Fig 2a. The author in [5] has demonstrated a double-loop Koch antenna which was optimized to operate from 0.9-2.4 GHz.

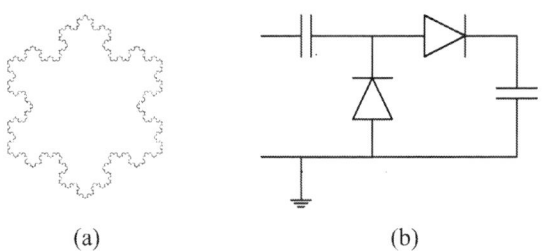

(a)   (b)

Fig. 2: Schematics of the (a) typical Koch fractal geometry and (b) single-stage voltage doubler.

B. RF Energy Harvesting Circuit

The RF energy harvesting circuit is composed of a matching network and a RF-DC converter. The biggest challenge when designing the RF-DC converter is to generate a suitably high DC output voltage from a very low input RF power. Diodes are typically used for rectification and they themselves as well as the circuit configuration can affect the conversion efficiency dramatically. Hence, in order to drive electronic devices, the circuit should be designed carefully.

Generally, there are two main categories for the RF-DC converter. One uses a single-ended topology; the other a differential topology [10]. Single-ended rectifiers, like commonly used voltage multipliers, have two basic schemes: Villard cascade (or Cockcroft-Walton) voltage multiplier and Dickson cascade voltage multiplier. Both schemes are based on the same "building block" that is shown in Fig. 2 (b); the Villard voltage multiplier uses series block connection, while Dickson voltage multiplier uses parallel block connection. Ideally, any voltage level can be achieved by adding enough stages.

A differential rectifier, usually realized in CMOS, is comprised of a cross-coupled diode bridge configuration driven by a differential RF input. This scheme exhibits a low activation threshold, which results in high conversion efficiency, especially when operating at low power levels [10].

In [11], the author proposed an improved rectifier circuit based on a conventional voltage multiplier mentioned above. The conversion efficiency has been optimized by introducing an extra inductor, which aims to store energy in a magnetic field during the negative cycle and discharge in the subsequent positive cycle. This can generate an opposition voltage polarity limiting current variation. As a result, the peak conversion efficiency of the proposed rectifier circuit is about 81.65% for 0 dBm input power at 868 MHz. Another design involving the differential rectifier can be found in [12], where an extra DC-DC converter was combined with a 3-stage differential rectifier to boost the output voltage without considerably jeopardizing efficiency. In addition, the start-up voltage of this scheme is low at 200 mV and the power consumption is less than 10 nW. This harvester has 68% peak efficiency at 900 MHz.

IV. CONCLUSION

This paper gives a brief review of RF energy harvesting techniques. The general structure of the RF energy harvesting system has been introduced in Section II. Both antenna and circuit designs have been classified and discussed in Section III. The gain of the antennas and conversion efficiencies of circuits can provide sufficient energy to power typical wireless sensor platforms and achieve self-sustainability.

REFERENCES

[1] Kim, Sangkil, et al. "Ambient RF energy-harvesting technologies for self-sustainable standalone wireless sensor platforms." Proceedings of the IEEE 102.11 (2014): 1649-1666.

[2] T. K. Sarkar, J. Zhong, K. Kim, A. Medouri, M. Salazar-Palma, "A survey of various propagation models for mobile communication," IEEE Antennas and Propagation Magazine, vol. 45, no. 3, pp. 51-82, June 2003

[3] J. Jose, S. George, L. Bosco, J. Bhandari, F. Fernandes, and A. Kotrashetti, "A review of RF energy harvesting systems in India," 2015 Int. Conf. Technol. Sustain. Dev., pp. 1–4, 2015.

[4] C. A. Balanis, Antenna theory: Analysis and design. 4rd ed. John Wiley & Sons, 2016.

[5] Volakis J L, O'Brien A J, Chen C. C. "Small and Adaptive Antennas and Arrays for GNSS Applications" Proceedings of the IEEE, 2016, 104(6): 1221-1232.

[6] A. Georgiadis, S. Member, G. Andia, and A. Collado, "Rectenna Design and Optimization Using Reciprocity Theory and Harmonic Balance Analysis for Electromagnetic (EM) Energy Harvesting," vol. 9, pp. 444–446, 2010.

[7] Hebelka V, Velim J, Raida Z. "Dual band Koch antenna for RF energy harvesting" 2016 10th European Conference on Antennas and Propagation (EuCAP). IEEE, 2016: 1-3.

[8] Simorangkir, R. B., Abbas, S. M., & Esselle, K. P. "A printed UWB antenna with full ground plane for WBAN applications" 2016 International Workshop on Antenna Technology (iWAT). IEEE, 2016: 127-130.

[9] Khan, H. A., Ullah, S., Afridi, M. A., & Saleem, S. "Patch antenna using EBG structure for ISM band wearable applications" 2016 International Conference on Intelligent Systems Engineering (ICISE). IEEE, 2016: 157-160.

[10] H. Dai, Y. Lu, M. K. Law, S. W. Sin, U. Seng-Pan, and R. P. Martins, "A review and design of the on-chip rectifiers for RF energy harvesting," 2015 IEEE Int. Wirel. Symp. IWS 2015, pp. 3–6, 2015.

[11] I. Chaour, S. Bdiri, A. Fakhfakh, and O. Kanoun, "Modified Rectifier Circuit for High Efficiency and Low Power RF Energy Harvester," 13th Int. Multi-Conferrence Syst. Signals Devices, pp. 619–623, 2016.

[12] D. Michelon, E. Bergeret, A. Di Giacomo, and P. Pannier, "RF energy harvester with sub-threshold step-up converter," 2016 IEEE Int. Conf. RFID, pp. 1–8, 2016.

Skew Control Methodology for Useful-Skew Implementation

SangGi Do
Electrical and Computer Engineering
UNIST
Ulsan, Korea
sanggido@unist.ac.kr

Seungwon Kim
Electrical and Computer Engineering
UNIST
Ulsan, Korea
kskyh002@unist.ac.kr

Seokhyeong Kang
Electrical and Computer Engineering
UNIST
Ulsan, Korea
shkang@unist.ac.kr

*Abstract— **Skew optimization** is an important stage of the physical design. Previous studies suggested various skew optimization algorithms [1-7]. However, many of them have only focused on the zero-skew optimization [1-3], and several recent studies focus on a useful-skew optimization [5-7]. In this paper, we propose a novel skew optimization method for useful-skew implementation. Our proposed method generates optimal skew values, and applies them to a clock tree without any buffer insertion.*

Keywords: Useful-skew, CTS, skew optimization, skew control

I. INTRODUCTION

A skew optimization has recently gained great concentration in a physical design flow [10]. In this paper, we verify the possibility of skew control for useful-skew implementation. The skew is the relative delay differences between different clock paths. As clock signal goes from source to sink, every path has different route, wire length and wire capacitance. This difference makes delay differences from source to sinks, and this is called a *clock skew*. Since the skews are the delay difference, it can be estimated from a delay calculation model. With the skew estimation, Edahiro's method [1] finds a zero-skew point between two different nodes. However, the useful-skew technique requires additional clock skews to optimize the design further. In this work, we propose a clock skew control methodology for the useful-skew implementation. The skew estimation is based on FLUTE [8] routing method with HSPICE simulation. We use π-delay model for wire delay calculation and visualize our experimental result by MATLAB.

A. Related work

Previous skew optimization studies mostly proposed zero-skew methods [1-3]. One of the well-known method is Edahiro's zero-skew optimization [1]. They proposed two different algorithms; nearest neighbor selection (NS) algorithm and clustering based (CL) algorithm. Both algorithms have find-center and embedding stages. The embedding stages determines the best position for each node found from the previous find-center stage. However, in case of find-center stage, NS uses the nearest neighbor method and CL uses node weighting method to get node set for embedding. For node weighting, they use distance between two nodes and ordering difference.

There exist several useful-skew optimization methods [5-7]. One of popular method is NOLO [5]. The NOLO algorithm predicts an optimal useful-skew before the placement stage and uses the predictive useful-skew for clock skew without any iteration. Because of no iteration, they reduce runtime of the clock skew implementation. In addition, the NOLO optimizes

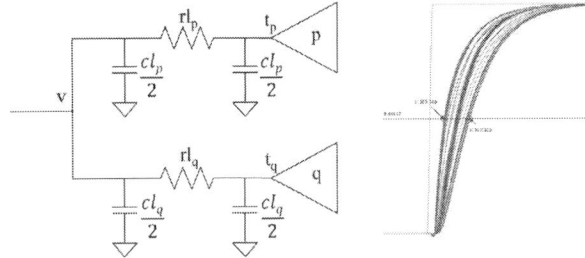

Figure 1.(a) π-delay model for our algorithm   (b) HSPICE simulation

data path delay since the predictive useful-skew can be used for placement, routing and CTS (Clock Tree Synthesis) stages. To obtain the predictive useful-skew, they use MMWC [9] (Maximum Mean Weight Cycle) method.

B. Our contribution

In this paper, we propose a novel clock skew control methodology. Our skew control is based on a topological CTS without any buffer insertion, which reduces area and power overheads from the clock buffers.

The rest of this paper is organized as follows. Section II describes the purposed skew control methodology and equations for delay calculation. Section III provides experimental environment and results. Section IV summarizes and concludes the paper.

II. SKEW CONTROL METHODOLOGY

The basic concept of our methodology is similar to NS algorithm of Edahiro's method [1]. In NS algorithm we have modified embedding stage to consider useful-skew. Our proposed algorithm calculates delay to generate intended skew for a useful-skew implementation. Algorithm 1 shows our proposed algorithm. First, we order S by target skew (Line 1) and search available point q from set S (Lines 3-7). Delta skew and maximum skew are calculated from Equation (2) and (3). Equation (1) calculates merging point of the previous point set (Line 8). These sequences are continued until the size of set S becomes one (Line 2). Figure 1 shows a delay calculation model in our experiments. r is the unit resistance per length, c is the unit capacitance per length, t_p is skew at p leaf, and t_q is skew at leaf, q. Point v is an optimal center point considering skews. l_p and l_q are the distance between v and p/q point.

EQUATIONS

$$t_v = rl_p(\, cl_p + C_p\,) + t_p = rl_q(\, cl_q + C_q\,) + t_q \tag{1}$$

$$\Delta t = rl_q(\, cl_q + C_q\,) - rl_p(\, cl_p + C_p\,) \tag{2}$$

$$\Delta t_{max} = rL(\, cL + C_{load}\,) \tag{3}$$

ALGORITHM 1: Build_Tree_NS
Input : Set of node S
Output : Tree TR
1   sort S by decreasing order of target skew
2   **while** size of $S > 1$
3   set p as most positive skew point from S
4   **for** point $q \in S$ without p
5   **if** $\Delta t < \Delta t_{max}$
6   set q and **break**
7   **end for**
8   find optimal point v of (p , q)
9   remove p and q from S
10   Insert r to S
11 **end while**
12 return TR

III. Experimental Result

Our first experiment is a target skew estimation. We generate four different simulation cases in terms of node layout. One is uniformly distributed and others are randomly distributed. All cases have 49 nodes and die size is $300um \times 300um$. Wire segments are divided by $1um$ unit wire and wire characteristics are same with Metal-1 of TSMC 65nm library. The routing tool is FLUTE [8]. Input signal transition time is 10ps, and VDD is 1.2V. Figure 1(b) shows the average skew result of four cases. From this result, we can see that target skew range is 40ps. The second experiment is to verify our skew control methodology. We run our algorithm with a uniformly distributed case generated from the previous experiment. The target skew values for each node are randomly generated between -20ps to +20ps. Figure 2 shows our algorithm results from each stage. Figure 2(a, b) represent find-center stage and Figure 2(c, d) represent embedding stage. In Figure 2, red and blue color areas show positive and negative target skew, respectively.

Figure 2 shows that our algorithm successfully controls the intended skew. As the algorithm continues from Figure 2(a) to Figure 2(d), each stage converts the node set to a new node. This converted node has a new target skew. From MATLAB visualization, we can figure out that each stage removes the red area; this means that intended skews are successfully generated for new merging nodes. Figure 3(a) and 3(b) show the final routing results from Edahiro's and our approach, respectively. From the results, total wire length of our result is 59% longer than Edahiro's result. However, our method completely generates the intended target skews for every node with the additional wire length.

IV. Conclusion

In this paper, we have successfully verified that clock skews can controlled with wire connections without any buffer insertion. Our proposed algorithm generates intended target skews, and it can successfully optimize clock skews. However, proposed algorithm increases the wire length compared with Edahiro's zero-skew optimization methods. The increased wire lengths can affect to a latency of the clock signal. However, in clock tree, the relative delay difference among clock paths is more important than the clock latency, and our approach can reduce the area and power overheads from additional clock

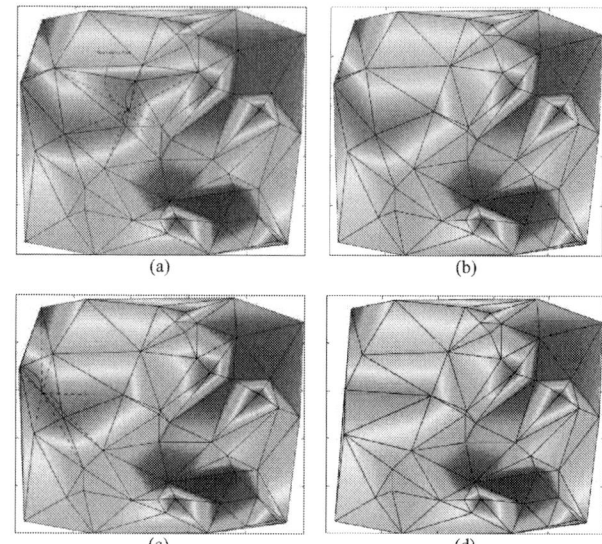

Figure 2. find center(a, b) and Embedding(c, d) stage

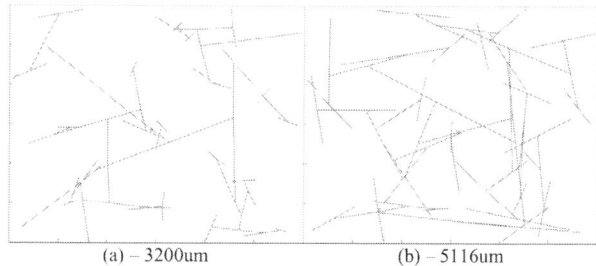

(a) – 3200um   (b) – 5116um

Figure 3. routing result of Edahiro's and Ours

buffers. As a result, our proposed method is effective for the skew optimization.

References

[1] M. Edahiro, "A Clustering-Based Optimization Algorithm in Zero-Skew Routings", *Proc. DAC*, 1993, pp. 612-616.

[2] K. D. Boese and A. B. Kahng, "Zero-Skew Clock Routing Trees with Minimum Wirelength", *Proc. ASIC*, 1992, pp. 17-21.

[3] T. H. Chao, Y. C. Hsu and J. M. Ho, "Zero Skew Clock Net Routing", *Proc. DAC*, 1992, pp. 518-523.

[4] D. J. Huang, A. B. Kahng and C. A. Tsao, "On the Bounded-Skew Clock and Steiner Routing Problems", *Proc. DAC*, 1995, pp. 508-513.

[5] T. B. Chan, A. B. Kahng and J. Li, "NOLO: A No-Loop, Predictive Useful Skew Methodology for Improved Timing in IC Implementation", *Proc. ISQED*, 2014, pp. 504-509.

[6] H. M. Chou, H. Yu and S. C. Chang, "Useful-Skew Clock Optimization for Multi-Power Mode Designs", *Proc. ICCAD*, 2011, pp. 647-650.

[7] S. Roy, P. M. Mattheakis, L. Masse-Navette and D. Z. Pan, "Clock Tree Resynthesis for Multi-Corner Multi-Mode Timing Closure", *IEEE Transactions on CAD*, 34(4) (2015), pp. 589-602.

[8] C. Chu and Y. C. Wong, "FLUTE: Fast Lookup Table Based Rectilinear Steiner Minimal Tree Algorithm for VLSI Design", *IEEE Transactions on CAD*, 27(1) (2008), pp. 70-83.

[9] C. Albrecht, B. Korte, J. Schietke and J. Vygen, "Maximum Mean Weight Cycle in a Digraph and Minimizing Cycle Time of a Logic Chip", *Discrete Applied Mathmatics* 123(1-3) (2002), pp. 103-127.

[10] A. B. Kahng, "New Game, New Goal Posts: A Recent History of Timing Closure", *Proc. DAC*, 2015, pp. 1-6.

Gap in pagination due to withheld paper.

Pages 223-224

μPnP-WAN: Wide Area Plug and Play Sensing and Actuation with LoRa

Fan Yang*, Gowri Sankar Ramachandran*, Piers Lawrence*, Sam Michiels*, Wouter Joosen*, and Danny Hughes*†

* iMinds-DistriNet, KU Leuven, 3001 Leuven, Belgium. Email: {first.last}@cs.kuleuven.be
† VersaSense NV, 3000 Leuven, Belgium. Email: danny@versasense.com

Abstract—The Internet of Things (IoT) is being applied in a wide variety of applications, which demand a range of networking support, including long range technologies. Unfortunately, emerging long range IoT platforms are difficult to deploy and configure for end-users who are not IoT specialists. This paper addresses this problem by introducing μPnP-WAN, which combines the ease of use of the μPnP peripheral system, with the long range LoRa network to realize the first true plug-and-play solution for long-range sensing and actuation. μPnP-WAN can achieve a range of up to 3.5 kilometers in ad-hoc suburban deployments and multi-year battery lifetime.

I. INTRODUCTION

The Internet of Things (IoT) has rapidly evolved from a concept to a broad marketplace of tangible products. There is a clear demand for monitoring physical phenomena at large scale and in a cost effective manner, to support application scenarios such as: smart buildings, smart cities, and smart factories. To minimise deployment complexity, IoT devices should automatically form a network, gather data from sensors and control actuators, while minimizing power consumption to guarantee a long battery lifetime.

To tackle the problem of plug-and-play integration of peripherals for embedded IoT devices, our prior work contributed μPnP [1]. μPnP features light-weight, low power hardware identification, platform independent driver software, and uses efficient multicast networking. As a technology, μPnP brings true plug-and-play peripheral integration, automatic driver installation, and extremely low power consumption. In [2], we extended μPnP with 802.15.4 mesh networking to provide reliable short range wireless sensing. In that paper we demonstrated battery lifetimes of over 6 years in realistic operating conditions.

This paper introduces the design of a new low-power, long range plug-and-play enabled platform for the IoT. The resulting system provides wide area plug-and-play sensing and actuation with multi-year battery lifetime and multi-KM range.

II. BACKGROUND

This section provides background on the two key technologies upon which μPnP-WAN is built: LoRa and μPnP.

A. Long range low power networking with LoRa

The LoRa Alliance technical workgroup defines both physical and data link standards for LoRa networking. At the physical layer, LoRa employs Chirp Spread Spectrum (CSS) to ensure robustness against interference and multi-path fading. The data rate of a LoRa network varies between 0.3 kbps and 50 kbps and has an inverse relationship with range. In an optimal deployment in free space, ranges of over 15KM have been reported.

The LoRaWAN specification [3] defines the frequency bands for LoRa communication. LoRa operates in the 868Mhz frequency band in Europe, 900MHz in Europe and 433MHz in Asia. In Europe, ETSI imposes strict guidelines about the use of various frequency bands. According to ETSI regulations, each end-device should follow a duty cycle between 0.1% and 10%. The FCC instead limits transmissions to 400ms in length.

LoRa end-devices are classified into: *Class A*, which support bi-directional communication, through short receive windows after an uplink transmission. *Class B* devices, which are operate using time-synchronised beacons from the gateway. *Class C* devices allow continuous reception of data due to its maximal receive slots. μPnP-WAN uses RN2483 in Class-A operational mode. The security of LoRa networks is ensured using AES-128 operating in CTR mode (RFC3602).

B. Plug and Play peripherals for the IoT: μPnP

In mainstream systems, the manual peripheral integration is no longer necessary thanks to the "plug-and-play" technologies, such as the Universal Serial Bus (USB), which auto-detect peripheral devices and automatically install drivers. However, mainstream approaches cannot be applied directly to embedded IoT devices due to excessive energy, computation and memory requirements.

μPnP [2] tackles the problem of plug-and-play IoT peripheral integration by providing a complete solution includes: 1) a low cost, low energy consumption hardware peripheral identification; 2) a platform independent software driver with a run-time environment that supports automatic driver downloading and installation; and 3) remote discovery and access based on IPv6 standard networking. μPnP achieves true plug-and-play peripheral integration, with a software stack that consumes less than 16kB of flash memory and 10 million times lower power than USB. In our previous work [2] we extended μPnP with support for IEEE-802.15.4 mesh networking, achieving 99.999% reliability and a battery life of over 6 years. Unfortunately, the μPnP

978-1-5090-3220-4/16 $31.00 © 2016 IEEE   225   ISOCC 2016

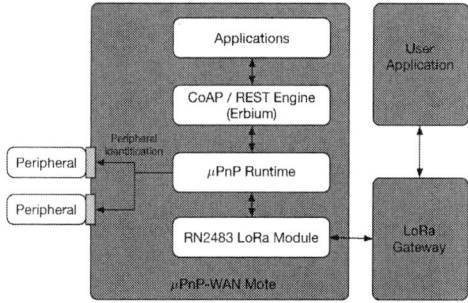

Figure 1.   Overview of the μPnP-WAN solution

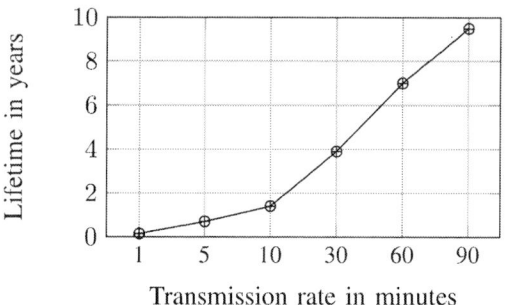

Figure 2.   Lifetime of μPnP-WAN mote varies with transmission frequency.

mesh solution offers limited range. This paper addresses this problem.

III.   μPNP-WAN

A. System Architecture

As can be seen in Fig. 1, there are three key elements in the μPnP-WAN system architecture: **Peripherals** are identified based upon four resistors that encode a 32-bit type identifier. Based on the type of peripheral, it is connected to the appropriate pins on the host micro-controller (μPnP currently supports ADC, I2C, SPI and UART peripherals [1]). Each **μPnP-WAN mote** therefore becomes a wireless hub for locally connected *peripherals*. The **LoRa gateway** receives the sensor data transmitted by the end-devices, which are arranged in a star topology. The LoRa gateway forwards the sensor data to the appropriate Internet application.

B. Low-power and long range WAN

μPnP-WAN uses Microchip's RN2483 module, which follows the LoRaWAN standard, realising a Class-A device. The interaction between μPnP and RN2483 happens via UART communication. The μPnP software stack realises a fully standards compliant network stack with IPv6 addressing using 6LowPAN (RFC4919) and application-level interaction using CoAP (RFC7952), which provides a lightweight REST implementation over UDP. The complete software stack is shown in Figure 1.

IV.   EVALUATION

We have built a production-ready prototype platform of μPnP-WAN using a 3v 3000mAh battery. The LoRA module was configured for maximum possible range (+14dBm tx power, spreading factor of 12).

A. Lifetime analysis

We calculate the battery lifetime of a μPnP-WAN device as shown in Figure 2 with varying transmission rates. The shelf life of a battery is 10 years. As can be seen from the graph, LoRa achieves the maximum possible battery lifetime when sending one message every 90 minutes and a battery lifetime of 5 years when sending a message once every 45 minutes.

B. Range and Reliability

We measured the coverage area and the packet delivery performance of μPnP-WAN in a suburban environment using indoor grade antennas, deploying μPnP-WAN devices in easy-to-reach locations in available buildings. Much longer ranges may be achieved by deploying motes on tall masts with very large antennas, but this frequently not possible.

In our tests, the coverage range of LoRa reached up to 3.5 kilometers, while maintaining a packet delivery reliability of over 95%, however it should be noted that range significantly in each direction due to the presence of tall buildings and interference sources. Our reliability results are far lower than μPnP-mesh (95% v 99.999%), but offers two order of magnitude greater range (3.5KM v 35M).

V.   CONCLUSION

In this paper we introduced the μPnP-WAN platform, which enables plug-and-play integration of peripherals such as sensors and actuators for the IoT, while supporting low power and long range communication through the LoRa communication protocol. Evaluation of our production-ready prototype demonstrates multi-year battery lifetime and multi-KM range.

ACKNOWLEDGMENTS

This research is partially funded by the Research Fund KU Leuven and the iMinds IoT program. The authors would like to thank VersaSense NV for the use μPnP hardware.

REFERENCES

[1] F. Yang, N. Matthys, R. Bachiller, S. Michiels, W. Joosen, and D. Hughes, "μPnP: Plug and Play Peripherals for the Internet of Things," in *European Conference on Computer Systems (EuroSys)*. Bordeaux, France: ACM, April 21-24 2015.

[2] N. Matthys, F. Yang, W. Daniels, S. Michiels, W. Joosen, D. Hughes, and T. Watteyne, "μpnp-mesh: The plug-and-play mesh network for the internet of things," in *Internet of Things (WF-IoT), 2015 IEEE 2nd World Forum on*, Dec 2015, pp. 311–315.

[3] *LoRaWAN Specification*, LoRa Alliance, 2015, rev. 1.0.

CAN FD Controller for In-Vehicle System

Jung Woo Shin, Jung Hwan Oh, Sang Muk Lee, and Seung Eun Lee[*]

Dept. of Electronic Engineering
Seoul National University of Science and Technology
Seoul, South Korea
*seung.lee@seoultech.ac.kr

Abstract— **In this paper, we propose a Controller Area Network with Flexible Data rate (CAN-FD) controller for communication network in automobile. The CAN FD is proper network protocol for in-vehicle and embedded communication which desires high reliability and data rate. We introduce our CAN FD controller which supports variable data length and faster data rate. The CAN FD controller receives data from virtual Electronic Control Unit (ECU) through Serial Peripheral Interface (SPI) and transmits the data to the system BUS. Our prototype has integrated along with processor for efficient control of the CAN FD communication.**

Keywords; CAN-FD; in-vehicle network; embedded network;

I. INTRODUCTION

Recently, real-time data communication with high data rate is the essential requirement for high performance system in the automobile and embedded system. Moreover, the internetworking among Electronic Control Units (ECUs) is getting more complicated due to the rapid variation of automotive and embedded devices. To keep pace with these variations, the in-vehicle network is becoming more important. The Control Area Network (CAN), which is a popular serial communications protocol, has been commonly used in the automotive industry over the past two decades due to high reliability and good real-time performance with low cost [1]. However, the increased use of electronic systems leads to inquire more bandwidth, performance, and low power consumption. In order to overcome these drawbacks of existing CAN protocol, a new communication protocol has come out, named CAN with Flexible Data rate (CAN-FD) [2]. The CAN-FD, which is based on CAN protocol, is easy to migrate from CAN. The CAN FD inherited admirable features of the existing CAN protocol such as arbitration, stuffing and acknowledge frame [3]. CAN FD also supports the extension of the data length and the accelerated data rate while transmitting data field over limitation of CAN (*1Mbps*).

In this paper, we present a CAN FD controller for in-vehicle network systems. One of the important features for in-vehicle and embedded network is the real-time data transfer with high throughput by adapting a faster data rate during the service of data transmission [4]. Thus, CAN FD not only provides variable data rate but also increases interconnection network performance [5]. In order to control the multiple ECUs of the in-vehicle system, we defines communication packet, which is transmitted through Serial Peripheral Interface (SPI) communication and converted to CAN FD frame. The CAN FD controller is implemented on FPGA and implementation demonstrates the feasibility of the CAN FD system.

II. PROPOSED SYSTEM

The Figure 1 illustrates the entire system structure. The proposed system consists of smart ECU and CAN FD controller. The CAN FD controller receive data from smart ECU in our SPI packet format. Then, CAN FD controller analyzes the data and converts to CAN FD frame. The CAN FD controller transmits the converted CAN FD data on the bus. Then, each CAN FD controller compares its own ID with data on the bus. The Figure 2 shows the organization of our SPI packet for in-vehicle network. The MSB 2 bits represent the *command* field. The *command* determines the frame type. The *mode* field indicates the type of the frame between CAN and CAN FD. The *extend* bit represents the extend mode. The *extend* field is '0' in base mode. Otherwise, *extend* bit is '1' in the extend format that supports 29 bits for ID length. The *length* field imply the data length to be transmitted right after the 4 fields.

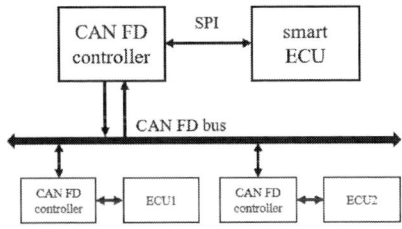

Figure 1. Entire system structure

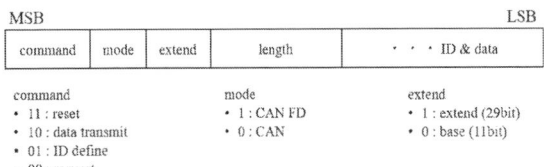

Figure 2. SPI packet format for our CAN FD controller

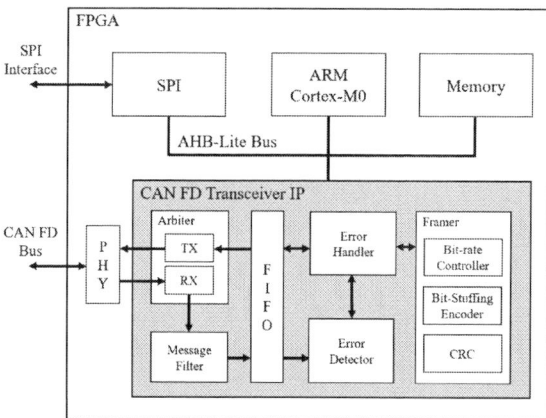

Figure 3. Block diagram of CAN FD controller

Figure 3 illustrates the hardware architecture of the proposed controller. The processor transmits SPI data to the CAN FD transceiver IP, and the data is transferred to CAN FD Bus. The CAN FD transceiver IP consists of AHB interface unit, arbiter, message filter, bit-stuffing encoder, error detector, error handler and framer. The arbiter monitors tx and rx bit on the bus and determines whether or not to send the data in order to solve data collision on the bus. Message filter compares own ID with the destination of the data. The framer includes 4 types of frame: *data*, *remote*, *error* and *overload* frame. The *data* frame is the frame for transmitting the data to the other node on the bus. Whenever the node requires to receive the data that another node possesses, the node sends the *remote* frame to request the desired data. When the error occurs, the node returns the *error* frame to receive correct data. Between *data* frames, delay is occasionally required. Because the CAN FD conducts continuous resynchronization to make sure whether the node received the message data correctly, *overload* frame is required in this circumstances. The stuffing bits are inserted after five successive bits of the same polarity in bit stuffing encoder. Reliability is one of the most important feature for automobile and embedded communication. Therefore, error detector and error handler conduct to detect and prevent errors for high reliable system.

III. IMPLEMENTATION

We implemented the CAN FD controller on FPGA. The figure 4 shows implemented hardware of CAN FD controller prototype and smart ECU. The CAN FD controller prototype is implemented into the Altera Cyclone III FPGA. We integrated ECU, which is emulated on Altera DE2-115 FPGA board, to demonstrate our design. The CAN FD controller receives SPI packet from SPI master, and transmits the data, suitable for the CAN FD frame. Another CAN FD controller receives data frame that matches with own ID on the CAN FD bus. Then, transmits the data to ECU. For real-time data analysis, the hardware scope monitors and analyzes the data frame on the bus. We verified that the data is successfully transferred on the CAN FD bus with using the hardware scope [6].

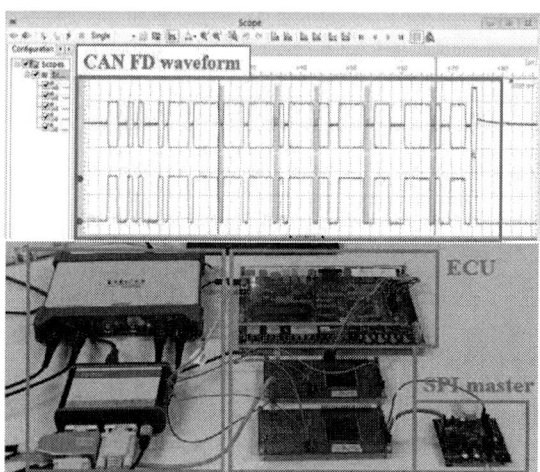

Figure 4. Implemented hardware of our CAN FD controller

IV. CONCLUSION

In this paper, we proposed CAN FD controller with high data rate for embedded and automobile network communication. Our system supports CAN FD data rate up to *2Mbps* and accomplished the data communication between the CAN FD controller and ECU successfully. We implemented CAN FD controller on FPGA. Experimental result demonstrates the feasibility of our proposal for in-vehicle and embedded communication network by using the hardware scope. We expect that our CAN FD controller is suitable for providing high reliable in-vehicle communication.

ACKNOWLEDGMENT

This study is supported by the Ministry of Trade, Industry & Energy (MOTIE, Korea) under Industrial Technology Innovation Program. No.10051106, 'Development of Band Type Wearable Device That Contains the Automotive Smart Key Function'

REFERENCES

[1] J. M. Flores-Arias, M. Ortiz-Lopez, F. J. Quiles-Latorre, V. Pallares and A. Chen, "Complete hardware and software bench for the CAN bus," 2016 IEEE International Conference on Consumer Electronics (ICCE), pp. 211-212, 2016.

[2] R. B. Gmbh, "CAN with flexible data-rat," Vector CANtech, Inc, Novi, Mi, USA, Specification Version 1.0, 2012.

[3] Jung Woo Shin, Jung Hwan Oh, Sang Muk Lee and Seung Eun Lee, "Live Demonstration: CAN FD Controller for In-Vehicle Network," *Circuits and Systems (APCCAS), 2016 IEEE Asia Pacific Conference on*, 2016.

[4] S. E. Lee and N. Bagherzadeh, "Increasing the Throughput of an Adaptive Router in Network-on-Chip (NoC)," *Proc. Int. Conf. Hardw./Softw. Codes. Syst. Synth.*, 2006, pp. 82-87.

[5] S. E. Lee and N. Bagherzadeh, "A variable frequency link for a power-aware network-on-chip," *Integr. VLSI J.*, vol. 42, no. 4, pp. 479–485, Sep. 2009.

[6] Vector Informatik GmbH, CANoe Manual (Version 9.0), http://vector.com, 2016.

Design of An Area-Efficient Hardware Filter for Embedded System

Ji Kwang Kim, Oh Seong Gwon, and Seung Eun Lee*
Dept. of Electronic Engineering
Seoul National University of Science and Technology
Seoul, Korea
*seung.lee@seoultech.ac.kr

Abstract— **In this paper, we propose an area-efficient hardware accelerated filter for embedded system. In order to minimize the area of hardware filter, the proposed filter architecture has a single multiplier. The filter operates by reusing the multiplier. In addition, we optimize the quantization bit length by analyzing the relationship between area and preciseness according to the quantization bit length. We verify the performance of the proposed filter by measuring frequency response in verification environment.**

Keywords; hardware filter; area-efficiency; quantization bit;

I. INTRODUCTION

Digital filters are generally used to process signal in embedded system such as healthcare device, audio device, and camera. As the performance of embedded system grow higher, the more amount of signal processing is demanded and it becomes more elaborate. However, in embedded system, the signal processing with the general-purpose processor has limitation in real-time and low-power consumption, because the signal processing requires many calculations [1]. For this reason, the signal processing is substituted by hardware accelerator [2]. In an embedded system, the area is important issues, because power consumption and production cost are restricted in the embedded system [3]. Therefore, the design of hardware accelerator for embedded system is required to possess low-area. In hardware digital filter, especially multiplier has a strong influence on area and power. To deal with these issues, there is a research which replaces multiplier to adder and shift operation for optimizing multiplier [4]. But, this strategy has limitation in that the filter uses constant coefficient and cannot be used flexibly.

In this paper, we propose single-multiplier IIR filter architecture in which single multiplier, multiplexers, and controller are disposed. In other words, the area of filter be reduced by reusing multiplier. And, it can process various types of signal by appropriately transforming coefficient. Furthermore, quantization bit length can be optimized in filter design in order to minimize area. Therefore, we optimize the quantization bit length of proposed filter by comparing the error rate and area according to quantization bit length.

This study is supported by the Ministry of Trade, Industry & Energy(MOTIE, Korea) under Industrial Technology Innovation Program. No.10060228, 'The Development of Low Power(mW) Mobile Health-Care SOC with Domestic CPU Core'

II. SINGLE-MULTIPLIER IIR FILTER

A. The area of IIR filter

Compared to Finite Impulse Response (FIR) filters, Infinite Impulse Response (IIR) filters require less delay, adder and multiplier. Therefore, the IIR filters provide a much better performance and less computational cost than FIR filters. [5][6]. Equation (1) is the expression of the *N*-order IIR digital filter. *x[k]* and *y[k]* indicate each input and output sample, b_k and a_k indicate coefficient and *k* indicates the level of delay. As you can see from Eq. (1), *2N+1* multiplications are required to process a sample of data. In implementation of hardware digital filter, the use of many multipliers increases area. The area of filter can be reduced by reducing the number of multiplier. Therefore, we propose single multiplier IIR filter in that the filter has a single multiplier. And, to reuse the multiplier, multiplexer and controller are disposed.

$$y[0] = \sum_{k=0}^{N} b_k x[k] + \sum_{k=1}^{N} a_k y[k] \tag{1}$$

B. Fixed-point filter and quantization error rate

Fixed-point digital filter has outstanding operation speed than floating-point digital filter [7] and has better performance about round-off noise in comparable bit length [8]. However, in fixed-point digital filter, quantization operation is inevitable as Eq. (2) in that round quantization operating is added to Eq. (1). Consequently, the fixed-point calculation degrades the precision of the filter, because of error caused by the loss of some point in quantization process. The quantization error can be reduced by increasing quantization bit length. However, the increase of quantization bit length increases the area. It is trade-off between the quantization error rate and the area. In that perspective, we measure the area and the error rate of cutoff frequency according to the quantization bit length and we analyze the relationship between both in order to optimize quantization bit length.

$$y[0] = \text{round}\left(\sum_{k=0}^{N} b_k x[k] + \sum_{k=1}^{N} a_k y[k] \right). \tag{2}$$

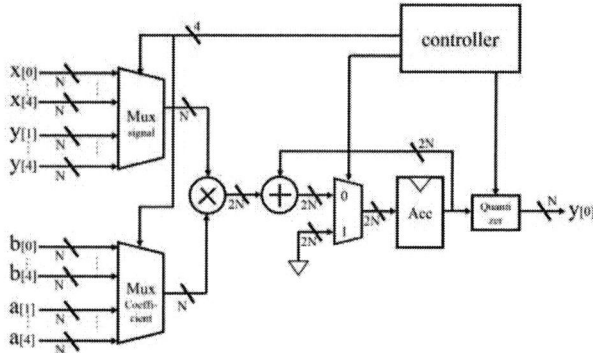

Figure 1. Microarchitecture of the designed filter

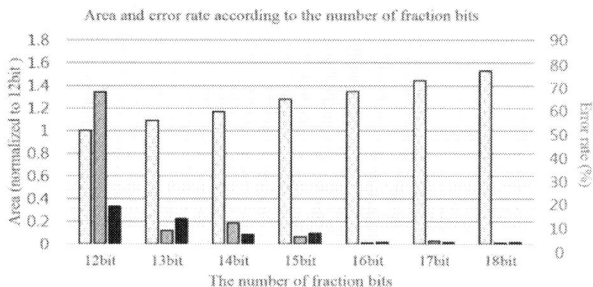

Figure 2. Area and error rate according to the number of fraction bits. Area represents a gate count of filter.

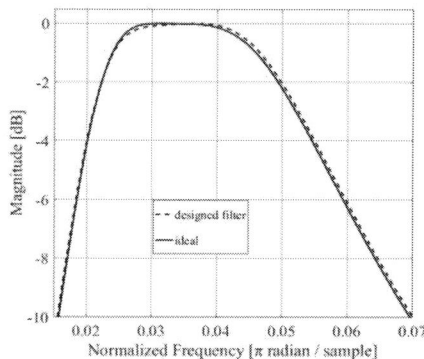

Figure 3. Frequency response of the designed filter

III. ANALYSIS AND IMPLEMENTATION

A. Design of Single multiplier IIR filter

Figure 1 is the structural block diagram of the designed 4-order IIR filter. First, the multiplexers transmit their input to multiplier as ordered pairs like $(x[0], b[0])$, $(x[1], b[1])$, ..., $(y[4], a[4])$. And the ordered pair multiplied each other in regular sequence like $(x[0]*b[0])$, $(x[1]*b[1])$, ..., $(y[4]*a[4])$. The multiplied data are accumulated in the accumulator. When the last multiplied data $y[4]*a[4]$ is added the accumulator, the IIR filter finishes the processing. Finally, the data of the accumulator is transferred to quantizer, the output of which is result.

B. Optimization and verification

Figure 2 shows the area and the error rate according to the different quantized bit size. In order to measure area and error rate, the bandpass-coefficient having a cutoff frequency lower about 0.7Hz and higher about 1.7Hz is used. As the quantized bit length increases, so the area increases linearly. The error rate shows the decay. The most efficient number of fraction bits for hardware implementation is the 16bit point, because after the 16bit point, the error rate decreases slightly.

Figure 3 indicates measured frequency response graph of designed filter, optimized quantization bit length to 16bit. We measure cutoff frequency error rate of the designed filter compare to the ideal filter in order to confirm precision of the designed filter. The result is under 1.4% error rate and that is same result that we analyzed before in Fig. 2.

IV. CONCLUSION

We proposed an area-efficiency hardware filter for embedded system. To minimize the area of filter, we disposed one multiplier. And, we reuse the multiplier at the entire multiplication in the digital filtering. Furthermore, we optimized the quantization bit length by measuring and analyzing the area and the error rate of cutoff frequency according to quantization bit length. The designed filter can reduce area of the total platform of embedded device and can process many types of signal for by transforming coefficient. We plan to verify availability of the designed filter by measuring power consumption of the designed filter in order to compare power consumption with other filter having same function.

REFERENCES

[1] P. Y. Chen, L. D. Van, H. C. Reddy and I. H. Khoo, "Area-efficient 2-D digital filter architectures possessing diagonal and four-fold rotational symmetries," Information, Communications and Signal Processing (ICICS) 2013 9th International Conference on, Tainan, 2013, pp. 1-5.

[2] Oh Seong Gwon, Ji Kwang Kim, Jung Woo Shin and Seung Eun Lee, "Live Demonstration: AHB based Digital Filter for Low Power Mobile Healthcare System," *Circuits and Systems (APCCAS), 2016 IEEE Asia Pacific Conference on*, 2016.

[3] N. Guan, M. Lv, Q. Deng and G. Yu, "A Real-Time Scheduling Algorithm with Buffer Optimization for Embedded Signal Processing Systems," Advanced Information Networking and Applications Workshops, 2007, AINAW '07. 21st International Conference on, Niagara Falls, Ont., 2007, pp. 772-777.

[4] M. Potkonjak, M. B. Srivastava and A. P. Chandrakasan, "Multiple constant multiplications: efficient and versatile framework and algorithms for exploring common subexpression elimination", IEEE Transactions on Computer-Aided Design of Integrated Circuits and Systems, vol. 15, no. 2, pp. 151-165, Feb 1996

[5] M. Sharifi and H. Mojallali, "Design of IIR digital filter using modified chaotic orthogonal imperialist competitive algorithm", Fuzzy Systems (IFSC), 2013 13th Iranian Conference on, Qazvin, pp. 1-6, Aug, 2013

[6] M. A. Sharifi and H. Mojallali, "A modified imperialist competitive algorithm for digital IIR filter design" Artificial Intelligence and Signal Processing (AISP), pp. 7-12, March, 2015

[7] A. Krukowski and I. Kale, "Two approaches for fixed-point filter design: "bit-flipping" algorithm and constrained downhill simplex method" Signal Processing and Its Applications, pp. 965-968 vol.2, Aug, 1999

[8] B. D. Van Veen and R. Baraniuk, "Matrix based computation of floating-point roundoff noise", IEEE Transactions on Acoustics, Speech, and Signal Processing, vol. 37, no. 12, pp. 1995-1998, Dec. 1989

A Network Architecture Design of Embedded System for Media Service in Bus

Sang Yub Lee, Duck Keun Park and Jae Jin Ko
Embedded Software Research Center
Korea Electronics Technology Institute
Seongnam, Republic of Korea
{syublee, parkdk, jaejini}@keti.re.kr

Jae Kyu Lee and Choul Jun Kang
Energy IT Convergence Research Center
Korea Electronics Technology Institute
Seongnam, Republic of Korea
{jae4850, kang-chouljun}@keti.re.kr

Abstract— recently, premium class bus is introduced to passengers. Premium class bus is equipped with multimedia service platforms. It enables to support the media contents to passengers while they are on the board. Existed media service in bus is consisted with Ethernet network and through IP based scheme, packet data transmission is executed. As you know, the method of packet data transmission based on IP addresses is hard to transmit the streaming contents such as music libraries without interruption. And structurally, Ethernet network topologies are needed to packet switching system and network hub system. It makes increase complexity and weight of wiring harness in bus [1]. In this paper, it is supposed the network architecture based on ring topology. Through the designed network service model, it served simple channel allocation and streaming data transmission for individual media services applied to bus seats. Compared with network properties, it is shown solution functionality, implementation and flexibility of proposed network system.

Keywords; network architecutre, media service, ring network topology, bus

I. INTRODUCTION

The premium class bus is developing from automotive suppliers. They are paving the way for bus operation companies to install Ethernet network service. A move toward Ethernet reflects the fact that in vehicle electronics are becoming more sophisticated to support autonomous driving, embedded displays and infotainment systems for multimedia service to passengers. In bus, Ethernet's greater bandwidth could, for instance, provide all passengers with turn-by-turn navigation and each backseat passenger watches videos on separate displays. Thus, Ethernet based system requires hardware and software combination of switching system and network hub and they support a broad range of onboard multimedia nodes for individual seats [2-3]. This system architecture causes the burden of data traffic and network delay for the media contents what passengers want to see. As described in Fig. 1, in order to be equipped multimedia system based on Ethernet service with bus [4], at least, 8 ports side hubs and 9 ports main hub are needed. Therefore, it needs the ring network based system which can fairly assign media data to all seat infotainment terminals in bus.

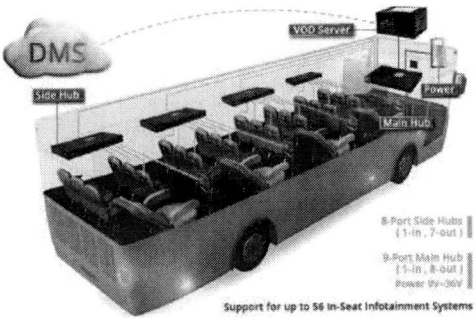

Figure 1 Ethernet based multimedia system in bus (source: funtro)

I. SYSTEM ARCHITECTURE

A. System model

As described in Fig. 2, the designed system architecture is composed of ring network devices. They are as a role of master and slave nodes in network. This is a point to multi-point data flow system, all devices shares a common system clock pulse derived from the data stream. They are in phase with each other and can transmit all data synchronously, which makes any mechanism for signal buffering and signal processing redundant. The system clock is generated by the master device depicted in Head Unit and all other nodes are synchronized onto this system clock pulse by means of a PLL connection and are referred to as slaves expressed in RSE Unit.

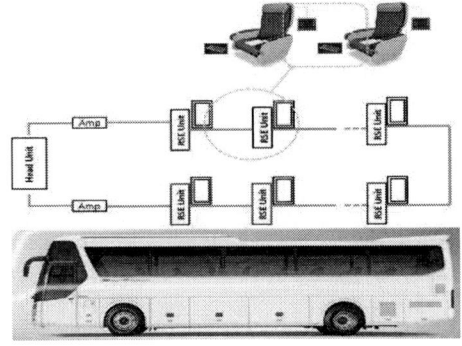

Figure 2 Proposed multimedia system in bus

B. Device Model

Server as a master device provides its functionalities by means of properties and methods of its function blocks used and controlled by an application. It requires partial system knowledge, which means that it must know the function blocks to be controlled. The Server is the interface to user of the system and thus presents the system function on a high abstraction level. It coordinates the various slaves. As a device model, application protocol procedure is introduced in Fig. 3.

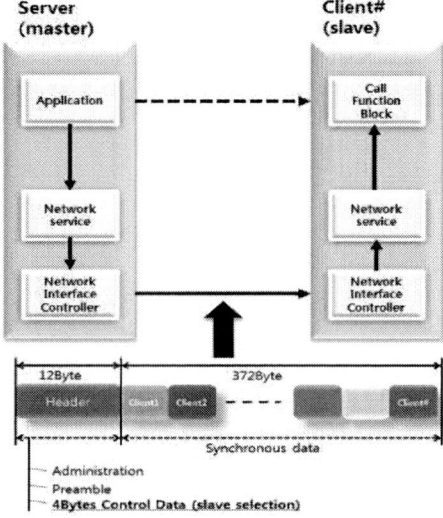

Figure 3 Application protocol in device model

II. IMPLEMENTATION OF DESIGNED ARCHITECTURE

System installation is applied to node position conception. The master always has node position 0x00, the following node 0x01. The maximum number of nodes in the ring is 64. By means of the node position, the delay between server and slave can be determined quite easily by the source informing the sink of its position and the sink determining the number of network nodes through which the signal has passed.

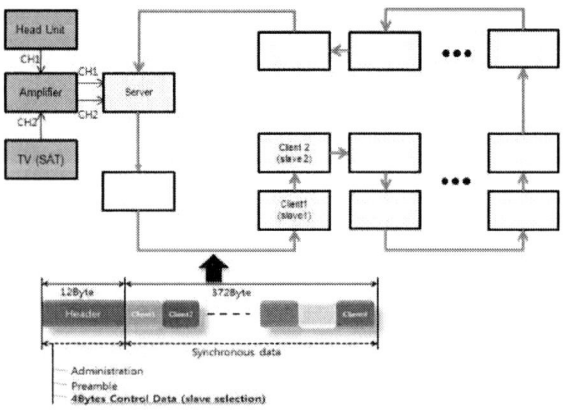

Figure 4 Ring network topology

A view of implementation of media system based on ring network topology is clear that streaming data are assigned to synchronous data of packet frame according to the node position shown in Fig. 3 and Fig. 4. Unlikely Ethernet based system, preoccupied channel for client number is operated transmitting the video and audio data continuously. When passengers request the video on demand service, selected contents will be transferred to the display on passenger's seat.

III. CONCLUSION

Through the proposed system model and device model for ring network based system in bus, simply and ease installation is enable to set media service platform up in the premium class bus. It is known that node position conception and application protocol method will be applied to implementation of network system. It enables multiple and simultaneous video on demand contents are transferred to passengers in bus. Ring network topology has advantages of system installation and little wiring harness. Furthermore, the side of maintenance, it has it over Ethernet based system.

승객석 인포테인먼트 시스템

Figure 5 Designed multimedia platform system in bus

ACKNOWLEDGMENT

This work was supported by the IT R&D program of MKE/KIAT [R0004937, Development of Lightweight VOD system for the premium class bus]

REFERENCES

[1] Microchip, "Audio video system in travel bus," 2015 (not published)

[2] Microchip, "Remote control system," 2014(not published).

[3] Andres Grzemba, "MOST25 to MOST150," MOST cooperation, Franzis, 2012.

[4] Stephan Kehrer, Oliver Kleineberg, Donal Heffernan, "A comparison of fault-tolerance concepts for IEEE802.1 Time Sensitive Networks," IEEE Emerging Technology and Factory Automation, 2014, pp.1-8.

A Study on Improvement of Recognition Accuracy by Applying Machine Learning Algorithms to the Vision-Based Traffic Condition Analysis System

Keonhee Lee*, Hyuntae Ju, Yong Mu Jeong and Soo-Young Min
Embedded.SW Research Center
Korea Electronics Technology Institute
Seongnam-si, Republic of Korea
{khlee, oskernel, ymjeong, minsy}@keti.re.kr

Abstract— **This paper proposes a method of applying a machine learning algorithm in order to improve the recognition rate of the video-based traffic information system. After applying the Error Backpropagation learning neural network algorithm to the traffic information it will be used for image recognition results. The training data is generated from the traffic information system, the noise of the generated data is removed by Gaussian smoothing. In this paper, we develop a machine learning based Traffic Condition analysis system was able to get an improved recognition rate than conventional vision-based system.**

Keywords; Error Backpropagation, Traffic condition recognition, machine learning

I. INTRODUCTION

To improve the stability, usability, operability, and accessibility for driver by the development of smart-car, software technic is more researched than hardware. Vision-based recognition system is one of field being studied from the smart-car technic. Representative systems such as LDWS(Lane Departure Warning System), PAS(Parking Assistance System), pedestrian recognition are commercialized.

In this paper, we studied to improve the accuracy of vision-based traffic condition analysis system. We propose the method using the machine learning and time-series data about traffic information for improving the recognition rate.

Representative time-series data learning method is SVM (Support Vector Machine). SVM is kind of supervised learning machine for data analysis and is able to learn regardless of linear/non-linear data. SVM has a demerit that learning time more increase as dimension of data increase. [1] To overcome this demerit of SVM, TSDCA(Time-series discriminant component analysis) introduce how to reduce the data dimension by using orthogonal transformation and posterior probabilities. TSDCA has an improved learning time and accuracy. [2] Also, ELM(Extreme Learning Machine) that is inverse matrix based learning machine is adopted the monthly drought index forecast method in Australia. Because ELM has a simple structure, it's result are faster and more exact than general ANN(Artificial neural network). However, as the input data more increases, ELM has the disadvantage of requiring a large hardware. [3]

The rest of this paper is organized as follows: An explanation about the Error Backpropagation learning algorithm in Section 2. Section 3 explains a proposed method, experiment, and traffic information used in input data. Section 4 concludes in this paper.

II. ERROR BACKPROPAGATION

Error Backpropagation(EBP) algorithm is one of the training algorithm for Multi-layer perceptron neural network(MLPNN). In order to train the neural network, Based on the output of the input data to update the weights of the network. Gradient descent method applied to the EBP is a representative algorithm for updating the network. Since EBP is directed learning algorithm, the input data will be specified the target. EBP is the learning proceeds until the convergence with the following conditions.

$$E = \frac{1}{2} \sum_{i=1}^{N_0} (t_j - O_j)^2 \qquad (1)$$

t_j is the target data, O_j is the output of the neural network. Learning procedure is repeated until the minimum value of E satisfies the condition. EBP network is represented by a multi-layer structure as shown in Fig 1.

The output of the k layer jth neuron of EBP network u_j^k is calculated by the following equation (2).

$$u_j^k = f\left(\sum_{i=0}^{M_{k-1}} (w_{ji} u_i^{(k-1)} + bias) \right) \qquad (2)$$

Output of next layer is determined by the product of the weights of the input of the current layer. Output of each layer is convert to the value of a certain range to activation function become input value of the next layer. The most often being used

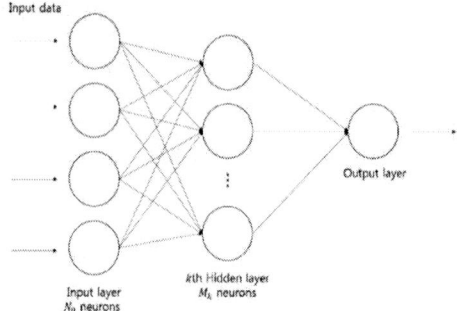

Figure 1. Architecture of Error Backpropagation

activation function is a logistic form sigmoid function (equation (3)).

$$f(x) = \frac{1}{1 + e^{-x}} \qquad (3)$$

Output computed until the end of the layer will be compare to the target of the input data. δ calculated by equation (4) will be obtained in every weight.

$$\delta_j = \begin{cases} (t_j - u_j)u_j(1 - u_j), Output layer \\ u_j(1 - u_j)\sum_k \delta_k w_{kj}, Otherwise \end{cases} \qquad (4)$$

Equation (5) updates the neural network on the amount of change in the weight calculated by applying the learning gain.

$$\Delta w_{ji} = \eta \cdot u_i \cdot \delta_j \qquad (5)$$

III. Improvement and Evaluation

Existing vision-based traffic condition analysis system receives real-time traffic video from vehicles to configure the database. The system via the received images to determine the road condition information, such as falling objects above the roadway, a collision accident, congestion.

We propose to apply a machine learning technique based on the traffic information in order to improve the recognition rate through the image. Road information data contains the accident information, traffic information such as time series, representative traffic information can be represented graphically as shown in Fig 2. The main data used in the experiments were configured to provide traffic information data system recognize the cumulative data from the Ministry of Land, Infrastructure, and Transport(MOLIT) as Time series data. All data are the Gaussian smoothing is applied to remove the noise. Performance of the algorithm is to measure the experimental results with 10 fold cross validation. The method 1 to test pieces after the learning 9 then divides the whole data to 10 to verify the performance.

Result of the experiment is as shown in Table 1, the image recognition applying the road traffic information can be obtained a better result. According to Table 1, the recognition rate for the four road condition information was to improve the recognition performance of the average of 9.69%.

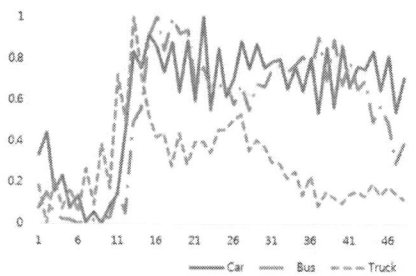

Figure 2. Traffic volume graph according to a type of the vehicle

TABLE 1. The Experiment results of applying the EBP

Methods	Recognition rate (%)			
	Traffic congestion	Accident	Obstacle	Normal
Video model	72.34	75.28	63.85	83.64
EBP mixed model	82.53	85.29	79.33	86.72

IV. Conclusion

Vision-based Traffic Condition Analysis System sends real-road status images to the server and judges the traffic condition. Because the vision-based recognition accuracy is depending on environment, we propose the learning method that uses the traveling area, time zone, area characteristic as an input data. The proposed method gives the weight to vision-based analysis result and improves the recognition rate. The proposed method has 9.69% higher accuracy than only vision-based traffic condition analysis system.

All traffic information recognized by the proposed method are stored in the database, and are able to use the prediction-based input data. To improve an accuracy of machine learning, we need a lot of database and researches about how to configure the data of image and traffic information.

Acknowledgment

The research was supported by the IT R&D program of The MOTIE/KEIT(Ministry of Trade, Industry and Energy/Korea Evaluation Institute of Industrial Technology) [10063329, "Cloud based Car Infotainment System with Realtime Traffic Condition"].

References

[1] N. I. Sapankevych, "Time Series Prediction Using Support Vector Machines: A Survey," IEEE Computational Intelligence Magazine, vol. 4, issue 2, pp. 24-38, May 2009

[2] H. Hayashi, T. Shibanoki, K. Shima, Y. Kurita and T. Tsuji, "A Recurrent Probabilistic Neural Network with Dimensionality Reduction Based on Time-series Discriminant Component Analysis," IEEE Trans, neural networks and learning systems, vol.26, issue. 12, pp 3021-3033, 2015

[3] D. C. Deo, M. Sahin, "Application of the extreme learning machine algorithm for the prediction of monthly Effective Drought Index in eastern Australia," Atmospheric Research, vol. 153, pp 512-525, Feb 2015

A Study on Improvement of Vision-Based Traffic Condition Analysis System by Comparing Feature Data of Images

Eunae Park*, Hyuntae Ju, Yong Mu Jeong and Soo-Young Min
Embedded.SW Research Center
Korea Electronics Technology Institute
Seongnam, Republic of Korea
{eunae, oskernel, ymjeong, minsy}@keti.re.kr

Abstract— **Despite the vision-based traffic condition analysis system provides better data quality than those of traditional UTS-based applications, it still have a problem of inefficiency because it handles commonly large size of video data. In this paper, we designed a feature-based image examining method to reduce the load of the vision-based application by measuring similarity of feature data before handling the original data.**

Keywords; feature data, optimization, image analysis, similarity

INTRODUCTION

Many applications that use traffic information obtains their necessary data from several established traffic information systems, but most of the source data may include somewhat imprecise information due to their technical limitations. Vision-based analyzing techniques have been designed in order to make up for the shortcomings, but they also have some technical weakness. Through this paper, we would like to discuss on a method to provide enhanced quality of traffic information while maintaining the efficiency.

VISION-BASED TRAFFIC CONDITION ANALYSIS

Existing traffic information systems as exemplified by UTIS(Urban Traffic Information System) are already able to produce traffic condition data, but most of them are not suitable for applications that require timely and accurate traffic information because they provide mostly imprecise data in long cycles.

A vision-based traffic condition analysis system provides improved quality of traffic information by analyzing complementary traffic status data by videos acquired from black boxes of subscribers' vehicles. Because the vision-based traffic condition analysis system leverages near real-time video data to make up for the weakness of UTIS, on the other hand, the system might not be able to cope with the enormous volume of network traffic and the tremendous storage usage if many(from a few to even many millions of) end devices try to upload their whole videos or images.

FEATURE EXTRACTION ALGORITHMS

An image data have several feature data such as contrasts, colors, edges and patterns. Generally a set of feature data has approximately a tenth part of the size of its original image. We can think of the methods in two categories to extract the features of the images.

A. Use of Keypoint (Interesting Point)

The term 'keypoint' means a point which can be identified easily because it could be distinguished clearly from its surroundings. In general, corners of objects or portions that contain relatively clear contrasts between light and shade may be selected as a keypoint.

The most widely used algorithms of extracting the keypoints include Harris Corner, FAST(Features from Accelerated Segment Test) and SIFT(Scale Invariant Feature Transform).

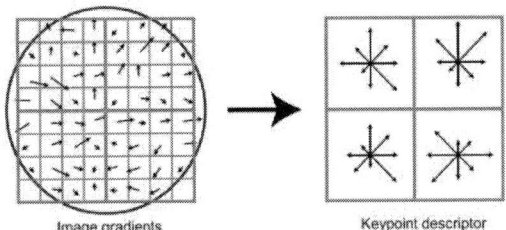

Figure 1. an example for keypoint descriptor of SIFT

Because the SIFT algorithm is less vulnerable to changes of size, rotation and point of sight, it is one of the strengths of the SIFT algorithm that it keeps good performance for feature matching[3]. The SIFT algorithm finds out 'extrema' by performing DOG(Difference of Gaussian) operations, then it figures out a keypoint and a descriptor vector using the extrema. A keypoint descriptor is a set of feature data that contain direction and magnitude of gradient of surrounding pixels.

B. Use of Patterns

The typical methods which extract feature data by comparing patterns for brightness and texture of the pixels include MCT(Modified Census Transform), LBP(Local Binary Pattern) and ULBP(Uniform Local Binary Pattern).

As an algorithm to calculate feature data from patterns of bit changes, the LBP stands up to changes of contrast and illumination[4]. With each pixel as the center, the LBP algorithm figures out feature data from each 3 x 3 matrix made up of surrounding pixels. It generates binary patterns by marking as 1 if the brightness value is greater than or equal to that of the central pixel, and marking as 0 otherwise.

$$LBP(x, y) = \sum_{n=0}^{7} B(p_c - p_n) \cdot 2^n, \ B(p) = \begin{cases} 1, & p \geqq 0 \\ 0, & p < 0 \end{cases} \quad (1)$$

The above formula represents the method to calculate histograms by LBP algorithm. In the formula, x and y denote the coordinates for current location. And also p_c and p_n mean the pixel at current position and the pixels around p_c respectively.

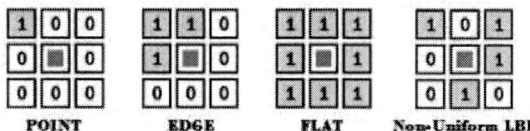

Figure 2.   an example for ULBP operation

As shown in figure 2, ULBP(Uniform Local Binary Pattern) algorithm is a method to utilize dots, edges and faces as feature data of an image.

SYSTEM DESIGN AND PERFORMANCE EVALUATION

By the feature-based image comparing technique which we designed in this paper, the whole new image will be uploaded only if the server does not have any images whose features are similar to the fresh one. To extract feature data from images, we employed MB-ULBP(Multi-scale Block Uniform Local Binary Pattern) and SURF(Speeded Up Robust Feature) in this paper. They verify the presence of similar feature data in the server by splitting the image into several sectors and comparing characteristics of each sector.

A. MB-ULBP(Multi-scale Block Uniform Local Binary Pattern)

Since MB-ULBP separates each sector into some blocks again and calculates ULBP histograms by the average value for characteristics of the blocks, it has the advantage of being possible to extract features from each image of various scales[6]. In this paper, we define 3 x 3 matrix of pixels as one block.

B. SURF(Speeded Up Robust Feature)

In order to accelerate the processing speed while preserving the advantages of SIFT, Herbert Bay designed SURF algorithm using integral images and Hessian Matrix. Integral images help the calculation of area of specific sector to be easier and faster. Since the Hessian Matrix operations adopted by SURF in order to detect objects are simpler than the SIFT's methods, SURF shows better performance than SIFT in general[7].

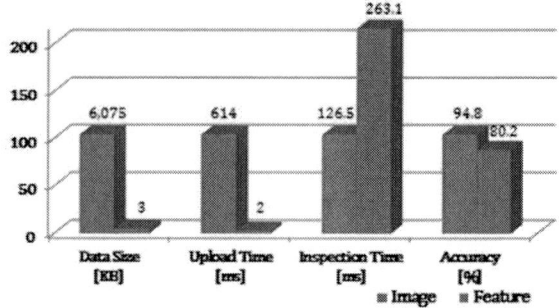

Figure 3.   Performance Evaluation

We use a 30 FPS and 20 seconds of full HD(1920 x 1080) video data for an experiment. The time value shown in figure 3 above is measured from calculation for feature data to decision of similarity. The computer specifications we use for this experiment as a server include Intel Core i7-2600K CPU and 24 gigabytes of RAM. In addition, we used automotive embedded devices as black boxes with quad-core application processor based on ARM Cortex-A9 at 1GHz clock speed.

CONCLUSION

Through the experiment above, we could confirm that the vision-based application systems can be improved in efficiency and performance by verifying feature data of images before handling its original data. We would like to continue the study for the further betterment of our feature-based image examining technique by removing unnecessary routines and performing some more optimizations. We especially hope to improve the inspection time and accuracy through future research.

ACKNOWLEDGMENT

The research was supported by the IT R&D program of The MOTIE/KEIT(Ministry of Trade, Industry and Energy/Korea Evaluation Institute of Industrial Technology) [10063329, "Cloud based Car Infotainment System with Realtime Traffic Condition"].

REFERENCES

[1] S. Kantawong, and T. Phanprasit, "Intelligent traffic cone based on vehicle accident detection and identification using image compression analysis and RFID system," ECTI-CON, pp. 1065-1069, May 2010.

[2] M. Jang, and D. C. Park, "Modified SURF-based Object Tracking System," Inter. J. of Computer Science and Electronics Engineering, vol. 3, pp. 319-323, 2015.

[3] D. G. Lowe, "Distinctive Image Features from Scale-Invariant Keypoints," Inter. J. Computer Vision, vol. 60, pp. 91-110, Nov. 2004.

[4] M. Topi, O. Timo, P. Matti, and S. Maricor, "Robust texture classification by subsets of Local Binary Patterns," Proc. 15th ICPR, vol. 3, pp. 935-938, 2000.

[5] G. Zhao, T. Ahonen, J. Matas, and M. Pietikaien, "Rotation-invariant image and video description with local binary pattern features," IEEE trans. on Image Processing, vol. 21, pp. 1465-1477, APRIL 2012.

[6] S. Liao, X. Zhu, Z. Lei, L. Zhang, and S. Z. Li, "Learning Multi-scale Block Local Binary Patterns for Face Recognition," ICB Proc. LNCS, vol. 4642, pp. 828-837, August 2007.

[7] H. Bay, A. Ess, T. Tuytelaars, and L. V. Gool, "Speeded up robust features (SURF)," J. Computer Vision and Image Understanding, vol. 110, pp. 346-359, June 2008.

A Study on river water level monitoring method in a debris barrier

Hyo Sub Choi
Embedded & Software Research Center
Korea Electronics Technology Institute
Seongnam-si, Rep. of Korea
hschoi@keti.re.kr

Deepak Ghimire
IT application Research Center
Korea Electronics Technology Institute
Jeonju-si, Rep. of Korea
deepak@keti.re.kr

Abstract— **In this paper, the river water level monitoring method in a debris barrier is presented. There are four steps in this proposed solution: Frame difference, Thresholding & Noise refinement, Candidate point detection, and Classification. The proposed method is able to calculate water flow occupancy value and monitor the change in water flow in a dam. This technique is very efficient to give warning in case of there is any abrupt change in water flow in the river.**

Keywords; water level monitoring; vision solution; frmae differencing; debris barrier

I. INTRODUCTION

A debris barrier is a small dam constructed across a swale, drainage ditch, or waterway to counteract erosion by reducing water flow velocity (1). Barriers require remote management as they are located in the mountains (2). Dams should be inspected every time that it is sited in the channel and after large storm (3). It is important that the change in water flow in the channel is monitored. Thus, in this paper, the river water level is extracted based on a segmentation using frame differencing. The proposed method is able to get a percentage of movement and track level of the water.

II. PROPOSED METHOD

In flowing river, there is movement of water in consecutive frames. It could be detected using optical flow or frame differencing. In this method, there are some problems. First, Segmentation becomes difficult if there is movement in non-water region, such was trees. Second, If the water is not flowing (eg. pound), there is no much movement in water surface. No information will be obtained using optical flow or frame differencing. We proposed a method for calculating percentage of movement by making a segmentation using frame differencing. While water level is normal, we can get percentage of movement in scene. In case if there is gradual increment in water flow in river, the percentage of movement in scene will increase. We can decide this is due to water flow change, because tree or other parts in scene is constant. If there is no wind, scene movement even decreases. The proposed river water level monitoring method is composed of four sub

parts: Frame difference, Thresholding & Noise refinement, Candidate point detection, and Classification.

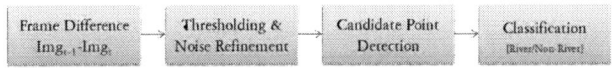

Figure 1. The steps of proposed method

A. Frame difference

In this solution, the term "frame difference" means difference between two consecutive frames in video sequence. In general frame difference gives the motion information in video sequence. If there is constant motion in some part of video sequence frame difference will give some information at that region even if the camera is still. In our case water is flowing in the river and it is creating wave or motion of water flow. The frame difference is calculated by using absolute difference of pixel value in gray domain in the corresponding position of two consecutive frames.

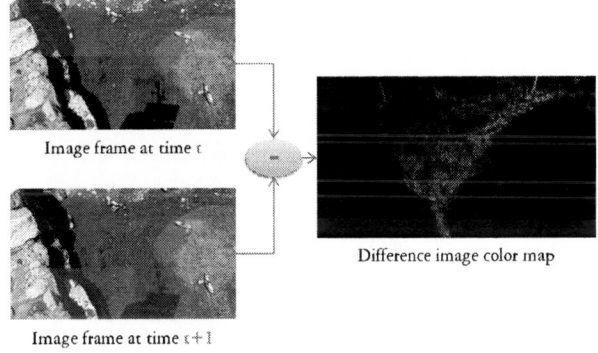

Figure 2. Frame difference

B. Thresholding & Noise refinement

After getting the frame difference by using two consecutive frames in video sequence, thus obtained response is thresholded to get large motions in video and remove noise and very small movements. Here we used threshold value 10, i.e. if the absolute difference is less than 10, response at that pixel

978-1-5090-3220-4/16 $31.00 © 2016 IEEE

position is removed. This is done to remove noise and small responses. Next step is to refine the thresholded response and remove the noise in the resultant image. We consider noise if in the 5x5 pixel region there is very low response of motion obtained using frame difference. Below first image is the response after thresholding

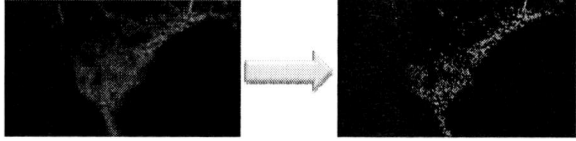

Figure 3.   Thresholding & Noise refinement

C. Candidate Point Detection

We take 5x5 window size. After frame differencing and thereholding, if the number of white pixel inside 5x5 window are more than given threshold we keep center pixel as candidate pixel of flowing river. Other considered as noise pixel and removed. Suppose Th = (5x5)/4 = 6 [i.e. 24%].

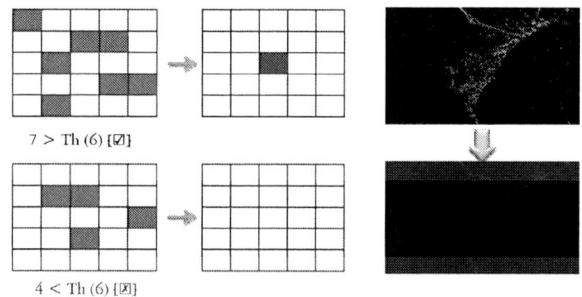

Figure 4.   Candidate Point Detection

D. Classification

After candidate water flow position detection it will be necessary to classify either the detected pixel positions actually refer to the flowing river or other region of the image such as waving tree and other objects. In many cases the motion response will be generated in waving trees and grasses other than flowing river waves. There are several information to differentiate river and other parts of the image. For example one clue is color of the background. Fig.5 below shows river detection results. In circular region shows wrong detection which needs to be removed using classification as river/non-river region.

Figure 5.   Classification

III.   RESULTS

Graphs below shows the percentage of candidate pixels that shows flowing river. The percentage is very low because we considered full image as river detection region at which only small region is corresponds to river [at least in test videos]. Fig. below shows that in one video sequence, the detected water flow occupancy value is around 4%, therefore we can set threshold as if it exceeds 10%, we can say that there is abrupt increase in water flow in river.

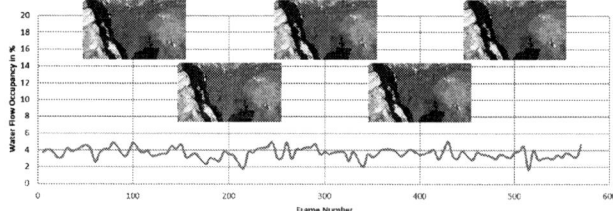

Figure 6.   Water Flow Occupancy

IV.   CONCLUSIONS

We have proposed a method to detect water flow based on percentage of movement. The obtained accuracy of river segmentation or water flow detection based on flow motion depends on the scene itself and intensity of water flow. In case of still river, it is hard to do segmentation of river/non-river region. In case of moving river, using motion information the river can be detected as water flowing region. The result is in best if the water flow intensity is high, i.e., with large motion and flow wave. This vision based technique is very efficient to give warning in case of there is any abrupt change in water flow in the river.

ACKNOWLEDGMENT

This work was supported by technical innovation R&D program of Small and Medium Business Administration and industry core R&D program of Ministry of Trade, Industry and Energy (Project Number : 10062934)

REFERENCES

[1] Marsh, William M. (2010). Landscape Planning: Environmental Applications (5th ed.). Danvers, MA: John Wiley & Sons, Inc. pp. 267–268.

[2] H.S. Choi and S.H. Jeong, "A study on water flow detection algorithm for monitoring a debris barrier," CICS2015, pp. 316–317.

[3] Ames, IA: Institute for Transportation at Iowa State University. Retrieved 28 October 2014.

Software Design for GUI Display in the Wearable Device

Gyutae Oh, Inhye Park, Sang-Yub Lee, and Jaejin Ko
Embedded & Software center
Korea Electronics Technology Institute
Seongnam-si, Gyeonggi-do, Republic of Korea
{gto, ine.park, syublee, jaejini}@keti.re.kr

Abstract— **This paper presents the software design for GUI(graphic user interface) display in the wearable device. For suitable operation of the wearable device, we design overall system structure including RTOS. Also, we design GUI routine and objective-oriented library format for GUI display. Through implementing the designed software in the wearable device platform, we confirmed that the proposed architecture is operated in wearable device.**

Keywords; Embedded software, RTOS, Embedded middleware

I. INTRODUCTION

To satisfy desires of users, the demand of small devices for the IoT(Internet of thing) including the wearable devices are increased. Users are increasingly want to lighter and slim devices, because of the convenience. We can get the result of a special research center, the shift of growth rate have been reduced gradually in the smartphone market [1]. The small devices have typical advantages of lower cost, smaller size and lower power consumption [2-3]. So, small devices have been using at many industrial aspects. Especially, the wearable device has been changing from simply supporting health services including heart rate and a step count to more advanced service. Gesture recognition using motion sensors of the wearable device is one of advanced operations [4-5]. Also requirements are increased to the visual display with other desires. Therefore, the suitable embedded software needs to user for the GUI display in wearable devices.

In this paper, we design software including RTOS(real-time operating system)[6] for GUI display in the wearable device. For the lower energy consumption and seamless operation, we design middleware and adjust RTOS. And we design GUI routine and objective-oriented library format for GUI display. This paper is organized as follows. Section 2 presents the system design including overall system architecture. And we describe software design for GUI display in the same section. At the Section 3, we implement proposed software in the wearable device platform. And then, the conclusion is expressed in last section.

II. SYSTEM DESIGN

A. System architecture

To design software of the wearable device, hardware component that we choose have to be defined. Our target

---

This research was supported by a grant from the IT Convergence program [10062943, Development wearable device and services for industrial convergence that support intelligent driver's ADAS system] (sponsored by MOTIE/KEIT)

---

device has several component that wearable device must contain, which is Bluetooth(BLE) module, user interface(e.g., LCD, Button), battery, and MEMS(Microelectromechanical systems) sensors. To operate all functions of wearable device at the same time, the light embedded OS have to be implemented. So we basically adjust RTOS as OS to manage resources and tasks. Else functions of the wearable device are implemented in the middleware that we customized.

In the middleware block, GUI converter, Event handler/Timer for interrupt from outside and inside, and Battery manager for the lower energy consumption. GUI converter supports GUI interface for application developers. GUI converter translate LCD device driver to objective-oriented library, and help converting image data. The Event handler/Timer support user interrupt from hardware and application both of all. Also, it module manage interrupts from inside kernel that generated by tasks(i.e., threads). Battery manager support low energy consumption of overall wearable device. For instance, Battery manager can reduce kernel copy that generated between application data and kernel data for out-forward communication.

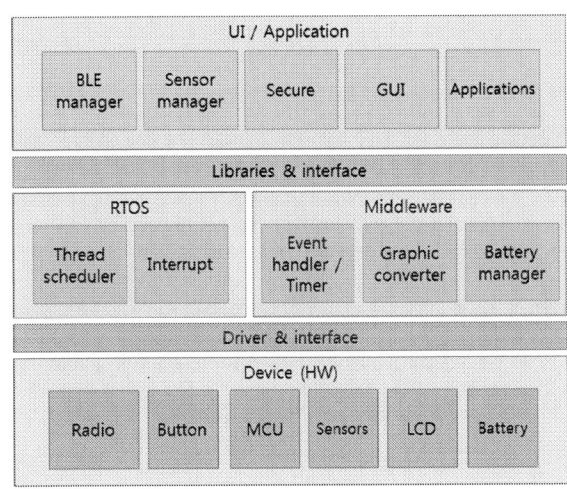

Figure 1 System architecture

At the UI/application block, many functions and applications for wearable device users are implementable. GUI for Graphic user display, sensor manager for MEMS 6-axis sensors, BLE manager for BLE connection, and applications including secure. The overall system architecture that we describe is represented in Fig 1. Shown in the Fig. 1, layered each block communicates through interfaces and library. And RTOS and Middleware place between hardware and application for users.

B. Software design for GUI display

The core development of the GUI is accessability to the display. The LCD display uses serial output through GPIO(general purpose input output). Therefore In the firmware or low level have to use character value information as a pixel unit. For the higher accessability, we design software Graphic converter in the middleware block. The GUI routine and block diagram is shown in the Fig. 2. After initializations, graphic converting and setting block are executed in the routine. In the graphic convert block, the image data convert to acceptable format of hardware such as RGB format. And the graphic data setting block reassigns image data from objective-oriented to hardware-oriented. After converting and data setting, serial forwarder and display block are sequentially executed to the LCD.

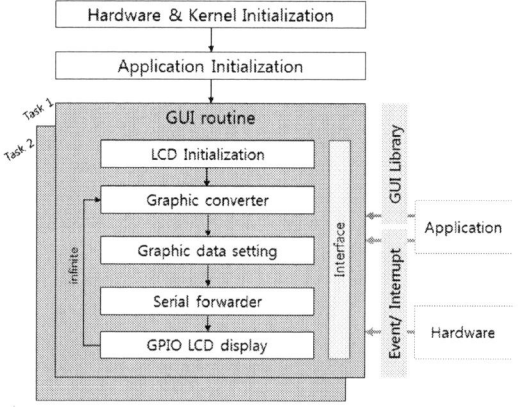

Figure 2 GUI routine block diagram

III. SOFTWARE IMPLEMENTATION

To confirm designed software operation, we implement proposed software architecture in the wearable device platform. Target wearable device platform equips Cortex-M3 MCU, MEMS sensor module, Bluetooth LE module, Button, and LCD module. We designed and implemented the at the designated development tool. Fig. 3 represents an example of low-level source code in the designed software. As shown in the Fig. 3, we set hardware pin number of specific chip to low-level block. The results that display the text is shown in the Fig. 4. As the text shown in the LCD display, the designed software is operated in the wearable device platform.

Figure 3 An example of source code

Figure 4 Software implementation in the wearable device platform

IV. CONCLUSION

This paper presents the software design for GUI(graphic user interface) development in the wearable device. For suitable operation of the wearable device, we designed overall system structure including RTOS. Also, we designed GUI routine and objective-oriented library format for GUI development. Through simple implementing the designed software in the wearable device platform, we confirmed that the proposed architecture is operated in wearable device.

REFERENCES

[1] E. Song, "The forecast of cellphone market at 2015 ," HI Investment & securities, Nov. 2014

[2] R. Noe-Nilssen and L. L. Helbostad, "Estimation of gait cycle characteristrics by trunk accelerometry," Elsevier Journal of Biomechanics, Vol. 37, pp. 121-126, 2004

[3] X. Yuan, S. Yu, S. Zhang, C. Liu, and S. Liu, "Modeling and Analysis of Wearable Low-Cost MEMS Inertial Measurement Units," In Proc. of IEEE 15th International Conference on Electronic Packaging Technology, 2014

[4] C. C. Yang, Y. L. Hsu, K. S. Shih, and J. M. Lu, "Real-Time Gait Cycle Parameter Recognition Using a Wearable Accelerometry System," Journal of Sensors, pp. 7314-7326, Nov. 2011

[5] M. Kusserow and O. Amft, "Modeling Arousal Phases in Daily Living Using Wearable Sensors," Journal of IEEE Transaction on Affective Computing Vol.4, No.1, pp. 93-105, Jan 2014

[6] FreeRTOS: http://www.freertos.org/, last access 7th July 2016

Resolution Tunable Ring Oscillator type TDC

Himchan Park, Zhang-Zhi Yu, Jinwoo Kim and Jinwook Burm*

Dept. of Electronics Engineering, Sogang University

, Sinsu-dong, Mapo-gu

Seoul 121-742, Korea

*burm@sogang.ac.kr

Abstract— **This paper presents a high resolution tunable ring oscillator type TDC (Time to Digital Converter). The proposed structure uses 2 ring oscillators composed of 8 differential inverters. Can be activated by selecting one of the three resolution in the structure, and can be selected in a ratio of 15/16, 13/32, 63/64, depending on fast and slow of the differential inverter delay. Also to increase the total conversion range of the TDC, since when added to the MSB 1bit added only one D-FF(D flip-flop), has the advantage also in the power consumption and chip area. The resolution of TDC is 1.54ps, maximum range is 50.6ns when the differential inverter delay ratio is 31/32. This paper is based on 28nm CMOS process.**

Keywords; formatting; ring oscillator, resolution tunable, TDC

I. INTRODUCTION

Previously studied TDC of has the features corresponding to each type. For fine time resolution in linear TDC, Vernier delay lines are generally considered [9], [10] where the time resolution is determined by the difference between two inverter delays. An excessive number of inverter stages are required to cover a large detection range, causing a long conversion time and a large power consumption [11]. The two-step TDC [12] to improve both the resolution and detectable range by amplifying the time residue after the coarse conversion for the fine conversion. However, the time amplification schemes [12], [14] utilize the delay increase by metastability and suffer from small input range and uncertainties of gain due to nonlinearity and PVT variations. Another type of TDC adopts a noise-shaping feature in TDC to improve effective resolution. By storing the time residue in the form of voltage on the capacitance of each node in the oscillator, the unconverted time residue is effectively added to the input of the next conversion, compensation of the current stage's quantization error in the following stages. However, since the noise shaping is based on the first order delta-sigma conversion with multiple sampling of input, high resolution is not guaranteed for fast-varying inputs. In addition, the multi-phase generation and multiple counters consume large power when used with a gated ring oscillator (GRO) for higher resolution [15]. This paper presents a different concept of resolution tunable TDC suitable for various applications. The proposed TDC achieves the minimum resolution of 1.54 ps, when the ratio of fast differential inverter delay/slow differential inverter delay is 31/32. Section II describes circuits of the proposed TDC. Section III concludes the results of this work.

II. THE PROPOSED TDC

A. Architecture of The Proposed TDC

A resolution tunable ring oscillator type TDC consist of 8 differential inverters and D-FFs. Fig 1 shows a simplified schematic of proposed TDC. Upper side is slow oscillator and bottom side is fast oscillator. If an input signal comes in both side, detecting the rising edge of input signal in the order of node number 5, 2, 7, 4, 1, 6, 3, 8. After the start signal enters into the fifth node of the slow oscillator, stop signal comes from the fifth node of the fast oscillator. The moment when the rising edges from slow and fast oscillators occur simultaneously is detected by the D-FF's, which will be the measure of the time difference.

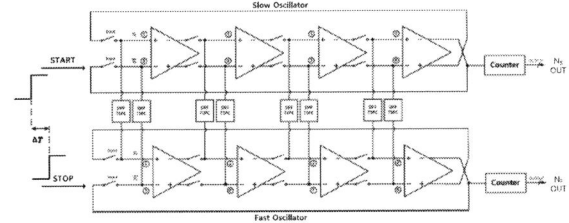

Figure 1.   simplified schematic of proposed TDC

This work was supported by the Industrial Core Technology Development Program (10049095, "Development of Fusion Power Management Platforms and Solutions for Smart Connected Devices") funded By the Ministry of Trade, industry & Energy. When the time difference between two input signals is ΔT, a method of translate the ΔT in digital code is as follows.

$$\Delta T = 8 \cdot N_S \cdot T_S + (8 \cdot N_F + B) \cdot (T_S - T_F) \qquad (1)$$

Where N_S and N_F are counting numbers of slow and fast oscillators. T_S and T_F are delay of slow, fast differential inverter. Fig. 2 shows the output is produced at the output of D-FF of the eight node (Fig. 1). For change the resolution, it is necessary to change the delay difference of each inverter in the slow and fast oscillators. MOS-capacitors are placed at each node of the oscillators for delay adjustment. By adjusting the capacitance value of each MOS capacitor, the delay of an inverter varies.

Fig 2 shows simulation results of slow and fast oscillators at fast/slow delay ratio is 32/31. Slow inverter delay is

978-1-5090-3220-4/16 $31.00 © 2016 IEEE   241   ISOCC 2016

49.43185ps and fast is 47.88881ps. Change the ratio of the delay, while maintaining the delay of slow inverter, to adjust the changing resolution of the delay of the fast oscillator inverter.

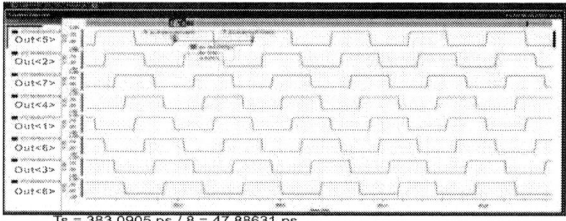

Ts = 383.0905 ps / 8 = 47.88631 ps

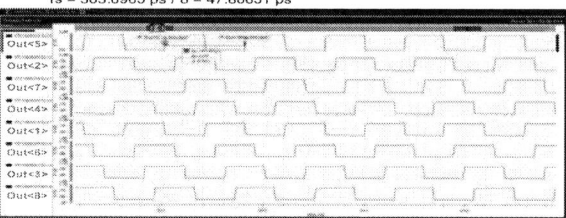

Ts = 395.451 ps / 8 = 49.43138 ps

Figure 2.   Simplified schematic of proposed TDC

The proposed TDC is designed using a 28-nm CMOS process. TDC occupies 348 X 80 μm² as shown in Fig 3. ENOB of proposed TDC is 15bit in delay ratio at 31/32. LSB resolution is 1.54 ps and maximum range is 50.6 ns. The performance of the TDC are summarized and compared with other TDCs Table 1. The proposed design shows a wide input range of 50.6 ns, while maintaining 1.54 ps time resolution. The power consumption of 0.9 mW is very low compared with other reports.

III.   CONCLUSION

A new TDC scheme is proposed to enlarge the input time range with low power consumption. Using two ring oscillators (fast and slow) with frequency difference, precise time resolution is achieved, which is the time delay difference between two oscillators. The input time range is 50.6 ns with 0.9 mW power consumption.

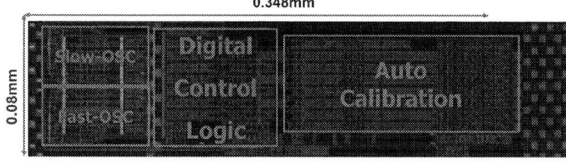

Figure 3.   Layout of proposed TDC

TABLE I
SUMMARY AND COMPARISON OF THE PERFORMANCE

	[10]	[12]	[13]	This work
Scheme	Vernier	Two-step	Noise shaping	Slow & Fast Oscillator
Technology	130nm	90nm	130nm	28nm
Supply	1.5V	1.5V	1.5V	1V
Area	0.26mm²	0.6mm²	0.04mm²	0.027mm²

	[10]	[12]	[13]	This work
1st resolution (N_S)	N/A	N/A	N/A	395.5ps (12.32ps x 2^5 = 395.5 ps)
2nd resolution (N_F)	N/A	N/A	N/A	12.32ps (1.54ps x 2^3 = 12.32 ps))
Min. resolution 3rd resolution (T_S-T_F)	8ps	1.25ps	1.2ps	1.54ps
Range	32ns	0.64ns	2.5ns	50.6ns
Power consumption	7.5mW	4.5mW	31.5mW	0.9mW

ACKNOWLEDGMENT

This work was supported by the Industrial Core Technology Development Program(10049095, "Development of Fusion Power Management Platforms and Solutions for Smart Connected Devices") funded By the Ministry of Trade, industry & Energy and by

REFERENCES

[1] V. Kratyuk et al., "A design procedure for all-digital phase-locked loops based on a charge-pump phase-locked-loop analogy," IEEE Trans. Circuits Syst. II, vol. 54, no. 3, Mar. 2007.

[2] T. Olsson and P. Nilsson, "A digitally controlled PLL for SoC applications,"IEEE J. Solid-State Circuits, vol. 39, pp. 751–760, May 2004.

[3] R. B. Staszewski, "Digitally controlled oscillator (DCO)-based architecture for RF frequency synthesis in a deep-submicrometer CMOS process," IEEE Trans. Circuits Syst. II, vol. 50, pp. 815–828, Nov. 2003.

[4] J. A. Tierno et al., "A wide power supply range, wide tuning range, all static CMOS all digital PLL in 65 nm SOI," IEEE J. Solid-State Cir.,vol. 43, no. 1, pp. 42–51, Jan. 2008.

[5] V. Kratyuk, "A digital PLL with a stochastic time-to-digital converter,"in Proc. Symp. VLSI Circuits, Jun. 2006, pp. 38–39.

[6] S.-Y. Yang et al., "A 7.1 mW 10 GHz all-digital frequency synthesizer with dynamically reconfigurable digital loop filter in 90 nm CMOS,"in IEEE Int. Solid-State Circuits Conf. Dig., Feb. 2009, pp. 90–91.

[7] R. B. Staszewski et al., "All-digital PLL and transmitter for mobile phones," IEEE J. Solid-State Circuits, vol. 40, pp. 2469–2482, Dec. 2005.

[8] R. B. Staszewski et al., "1.3 V 20 ps time-to-digital converter for frequency synthesis in 90-nm CMOS," IEEE Trans. Circuits Syst. II, vol. 53, no. 3, pp. 220–224, Mar. 2006.

[9] P. Dudek et al., "A high-resolution CMOS time-to-digital converter utilizing a vernier delay line," IEEE J. Solid-State Cir., vol. 35, no. 2, pp. 240–247, Feb. 2000.

[10] J. Yu et al., "A 12-bit vernier ring time-to-digital converter in 0.13 μm CMOS technology," in IEEE Symp. VLSI Circuits Dig., Jun. 2009, pp. 232-233.

[11] J. Lin et al., "A PVT tolerant 0.18 MHz to 600 MHz self-calibrated digital PLL in 90 nm CMOS process," in IEEE Int. Solid-State Circuits Conf., Feb. 2004, pp. 488–489.

[12] M. Lee et al., "A 9 b, 1.25 ps resolution coarse-fine time-to-digital converter in 90 nmCMOSthat amplifies a time residue," in IEEE Symp. VLSI Circuits Dig., Jun. 2007, pp. 168–169.

[13] M. Z. Straayer et al., "An efficient high-resolution 11-bit noise-shaping multipath gated ring oscillator TDC," in IEEE Symp. VLSI Circuits Dig., Jun. 2008, pp. 82–83.J. Clerk Maxwell, A Treatise on Electricity and Magnetism, 3rd ed., vol. 2. Oxford: Clarendon, 1892, pp.68–73.

[14] A. M. Abas et al., "Time difference amplifier," IEE Electronics Lett.,vol. 38, no. 23, pp. 1437–1438, Nov. 2002.

[15] Seon-Kyoo Lee "A 1 GHz ADPLLWith a 1.25 ps Minimum-Resolution Sub-Exponent TDC in 0.18 μm CMOS' in IEEE JournalL of Solid-State Circuits, Vol. 45, No. 12, December 2010.

Novel 8-T CNFET SRAM Cell Design for the Future Ultra-low Power Microelectronics

YoungBae Kim, Qiang Tong, Ken Choi
Electrical and Computer Engineering Department
Illinois Institute of Technology
Chicago, USA
{ykim102, qtong, kchoi12}@hawk.iit.edu

Yunsik Lee
School o Electrical and Computer Engineering
Ulsan National Institute of Science and Technology
Ulsan, Korea
leeys@unist.ac.kr

Abstract— In deep sub-micron technology, leakage power consumption has become a major concern in VLSI circuits, especially for SRAM, which is used to build the cache in System-on-Chip (SOC). In this paper, a low power 8-T SRAM cell, based on carbon nanotube field effect transistor (CNFET), is proposed to circumvent the leakage power issue. Experiment datas show that the proposed SRAM cell can save 97.94% static power consumption compared to existing 6T CNFET SRAM cell. In case of writing, the proposed SRAM cell comsumes 39.27% less power than the traditional SRAM cell for writing 0 and 58.79% less for writing 1. Also, because of the adoption of a colaborated voltage sense amplifier and independent read component, our 8T SRAM shows much improved dealy performance, the delay is observed to reduce by approximate 30% in write operation and approximate 90% in read operation.

Keywords; CNFET; low power SRAM; 8-T SRAM; SRAM; carbon nanotube filed effect transistor;

I. INTRODUCTION

In field of VLSI system design and microprocessor design, the power consumption has become overwhelmingly critical, especially for SRAM design. Three reasons contribute to this situation: 1) Most System-on-Chip (SOC) in modern VLSI design integrates large portion of cache, which is built by SRAM cell; 2) in deep sub-micron technology, the area of die becomes smaller and smaller, the smaller area raises the temperature issue in VLSI circuits, and hence reducing power consumption is necessary to limit the temperature in the chips; 3) the popularity of portable devices which are power by batteries makes the lower power VLSI circuits more competitive in consumer electronics market. For these reasons, many researchers and SRAM designers have been trying to find method to achieve low power dissipation and high power efficiency in SRAM. Among those existing SRAM cell designs, carbon nanotube filed effect transistor (CNFET) is a promising substitution of traditional MOSFET in SRAM cell design due to its power efficiency and variation resistance. Carbon nanotube filed effect transistor (CNFET) is considered to be a promising candidate for the future nano-electronic devices. It has excellent electrical properties, such as High-speed, high-K compatibility, and chemical stability [2]. In this paper, we propose a novel 8-T SRAM cell based on carbon nanotube field effect transistor (CNFET) to reduce static and dynamic power.

II. PROPOSED 8-T SRAM CELL DESIGN

To construct the CNFET 8-T SRAM circuit, we used a Stanford CNFET HSPICE model [6]. In order to reduce switching power consumption, we separated reading part and writing part as shown in Figure 1. Unlike conventional 6T SRAM, the proposed 8-T cell uses only two driver transistor M1 and M2. During the write operation, the circuit does not include a race between the latch and the access transistors. This in turn enhances the write speed and saves a lot of power consumption. In terms of reading part, our proposal offers better low-voltage operation performance due to the stacking of devices in the read port, though it suffers a commensurate area penalty.

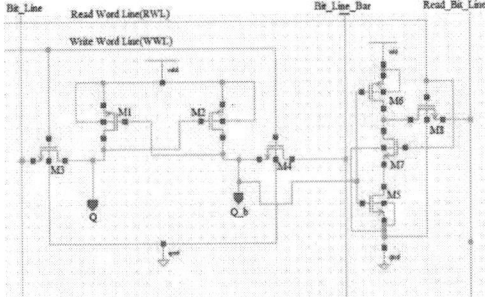

Figure 1. Proposed nevel 8T CNFET SRAM Cell

Such alternative SRAM cell designs successfully cope with the difficulty of maintaining proper operation at high yield constraints in the subthreshold operating region, and offer promising characteristics for realizing reliable cache.

TABLE I. SIZE OF THE CNFET

No	Number of the CNFET	Size(No. of tubes)
1	M1	2
2	M2	2
3	M3	6
4	M4	6
5	M5	2
6	M6	2
7	M7	2
8	M8	2

978-1-5090-3220-4/16 $31.00 © 2016 IEEE

III. SIMULATION AND RESULTS

For the simulation, we used a Stanford CNFET HPICE model [6] and set the Vdd to 0.9v, and all the CNFET transistor of 4T PLNA cell are used minimal size as shown in Table 1. To simulate the circuit, we built the whole circuit for 8-T CNFET SRAM. Before the write operation, both Bit_Line and Bit_Line_Bar have pre-discharged to ground. We used a pseudo-random sequence of 0110100 to operation of writing and reading.

TABLE II.   POWER CONSUMPTION COMPARISION

Power(W)	Traditional 6T CNFET	Proposed	Improvement
Static	1.816E-09	3.725E-11	97.94%
Write 0	1.591E-06	9.662E-07	39.27%
Write 1	4.701E-06	1.937E-06	58.79%
Read 0	4.688E-09	1.019E-10	97.82%
Read 1	2.485E-09	1.026E-10	95.87%

As seen in Table 2, due to the characteristic of CNFET and elimination of race condition between the latch and the access transistors, our proposed SRAM reduce static power by 97.94% when the SRAM is in standby mode. During a write operation, the dynamic power consumption varies drastically with the cell data. During writing '1' operation, the dynamic power consumption of the proposed 8-T cell is reduced by 58.79%, compared to traditional 6-T cells. And for writing '0' operation, the dynamic power consumption of the proposed 8-T cell is reduced by 39.27%. Since the reading part of our design can be shared for all the cells in a SRAM column, the average transistors used in each SRAM cell of our proposed design is around 4, so our proposed SRAM design will also have area improvement compared to the traditional 6-T SRAM design. In summary, our proposed SRAM cell has both area and power advantage over the traditional 6-T SRAM design.

TABLE III.   DELAY COMPARISION

Delay(s)	Traditional 6T CNFET	Proposed	Improvement
Write 0	1.386E-08	1.002E-08	27.71%
Write 1	8.117E-09	5.610E-09	30.88%
Read 0	7.4E-07	1.10E-07	85.13%
Read 1	4.61E-07	1.645E-08	96.43%

As shown in Table3, Because of the optimized 4T PLNA cell and size, the speed of write operation is improved by approximate 30%. Both the independently read access and speed of sense amplifier(Figure 2) are going to be affected the read latency. So, we could improve 85% when reading 0 and 96% when reading 1. Since the our separated reading section from internal node of latch, one side the internal node can be changed the semi output node without any voltage change. Thus, our 8T SRAM cell removed the problem of read SNM by buffering the stored data during the reading access.

IV. CONCLUSION

In this paper, we proposed an 8-T SRAM cell based on CNFET in 32nm technology by using HSPICE simulations and Stanford CNFET models And compared with traditional 6-T CNFET SRAM cell, the new 8T SRAM cell can reduce substantial static and dynamic power. Especially for the static power consumption, experimental result shows that our proposed design reduces static power consumption by 97.94%, compared to traditional SRAM cell. Also, due to the combination of independent reading access component and voltage sense amplifier, the read speed of the our 8T SRAM is improved by 96% for reading 1 and 85% for reading 0. As of future work, we may include further improvement of SNM.

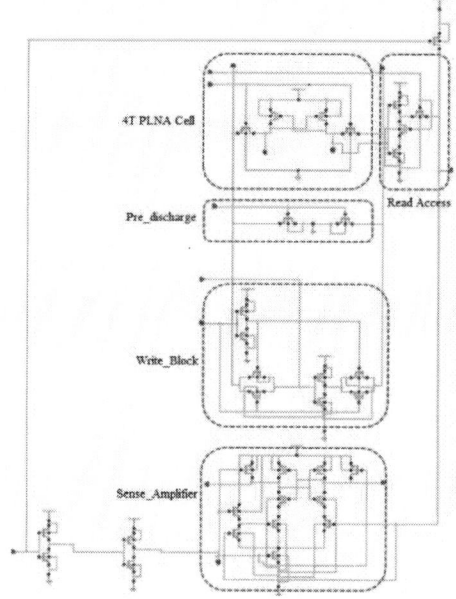

Figure 2.   Novel 8-T SRAM cell operation

REFERENCES

[1]   Yinhui Chenn, Zhiyuan Yu, Hailing Nan, and Ken Choi, "Ultralow Power SRAM Design in Near Threshold Region using 45nm CMOS Technology

[2]   J.Appenzeller "Carbon Nanotubes for High-performance Electronics-Progress and Prospect," Proc, IEEE, Volume 96, Issue 2, pp.201 – 211, Feb.2008

[3]   K.Noda, K. Matsui, K.Takeda, N. Nakamura," A loadless CMOS four-transistor SRAM cell in a $0.18\mu m$ logic technology," IEEE Trans Electron Devices, Vol48, no.12,pp2851-2856, Dec.2001

[4]   J. Deng and H.-S.P. Wong, "A compact SPICE Model for Carbon-nanotube Field-Effect Transistors Including Nonidealities and Its application-Part I:Model of the intrinsic Channel Region," IEEE Transactions on Electron Devices, Vol.54,2007, pp. 3186-3194

[5]   P.L. McEuen and M.S. Fuhrer, "Single-walled carbon nanotube electronics," IEEE Transactions on Nanotechnology, Vol. 1, Mar.2002, pp. 78-85.

[6]   Albert Lin, Gordon Wan, Jie Deng, and H-S philip Wong "A Quick user guide on stanford University Carbon Nanotube Field Effect Transistor(CNFET) HSPICE model,"2008

Gap in pagination due to withheld paper.

Pages 245-246

A MDLL-based Multi-Phase Clock Multiplier

Junsub Yoon and Jongsun Kim
Electronic and Electrical Eng., Hongik University
Seoul, Korea
js.kim@hongik.ac.kr

Abstract— **A multiplying delay-locked loop (MDLL)-based multi-phase clock multiplier is presented. The proposed clock multiplier provides 8-phase output clocks and achieves a frequency range of 0.6–1.0 GHz with programmable fractional multiplication ratios of N/M, where N = 4, 5, 8, 10 and M = 1, 2, 3. The proposed clock multiplier is implemented in a 65 nm CMOS process and occupies an active area of 0.01 mm^2. It achieves an effective peak-to-peak jitter of 5 ps and dissipates 3.4 mW from a 1.0 V supply at 1 GHz.**

Keywords; clock multiplier, clock generator, MDLL

I. INTRODUCTION (HEADING 1)

Phase-locked loop (PLL)-based multi-phase clock generators and multipliers have been widely used for various integrated circuit design [1]. PLLs, however, require a large loop capacitor and exhibit jitter accumulation problems. Multiplying delay-locked loops (MDLLs) [2, 3, 4] have been have been introduced to provide integer-ratio (×N) frequency multiplication, and fractional-ratio frequency multiplying DLL has been introduced, which can provide ×N/M multiplication [5]. However, [5] can generate only single-phase multiplied clock. This limits its application scope because it cannot be used in instances where parallel data has to be serialized using multi-phase clocks or when a sub-sampling ADC requires multi-phase clocks. In this paper, a new MDLL-based multi-phase clock multiplier is presented. To achieve low-jitter fractional-ratio clock multiplication with 8-phase output clocks, the proposed architecture adopts a 4-stage differential voltage-controlled delay line.

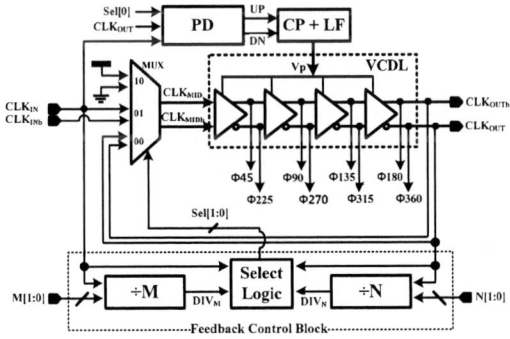

Figure 1.   Proposed multi-phase clock multiplier

II. PROPOSED MULTI-PHASE CLOCK MULTIPLIER

The Fig. 1 shows the block diagram of the proposed multi-phase clock multiplier. It consists of a phase detector (PD), a charge pump (CP) and a loop filter (LF) combination, a differential 3-to-1 multiplexer (MUX), a voltage controlled delay line (VCDL), and a feedback control block. The feedback control block consists of a select logic, a divide-by-N divider (÷N), and a divide-by-M divider (÷M). With the placement of two dividers (÷N, ÷M), this clock multiplier can provide multiplied fractional-ratio output clock $f_{clk\_out} = f_{clk\_in} \times$ (N/M), where N and M are integer and programmable and f_{clk_out} and f_{clk_in} are the frequencies of the output and input clocks. With this scheme, no skew error is generated between the input and output clocks. Unlike the architecture of [5] which generates only single-phase output clock, the proposed architecture provides equally spaced 8-phase output clocks ($\Phi 45°$, $\Phi 90°$, $\Phi 135°$, $\Phi 180°$, $\Phi 225°$, $\Phi 270°$, $\Phi 315°$, $\Phi 360°$) by adopting a differential 4-stage VCDL. The select logic generates the Sel[1:0] signal to select the MUX's output among the three differential MUX input signals: CLK$_{IN}$/CLK$_{INB}$, CLK$_{OUT}$/CLK$_{OUTB}$, and Supply/Ground. When Sel[1:0] = [0, 0], the MUX selects CLK$_{OUT}$/CLK$_{OUTB}$ and the VCDL operates as a ring oscillator. When Sel[1:0] = [1, 0], the MUX selects Supply/Ground and the delay of the VCDL is extended without the need for additional delay cells. When Sel[1:0] = [0, 1], the MUX selects CLK$_{IN}$/CLK$_{INB}$ and the integrated jitter is eliminated through the injection of the clean reference clock. This MUX switching operation is repeated at every M reference input clock cycle until the VCDL adjusts its delay to achieve $f_{clk\_out} = f_{clk\_in} \times$ (N/M).

III. EXPERIMENTAL RESULTS

Fig. 2 displays the chip layout and the test chip-on-board (CoB) of the proposed multi-phase clock multiplier which is fabricated in a 65nm TSMC CMOS process. It occupies an active area of only 0.01 mm^2 and achieves an output frequency range of 0.6–1.0 GHz with programmable multiplication ratios of N = 4, 5, 8, 10 and M = 1, 2, 3. It consumes 3.4 mW when N/M = 10/3 from a 1.0 V supply at 1 GHz. Fig. 3 shows the measured locking process when $f_{clk\_out} = 1.0$ GHz, $f_{clk\_in} = 300$ MHz, and N/M = 10/3. After locking, the FMDLL's (N+1)th rising edge of CLK$_{OUT}$ is exactly aligned with the (M+1)th rising edge of CLK$_{IN}$, which clearly confirms the deskewed multiplication of N/M = 10/3. Fig. 4 shows the simulated 8-phase output clock signals at 1 GHz. Ideally, two adjacent

978-1-5090-3220-4/16 $31.00 © 2016 IEEE   247   ISOCC 2016

phases are 125 ps apart, and the simulated maximum phase difference error is only 0.67 ps. It achieves a measured peak-to-peak (p-p) output clock jitter of 20 ps (with p-p input clock jitter of 15 ps at f_{clk_IN} = 500 MHz) and root-mean-square (RMS) jitter of 2.78 ps at 1 GHz when N/M = 4/2.

Table I summarizes the performance of the proposed multi-phase clock multiplier and compares it with the prior art. It can be seen that the proposed architecture achieves lower power and smaller area while maintaining the ability to supply deskewed fractional-ratio multi-phase output clocks.

IV. CONCLUSION

A MDLL-based multi-phase clock multiplier is presented. The proposed fractional-ratio clock multiplier operates over a frequency range of 0.6–1.0 GHz with programmable multiplication ratios of N/M, where N = 4, 5, 8, 10 and M = 1, 2, 3. It provides 8-phase output clocks and achieves an effective p-p jitter of 5 ps and dissipates 3.4 mW from a 1.0 V supply at 1 GHz. It can be applied to low-jitter, low-power, and small-area clock generators that require flexible frequency multiplication with deskewed multi-phase output clocks.

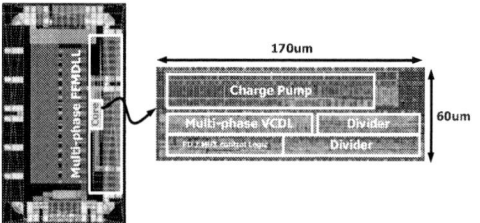

Figure 2. Chip layout and test Chip-on-Board (CoB) of the proposed multi-phase clock multiplier

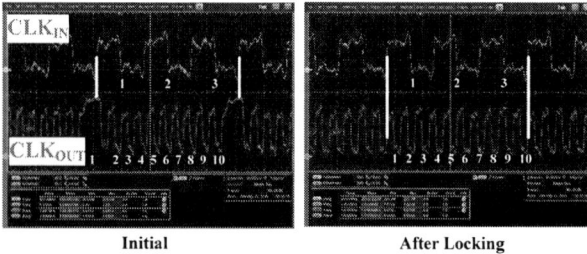

Figure 3. Measured locking process of the proposed multi-phase FFMDL when f_{clk_out} = 1.0 GHz, f_{clk_in} = 300 MHz, and N/M = 10/3

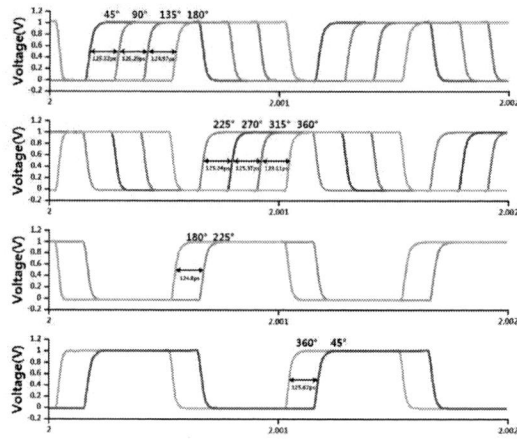

Figure 4. Simulated 8-phase output clocks at 1 GHz

TABLE I. PERFORMANCE SUMMARY AND COMPARISON

	[2]	[3]	[5]	This Work
Architecture	MDLL	MDLL	MDLL	MDLL
Process	0.18 μm	0.18 μm	0.13 μm	65 nm
Supply (V)	1.8	1.8	1.2	1.0
Frequency range (GHz)	0.2–2	0.9–2.9	0.85–1.8	0.6–1.0
Multiplication factor (N)	4,5,8,10	13–20	4,5,8,10	4,5,8,10
Division factor (M)	x	x	1, 2, 3	1, 2, 3
DLL mode support	x	x	x	O
Multi-phase output	x	x	x	O
Effective p-p jitter (pS)	13.11 @2GHz	12.9 @2.1GHz	7.5 @1GHz	5.0 @1GHz, N/M=4/2
RMS jitter (pS)	-	-	-	2.78 @1GHz
Phase noise (dBc/Hz)	-	-	-	-123.3 @100KHz
Power (mW)	12 @2GHz	19.8 @2GHz	9.0 @1.5GHz	3.4 @1.0GHz
Active area (mm²)	0.05	0.07	0.018	0.01

ACKNOWLEDGMENT

This work (C0396252) was supported by Business for Cooperative R&D between Industry, Academy, and Research Institute funded Korea Small and Medium Business Administration in 2016. This work was also supported by the Korea Institute for Advancement of Technology (KIAT) grant funded by the Korean government (Motie: Ministry of Trade, Industry & Energy, HRD Program for Software-SoC convergence) (No. G02N04500000301).

REFERENCES

[1] M. Demirkan, et al, "A pulse-based ultra-wideband transmitter in 90-nm CMOS for WPANs," IEEE J. Solid-State Circuits, 43, No.12, pp.2820-2828, 2008.

[2] R. Farjad-Rad, et al, "A low-power multiplying DLL for low-jitter multigigahertz clock generation in highly integrated digital chips," IEEE J. Solid-State Circuits, 37, no. 12, pp. 1804–1812, 2002.

[3] Q. Du, et al, "A low-phase noise, anti-harmonic programmable DLL frequency multiplier with period error compensation for spur reduction," IEEE Trans. Circuits Syst. II, Vol. 53, pp. 1205-1209, 2006.

[4] G. Park, H. Kim, and Jongsun Kim, "A reset-free anti-harmonic anti-harmonic programmable MDLL-based frequency multiplier", J. Semiconductor Technology and Science, Vol. 13, no. 5, pp. 459-464, Oct. 2013.

[5] S. Han, J. Kim, and Jongsun Kim, "Programmable fractional-ratio frequency multiplying clock generator", IET Electronics Letters, Vol. 50, no. 3, pp. 163-165, 2014.

Proposal for sensitive frequency demodulator for 10-Gb/s transmission labeling signal system

Natsuyuki Koda, Kosuke Furuichi, Hiromu Uemura, Hiromi Inaba and Keiji Kishine

The University of Shiga Prefecture

2500, Hikone Hassaka, Shiga, 522-8533, Japan

Email: ot23nkouda@ec.usp.ac.jp

Abstract—**In this paper, we propose a frequency demodulation circuit for realizing multiplex communication called a "labeling signal system." The proposed circuit configuration consists of delay circuits and logic circuits. In this work, a frequency demodulator designed with discrete devices detects information with an FPGA. Compared with the conventional delay detection circuit, the proposed circuit successfully increased the voltage change in accordance with the frequency shift. We also confirm its operating characteristics and advantages from the experimental results.**

I. INTRODUCTION

Recent communication systems transmit and receive a large amount of information. Therefore, we previously proposed a system for transmitting frequency-modulated data frames to communicate more information, as shown in Fig. 1[1]. In this system, a transmitter modulates the frequency of a flame signal to add information in accordance with an additional signal called a "labeling signal." The receiver has a clock and data recovery (CDR) and frequency demodulation circuit. The CDR recovers a modulated flame signal, and the frequency demodulation circuit detects the labeling signal. A delay detection circuit [Fig.2(a)] is used for the demodulation circuit in the proposed system. But the issue is that this circuit causes a slight voltage change in accordance with the frequency shift. Therefore, we have proposed a circuit configuration [2][Fig. 2(b)] so that the proposed system can sensitively change the voltage in accordance with the frequency shift by using transient response characteristics. The proposed circuit consists of delay and AND circuits following the conventional circuit. In this paper, we show the detailed comparison of the conventional and proposed circuits. In addition, it is shown that the experimental results by using discrete devices and an FPGA at around 10-Gb/s data transmission. From the results, the proposed circuit could detect about twice the detection data than could the conventional circuit.

II. DETAILED COMPARISON OF THE CONVENTIONAL AND PROPOSED DEMODULATORS

We have proposed a sensitive demodulator for the proposed system. We show the detailed comparison of the conventional and proposed circuits below.

A. Conventional delay detection circuit

Figure 2(a) shows the conventional circuit configuration. The delay module of this circuit is designed so that the phase

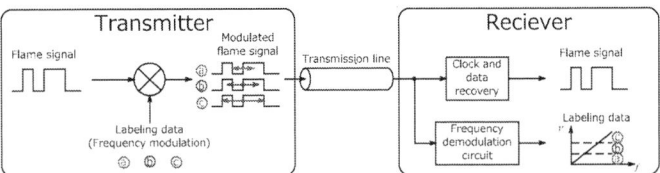

Fig. 1. Overview of labeling signal transmission transmission system

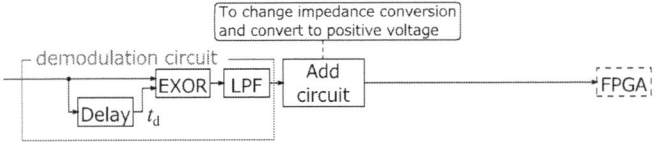

(a) Conventional delay detection circuit configuration

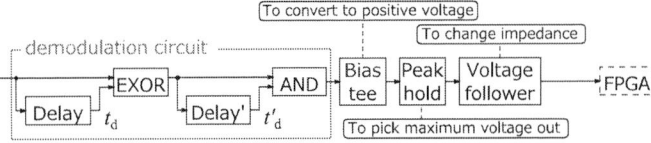

(b) Proposed circuit configuration

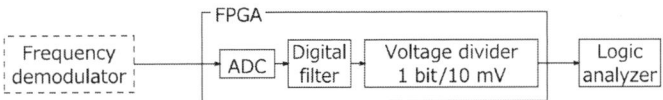

(c) Detection system configuration of frequency demodulation

Fig. 2. Configuration of frequency demodulation circuit

difference is $\pi/2$[rad] (t_d=50ps) at the time of the reference bit rate (T_{ref}). In the EXOR output, while the high-level period is the delay time in the delay module, the low-level period changes in accordance with the transmission rate of the input signal [Fig. 3(a)]. The duty rate change is detected as the average voltage change with the low pass filter (LPF), as shown in Fig. 3(c). The LPF voltage of the conventional circuit V_E is obtained with the following equation.

$$V_E = \frac{t_d}{T_{ref} - \Delta T} V_{Emax} = \frac{T_{ref}}{2(T_{ref} - \Delta T)} V_{Emax} \quad (1)$$

ΔT shows the period change of the modulated data frequency and V_{Emax} is the maximum amplitude of the EXOR circuit.

B. Proposed circuit configuration

Figure 2(b) shows the proposed circuit configuration. When the delay time of each delay module is set to be around 40ps, the AND circuit can detect the period-change of the EXOR

978-1-5090-3220-4/16 $31.00 © 2016 IEEE

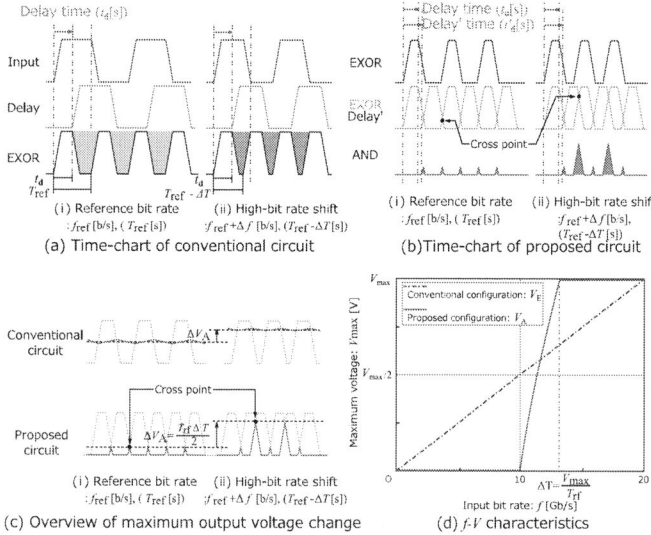

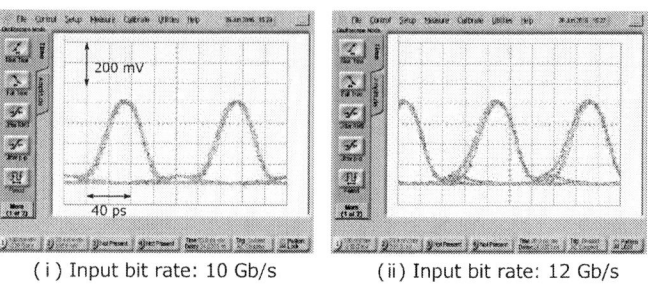

(ⅰ) Input bit rate: 10 Gb/s   (ⅱ) Input bit rate: 12 Gb/s

(a) Waveform of conventional circuit

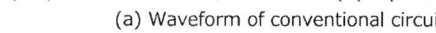

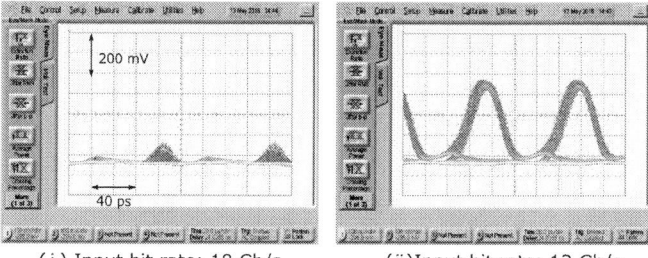

(ⅰ) Input bit rate: 10 Gb/s   (ⅱ)Input bit rate: 12 Gb/s

(b) Waveform of proposed circuit

Fig. 4. Waveforms of experimental results

Fig. 3. Time chart and f-V characteristics

low-level that depends on the modulated frequency. By using the gradient of the rise and fall time of the EXOR output waveform (T_{rf}), the output voltage of the proposed circuit V_A is changed by the crosspoint of the EXOR and Delay' voltage [Fig. 3(c)]. The output voltage of the proposed circuit V_E is obtained with the following equation.

$$V_A = \begin{cases} 0 & (\Delta T \leq 0) \\ \dfrac{T_{rf}\Delta T}{2} & (0 < \Delta T \leq \dfrac{V_{Amax}}{T_{rf}}) \\ V_{Amax} & (\dfrac{V_{Amax}}{T_{rf}} < \Delta T) \end{cases} \quad (2)$$

V_{Amax} is the maximum amplitude of the AND circuit. Figure 3(d) shows V_A and V_E calculated by (1),(2) when it is defined that the T_{rf} is 1.75×10^{10} and the V_{max} is 400 mV. In the proposed circuit, while a frequency band becomes narrow, this circuit can detect the frequency shift sensitively.

III. EXPERIMENTAL RESULTS

To investigate the output characteristics of each circuit, a test environment with an FPGA and discrete devices was designed, as shown in Fig. 2. The input signal is generated by a pulse-pattern-generator (PPG). The PPG output signal was PRBS ($2^7 - 1$) signal, and the input bit-rate of each demodulator was changed from 9.5 Gb/s to 12.5 Gb/s. In the FPGA [Fig. 2(c)], the voltage during 10-Gb/s transmission was set as the standard voltage, and every time the binary detection data changed 10mV, it was regarded as a frequency shift.

Figures 4 show the output waveforms of the conventional EXOR and the proposed AND circuit, respectively. For each waveform, it was confirmed that the desired movement was performed as shown in Section II. Figure 5(a) shows the output voltage change of each circuit before being input into the FPGA. While the output voltage of the conventional circuit could change the output only about 60 mV, the proposed circuit could it about 200 mV. Therefore, the proposed circuit could sensitively change the voltage in accordance with the frequency shift. Figure 5(b) shows the change of the FPGA detection data in accordance with the frequency shift. A peculiar point appeared in the result around the 11-Gb/s point in the

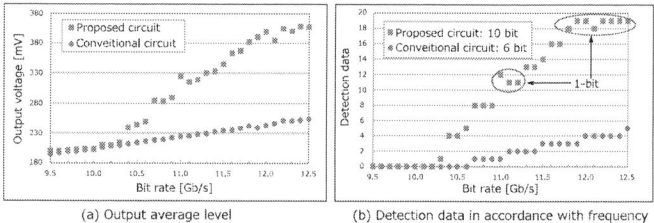

(a) Output average level   (b) Detection data in accordance with frequency

Fig. 5. Frequency characteristics of experiment results

proposed circuit. When this point was converted together by the FPGA cord description in the proposed circuit, the different values that can be detected was 6 with the conventional circuit and 10 with the proposed one. This confirms that the detection data that can be detected becomes about double.

IV. CONCLUSION

We have proposed a frequency demodulation circuit for a labeling signal system. The maximum output voltage of this circuit can change in accordance with the input bit rate sensitively. In this paper, we made a comparison of the conventional and proposed circuits. In addition, we investigated the characteristics of the proposed circuit by using discrete devices and an FPGA. The proposed circuit successfully detected about twice the detection data than did the conventional circuit.

ACKNOWLEDGMENTS

A part of this research was supported by Grants-in-Aid for Scientific Research from the Japan Society for the Promotion of Science.

REFERENCES

[1] D. Omoto, et. al., "Simple routing control system for 10-Gb/s data transmission using a frequency modulation technique" International Conference on Electronics, Information, and Communication,2016.

[2] N. Kouda, et. al.,"Simplification of Labeling Signal Detecting Circuit Using a Delay Detection Circuit" The Institute of Electronics, Information and Communication Engineers,p.19,2016.

A Transient Enhanced External Capacitor-Less LDO With A CMOS Only Sub-Bandgap Voltage Reference

Chang-Bum Park, Chan-Kyeong Jung
Dept. Electronics and Computer Engineering
Seokyeong University
Seoul, Korea
wdrqq1@skuniv.ac.kr, ckjung@skuniv.ac.kr

Shin-Il Lim
Dept. Electronics Engineering
Seokyeong University
Seoul, Korea
silim@skuniv.ac.kr

Abstract— **This paper presents an external capacitor-less low-dropout(LDO) regulator with a voltage spike detection circuit for the enhanced transition response and with a CMOS only sub-bandgap voltage reference(BGR) operated in subthreshold region. CMOS sub-BGR adopted a weighted Vgs structure and a body bias technique for reducing the variations from process, voltage and temperature (PVT). The proposed LDO achieved the PSRR of -96dB at DC and -34dB at 1MHz by using 3-stage configuration. The proposed LDO operates with the reference voltage of 283mV from the Sub-BGR and provides the output voltage of 1.5V. Simulated results shows that overshoot and undershoot of output voltage were reduced to 62mV and 56mV respectively when the load current changes from 0 to 50mA. Total power consumption was 60μA and the chip area was 0.03358mm² with 0.18μm CMOS process.**

Keywords : Bandgap reference, sub-BGR, LDO, regulator

I. INTRODUCTION

The LDO plays a key role for minimizing the ripples of supply voltage that considerably reduce the performance of systems-on-chip(SoC) and for reducing the power consumption. Moreover, the capacitor-less LDO with high power supply rejection ratio (PSRR) has been attractive to reduce the size of printed circuits board (PCB) and to reduce the noise in analog-front-end circuits. According to this trend, an external capacitor-less LDO which has 2-stage error amplifier with internal zero compensation techniques was published [1]. However, the LDO suffers from large voltage spike due to the abrupt change of load currents when the external capacitor is removed. To solve this problem, in this paper, we adopted a transition enhancement circuit [2] and also the CMOS only sub bandgap voltage reference with a body bias technique is proposed to produce the reference voltage which is robust to process, voltage and temperature (PVT) variations.

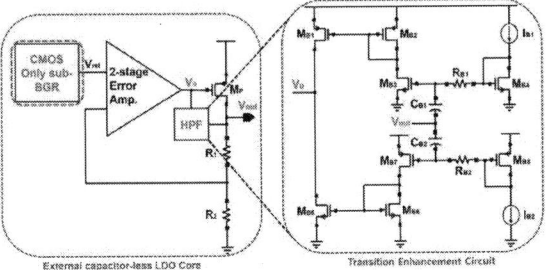

External capacitor-less LDO Core / Transition Enhancement Circuit

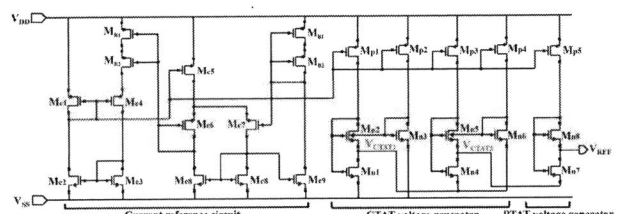

Figure 1. Structure of proposed LDO with transition enhancement circuit

Figure 2. Schematic of proposed sub-BGR

II. PROPOSED CIRCUITS

A. Proposed LDO

Fig. 1 shows the structure of the proposed LDO regulator and the schematic of the transition enhancement circuit. Total 3-stage including the power MOS transistor (Mp) with 2-stage error amplifier is used for achieving high PSR [1]. And to achieve enough phase margin for stability with a small internal capacitor, cascode and current buffer compensation techniques are applied in the proposed LDO [1]. Reference voltage is provided by the CMOS only sub-BGR circuits. Also, the transition enhancement circuits which produce a function of a high pass filter in feedback network are adopted [2] to reduce the voltage spike during the significant change of the load current. If Vout is lowered due to the abrupt increase of the load current, the gate voltages of M_{B3} and M_{B7} will be lowered. This in turn lowered the gate voltage of driver transistor M_P. This lowered gate voltage of M_P makes the output voltage of Vout to increase. As a result, we can achieve the regulation.

B. Proposed sub-BGR

Fig. 2 shows the schematic of proposed sub-BGR which is consist of current reference circuits [3] [4], complementary to absolute temperature (CTAT) voltage generator and proportional to absolute temperature (PTAT) voltage generator [5]. The MOS transistors of Mn3 and Mn6 are used to provide a body voltage to the Mn2 and Mn5 for reducing their threshold voltage. As a result, Mn2 and Mn5 have a lower threshold voltage than Mn1 and Mn4. The threshold voltage difference of Mn1 and Mn2 allows V_{CTAT1} to have CTAT voltage. PTAT voltage is created by the MOS transistors of Mn7 and Mn8. Assuming these two MOS transistors have the same threshold voltage, PTAT voltage is determined by the

978-1-5090-3220-4/16 $31.00 © 2016 IEEE   251   ISOCC 2016

W/L ratio of Mn7 and Mn8. All transistors are operated in the sub-threshold region to reduce the power consumption.

III. SIMULATION RESULTS

The proposed sub-BGR and LDO were implemented with the Magna 0.18-μm CMOS technology. The output reference voltage of sub-BGR was 283mV as shown in Fig. 4. The voltage variation at each process corner is 20mV. As shown in Fig. 3, the simulated temperature variation was 26ppm/°C at typical-typical. Fig. 5 shows the load transition results of proposed LDO when the output load current has the change of 50mA. With the transition enhancement circuits, the undershoot and overshoot of LDO output voltage were reduced from 276mV to 62mV and from 163mV to 56mV respectively when the load current changes from 0 to 50mA and also from 50mA to 0. The total active area was 230μm by 146μm. The performance summary was represented in table 1.

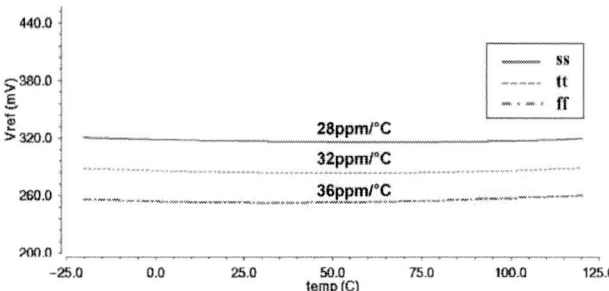

Figure 3.   Temperature and process variations of proposed sub-BGR

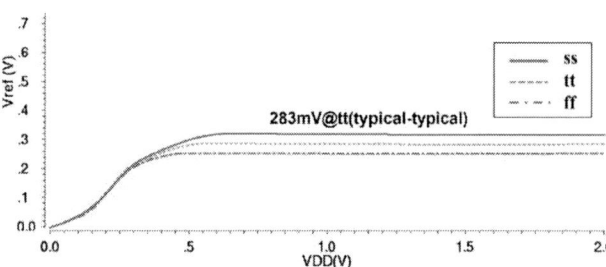

Figure 4.   Voltage and process variations of proposed sub-BGR

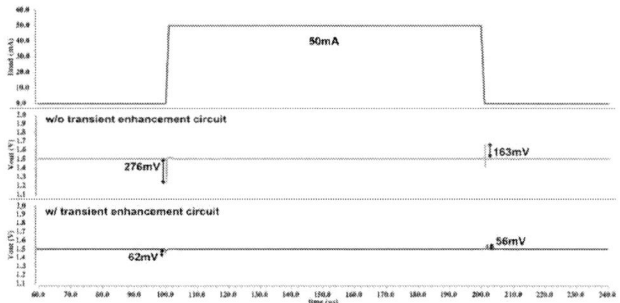

Figure 5.   Load transition of proposed LDO

IV. CONCLUSION

We proposed the external capacitor-less high PSRR LDO with a transition enhancement circuits. Moreover, the full CMOS sub-BGR without any BJT and resistors was proposed. To guarantee the stability of LDO, the cascode and current buffer compensation based on Miller compensation technique were adopted. With the transition enhancement circuits, the voltage spike was reduced to under 1/3 when the load current rapidly changes.

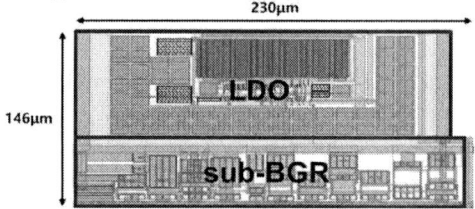

Figure 6.   Layout

TABLE I.   PERFORMANCE SUMMARY

Parameter	*This work*
Technology	0.18um(Magna)
Vin/Vout	1.8V/1.5V
Reference voltage	283mV
Temperature range	-40°C ~ 120°C
Temperature variation	26ppm/°C
PSRR	-96dB@DC, -34dB@1MHz
Load Max. current	50mA(Extendable)
Line regulation	0.52mV/V(with 50mA)
Load regulation	2.1uV/mA
Load transition	62mV(undershoot) 56mV(overshoot)
Power consumption	60μA
Chip Area(mm²)	0.03358

ACKNOWLEDGMENT

This research was supported by the Industrial Core Technology Development Program (10049009) funded by the Ministry of Trade and also supported by the MSIP(Ministry of Science, ICT and Future Planning), Korea, under the ITRC(Information Technology Research Center) support program(IITP-2016-H8501-16-1010) supervised by the IITP(Institute for Information & communications Technology Promotion). The CAD tools were supported by IC Design Education Center (IDEC).

REFERENCES

[1] Jin Woo Kim, Shin-Il Lim "An Output-Capacitorless High PSRR LDO for wide Frequency Range" *International Conference on Electronics, Information and Communication (ICEIC).* Jan, 2013.

[2] Or, Pui Ying, and Ka Nang Leung. "An output-capacitorless low-dropout regulator with direct voltage-spike detection." *Solide-State Circuit, IEEE Journal of.* Vol.45, no. 2 pp. 458-466. Feb. 2010.

[3] Y. Osaki and T. Hiroseet, "1.2-V Supply, 100-nW, 1.09-V Bandgap and 0.7-V Supply, 52.5-nW, 0.55-V Subbandgap Reference Circuits for Nanowatt CMOS LSIs", *Solid-State Circuits, IEEE Journal of* 48.6, pp. 1530-1538, 2013

[4] T.Hirose, Y. Osaki, N. Kuroki, and M. Numa, "A nano-ampere current reference circuit and its temperature dependence control by using temperature characteristics of carrier mobilities," in *Proc. Eur. Solid-State Circuits Conf.*, 2010, pp. 114-117.

[5] A.-J. Annema, "Low-Power bandgap references featuring DTMOST's," *IEEE J.Solid-State Circuits*, vol. 32, no.7, pp.949-955, Jul. 1999.

A Fast-Locking Clock Multiplying DLL

Jongsun Kim and Bongho Bae
Electronic and Electrical Eng., Hongik University
Seoul, Korea
js.kim@hongik.ac.kr

Abstract— A fast-locking clock multiplying delay-locked loop (MDLL) for fractional-ratio frequency multiplication is presented. A new phase detecting controller (PDC) has been adopted to resolve the long locking time problem of conventional MDLLs. The proposed FMDLL was implemented in 65-nm CMOS process and occupies an active area of 0.015 mm². It operates over a frequency range of 2.0–4.0 GHz with a frequency multiplication factor of N/M, where N = 4, 5, 8, 10 and M = 1, 2, 3. It achieves a peak-to-peak output clock jitter of 13.5 ps at 4 GHz while reducing locking time of about 75%.

Keywords; clock multiplier, clock generator, MDLL

I. INTRODUCTION (HEADING 1)

In recent years, multiplying delay-locked loops (MDLLs) have been introduced as an alternative to conventional PLLs [1-4]. An MDLL-based clock multiplier achieves better phase noise performance by periodically injecting a clean reference input clock to the delay line [1-6]. However, conventional MDLL architectures have an unnecessarily long phase detecting period that causes longer locking time and even lock-in fail problem. The phase detecting and selecting structure of conventional MDLL [2] causes undesired charge pump current, resulting in increased locking time and even logical fails. In this paper, a fast-locking MDLL with a new phase detecting controller (PDC) is presented to achieve a shorter locking time and eliminate lock-in fail problems.

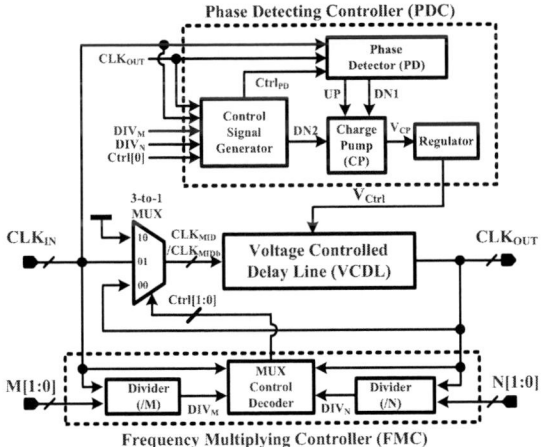

Figure 1.   Proposed fast-locking MDLL

II. PROPOSED MULTI-PHASE CLOCK MULTIPLIER

Fig. 1 shows a block diagram of the proposed fast locking FMDLL with de-skewed fractional-ratio frequency multiplication capability. It consists of a 3-to-1 MUX, a VCDL, two programmable frequency dividers (/M and /N), a MUX controller, and a new phase detecting controller (PDC). The PDC includes a PD, a CP, a regulator, and a new PD/CP controller. The PDC is used to define a new t_{PDP} window and boost the CP discharge current for fast locking.

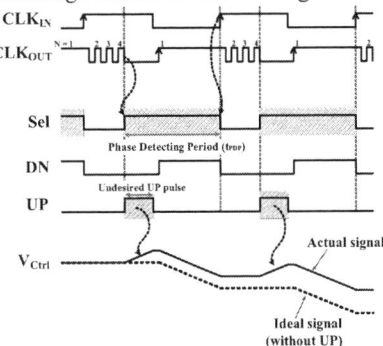

Figure 2.   Locking process of conventional MDLL [2]

Fig. 2 shows the locking process and phase detecting period of conventional MDLL [2] with N = 4. The phase detecting period (t_{PDP}) is defined by the Sel signal which is generated by select logic and activated high from the Nth falling edge of CLK_{OUT} to the rising edge of the next CLK_{IN}. Since the MDLL starts with its minimum delay, the problem is that the t_{PDP} window can become longer than half of the CLK_{IN} period. In this situation, the PD generates two CP control signals (UP and DN) simultaneously, where UP is used to charge up and DN is used to discharge the CP in [1]. In Fig. 2, the VCDL delay should be increased by lowering the V_{Ctrl} voltage for proper locking. However, the undesired UP pulses increase the V_{Ctrl} voltage (Actual signal), which increases the locking time seriously and even causes lock-in fails. Therefore, this kind of phase detecting and CP control scheme is not proper to use in applications where fast locking is required for frequency multiplication.

Fig. 3 shows the locking process of the proposed fast-locking MDLL, where N/M = 8/3 for an example. Similar to [1], the proposed MDLL has three operation modes for N/M frequency multiplication: ring oscillator (RO), reference

978-1-5090-3220-4/16 $31.00 © 2016 IEEE   253   ISOCC 2016

injection (RI), and supply injection (SI). Unlike the previous phase detecting scheme with a large t_{PDP} window, the proposed PDC generates a Sel$_{PD}$ signal to define a new t_{PDP} window and eliminates the undesired UP pulses at the beginning of the operation. It also generates the DN2 signal to provide fast discharging period of the charge pump. As a result, the proposed MDLL achieves fast locking operation and eliminates the lock-in fail problem.

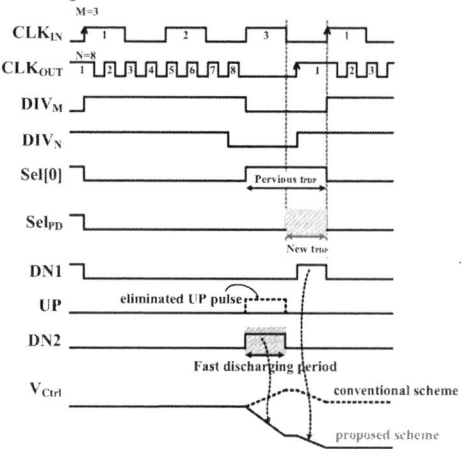

Figure 3.   Locking process of the roposed fast-locking MDLL

III.   EXPERIMENTAL RESULTS

Fig. 4 displays the chip layout of the proposed MDLL which is fabricated in a 65nm TSMC CMOS process. It occupies an active area of only 0.015 mm² and achieves an output frequency range of 2.0–4.0 GHz with programmable multiplication ratios of N = 4, 5, 8, 10 and M = 1, 2, 3. It achieves a peak-to-peak output clock jitter of 13.5 ps at 4 GHz. Compared with the previous phase detecting architecture, the locking time has been reduced about 75 %. Fig. 5 shows the measured input and output clocks when f_{clk_out} = 2.0 GHz and 4.0 GHz with N/M = 4/2 and 10/2, respectively.

Table I summarizes the performance of the proposed MDLL and compares it with the prior art. It can be seen that the proposed architecture achieves higher frequency operation with fast-locking capability.

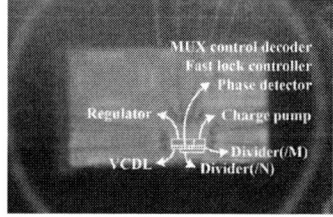

Figure 4.   Chip layout of the proposed MDLL

IV.   CONCLUSION

A fast-locking multiplying DLL is presented. The proposed MDLL operates over a frequency range of 2.0–4.0 GHz with programmable multiplication ratios of N/M, where N = 4, 5, 8, 10 and M = 1, 2, 3. It adopts a new PDC to resolve the long locking time problem of conventional MDLLs. Compared with the previous scheme, the locking time has been reduced about 75%. The proposed MDLL achieves an effective p-p jitter of 13.5 ps from a 1.2 V supply at 4 GHz.

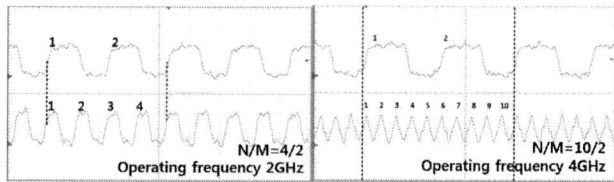

Figure 5.   Measured input and output clocks when f_{clk_out} = 2.0 GHz and 4.0 GHz with N/M = 4/2 and 10/2, respectively.

TABLE I.   PERFORMANCE SUMMARY AND COMPARISON

	[1]	[2]	This work
Process	0.13μm	0.18 μm	65nm
Supply (V)	1.2	1.8	1.2
Fractional-ratio frequency multiplication capability	Yes. N=4, 5, 8, 10 and M=1, 2, 3	No. N=4, 5, 8, 10 (Integer only)	Yes. N=4, 5, 8, 10 M=1, 2, 3
Frequency range (GHz)	0.85-1.8	0.2-2	2-4
Effective pk-pk jitter (ps)	27.5@1 GHz	13.11@2GHz	13.5@4GHz
Power (mW)	9@1.5GHz	12@2GHz	6.7@2GHz
Active area (mm²)	0.018	0.05	0.015

ACKNOWLEDGMENT

This work (C0396252) was supported by Business for Cooperative R&D between Industry, Academy, and Research Institute funded Korea Small and Medium Business Administration in 2016. This work was also supported by the Korea Institute for Advancement of Technology (KIAT) grant funded by the Korean government (Motie: Ministry of Trade, Industry & Energy, HRD Program for Software-SoC convergence) (No. G02N04500000301).

REFERENCES

[1]   S. Han, J. Kim, Jongsun Kim, "Programmable fractional-ratio frequency multiply-ing clock generator," Electronics letters, vol. 50, No. 3, pp. 163-165, 2014.

[2]   R. Farjad-Rad, et al, "A low-power multiplying DLL for low-jitter multigigahertz clock generation in highly integrated digital chips," IEEE J. Solid-State Circuits, 37, no. 12, pp. 1804–1812, 2002.

[3]   Q. Du, et al, "A low-phase noise, anti-harmonic programmable DLL frequency multiplier with period error compensation for spur reduction," IEEE Trans. Circuits Syst. II, Vol. 53, pp. 1205-1209, 2006.

[4]   G. Park, H. Kim, and Jongsun Kim, "A reset-free anti-harmonic anti-harmonic programmable MDLL-based frequency multiplier", J. Semiconductor Technology and Science, Vol. 13, no. 5, pp. 459-464, Oct. 2013.

[5]   Ali. Et all., "A 4.6GHz MDLL with -46dBc reference spur and aperture position tuning," ISSCC Dig. Tech. Papers, pp. 466-467, 2011

A CMOS 10-bit SAR ADC with Threshold Configuring Comparator for 5 MSBs

Sang Heon Lee
Department of Electronic Engineering
Inha University
Incheon, Korea
lsh6032@naver.com

Kim Jong Gu
Department of Electronic Engineering
Inha University
Incheon, Korea
kjg8912@naver.com

Seong Jae Hyeon
Department of Electronic Engineering
Inha University
Incheon, Korea
sungjaehyeon@naver.com

Kwang Sub Yoon
Department of Electronic Engineering
Inha University
Incheon, Korea
ksyoon@inha.ac.kr

Abstract—**This paper describes a CMOS 10-bit Successive Approximation Register (SAR) Analog to Digital Converter (ADC) using TCC(Threshold Configuring Comparator) for the 5 MSBs. This architecture enables SAR to simplify C-DAC and reduce power consumption. The proposed SAR ADC is fabricated in 180nm CMOS and occupies a core area of 750um x 700um. It consumes 53uW and achieves an ENOB of 9.7 bits at sampling frequency 10MS/s, power supply of 1.8V, and reference of 1.2V. The Figure of Merit (FOM) is simulated to be 6.37fJ/step.**

Keywords; ADC, SAR, TCC, C-DAC

I. INTRODUCTION

Advances in integration and process technology of the circuit design evolve telecommunications industry over the low power consumption and miniaturization of electronic devices. For these reasons, these developments accelerated the portability of the device. Recently, the low power circuit designs of SAR ADCs with enhancement of operation speed have become popular. Many researches made progresses in the

direction to meet the low power and high sampling speed [1-3]. This paper proposes a SAR ADC architecture to adapt TCC for the 5 MSBs to reduce the size of C-DAC. This architecture allows TCC to determine the 5 MSBs and uses a conventional C-DAC to decide the 5 LSBs.

The paper is organized as follows. Section II describes the proposed architecture and the operational principle of the proposed SAR ADC with simulation results. Section III draws conclusions.

II. THE PROPOSED ARCHITECTURE

The proposed SAR ADC consists of conventional C-DAC for the 5 LSBs, SAR Logic, shift registers, TCC control logic, clock generator, and TCCs for 5 MSBs, as presented in Fig. 1. This architecture employs TCC with control logic to determine the 5 MSBs. As errors are caused by process mismatches in using TCC, it results in code error in determining 5 LSBs. Due to this reason, the proposed architecture employs a conventional C-DAC to determine 5 LSBs.

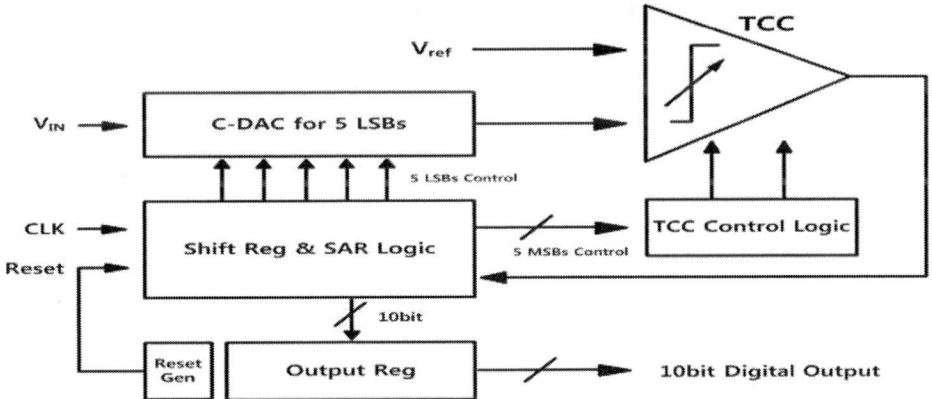

Figure 1. The block diagram of the proposed 10-bit SAR ADC.

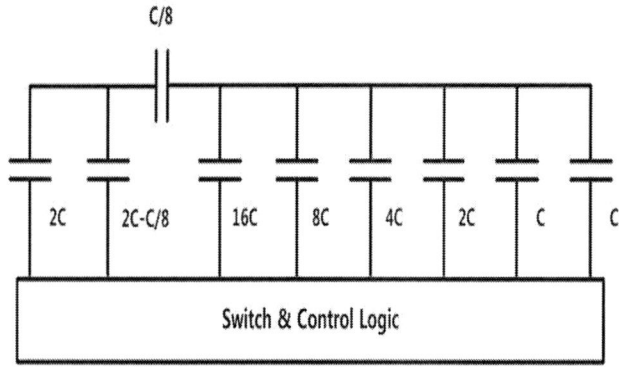

Figure 2. The block diagram of proposed C-DAC for 5 LSBs.

The circuit schematic of the proposed 5 LSBs capacitor DAC is presented in Fig. 2. While conventional 10-bit C-DAC requires 2^{10} unit capacitors, the proposed C-DAC only requires only 36 unit capacitors. Due to the reduction of the size of the capacitors, the proposed architecture allows ADC to be able to reduce the switching power of C-DAC. It reduces the die size of the layout.

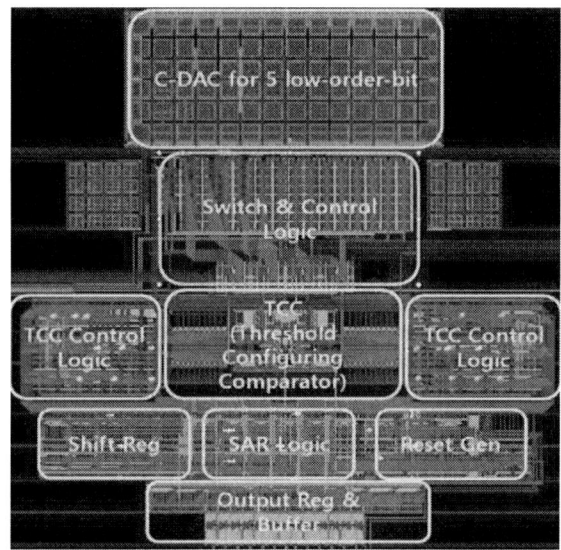

Figure 3. The layout of proposed SAR ADC

The proposed SAR ADC is designed with a 0.18um CMOS process and the layout of the designed circuit is presented in Fig. 3. It occupied a core area of 750um x 700um. Power supply voltage is 1.8V and the reference voltage is 1.2V.

Figure 4 shows the FFT post simulation result with a 10MS/s sampling frequency at input signal of 1.5kHz. The SNDR, SFDR, and ENOB are simulated to be 60.12dB, 63.7dB, and 9.7 bits, respectively. The power consumption and figure of merit (FOM) of the ADC are 53uW and 6.37fJ/step, respectively.

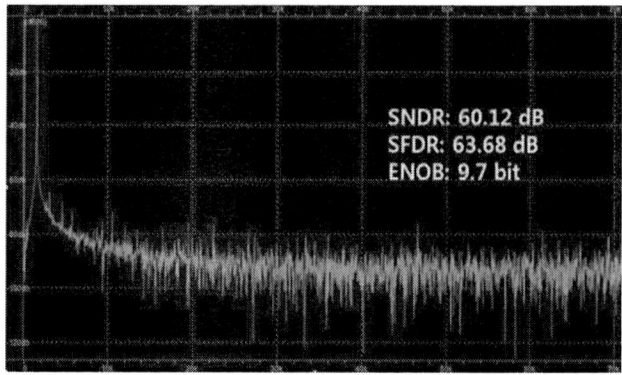

Figure 4. FFT spectrum simulation result.

III. CONCLUSION

The proposed SAR ADC was implemented in 0.18um CMOS process with a power supply of 1.8V and reference voltage of 1.2V. The design technique to employ TCC for the 5 MSBs resulted in the reduction of power consumption and chip area. The simulated SNDR and FOM were 60.12dB and 6.37fJ/step, respectively. The proposed ADC can be utilized not only for the portable devices, but for the bio-signal signal processing systems.

ACKNOWLEDGMENT

THIS RESEARCH WAS SUPPORTED BY THE MSIP(MINISTRY OF SCIENCE, ICT AND FUTURE PLANNING), KOREA, UNDER THE ITRC(INFORMATION TECHNOLOGY RESEARCH CENTER) SUPPORT PROGRAM (IITP–2016-H8601–16–1003) SUPERVISED BY THE IITP(INSTITUTE FOR INFORMATION & COMMUNICATIONS TECHNOLOGY PROMOTION)

This work was supported by the National Research Foundation of Korea(NRF) grant funded by the Korea government(2015R1B1A1A01058603)

This work was supported by the IDEC

REFERENCES

[1] Kentaro Yoshioka, Akira Shikata, Ryota Sekimoto, Tadahiro Kuroda, Hiroki Ishikuro, "An 8 bit 0.3–0.8 V 0.2–40 MS/s 2-bit/Step SAR ADC With Successively Activated Threshold Configuring Comparators in 40 nm CMOS", IEEE Transactions on VLSI, Vol. 23, No. 2, 2014, pp. 356-368.

[2] Hyeok-Ki Hong, Wan Kim, Sun-Jae Park, "A 7b 1GS/s 7.2mW Nonbinary 2b/cycle SAR ADC with Register-to-DAC Direct Control, " Custom Integrated Circuits Conference (CICC), 2012, pp.1-4.

[3] Cao, Zhiheng, Shouli Yan, and Yunchu Li, "A 2mW 1.25 GS/s 6b 2b/step SAR ADC in 0.13 um CMOS", IEEE Journal of Solid-State Circuits , vol. 44 ,2008, pp.862-873.

A Low-Power 10-bit Single-Slope ADC Using Power Gating and Multi-Clocks for CMOS Image Sensors

Byoung-Kwan Jeon, Seong-Kwan Hong, and Oh-Kyong Kwon

Dept. of Electronic Engineering, Hanyang University, Seoul, Korea
okwon@hanyang.ac.kr

Abstract—**This paper proposes a low power 10-bit single-slope analog-to-digital converter (SS-ADC) for CMOS image sensors (CISs) with a column-parallel readout structure. The power consumption of the proposed SS-ADC is reduced by using a power gating scheme for the comparator and multi-clocks having different frequencies. The proposed SS-ADC was designed using a 0.13-μm CIS process technology. The simulation results show that the power consumption of the proposed SS-ADC is 9.7 μW, which is 59.4 % less than that of the conventional SS-ADC.**

Keywords- single-slope ADC; CMOS image sensor; counter, comparator; shot noise.

I. INTRODUCTION

Recently, single-slope analog-to-digital converter (SS-ADC) has been widely adopted for CMOS image sensors (CISs) with a column parallel readout structure because of its high linearity and small area. In the CIS, the SS-ADC, which consists of a comparator and a counter, is implemented inside of each readout channel, whereas a ramp generator, which provides a ramp signal to the SS-ADCs, is implemented outside of the readout circuit. As the number of pixels increases in the high-resolution CIS, the power consumption of the SS-ADCs increases, which occupies up to 60% of that of the entire CIS [1].

Since the conventional SS-ADC in [2] always turns on the comparator during the A/D conversion, it consumes large static power. In addition, as the light intensity increases, the shot noise of the CIS induces more noise into the ADC input voltage. Therefore, the quantization step size of the SS-ADC can be increased at the high light intensity [1]. However, since the conventional SS-ADC uses a clock having a fixed frequency, it has a constant quantization step size regardless of the light intensity, thereby consuming unnecessary dynamic power at the high light intensity.

In this paper, we propose a low power 10-bit SS-ADC for the CISs. To reduce the static and dynamic power consumptions, the proposed SS-ADC, which uses three clocks having different frequencies, turns off the comparator right after the comparator output changes from low to high. Sections II and III describe the operating principle and circuit implementation of the proposed SS-ADC, respectively. In Section IV, the simulation results of the proposed SS-ADC are analyzed. Finally, the conclusions are given in Section V.

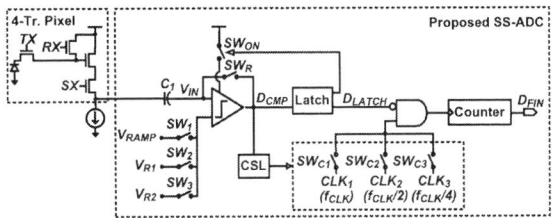

Figure 1. Block diagram of the proposed SS-ADC with 4-Tr. pixel.

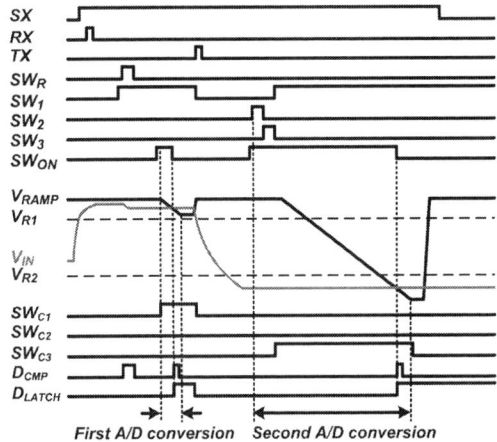

Figure 2. Timing diagram of the proposed SS-ADC and 4-Tr. pixel.

II. OPERATING PRINCIPLE OF THE PROPOSED SS-ADC

Fig. 1 shows the block diagram of the proposed SS-ADC with 4-Tr. pixel. The SS-ADC input is connected to the pixel output via capacitor, C_1, and its voltage, V_{IN}, is determined by the pixel control signals, SX, RX, and TX, and the comparator reset signal, SW_R [2]. The SS-ADC consists of a comparator, a latch, a clock selection logic (CSL), and a counter. The comparator compares V_{IN} with the reference voltages, V_{R1} and V_{R2}, and the ramp signal, V_{RAMP}, according to the control signals, SW_1, SW_2, and SW_3, respectively. The latch, which stores the comparator output, D_{CMP}, generates the latch output, D_{LATCH}, and the turn-on signal of the comparator, SW_{ON}. The CSL generates the control signals, SW_{C1}, SW_{C2}, and SW_{C3} to select one of the three clocks, CLK_1, CLK_2, and CLK_3, having different frequencies of f_{CLK}, $f_{CLK}/2$, $f_{CLK}/4$, respectively. The counter counts the number of selected clocks until D_{LATCH} changes from low to high.

978-1-5090-3220-4/16 $31.00 © 2016 IEEE   257   ISOCC 2016

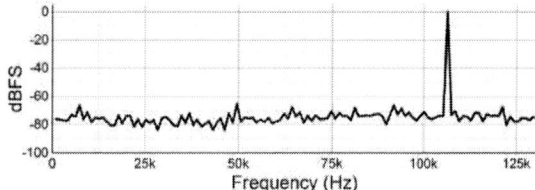

Figure 3. Simulated FFT spectrum of the proposed SS-ADC.

Fig. 2 shows the timing diagram of the proposed SS-ADC and 4-Tr. pixel. In the first A/D conversion, V_{RAMP} decreases. When V_{IN} is equal to V_{RAMP}, D_{CMP} changes from low to high. Then, since V_{RAMP} is always lower than V_{IN}, the latch keeps D_{LATCH} high and turns off the comparator by using SW_{ON} to reduce the static power consumption of the SS-ADC.

As the light intensity increases, V_{IN} decreases, whereas the noise of V_{IN} increases due to the shot noise [1]. In the second A/D conversion, V_{R1} and V_{R2} are sequentially compared with V_{IN} according to SW_2 and SW_3, respectively. After that, CSL keeps SW_{C1}, SW_{C2}, or SW_{C3} high when V_{IN} is higher than V_{R1}, when V_{IN} is between V_{R1} and V_{R2}, or when V_{IN} is lower than V_{R2}, respectively. And then, V_{RAMP} is compared with V_{IN}. Since three clocks have different frequencies, the second A/D conversion result is obtained by multiplying the counter output by 1, 2, or 4, when CLK_1, CLK_2, or CLK_3, is selected, respectively. Therefore, at the high light intensity, as the quantization step size of the SS-ADC increases according to the shot noise, the number of internal signal transitions of the counter decreases, and thereby the dynamic power consumption of the SS-ADC is reduced. To reduce static power consumption, the comparator operates in the same manner as the first A/D conversion.

To reduce the fixed pattern noise of the CIS, the counter calculates the final A/D conversion result of the SS-ADC, D_{FIN}, by subtracting the first A/D conversion result from the second one. [2-3].

III. CIRCUIT IMPLEMETATION OF THE PROPOSED SS-ADC

A 0.13-μm CIS process technology was used to design the proposed SS-ADC with 10-bit resolution, which is applicable for 60 frames/s 8 Mpixels CIS. The proposed SS-ADC uses supply voltages of 3.3 V and 1.2 V, and in addition, employs the comparator in [4], the counter in [3], and three clocks having frequencies of 400 MHz, 200 MHz, and 100 MHz. Considering the shot noise of the CIS in [1], V_{R1} and V_{R2} are chosen to be 1.875 V and 1.250 V, respectively, and V_{RAMP} varies from 1 V to 2 V.

IV. SIMULATION RESULTS

Fig. 3 shows the simulated fast Fourier transform (FFT) spectrum of the proposed SS-ADC at an input frequency of 105 kHz. The proposed SS-ADC uses a clock having a frequency of 400 MHz for evaluating the accuracy in the full ADC input range. The simulated signal-to-noise-distortion ratio and effective number of bits are 58 dB and 9.4-bit, respectively.

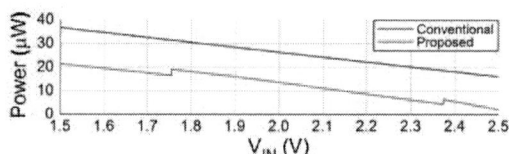

Figure 4. Power consumptions of the proposed and conventional SS-ADCs.

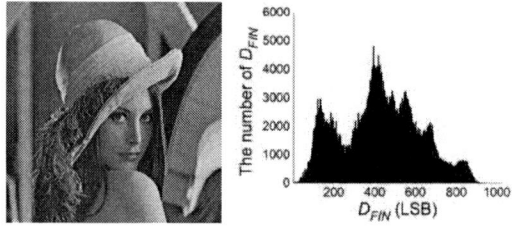

Figure 5. LENA image and its histogram of D_{FIN}.

Fig. 4 shows the power consumptions of the proposed and conventional SS-ADCs according to V_{IN}. Since the proposed SS-ADC, which uses three clocks, turns off the comparator when the comparator output changes from low to high, its power consumption decreases as V_{IN} increases.

Fig. 5 shows the LENA image, which is widely used for evaluating the performance of the CIS, and its histogram of D_{FIN}. Since the power consumption of the proposed SS-ADC is determined by D_{FIN}, the LENA image was used to estimate the power consumption. The simulated power consumptions of the proposed and conventional SS-ADCs are 9.7 μW and 24.0 μW, respectively. These simulation results show that the proposed SS-ADC reduces the power consumption by 59.4 % compared with the conventional SS-ADC.

V. CONCLUSIONS

In this paper, a low power SS-ADC is presented. The static and dynamic power consumptions of the proposed SS-ADC are reduced by turning off the comparator right after the comparator output changes from low to high and using multi clocks, respectively. The simulation results show that the proposed SS-ADC reduces the power consumption by 59.4 % compared with the conventional SS-ADC. Therefore, the proposed SS-ADC is suitable for the low power CIS applications.

VI. REFERENCES

[1] M. F. Snoeij, P. Donegan, A. J. P. Theuwissen, K. A. A. Makinwa, and J. H. Huijsing, "Multiple-ramp column-parallel ADC architectures for CMOS image sensors," *IEEE J. Solid-State Circuits*, vol. 42, no. 12, pp. 2968–2976, Dec. 2007.

[2] S. Yoshihara, et al., "A 1/1.8-inch 6.4Mpixel 60 frames/s CMOS image sensor with seamless mode change," *IEEE J. Solid-State Circuits*, vol. 41, no. 12, pp. 2998–3006, Dec. 2006.

[3] D. Lee, and G. Han, "High-speed, low-power correlated double sampling counter for column-parallel CMOS imagers," *Electron. Lett.*, vol. 43, no. 24, pp. 1362–1364, Nov. 2007.

[4] Takayuki Toyama, "Physical Quantity Distribution Detecting Apparatus And Imaging Apparatus" U.S. Patent 7,755,686 B2, Jul. 13, 2010.

A 200-Mb/s to 3-Gb/s Wide-band Referenceless CDR Using Bidirectional Frequency Detector

Nguyen Huu Tho, Kyung-Sub Son, Kyongsu Lee, and Jin-Ku Kang

Dept. of Electronics Engineering, Inha University
100 Inha-ro, Nam-Gu, Incheon 402-751, Republic of Korea
E-mail: thonh@inha.edu, jkang@inha.ac.kr

Abstract—This paper presents a 200-Mb/s to 3-Gb/s half-rate referenceless clock and data recovery (CDR) circuit in 180nm CMOS process. A bidirectional frequency detector (FD) is proposed to eliminate the harmonic locking issue and reduce the frequency acquisition time. A frequency band selector for wide-range the voltage-control oscillator (VCO) is also presented to select an exact frequency band of the VCO. The simulation shows the CDR achieves 10-ps peak-to-peak jitter at 3Gb/s and the frequency acquisition time of 12.9 µs.

Keywords—*Clock and data recovery, bidirectional frequency detector, referenceless, wide-band*

I. INTRODUCTION

The referenceless CDR plays a very important role in high-speed interface systems to extract the data and clock signals from the received signal. The most significant challenge of referenceless CDR is the harmonic locking issue. To solve this problem, several wide-range unilateral FDs are presented in literature. The unilateral FDs always start from minimum frequency [3] or maximum frequency [4] of the VCO for the frequency acquisition process, thus increasing frequency acquisition time. To overcome this problem, bidirectional FDs are presented in [2], [5]. The reference [2] assumes that run-length of the input data is fixed and can work by interaction between phase detector (PD) and FD because of absence of the lock detector. In [5], an additional quadrature divider is required for the frequency decrement acquisition, thus complicating the FD design. This paper proposes a simple bidirectional FD for wide-band referenceless CDR circuits. This proposed FD is free from harmonic locking issue, does not depend on run-length of the input data.

II. CIRCUIT DESCRIPTION

The block diagram of the proposed half-rate CDR circuit is shown in Fig. 1. It consists of the proposed bidirectional FD, a frequency lock detector (FLD), a half-rate bang-bang phase detector [1], two-band VCO, a frequency band selector (FBS), and two charge pump (CP) circuits.

In Fig. 1, when the signal *EN* is activated, switch S1 turns on and S2 turns off. Based on bit rate of the input data, the proposed frequency band selector selects an exact frequency band from two-band VCO by updating the control bit D0. Then, the proposed FD tracks the frequency error between the input data and the output clocks. Once the frequency error is small enough, the frequency lock detector triggers the *LOCK*

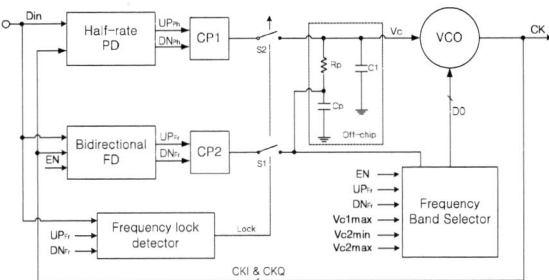

Figure 1. The block diagram of the proposed wide-band referenceless CDR

signal to turn off S1 and turn on S2. After that, the PD takes over the acquisition process. The frequency band selection algorithm for the wide-band VCO is shown in Fig. 2.

Fig. 3 shows the proposed bidirectional FD. This proposed FD consists of two unilateral FDs, a D-type flip-flop (D-FF) and a multiplexer. We use the *STOP* signal to control *DN* of the frequency decrement acquisition. Initially, the external pulse signal *EN* keeps low to reset *STOP*. When the clock is faster than the data, as long as *STOP* keeps low, then signal *DN* goes to high to discharge the loop filter to decrease the clock frequency. When the data rate is faster than the clock, signal *UP* goes high. Then, signal *STOP* is activated to disable *DN* and stop decreasing the clock frequency. After that, the frequency acquisition process can be accomplished by the frequency increment acquisition FD. The proposed FD uses two phases of the clock, i.e., *CKI* and *CKQ*.

The frequency decrement acquisition circuit is implemented with only two D-FFs, which is simpler than [2] and [5]. The FD counts the number of consecutive transition edges of the clock in sequence "010" of the data. As long as the data *Din* encloses pulse of clock *CKI/CKQ*, signal *DN1* is generated. In addition, the proposed FD work well with long-run of data.

To reduce frequency acquisition time, instead of counting the rising edges of input data during one clock period of *CKI/CKQ* [3], we propose to count the number of consecutive transition edge of the data in the half of clock period. Furthermore, the FD in [5] works by detecting a specific pattern "101" of the data in a half of clock period and use only a phase of clock. Our proposed bidirectional FD detects both specific pattern "101" and "010" of the data and use two phases of the clock *CKI*, *CKQ*. Therefore, this FD can decrease frequency acquisition time. As long as the pulse of clock *CKI/CKQ* encloses a single bit of data D_{in}, signal *UP* is generated.

978-1-5090-3220-4/16 $31.00 © 2016 IEEE   259   ISOCC 2016

III. SIMULATION RESULT

The proposed half-rate CDR circuit is implemented in a 180 nm CMOS process. The simulation results show that the circuit successfully recovery with 2^7-1 PRBS data from 200Mb/s to 3Gb/s. The operating data range of the proposed FD is unlimited. It can be extended by adjusting the frequency range of the VCO.

Fig. 4 shows the frequency and phase acquisition process for the CDR when the data rate is 3Gb/s. The acquisition process of the CDR is divided by three periods. At the start, the frequency band selector operates to search the true frequency band of the VCO. Then, the frequency acquisition process can be accomplished by the proposed FD. The frequency lock detector toggles the signal *LOCK* to the high state when frequency error is small enough. After that, the CDR transfers the loop control to the PD. The performance comparisons with other referenceless CDRs are shown in Table I and the layout is shown in Fig. 5, respectively.

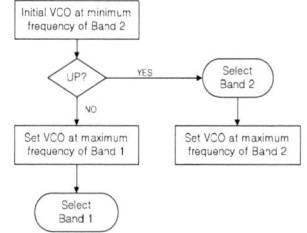

Fig. 2. Frequency band selection algorithm for the wide-range VCO

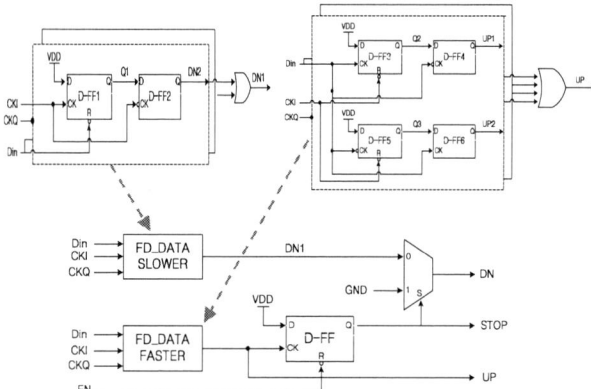

Fig.3. The block diagram of the proposed bidirectional FD

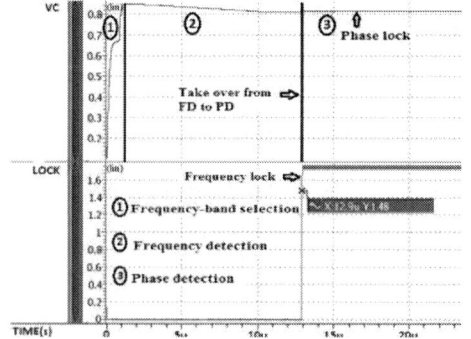

Fig. 4. Simulation result of frequency and phase acquisition process

Fig. 5. Layout of the proposal CDR

TABLE I. PERFORMANCE COMPARISION OF WIDE-BAND REFERENCELESS CDR

	[2]	[3]	[5]	This work
Technology(nm)	180 CMOS	350 BiCMOS	130 CMOS	180 CMOS
Supply (V)	1.8	3.3	1.5	1.8
Data rate (Gb/s)	0.15552-3.125 Full-Rate	0.0125-2.7 Full-Rate	1-16 Half-Rate	0.2-3 Half-Rate
FD type	Bidirectional	Unilateral	Bidirectional	Bidirectional
Divider in FD	Yes	Yes	Yes	No
Jitter$_{p-p}$ (ps)	82.2@1.2Gb/s	N/A	146@1Gb/s	10@3Gb/s
Freg.acquisition time (µs)	100	1000	1000	12.9
Area (mm^2)	0.88 (withoutFBS)	9 (with FBS)	0.134 (without FBS)	0.345 (with FBS)
Power (mW)	95	750	160	37.8

IV. CONCLUSION

A referenceless 200Mb/s to 3Gb/s rate CDR circuit is implemented in 180nm CMOS process. The proposed bidirectional FD is free from harmonic locking issue, and achieves an unlimited frequency acquisition range. It has the short frequency acquisition time of 12.9 µs in simulation with a 37.8mW power consumption. In addition, a frequency band selection algorithm for wide-range VCO is also presented.

ACKNOWLEDGMENT

This research was supported by NRF(2010-0022670) and the MSIP, Korea, under the ITRCsupport program (IITP-2015-H8501-15-1010) supervised by the IITP.

REFERENCES

[1] B. Razavi, *Design of Integrated Circuits for Optical Communication Systems*. New York: McGraw-Hill, 2003.

[2] R.J. Yang, K.H. Chao, S.C. Hwu, C.K. Liang and S.I. Liu, "A 155.52 Mbps-3.125 Gbps Continuous-Rate Clock-and-Data-Recovery Circuit," *IEEE J. Solid-State Circuits*, vol. 41, no. 6, pp.1380-1390, Jun. 2006.

[3] D. Dalton, K. Chai, E. Evans, et. al., "A 12.5-Mb/s to 2.7-Gb/s Continuous-Rate CDR with Automatic Frequency Acquisition and Data-Rate Readback," *IEEE J. Solid-State Circuits*, vol. 40, no. 12, pp. 2713-2725, Dec., 2005.

[4] M.S. Hwang, S.Y. Lee, J.K. Kim, S. Kim and D.K. Jeong, "A 180-Mb/s to 3.2-Gb/s, Continuous-Rate, Fast-Locking CDR Without Using External Reference Clock," *IEEE Asian Solid-StateCircuits Conf.*, pp. 144-147, Nov. 2007.

[5] C.L. Hsieh and S.I. Liu, "A 1–16-Gb/s Wide-Range Clock/Data Recovery Circuit With a Bidirectional Frequency Detector," *IEEE Transactions on Circuits and Systems-II*, vol. 58, no. 8, Aug. 2011.

Design of High-Linearity Delay Detection Circuit for 10-Gb/s Communication System in 65-nm CMOS

Kosuke Furuichi, Hiromu Uemura, Natsuyuki Koda, Hiromi Inaba and Keiji Kishine

University of Shiga Prefecture

2500, Hikone Hassaka, Shiga, 522-8533, Japan

Email: oo23kfuruichi@ec.usp.ac.jp

Abstract—**We have proposed a transmission system of the additional data by using frequency modulation technique. In a receiver, demodulation characteristics deteriorate according to data speed. In the previous work, we showed the emphasis technique can reduce the degradation in demodulated signal. In this paper, we designed the delay detection circuit on the basis of the detailed analysis. To investigate the proposed circuit, we fabricated the delay detection circuit with emphasis using 65-nm CMOS process. We confirmed the improved linearity in demodulated signal, which indicate our proposed circuit can be applicable to the 10-Gb/s demodulating systems.**

I. INTRODUCTION

Recently, communication systems transmit and receive a large amount of information. In previous work, we proposed a communication system which can embed additional data without altering the data frame[Fig. 1(a)][1]. In the system, additional data is transmitted by frequency modulation on the data frame. The demodulation characteristics strongly depend on the data speed. The key issue in the system is how to reduce the linearity deterioration on the demodulated signal when the data speed gets higher[Fig. 1(b)]. To obtain the higher-speed operation, we have proposed the emphasis technique for the receiver circuits[Fig. 1(c)][2]. In this paper, we show the detailed design of the delay detection circuit using the emphasis technique, which provide the higher speed operation. In implementing the design, we fixed the circuit parameters by using a small-signal equivalent circuit. In addition, we fabricated delay detection circuit using 65-nm CMOS process. The IC achieved a 10% higher detection-characteristic linearity than a receiver without emphasis.

II. ANALYSIS OF DELAY DETECTION CIRCUIT

A delay detection circuit with low-pass filter is used as the demodulator[Fig. 1(c)]. In the demodulator, the degradation of an EXOR output signal[Fig.2(a)] affects the linearity of the detection characteristics. As the data speed gets higher, the linearity degradation in the output signal becomes larger[Fig. 1(b)]. The detection-characteristic linearity degradation is caused by reduction in the average voltage of the output signal depending on the data speed. To avoid this, we use a high-pass filter(HPF) as an emphasis circuit[Fig.2(b)] in the Emphasis_path. The Emphasis_path generates high-frequency components in the signal. A signal generated in the Emphasis_path is combined with the signal in the main_path to compensate for the degradation. The appropriate amount of emphasis at the target frequency must be designed according to the degradation. Therefore, it is necessary to determine

the appropriate circuit parameter in emphasis circuit. So we calculate the average voltage using a small-signal equivalent circuit[Fig. 3] for determining the appropriate circuit parameter against the linearity degradation. The average voltage can be calculated from transient response of the output signal. Therefore, we calculated a gain of the EXOR circuit and the emphasis circuit(G_{exor} and G_{emp}). In addition, we obtained each of the transient responses[$v_{\text{exor}}(t)$ and $v_{\text{emp}}(t)$] using the inverse Laplace transform on G_{exor} and G_{emp}.

$$G_{\text{exor}}(\omega) = -\frac{g_{m1} R_D}{(1 + j\omega C_p R_D)} \tag{1}$$

$$G_{\text{emp}}(\omega) = -\frac{g_{mC} R_D}{1 + j\omega C_p R_D} - \frac{j\omega T g_{mE} R_D}{(1 + j\omega T)(1 + j\omega C_p R_D)} \tag{2}$$

$$v_{\text{exor}}(t) = g_{m1} R_D (1 - e^{-\frac{t}{C_p R_D}}) v_{in}(t) \tag{3}$$

$$v_{\text{emp}}(t) = g_{mC} R_D (1 - e^{-\frac{t}{C_p R_D}}) v_{in}(t) + \frac{T g_{mE} R_D}{T - C_p R_D} (e^{-\frac{t}{T}} - e^{-\frac{t}{C_p R_D}}) v_{in}(t) \tag{4}$$

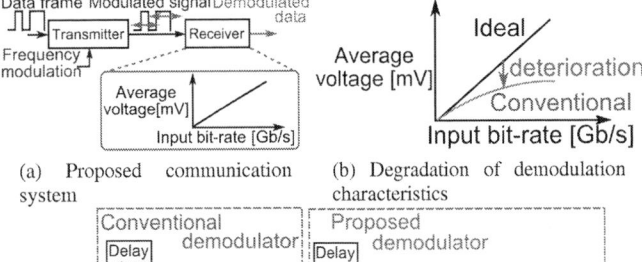

(a) Proposed communication system
(b) Degradation of demodulation characteristics

(c) Demodulator in proposed system

Fig. 1. Overview of proposed system

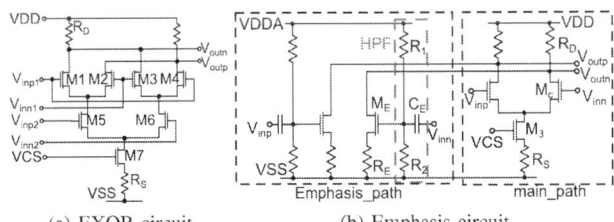

(a) EXOR circuit
(b) Emphasis circuit

Fig. 2. Detailed view of the configuration circuit

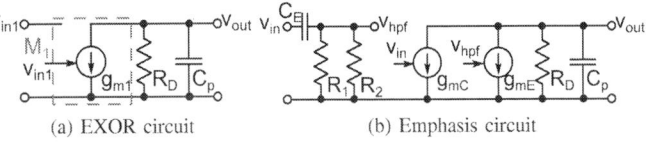

(a) EXOR circuit
(b) Emphasis circuit

Fig. 3. Small-signal equivalent circuit

978-1-5090-3220-4/16 $31.00 © 2016 IEEE

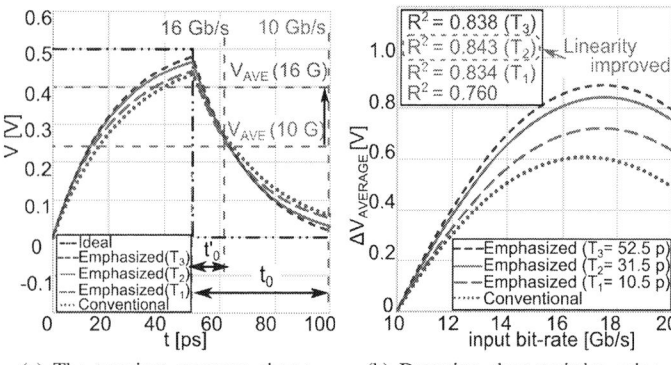

(a) The transient response characteristics

(b) Detection characteristics using theoretical formulas

Fig. 4. The detection characteristics led from the theoretical formula

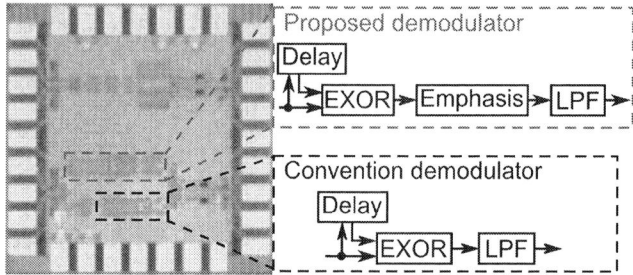

Fig. 5. Fabricated IC and implementation in circuit

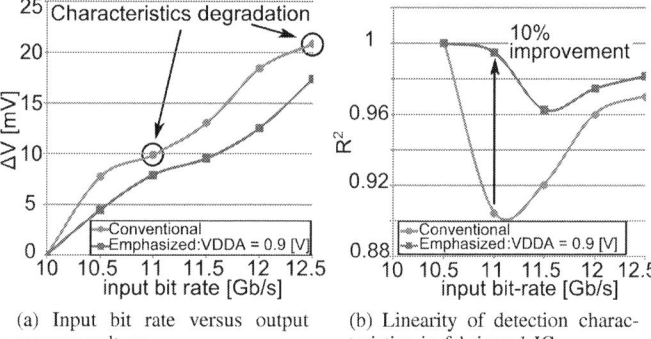

(a) Input bit rate versus output average voltage

(b) Linearity of detection characteristics in fabricated IC

Fig. 6. Detection characteristics of fabricated IC

In the formula(1), (2), (3) and (4), the time constant of the high-pass filter(T) is $\frac{C_E R_1 R_2}{R_1 + R_2}$. The emphasis characteristics depends on the T. Here, the emphasis characteristics are emphasis amount and emphasis frequency band. The emphasis amount can be controlled by using an external terminal(VDDA). Therefore, we designed T by focusing on the emphasis frequency band. Figure 4(a) shows the transient responses of the EXOR circuit output and the emphasis circuit outputs. In Fig. 4(a), we compare several $T(T_1, T_2$ and $T_3)$ in the emphasis circuit. Figure 4(b) shows the average-voltage of the output signal obtained from the $v_{exor}(t)$ and $v_{emp}(t)$. As a verification of the detection-characteristic linearity, we compare the linearity of the average voltage in Fig. 4(b). Here, we used the coefficient of determination(R^2) as an index of linearity[3]. The R^2 is 1, when there is no deterioration in the detection characteristics. As the detection characteristics deteriorate, the R^2 reduces from 1. From the value of the R^2, the linearity of the detection characteristics are higher than that of a demodulator without emphasis[Fig. 4(b)]. Next, we determined that the appropriate parameter of the T. When T is lower than the appropriate value(T_1), the emphasis frequency band becomes high. In this case, the emphasis frequency band becomes higher than the deteriorated frequency band in the EXOR circuit. Therefore, the detection characteristics remains deteriorated. In contrast, when T is larger than the appropriate value(T_3), the emphasis frequency becomes low. In this case, the linearity of the detection characteristics remains deteriorated. From the value of the R^2, the linearity of the detection characteristics become maximum when T is T_2[Fig. 4(b)]. Therefore, we fixed that the appropriate parameter is T_2.

III. EXPERIMENTAL RESULTS

To confirm the advantages of the proposed delay detection circuit, we fabricated an IC with a delay detection circuit using the 65-nm CMOS process[Fig. 5]. The time constant of the high-pass filter[T] was 31.5 ps in the IC. When measuring the fabricated IC, the input signal was generated by using a pulse-pattern-generator (PPG). The PPG output signal was a PN7, PRBS signal. We changed the bit-rate of the input signal from 10 Gb/s to 12.5 Gb/s. In this way, we changed the frequency of the input-signal as a substitute for frequency

modulation. The experimental results and the linearity of the detection characteristics are shown in Fig. 6. The IC achieved a 10% higher-linearity than a receiver without emphasis by applying the appropriate VDDA.

IV. CONCLUSION

In this paper, we described the performance improvement of a delay detection circuit used as a receiver. We improved the linearity of the detection characteristics of the receiver using an emphasis circuit. In implementing the design, we determined the circuit parameters using a small-signal equivalent circuit. We fabricated an IC by using the 65-nm CMOS process. From the experimental results, we confirmed that the linearity of the detection characteristics was improved by determining the circuit parameters on the basis of theoretical formulas and applying an appropriate voltage to the VDDA. The IC achieved a 10% higher detection-characteristic linearity than a receiver without emphasis at 10-Gb/s.

ACKNOWLEDGMENTS

A part of this research was supported by Grants-in-Aid for Scientific Research from the Japan Society for the Promotion of Science.

REFERENCES

[1] D. Omoto,, et. al., "Simple routing control system for 10-Gb/s data transmission using a frequency modulation technique" International Conference on Electronics, Information, and Communication,2016.

[2] K. Furuichi, et al., " Bandwidth of the Delay Detection Circuit according Emphasis Circuit," The Institute of Electronics, Information and Communication Engineers, p. 25, 2015.

[3] W. Mendenhall, " Introduction to Probability and Statistics," p.518, 2008.

Design of Pseudo-Random Bit Sequence Generator with Adjustable Sinusoidal Jitter

Hong-Jhih Chen, Jau-Ji Jou, and Tien-Tsorng Shih
Department of Electronics Engineering
National Kaohsiung University of Applied Sciences
Kaohsiung City, Taiwan
sda26840704@gmail.com, jjjou@kuas.edu.tw

Abstract—**In this paper, a 2^7-1 pseudo-random bit sequence (PRBS) generator with built-in clock is designed in 0.18-μm CMOS technology. The MOS current mode logic (CML) is used in the PRBS generator for high-speed operation. The clock circuit is realized using a three-stage ring voltage controlled oscillator (VCO). The frequency of the clock can be from 3.3 to 2.7 GHz, while the controlled voltage of the VCO is varied from 0 to 1.1 V. The periodic jitter of the clock can be also controlled, while a sinusoidal wave is added in the controlled voltage of the VCO. The maximum sinusoidal jitter amplitude of our 3-Gb/s PRBS signal can be adjusted to about 0.9-UI.**

Keywords: pseudo-random bit sequence (PRBS) ; voltage controlled oscillator (VCO); sinusoidal jitter

I. INTRODUCTION

In high-speed digital circuit and transmission system, the bit-error-rate (BER) is an important indicator of signal quality. A pseudo-random bit sequence (PRBS) signal is injected into the transmitter (Tx), and the BER is tested at the receiver (Rx) end through a device under test (DUT) or a transmission path. However, the measurement equipment of BER test is very expensive, and the test time usually requires a longer time. Therefore, a built-in self-test (BIST) system with BER test on chip can reduce test cost and time [1]. In serial link applications, the most important components of BIST system are PRBS generator and checker circuits.

Moreover, a receiver of high-speed digital data has to tolerate a certain amount of time jitter. The jitter tolerance mask is taken from some serial transmission standards. The intention for such a jitter mask is to ensure margin for all types of frequency jitter, wander, noise, crosstalk and other variable system effects. Under these conditions, a receiver must run properly for a very low specific BER. Jitter tolerance tests are commonly performed on receiver circuits using sinusoidal jitter (i.e. periodic jitter). In addition to sinusoidal jitter, deterministic jitter and random jitter have also been proposed to use in jitter tolerance tests. However, only sinusoidal jitter can provide the worst-case jitter to the transmission test system [2].

The block diagram of BER and jitter tolerance tests is shown in Figure 1. The clock with sinusoidal jitter is a phase modulated clock, as below expression [2],

$$Clcok(t) = A_c \cos\left[\omega_c t + J_{pp}\pi \sin(\omega_m t)\right] \quad (1)$$

This paper is supported by the Taiwan National Chip Implementation Center and the Taiwan Ministry of Science and Technology (MOST 105-2221-E-151-002).

where A_c is amplitude of clock, ω_c is angular frequency of clock, J_{pp} is sinusoidal jitter amplitude (UI), and ω_m is sinusoidal jitter angular frequency. In this paper, we will design a PRBS generator with adjustable sinusoidal jitter for jitter tolerance BIST.

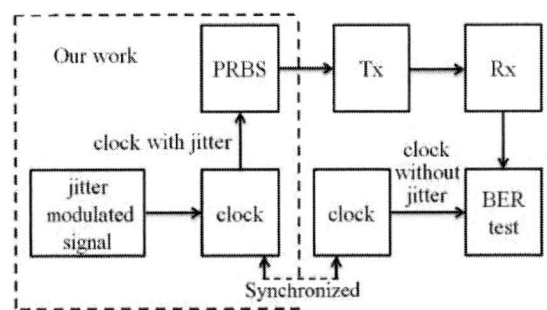

Figure 1. Block diagram of BER and jitter tolerance tests.

II. DESIGN OF CIRCUIT

Figure 2 shows a PRBS generator with a built-in clock. The PRBS generator is based on a linear feedback shift register (LFSR), and comprises seven D-type flip-flops (DFF) and a XOR gate. The clock circuit is a voltage controlled oscillator (VCO). The clock buffers are added between VCO and DFFs, and the output buffer is added at the PRBS output end.

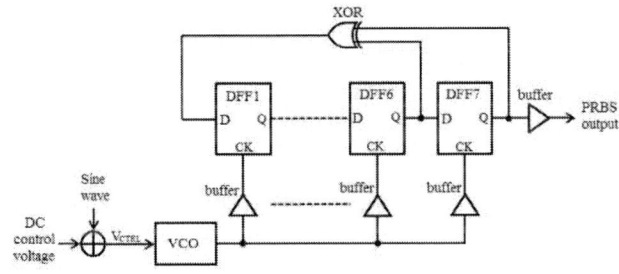

Figure 2. PRBS generator with a built-in clock.

Current mode logic (CML) is a popular logic style for high-speed circuits. Our PRBS generator circuit uses the CML style. The schematic of a CML master-slave DFF including a master latch and a slave latch is shown in Figure 3. The operation mode (sampling mode or storage mode) of the latch is controlled through the clock signal [3].

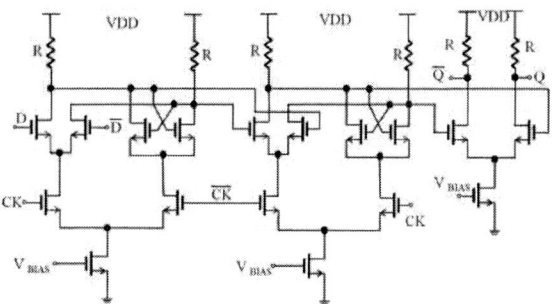

Figure 3. CML D-type flip-flop.

Our clock circuit is realized using a three-stage multiple-loop ring oscillator [4], as shown in Figure 4. The primary loop (black line) works as a normal differential ring oscillator. The secondary loop (thin line) can help to decrease the slew time of the output nodes.

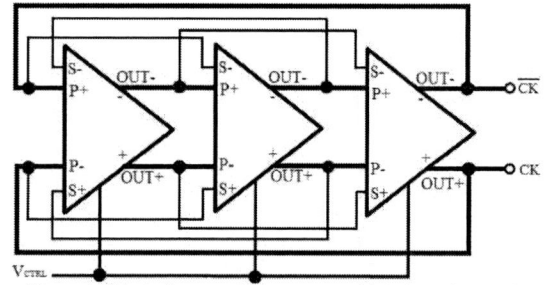

Figure 4. Block diagram of three-stage multiple-loop ring VCO.

The circuit schematic of the proposed delay cell is shown in Figure 5. M1 and M2 form the input pair for the primary loop, while M5 and M6 serve as the input pair of the secondary loop. The operating frequency is controlled by the gate voltage of M3 and M4.

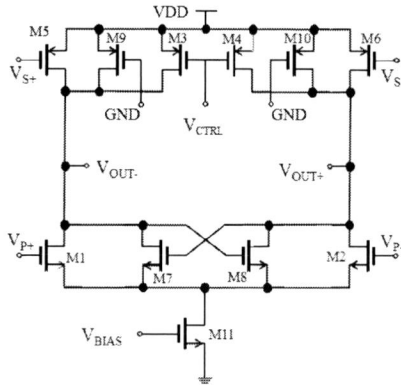

Figure 5. Proposed delay cell for multiple-loop ring VCO.

The control voltage of the VCO can be added a sine wave, and the oscillated frequency will be varied by the sine wave. The frequency modulated VCO clock can be express as

$$Clock_{VCO}(t) = A_c \cos\left[\omega_c t - \frac{2\pi A_m G_{VCO}}{\omega_m}\cos(\omega_m t)\right] \qquad (2)$$

where A_m is amplitude of the modulated sine wave and G_{VCO} is VCO gain.

III. SIMULATION RESULTS AND DISCUSSION

The PRBS generator with built-in clock is designed in TSMC 0.18-μm CMOS technology. The power consumption of the circuit is about 141-mW and the chip area is 567-μm×558-μm. While the control voltage of the VCO is from 0 to 1.1-V, the oscillated frequency is from 3.3 to 2.7-GHz, as shown in Figure 6(a). The VCO gain is about 0.6-GHz/V. At 0.5-V control voltage, the 3-Gb/s PRBS signal without sinusoidal jitter can be obtained, as shown in Figure 6(b).

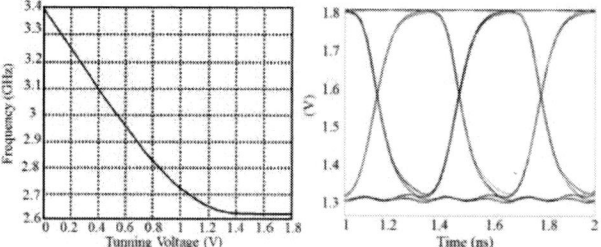

Figure 6. (a) VCO oscillation frequency vs. control voltage and (b) eye diagram of 3-Gb/s PRBS signal without sinusoidal jitter.

Figure 7 shows 3-Gb/s PRBS signals with 0.1UI and 0.9UI sinusoidal jitter amplitude and 100-MHz jitter frequency. Generally, in low-frequency jitter tolerance test, the large sinusoidal jitter amplitude is required. However, our VCO gain is not enough, so the maximum sinusoidal jitter amplitude is just about 0.9UI.

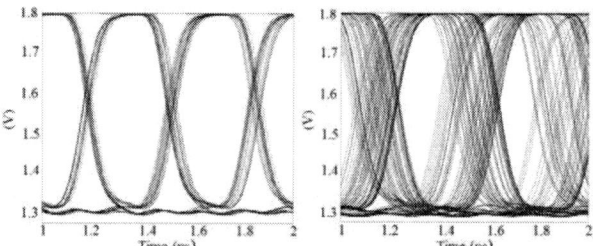

Figure 7. Eye diagrams of PRBS signals with (a) 0.1UI and (b) 0.9UI sinusoidal jitter.

IV. CONCLUSION

A 2^7-1 PRBS generator with built-in clock has been designed in 0.18-μm CMOS technology. A 3-Gb/s PRBS signal with 0.9UI sinusoidal jitter can be generated to apply in jitter tolerance test.

REFERENCES

[1] S. W. Kang, J. H.Chun, Y. H.Jun, and K. W. Kwon, "A study on accelerated built-in self test of multi-Gb/s high speed interfaces," IEEE 2nd International Conference on Networked Embedded Systems for Enterprise Applications, pp. 1-4, Dec. 2011.

[2] M. Shimanouchi, "Periodic jitter injection with direct time synthesis by SPPTM ATE for SerDes jitter tolerance test in production," International Test Conference, pp. 48-57, 2003.

[3] P.Heydari and R. Mohanavelu, "Design of ultrahigh-speed low-voltage CMOS CML buffers and latches," IEEE Transactions on Very Large Scale Integration (VLSI) Systems, Vol. 12, pp. 1081-1093, 2004..

[4] H. Q. Liu, W. L. Goh, L. Siek, W. M. Lim, and Y. P. Zhang, "A low-noise multi-GHz CMOS multiloop ring oscillator with coarse and fine frequency tuning," IEEE Transactions on Very Large Scale Integration (VLSI) Systems, Vol. 17, pp. 571-577, 2009.

A study of the referenceless CDR based on PLL

JiHoon Kim, YoungJu Hwang and Yong Moon

School of Electronic Engineering, Soongsil University, Seoul, Korea

lixc1988@naver.com, ghkddudwn89@naver.com, moony@ssu.ac.kr

Abstract— Clock data recovery (CDR) circuit is an essential component for serial data communication. S/PDIF which is one of data coding is used. The CDR based on PLL recovers clock and data of 2.8224 ~ 24.576MHz and was designed with the frequency detector (FD) to detect the frequency by using the preamble. The PLL, frequency detector (FD) and the reset circuits were used to design the refernceless CDR based on PLL. 65nm CMOS process is used in this study.

Keywords CMOS, CDR , SPDIF , PLL , Frequency Detector, CDR based-on PLL

I. INTRODUCTION

Clock data recovery (CDR) circuit is an essential component for serial data communication. The clock information is included in digital coded data. For this reason, CDR circuits are important to the clock and data restoration of digital coded data. Recovery circuit could be designed in several different ways. Most of CDR circuits are based on PLL. CDR based on PLL is classified as reference CDR and referenceless CDR. Referenceless CDR is better than reference CDR in cost. Because it does not use of crystal oscillator in receiver [1] [2].

This paper uses the S/PDIF signal which is one of data coding. SPDIF is the acronym of Sony / Philips Digital Interface Format, data protocol for transmitting digital audio signals. SPDIF digital data stream is encoded using the BMC (Bi-phase Mark Code). SPDIF signal has 192 frames in one block. One frame has two sub-frames. The preamble signal of 8bits exists in each the sub-frame. In addition, other S/PDIF features follow the standard IEC-60958 [3].

The proposed CDR could cover the full range of SPDIF. The frequency detector is designed to detect the sampling rate change in S/PDIF. Reset circuit is designed and described from the following chapters.

II. ARCHITECTURE

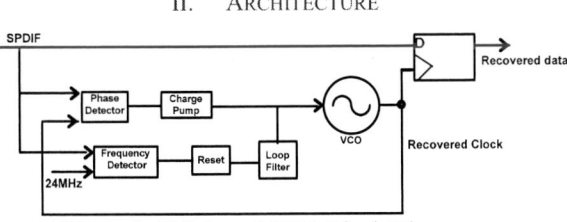

Figure 1. PLL-based CDR Block Diagram

Figure 1 is the block diagram of CDR based on PLL using reset signal. The operation of CDR is described in detail.

S/PDIF is the input signal. The phase locked loop (PLL) block recovers clock in SPDIF signal. PLL block is locked to the recovered clock. When the sampling rate changes in S/PDIF, frequency detector (FD) are designed to detect the frequency change. When the change in frequency is detected, FD generates 3 bit signals.3bit signals generate the RESET signal. If the RESET again signal is detected, PLL block find the target frequency.

A. Frequency Detector design

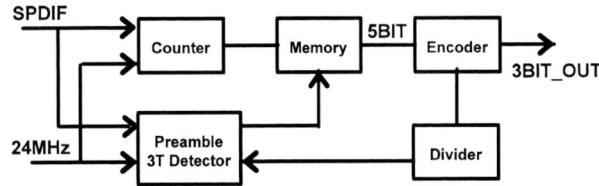

Figure 2. Frequency Detector Block Diagram

Figure 2 is the block diagram of the frequency detector. Frequency detector sample is SPDIF signal according to 24MHz clock. If frequency detector detects the preamble, stores the count value in the memory block. The frequency detector can detect the frequency using the preamble. After the detection of the frequency from preamble, frequency detector outputs 3bit signals corresponding to the detected frequency.

B. The PLL-based CDR

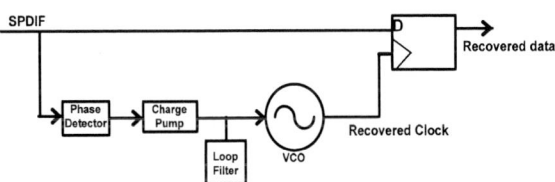

Figure 3. The CDR based on PLL Block Diagram

Figure 3 is the block diagram of the PLL-based CDR. The Voltage Controlled Oscillator (VCO) of the CDR based on PLL is composed of the ring oscillator 5 stage inverter with feedback loop. The oscillation frequency of VCO is determined by the amount of current passing through the inverters. CDR should use support the full range of SPDIF. The input signal of CDR circuit is random data. Phase frequency detector (PFD) of the conventional PLL could not be used, because the output value of PFD follows CDR follows random data, so CDR loses the lock condition. Therefore, the hogge Phase Detector (PD) is used [4]. The

978-1-5090-3220-4/16 $31.00 © 2016 IEEE   265   ISOCC 2016

pumping current of charge pump circuit is 60uA. D flip flop is master and slave type. The third order loop filter used.

C. Reset circuit

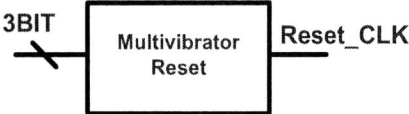

Figure 5. The Multivibrator Reset Block Diagram

Fig 5 is the block diagram of multivibrator reset circuit .When sampling rate changes, 3 bit signals show the transition. The RESET signal duration is generated by monostable multivibrator circuit.

III. MEASUREMENT AND CONCLUSION

The simulation results are as follows

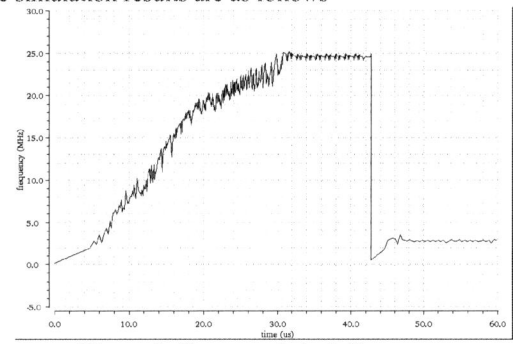

Figure 6. Recovered clock Frequency

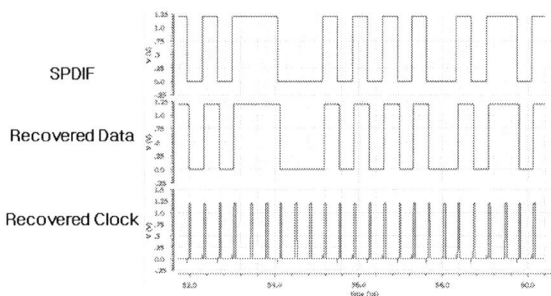

Figure 7. SPDIF/ Recovered clock / Recovered data Graph

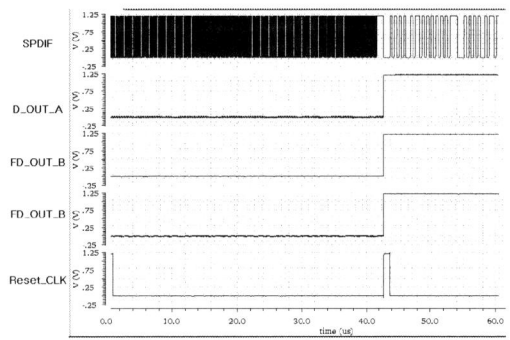

Figure 8. SPDIF / FD OUTPUT_A,B,C / Reset_CLK

Figure 9. PLL-based CDR Block Layout

Figure 6 shows the post simulation result of the recovered frequency of the SPDIF using the CDR. This shows the change in frequency from 24.576MHz to 2.8224MHz.

Figure 7 shows the transient simulation result. Recovered clock, recovered data and S/PDIF signals are shown in plot.

Figure 8 is shows FD OUTPUT_A, B, C and RESET_CLK signal. When the sampling rate changes in S/PDIF, PD operates. 3 bit signals of FD simulation are result. RESET_CLK is generated after 3 bits has passed Multivibrator Reset circuit.

Clock and data is recovered according to S/PDIF change. The CDR circuit has been designed in a 65nm CMOS technology. The layout is shown in Figure 9. CDR layout size occupies the area of 502 um x 581um.

IV. ACKNOWLEDGMENT

This work was supported by the Human Resources Development program (No.20144030200600) of the Korea Institute of Energy Technology Evaluation and Planning (KETEP) grant funded by the Korea government Ministry of Trade, Industry and Energy.

V. REFERENCES

[1] Hong Jun Park, "CMOS Analog Integrated circuit Design"

[2] Xuehui Chen, Yingmei Chen," A 9.95-11.5Gb/s Full Rate CDR with Jitter Attenuation PLL in 65-nm CMOS Technology" 2011 International Symposium on, April 2011

[3] IEC, IEC-6095

[4] Behzad Razavi, "Challenges in the Design of High-Speed Clock and Data Recovery Circuits", IEEE Communications Magazine, August 2002.

A design of NFC Analog Front-End with the Frequency Selector

Jin-ho Kim and Yong Moon

Department of Electronic Engineering, Soongsil University, Korea

Email : jh4747h2@ssu.ac.kr, moony@ssu.ac.kr

Abstract— We have designed the NFC analog front-end including the frequency selector using 0.18μm CMOS Process. The NFC analog front-end satisfied the ISO/IEC-14443A/B. It consists of the power supply block and the data transmitter block. The power supply block has the DC rectifier, voltage multiplier, bandgap reference, and regulator. The data transmitter block has the demodulator and the load modulator. The frequency selector distinguishes NFC signal which satisfies the ISO/IEC-14443A standard and the wireless power transfer signal which satisfies Rezence standard. If the frequency selector receives the frequency signal of 6.78MHz, output becomes 1V. If the frequency selector receives the frequency signal of 13.56MHz, output becomes 0V and the data transmitter block of NFC analog front-end is turned on.

Keywords; NFC , Rezence, wireless power trasnfer tag.

I.   INTRODUCTION

NFC (Near Field Communication) uses 13.56MHz of high frequency band and communicates the distance of 10cm between reader and tag. It also satisfied the ISO/IEC-18092 and ISO/IEC-14443A/B standards. It has been applied to the Smart Card like a transportation card, access control and electronic cash. WPT (Wireless Power Transfer) for wireless charging has the two methods in wireless charging market. One is inductive method, and the other is resonant method. Recently inductive method takes the lead in wireless charging market. Inductive method has the Qi Standard of WPC (Wireless Power consortium) and PMA (Power Matters Alliance) standard. A4WP (Alliance for Wireless Power) founded in 2012 uses resonant method and named the standard as Rezence. Presently A4WP and PMA incorporated Air fuel Alliance to lead the wireless charging market in the near future. The proposed NFC analog front-end with the frequency selector satisfied the standard of ISO/IEC-14443A and the Rezence of A4WP. The proposed circuit shares the supply block of NFC analog front-end. The supply block is operated according to the input frequency signal. If the input frequency signal is 13.56MHz, supply block output is connected to demodulator for NFC data transmit to send the demodulated data. If the input frequency signal is 6.78MHz, supply block output is connected to the wireless battery charging block.

TABLE I.   DESIGN SPECIFICATION

	NFC	Wireless Charging
Standard	ISO/IEC-14443A	Rezence
Frequency band	13.56MHz	6.78MHz
Operation Mode	Passive	.
Supply Voltage	1V	

II.   THE NFC ANALOG FRONT-END WITH THE FREQUENCY SELECTOR

A.   The NFC analog front-end

The block diagram of the proposed NFC analog front-end with the frequency selector is shown in Figure 1.

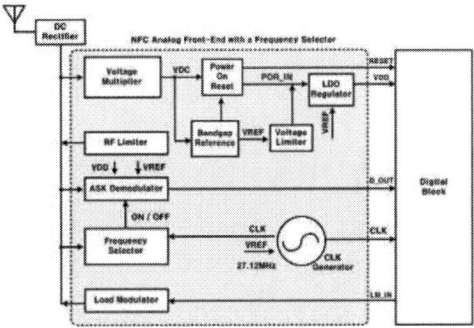

Figure 1.   The block diagram of the proposed NFC analog front-end with the frequency selector

The NFC analog front-end consists of the power supply block and the data transmitter block. The power supply block supplies stable DC Voltage to the ASK demodulator, the clock generator and the digital block. For generating the stable DC voltage, the power supply block has the voltage multiplier, the bandgap reference and the regulator. At first, the voltage multiplier generates large DC voltage due to the multi-Stage structure. And the voltage limiter is required because large DC voltage may cause damage to the internal circuitry. The DC voltage generated by the voltage multiplier is unstable. So the bandgap reference and the regulator generate stable DC voltage to supply the VDD for other circuitry. The data transmitter block has the ASK demodulator and the load modulator. The ASK demodulator transmits data to the digital block. The power of the circuit is supplied from the power supply block. The load modulator responds data from the digital block. The clock generator supplies the constant clock to digital block for the synchronization of input signals.

B.   The frequency selector

The proposed NFC analog front-end has two functions. At first, it transmits data to the digital block for NFC communication. At second, it supplies DC voltage to the wireless battery charging block. So the NFC analog front-end has to distinguish two frequency bands and the frequency selector performs this function. Figure 2. shows the proposed frequency selector circuit.

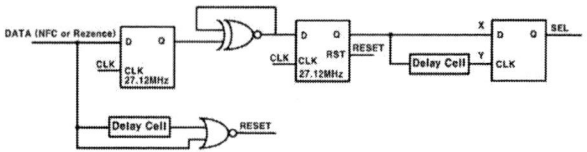

Figure 2.   The proposed frequency selector circuit

The first Flip-Flop performs the synchronization between the internal clock (output of the clock generater) and the input signal. We used the internal clock of 27.12MHz. This frequency is 4 times of the Rezence standard frequency period and twice of the NFC standard frequency period. After the synchronization, the input signals are compared with the internal clock. If the frequency selector receives the frequency signal of 6.78MHz (wireless charging signal), output becomes 1V. If the frequency selector receives the frequency signal of 13.56MHz (NFC), output becomes 0V. Because the reader sends the ASK 100% signal including the 0V data, the output of the frequency selector data have to be 0V.

III. SIMULATION RESULTS

A. The NFC analog front-end

NFC Analog Front-End Simulation Result

Figure 3.   The NFC analog front-end simulation result

Figure 3. shows the simulation result of the NFC analog front-end consists of the power supply block and the data transmitter block. The ASK signal using modified Miller coding from the reader goes into the voltage multiplier. It makes the output voltage, VDC. The voltage limiter did not work because the voltage of the ASK signal which is entering is low at this simulation. The bandgap reference supplied the output voltage of the DC rectifier (VDC). The bandgap reference outputs the constant voltage (VREF). After the regulator receives VREF from the bandgap reference, VDD can be obtained through the error amplifier operation of the regulator. It is used as the supply voltage (VDD) of the NFC

analog front-end. D_OUT_HF shows the operation of demodulation. The digital block received the data through the operation of demodulation and sends the response signal (DIGTAL RESPONSE).

The response signal based on Manchester coding has the carrier frequency of 847 KHz, the amplitude of the input signal changing at the rate of 847 KHz by receiving the response signal.

B. The frequency selector

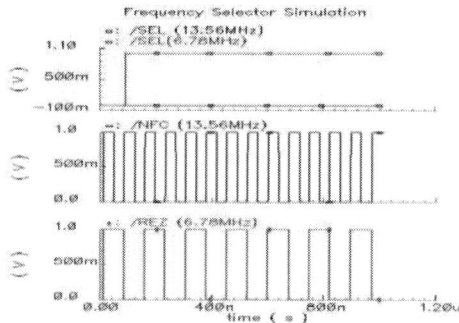

Figure 4.   The frequency selector simulation result

Figure 4. shows the simulation result of the frequency selector. NFC signal has the frequency of 13.56MHz, REZ (Rezence standard signal) has the twice period of the NFC signal. Two signals are compared with the internal clock of 27.12MHz. If the input signal has the frequency of 6.78MHz, output voltage becomes 1V. If the input signal has the frequency of 13.56MHz, the output voltage becomes 0V. SEL is the output voltage of the frequency selector in this simulation

IV. CONCLUSION

We have designed the NFC analog front-end including the frequency selector using 0.18μm CMOS Process. The frequency selector circuit has two functions. If the input signal operates at the frequency of NFC, the frequency selector output becomes 1V and the NFC analog front-end internal circuit is turned off. If the input signal operates at the frequency of the wireless charging signal (Rezence), the frequency selector output becomes 0V and NFC analog front-end block is turned on.

ACKNOWLEDGMENT

"This research was supported by the MSIP(Ministry of Science, ICT and Future Planning), Korea, under the ITRC(Information Technology Research Center) support program (IITP-2016-H8501-16-1010) supervised by the IITP(Institute for Information & communications Technology Promotion)"

REFERENCES

[1] Hyun-Chul Shim, Chung-Hyun Cha, Jong-Tae Park, and Chong-Gun Yu, "Design of a Low-Power CMOS Analog Front-End Circuit for UHF Band RFID Tag Chips,"Journal of The Institute of Electronics Engineers of Korea (IEEK), Vol.45, No.6, pp.28-36, Jun. 2008.

Speed-Adaptive Ratio-Based Lane Detection Algorithm for Self-Driving Vehicles

Seongrae Kim, Junhee Lee, and Youngmin Kim
Department of Computer Engineering
Kwangwoon University
Seoul 01897, Republic of Korea
{ksr8601, junhee1991, youngmin}@kw.ac.kr

Abstract— **Lane detection algorithm using a vision sensor or a camera would be more effective for self-driving vehicles to keep in lane, if it is possible to derive a distance ratio between a vehicle and left-right lanes. However, a dangerous situation may occur if the performance of the camera (e.g., frame/sec.) and the real-time speed of the vehicle are not considered properly because of the huge distance difference among frames for a fast moving vehicle with a low-speed camera. In this study, we propose a simple method to anticipate the relative position of the vehicle in the following frame from the current frame image. The expected ratio between a vehicle and the left-right lanes can be obtained by using of the speed of a vehicle and the frame speed of a camera. Experiment results show that less than 5.28% error occurs by the proposed algorithm for various cars and cameras.**

Keywords - Self driving, Lane detection, Lane keeping, Speed-adaptive, Lane ratio, Frame speed of camera

I. INTRODUCTION

Self-driving is one of the most interesting topics on the numerous automotive applications-related studies recently. Among them, many researches are being conducted in lane detection area using a vision processing system.

Main focus of the previous research is concentrated on two topics. One of them is detecting the lanes effectively by using various image processing methods [1]. The other is measuring lane departure for LDWS (Lane Departure Warning System) [2]. Because eventual purpose of the lane detection is reliable self-driving, close interaction between a vision process system and a control system is necessary. In addition, lane detection algorithm using a vision sensor should be able to derive practical parameters that can be used in control system. Therefore, it will be effective if the lane detection system can derive how much the vehicle is one-sided from the center of lane through a ratio of the distance from the left and right lanes [3].

However, the vehicle moving distance between frames will vary according to the moving speed of the vehicle and the performance of the vision system (i.e., frame per second). If the vehicle is not properly controlled due to huge difference between images of the current frame and the next frame, a dangerous situation may occur because it is not guaranteed that the vehicle is located on center of lane in the next frame. So, the previously mentioned lateral position ratio of the vehicle becomes more safe if we can derive the value in advance that the vehicle will

be located in the next frame rather than in the current frame. The expected lateral position ratio is then transferred to the control system one frame ahead for more efficient and safe self-driving.

In this paper, we propose a simple algorithm that can derive the lateral position ratio between the vehicle and left-right lanes in the next frame based on the relation between the vehicle speed and the performance of the camera equipped with the vehicle.

II. PROPOSED SPEED-ADAPTIVE RATIO-BASED ALGORITHM

The flow chart of the proposed speed-adaptive ratio-based algorithm is shown in Fig. 1. In order to apply our proposed idea, only lanes have to be detected in an image first. The lane detection is a well-known problem and a conventional method can be used [1], [4]. When lanes are identified through several image processing algorithm as shown in red lines in Fig. 1, a virtual longitudinal line which indicates the center of the vehicle (or the center of the image) can be obtained. Then, the lateral position ratio can be calculated by the distances from the longitudinal line to left-right lanes. The vehicle speed and frame speed of the camera can be exploited to anticipate how far the vehicle moves longitudinally at the next image frame from the current image.

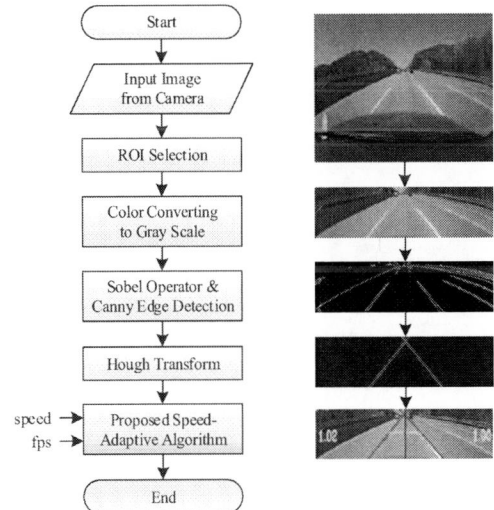

Figure 1. Flow chart of the proposed lane detection algorithm

The actual moving distance of the vehicle between frames, D_A (in meter) can be approximately calculated as follows;

$$D_A \approx \frac{0.28 \times V}{F} \qquad (1)$$

Where, V is the current vehicle speed (km/h) and the F is the frames per second (fps) of the camera. The V is a variable and real-time parameter which is provided from the speedometer and the F is a fixed parameter indicating the performance of the camera. For each frame, when the values of F and V are entered in equation (1), the value D_A, which means the expected actual distance of the vehicle between the current frame and the next frame will be derived. Then, we can obtain the lateral position ratio of the vehicle by using the point that the vehicle will be located in the next frame and the x-coordinates of the left-right lanes in the current image.

However, the actual distance D_A decreases exponentially as moving from the bottom to top of an image due to the geometric characteristic of the input image. Fig. 2 shows the relationship between the average value of the actual distance and the longitudinal percentage (F_P) of each point in the image of region of interest (ROI) measured from experiment with three cars and 10 different camera angles. The relationship is also plotted in Fig. 3 and the generalized model can be extracted as follows through a curve fitting.

$$F_P = -65.81 e^{D_A/-3.07} + 65.98 \qquad (2)$$

Thus, the value D_A is obtained by equation (1) with real-time speed information and the F_P can be calculated by the D_A with the empirical model as shown in equation (2). If the pixel points of $x_{max}/2$-coordinate (P_x) and y-coordinate (P_y) at the extracted percentage line (F_P) in a current image are determined as shown

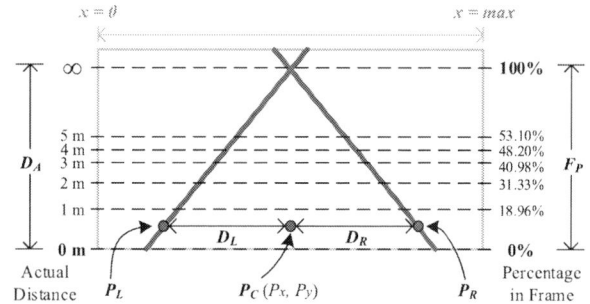

Figure 2. Relation between actual distance and percentage in ROI image with extracted lanes

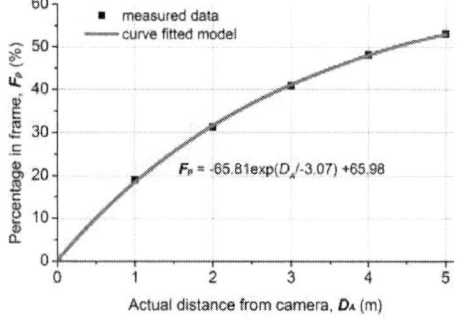

Figure 3. Curve fitting to model the relation between D_A and F_P in ROI image

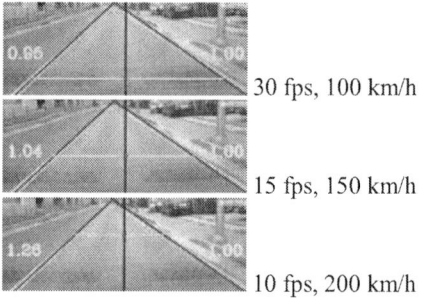

30 fps, 100 km/h

15 fps, 150 km/h

10 fps, 200 km/h

Figure 4. Experiment results of three cases of vehicle speed and fps of camera

in Fig. 2, the lateral vehicle position ratio (R) which is defined as below can be obtained.

$$\text{Lateral Vehicle Position Ratio } (R) = D_L : D_R \qquad (3)$$

Where, D_L and D_R are pixel-wise distances from the expected vehicle location point (P_C) to the left (P_L) and right (P_R) lanes, respectively.

III. EXPERIMENT RESULTS

All of the image processing and the proposed algorithm are implemented by C++ using OpenCV library. The resolution of the captured video image is 720×480 pixels. Experiment results are shown in Fig. 4 for three different cases of vehicle speed and frames per second of camera. The red lines indicate the extracted left and right lanes through image processing. The blue vertical line is the virtual center line and the yellow horizontal line is the calculated F_P line with speed and fps information for measuring the lateral position ratio. As you can see, the yellow line and the ratio measuring points are far from the vehicle as the speed is faster and the camera speed is lower. Measurements with five speeds and five cameras for three different cars have been conducted to verify the correctness of the proposed method. As a result, the error between the longitudinal real distance and P_C in the image is 5.28% of the ROI, and the discrepancy between the real ratio and the ratio calculated by the proposed algorithm is 0.07 at maximum.

IV. CONCLUSION

In this study, a simple method to anticipate where the vehicle will be located laterally in the next image frame is proposed. The proposed algorithm can predict the speed-adaptive lateral ratio between left and right lanes by using of the vehicle speed and camera frame information. It will be helpful in lateral control for lane keeping when the ratio is provided in the control system of the self-driving cars. Experiment results verified the efficiency and correctness of the proposed idea. Also, the proposed idea is possible to implement in a hardware system because the operations are simple and easy.

ACKNOWLEDGMENT

This work was sponsored by NRF Grant #2014-R1A1A2057715.

REFERENCES

[1] Y. Wang, E. Teoh, and D. Shen, "Lane detection and tracking using B-snake," *Image Vis. Comput.*, vol. 22, no. 4, pp. 269–280, Apr. 2004.

[2] J. He et al., "A lane detection method for lane departure warning system", in *Proc. ICOIP*, 2010, pp. 28-31.

[3] D. C. Tseng and C. W. Lin, "Versatile lane departure warning using 3D visual geometry," *Proc.* ICIC, vol. 9, no. 5, May 2013.

[4] D. Ding, C. Lee, and K. Lee, "An adaptive road ROI determination algorithm for lane detection," in *Proc. TENCON*, Oct. 2013, pp. 1–4.

A Design of Tunable Component for Font End Module

Suk-Hui Lee, Ki-Jin Kim, K.H. Ahn
Convergence Communication Components Research Center
Korea Electronics Technology Institute
Seongnam-si, South Korea
sshviole@keti.re.kr

Sung-il Bang
Department of Electronics and Electric Engineering
Dankook University
Yongin-si, South Korea
bang@dku.edu

Abstract— **This paper describes VCO designs based on varactor tuned architecture. The oscillators of FMCW generator have been designed in a 0.13μm SiGe BiCMOS technology, thus targeting automotive Radar and millimeter-wave applications. Their tuning ranges are 2.98 GHz. The VCO chip achieve low phase noise characteristics -102.68 dBc/Hz at 1 MHz offset from the tunable frequency.**

Keywords; BiCMOS, millimeter wave, Automotive Radar, Oscillator, Driving amplifier

I. INTRODUCTION

A millimeter-wave radar utilizes radio waves in the millimeter-wave band, and the operation on electromagnetic waves makes it less affected by climatic elements such as rain, snow and fog as well as a vehicle's rear lights than image sensors and laser radars, which have used visible light or infrared so far. There are 77 GHz-band and 24 GHz-band radars already commercialized for in-vehicle applications, but higher performance is in demand as the systems continue to become more sophisticated as in driving safety support and automatic driving, which require wider detection areas and enhanced resolution to be operable in more complex environments such as on urban roads. To enhance the resolution requires wider bandwidth. In Japan, the available bandwidth is limited to 0.5 GHz by law, and thus the existing 77 GHz-band millimeter wave radar cannot have the resolution necessary for automatic driving. This is why there are high expectations for the development of 79 GHz-band radars, which are capable of achieving high resolution in the ultra-wide 4 GHz bandwidth. The driving safety support system includes Adaptive Cruise Control (ACC) and Autonomous Emergency Braking (AEB), as well as lane-changing assistance and automatic collision avoidance[1-2]. In order to meet the requirements of the above-mentioned systems, an Advanced Driver Assistance System (ADAS) based on radar system as shown in Figure 1 must be developed. Long- and mid-range radars are installed in the front and rear of a vehicle, and short-range radars are mounted on both sides. Currently we are considering the use of multi-purpose radar for 79 GHz-band high resolution millimeter-wave radar single module. The 79 GHz-band millimeter-wave radar enhances the range resolution through wider bandwidth, allowing for downscaling and lower costs.

This work was supported by ICT R&D program of MSIP/IITP. [B010-16-1353 , Development of high efficiency, fully integrated Multi-mode Multi-band RF Module for Smart hand held Devices]

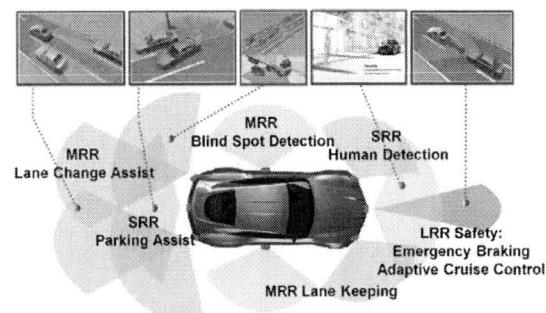

Figure 1. Radar system structure

II. FMCW GENERATOR STRUCTURE

In general, FMCW generator have two structures, such as direct VCO control and mixer with high speed DAC. A direct VCO control FMCW generator is operated by voltage waveform from DAC. This structure is limited by process variation and non-linearity of VCO. To solve the VCO issues, the other FMCW generator use mixer with high speed DAC. A FMCW waveform generate from high speed DAC then it pass through the up conversion by mixer. But 79 GHz-band ADAS radars must generate broadband signal on 4 GHz bandwidth, the performance of DAC and linearity of mixer decrease the total system performance.

We design FMCW generator with PLL. The PLL is negative feedback system, it generate output frequency which is multiplied F_{REF} by ratio. A figure 2 shows suggested structure. A VCO operate at 40 GHz-band, it is easy to generate 80 GHz-band signal by doubler. Also broadband 80 GHz-band divider design is difficult than 40 GHz-band divider.

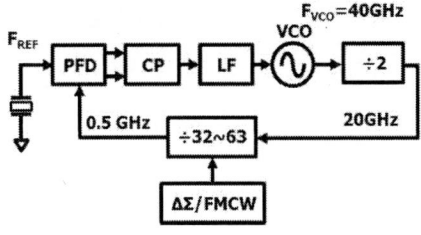

Figure 2. FMCW generator with PLL

III. DESING OF VCO

In this paper, a VCO core is proposed and designed in a 0.13μm SiGe BiCMOS process. The main targeted performances are significant in term of low phase and low power consumption.

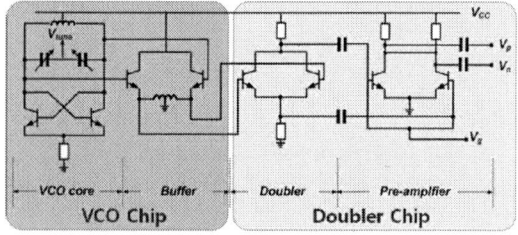

Figure 3. Sturucture of 79GHz VCO

Figure 3 shows the architecture of our 79GHz VCO structure. It consists of a 39.5GHz VCO chip and doubler chip. The VCO core includes a cross coupled VCO and a buffer. The gate and drain of the cross-coupled nFETs in the VCO core are DC-coupled to simplify layout and to minimize parasitic. A MOS varactor and an inductor form the default LC tank; A λ/4-stub is placed at the common source to achieve high impedance at $2\omega_0$ for reduced PN similar to LC-filtering. The buffer is a simple source-follower. The doubler topology is adopted from SiGe BiCMOS chipset [3-4] and provides differential output. Also, it is DC-coupled to the buffer output, i.e., doubler gates are grounded, which loads the VCO slightly. A common-source, differential pre-amplifier is then used to recover the doubler loss and enable the VCO to be readily integrated. There are various advantages of this modular architecture. The lower frequency in the fundamental VCO core results in higher Q factor for the LC tank, which in turn means lower phase noise and smaller nFETs to provide the negative resistance; the smaller nFETs result in less parasitics and hence, higher TR. Furthermore, the larger varactor capacitance for the lower fundamental frequency reduces the relative impact of parasitic capacitances, and thus increases TR. Finally, the modular architecture enables optimizations of the VCO core as needed. Figure 4 shows the VCO chip layout of our VCO core and buffer stage with PTAT current generator.

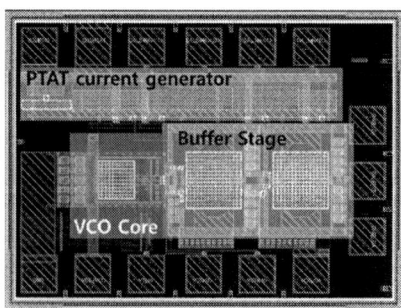

Figure 4. Layout of VCO core and Buffer

Figure 5 shows the simulated oscillation frequency (f_{osc}) versus tuning voltage (V_{tune}) with 2.5V supply. The f_{osc} is tunable from 37.88 GHz to 40.86 GHz. The tunable bandwidth of VCO is about 2.98 GHz. Figure 6 shows the simulated phase noise (PN) versus V_{tune}. The PN is lower than -102.68 dBc/Hz at 1 MHz offset for 37.88 GHz and 40.86 GHz.

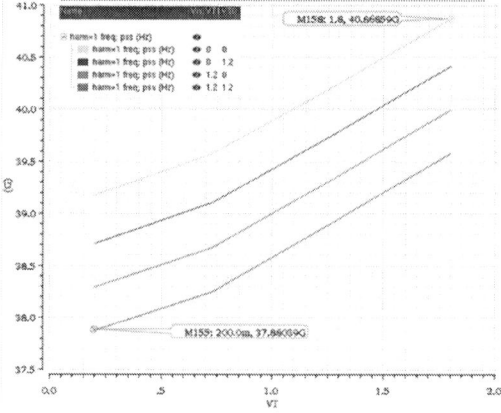

Figure 5. Simulation result of f_{osc} versus V_{tune}

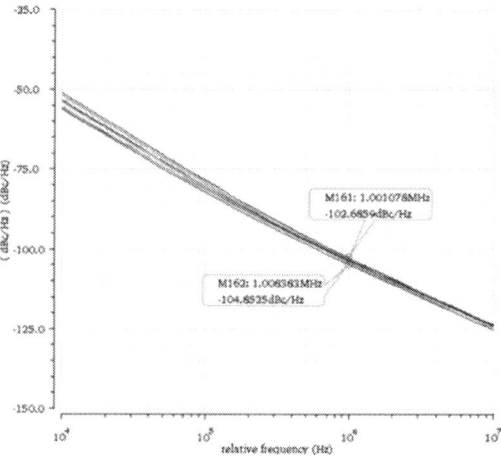

Figure 6. Simulation result of phase noise versus V_{tune}

IV. CONCULSION

In this paper a 39.5 GHz VCO chip for FMCW generator has been designed. The simulated performances exhibit a minimum phase noise of -102.68 dBc/Hz at 1 MHz offset and broad tuning bandwidth of 2.98 GHz. Such results in a 0.13μm SiGe BiCMOS process may apply to the development of low cost ADAS base on radar system.

REFERENCES

[1] N. Shima, O. Isaji, M. Kishida, N. Okubo and S. Yamano,"High Resolution Long Range Radar with Small Aperture," IS:04-24, Media Interactive Sessions, ITS-WC 2010.

[2] H. Asanuma, Y. Sekiguchi and M.Kishida, "Side Forward Looking Millimeter Wave Radar for Front Diagonal Pre-Crash Safety System," IS:3167, TS102,Technical Session, ITS-WC 2009.

[3] S. Trotta, et al., "A multi-channel Rx for 76.5GHz automotive radar applications with 55dB IF channel-tochannel isolation," *EuMIC 2009*, pp. 192-195.

[4] Forstner, H. P., et al. "A 19GHz VCO downconverter MMIC for 77GHz automotive radar frontends in a SiGe bipolar production technology." *Microwave Conference, 2007. European*. IEEE, 2007.

Efficient and Real-time Stereo Matching Hardware Architecture for High-resolution Image

Haengson Son, Seonyoung Lee, and Kyoungwon Min

Mobility Platform Research Center
Korea Electronics Technology Institute
Seongnam, Republic of Korea
{hsson, drleesy, minkw}@keti.re.kr

Abstract—**In this paper, we propose an efficient and real-time stereo matching hardware architecture for a high resolution image. Disparity estimation algorithm must be operated at a real-time to be of practical use for applications such as an autonomous driving. However, they generally require large computational efforts and high memory capacities in the embedded processor-based systems. To solve this problem, we studied the real-time stereo matching hardware architecture and implemented in hardware system. Our architecture was implemented using Verilog HDL. Our circuit uses 95% LUT, 92% FF and 80% BRAM of Zynq XC7Z020 FPGA. Also, our hardware circuit can generate the depth data for the high-resolution images which receive from cameras without delays in the real time.**

Keywords-component; stereo matching; high-resolution; hardware architecture; real-time

I. INTRODUCTION

The stereo matching has been used in the fields of automobiles, and robots as a pre-processing step for detecting the distance and background surface. Intelligent vehicles should detect an object and recognize the distance of the object for the safe autonomous driving. Recently, applications of the stereo matching has been expanded to wearable devices and smart security systems. Stereo matching is a traditional method to acquire 3-D (dimensional) information from image pair and has some merits over other 3-D sensing methods in terms of operating condition and reliability [1]. Stereo matching shows the best performance to detect objects and estimate distance the detected objects. So, many algorithms have been developed, such as global matching and local matching. Global matching stereo algorithm [2] shows a good performance to generate disparity values. However, it exhibits a high sensitivity to adjust parameters and generally requires large computational efforts with high memory capacities. Local matching stereo algorithm [3] has a fast operation time, but requires an adequate choice of window size. Since depth errors increase quadratically with the far distance [4], high-resolution images are needed to obtain accurate 3D representations. However, it is required to detect an object such as cars and pedestrians, and calculate the object distance for a short time. In the embedded processor-based systems, it is difficult to process in real-time for a high resolution image because these stereo matching algorithms are necessary much calculation complexity.

In this paper, we propose an efficient and real-time stereo matching hardware architecture for high resolution images. First, we analyzed the statistical characteristics of the depth image using KITTI dataset to overcome the disadvantage of local matching algorithm. The experiment has shown us that the surrounding pixels are similar to the variation distribution in the horizontal/vertical direction. Based on this characteristic, we developed some window types suitable for the road environment when implementing in the hardware. Also, our real-time stereo matching architecture for high resolution image was implemented using Verilog HDL (hardware description language). Our stereo matching logic implemented in hardware was tested its operation using Xilinx's Zynq-7000 FPGA (XC7Z020) device. The synthesis results show that 95% LUT, 92% FF and 80% BRAM utilization. Our circuit operates on 150 MHz clock frequency and calculates the depth data at 30 frames/sec for HD (1280x672) images.

II. REAL-TIME STEREO MATCHING HARDWARE ARCHITECTURE

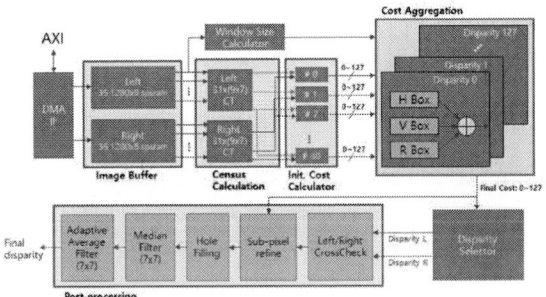

Figure 1. Proposed hardware architecture for real-time stereo matching.

Our proposed hardware architecture for efficient and real-time stereo matching is configured as shown in Fig. 1. 'DMA IF' module receives the stereo image and transmits the disparity data through AMBA AXI bus. 'Image Buffer' module stores the image of the 36 lines left/right image data into single-port SRAM for the real-time stereo matching. 'Census Calculation' module performs the census transform for each of the left/right image. 'Init. Cost Calculator' module calculates the initial cost for the 31 lines and 128 disparities. 'Cost Aggregation' module computes a 128 disparity information through the aggregation process. This module performs aggregation operations on the three type windows. 'Disparity Selector' module selects the minimal disparity values from the cost aggregated information. 'Post-processing' module performs a Left/Right cross check (consistency check), Sub-pixel refine, Hole filling (interpolation) and depth map improvement (median filter and adaptive average filter). 'Window Size Calculator' module determines the window size for the cost aggregation of the horizontal direction. Local matching algorithm has a problem that it is not easy to implement in a hardware because of the variable window size. To overcome this problem, the statistical characteristics of depth images were analyzed using KITTI dataset. Figure 2 (a) shows the distribution variation of the pixel value with the surrounding pixels in a histogram KITTI dataset. The center point is the pixel location to estimate the current depth. X-axis and Y-axis represents the frequency of pixels having the same variation of values when the position is far away from the pixel of interest. This figure shows that a pixel having a similar disparity exists much in the horizontal/vertical. We decided the three types of window for the cost aggregation through the histogram distribution, as shown in Fig. 2 (b). Based on these results, we developed three window types suitable for the road environment shown in Fig. 2 (c). Figure 3 shows the block diagram of an overall implemented real-time stereo matching system. Our architecture is composed of 'Stereo Camera Interface', 'Stereo Matching Engine', 'Interface Timing Generator', 'System Control Registers' and 'Cam Reset' logics. 'Stereo Camera Interface' module receives data input from the stereo camera and aligns two images to get a good depth results. 'Stereo Matching Engine' module extracts the disparity map from the rectified images. 'Interface Timing Generator' module generates the camera interface signal of ITU-T 601 format and transmits the depth information. 'System Control Registers' and 'Cam Reset' module controls the stereo matching system and stereo camera.

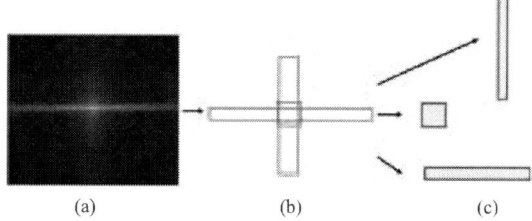

(a)   (b)   (c)

Figure 2.   Window types for cost aggregation.

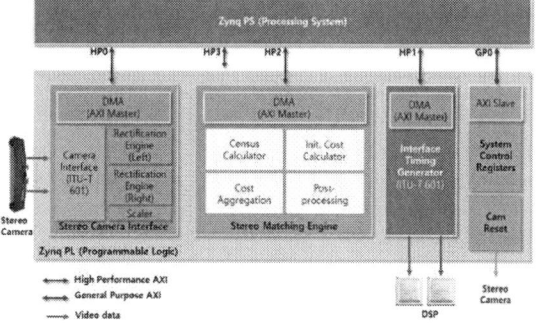

Figure 3.   Overall architecture of implemented stereo matching system.

III.   IMPLEMENTATION RESULTS AND CONCULSTION

Our real-time stereo matching architecture was described into RTL (register-transfer level) circuit using Verilog HDL. We tested its operation in the Xilinx's Zynq-7000 FPGA (XC7Z020) device. Our circuit showed 95% LUT, 92% FF and 80% BRAM utilization. Table 1 represents a summary of the features for the real-time stereo matching architecture. Our circuit operates on 150 MHz clock frequency and calculates the depth information at 30 frames/sec for HD (1280x672) images with 128 disparity range. Upper image of Fig. 4 shows the original image from camera and lower image of Fig. 4 shows the color disparity map for the top image using our implement circuit.

TABLE I.   FEATURES OF PROPOSED STEREO MATCHING SYSTEM

Device	Xilinx Zynq-7000	
Camera Input	HD Stereo images, 60Hz	
Output Results	Disparity (128 range) Rectified HD color image	
Speed of output data	30 frame/sec	
Operating Frequency of System	ARM Cortex-A9	667 MHz
	DDR3	533 MHz
	Camera IF	74.5 MHz
	Stereo Matching IP	150 MHz

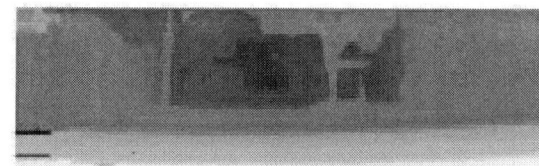

Figure 4.   Results of test image: (upper) Original camera image, (lower) color disparity map calculated by implemented circuit.

ACKNOWLEDGMENT

This work was supported by the Knowledge & Economy Technology Innovation program (10052731) funded by the Ministry of Trade, Industry & Energy (MOTIE), Korea.

REFERENCES

[1] J. I. Woodfill, G. Gordon, and R. Buck, "Tyzx DeepSea high speed stereo vision system," in Proc. IEEE Comput. Soc. Workshop Real-Time 3-D Sensor Use Conf. Compt. Vision Pattern Recog., Washington D.C., 2004, pp. 41-46.

[2] Felzenszwalb, P., Huttenlocher, D.: Efficient belief propagation for early vision. International Journal of Computer Vision 70 (2006) 41–54.

[3] Weber, M., Humenberger, M., Kubinger, W.: A very fast census-based stereo matching implementation on a graphics processing unit. In: IEEE Workshop on Embedded Computer Vision. (2009).

[4] D. Gallup, J.M. Frahm, P. Mordohai, and M. Pollefeys, Variable baseline/resolution stereo. In: CVPR. pp. 1–8, 2008

A Low-Power, Low-Noise Neural Recording Amplifier for Implantable Biomedical Devices

Hyung Seok Kim and Hyouk-Kyu Cha

Department of Electrical and Information Engineering
Seoul National University of Science and Technology
Seoul, Korea
hkcha@seoultech.ac.kr

Abstract— **This paper presents a low power and low noise neural amplifier IC for processing both action potential and local field potential signals in neural implant devices. Based on a capacitive-feedback topology, the core operational transconductance amplifier utilizes a two-stage structure with current buffer achieving wide bandwidth, large output swing, and small area. The proposed neural amplifier is designed using 0.18μm CMOS process and achieves 46 dB gain, a bandwidth of 0.9 Hz-13.8 kHz, and integrated input-referred noise of 5 μV$_{rms}$ in the range of 1 Hz to 10 kHz. The noise efficiency factor of the designed neural amplifier is 2.6 and consumes 2 μA of current from a 1.2 V supply with an area of 0.136 mm^2.**

Keywords–neural amplifier; neural recorder; capacitive feedback; operational transconductance amplifier

I. INTRODUCTION

A neural recorder is an implantable medical device used to collect neural signals for observing and diagnosing certain symptoms of various diseases associated with the brain, and to analyze the signals of the brain related to physical activity [1]. In particular, neural signals known as action potential (AP) which reside in the frequency band from 300 Hz to around 7 kHz, and local field potential (LFP) which is observed around 1 Hz to 300 Hz frequencies, are of interest. These are very weak signals with amplitudes ranging from 100 μV$_{pp}$ up to 1 mV$_{pp}$. Thus, it is important to provide sufficient amplification prior to further processing of the signal, and to minimize the addition of noise at the same time. To interface the neural recording IC with multi-array probe for high-resolution processing, power consumption is also important as it may cause temperature rise and lead to tissue damage. The neural amplifier is the first block in the neural recording chain and is critical in deciding the overall performance of the recording device. This paper presents a low-power and low-noise amplifier for neural recording systems.

II. CIRUIT DESIGN

Figure 1 shows the block diagram of the designed neural amplifier. It utilizes an operational transconductance amplifier (OTA) with capacitive-feedback. The capacitive-feedback topology enables accurate gain setting, where the mid-band gain A_M is determined by the ratio C_{IN}/C_F. Also, AC coupling is used in order to remove the large DC offset voltage, caused by the chemical reaction at the electrode-neuron interface. In this

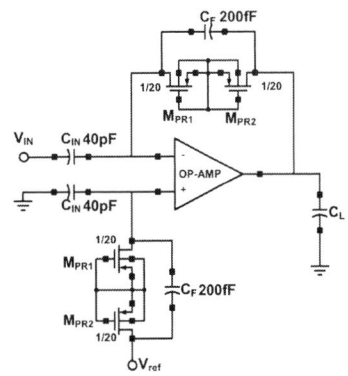

Figure 1. Block diagram of the designed neural amplifier

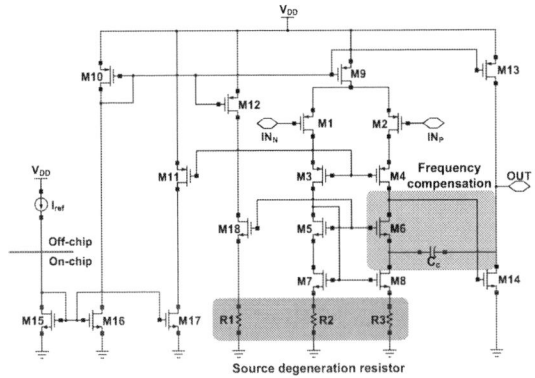

Figure 2. Schematic of operational transconductance amplifier

design, the gain is decided to be around 46 dB using the values of C_{IN}=40 pF and C_F=200 fF. The low-pass bandwidth of the amplifier is approximately equal to $G_m/(A_M \cdot C_L)$ where G_m is transconductance of the OTA. The low-pass cutoff must be set to be high enough to support AP signals. The sub-Hertz high-pass cut-off frequency is determined by $1/(2\pi \cdot R_P \cdot C_{IN})$, where R_P is a large resistance consisting of M_{PR1} and M_{PR2} acting as a pseudo-resistor. Figure 2 presents the circuit schematic of the proposed OTA. Among several topologies available in the design of the OTA, this work utilizes a two-stage

978-1-5090-3220-4/16 $31.00 © 2016 IEEE   275   ISOCC 2016

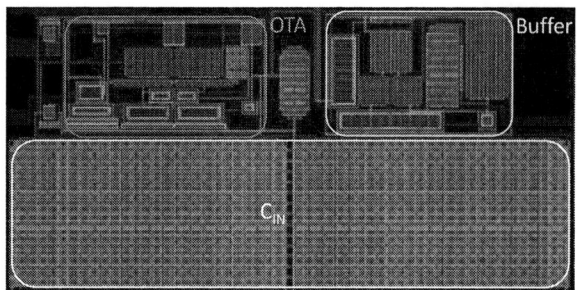

Figure 3. Layout capture of neural amplifier and unity-gain test buffer

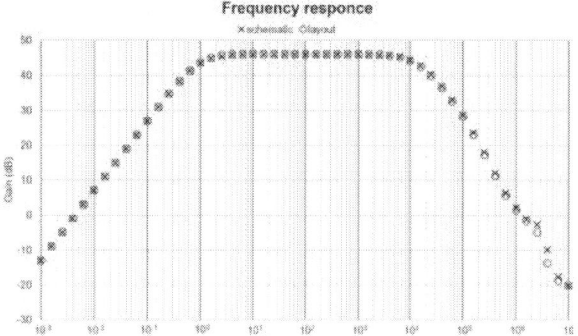

Figure 4. Frequency response of neural amplifier

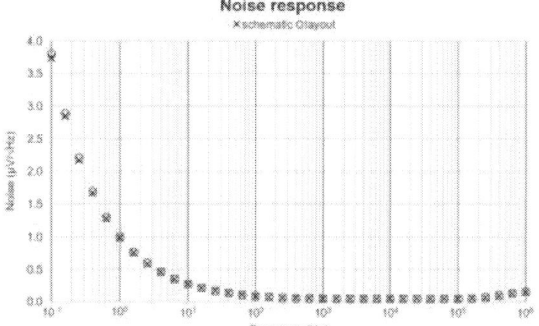

Figure 5. Noise performance of neural amplifier

architecture for high open-loop gain and large output swing. The first stage uses a telescopic structure with PMOS input transistors biased in subthreshold for better flicker noise performance and maximized g_m/I_D ratio, respectively. In addition, source degeneration resistors are used in the current mirror loads for improved noise performance [2]. Frequency compensation is done employing Miller capacitor C_C with the common-gate current buffer [3]. This strategy is used to reduce the required value of C_C and improves the bandwidth of the OTA.

III. SIMULATION RESULTS

The proposed neural amplifier is designed using 0.18μm CMOS process. Figure 3 shows the layout of the neural amplifier and the area is 0.136 mm². Figure 4 shows the schematic and post-layout frequency response of the designed neural amplifier. A mid-band gain of 46 dB is achieved with

high-pass and low-pass cutoff frequencies at 0.89 Hz and 13.8 kHz, respectively. Figure 5 presents the schematic and post-layout input noise density plot. The integrated input noise from 1 Hz to 10 kHz is 5 μV$_{rms}$. Table I shows the performance summary of the designed neural amplifier. The proposed neural amplifier consumes very low power while achieving comparable performance to other recent similar works.

TABLE I.   COMPARISON OF THE PROPOSED NEURAL AMPLIFIER WITH THE PREVIOUS WORKS

Parameter	This work*	[4]	[5]
CMOS Technology (μm)	0.18	0.18	0.18
Supply voltage (V)	1.2	1.8	1
Supply current (μA)	2	5.6	2.52
Gain (dB)	46.2	48 / 60	54.2
Bandwidth (kHz)	13.8	0.03-0.29/9	5.7
Low-frequency cutoff (Hz)	0.89	0.3-900	200
Input referred noise (μV$_{rms}$)	5 (1Hz-10 kHz)	5 (1Hz-8kHz)	3.83 (200 Hz-10kHz)
Noise efficiency factor	2.6	4.6	3.07
CMRR (dB)	>66	48	65
PSRR (dB)	>74	55	67
Area (mm²)	0.136	0.065	0.109

*Simulation results

IV. CONCLUSIONS

A low-power capacitive-feedback neural amplifier using two-stage OTA with current buffer based compensation and source degeneration for implantable neural recording device is designed using 0.18μm CMOS process. The neural amplifier achieves 46 dB gain and integrated noise performance of 5 μV$_{rms}$ while consuming 2 μA from a 1.2-V supply.

ACKNOWLEDGMENT

This research was supported by Basic Science Research Program through the National Research Foundation of Korea (NRF) funded by the Ministry of Science, ICT, and Future Planning, (NRF-2015R1A2A2A01006502). This work was supported by IDEC for EDA Tool and MPW support.

REFERENCES

[1] R. R. Harrison and C. Charles, "A low-power low-noise CMOS amplifier for neural recording applications," IEEE J. Solid-State Circuits, vol. 38, no. 6, pp. 958–965, Jun. 2003.

[2] W. Wattanapanitch, M. Fee, and R. Sarpeshkar, "An energy-efficient micropower neural recording amplifier," IEEE Trans. Biomed. Circuits Syst., vol. 1, no. 2, Jun. 2007

[3] G. Palmisano and G. Palumbo, "A compensation strategy for two-stage CMOS opamps based on current buffer," IEEE Trans. Circuits Syst. I, vol. 44, no. 3, pp. 257–262, Mar. 1997

[4] P. Kmon and P.gryboś, "Energy efficient low-noise multichannel Neural amplifier in submicron CMOS process," IEEE Trans. Circuits Syst. I, Reg. Papers, vol. 60, no. 7, pp. 1764–1775, Jul. 2013

[5] Yi Chen, A. Basu, L. Liu, X. Zou, R. Rajkumar, G. S. Dawe, and M. Je "A digitally assisted, signal folding neural recording amplifer," IEEE Trans. Biomed. Circuits Syst., vol. 8, no. 4, pp. 528–542, Aug. 2014

Design of Emotion Lighting Control System on the Power Spectrum Algorithm

Su-Jeong Yun
School of Computer
Semyung University
Jecheon-city Chung-buk, Korea
sjbaby79@naver.com

Chi-Ho Lin
School of Computer
Semyung University
Jecheon-city Chung-buk, Korea
ich410@semyung.ac.kr

Abstract— Modern society is the emotional center of this design, each company has a number of factors are associated with emotions such as color and shape to fit the emotional paradigm focused to meet consumer sensibility. In this paper, the power spectrum algorithm proposed an algorithm for controlling the extracted emotional lighting a biological signal. LED lamp has been in use for emotional lighting is eco-friendly and has a high efficiency and prolonged lifetime. In particular, LED lighting has the advantage also possible that various color representation single light port. And it was used as a biological brain wave signals to determine more accurately the human emotions, analyze the collected brain waves using EEG equipment was considered an emotion.

Keywords—Emotion;Light;LED; BCI; EEG

I. INTRODUCTION

Full-Color LED is possible to control the light source, etc. Unlike conventional color, color temperature, brightness. Controlled variation of the attributes of the light has a study on the correlation between the light and human emotion with the development of LED due to exert a great influence on the sensitivity of human attention. The final goal of such a study can be said to provide a light environment optimized for the human life environment. [1-2]

Using the mechanisms of the brain varies according to the latest thinking brain-computer interface (BCI) technology to control the computer and communication has emerged. And EEG is possible to implement compact and lightweight, called brain waves easy portability and low cost compared to other sensors, it can be said that the real-time measurement is possible, the most suitable for the BCI system implementation. Emotion-based intelligent system that the computer understands the human emotions, and to actively perform the human requirements are necessary in order according to advanced technology development to provide a light environment optimized using BCI technology with the best way to determine the sensitivity [3-6]

In this paper, the user's emotional state classification based on the bio-signal, and propose an efficient LED emotional lighting control system to control the LED light itself. In addition, we have designed and implemented using the hardware Openeeg to measure the bio-signal. [7]

II. A EFFICIENT LED EMOTIONAL-LIGHTING CONTROL ALGORITHM

Emotion LED lighting control system proposed in this paper is to control the LED light itself by using the emotion states determine the user's emotional state, and determine by using the bio-signal. Bio-signal is used the EEG signal. A method for measuring the EEG signal with the board was classified Openeeg a sleepy state, focus state, the emotional state of the user in a strained state, and excited to measure the EEG signal. Also it applies to the emotional lighting control system for controlling the LED light with an emotional state classification.

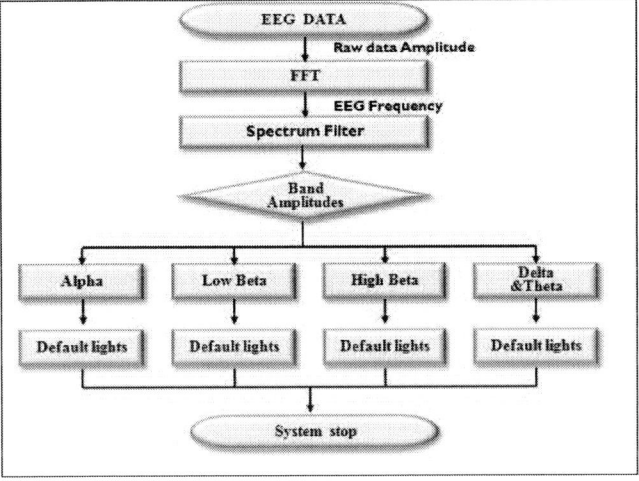

Fig.1. Emotion lighting control algorithm

Figure 1 shows a lighting control algorithm for controlling emotion on a PC using the EEG data measured at the analog circuits. In the light control sensitivity algorithm was changed to the first digital signal is a discrete process for continuous EEG signals from the computer to enable. It is typical of a variety of methods to analyze the EEG power spectrum analysis. Power spectrum analysis is commonly used for this method and FFT correlation function method, a case can not be obtained sufficiently long the length of the data has to be used usually linear predictive model method. In the FFT method, without estimating the correlation function, the power spectrum directly from the Fourier transform of the observed data were estimated. From the measured data by using only, $x_0, x_1 \, x_{N-1}$, of the data was used as an expression (1) to estimate the power spectrum.

$$\hat{P}_{PER}(f) = \frac{1}{N \triangle t} \left| \triangle t \sum_{n=0}^{N-1} x_n \exp(-j2\pi f n \triangle t) \right|^2 \tag{1}$$

At this time, the band-dependent power generated in the electrode of the scalp as an absolute power, and a value obtained by calculating the absolute power of each frequency band at a rate based on the absolute power of the entire frequency band that the relative power of each frequency band. Relative power spectral analysis Study on the brain because of the absolute can compare to reduce the measured variables such as the thickness of the skull with the difference, the electrical condition of the scalp during the measurement, tension on the power spectrum analysis, showing the relevance of the cognitive function and brain waves well from being widely used

III. RESULTS

In this paper, by utilizing Openeeg board can measure the brain wave data 2CH designed a board that has a resolution of 10 bits. An electrode for brain wave measurement was attached to the prefrontal cortex (Fp1, Fp2) by a batch method in international electrode International 10-20 System. A reference electrode was attached to the right behind the earlobe, it was attached to the ground electrode after the left earlobe.

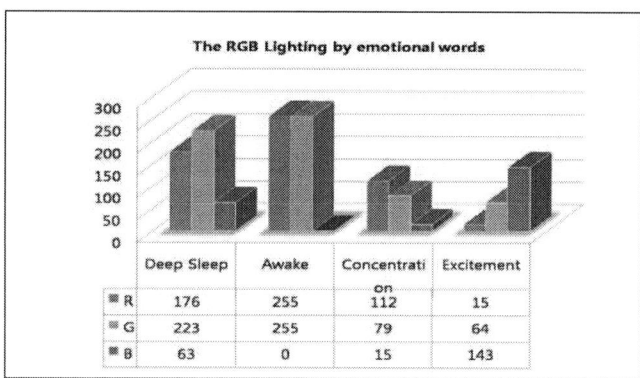

Fig. 2. The RGB Lighting by Emotional Words

Figure 2 shows the RGB values stand set of emotional vocabulary star light. The emotional word choice was determined after a study found out the color image scale and the adjective image scale infer the emotional vocabulary to three primary colors and color distribution of the reasoning used in the emotional final engineering chromaticity coordinates with. Through the study material and free

association, etc. Emotional engineering were collected primary emotional vocabulary, select the 56 second emotional vocabulary words around one trillion people involved were selected final emotional vocabulary.

Show the EEG data by applying a band-specific emotional lighting control algorithm proposed in this paper, the EEG signal measured from a user, relative to the frequency band as shown in Table 1. Set the RGB values for sensitivity to find words to express the general emotional state that each brainwave band has lighting and lighting control to be applied.

TABLE I. RGB lighting value of the EEG classification

Indicator	Frequency Definition	State	R	G	B
Delta wave & Theta wave	0.5~7Hz	Deep Sleep, Sleep	176	223	63
Alpha wave	8~12Hz	Awake	255	255	0
Low beta wave	13~20Hz	Concentration, Activity	112	79	15
High beta wave	21~30Hz	Tension, Excitement	15	64	143

In this paper, the user's emotional state classification based on the biometric signal, and proposed an efficient LED emotional lighting control system to control the LED light itself. Therefore, the accuracy of the difference signal to the measurement environment, but emotional lighting control algorithm could be sufficient to extract the brain wave signal band by using the brain wave signal to the control signal applied to the emotional lighting control system. In the future, if you want to use the less it receives EEG extraction algorithm, the impact of environmental apply emotional lighting control algorithm, will be able to get the exact biological signals, lighting appropriate to the individual user the convergence of LED lighting and IT technology that is currently receiving attention It will be able to provide an environment.

REFERENCES

[1] S. H. Baik, J. C. Kim, I. Y. Jeong, J. T. Kim, "Preferences of Work-plane by Correlated Color Temperature of LED Light sources", *Architecture&Urban Research Information Center*, Vol. 8, No. 2, 2008.

[2] S. D. Jee, S. H. Lee, K. J. Choi, J. K. Park, C. H. Kim, "Sensibility Evaluation on the Correlated Color Temperature in White LED Lighting", *Journal of Korean Institute of Illuminating and Electrical Installation Engineers*, Vol. 22, No. 2, 2008.

[3] L. Kirkup, A. Searle, A. Craig, P. Mclsaac, and P Moses, "EEG-based system for rapid on-off switching without prior learning", *Medical and Biological Engineering and Computing*, vol. 35, pp. 504-509, 1997.

[4] B. Blankertz, F. Losch, M. Krauledat, G. Dornhege, G. Curio, K. R. Muller, "The Berlin Brain-Computer Interface: Accurate Performance From First-Session in BCI-Nave Subjects," *Biomedical Engineering, IEEE Transactions on*, Vol. 55, No. 10, pp. 2452-2462, October 2008.

[5] A. Erfanian and A. Erfani, "ICA-based classification scheme for EEG-based brain-computer interface: the role of mental practice and concentration skills," in Proc. of the 26th Annual International Conf. on of the IEEE EMBS, pp. 235-238, September 2004.

[6] H. G. Yeom, J. S. Kim, C. K. Chung, "Estimation of the velocity and trajectory of three-dimensional reaching movements from non-invasive magnetoencephalography signals," *JNeural Eng.* Vol. 10, No 2, February 2013..

[7] Openeeg, Retrieved April., 30, 2015. from http:// openeeg. sourceforge.net/doc

A Low-Power Capacitive-Feedback CMOS Neural Recording Amplifier for Biomedical Applications

Hyung Seok Kim and Hyouk-Kyu Cha
Department of Electrical and Information Engineering
Seoul National University of Science and Technology
Seoul, Korea
hkcha@seoultech.ac.kr

Abstract—A low-power capacitive-feedback amplifier IC for neural SoCs in medical implant devices using 0.18μm CMOS technology is presented. The proposed neural amplifier, which is based on the source-degenerated folded-cascoded OTA, achieves 40 dB of voltage gain and integrated input-referred noise of 4.3 μV$_{rms}$ in the range of 1 Hz to 10 kHz while dissipating 2.2 μA of current from a 1 V supply. The designed amplifier IC achieves the noise efficiency factor of 3.07 and occupies 0.136 mm^2 of area.

Keywords–neural amplifier; capacitive-feedback; folded-cascode operational transconductance amplifier

I. INTRODUCTION

Due to the need to monitor, record, and analyze the activity of neurons in human brain to enable and control various neural prosthetic medical devices, much research and development of neural recording ICs have been on-going for the past few years [1],[2]. A neural amplifier is needed to amplify weak neural signals residing in the band of 1 Hz to around 5 kHz, while minimizing the addition of noise. As many number of amplifiers are used to process neural signals from the micro electrode array (MEA), the power consumption of each neural amplifier should be low considering the limitation in wirelessly transferring sufficient power from the external device. This paper presents a low-power, low-noise recording amplifier suitable for multi-array neural system-on-chips (SoCs).

II. CIRUIT DESIGN

The overall system block diagram of a general neural recording SoC is shown in Fig. 1. This work includes the single-channel low-noise neural amplifier which is the first block interfacing the MEA in a neural SoC. Fig. 2 shows the schematic of the proposed capacitive-feedback neural amplifier. The input capacitors C_{INP} and C_{INN} are 20 pF mim-capacitors used to block DC offset. The midband gain is set by C_{INN}/C_F ratio. To provide a voltage gain of 40 dB, the size of C_F is 200 fF. Along with C_F, the resistor R_F is used to set the high-pass filter cut-off frequency to be around 870 mHz. The resistances are implemented using pseudo-resistors to achieve large resistance with small area. The high-frequency low-pass filter cut-off is set to be around 7.8 kHz which is dominantly decided by the core OTA of the neural amplifier. The design approach of the core OTA is focused on noise performance while considering low power consumption. A folded-cascode topology is chosen for its large open-loop gain with large-size

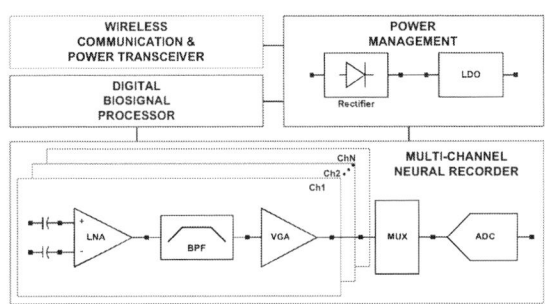

Figure 1. Block diagram of a neural recording SoC

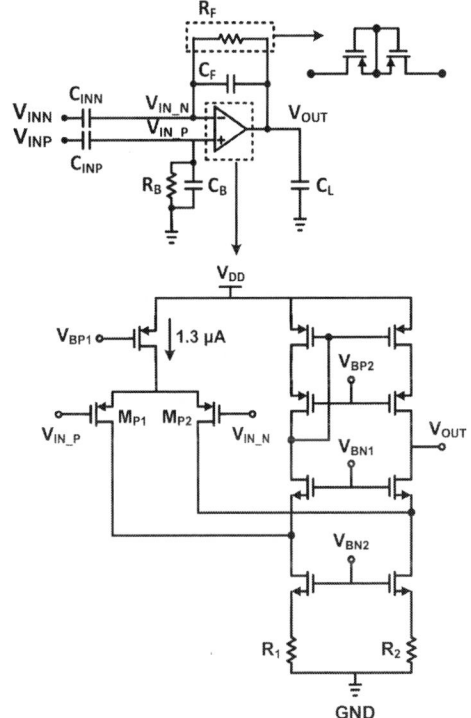

Figure 2. Schematic of proposed neural amplifier IC

PMOS input transistors to reduce 1/f noise. The input transistors are biased in weak inversion to achieve large g_m/I_D

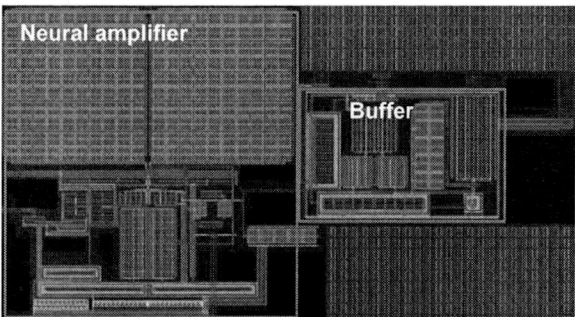

Figure 3.   Layout capture of designed neural amplifier IC

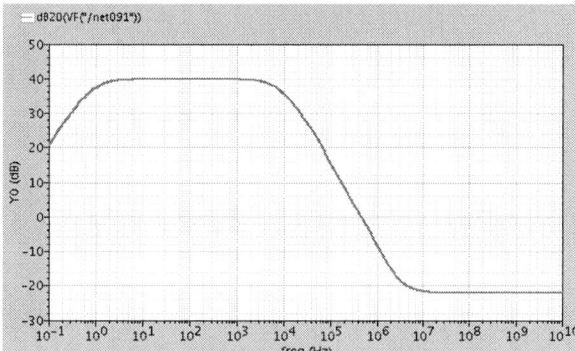

Figure 4.   Gain response of neural amplifier IC

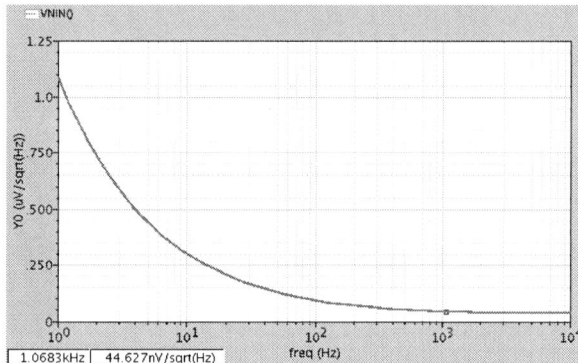

Figure 5.   Input noise response of neural amplifier IC

ratio. Careful optimization in sizing of the transistors is done to maximize the overall transconductance of the OTA for a given total current. In order to improve the thermal noise performance, source degeneration resistors are used [2]. The resistance values are optimized considering the resulting noise performance and the voltage headroom margin.

III.   SIMULATION RESULTS

Figure 3 shows the layout capture of the neural amplifier and the area is 0.097 mm². Figure 4 shows the simulated gain response of the proposed neural amplifier. A mid-band gain of 40 dB is achieved with high-pass and low-pass cutoff frequencies at 0.89 Hz and 7.8 kHz, respectively. Figure 5 presents the simulated input noise density plot. The integrated

input noise from 1 Hz to 10 kHz is 4.67 μV_{rms}. The noise efficiency factor (NEF) used as a performance indicator in the neural amplifier design is

$$NEF = V_{ni,rms}\sqrt{\frac{2I_{tot}}{\pi \cdot U_T \cdot 4kT \cdot BW}} \qquad (1)$$

where $V_{ni,rms}$ is the total input-referred noise, I_{tot} is the total supply current, U_T is the thermal voltage, k is the Boltzmann's constant, T is the absolute temperature, and BW is the –3 dB bandwidth of the amplifier, respectively. The NEF is calculated to be 3.07. Table I summarizes the performance of the designed neural amplifier. The proposed low-power neural amplifier achieves favorable performance in comparison to other recent similar works using 0.18 µm CMOS process.

TABLE I.   COMPARISON OF THE PROPOSED NEURAL AMPLIFIER WITH THE PREVIOUS WORKS

Parameter	This work*	[2]	[3]
CMOS Technology (µm)	0.18	0.18	0.18
Supply voltage (V)	1	1.8	1.8
Supply current (µA)	2.2	5.6	4.4
Gain (dB)	40	48 / 60	39.4
Bandwidth (kHz)	7.8	0.03-0.29/9	7.2
Low-frequency cutoff (Hz)	0.87	0.3-900	10
Input referred noise (μV_{rms})	4.67 (1Hz-10 kHz)	5 (1Hz-8kHz)	3.5 (10Hz-100kHz)
Noise efficiency factor	3.07	4.6	3.35
Area (mm²)	0.097	0.065	0.0625

*Simulation results

IV.   CONCLUSIONS

A low-power neural amplifier using a source degenerated folded-cascode OTA for biomedical applications is designed using 0.18µm CMOS process. The neural amplifier achieves 40 dB of voltage gain and integrated noise performance of 4.67 μV_{rms} and dissipates 2.2 µA from a 1-V supply.

ACKNOWLEDGMENT

This research was supported by Basic Science Research Program through the National Research Foundation of Korea (NRF) funded by the Ministry of Science, ICT, and Future Planning, (NRF-2015R1A2A2A01006502). This work was supported by IDEC for EDA Tool and MPW support.

REFERENCES

[1]   X. Zou, L. Liu, J. H. Cheong, L. Yao, P. Li, M.-Y. Cheng, W. L. Goh, R. Rajkumar, G. S. Dawe, K.-W. Cheng, and M. Je, "A 100-channel 1-mW implantable neural recording IC," *IEEE Trans. Circuits and Systems I.*, vol. 60, no. 10, pp. 2584–2596, Oct. 2013.

[2]   P. Kmon and P.Gryboś, "Energy efficient low-noise multichannel Neural amplifier in submicron CMOS process," *IEEE Trans. Circuits Syst. I. Reg. Papers*, vol. 60, no. 7, pp. 1764–1775, Jul. 2013

[3]   V. Majidzadeh, A. Schmid, and Y. Leblebici, "Energy efficient lownoise neural recording amplifier with enhanced noise efficiency factor," *IEEE Trans. Biomed. Circuits Syst.*, vol. 5, pp. 262–271, Jun. 2011

Development of an IoT-based Visitor Detection System

Hyoung-Ro Lee[1,2]

School of Computer
Semyung University[1]
Chungbuk, Korea
lhr60127@etri.re.kr, ich410@semyung.ac.kr

Chi-Ho Lin[1]

Won-Jong Kim[2]

SW-SoC R&BD Center
ETRI[2]
Gyeong-gi, Korea
wjkim@etri.re.kr

Abstract— **In this paper, we proposed an IoT-based visitor detection system. It uses an IR sensor to detect human body and two ultrasonic sensors to locate visitor servo motor under the position. When a visitor is detected it drives camera module to locate the visitor. Recoded video and sensor data are stored in the Database. Saved data can see via the PC and Smart device. We developed the system using Raspberry Pi2 and sensor modules to verify the concept. It can track the visitor moving route and minimize the blind spots of the camera. And sensor data and recoded video are checked internet possible all remote location.**

Keywords; IR sensor, Ultrasonic sensor, IoT, Raspberry Pi

I. INTRODUCTION

A variety of digital smart home services have recently been provided as technology develops securing a resident's convenience. Moreover, as more attention has been paid to smart home security, various studies are being conducted on home security, particularly CCTV monitoring systems and home monitoring systems. [1-3]

However, most of the existing systems use fixed cameras which may have blind spots. [5-7]

In this paper, we suggests an IoT-based visitor detection system to minimize blind spots and to check all remote location. The proposed system is designed based on Raspberry Pi2 with a Web server, IR sensor, two ultrasonic sensor, and a camera module with a driving servo motor.

II. IoT-BASED VISITOR DETECTION SYSTEM

The IoT-based visitor detection system proposed in this paper used Raspberry Pi2 as controller, and IR sensor to detect a visitor. In addition, two ultrasonic sensors are used to locate the position of the visitor. The camera module was equipped with a servo motor to change the direction of the camera to the visitor. A web server is used to provide visitor information and sensor data to any internet-enabled remote location.

Figure 1 shows configuration of the proposed IoT-based visitor detection system.

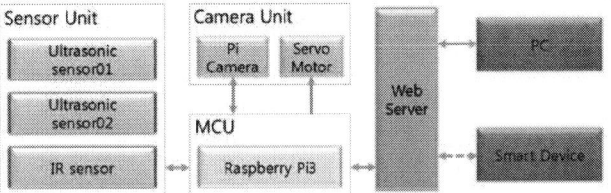

Figure 1. System configuration

When the detection system starts, it initializes the IR sensor and two ultrasonic sensors. The system determines the presence of a visitor from the data of the IR sensor. When a visitor is detected, the two ultrasonic sensors are activated to spot the location of the visitor. The system identifies the location of the visitor with the data from the ultrasonic sensors. And then, the camera moves the direction to the visitor by driving the servo motor where the camera is attached. The system controls the servo motor by supplying PWM current. The system stores recorded video and sensor data in database. The saved data can be seen from any remote location via internet.

Figure 2 shows flow chart of the IoT-based visitor detection system.

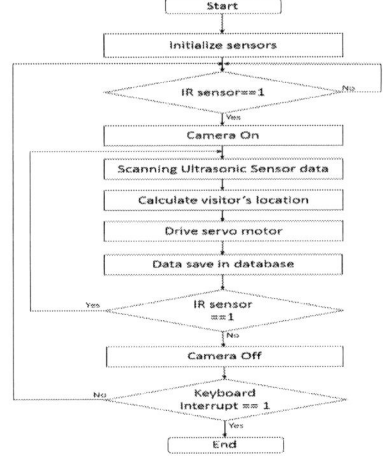

Figure 2. Flow chart of the Visitor Detection System

III. EXPERIMENTAL RESULTS

To test the proposed visitor detection system in an actual situation. We developed an experimental system as shown in Figure 3. We tested IR sensor, Ultrasonic sensors, the camera module, and the servo motor separately to validate each device. In addition, the monitoring to web and smart device was tested.

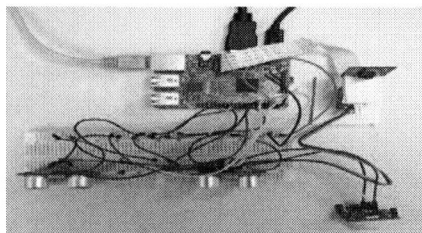

Figure 3.   Experimental environment

Next we used the Python language to implement the algorithm shown in Figure 2. Raspberry Pi2.

Table 1 shows the data values of the two ultrasonic sensors for each case. Case I is for visitor detection in ultrasonic sensor 1 side, and case II is for visitor detection in ultrasonic sensor 2side.

TABLE I.   DATA VALUES OF THE ULTRASONIC SENSORS

Each case	Ultrasonic data value	
	Ultrasonic sensor 1	*Ultrasonic sensor 2*
Case 1 : Initialized data	50 cm	55 cm
Case 2 : Detected by Ultrasonic sensor 1	45 cm	49 cm
Case 3 : Detected by Ultrasonic sensor 2	47 cm	42 cm

Figure 4 shows the images of each situation. Figure 4 (a) is a screen shot after locating the visitor by using the system. Figure 4 (b) is a screen shot of same situation on smart device.

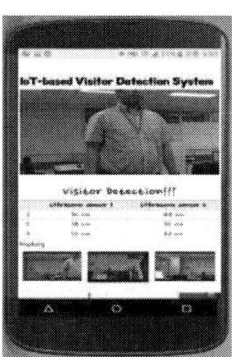

(a)   (b)
After locating the visitor using   same situation on smart
ultrasonic sensors   device

Figure 4.   Recorded images

IV. CONCLUSION

We developed an IoT-based visitor detection system using IR sensor and two ultrasonic sensors to locate the visitor. It controls the servo motor to change the director of the camera for correct recording of visitors. It can also track the moving visitor to minimize the blind spots of the camera. And the recorded video and sensor data can be checked from any remote location using web browser.

For future works, we can extend this study for more than two ultrasonic sensors and an infrared camera to locate a visitor more accurately.

ACKNOWLEDGMENT

This work was supported by the ICT R&D program of MSIP/IITP 16GS1700, Pangyo TechnoValley Related Open SW-SoC Convergence Platform Development.

REFERENCES

[1]   Ye-Jin Jang, Young-Tae Chun, "Technology trend of Smart-home Security System", Korean Security Science Review, no.30, pp.119-138, 2012.

[2]   Il-Sik Chang, Hyun-Hee Cha, Goo-Man Park, Kwang-Jik Lee, "A study of Scenario and Trends in Intelligent Surveillance Camera", The Journal of The Korea Institute of Intelligent Transport Systems, Vol.8, No.4, pp.93-101, August 2009.

[3]   V S Rakesh, P R Sreesh, Sudhish N George, "An improved real-time surveillance system for home security system using BeagleBoard SBC, Zigbee and FTP webserver", IEEE Conference Publications, 2012 Annual IEEE India Conference (INDICON), pp.7-9, January 2013.

[4]   Mrutyunjaya Sahani, Chiranjiv Nanda, Abhijeet Kumar Sahu, Biswajeet Pattnaik, "Web-based online embedded door access control and home security system based on face recognition", IEEE Conference Publications, Circuit, Power and Computing Technologies (ICCPCT), 2015 International Conference on, pp.19-20, March 2015.

[5]   Wang Zai-ying, Chen Liu, "Design of mobile phone video surveillance system for home security based on embedded system", IEEE Conference Publications, The 27th Chinese Control and Decision Conference (2015 CCDC), pp.5856-5859, July 2015.

[6]   Jung-Hwan Ko, Jung-Suk Lee, Young-Hwan, "Autonomous Surveillance-tracking System for Workers Monitoring", Journal of The Institute of Electronics Engineers of Korea, pp. 38-46. June 2010.

[7]   Yun-Kyu Choi, Woo-Soo Choi, Jae-Bok Song, "Obstacle Avoidance of a Mobile Robot Using Low-Cost Ultrasonic Sensors with Wide Beam Angle", Journal of Institute of Control, Robotics and Systems, Vol.15, No.11, pp.1102-1107, November 2009.

Current Mode Four-Quadrant Multiplier Design Using CNTFET

Gyunam Jeon[1], Minsu Choi[2], Kyung Ki Kim[3], Yong-Bin Kim[1]

[1]Dept of ECE, Northeastern University, Boston, MA, USA, {gjeon, ybk}@ece.neu.edu
[2] Dept of ECE, Missouri Univ of Science & Technology, Rolla, MO, USA, choim@mst.edu
[3] Dept of Electronic Eng., Daegu University, Gyeongsan, Korea, kkkim@daegu.ac.kr

Abstract— **This paper proposes presents a low power and high-speed four-quadrant analog multiplier in the current mode based on dual translinear loops using 32nm *CMOS* and 32nm *CNTFET* technologies to investigate and compare the performance differences of the analog circuits on CNTFET technology and CMOS 32nm technology nodes. All the simulations were performed using *hspice* with 32nm *CMOS* from PTM library and 32nm *CNTFET* from Stanford University technologies at the same power supply level. CNTFET based multiplier shows a wider linearity over considerable range of outputs (-10μA to +10μA) while the CMOS based multiplier shows (-7μA to +7μA) and the 3db frequency of the CNTFET based multiplier is 110GHz while the 3dB frequency of the CMOS based multiplier is only 2.45GHz.**

Keywords; Carbon Nano Tube FET(CNTFET), multiplier design, low power circuits, emerging technology

I. INTRODUCTION

As technology scales down to 90 nm and below, the bulk CMOS technology has approached the scaling limit due to the increased short-channel effects, increased leakage power dissipation, severe process variations, high power density, and so on. To overcome this scaling limit, different types of materials have been experimented. Si-MOSFET-like carbon nanotube FET (hereafter called CNFET) devices have been evaluated as one of the promising replacements in the future nanoscale electronics. The reason that makes CNFETs a promising device is that they are compatible with high dielectric constant materials and a unique 1-D band-structure that restrains back-scattering and makes near-ballistic operation a realistic possibility. Using this CNFET, a high-k gate oxide can be deployed for lower leakage currents while keeping the on-current drive capability (compared to Si-MOSFET). CNFET has lower short-channel effect and a higher subthreshold slope than Si- MOSFET [1].

Digital circuits design using CNTFET have been reported in wide range of articles. However, analog circuits design using CNTFET have been rarely reported so far. Optimizations of the elements of analog circuits can improve their features such as linearity, bandwidth and power. Since CNTFET is regarded as the best option among the emerging technologies due to its better features such as current density, transconductance and lower parasitic capacitance [2], it will be a technology of choice for analog integrated circuits design in near future. One of the basic blocks in analog circuits is a multiplier and it is the most important part of adaptive filters, frequency doublers, modulators, signal processing circuits and fuzzy logic

controllers. An ideal multiplier produces a linear output signal, which is obtained from two linear input signals with a constant designated as *k* [3]. The analog multiplier can be categorized into two groups, voltage mode and current mode.

This paper presents a low power and high-speed four-quadrant analog multiplier in the current mode based on dual translinear loops using 32nm CMOS and 32nm CNTFET technologies to investigate and compare the performance differences of the analog circuits on CNTFET technology and CMOS 32nm technology nodes.

II. CARBON NANOTUBE FET

Carbon nanotube FETs employ semiconducting single-wall carbon nanotubes to assemble electronic devices, and the single walled CNFET is obtained by replacing the channel of a conventional MOSFET with carbon nanotubes (a 1-D conductor obtained by rolling a sheet of graphite) as shown in Fig. 1 [4]. The nanotubes can be either a metallic (conductor) or a semiconducting (semiconductor) depending on the angle (represented as a chirality integer vector (n, m)) of the atom arrangement along the nanotube. The nanotube is metallic if (n = m) or (n − m = "a multiple of 3"); otherwise, the tube is semiconducting.

The CNFET device has four terminals (drain, gate, source, and back-gate); a dielectric film is wrapped around a portion of the un-doped nanotube in the intrinsic region, and a metal gate surrounds the dielectric while the other nanotube regions are heavily doped for a low series resistance during the on-state. As shown in Fig. 1(a), the top-gated CNFETs are fabricated on an oxidized Si substrate that can be used as a back-gate in the CNFET. In the early 1990s, most CNFETS studied had adopted a back-gate top-contact structure [5], in which the nanotubes are grown on a conducting substrate covered by an insulating layer. Two metal contacts are deposited on the nanotube to serve as source and drain electrodes while the conducting substrate is the gate electrode in the three-terminal device.

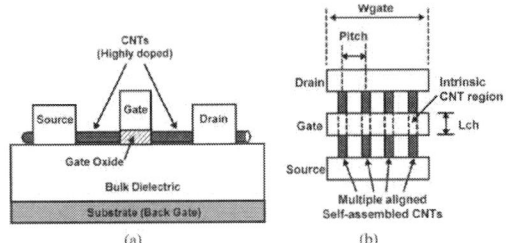

Figure 1. *CNFET* structure (a) Cross-sectional view (b) Top View

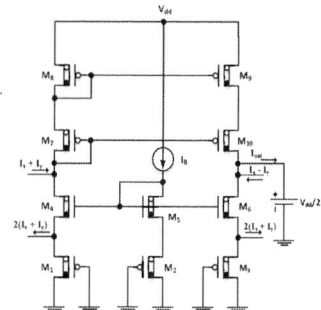

Figure 2. A *CNTFET* based four-quadrant analog multiplier [22]

The currents of the CNFET device are controlled by adjusting device parameters such as gate length (L_{ch}), the number of nanotubes, chirality vector, and pitch between nanotubes [6].

III. FOUR-QUADRANT CURRENT MODE ANALOG MULTIPLIER DESIGN

One of the most important fundamental blocks in analog integrated circuits is multiplier. A CMOS based current mode four-quadrant analog multiplier is one of the last multiplier design to be optimized. Fig. 2 shows the multiplier. Two inputs are determined by (1) and (2), and the output of the multiplier is written on (3).

$$I_{in1} = I_x + I_y \qquad (1)$$
$$I_{in2} = I_x - I_y \qquad (2)$$
$$(I_{in1})^2 - (I_{in2})^2 = kI_xI_y \qquad (3)$$

Square and subtraction functions are used to implement (3) at the dual transistor multiplier, which implement the equations, as shown in Fig 2. In addition, dual translinear loops are *m1, m2, m4, m5 and m2, m3, m5, m6*. KVL is applied in the first translinear loop and it can be written by the following (4).

$$V_{gsm1} + V_{gsm4} - V_{gsm2} + V_{gsm5} \qquad (4)$$

Due to the reason ignoring the body effect of the transistor, *CMOS*-based transistors used in the multiplier can be represented as an inverse square relationship between source-drain current and gate-source voltage. Thus, the Eq. 5 can be rewritten from (4).

$$\sqrt{I_{m1}} + \sqrt{I_{m4}} = \sqrt{I_{m2}} + \sqrt{I_{m5}} \qquad (5)$$

In the second translinear loop, *m2, m3, m5 and m6* the same equation also can be written as shown in (6).

$$\sqrt{I_{m3}} + \sqrt{I_{m6}} = \sqrt{I_{m2}} + \sqrt{I_{m5}} \qquad (6)$$

Moreover, application of KCL in the input node and another node yields the following current equations: $I_{in1} + I_{m7} = I_{m4}$, $I_{m4} = 2I_{in1} + I_{m1}$. The analyzes of a different drain-source current of transistors, which is shown in (5) and (6) can help to write the output signal. The uses of Eq. 3 and Eq. 4 can help to write I_{out} as shown in (7) below.

$$I_{out} = \frac{(Ix+Iy)^2 - (Ix+Iy)^2}{4Ib} \qquad (7)$$

Finally, I_{out} is directly proportional to I_x and I_y as shown in (8).

$$I_{out} = \frac{Ix\,Iy}{Ib} \qquad (8)$$

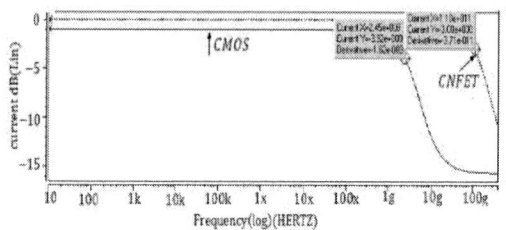

Figure 3. Simulation results for -3*db* bandwidth

IV. SIMULATION RESULTS

The 32nm *CMOS* and 32nm *CNTFET* of CML multipliers at a supply voltage, 3.3V were simulated by the *Hspice* simulator. Simulation results show that the dc transfer characteristics of differently modeled multipliers with 32nm *CNTFET* and 32nm *CMOS*. The results also show more linearity on the *CNTFET* based multiplier over a considerable range of inputs (-10μA to 10μA), which have a maximum value that is equal to $\pm I_b$. The output signal range of a multiplier using 32nm CMOS is between -7μA and 7μA, but the output signal range of a multiplier that uses CNTFET shows a higher range of output signal than the CMOS based multiplier because a CNTFET based multiplier possesses more linearity. Fig. 3 shows the -3*db* bandwidths of the multiplier using CNTFET and CMOS technologies. The -3db bandwidth of the *CNTFET* based multiplier is 110GHz of bandwidth, which is much higher than the bandwidth of the CMOS based multiplier, 2.45GHz. The CNTFET multiplier does not have any attenuation beyond the cut off frequency of the CMOS multiplier, which has a 5db attenuation as can be seen in the Fig. 3.

V. CONCLUSION

This paper presents a current mode four-quadrant analog multiplier using *CNTFET* and and *CMOS* technologies. At the same conditions of supply voltage, the simulations are measured using *Hspice*. A *CNTFET* based multiplier circuit can be said it is the viable solution to overcome the *CMOS* based multiplier's disadvantages, such as relatively low speed, low linearity, high power consumption in transient and dc simulations.

REFERENCES

[1] A. Javey, Q. Wang, W. Kim, and H. Dai, "Advancements in complementary carbon nanotube field-effect transistor," in Proc. 2003 IEEE Int. Electron Devices Meeting, pp. 31.2.1 –31.2.4.

[2] S. Hasan, S. Salahuddin, et. al., "High frequency performance projections for ballistic carbon-nanotube transistors," *IEEE Trans. Nanotechnol.*, vol. 5, no. 1, pp. 14–22, Jan. 2006.

[3] Gilbert, B., "A precise four quadrant multiplier with sub-nanosecond response", IEEE J Solid-State Circuits 1986, Vol. 21. 430–5.

[4] H. Hashempour and F. Lombardi, "Device model for ballistic CNFETs using the first conducting band," IEEE Design Test of Computer., vol. 25, no. 2, pp. 178–186, Mar./Apr. 2008.

[5] P. L. McEuen, M. S. Fuhrer, and H. Park, "Single-walled carbon nanotube electronics," IEEE Trans. Nanotechnology., vol.1 , no. 1, pp. 78–85, Mar. 2002.

[6] J. Deng and H.-S . P. Wong, "A compact spice model for carbon-nanotube field-effect transistors including nonidealities and its application," IEEE Trans. Electron. Devices, vol. 54, no. 12, pp. 3186–3194, Dec. 2007

A Flexible MCMC Detector ASIC

Dominik Auras, Sebastian Birke, Tobias Piwczyk, Rainer Leupers, and Gerd Ascheid

Institute for Communication Technologies and Embedded Systems

RWTH Aachen University, Aachen, Germany

Abstract— **This paper presents a flexible Markov Chain Monte Carlo based MIMO (multi-antenna) detector ASIC that efficiently uses the available computing resources by leveraging inherent chain-level and symbol-level parallelism offered by the underlying detection algorithm. Compared to a reference architecture from literature, we roughly double the throughput on average, while achieving a speedup of up to 4.5x for certain cases. The implementation layouted for a 90nm CMOS technology requires 1.3mm² and runs at 397 MHz.**

MIMO detection, Markov Chain Monte Carlo; Multi-antenna

I. INTRODUCTION

The trend in wireless broadband communications goes to very high transmission rates beyond 1 Gbit/s. Recent standards, such as Wi-Fi or LTE, adopt amongst others multi-antenna (MIMO) technologies to achieve this objective. They use multiple spatial data streams to convey more information in the same radio channel bandwidth, known as spatial multiplexing. These standards employ bit-interleaved coded modulation (BICM). Iterative detection and decoding is another promising technique to increase the data rates, also known as BICM-ID. It improves the receiver systems' spectral efficiency, i.e., how many bits per channel use we can transmit.

We present a Markov Chain Monte Carlo (MCMC) based MIMO detector, which recovers the information bits from the received signals, for BICM-ID receiver systems. It is an extension of [1,2] in prospect to mitigate the relatively low throughput. We double the number of processing units (GS/M circuits) and support the simultaneous processing of up to four inputs. Furthermore, the new flexible architecture allows a more fine-grained control of the algorithm's run-time parameters.

II. BACKGROUND

Markov Chain Monte Carlo based MIMO detection performs random sampling of the code bit's posterior likelihood function in order to compute the extrinsic log-likelihood ratios (LLRs) according to

$$\lambda_i^e = \ln \frac{P(c_i = 1)}{P(c_i = 0)} - \lambda_i^a \qquad (1)$$

given the received signal, channel transfer function observation, noise density etc. The likelihood function is unknown, but can be evaluated at distinct points. The samples are used to estimate the function.

Basically, the detector first draws several random code bit vectors as starting points for the MCMC chains. Then, it walks around by randomly accepting or rejecting transitions to close neighbors (one bit flipped). The code bit likelihood function is evaluated at every point. All values are finally used in (1) as estimate of the true likelihood function.

The transition probability depends on the current state of the MCMC chains, which summarizes all previously evaluated samples, and converges towards the desired likelihood function, which means that the chains step by step approach regions of interest with high likelihood values.

III. ARCHITECTURE

The proposed architecture, depicted in Fig. 1, bases on [1,2]. We process up to four inputs at once. The data distribution stage assigns an input to an idle FEP/Cluster combination. There are 16 GS/M circuits, executing the MCMC detection, organized into four clusters. Each cluster runs four parallel MCMC chains. Four parallel front-end processing (FEP) units preprocess the input data. Each FEP feeds only one cluster. This avoids layout-level signal congestion at the expense of energy consumption due to redundant computations. The LLR circuits support three cases: i) four clusters work on the same input, i.e., we combine 16 parallel chains to produce the output LLRs (use L16 in Fig. 1), ii) four clusters work on two inputs, then L16 and L8 produce the output, and finally iii) with four inputs, we use all four LLR circuits, one per cluster.

The detector implementation supports the two symmetric MIMO antenna configurations 2x2 and 4x4. Its hand-shaking based interface accepts one symbol per clock cycle and stalls appropriately if busy. The output interface shifts out one extrinsic LLR value per clock cycle.

The new flexibility of the architecture is to support multiples of four chains per symbol (finer granularity, [1] supports eight) while making use of all available GS/M circuits, by processing up to four symbols in parallel ([1] uses no symbol-level parallelism). Also, we doubled the number of GS/M circuits ([1] has eight).

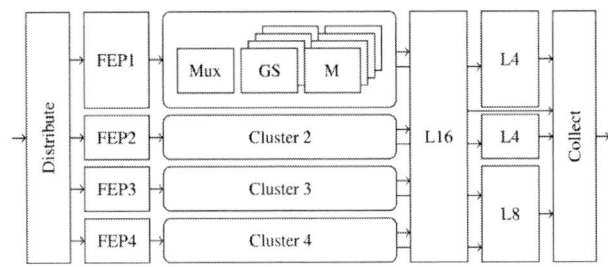

Figure 1: Architecture diagram

TABLE I.   IMPLEMENTATION RESULTS

	This Work	[1]
Area	1.3 mm²	0.47 mm²
Gates	367 kGE[a]	149.5 kGE
Frequency	397 MHz	479 MHz
Chip density	88,6%	n/a
Code bit throughput	87.4[b] Mbit/s	52.0 [b] Mbit/s

a. One gate equivalent (GE) corresponds to the area of one 2-input drive-1 NAND gate.
b. Throughput for 4x4 64-QAM mode and 8 chains with 8 samples per chain.

The proposed detector accepts a new symbol every

$$n_{det} = \max \begin{pmatrix} KN_t, \\ ((N_s + 1)KN_t + 2)N_q/16, \\ (N_t^2(N_t + 2)/2 + 5)N_q/16 \end{pmatrix} \quad (2)$$

cycles on average in the steady state, where N_s denotes the number of samples per chain, N_q is the number of chains per symbol, and KN_t is the number of code bits per symbol. This considers the execution times of the interface, the four FEP circuits, and the four clusters with four GS/M circuits each. The implementation can currently run in parallel four 4-chains, two 8-chains, or one 16-chains operation.

The original MCMC architecture requires

$$n_{old} = \max \begin{pmatrix} KN_t, \\ \lceil N_q/8 \rceil (N_s + 1)KN_t + 5, \\ N_t^2(N_t + 2)/2 + 3 \end{pmatrix} \quad (3)$$

cycles per symbol. If N_q is not a multiple of 8, some GS/M circuits are disabled. For more than 8 chains, the detector runs them sequentially, e.g. a 16-chains detection is basically two sequentially chained 8-chains detections.

IV.   IMPLEMENTATION RESULTS

The detector's RTL design has been synthesized with Synopsys Design Compiler in topographical mode, and a layout was obtained with the Cadence SoC Encounter 13.1 Foundation Flow, using a 1.0V standard-performance standard-cell library for the UMC 90nm SP-RVT LowK CMOS process. Tbl. I compares this work to [1]. The chip layout is shown in Fig. 2. One cluster is larger than a FEP or LLR circuit. Clusters and FEP/LLR-circuits each occupy roughly 50% of the total chip area. Similar to [1,2], the critical path is in the clusters.

We define the speedup over [1] as

$$S = \frac{n_{old}}{n_{det}} \times \frac{397 \text{ MHz}}{479 \text{ MHz}}. \quad (4)$$

Considering all possible combinations of $K = \{2, 4, 6\}$, $N_t = \{2, 4\}$, $N_q = \{4, 8, 16\}$, $N_s = \{2, 4, 8, 16\}$, we observe

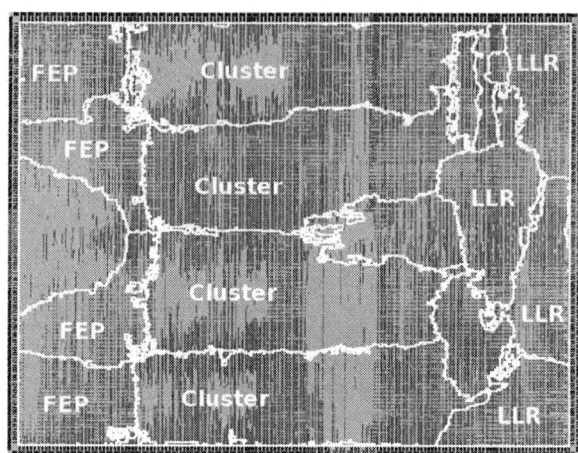

Figure 2:   Chip Layout

that the maximum speedup is 4.5x. For one single case ($K = 2$, $N_t = 4$, $N_q = 16$, $N_s = 2$) the proposed ASIC is 20% slower than [1]. In all other cases, we are at least 30% faster, achieving 2.2x on average.

As expected, the architecture's additional flexibility leads to an increased area consumption and a lower operation frequency compared to [1]. This ultimately limits the implementation to achieve the full expected speedup given that we doubled the number of GS/M circuits. However, the more fine-grained run-time parameter control actually allows for a more efficient trade-off between communications performance and throughput, as analyzed in [2]. Due to the changed granularity, we are not directly comparable to [1,2]. A detailed communications performance is necessary to assess the benefit of the flexibility.

V.   CONCLUSIONS & OUTLOOK

The proposed flexible architecture leverages several independent parallelism degrees in order to efficiently use the available resources: it computes independent MCMC chains in parallel, and process several input data vectors simultaneously. Its more fine-grained run-time parameter control compared to [1,2] allows a more efficient trade-off between communications performance and throughput.

As future work, we propose to share major parts of the LLR circuits, i.e., use the compare-select tree from the L16 circuit for all LLR computations. Furthermore, some chain samples can be processed simultaneously, which could possibly double the throughput, by combining neighboring GS/M circuits on-demand at run-time.

REFERENCES

[1] D. Auras, U. Deidersen, R. Leupers, and G. Ascheid, VLSI-SoC: Internet of Things Foundations: 22nd IFIP WG 10.5/IEEE International Conference on Very Large Scale Integration, VLSI-SoC 2014, Playa del Carmen, Mexico, October 6-8, 2014, Revised Selected Papers. Cham: Springer International Publishing, 2015, ch. A Parallel MCMC-Based MIMO Detector: VLSI Design and Algorithm, pp. 149–169.

[2] D. Auras, U. Deidersen, R. Leupers, and G. Ascheid, "VLSI design of a parallel MCMC based MIMO detector with multiplier-free gibbs samplers," in Very Large Scale Integration (VLSI-SoC), 2014 22nd International Conference on, Oct 2014, pp. 1–6.

Throughput Enhancemnet with Optimal Fragmented MSDU Size for Fragmentation and Aggregation Scheme in WLANs

Eunbi Ku, Chulho Chung, Byungcheol Kang, and Jaeseok Kim

Department of Electrical & Electronics Engineering

Yonsei University

Seoul, Korea

Email: eeeb9003@yonsei.ac.kr

Abstract—**The fragmentation and aggregation (F&A) algorithm improves the throughput while ensuring the reliability in the MAC layer. To determining the optimal fragment size is a key issue of effectively operating the F&A algorithm. In this paper, the proposed method selects the optimal fragmented frame size and uses aggregation scheme based on various channel conditions. As a result the throughput can be further increased.**

Keywords—MAC; Fragmentation scheme; Aggregation scheme; Fragment size

I. INTRODUCTION

In the medium access control (MAC) layer, a variety of the latest technology to improve the throughput and ensure the reliability has emerged based on the WLAN standard. Among them, the fragmentation and aggregation (F&A) algorithm combines the fragmentation technique in the initial IEEE 802.11 and the aggregate-MPDU (A-MPDU) technique, the opposite way to pursue the fragmentation technique, in the IEEE 802.11n. When using F&A algorithm, the network throughput can be further improved.

In this paper we propose an optimal fragment size for the F&A algorithm, which can maximize the throughput by selecting the optimized fragment length based on various channel conditions.

II. F&A ALGORITHM BACKGROUND

The F&A algorithm in Fig. 1 combines the MAC service data unit (MSDU) fragmentation technique to enhance the reliability and the MAC protocol data unit (MPDU) aggregation technique to increase the throughput. When using F&A algorithm, the network throughput can be enhanced ensuring the reliability.

The fragmentation technique is a method to create MSDUs smaller than the original MSDU to increase the probability of successful data transmission of the MSDU in the 802.11 MAC specifications [1], [5]. This technique to reduce the size of the frame improves the reliability in the channel environment that is difficult to receive a frame by

increasing the transmission success probability. On the other hand, if there are a large

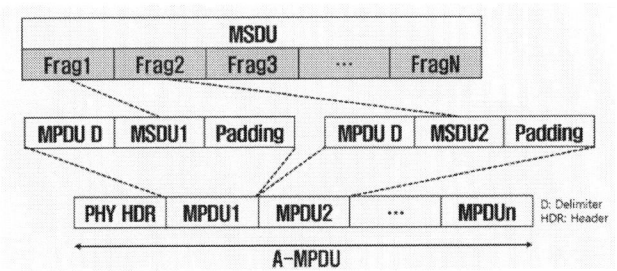

Fig. 1. Frame format of the F&A algorithm

number of fragmented frames, the divided frames are undergoing a separate process about each received response and attached the overhead caused by the frame format. Eventually it results in the throughput reduction [1], [2]. The maximum length of the MSDU received from the LLC layer is limited to 2304 bytes [5]. Thus, in the high-speed transmission rate, the technique improving the throughput is limited through increasing the amount of the data contained in the frame. In order to solve this problem the frame aggregation technique has been introduced in the IEEE 802.11n, where multiple MPDUs are transmitted in one A-MPDU. The aggregation technique is divided into the two types of the aggregate-MSDU (A-MSDU) and the A-MPDU. In this paper, we use the A-MPDU which is robust against errors due to the presence of individual CRC per MPDU [2]. The A-MPDU technique accepts the data frame from the MAC layer and aggregate it in the PHY layer [3]. In this scheme, the throughput improves by reducing the time loss due to the redundant protocol overhead caused by the header, the FCS, acknowledgments, the IFS attached each frame and MAC coordination.

III. OPTIMAL FRAGMENT LENGTH

To determining the optimal frame size is a key issue of effectively operating the F&A algorithm. Moreover, there is always a tradeoff between reliability and throughput for WLAN. We study this tradeoff and propose an optimal

978-1-5090-3220-4/16 $31.00 © 2016 IEEE

fragment size for F&A algorithm, which can maximize the throughput by selecting the optimized fragment length based on different channel conditions. Simulation results are

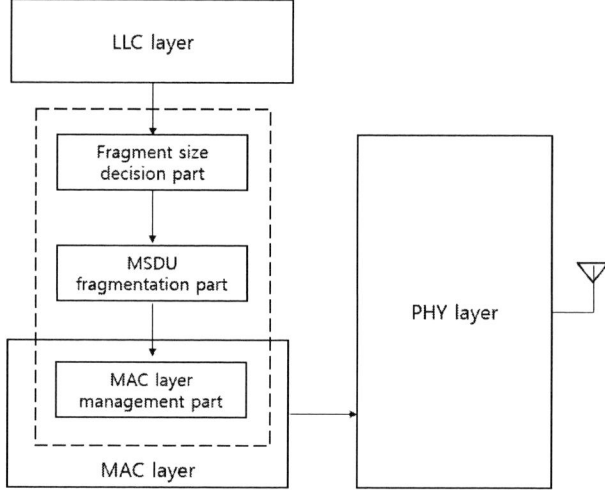

Fig. 2. Architecture block diagram

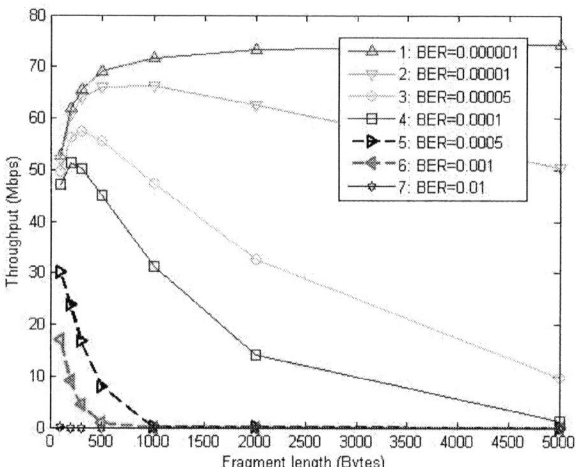

Fig. 3. Throughput result in F&A scheme

presented demonstrating that the throughput gradually increases when transmitting the MSDU frame with optimal fragment size.

Figure 3 shows the functional block diagram of the proposed method. In the size decision part, we determines the optimal MSDU fragment size using the transmission time algorithm during the transmission opportunity (TXOP) limit time of STAs. In the MSDU fragmentation part, we fragment the MSDU to the determined size and generate numerous MPDUs. In the MAC layer management part, we constitute pad bits and a delimiter to identify plenty of MPDUs. Applying the aggregation technique, we generate the A-MPDU. Accordingly we can guarantee the reliability and improve the throughput at the same time by delivering the generated A-MPDU to the PHY layer.

IV. PERFORMANCE EVALUATION

The parameter setting for the performance analysis was based on the IEEE 802.11ac networks. The Simulation was made correspondingly using MATLAB. We compared the performance of a high rate when the transmission rate is 780 Mbps in one antenna. The MSDU length received from the upper layer was set 5000 bytes. We have simulated a variety of channel conditions, and found that the throughput improvement by adopting optimal fragment length is convincing over various channel conditions. The curves in Fig. 4 illustrated effective throughput performance of the MAC layer with a wide range of fragmented MSDU sizes as bit error rate (BER). These performance provided us with the optimal fragment length for given BERs. For example, the optimal fragment size was 2000 bytes or 300 bytes when BER is 10^{-6} or 10^{-4} respectively. It could also be seen that in the case of fine channel condition with BER less than 10^{-6}, the throughput increases hardly noticeable with the fragmented MSDU length when this length is more than 500 bytes. Namely, in the good channel environment we aggregated just the MSDU fragment of more than 500 bytes. When the BER was poorer than 10^{-6}, the enhancement by adopting an optimal fragment size is significantly crucial result.

V. CONCLUSION

In this paper, we analyzed the optimal fragment size of MSDU in the F&A algorithm, which combines the fragmentation and aggregation scheme together, in order to further optimize network throughput in variety of channel conditions. We showed that the optimal fragment length in the F&A schemes increases the throughput for high data rates through simulation result.

ACKNOWLEDGMENT

This work was supported by the National Research Foundation of Korea(NRF) grant funded by the Korean government(MSIP) (No. NRF-2015R1A2A2A01004883), and was also supported by IDEC(IPC, EDA Tool, MPW)

REFERENCES

[1] A. Saif, M. Othman, S. Subramaniam, "Frame Aggregation in Wireless Networks: Techniques and Issues," IETE Technical Review, vol 28 issue 4, Jul-Aug 2011.

[2] Sidelnikov, Alexey, Jeonggyun Yu, and Sunghyun Choi. "Fragmentation/aggregation scheme for throughput enhancement of IEEE 802.11 n WLAN," Network, 2002.

[3] Chulho Chung, Taewook Chung, Byungcheol Kang, Jaeseok Kim, "A-MPDU using fragmented MPDUs for IEEE 802.11ac MU-MIMO WLANs," TENCON 2013 - 2013 IEEE Region 10 Conference, pp.1-4, Oct 2013.

[4] "IEEE Standard for Information technology – Telecommunications and information exchange between systems Local and metropolitan area networks–Specific requirements Part 11: Wireless LAN Medium Access Control (MAC) and Physical Layer (PHY) Specifications," IEEE Std 802.11a, 1999.

[5] Xin He, Frank Y. Li, and Jiaru Lin, "Link Adaptation with Combined Optimal Frame Size and Rate Selection in Error-Prone 802.11n Networks," IEEE ISWCS, October, 2008.

Design of NFC transceiver for automotive applications

Yeong-Gyo Gim[1], Shiho Kim[1,2]

Yonsei Insititute of Convergence Technology[1] and School of Integrated Technology[2]
Yonsei University
85, Songdo-Gwahak-ro, Incheon, 21893, Republic of Korea
Phone: +82-32-749-5872, Fax: +82-32-818-5801, E-mail: {kyo.kim, shiho}@yonsei.ac.kr

Abstract

We present a design and application of NFC (Near Field Communication) transceiver for the using an inductive powered receiver to enable all type of NFC communication modes such as the passive, active, RW, and RFID modes. NFC technology has been ready massively adopted in the automotive applications. We proposed a system architecture of NFC Automotive Applications with Trused Execution Envinonment to enforce protection, confidentiality, integrity and access rights of the resources and data.

Keywords-component; NFC(near field communication); automotive applications; TEE(Trusted Execution Environment); SE(Secure Element)

Introduction

Recently, many mobile devices and automotive applications adopting the NFC (near field communication) technology are emerging rapidly. Especially, new automotive service trend such as car sharing, corporate fleet management is a driving force of the automotive NFC market driven by combined forces of the NFC-enabled mobile devices, the demand for personalization as well as Wi-Fi and/or Bluetooth pairing inside a cabin[1]. All of these applications can potentially make good use case of NFC-enabled smart mobile devices. NFC is a standards-based, short-range (upto 10 centimeters) half duplex wireless connectivity technology that enables simple but safe interactions between cards or electronic devices with just an tab [2]. According to the NFC Forum[3], the following applications are most likely into the market over the next few years,

A. Car access using NFC-enabled devices as a key to lock/unlock doors and Ignition control, etc.
B. Personalized settings inside the car such as Climate control settings as well as automatic seat and mirror adjustments, etc.
C. Pairing hands-free interface or infotainment systems connections.
D. In-car payment for tolls, drive-through, etc.
E. Acquiring vehicle information and Vehicle management, or Vehicle diagnostics through the Wi-Fi or Bluetooth personal devices.
F. Alert the owner of pending service requirements.

The NFC technology was suggested for data transfers between very close proximity devices using a 13.56MHz RFID band [2, 4]. The standard of NFC is defined by ISO

18092 and EMCA 340 . The communication methods and key parameter of standards are completely similar with ISO14443, ISO15693 and ISO18000-3 as well. Therefore, NFC devices can easily support multiple standards of 13.56MHz RFID in many applications. However, it is highly desirable that RFID operation is possible when battery of the device has been fully exhausted [4]. In this context, we have focused on achieving operation of all types of communication modes on NFC devices, adding only a few additional circuits while maintaining a simple design for an inductive powered receiver. The rest of this paper is organized as follows. Section II described proposed structure of transceiver for NFC. Section III presents the architecture of NFC enabled smartphone service platform with secure element and trusted execution environment(TEE). Finally, we conclude this paper with a discussions of a secure architecture.

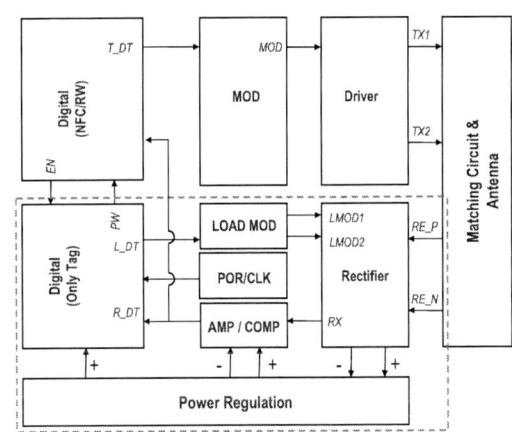

Figure 1. A functional block diagram of NFC controller.

Functional Block and Structure of an NFC Transceiver

ISO/IEC 18092 NFC standard specifies a behavior of NFC devices that makes use of an Initiator and a Target. The Initiator is the device that initiates and controls the exchange of data. The target is the device that answers the requests from the Initiator. Also, the communication mode of NFC distinguishes between the active which generate their own RF field and passive which is only initiator generates RF field of communication mode. When the device operates as the target of the passive communication mode, the physical link is similar with 13.56MHz passive RFID as well. The transceiver shown in Fig. 1 consists of a digital controller to support multiple standards for 13.56MHz including the NFC protocol,

978-1-5090-3220-4/16 $31.00 © 2016 IEEE

antenna driver to transmit RF signal through coil antenna, rectifier to convert the RF signal into a DC signal, receiver to demodulate a command signal of 106Kbs and 212/424Kbps or even 847KHz subcarrier signal from NFC passive communication mode target or passive 13.56MHz RFID tags. Some of the circuits outside the red boundary line may be powered from a battery.

Achitecture of NFC enabled smartphone

Three types of operating modes, reader/writer, peer-to-peer, and RFID card emulation are involved in NFC communication as shown in Fig. 2. The USIM Secure element(SE) of mobile phone enables secure storage and secured transactions among in smartphones. Recently, SE alternatives for NFC transactions is Host Card Emulation (HCE) implemented as a software in mobile phone operating system to secured transactions among NFC devices in NFC enabled smartphones. The latest versions of the Android operating system support HCE as a software based virtual card technique [5]. Required technique to protect security of data and transaction must be implemented on top of the HCE implementation, because HCE does not provide any security or specify any security techniques. For automotive applications, careful risk analysis is required to design the most appropriate security solution for each case.

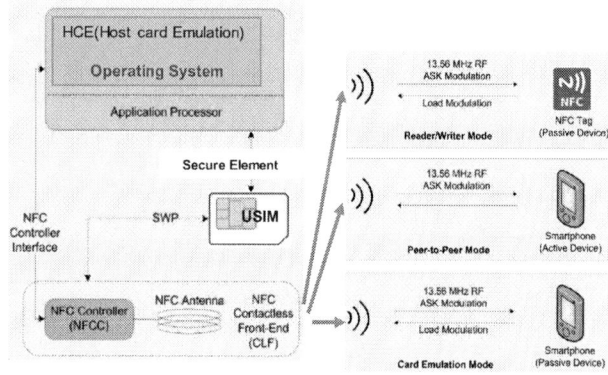

Figure 2. Architecture of NFC enabled smartphone service platform with secure element and Host card emulation (SWP: single wire Protocol).

NFC technology has emerged as a strong contender for replacement of the current transaction and mobile payment techniques due to its security and connectivity features. Similarly, the hardware architecture with considerable security promised by NFC is also very attractive to the automotive industry. The TEE (Tresuted Execution Environment) is an independent execution environment that runs alongside operating system of a mobile device (Rich OS) and to offer an execution space that enables secure stroing of memory data, storage and transaction record of applications. The international organization, Global Platform[6], has developed a set of standards for the TEE's APIs and security services. Because the TEE delivers a greater security than the Rich OS environment with considerably lower cost than using the external SE like USIM, it can provide perferable system architecture of NFC automotive applications as shown in Fig. 3. The Trusted Execution Environment (TEE) protects security

functions such as trusted applications or trusted boot from software attack. The TEE Kernel provides safe execution of authorized trusted function "trusted applications" such as payments, Credentail, and ignition control. Inside the TEE, each trusted applications is independent from the others and one trusted application can not access trusted resources.

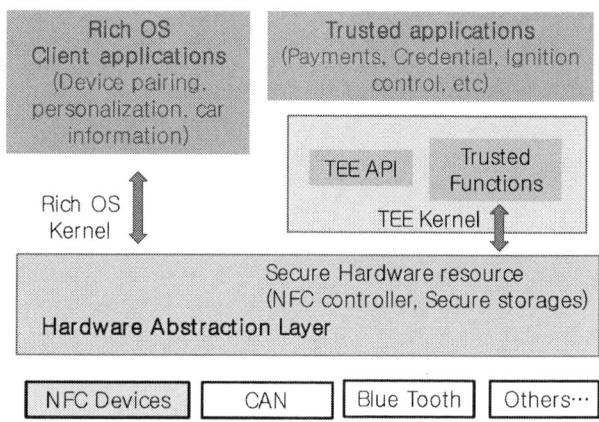

Figure 3. System Architecture of NFC Automotive Applications with Trused Execution Envinonment.

Discussions & Conclusions

NFC technology has been ready massively adopted in the automotive applications. Many experts project that NFC-enabled intelligent vehicles will grow significantly in the next 10 years. However, except for a few early adopters, many automotive OEMs are still reluctant to take on NFC solution and are still in favor of other alternatives such as Bluetooth-enabled digital car keys. In this paper, the structure of NFC transceiver which can support all types of NFC communication modes such as Active, Passive, RW, and RFID modes using inductive powered receiver has been presented. We present System Architecture of NFC Automotive applications with HCE and Trused Execution Envinonment.

Acknowledgment

This work was supported by the MSIP, under the "IT Consilience Creative Program"((IITP-R0346-16-1008) supervised by the IITP. The authors thank the LG Electronics R&D group for research grant to the IT Consilience Creative Program.

References

[1] S. Rainer, et al. "Near field communication (NFC) in an automotive environment." International Workshop on NFC 2010.

[2] Coskun, et al. "A survey on near field communication (NFC) technology." Wireless personal communications (2013): pp. 2259-2294.

[3] White paper, "NFC Non-Payments Use Cases" smart card alliance (2015)

[4] J. Cho et al. "An NFC transceiver using an inductive powered receiver ". Proceedings of ISOCC 2009, pp. 456–459.

[5] B. Lepojevic, et al. "Implementing NFC service security–SE VS TEE VS HCE."

[6] https://www.globalplatform.org/ last visited 9-September-2016.

Possibility Verification of Drone Detection Radar based on Pseudo Random Binary Sequence

Sung Jun Lee, Jae Ho Jung, and Bonghyuk Park

Mobile RF Research Section

Electronics and Telecommunications Research Institute (ETRI)

218 Gajeongno, Yuseong-gu, Daejeon, 34219, Korea

Email: {sea0310, jhjung, bhpark}@etri.re.kr

Abstract—**This paper presents the possibility verification of a drone detection radar based on pseudo random binary sequence (PRBS). It contains the design of a probing signal and resultant specifications of the radar, measurement setup using commercial signal generator/analyzer and antennas, and laboratory/outdoor measurements. Measurements in the 2 GHz band with 6 dBm transmit condition show the drone detection at the range over 100 m.**

Keywords; moving target detection radar; pseudo random binary sequence (PRBS)

I. INTRODUCTION

Recently, the need of drone detection has been increased due to the noticeable growth of drone usage. Radar may be one of the means for the drone detection. To verify directly the possibility of drone detection by radar, it is decided to utilize radar based on pseudo random binary sequence (PRBS). This is because the fast verification is possible owing to the author's familiarity with a PRBS channel sounder [1], and performance of PRBS radar reported in [2] gives the above verification enough significance.

According to verification steps, the design of a probing signal and resultant specifications in terms of signal processing are explained in section II. And, the measurement setup using commercial RF instruments and components is described in section III. Then, a rural area measurement for verifying drone detection is discussed in section IV.

II. PROBING SIGNAL & RESULTANT SPECIFICATIONS

Thanks to the well known correlation property of the PN sequence [3], the sliding correlation of received sequences by a prepared PN sequence can give a time-domain delay profile from a transmitted signal to a received signal. In radar case, this delay profile is related to a time-domain target range profile through the light velocity. With range profiles acquired at different instants, time variations of a range profile can be observed, which gives target velocity estimation. These are the basic principles of radar that utilizes a PRBS based probing signal. The designed probing signal is shown in Fig. 1.

Data acquisition in a receiver will be synchronized with the start of any PN iteration block (PNI) shown in Fig. 1. In the duration of each PNI, except the first and last PN duration, data

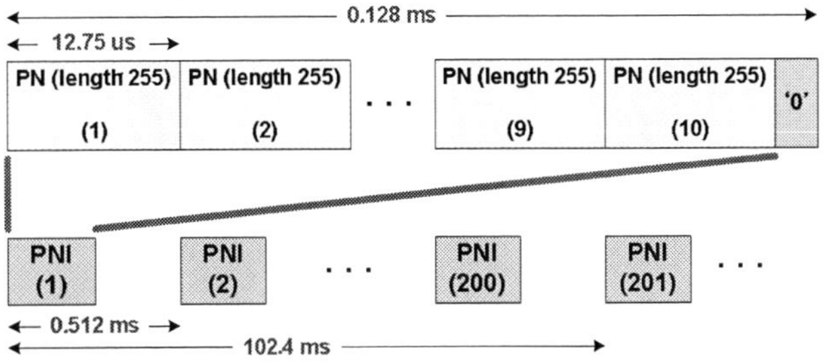

This work was supported by ICT R&D program of MSIP/IITP [B0117-16-1007, The Study of Future Original Technologies of Hyper-connected Networking].

Figure 1. The designed probing signal utilizing a PN sequence of length 255

over 8 PN durations will be averaged to enhance signal to noise ratio (SNR). The exceptions are due to a time delay between transmitted and received signals. Data in the first and last PN duration does not correspond to a complete PN sequence. After a duplication of the averaged signal, a sliding correlation with a prepared PN sequence give a time-domain range profile at each instant. Range resolution corresponds to the 50 ns chip duration of the PN sequence, maximum unambiguous range to 12.75 us period of the sequence, and dynamic range to 255 length of the sequence. To observe time variations of a range profile, 200 range profiles will be used. Discrete Fourier transform (DFT) for any same range in the 200 range profiles gives a spectral-domain range profile. Frequency in the spectral-domain is a Doppler frequency. Sample spacing and meaningful range in the spectral-domain is determined from the DFT definition [4]. The corresponding radar specifications are listed in Table I.

III. MEASUREMENT SETUP

A Keysight MXG (N5182B) plays a role of a transmitter. Details are as below. The 0.512 ms including one PNI shown

TABLE I.   THE RESULTANT RADAR SPECIFICATIONS

Parameters	Values
Range Resolution	7.5 m
Maximum Unambiguous Range	~ 1.9 km
Velocity Resolution	~2 km/h
Maximum Detectable Velocity	~ 200 km/h
Dynamic Range @ Time-Domain Range Profile	48.1 dB

in Fig. 1 is generated in a personal computer (PC), in which 4-oversampling and a root raised cosine (RRC) filtering to limit bandwidth and avoid inter symbol interference are included. The generated baseband signal is downloaded into the MXG. With an arbitrary waveform function of the MXG, the designed probing signal is up-converted to 2 GHz band.

A receiver is composed of a commercial cavity band pass filter (BPF) and a low noise amplifier (LNA) followed by a Keysight PXA (N9030A). The insertion loss of the BPF is under 1.0 dB, and the gain and noise figure (NF) of the LNA are 30 dB and under 2.0 dB respectively. The NF of the PXA is about 5 dB in case of the 0 dB inner mechanical attenuator, and the signal to quantization noise ratio (SQNR) is 98 dB [5].

Two of a commercial antenna are used as a transmit/receive antennas. The gain, 3-dB beamwidth, and front-to-back ratio of the antenna are 13 dBi, around 40 degree, and over 20 dB.

IV. MEASUREMENTS

For B2B calibration, the MXG is directly connected to the PXA with about 0.5 m coaxial cable. Data acquisition of the PXA is triggered with a signal synchronized with the start of a PNI in the MXG. And, a normalized spectral-domain range profile is obtained from a post-processing of the acquired data. The RRC filtering, averaging, sliding correlation and DFT are included in the post-processing. The ranges of the maximum intensity in several B2B measurements are 45~52.5 m, which may be due to a filter delay of PXA in VSA (Vector Signal Analyzer) mode. It is found that above time difference can be calibrated by stating data acquisition in the PXA after 350 ns from the receiving of the above trigger signal.

The measurement situation in a rural area is shown in Fig. 2. The used drone is 'a phantom 3 of the DJI', the diagonal length of which except a propeller is 350 mm and thus corresponding to about 3-wavelengths. The spectral-domain range profile with 6 dBm transmit (0 dBm setting in MXG) is shown in Fig. 3. A crosstalk between a transmitter and a receiver and returns from surroundings such as mountains are in evidence. Also, a faint detection marked in white circle is found, range and velocity of which are 217.5 m and 37.52 km/h respectively. Through the cut-view shown in Fig. 4, it is apparent that this faint detection is the drone return. The range of the drone is quite sufficiently over 100 m, even if coaxial cables between transmitter/receiver and antennas, space between antennas, and the 50 m found in the B2B calibration are considered by a wide margin.

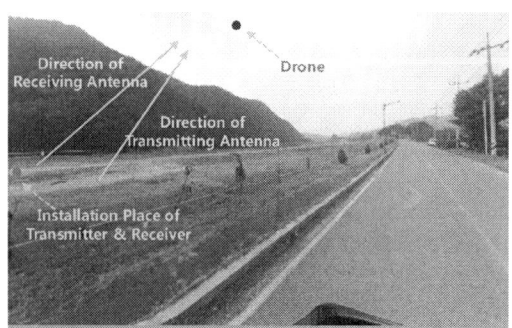

Figure 2. Measurement situation in a rural area (Nonsan, Chungnam, Korea)

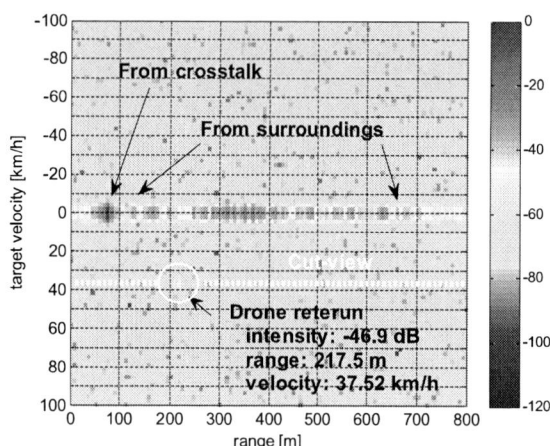

Figure 3. The spectral-domain range profile observed in the measurements

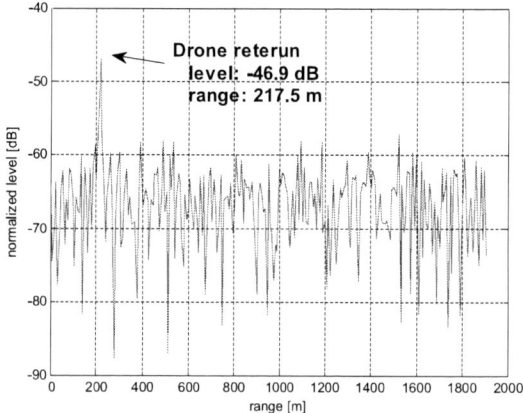

Figure 4. The cut-view with the velocity of 37.52 km/h

V. CONCLUSION

The drone detection by radar was verified using a PRBS based probing signal and commercial measuring instruments and components, which focuses on fast verification. The result shows the drone detection over 100 m range in the 2 GHz band with 6 dBm transmit condition, which confirms the possibility of drone detection radar.

REFERENCES

[1] J. Kivinen et al., "Wide-band radio channel measurement system at 2 GHz," IEEE Trans. Instrumentation and Measurement, vol. 48, no. 1, pp. 39-44, Feb. 1999.

[2] H. Ng, R. Feger, and A. Stelzer, "A Fully-Integrated 77-GHz UWB Pseudo-Random Noise Radar Transceiver With a Programmable Sequence Generator in SiGe Technology," IEEE Trans. Circuits. Syst. I, vol. 61, no. 8, pp. 2444-2455, Aug. 2014.

[3] S. Haykin, Communication Systems, John Wiley & Sons, 2001.

[4] G. E. Carlson, Signal and Linear System Analysis, Houghton Mifflin Company, 1992.

[5] Kesight technol., datasheet, "PXA X-Series Signal Analyzer N9030A."

Design of Low Latency Successive Cancellation Decoder for Polar Codes

Zheyan Piao and Jin-Gyun Chung
Div. of Electronic Engr.
Chonbuk National University
Jeonju, South Korea
{capark, jgchung}@jbnu.ac.kr

Abstract—**Polar codes have recently become increasingly popular due to their simple structure and low decoding complexity. However, polar codes are still not suitable for real-time applications because of the long decoding latency. In this paper, by analysis of the conventional architecture of SC decoder, a low latency SC decoder architecture is proposed. Using the proposed architecture, the decoding latency can be reduced by 25% compared with conventional architectures**

Keywords;polar codes,SC decoder, latency reduction.

I.   INTRODUCTION

Polar codes, introduced by Arikan [1], have received a lot of attentions due to its low encoding and decoding complexities. In general, to achieve good performance of polar codes, code length N needs to be larger than 2^{10}. But for the successive cancellation (SC) decoding scheme, large code length leads to a long decoding latency. In this paper, we present a low latency SC decoder architecture.

II.   PROPOSED SC DECODER ARCHITECTURE

Fig. 1 shows the pre-computation look-ahead SC (PCLASC) decoder architecture with $N = 8$ [2]. TABLE I shows the decoding schedule. As can be seen in TABLE I, the PCLASC decoder needs total $N - 1$ clock cycles. By merging type I and type II processing elements, the PCLASC decoder can reduce 50% decoding latency compared with conventional SC decoder architecture.

One of the reasons leading to long decoding latency of polar codes is the successive cancellation decoding scheme. The estimation of the current bit must depend on the value of the previous decoded bits. Therefore, we can improve the existing SC decoder architecture by decoding more bits in the same clock cycle.

In [3], an algorithm called merged processing element sharing (MPES) is introduced to reduce the number of merged processing elements (MPEs) in list SC decoders. MPES algorithm is applied to stages 1 and 2, since these stages contain approximately 75% of all the MPEs used. As can be seen in Fig. 2, by using five MPE blocks in the second stage, all kinds of output data are computed in second stage in one clock cycle.

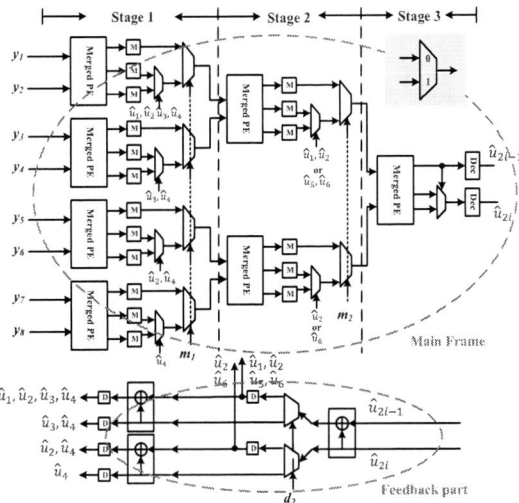

Fig. 1. Architecture of PCLASC decoder.

TABLE I. Decoding schedule of the 8-bit SC decoder.

Clock / Stage	1	2	3	4	5	6	7
1	4 MPEs						
2		2 MPEs			2 MPEs		
3			1 MPE	1 MPE		1 MPE	1 MPE
Output			$\hat{u}_1, \hat{u}_2$	$\hat{u}_3, \hat{u}_4$		$\hat{u}_5, \hat{u}_6$	$\hat{u}_7, \hat{u}_8$

In the proposed architecture, to reduce the decoding latency, MPES algorithm is applied to the last two stages instead of the first two stages. Fig. 3 shows the proposed 8-bit SC decoder architecture.

By using MPES algorithm to replace the last two stages, the decoder allows to decode four bits in the same clock cycle as shown in the rightmost part of Fig. 3. Therefore, the total decoding latency can be reduced as follows:

$$T_{latency} = \frac{3}{4}N - 1. \qquad (1)$$

In successive cancellation decoding scheme, first two decoded bits are used to select next two decoded bits. For example, in PCLASC decoding schedule, after decoding $\hat{u}_1$

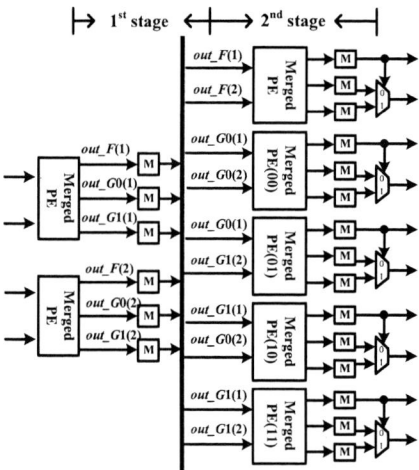

Fig. 2. Architecture of MPES algorithm.

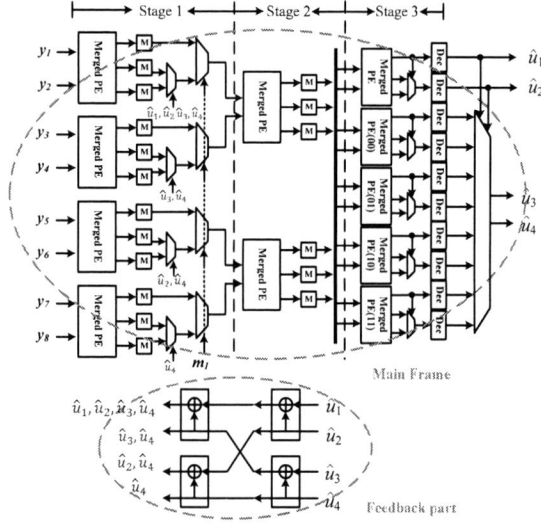

Fig. 3. Proposed 8-bit SC decoder architecture.

TABLE II. Truth table for selecting signals in last stage.

Decoded bits		Selecting values for last stage		Decoded bits selected	
$\hat{u}_{4i+1}$	$\hat{u}_{4i+2}$	$\hat{u}_{4i+1} \oplus \hat{u}_{4i+2}$	$\hat{u}_{4i+2}$	$\hat{u}_{4i+3}$	$\hat{u}_{4i+4}$
0	0	0	0	Merged PE(00)	
0	1	1	1	Merged PE(11)	
1	0	1	0	Merged PE(10)	
1	1	0	1	Merged PE(01)	

and $\hat{u}_2$, selecting values $\hat{u}_1 \oplus \hat{u}_2$ and $\hat{u}_2$ in feedback part are used to decode $\hat{u}_3$ and $\hat{u}_4$.

In the proposed architecture, we need to use a multiplexer to select decoded bits from pre-computed MPE blocks in the last stage. The truth table in TABLE II shows how we select

TABLE III. Comparison of SC decoders with different designs.

	PCLASC	[1]	[4]	[5]†	Proposed
# of MPEs	N-1	N-1	N-2	N+17	N+3
Latency	N-1	$2N$-2	N-2	N-1	$3/4N$-1
Throughput‡	2	1	2	2	2.67

† Results obtained with k_0=2.
‡ Normalized results are compared.

the MPE blocks in the last stage based on the first two decoded bits.

Because of the five MPEs in the last stage, the number of MPEs is slightly increased as follows:

$$Num_{MPE} = N - 2 + 5 = N + 3. \qquad (2)$$

III. COMPARISON AND DISCUSSION

TABLE III compares complexity, decoding latency and throughput of different designs. For the number of MPEs, all the listed designs are nearly the same. For N =1024, compared with other designs, decoding latency is reduced at least by 25% by the proposed architecture.

In order to achieve further decoding latency reduction, we can modify the last three stages using MPES algorithm. In this case, though more decoding latency can be reduced by decoding eight output bits in the same clock cycle, the number of MPEs will also be increased.

IV. CONCLUSIONS

In this paper, a latency reduced SC decoder was proposed. It was shown that, by applying MEPS algorithm to the last two stages, the decoding latency can be reduced by 25% compared with conventional architectures.

ACKNOWLEDGMENT

This work was supported by the Brain Korea 21 PLUS Project, National Research Foundation of Korea.

REFERENCES

[1] E. Arikan, "Channel polarization: a method for constructing capacity-achieving codes for symmetric binary-input memoryless channels," *IEEE Trans. on Inf. Theory*, vol. 55, no. 7, pp. 3051-3073, July 2009.

[2] C. Zhang and K. K. Parhi, "Low-latency sequential and overlapped architectures for successive cancellation polar decoder," *IEEE Trans. Signal Process.*, vol. 61, no. 10, pp. 2429-2441, May 2013.

[3] Zheyan Piao and Jin-Gyun Chung "Design of low area list successive cancellation decoder for polar codes", in *ISOCC*, pp: 35-36, Nov. 2015

[4] Xing Liu, Jin Sha, Chuan Zhang and Zhongfeng Wang, "A stage-reduced low –latency successive cancellation decoder for polar codes," in *IEEE Int. Conf. on DSP*, pp. 258-262, July. 2015.

[5] Dan Le, Qiong Li, Lu Shan and Xiamu Niu, "A new reduced-latency SC decoder for polar codes," in *IIH-MSP*, pp. 214-218, Sept. 2015.

Adaptive Approximate Adder (A^3) to Reduce Error Distance for Image Processor

Sunghyun Kim and Youngmin Kim

Dept. of Computer Engineering,
Kwangwoon University, Nowon-gu, Gwangun-ro 20
Seoul 01897, Republic of Korea
nayam43@kw.ac.kr

Abstract— **Approximate computing is a solution for energy-efficient designs providing trade-off between accuracy and power. Especially vision-related processors are suitable applications to use approximation. So both error distance and error rate, which are important for output quality, are essential metrics in approximate computing logics. In this study, we propose an adaptive approximate adder by using of a modified XNOR-based adder with an adaptive method to configure the approximation bits during runtime. Simulation results of 16-bits adders show that both error distance and error rate are significantly improved by the proposed adder compared to other approximation adders.**

Keywords; Approximate Computing, Adaptive Adder, Configurable Adder, Error Distance, Error Rate, Output Quality

I. INTRODUCTION

Nowadays reduction of power consumption has been a critical issue for mobile, wearable device, and bio-related integrated circuit (IC) designs. Approximate computing, in which computation returns inaccurate but acceptable results rather than accurate results, provides a low-power and small size design by reducing hardware overhead and complexity [1]. Especially vision-related processors can exploit the trade-off between accuracy and power saving of the approximate computing, since in image processing, the error significance (i.e., error distance [2]) is one of output quality metrics [3].

Many researches have been conducted in approximate adder designs to optimize the trade-off between power consumption and precision. Lower-part-OR Adder (LOA) [4] is proposed to be used in bio-related circuits. LOA is an adder consists of a precise adder for higher bits and a simplified approximate adder for the lower bits with OR gates and an extra AND gate. In [5], modeling of a timing-starved adder and synthesis of conditional bounding logics are introduced to optimize the trade-off between quality of image and power saving. In [6], several approximate mirror adders are proposed to save power consumption of digital signal processing (DSP) architectures (e.g., discrete cosine transform). Accuracy-configurable adder, in which the accuracy of results can be configurable during runtime based on accuracy requirement, is proposed in [7]. Recently, a modified Vdd-connected approximate XNOR-based adder (VAXA) is proposed to improve an error rate and 50% reduction in the error rate is achieved [8].

Both error distance and error rate are metrics for evaluating the performance of the approximate adder [2]. In addition, these metrics can be used to measure image and video quality by compression [3]. Power consumption of the LOA is less than the accurate adder and the reduction can be further improved by increasing the bits of approximation. Even though the error distance of the LOA is better than other approximate adders (e.g., XNOR-based or modified XNOR-based), the error rates are naturally too high even when small number of bits (e.g., 1bit or 2bits) are allocated to the lower-part approximate adder because it consists of OR gates without carry chain [4][8]. Therefore, a novel approximate adder architecture, which improves output quality by reducing both error rate and error distance in the image processors is required.

In this study, we propose an adaptive approximate adder (A^3) by using a modified XNOR-based adder [8] and an adaptive concept to configure the approximation bits during runtime for reducing overall error distance occurred by inexact addition. Simulation results show that the average error distance (AED) is 40 for nine-bits approximation out of total 16-bits addition. This result is less than 58% compared to that of LOA for same bit approximation. In addition, the error rate is also smaller than other approximate adders for various approximation bits.

II. ADAPTVE APPROXIMATE ADDER (A^3)

The proposed adaptive approximate adder (A^3) is shown in Fig. 1. It consists of three parts; a precise adder for the higher bits (upper-part), a VAXA-based approximate adder for the lower bits (lower-part), and a XOR-based lower-part length (LPL) selector to configure the number of lower bit of approximation (adaptive-part). The upper m-bits precise adders in total p-bits adders can be made by using of any accurate adders to obtain accurate results. And the lower n-bits

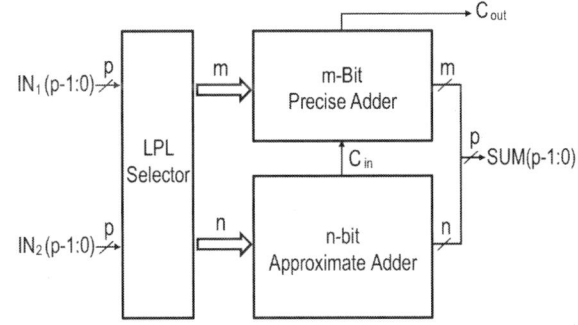

Fig. 1. Proposed Adaptive Approximate Adder (A^3)

978-1-5090-3220-4/16 $31.00 © 2016 IEEE

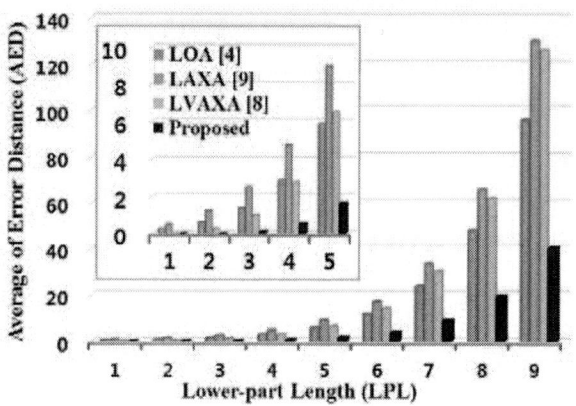

Fig. 2. Average Error Distance (AED) comparison

Fig. 3. Error rate or probability of error comparison

approximate adders use VAXAs. Both the 'm' of upper-part and the 'n' of lower-part are decided using the input patterns during runtime by the LPL selector. The LPL selector sequentially compare both input streams of IN_1 and IN_2 from the maximum low bit length (9th bit in this study) to the first bit position by XOR operations. During this, if XOR result is equal to zero (i.e., same input bit patterns), 'n' becomes the current bit position plus one and the 'm' becomes 'p' minus 'n'. For example, when p is 6, the maximum low bit length is 4, IN_1 is '110111', and IN_2 is '111011', XOR generates zero at the 2nd bit position at first. Thus, 'n' is 3 ($= 2 + 1$) and 'm' is 3 ($= 6 - 3$), respectively.

In this way, at least, the most significant bit (MSB) of the approximate adder can guarantee an accurate result for any inputs as explained by the error cases of VAXA in [8]. So the maximum error distance (ED) of the proposed adder is less than that of the LOA which is 2^{n-1}. Furthermore, error rate of the proposed adder will be improved over the LOA and other approximate adder architectures because it utilizes advantage of the VAXA adder adaptively based on input patterns.

III. SIMULATION REULT

Simulation is conducted using a logic simulator, NC-Verilog. Total number of bits of adders is fixed to 16 bits. Four adders, LOA, Lower-part AXA (LAXA), Lower-part VAXA (LVAXA), and the proposed adder are designed and compared. In addition, the lower-part length (LPL) is varied from 0 (i.e., all accurate adders) to 9 bits (i.e., approximate adders for the low 9 bits and accurate adders for the high 7 bits). The average error distance (AED) are compared to identify the amount of errors in the approximate adders as shown in Fig. 2. The inset in Fig. 2 shows the enlarged chart of LPL from 0 to 5 for better comparison. As shown, the AED improves by the proposed adder significantly over others. For example, the AED for 5-bit LPL of the proposed adder is 1.7, which is almost equal to the AED for 3-bit LPL of the LOA. When LPL is the maximum (9 in this study), the AED of the proposed adder is 40, which is 58% less than that of the LOA. The probability of errors (PE), or error rate is presented in Fig. 3. As shown, no errors are obtained up to 2-bits LPL in the proposed adder and the error rates are reduced for larger LPL values. For example, less than 60% error rate can be achieved up to nine lower-bits approximation by using of the proposed adder. Furthermore, the error rate for 9-bit

LPL by the proposed adder is almost equal to that of LVAXA with 7-bit LPL or that of the LOA only with 3-bit approximation. Lastly, for maximum approximation (or when LPL is 9), the PE of the proposed adder is 40% and 20% smaller than the LOA and the LVAXA, respectively.

IV. CONCLUSION

In this paper, we propose an adaptive approximate adder (A^3) architecture to configure the bit length of the approximate adder during runtime. Simple XOR operation can be used to decide the bit lengths for the accurate adder and the approximate adder. Simulation results show that significant improvement in both error distance and error rate can be achieved by the proposed adder. Therefore, the proposed adder can be exploited to design image processors providing low power with improved output quality.

ACKNOWLEDGMENT

This work was sponsored by NRF Grant #2014-R1A1A2057715. EDA tools were partially supported by IDEC at KAIST.

REFERENCES

[1] J. Han and M. Orshansky, "Approximate computing: an emerging paradigm for energy-efficient design," in Proc. *European Test Symposium (ETS)*, pp. 1-6, May 2013.

[2] J. Liang, J. Han and F. Lombardi, "New metrics for the reliability of approximate and probabilistic adders", *IEEE Trans. on Computer*, vol. 62, no. 9, pp. 1760-1771, June 2012.

[3] I.S. Chong, et al., "New quality metric for multimedia compression using faulty hardware", in Proc. *Int'l Workshop on Video Processing and Quality Metrics for Consumer Electronics*, 2006.

[4] H.R. Mahdiani, et al., "Bio-inspired imprecise computational blocks for efficient VLSI implementation of soft-computing applications," *IEEE Trans. on Circuits and Systems I: Regular Papers*, vol. 57, no. 4, pp. 850-862, April 2010.

[5] J. Miao, K. He, A. Gerstlauer and M. Orshansky. "Modeling and synthesis of quality-energy optimal approximate adders", in Proc. *International Conference on Computer-Aided Design (ICCAD)*, pp. 728-735, 2012.

[6] V. Gupta, D. Mohapatra, A. Raghunathan and K. Roy, "Low-power digital signal processing using approximate adders," *IEEE Trans. on CAD of Integrated Circuits and Systems*, vol. 32, no. 1, pp. 124-137, Jan 2013.

[7] A.B. Kahng and S. Kang, "Accuracy-configurable adder for approximate arithmetic designs," in Proc. *DAC*, pp. 820-825, 2012.

[8] S. Kim and Y. Kim, "Energy-efficient hybrid adder design by using inexact lower bits adder", in Proc. *APCCAS*, 2016.

[9] Z. Yang, et al., "Approximate xor/xnor-based adders for inexact computing," in Proc. *Nanotechnology (IEEE-NANO)*, pp. 690-693, 2013.

Artificial Neural Network Implementation in FPGA: A Case Study

Shuai Li, Ken Choi
Department of Electrical and Computer Engineering
Illinois Institute of Technology, USA
sli97@hawk.iit.edu, kchoi@ece.iit.edu

Yunsik Lee
School of ECE
UNIST, Ulsan, Korea
leeys@unist.ac.kr

Abstract— **Artificial Neural Network (ANN) is very powerful to deal with signal processing, computer vision and many other recognition problems. In this work, we implement basic ANN in FPGA. Compared with software, the FPGA implementation can utilize parallelism to speedup processing time. Additionally, hardware implementation can save more power compared with CPU/GPU. Our ANN in FPGA has a high learning ability, for logical XOR problem, which reduced the error rate from 10^{-2} to 10^{-4}.**

Keywords; artificial neural network, parallelism, back propagation, LReLU, FPGA

I. INTRODUCTION

Artificial neural network (ANN) is a model that inspired by the biological neural network of human brain. ANN based on a set of algorithms that attempt to model high-level abstraction in data by using multiple processing layers, which has the ability to automatically infer rules for expected results [1-6]. Furthermore, the network can improve the accuracy by increasing the training times. This paper discusses various aspects of hardware implementation of ANN, e.g., generic architecture, back propagation, precision, etc. One of the most arguments in hardware is the exploitation of the parallelism in neural network, which can be very fast, especially for well-defined signal processing usages. The rest of the paper is organized as follows. Section II describes the proposed method. The experimental results are given in section III. Finally, section IV concludes this paper.

II. THE PROPOSED METHOD

A. Generic Architecture of ANN

For the implementation of ANN algorithm, we have the following architecture: As shown in figure 1, basically, there are three types of layers in the feedforward structure. The leftmost layer in network is called input layer. The rightmost layer is output layer, the middle layer(s) are hidden layers since they act as a black box to the input and output layers. We use w_{ij}^n to denote the weight from j^{th} neuron in $(n-1)^{th}$ layer to the ith neuron in n^{th} layer, b is bias vector. Bias is influences the output without interacting with the actual input data. We have:

$$a_i^n = \sigma\left(\sum_j w_{ij}^n a_j^{n-1} + b_i^n\right) \qquad (1)$$

Where a_i^n is the activation function of the ith neuron in n^{th} layer. $\sigma(\blacksquare)$ denotes elementweise application of the function.

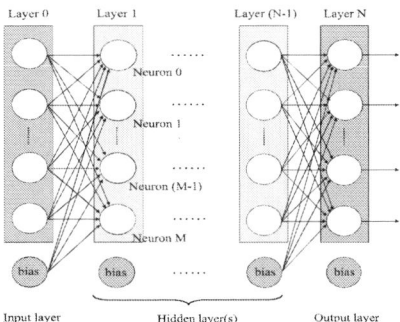

Figure 1. Architecture of feedforward ANN.

B. Back Propagation and Parallelism

Back propagation is to update weights and biases. The purpose is to minimize the difference between target value (ground value) and trained output value. The algorithm of back propagation is shown in Figure 2.

Algorithm: Back Propagation
1) Staring from output layer towards the input layer, calculate the local gradients.

$$\varepsilon_i^n = \begin{cases} O_i - a_i^n & n = N \\ \sum_{j=0}^{M_{n+1}} w_{ij}^{n+1}\delta_j^{n+1} & n = 1,2,\dots,N-1 \end{cases} \qquad (2)$$

Where ε_i^n is the error of ith neuron in n^{th} layer. O_i is target value. δ_j^{n+1} is local gradient of jth neuron in $(n+1)^{th}$ layer. Specifically,

$$\delta_j^n = \varepsilon_j^n f'(a_j^n) \qquad (3)$$

where $f'(a_j^n)$ is derivative of activation function.
2) Calculate weight changes.

$$\Delta w_{ij}^n = \eta \delta_i^n a_j^{n-1} \qquad (4)$$

Where η is learning rate.
3) Update the weights.

$$w_{ij}^n(n+1) = w_{ij}^n(n) + \Delta w_{ij}^n(n) \qquad (5)$$

Figure 2. Algorithm of back propagation.

```
for layer n from 0 to N
    for neuron i from 0 to M
        for each weight j (from 0 to M) in nth layer
            Update weight in (n+1)th layer: wij^n(n+1)
        endfor
    endfor
endfor
```

Figure 3. Updating weight by using 3 loops in software.

A typical ANN has three components: layers, neurons and weights. In software, we implement the function of updating weights by using three loops as shown in Figure 3. In Figure 4, there are three loops, we need at least N*M*M cycles to finish updating the weights. On the contrary, in hardware implementation, there are 3 different ways of achieving parallelism to save the processing cycles dramatically.

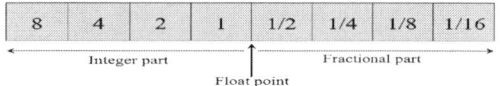

Figure 4. Parallelism used in hardware optimization.

C. Precision and Activation Function

We use floating numbers in ANN, however, in FPGA all the operations are bitwise and integers. Back propagation is very sensitive to numerical precision. The numerical precision and dynamic range such as multiply and add operations should be studied with care. A reduction of precision will have a negative impact on the training effect. To solve the problem, we use the method in Figure 5.

8	4	2	1	1/2	1/4	1/8	1/16

Integer part — Float point — Fractional part

Figure 5. Representation of 8-bit floating number.

As shown in Figure 5, it is an 8-bit floating number, the least significant 4 bits are fractional part and the most significant 4 bits are integer part. The precision is $1/16=0.0625$. In our work, we use a 16-bit to represent the fractional part, the precision is $1/2^{16}$ which can meet the requirement. There are different kinds of activation functions, such as tanh and Leaky Rectified Linear Unit (LReLU) function, etc.

$$\tanh(x) = \frac{e^x - e^{-x}}{e^x + e^{-x}} \qquad (6)$$

For LReLU, we have

$$f(x) = \begin{cases} x & if \ x > 0 \\ 0.01x & otherwise \end{cases} \qquad (7)$$

In back propagation, the derivative function of tanh and LReLU are:

$$\frac{d}{dx}\tanh(x) = 1 - tanh^2(x) \qquad (8)$$

$$\frac{d}{dx}f(x) = \begin{cases} 1 & if \ x > 0 \\ 0.01 & otherwise \end{cases} \qquad (9)$$

The ReLU function has a significant contribution to reduce the error rate, which will be evaluated in section III.

III. EXPERIMENTAL RESULTS

The logical XOR is typical benchmark to verify the learning ability of ANN. We set the training times for 2000 epochs, and learning rate η is 0.15. The initial weights and biases are random numbers between 0 and 1. As shown in figure 6, both of the inputs are 1, when epoch = 3, the output value $(0.1011101110011101)_2 = (0.732864)_{10}$; when epoch = 1950, the output value $(0.0000000000000100)_2 = (0.000061)_{10}$, the result is very close to the target value 0. To evaluate the

performance of ANN, we use error rate vs. epoch times to show the computational accuracy. Equations (10)-(12) compute the error rate.

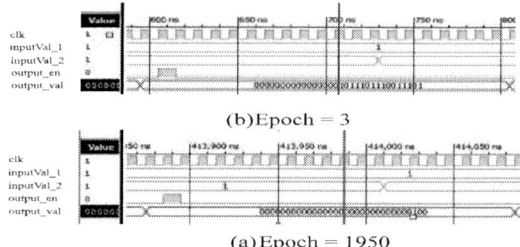

(b) Epoch = 3

(a) Epoch = 1950

Figure 6. Test results of ANN.

$$\varepsilon_i^n = O_i - a_i^n \qquad n = N, i = 0,1,2,\dots,M \qquad (10)$$

$$\xi_k = \sqrt{\frac{\sum_{i=0}^{M}(\varepsilon_i^n)^2}{M}} \qquad (11)$$

$$\bar{\xi}_k = \frac{\bar{\xi}_{k-1}\beta + \xi_k}{\beta + 1} \qquad (12)$$

Where M is the number of neurons in output layer, ξ_k is the error rate in epoch k, $\bar{\xi}_k$ is current average error rate, β is the soothing factor, here $\beta = 100$.

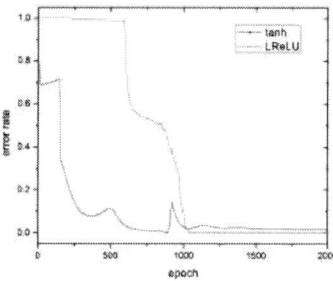

Figure 7. Error rate vs. epoch.

Figure 7 demonstrates the changes of error rate during 2000 epochs. After 2000 times training, the error rate of LReLU is as low as 10^{-4}, which is superior compared with tanh function which with the error rate of 10^{-2}.

IV. CONCLUSION

This paper has presented the implementation of ANN in hardware. This work is a part of project, and the system is targeted for detecting objects with small hardware resources and low power consumption.

REFERENCES

[1] L. Deng, D. Yu, "Deep Learning: Methods and Applications", Foundations and Trends in Signal Processing 7: 3–4, 2014.

[2] Y. Bengio, "Learning Deep Architectures for AI", Foundations and Trends in Machine Learning 2 (1): 1–127, 2009.

[3] Y. Bengio, A. Courville, P. Vincent, "Representation Learning: A Review and New Perspectives", IEEE Transactions on Pattern Analysis and Machine Intelligence 35 (8): 1798–1828, 2013.

[4] J. Schmidhuber, "Deep Learning in Neural Networks: An Overview", Neural Networks 61: 85–117, 2015.

[5] Y. Bengio, Y. LeCun, G. Hinton, "Deep Learning", Nature 521: 436–444, 2015.

[6] I. Arel, D. C. Rose, T. P. Karnowski, "Deep Machine Learning – A New Frontier in Artificial Intelligence Research", IEEE Computational Intelligence Magazine, 2013.

Resource-Efficient FPGA Architecture of Canny Edge Detector

Yunseok Jang , Junwon Mun and Jaeseok Kim School of Electrical and
Electronic Engineering
Yonsei University
Seoul, Korea 120-749
Email: {yunejang, mjw5554 and jaekim}@yonsei.ac.kr

Abstract—Edge detection is one of the key stages of image processing and feature extraction. The Canny edge detector is the most popular edge detector because of its ability to detect edges in noisy images. However, it is a time and resource consuming algorithm which contain many stages. So we need to reduce the size of the Canny edge detector. In this paper, a hardware architecture for Canny edge detector is proposed. A 5 by 5 sliding window is adopted to conduct image smoothing and get gradient at the same time. By using same divider value twice, the angular value for all edges with one degree resolution is obtained. Synthesis and simulation results are presented.

Index Terms—Canny edge detection, real time, FPGA architecture.

I. INTRODUCTION

Edge detection is essential and first step for many image processing algorithms. And Canny edge detection[1] is most popular algorithm among them due to good performance and robust to noise.

Many algorithms realization Canny algorithm on FPGA have been proposed. In [2], a self-adapt threshold calculation method of Canny is proposed, it is robust to environment change but show low frame rate. In [3], a 4 pixel parallel computation which can achieve high frame rate is proposed. In [4], Canny architecture which use improved median filter and shifting LUT has been proposed which can also achieve high frame rate.

The remainder of the paper is organized as follows. The Canny edge detection algorithm is presented in Section II. Our hardware architecture is presented on III. And then, Section IV shows our synthesis result. Lastly, Section V summarizes the conclusions.

II. CANNY EDGE DETECTION

The block diagram of the Canny edge detection is demonstrated in Fig. 1. First smooth the image by Gaussian smoothing filter to reduce the noise contained on the image. A 5x5 mask is a common choice for a size of a Gaussian filter.

The next step is to calculate all pixels direction and magnitude. By obtain vertical and parallel gradient by 3x3 Sobel operator. Gradient magnitude and orientation is obtained by following equations.

$$|G| = |G_x| + |G_y| \qquad (1)$$

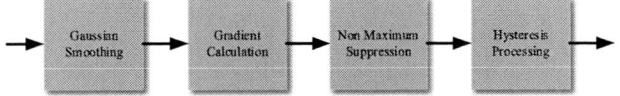

Fig. 1. Canny edge detection block diagram

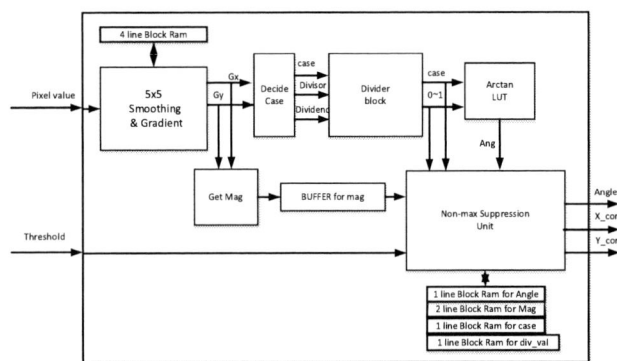

Fig. 2. Hardware architecture of Canny edge detection

$$\theta = \arctan(G_x/G_y) \qquad (2)$$

The following step is non maximum suppression, which is used for thinning the edge. However, after this step, there is still false edges, so we need to eliminate those by hysteresis thresholding. In this step two thresholds are set and linking or eliminating edges by thresholded binary image.

III. HARDWARE ARCHITECTURE

The block diagram for our hardware architecture is shown in Fig. 2. We use 5x5 filter to smooth image and get gradient simultaneously. And we use divider value on both arctangent LUT and non maximum suppression unit to reduce hardware size. We use 4 block rams for 5x5 filtering and 5 block rams for non maximum suppression.

A. 5x5 smoothing and gradient filter

We made 5x5 filter by convoluting 3x3 Gaussian blur filter and 3x3 Sobel filter to get x-direction and y-direction gradient value. Four block rams are sufficient for our implementation.

978-1-5090-3220-4/16 $31.00 © 2016 IEEE   299   ISOCC 2016

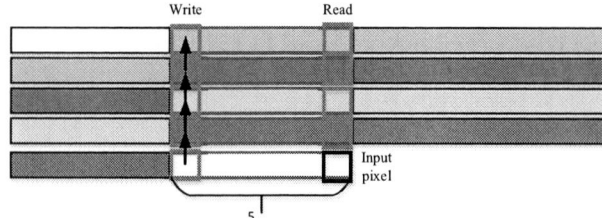

Fig. 3. Block ram operation in 5x5 filtering

TABLE I
CLASSIFY DIVIDER CASE ACCORDING TO GRADIENT VALUE

Condition	Case	Divide
$G_y \geq 0, G_x \geq G_y$	1	G_y/G_x
$G_x \geq 0, G_x < G_y$	2	G_x/G_y
$G_x < 0, G_y \geq -G_x$	3	G_x/G_y
$G_y \geq 0, G_y < -G_x$	4	G_y/G_x
$G_y < 0, G_x \leq G_y$	5	G_y/G_x
$G_y \leq 0, G_x > G_y$	6	G_x/G_y
$G_x > 0, G_y \leq -G_x$	7	G_x/G_y
$G_y \leq 0, G_y > -G_x$	8	G_y/G_x

Fig. 3. shows implementation of filter. Blue box indicate registers for 5x5 operation and value in red box indicate pixel to write or read to block ram and value in black box indicate current input pixel.

B. Divider to non maximum suppression

After 5x5 filtering, we classify pixel according to gradient values like TABLE I. Then we divide the gradient value which quotient to have range between 0 to 1. We use this quotient in both arctangent LUT block and non maximum suppression unit. The non maximum suppression is operated by interpolating near two pixel as shown in Fig. 4.

IV. EXPERIMENTAL RESULT

The above Canny edge detector design was synthesized with Xilinx zynq7020 which has small logic size by Xilinx ISE Design Suite 14.7. The result is presented in Table II. Even in small FPGA we use, our implementation achieves an operating frequency of 130MHz. The total block ram use was 20kbytes and we use only 14% of the FPGA logic.

We used 640x480 VGA image to verify our proposed architecture and Fig.5. Shows our result.

V. CONCLUSION

In this paper a 5x5 filter and divider shared resource efficient Canny implementation is presented. By sharing divider on arctangent LUT and non maximum suppression unit we can reduce logic size. And our design was synthesized for low-end Xilinx zynq7020 FPGA, The result show that our design

TABLE II
SYNTHESIS RESULT

Synthesis Result	Slices	B-Ram	Frequency
Zynq7020	7894(14%)	20kbytes(3%)	132.4MHz

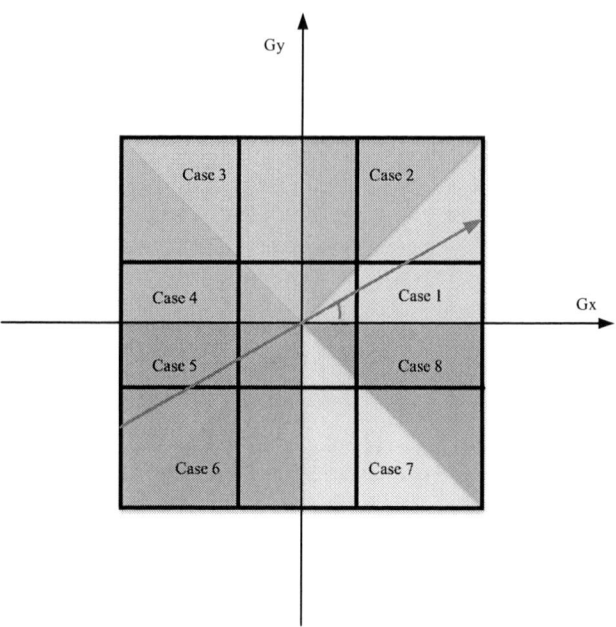

Fig. 4. Non max suppression operation

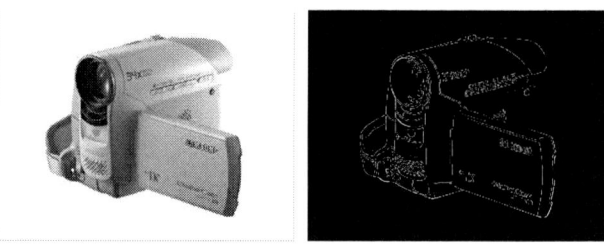

Fig. 5. Canny edge detection result

achieve operating frequency of 130MHz, only occupy 14% of entire logic and use 3% of Block Ram.

ACKNOWLEDGMENT

This work was supported by the National Research Foundation of Korea(NRF) grant funded by the Korean government(MSIP) (No. NRF-2015R1A2A2A01004883), and was also supported by IDEC(IPC, EDA Tool, MPW)

REFERENCES

[1] Canny. J. "A Computational Approach to Edge Detection" *IEEE Trans. Pattern Anal. Mach. Intell.* vol. 8, no. 6, pp. 679–698, NOV. 1986.
[2] Wenhao. H. Kui. Y "An improved Canny edge detector and its realization on FPGA" *World Congress on Intelligent Control and Automation* pp. 6561–6564, JUN. 2008.
[3] Gentsos. C, Sotiropoulou. C. L "Real-time Canny edge detection parallel implementation for FPGAs" *IEEE Int. Conf. Electronics, Circuits, and Systems.* pp. 499–502, DEC. 2010.
[4] Xiaoyang. L, Jie. J, Qiaoyun. F "An improved Real-time Hardware Architecture for Canny Edge Detection Based on FPGA" *International Conference on Intelligent Control and Information Processing* pp. 445–449, JULY. 2012.

978-1-5090-3220-4/16 $31.00 © 2016 IEEE

Standing Wave Oscillator Based Clock Distribution

Wei Zhang, Youde Hu, Keji Cui, Dongxuan Bao,Dashan Pan,Lebo Wang and Lirong Zheng

Shool of Information Science and Technology

Fudan University

Shanghai, China

Email: {wei_zhang, youde_hu, kcui13, dbao10,12110720039,lebowang, lrzheng}@fudan.edu.cn

Abstract—**This paper introduces a two-level bufferless standing wave oscillator based clock distribution network to minimize the clock power consumption. The global clock network is a serpentine architecture of many coupled standing wave oscillators. The local clock network is a group of fishbone architectures connected to the standing wave oscillators. The experimental result using simulation demonstrates that this architecture reduces 30% local clock power consumption.**

Keywords: Standing wave oscillator; Clock distribution

I.   INTRODUCTION

In modern high-performance chips, the on-chip clock distribution network (CDN) consumes in excess of 35% of total chip power and occasionally as much as 70%. Reducing the power consumption of the clock signal can reduce overall chip power dramatically. Recently proposed resonant clocking can recycle the energy present in clock switching and can improve the power efficiency during peak activity.

LC tank clock has been used in global clock distribution [1,2], but most clock power is dissipated in local clock network. Rotary clock and standing wave clock can be constructed to a bufferless grid architecture to reduce clock power[3,4]. However, it is a gargantuan task to develop automation design flows. In this paper, we present a standing wave oscillator (SWO) based mixed interconnection of a serpentine network and many fishbone architectures for three purposes. The first is a resonant global clock network suitable for being integrated in nowdays digital chip floorplan. The second is a bufferless resonant local clock network to reduce clock power. The last is minimal change to existing design methodologies.

II.   CLOCK DISTRIBUTION ARCHITECTURE

A.   Global Clock Network

As widely accepted, the grid architecture can distribute a low skew clock signal in the whole chip. However, a great deal of metal resource is required in this architecture. In this paper, we use a serpentine architecture to distribute many coupled SWOs all over the chip area. Each SWO consists of a cross coupled inverter pair (CCIP) and the differential transmission line described in [5]. The coupling architecture of SWOs is described in [4]. Figure 1(a) shows the serpentine network which distributes locked low skew standing wave clock across the whole chip. This architecture uses less metal resource and provides channels to route the global signals.

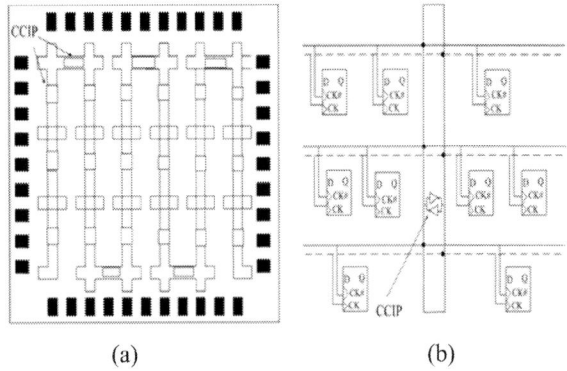

(a)   (b)

Figure 1.   (a) Global serpentine network   (b) Local fishbone network.

B.   Local Clock Network

Clock mesh is widely used with low skew and high robustness because the redundant wires reduce the impact of variation. It also has more power consumption because of the increasing capacitance. Fishbone architecture is similar with mesh architecture, and has only horizontal clock wires. This architecture has more skew than mesh but less power and metal resource uasge. The fishbone architecture used for the SWO is shown in Figure 1(b). This architecture has horizontal wires connected to SWO's differential transmission line.

C.   Working Conditions

In order to provide high quality clock signal, all coupled SWOs should have nearly the same resonant capacitance to decrease skews. A certain amount of load capacitance should be selected for a prefered frequency, including the capacitance of fishbone wires, vias and flip-flops. The load capacitance should also be distributed evenly along the transmission line.

III.   DESIGN METHODOLOGY

A.   Design of Flip-flop

Normally a special buffer is required for SWO to convert the low-swing, differential sinusoids to digital levels to interface with a conventional digital clock network. However, the utilization of buffers decreases clock power savings due to the fact that most power is dissipated in local clock distribution. Hence, to increase energy efficiency, the clock sinks should be directly driven by the SWO.

978-1-5090-3220-4/16 $31.00 © 2016 IEEE

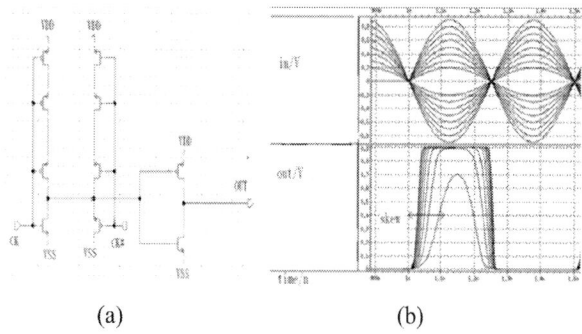

(a)   (b)

Figure 2. (a) Flip-flop architecture (b) Skews caused by voltage difference.

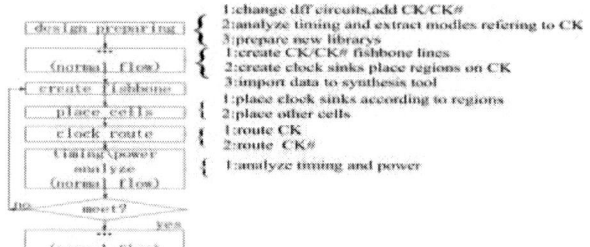

Figure 3. Synthesis flow of SWO based fishbone architecture

We replace the clock driver in a flip-flop with the buffer shown in Figure 2(a) which has nearly the same input capacitance. This buffer accepts the differential input signals named CK and CK#, and outputs the digtal clock signal named OUT. While connected on different places of transmission line, the different voltage amplitudes cause different delays on the output of the new buffer. As shown in Figure 2(b), when the voltage amplitude varys from 0.9V to 0.3V, the delay varays from 29.4ps to 66.1ps. We choose a proper connecting region on transmission line to realize low skew.

B. Synthesis Flow

All EDA synthesis flows have realized the design automation of clock mesh architecture. As shown in Figure 3, a similar flow can be used for the synthesis of a SWO based fishbone network. In design preparing step, all flip-flops should be redesigned, simulated and extracted. When extracting new timing library, CK and CK# will be treated as two clock signals, but timing check is according to CK only. In the following steps, we can use the normal EDA tool functions to realize synthesis according to CK signal, including creating clock regions and placing cells.

IV. EXPERIMENT RESULT

In this paper, a synchronous module of 6 parallel pipelined floating-point fused multiply-add (FMA) is designed to analyze the power efficiency under the proposed fishbone clocking network. The module has 7 gated clocks of 2.6GHz, and involves 3984 flip-flops. The module is realized using normal design flow under 28nm CMOS process. The layout is 320um*230um, and has 7 fishbone clock wires of 200um. The total capacitance of these 7 clocks is 6740.2ff including drivers, wires and flip-flops. The power of these clocks is 14.19mw.

TABLE I.   SWO SIMULATION PARAMETERS

Parameters	Value
Transmission line layer	metal 9
Transmission line length	1600um
Transmission line width	12 um
Transmission line space	1.5 um
nmos/pmos width of CCIP	151.2 um
nmos/pmos length of CCIP	30nm

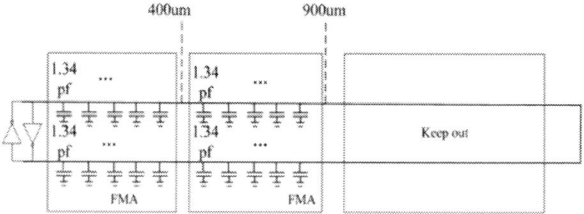

Figure 4. Power simulation architecture of SWO based fishbone clock.

We use the SWO simulation architecture described in [5] and simulate the clock power of FMA under this SWO architecture. The parameters of this SWO are shown in TABLE I. Figure 4 shows that two modules are connected to SWO clock architecture. In each module, we suppose using 5 pairs of fishbone wires of 200um to connect the new flip-flops. The capacitance of double 6740.2ff is distributed equally on these 5 pairs of wires. The simulated power of this architecture is 20mw which is approximately 70% of the sum of these two module's traditional clock power.

V. CONCLUSION

In this paper, we present a two-level hierarchical interconnection architecture for bufferless standing wave clock distribution. This architecture consists of a serpentine global network and many fishbone local networks, and connects flip-flops directly without any buffers. The simulation result shows that this fishbone local network decreases more than 30% clock power.

REFERENCES

[1] S.C. Chan, P.J. Restle, T.J. Bucelot, J.S. Liberty, S. Weitzel, J.M. Keaty, B. Flachs, R. Volant, P. Kapusta and J.S. Zimmerman, "Resonant global clock distribution for the cell broadband engine processor," in Journal of *Solid-State Circuits*, vol.44, no 1, pp.64-72, 2009.

[2] V.S. Stathe, S. Arekapudi, A. Ishii, C. Ouyang, M.C. Papaefthymiou and S. Naffziger, "Resonant-Clock Design for a Power-Efficient, High-Volume x86-64 Microprocessor," in *IEEE Journal of Solid-State Circuits*, vol.48, no 1, pp.140-149, 2013.

[3] Hu Xuchu and M.R. Guthaus, "Distributed resonant clock grid synthesis (ROCKS)," in *48th ACM/EDAC/IEEE Design Automation Conference (DAC)*, pp.516-521, 2011.

[4] F. O'Mahony, C.P. Yue, M.A. Horowitz and S.S.Wong, "A 10 GHz global clock distribution using coupled standing wave oscillators," in *IEEE Journal of Solid-State Circuits*, vol.38, no 11, pp.1813-1820, 2003.

[5] Wei ZHNAG, You-de Hu, Li-romg Zheng, " Design and simulation of a standing wave oscillator based PLL," in Frontiers of Information Technology & Electronic Engineering, vol.17, no 3, pp.258-264, 2016.

Area-efficient and High-speed Binary Divider Architecture for Bit-Serial Interfaces

Yunho Park, Jonghyuk Kwon and Youngjoo Lee

Department of Electronic Engineering
Kwangwoon University
20, Gwangun-ro, Nowon-gu, Seoul 01897, Republic of Korea
Email: yhpark.ims@gmail.com, jhkwon.ims@gmail.com, yjlee@kw.ac.kr

Abstract— This paper proposes a low-complexity binary divider for high-speed serial interfaces. In contrast that the previous parallel dividers require a large amount of hardware costs, the proposed divider achieves a low-complexity division by targeting the serial interface that is widely used for the recent digital communications. By modifying the linear feedback shift register (LFSR) architecture for polynomial division, the high-speed processing element is newly introduced to handle the serialized dividend input. As a result, the proposed serialized architecture reduced the area-time complexity by 75 times compared to the previous slow parallel divider.

Keywords; Arithmetic Circuits; Binary Divider; Serial Interface;

I. INTRODUCTION

In digital systems, the binary divider is one of the most complicated arithmetic circuits due to the numerous subtractors and registers for managing the parallel dividend inputs [1]–[3]. In many applications, however, the data is serially transferred to reduce the interconnection overheads [4]. Before activating the division, in the conventional parallel-in/parallel-out divider architecture, all the bits of dividend are received first, requiring a large size of buffers. Targeting the serial data interfaces, we present in this paper an area-efficient serial-in/serial-out binary divider based on the linear feedback shift register (LFSR). As the proposed architecture splits the division operation into several steps, in addition, the critical delay is greatly reduced.

II. PROPOSED SERIAL-IN/SERIAL-OUT DIVIDER

A. Interface and Timing Diagram

Fig. 1(a) shows the conventional n-bit parallel-in/parallel-out divider interface. As shown in the figure, an n-bit dividend and an n-bit divisor, which are represented as $A = (a_{n-1}, a_{n-2}, \ldots, a_0)$ and $D = (d_{n-1}, d_{n-2}, \ldots, d_0)$, respectively, are issued to the divider at the same time. The combinational logic computes an n-bit quotient $Q = (q_{n-1}, q_{n-2}, \ldots, q_0)$ and an n-bit remainder $R = (r_{n-1}, r_{n-2}, \ldots, r_0)$. We assume that the divisor is already prepared where the dividend is serially transferred in descending order, which is an acceptable scenario for many applications [4], [5].

Considering the serial data communication, the interface of the proposed serial-in/serial-out divider is illustrated in Fig. 1(b). Note that there is only a single input port for the dividend, which is denoted as a_x, so that all the bits of n-bit dividend are serially issued to the divider in descending order, which is identical to the transmission sequence. After receiving the first bit of dividend, i.e., a_{n-1}, the proposed divider provides the n

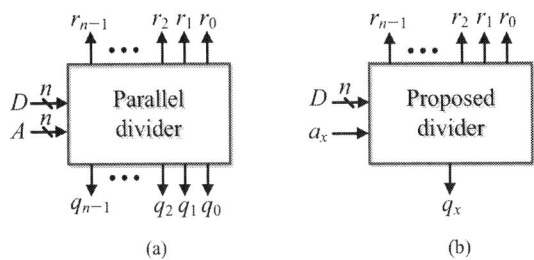

Figure. 1 Interfaces of (a) the parallel divider and (b) the proposed divider.

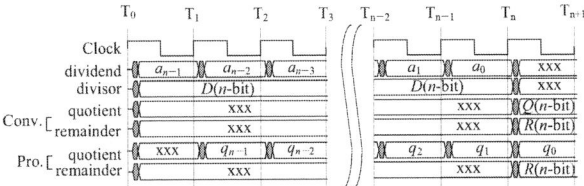

Figure. 2 The timing diagram of division operations.

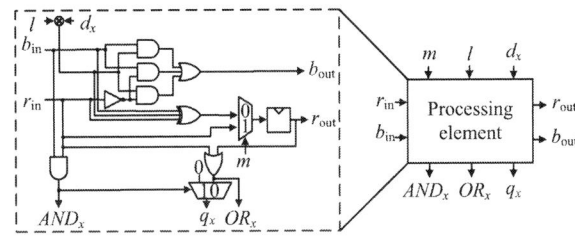

Figure. 3 The detailed architecture of the processing element.

quotient bits q_x in descending order. Similar to the previous divider, the remainder R is calculated in parallel.

Fig. 2 shows the timing diagram of two dividers. Including the transfer latency, the previous divider uses $n+1$ cycles for finishing the division. Note that the serial divider requires the same cycles as it generates quotient bits during the transmission time. Moreover, the proposed architecture eliminates the input buffer in the parallel divider, which is used for capturing the serially issued dividend to construct the parallel input.

B. Processing Element

For the serial computation, the proposed divider uses the processing element (PE) as depicted in Fig. 3. Each PE contains a flip-flop and corresponds to a bit of remainder similar to the

---

This work was supported by the National Research Foundation (NRF) grant funded by the Korea government (MSIP) (No. 2016R1C1B1007593).

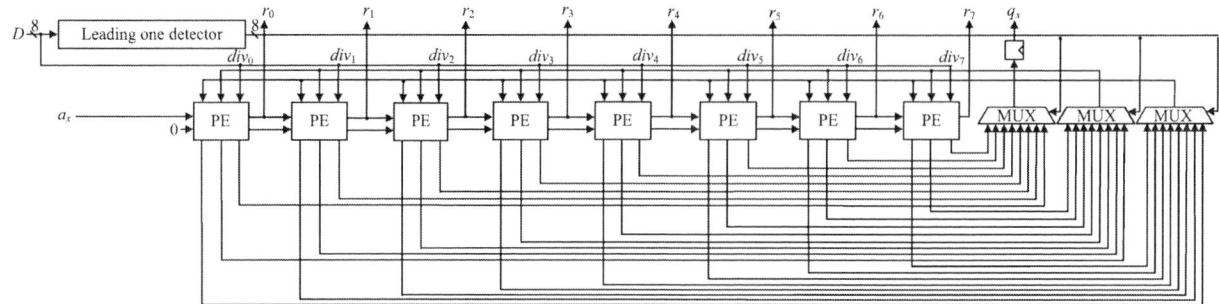

Figure 4. The proposed 8-bit binary divider.

TABLE I. IMEPLEMENTATION RESULTS

Bitwidth	Conventional Parallel [3]		Proposed Serial	
	Area (μm^2)	Delay (ns)	Area (μm^2)	Delay (ns)
8bit	10209.72	2.92	2622.36	1.86
16bit	63990.96	4.74	5839.12	2.59
32bit	331721.82	9.59	11534.35	3.69
64bit	1479082.32	23.25	17038.62	5.94

LFSR for polynomial division. A PE mainly generates two signals for realizing the binary division, i.e., the remainder output r_{out} and the borrow output b_{out}. These two outputs are directly connected to the same inputs of the next PE. Each PE requires the corresponding divisor bit d_x and generates the x-th quotient bit, i.e., q_x. In addition, control signals are introduced for calculating the quotient in serial manner. For the input control signals, more precisely, m denotes the comparison result between the divisor and the current remainder, and l stands for subtracting the divisor from the current remainder. Note that m is determined by the AND_x signals from all the PEs where l is dependent on all the OR_x signals. Basically, these control input/output signals are necessary for supporting an arbitrary divisor by checking the leading one position of D.

C. Divider Architecture

Fig. 4 exemplifies the 8-bit divider based on the proposed PE architectures. Note that 8 PEs are serially connected to each other for realizing the serial-in/serial-out operations. For the general division, 8-bit divisor is tested by the leading-one detector for selecting the proper control signals among the outputs of 8 PEs. Since each PE contains only few logic gates, as depicted in Fig. 3, the complexity of the proposed divider is naturally relaxed compared to the conventional parallel divider having long and complicated gate networks [3].

III. IMPLEMENTATION RESULTS

For fair comparisons, we implement two different binary dividers; the parallel structure and the proposed serial one. Both architectures are equally synthesized in 130nm CMOS process. Table I summarizes and compares two divider architecture for various bit-widths. Compared to the conventional architecture, it is noticeable that the proposed divider always enhances the area costs as well as the critical delay. For the case of 32-bit

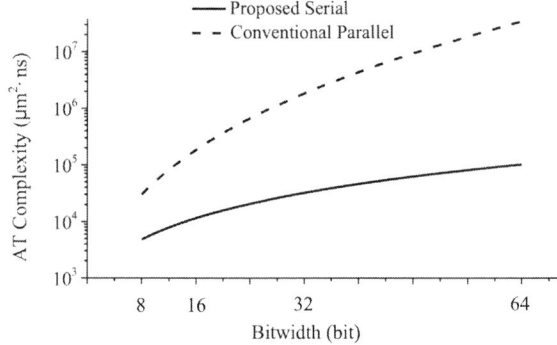

Figure. 5 AT complexities of two different dividers.

division, for example, the proposed divider reduces the silicon area and the critical delay by 29 and 2.6 times, respectively. Fig. 5 shows the area-time (AT) complexity of each divider, where AT complexity can be obtained by multiplying the area by the delay. Note that the proposed divider is superior to the parallel architecture, leading to the cost-effective division operation for serial interfaces.

IV. CONCLUSION

In this paper, we have presented a cost-effective binary divider, which targets the high-speed serial interfaces. Based on the proposed serialized PE-based architecture, the proposed divider relaxes both the hardware complexity and the critical delay compared to the conventional parallel structures.

REFERENCES

[1] F. Hassan, J. L. Magalini, V. C. Pentea, and J. E. Carletta, "A booth-like modulo operator," in *Proc. IEEE Midwest Symp. Circuits Syst. (MWSCAS)*, 2015, pp. 1–4.

[2] P. Saha, A. Banerjee, P. Bhattacharyya, and A. Dandapat, "Vedic divider: Novel architecture (ASIC) for high speed VLSI applications," in *Proc. Int. Symp. Electron. Syst. Design (ISED)*, 2011, pp. 67–71.

[3] I. Koren, *Computer arithmetic algorithms*, 2nd ed., A. K. Peters, Natick, MA, 2002.

[4] J. H. Lee and M. H. Sunwoo, "Low-complexity first-two-minimum-values generator for bit-serial LDPC decoding," *IEEE Trans. Circuits Syst. II, Exp. Briefs*, vol. 63, no. 5, pp. 483–487, May 2016.

[5] P. A. Marshall, V. C. Gaudet, and D. G. Elliott, "Deeply pipelined digit-serial LDPC decoding," *IEEE Trans. Circuits Syst. I, Reg. Papers*, vol. 59, no. 12, pp. 2934–2944, Dec. 2012

Hardware Design Exploration of Fully-Connected Deep Neural Network with Binary Parameters

Jinkyu Kim, Juyeob Kim, Byungjo Kim, Miyoung Lee and Joohyun Lee
Intelligent SoC Research Department
Electronics and Telecommunications Research Institute
Gajeong-ro, Yuseong-gu, Daejeon, 305-700, South Korea
kimjk@etri.re.kr, juyeob@etri.re.kr, kimbj@etri.re.kr, sharav@etri.re.kr, juehyun@etri.re.kr

Abstract— **This paper describes the exploration and analysis to design hardware of the fully connected deep neural network with binary weight value. The fully connected deep neural network is a promising reference model in order to implement fully hardwired classifier in mobile and IoT (Internet of Things) device. So, we analyzed its learning accuracy according to the number of layers and nodes through environment of reference simulation. And we analyzed hardware complexity and usage in terms of FPGA. We used Caffe framework to extract parameter and accuracy as reference model. We used Xilinx Vivado 2015.2 as synthesis tool for hardware design exploration.**

Keywords; DNN; FPGA;

I. INTRODUCTION

DNN (Deep Neural Network) is popular theme to provide image classification, segmentation and object detection for input data of sensor device. But its complexity and memory usage are excessive for mobile and IoT (Internet of Things) devices with the limitation of computation resource and memory capacity. The number of weight value as learning parameter in DNN model is beyond the millions in case of high accuracy CNN (Convolution Neural Network), such as VGG Net and AlexNet [1, 2]. Recently, several models were reported as train and test scheme to decrease computation power and memory access, such as binarization and quantization of weight value [3, 4]. These methods and algorithms should be provided with dedicated hardware in order to achieve efficient operation. So, we analyzed to implement a dedicated hardware with binary weight parameters stored on hardware. We followed the algorithm and the data processing flow of [3] as train and test model. And we applied Caffe framework in order to obtain reference model and test-vector of hardware design.

II. DNN WITH BINARY WEIGHT VALUE

A. DNN Data Flow

DNN operation can be divided into train and inference mode. We deal with the hardware design and architecture about the inference mode among the two modes. In the inference mode, main operation is multiplication and accumulator (MAC) between feature data and weight value in one node unit. Generally, MAC operations are present as the number of node connection between layers. In order to complete overall operation, multipliers are exhaustively placed in its hardware design when we use a formal data flow. It is not appropriate to

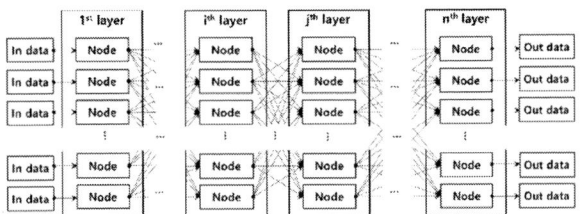

Figure 1. DNN Data Flow

simultaneously operate massive multiplier because of a constrained hardware area and power dissipation. Limitation of external memory access should be particularly avoided by decreasing of quantization bit width of weight value. Feature data can be expressed on bit-width of less than 16 bits without loss of accuracy even though it is a single-precision value that is depended on input data of DNN. Weight values are generated by train procedure and it is not variable in inference phase as Figure 1. Actually, it is difficult to shrink its bit-width in terms with input data, but [3] has shown that weight can be reduced by 1 bit without critical loss of accuracy with their novel learning procedure.

B. Data Flow with Binary Weight

When weight value has binary type, DNN operation can achieve 16 x area reduction in terms of weight value. And it can reduce power dissipation by memory access. Operation of each layer with binary weight value can usually be expressed by (1).

$$O_l = BatchNorm(I_l * W_l) \qquad (1)$$

I_l is input vector of each layer can be performed convolution operation, $*$, with W_l. After that, batch normalization operation should be performed with mean and variance value that are fixed in train phase. Output data vector, O_l, of lth layer should be issued as input data vector for next layer.

III. HW ARCHITECTURE EXPLORATION

A. Reference Model

We analyzed network structure for the fully connected deep neural network from 1 hidden layer to 3 hidden layers with 2048 nodes. We traced learning accuracy with MNIST dataset [5]. So, input layer has 784 nodes in order to process MNIST dataset format that has 28x28 image and gray scale, and output

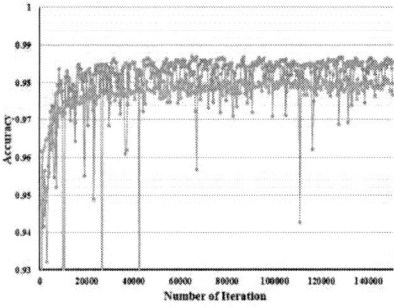

Figure 2. Accuracy in terms with the number of layer

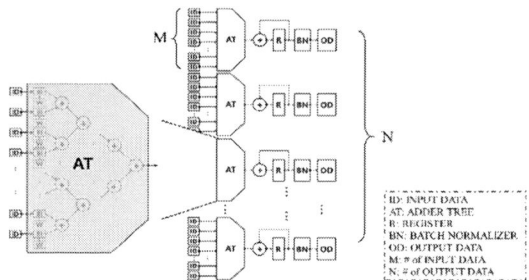

Figure 3. Accuracy in terms with the number of layer

layer has 10 nodes for classification with 0 ~ 9 digits. We utilized the Caffe framework as train tool, a reference model and a software tool for extracting a test vector for hardware design. It was a task to add the training flow of the binary neural network and the regularization of quantization level in part of pre-defined layer model of Caffe framework. So, we got results about loss of accuracy in case of the number of hidden layer as Figure 2. The accuracy of 2 hidden layers and 3 hidden layers are almost 98.5%. The accuracy of 1 hidden layer is almost 97.6 % through 150000 iterations. Generally, in case of MNIST dataset, the accuracy should be achieved by almost 99% in DNN model with single-precision value weight.

B. Design and Architecture

Multiplication between 16 bits input data and 1bit weight value is performed by two cases, which are separated by '-1' and '1'. If the weight value is '1', multiplication can be skipped. In case of weight value '-1', the sign conversion of input data should be performed as Figure 3. The sign conversion is described by the bitwise inverting and one adder. Especially, it is necessary to set up batch normalization operation in data flow of binary neural network. So, we can apply the count number of one addition for sign conversion into the mean and variable value of batch normalization operation. When one adder tree module finishes accumulation of input data as the number of pre-layer node, one node operation of current layer is completed. This operation is repeatedly performed by the number of current layer node.

C. Hardware Synthesis

We synthesized hardware for operating on Xilinx FPGA XZ7045 with system-verilog. We can try to minimize the multiplier in the hardware of the fully connected deep neural network with binary weight value. But, the HW resources, such as look-up table and flip-flop, are not enough to achieve full parallelism even if the number of multiplier can be reduced.

TABLE 1. HW Resource Usage

Input X Output (M × N)	Slice LUT (Available: 218,600)	Slice Register (Available: 437,200)
16 × 16	3, 664 (1.68%)	4.915 (1.12%)
32 × 32	13,951 (6.38%)	17,987 (4.11%)
64 × 64	52,544 (24.04%)	68,805 (15.74%)
128 × 128	203,392 (93.04%)	268,690 (61.46%)

So, in order to get maximum parallelism case for a target FPGA, we conducted to synthesize hardware module according to the complexity of layer with the synthesizer of the Xilinx Vivado 2015.2. Its limitation is the hardware module of 128 × 128 dimension, its operating frequency is almost 250MHz.

IV. CONCLUSION

We analyzed learning accuracy and hardware complexity in terms of the fully connected deep neural network with binary weight value. We checked environment for confirmation of learning accuracy according to quantization in Caffe framework. We conducted hardware design exploration with scalable hardware module.

V. FUTURE WORKS

We will implement the hardware of CNN (Convolution Neural Network) for classification of image. We will utilize hardware of classifier in back-end of CNN model. Especially, in case of VGG Net, it has the weight value of 138 million. Most of weight value is required to perform the fully connected deep neural network as classifier. Therefore, we expect to provide effective approach as candidate model.

ACKNOWLEDGMENT

This work was supported by Institute for Information & communications Technology Promotion (IITP) grant funded by the Korea government (MSIP) (No. R7117-16-0166, Brain-Inspired Neuromorphic Perception and Learning Processor)

REFERENCES

[1] Karen Simonyan and Andrew Zisserman, "VERY DEEP CONVOLUTIONAL NETWORKS FOR LARGE-SCALE IMAGE RECOGNITION," CoRR (2015)

[2] Krizhevsky, A., Sutskever, I., and Hinton, G. E, "ImageNet classification with deep convolutional neural networks," NIPS, pp. 1106-1114, 2012

[3] Courbariaux. M, Bengio. Y, David. J.P, "Binaryconnect: Training deep neural networks with binary weights during propagations," In: Advances in Neural Information Processing Systems. (2015) 3105–3113

[4] Mohammad Rastegari, Vicente Ordonez, Joseph Redmon, Ali Farhadi, "XNOR-Net: ImageNet Classification Using Binary Convolutional Neural Networks," CoRR (2016)

[5] Yann LeCun, Corinna Cortes and Chistopher J.C. Burges, "THE MNIST DATABASE of handwirttern digits,"http://yann.lecun.com/exdb/mnist

A Pre-characterization Method for Multiple Single-Event Transient Analysis in Cell-based Designs

J. K. Park, *Member IEEE*, J.-S. Go and J. T. Kim

School of Electronic and Electrical Eng., College of Information and Communication Eng.

Sungkyunkwan Univ.

Suwon, Rep. of Korea

{jkpark1, jtkim}@skku.edu

Abstract—**This paper briefly presents an efficient pre-characterization method of logic cells for single event transient analysis. Based on the current source model and I-V conversion, it mainly estimates the electrical property of multiple single-event transients (SETs). A combination of multiple SETs can be freely chosen at the analysis stage. This effectively reduces the number of pre-characterizing elements. The accuracy of generating multiple SETs shows 8% mean errors for SPICE-level simulation results in typical CMOS cell libraries.**

Keywords; single event transient; soft error; multiple faults; characterziation; gate-level design

I. INTRODUCTION

Multiple soft errors or soft faults originating from a high-energy particle are increasingly concerning for reliability issues in highly integrated circuits (ICs) with deep sub-micron technologies and very low power devices [1,2]. More frequent and untraceable failures deteriorate product quality in commercial off-the-shelf ICs. For a cell-based design flow, it is common that logic and sequential cells, such as NAND and d-type flip-flops, are pre-characterized by a SPICE simulation prior to the static path analysis [3,4]. The electrical properties, such as the width and amplitude of the target single-event transient (SET) can be picked up or interpolated from these pre-defined values. For a single fault or a SET, only a faulty site can be chosen at any step of the analysis. For cases with multiple faults inside the logic cell, a sensitized output contains the aggregated voltage waveform which is generated by all faulty transistor sources. In principle, every possible combination of these sources should be defined before the analysis of SET.

Our motivation for this work is to exploit advantages of the current source model [5] in characterizing concurrent and multiple SETs. Existing SET sources [3,4] can be modeled by current sources with collected charges. In addition, the resultant output waveform can be generated by adding up each faulty current waveform. In this way, the number of characterizing elements is only proportional to the number of transistors, which correspond to the same order in the conventional single fault analysis [4]. The method is able to every combination of internal SETs. Experiments show that it can reduce 95% of look-up table sizes in full-combination cases while the estimation errors are limited to 8% of SPICE simulation results on average.

This research was supported by the Basic Science Research Program through the National Research Foundation of Korea (NRF) funded by the Ministry of Education (2013R1A1A2060954).

II. MODELING MULTIPLE SETs

A. Characterization of individual SETs

Fig. 1 shows a basic structure of the two-input NAND gate, including three possible SET sources. For a given load capacitance and the input condition of the NAND gate, the fault waveforms $1\rightarrow0\rightarrow1$ and $0\rightarrow1\rightarrow0$ can be generated from the NMOS and PMOS drain regions, respectively. The current source for each faulty site can be defined by:

$$I(q,t) = 2q/\tau\sqrt{\pi}\sqrt{t/\tau} \cdot e^{-t/\tau} \qquad (1)$$

where q, t, and τ are the collected charge, time instant, and the time constant, respectively, for the given technology. The charge q can be determined by the charge distribution that belongs to the given faulty site of the transistor. This information can be obtained by a cell placement [1,2], a design budget or, a constraint. Since the excessive charge distribution for a high-energy particle obeys one of the Poisson distributions [6,7], the charge collection and resulting SET waveform can be generated by sampling and accumulating the total charges involved in the effective area of each faulty site. This can occur once the total generating probability for the specific distribution is determined. Therefore, it is sufficient to separately store the characterized SETs for each transistor without considering the variety of combinations of multiple faults at this stage. The only difference to the conventional

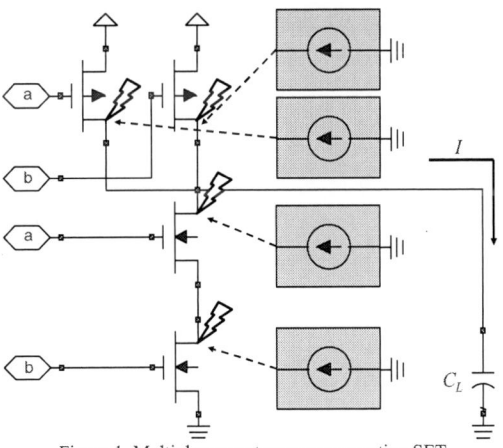

Figure 1. Multiple current sources generating SET

method is that the current samples, not voltage samples, should be stored on the look-up table in the cell library. Depending on the transfer model, the current samples can be further utilized in propagating SETs at an accuracy near SPICE-level.

B. Estimating the output waveform from multiple SETs

As shown in Fig.1, the time-domain current output is combined by the individual SET sources, which are defined by (1). To employ the current sample as an input of the conventional SET propagation procedure [1-4], it should be converted to the voltage domain signals. The following steps summarize the estimation of the output waveform resulting from multiple SETs in the proposed method.

i) Charge distribution and collected charges q_i for individual SETs are generated for the given logic cell and the input bias condition. The corresponding generation probability will also be given by the probability density function of the collected charge q, as in [3].

ii) Obtaining individual SET output waveforms by looking up the pre-stored values presented in Section II.A. Two-dimensional interpolation will also be required to estimate each output for the given load capacitance and q_i.

iii) Summing up all of the excessive current samples obtained in step ii) along the simulation time.

iv) In the case of applying V-to-V transfer models during the SET propagation, the current samples should be converted to the voltage samples for the given load capacitance.

The time complexity for step i) through to step iv) is proportional to the number of current samples. This number is usually limited due to the specified clock period of a sequential design and time resolution; consequently, O(1) is achieved for the above estimation procedure.

III. RESULTS AND DISCUSSION

To validate the present method, we chose several logic cells

in 45nm and 130nm cell libraries. Fig. 2 shows an example of $1 \rightarrow 0 \rightarrow 1$ multiple SETs estimated by the proposed technique. Differences between the estimation and SPICE results are due to calculation errors and linear interpolation of discrete q_i and capacitance samples. The plots with a voltage characterization where two waveforms are synthesized result in a large amount of errors. By investigating 12 different logic cells, 8% mean errors were observed compared to the SPICE result. In Fig. 3, we compared the size of different SET combinations with the case of straightforward conversion in a given logic cell. Note that the NAND_X5 cell has a smaller number of transistors than NAND_X4. This reduces the number of SET instances, accordingly. As the driving strength and the number of transistors increase, the required table size can be reduced by up to 95%.

IV. CONCLUSION

This paper presents an efficient characterization technique for multiple SETs based on the current source model. In conjunction with the existing composite current source models, the proposed method can give an accurate electrical estimation for SETs under gate-level designs compared to the results of SPICE simulation. Memory requirement and time complexity of the estimation procedure are proportional to the number of transistors and time samples, respectively.

REFERENCES

[1] M. Ebrahimi, H. Asadi, R. Bishnoi and M.B. Tahoori, "Layout-Based Modeling and Mitigation of Multiple Event Transients," IEEE Trans. on CAD of Integrated Circuits and Systems, Vol. 35, No. 3, pp. 367-379, 2016.

[2] H.-M. Huang and C. H.-P. Wen, "Layout-Based Soft Error Rate Estimation Framework Considering Multiple Transient Faults—From Device to Circuit Level," IEEE Trans. on CAD of Integrated Circuits and Systems, Vol. 35, No. 4, pp. 586-597, 2016.

[3] J.K. Park, M. Kim and J.T. Kim "Cascaded Propagation and Reduction Techniques for Fault Binary Decision Diagram in Single-Event Transient Analysis," JSTS, accepted for publication, 2016.

[4] B. Zhang, W. Wang and M. Orshansky, "FASER: Fast Analysis of Soft Error Susceptibility for Cell-Based Designs," Proc. of 7th Int'l. Symp. on Quality Electronic Design, pp. 755-760, 2006.

[5] Synopsys, "CCS Noise Technical White Paper," 2005.

[6] F. B. McLean and T.R. Oldham, "Charge funneling in N- and P-type Si substrates," IEEE Trans. Nucl. Sci., Vol. 29, pp. 2018-2023, 1982.

[7] K. M. Warren, "Sensitive Volume Models for Single Event Upset Analysis and Rate Prediction for Space, Atmospheric, and Terrestrial Radiation Environments," Dissertation, Vanderbilt Univ., 2010.

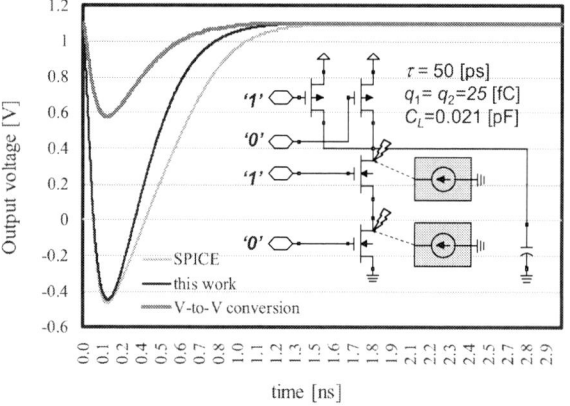

Figure 2. An estimated waveform by the proposed technique

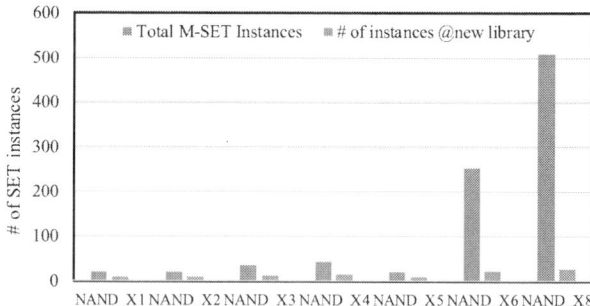

Figure 3. The number of elements required for multiple SETs

Hardware implementation of fast high dynamic range processor for real-time 4K UHD video

Sang-Seol Lee, Eunchong Lee, Youngbae Hwang and Sung-Joon Jang

Korea Electronics Technology Institute
Daewangpangyo-ro 712beon-gil, Bundang-gu, Seongnam-si
Gyeonggi-do, South Korea
{sslee81, elee, ybhwang, sjjang0626}@keti.re.kr

Abstract— **The high dynamic range (HDR) has become very important because of the rapid increase in demand for a variety of applications. However, most of them were implemented by expensive systems due to the high complex computation for processing the real-time 4K UHD video. In the proposed hardware, the non-linear camera response function (CRF) with the area optimization of logarithmic computations has been applied to improve HDR quality. And, for embedding in Field Programmable Gate Array (FPGA), we implement a dedicated hardware using 4006 lookup table (LUT) and 21KB sized internal memory. The proposed architecture enables a real-time HDR processing with pipelining for a UHD video (8 Mega pixels) at 30 frames per second.**

Keywords; High Dynamic Range(HDR); Radiance Recover; Tone-mapping; UHD ISP;

I. INTRODUCTION

The dynamic range (DR) is a widespread in a natural environment. Typical camera has a low dynamic range (LDR) due to the analog to digital converter (ADC) noise level and the 8 to 14bit output range limitation. In recent years, the importance of the HDR video at the same level as the human eye has been increased for surveillance cameras, robot vision and automotive applications [1].

There are two well-known approaches for creating LDR to HDR image. The first method is to develop a sensor capture technic such as self-reset, time-to-saturation and well-capacity adjusting [2]. Second, combine more than two different exposure images to make a single image [3]. Multiple images are captured under various exposure time then taking a shadow area in the over exposure and highlight area in under exposure to combine into a single HDR image which has proper exposure. The HDR rendering method, weighted average [4] and restoring the real radiance from CRF estimation [5] are used. The method of weighted average, there is an advantage that is easy to implement hardware. However, there is a drawback that contrast is broken down because of overall image histogram is biased towards to the middle. The CRF estimation with logarithmic approach needs high complexity and GPU acceleration cannot achieves real-time processing. Therefore, the CRF estimation method has chosen for HDR hardware implementation with the low complexity optimization.

After restoring the real radiance, compression process as tone-mapping is required to fit the output range of the display device. Thus tone-mapping architecture with reasonable complexity include in the proposed hardware.

II. PROPOSED ARCHITECTURE

The main algorithm of proposed architecture is based on the Debevec & Malik [5]. The overall hardware structure is shown at Fig. 1. There are five main blocks in the proposed HDR hardware architecture. The LUT of CRF control and weight generator are configured in parallel to handle two different exposure images at the same time. The radiance calculator that perform complex operation increase operating frequency by the pipeline technique. The LUT size reduction is applied to log lookup table and tone-mapping. Detailed architecture for proposed hardware architecture will explain in this section.

A. CRF(Camera Response Function) Estimation

The CRF estimation is needed for radiance recover because the digital camera image sensor has a non-linear response curve. The CRF values are calculated as estimation that the value to find minimum error by the singular value decomposition from the same scene of 10 sample images having different exposure. Due to the large computational complexity, the CRF estimation was calculated from off-chip in matlab tool. The calculated CRF values are converted to 21-bit (1-bit sign, 3-bit integer, 17-bit fraction) data and stored in 4096x21bit LUT.

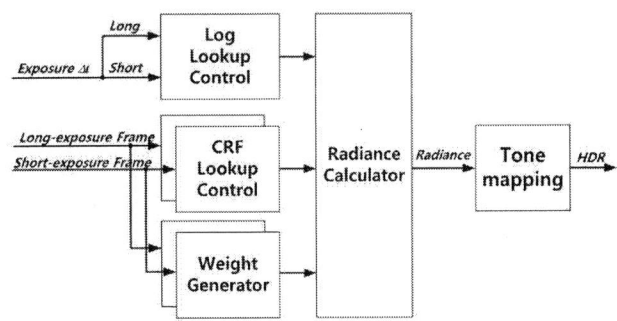

Figure 1. Top module architecture

B. Logarithmic Exposure Time Calculation

For the radiance recover, it is necessary to convert the long and short exposure time to logarithmic scale. To reduce the hardware complexity, the logarithmic conversion was replaced by the LUT. Also, only 6 kind of selected shutter speeds are stored in the LUT which is mainly used in commercial camera as 1/30, 1/60, 1/125, 1/250, 1/500 and 1/1000 sec. For real-time image processing, individual log exposure LUT used for each shutter speed [6]. However the exposure time is never changed within the same frame so, only one LUT is used for configuration of long and short exposure time in our hardware.

C. Tone-mapping

To display the HDR image, the recovered radiance need to be a compression process to fit into output range of display device. Both radiance calculator output and tone-mapping input are the natural logarithmic value, since it does not require pre-processing process, we selected [7] as tone-mapping algorithm (1).

$$(\ln E - \ln I_{min})/(\ln I_{max} - \ln I_{min}) = HDR \qquad (1)$$

I_{min}, I_{max} denotes the minimum maximum values of pixel within the frame, $\ln E$ is a recovered radiance. Assuming only little scene is changed in the connected sequence of video, the value of I_{min}, I_{max} is replaced from previous frame for the real-time video processing. In Fig. 2, to reduce hardware complexity, previously calculated reciprocal of the denominators are stored in the LUT. Moreover, we reduce the size of LUT that the value of 0 to 4095 was converted to 64 levels.

III. EXPERIMENTAL RESULTS AND IMPLEMENTATION

The proposed hardware was implemented in Verilog-HDL and was simulated with 12bit Bayer raw data, captured by digital camera. The simulation source and result are shown at Fig. 3. In this paper, we propose an area-optimized and high performance HDR hardware that includes radiance recover and tone-mapping. Finally, a dedicated hardware IP is synthesized by the place and route tool using Xilinx ISE for xc6vlx7601 FPGA. The proposed hardware is working on maximum frequency at 283 MHz. It is enough to generate the real-time HDR 4K image 30 frames per second on operating frequency at 258 MHz.

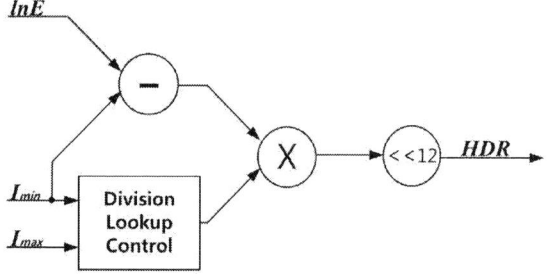

Figure 2. Tone-mapping architecture

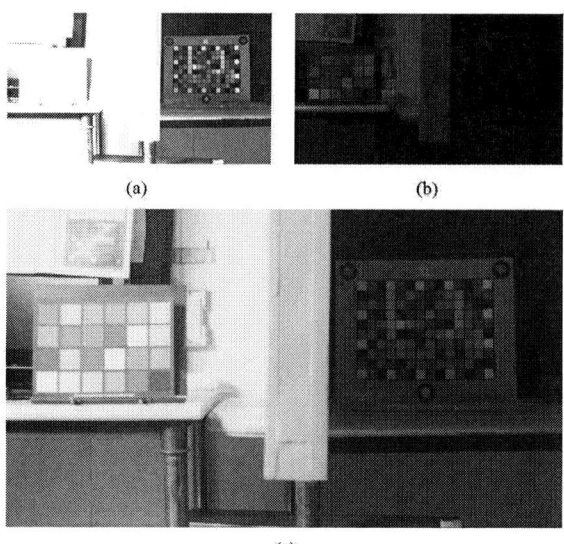

Figure 3. HDR process;
(a) Long Exposure image, (b) Short Exposure image, (c) HDR Results

TABLE I. IMPLEMENTATION RESULTS

Logic utilization	Used		Utilization
Number of Occupied Slices	1194		1.0%
Number of FIFO/BRAMs	5		0.7%
Complexity distribution	LUT	FF	Memory
Radiance Calculator	3627	2493	-
Log Lookup Control	60	23	
CRF Lookup Control	-	-	21KB
Weight Generator	152	96	-
Tone-mapping	231	147	-

ACKNOWLEDGMENT

This work is supported by the Industrial Core Technology Development Program of MOTIE/KEIT, KOREA. [10060387, Development of realtime processing ADAS algorithm IPs based on HD Camera image for Autonomous driving car] .

REFERENCES

[1] A. Darmont, High Dynamic Range Imaging; Sensors and Architectures, SPIE PRESS, Bellinham, WA, USA, 2012.

[2] S. Kavusi, A. Gamal "A quantitative study of high dynamic range image sensor architecture," SPIE. Proc. Electronic Imaging. vol. 5301, pp. 264–275, 2004.

[3] S. Mann, R. W. Picard "On being 'undigital' with digital cameras: extending dynamic range by combining differently exposed pictures," Proc. S&T 46th annual conference, pp. 422–428, 1995

[4] K. Yamada, T. Nakano, S. Yamamoto "A Vision sensor having an expanded dynamic range for autonomous vehicles," IEEE. Trans. Vehicular Technology. vol. 47, no. 1, pp. 332–341, February 1998.

[5] P. E. Debevec, J. Malik "Recovering high dynamic range radiance map from photographs," Proc. 24th annual conference ACM SIGGRAPH, pp.369–378, 1997.

[6] P. Lapray, B. Heyrman, D. Ginhac "HDR-ARtiSt: an adaptive real-time smart camera for high dynamic range imaging," Real-Time Image Processing. pp.1-16, 2014.

[7] J. Duan, M. Bressan, C. Dance, G. Qiu "Tone-mapping high dynamic range images by novel histogram adjustment," Pattern Recognition. vol. 43, issue. 5, pp.1847-1862, 2010.

A low power, high speed FinFET based 6T SRAM cell with enhanced write ability and read stability

Rahaprian Mudiarasan Premavathi, Qiang Tong, and
Ken Choi

Department of Electrical and Computer Engineering
Illinois Institute of Technology, Chicago, USA
kchoi@ece.iit.edu

Yunsik Lee

School of ECE
UNIST, Ulsan, Korea
leeys@unist.ac.kr

Abstract— **This paper presents a FinFET based 6T SRAM cell, with separate read access path and write path, designed by combining the advantages of conventional single ended 5T and the conventional 8T SRAM cells. The proposed SRAM cell achieves 70% and 55% of write performance improvement in terms of Power delay product (PDP) than 8T (also conventional 6T) and 5T SRAM cells respectively. Proposed cell achieves 78% of hold 1 and 40% of hold 0 static power reduction than the conventional 5T, 6T and 8T cells. The proposed cell is read SNM free and also achieves better hold SNM and write ability than 5T and 8T SRAM cells.**

Keywords — Read SNM free, Power Delay Product, Write Trip Point.

I. INTRODUCTION

Leakage power, dynamic power, read and write stability have become the major concern of SRAM design now a days. Major portion of the dynamic power is consumed during write operation as the write bit lines are confined to have full voltage swing between Vdd and 0. A single-ended asymmetric 5T SRAM cell overcomes this problem and is write stable [1] [2]. The read stability of these cells decline compared to conventional 6T cell. A conventional 8T SRAM cell is read SNM free, but lacks in write stability and performance [9]. The aim of this paper is to offer improved read, write stability and low power altogether.

A tied-gate FinFET based 6T SRAM cell with separate read and write paths has been proposed in this paper. This proposed SRAM cell offers low power with enhanced write ability and is read SNM free. Static power is reduced by imposing negative word line scheme over both read and write word lines [3].

The rest of the paper is organized as follows. Section II introduces the proposed single-ended 6T SRAM cell. Section III presents the simulation and comparison results of the proposed 6T SRAM cell with the conventional single ended 5T SRAM [2], conventional 8T SRAM [4] and conventional 6T SRAM. Section IV provides the conclusion of the work.

II. PROPOSED 6T SRAM CELL

The Fig 1. shows the proposed SRAM cell. The Proposed cell uses only one driver transistor (M3), unlike conventional one. Minimum sizing (1 fin) is followed all over except M2 (4 fins). M0, M1, M2, M3 and BL (write bit line) form the write path whereas M4, M5, M6 and RBL (Read BL) form the read path. BL is pre-discharged before each write operation.

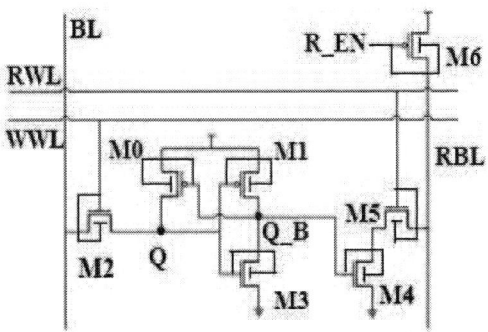

Fig. 1.   Proposed Single-ended tied-gate FinFET based 6T SRAM cell.

A. Write Operation

Due to large number of fins in M2 the write margin and write speed increases. It has only one write bit line (BL) to avoid the full voltage swing of the BL during each write operation, so as to reduce the dynamic power. For write operation, initially BL is pre discharged to 0V. Then if the data to be written is 1, bit line BL is charged to Vdd via write block. Power consumption is very less for write 0, as BL is already pre discharged to 0V.

B. Read Operation

Initially during off mode the Read Enabled Signal (R_EN) is enabled and hence read bit line (RBL) is charged to 0.4V. During read mode M5 is switched on by asserting the read word line (RWL). Depending on the value of Q_B, M4 is either switched on or off, thus discharging or retaining the RBL accordingly. Then depending on RBL value the current sense amplifier senses the Q value and outputs it. RWL and WWL are maintained at negative bias (-0.3V), during write (to reduce the leakage through read path) and read operation (to reduce the leakage in write path) respectively. Proposed cell is read SNM free it has separate read and write paths.

III. SIMULATION RESULTS

The proposed and conventional 5T, 6T, 8T SRAM cells are simulated with Predictive Technology Model FinFET HP device models at room temperature of 25°C and Vdd as 0.4V.

A. Write Results

Fig 2. shows the write waveform of proposed cell. Din is the data sequence (111000). Table I and Table II gives Write Dynamic Power and delay Comparison

978-1-5090-3220-4/16 $31.00 © 2016 IEEE   311   ISOCC 2016

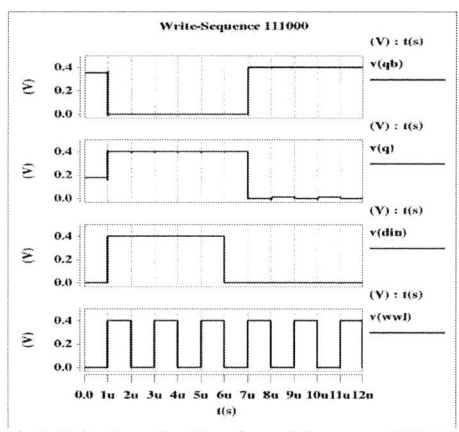

Fig 2. Write Operation Waveform of the proposed SRAM

TABLE I   WRITE DYNAMIC POWER COMPARISON

Operation	Conventional 8T/ 6T (nW)	5T SRAM (nW)	Proposed SRAM (nW)
Write1(on0)	396.2	395.7	115.8
Write1(on1)	2.443	1.438	5.572
Write0(on1)	164.5	117.7	118.3
Write0(on0)	2.484	2.872	1.600

TABLE II   WRITE DELAY COMPARISON

Operation	8T/ 6T (ps)	5T (ps)	Proposed (ps)
Write1	51.22	12.45	54.09
Write0	429.2	344.0	282.7

The proposed cell consumes very less power for writing 0 as pre-discharging the BL avoids full voltage swing of bit line unlike other cells. The proposed cell writes 0 much faster due to its pre-discharge scheme. Table III and Table 1V give the static power (hold 0 and hold 1) and Power delay product comparisons respectively.

TABLE V   STATIC POWER COMPARISON

Operation	8T (nW)	5T (nW)	Conventional 6T (nW)	Proposed (nW)
Hold 1	2.872	2.874	2.872	0.6052
Hold 0	2.864	2.873	2.864	1.701

The use of negative bias technique for RWL and WWL and decrease in number of connection paths to ground are the main reasons for reduction in static power of the proposed cell.

TABLE IV   POWER DELAY PRODUCT COMPARISON

Operation	8T (e-17 Ws)	5T (e-17 Ws)	Conventional 6T (e-17 Ws)	Proposed (e-17 Ws)
Write	3.396	2.306	3.396	1.015

Thus the proposed cell achieves better write performance (PDP). This is due to the use of Single-ended write scheme and separate read and write paths.

B. Hold SNM Analysis

The Hold SNM of the proposed (Fig. 3) and 8T SRAM cells, obtained from the butterfly curve at 25°C is 0.195 V and 0.155 V respectively which means that the proposed SRAM cell offers better hold SNM than 8T SRAM cell.

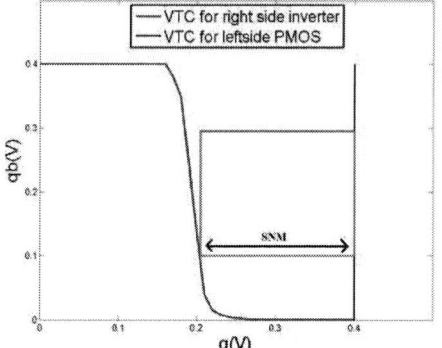

Fig 3. Hold SNM Butterfly curves for the proposed cell

C. Write Ability Analysis

Write ability is measured using Write Trip Point (WTP) analysis. On increasing the BL voltage from 0 to Vdd, the BL voltage where Q and QB flip is WTP. The write trip point of the proposed (Fig 4) and the 5T SRAM cells are 0.094V and 0.129V respectively. Thus the proposed cell is more stable (write).

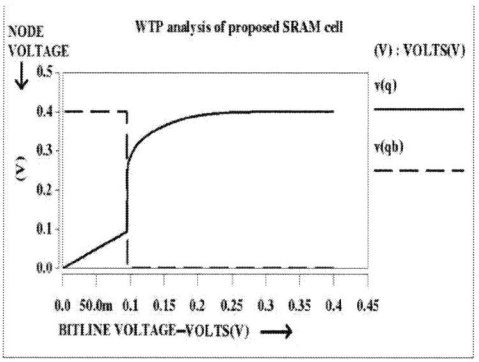

Fig 4. Write Trip Point analysis of the proposed 6T SRAM

IV.   CONCLUSION

In this paper a low power, highly stable 6T FinFET based SRAM operating in sub-threshold region is presented. The proposed cell achieves lower dynamic and leakage power due to single-ended write bit line scheme and the presence of only one driver transistor respectively. The proposed cell also achieves better hold SNM, write ability and is read SNM free due to separate read and write paths and greater pull up ratio.

REFERENCES

[1] Y. Chang, F. Lai, C. Yang, "Zero-aware asymmetric SRAM cell for reducing cache power in writing zero," IEEE Trans. Very Large Scale Integr. (VLSI) Syst., vol. 12, no. 8, pp. 827-836, Aug. 2004

[2] I. Carlson et al., "A high density, low leakage, 5T SRAM for embedded caches," European Solid-state Circuits, 2004, ESSCIRC 2004. Conference on, pp. 215-218, Sept 2004.

[3] C. Wang, C. Lee, W. Lin, "A 4-kb low-power SRAM design with negative word-line scheme," IEEE Trans. Very Large Scale Integr. (VLSI) Syst., vol. 54, no. 5, pp. 1069-1076, May. 2004

[4] M. Turi, J. Delgado-Frias, " An evaluation of 6T and 8T FinFET SRAM cell leakage currents," in Proc. IEEE 57th Int. Midwest Symp. Circuits Syst., pp. 523-526, Aug 2014.

A Dual-Retention Time Architecture towards Secure and High Performance STT-RAM Main Memory Subsystem

Taemin Lee
Department of Computer Science and Engineering
Seoul National University
Seoul, Korea

Sungjoo Yoo
Department of Computer Science and Engineering
Seoul National University
Seoul, Korea

Abstract— **Spin-transfer torque RAM (STT-RAM)-based main memory can suffer from a security problem due to its non-volatility. Data encryption, which can be applied to resolve this problem, can give significant performance degradation due to the additional latency of decryption/encryption in main memory accesses. We propose an STT-RAM architecture where two regions have different retention times. A small short retention-time region, which works as a write-through cache, keeps data unencrypted thereby avoiding the additional latency. The short retention time also makes the memory chip secure by decaying data fast in case of sudden power off. The large long retention-time region contains the original version of main memory data in the encrypted form thereby providing full security. Experimental results, with SPEC2006 benchmarks, show that the proposed method offers 17.1%~36.3% improvement in memory access latency. For 10 memory-intensive programs, we can obtain average 15.3% improvement in total execution cycles.**

Keywords; STT-RAM; main memory; security

I. INTRODUCTION

Spin-transfer torque RAM (STT-RAM) is considered to be a strong candidate of large capacity cache and main memory as DRAM faces its scaling limit. However, STT-RAM suffers from security problems due to its non-volatility. For instance, assume that a stolen mobile device has STT-RAM main memory. In such a case, malicious users can detach the memory chip from the original board and install it on their own board. Then, they can retrieve all the confidential data resident in the chip. In order to resolve the security problem, we can make the data in the STT-RAM encrypted all the time.[1]

The encryption-based solution can suffer from a significant performance overhead because every access to the STT-RAM main memory. In order to resolve the above security-related performance problem, we propose a dual-retention time STT-RAM architecture. The proposed memory architecture consists of two types of memory regions each having a different retention time. The short retention time region (called SR region) plays the role of write-through cache for the long retention time region (called LR region).

The concept of multi-retention time STT-RAM is not new [1]. However, to the best of our knowledge, our proposed method is the first approach to apply the concept of multi-retention time to resolve the security-related performance

---

[1] There are two methods of encrypting main memory contents, during runtime or before system shutdown. In the security attack scenario, we cannot apply the second method due to sudden power off.

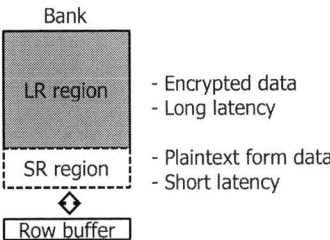

Figure 1 Dual-retention STT-RAM

problem in the STT-RAM-based main memory while the existing work aims at reducing the overhead of high write power and latency in STT-RAM cache.

II. PROPOSED IDEAS

A. Preliminary: Encryption for Security

A symmetric key encryption method, Advanced Encryption Standard (AES) is widely used for memory/storage security [2]. AES generates encrypted data for a 512b cache block in the cipher block chaining mode. AES is typically applied at the granularity of 128b data. Thus, a cache block is first decomposed into four chunks of 128b. The encrypted data of the previous data chunk is used as an input to the encryption of the next data chunk. After four runs of AES algorithm, the four encrypted data chunks are obtained and stored in the main memory.

Each AES algorithm run takes multiple (10~16) rounds to generate an encrypted data. The number of rounds varies depending on the length of cipher key (128b, 192b or 256b). Each round consists of four (or three in the final round) transformation steps of SubBytes, ShiftRows, MixColumns, and AddRoundKey. It is reported that a step of AES decryption run takes about 64ns of decryption latency in the hardware implementation of encryption engine at 0.13μm technology. Thus, the total decryption latency of a cache block can amount to 256ns. Considering the state-of-the-art technology (32nm or 28nm), the decryption latency (4x less than 256ns, i.e., 64ns) is larger than the typical memory access latency (15~30ns in DRAM).

B. Dual-retention STT-RAM

Our basic idea to resolve the performance problem is to prepare a small SR region on the STT-RAM main memory

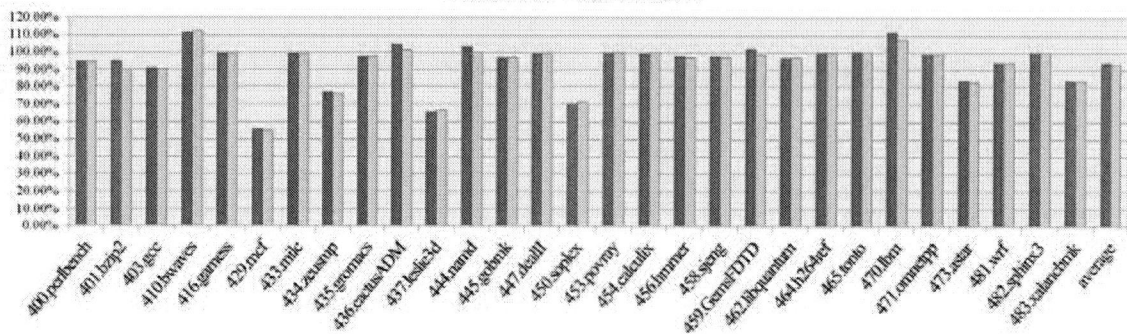

Figure 2 Total execution cycle (SR region size=256MB)

and keep decrypted data on the SR region while the large LR region keeps all the data encrypted.

Figure 1 illustrates the proposed dual-retention STT-RAM structure having both SR and LR regions. We assume a fraction of STT-RAM cells (a subset of rows in an STT-RAM bank) are fabricated for short retention time while keeping most of existing architecture, circuit and layout unchanged. STT-RAM cells with a short retention time can be implemented in a similar way to the case of quasi-volatile STT-RAM cache. e.g., by controlling the height of tunneling layer in the magnetic tunneling junction. Note that each STT-RAM bank consists of LR and SR regions. Thus, both regions share a per-bank row buffer

In order to realize the proposed dual-retention STT-RAM, we need to address two issues, refreshing data in the SR region and reducing the overhead of accessing the LR region. The size of SR region needs to be small because too large SR region decreases the effective capacity of main memory chip. However, the small SR region will not be effective in reducing average memory access latency, especially, for large-footprint programs which can incur high miss rates on the SR region. In order to address the two issues, we propose the following ideas.

- Speculative decryption to reduce average read latency
- Selective and runtime adaptive refreshes of SR region data to adjust the level of speculative decryption and reduce the refresh overhead

III. EXPERIMENTS

We enhanced a multi-core simulation environment called McSim [3] to incorporate our model of STT-RAM main memory subsystem. Table 1 shows the architectural parameters used in the experiments. The STT-RAM has the same parameters as a typical DDR2 DRAM. The energy model includes the energy consumption in tag access (obtained from the 32nm LOP model in CACTI) as well as STT-RAM energy consumption [4]. Note that, in SR region, we use the data units of 256B (1/4 size of row in LR region) through a sensitivity analysis of performance on the data unit size. We run SPEC2006 benchmarks on the simulation model

Figure 2 gives the comparison of total execution cycles normalized to the baseline when the SR region size is 256MB. Our proposed method (the dynamic version) gives average 6.7% improvement compared with the baseline. However, regarding 10 memory-intensive programs, performance improvement amounts to average 15.3%.

Our proposed method incurs additional energy consumption in keeping data in SR region. The dynamic version of proposed method gives average 14.6% overhead in energy consumption compared with the baseline. The additional energy consumption results from additional copy between LR and SR regions and refresh operations.

IV. CONCLUSION

In this paper, we proposed a dual-retention STT-RAM which can resolve the security-related performance problem. We also proposed speculative decryption and selective refresh.

V. ACKNOWLEDGEMENT

This work was supported by Research Resettlement Fund for the new faculty of Seoul National University and the IT R&D program of MKE/KEIT (No. 10041608, Embedded System Software for New Memory based Smart Devices).

REFERENCES

[1] Z. Sun, et al., "Multi Retention Level STT-RAM Cache Designs with a Dynamic Refresh Scheme," Proc. MICRO, 2011.

[2] N. I. of Science and Technology, FIPS PUB 197: Advanced Encryption Standard (AES), November 2001.

[3] Manycore simulation infrastructure, available at http://cal.snu.ac.kr/mediawiki/index.php/McSim.

[4] C. W. Smullen, et al., "The STeTSiMS STT-RAM Simulation and Modeling System," Proc. ICCAD, Nov. 2011.

[5] A. Jog, et al., "Cache Revive: Architecting Volatile STT-RAM Caches for Enhanced Performance in CMPs," Proc. DAC, 2012.

Table 1 Architectural parameters

Core	x86, 2GHz, in-order
L1 cache	32KB/32KB I/D cache, 4-way, 1 cycles
L2 cache	512KB, I/D shared, 16-way, 6 cycles, 64B line
STT-RAM	2GB, DDR2-667, 16b, 8 banks, t_{RCD}=t_{RP}=6, t_{CL}=5 cycles
	LR region: E_{read}=1.27pJ/b, E_{write}=1.40pJ/b [4]
	SR region: E_{read}=1.27pJ/b, E_{write}=0.47pJ/b [5][2]
	SR region size=32MB~256MB
	Retention time (RP) in SR region=64ms
	Data unit size of SR region=256B
	Prefetch method=strided prefetch

---

[2] We assumed that SR cells have 1/3 of write power of LR cells based on [5] while both types of cells have the same read energy since read energy is dominated by memory array structure.

Selective Refresh to Avoid Read Disturb Errors in STT-RAM Main Memory

Taemin Lee
Department of Computer Science and Engineering
Seoul National University
Seoul, Korea

Sungjoo Yoo
Department of Computer Science and Engineering
Seoul National University
Seoul, Korea

Abstract— **Spin-Transfer Torque RAM (STT-RAM) has read disturb error problems where read operations can flip bits in the cell. The problem is expected to become critical as the write current is being significantly reduced and becomes comparable to the read current. In this work, we propose a method called selective refresh which corrects read disturb errors and writes back the correct data only when the number of row activations reaches a threshold. We also propose a low cost implementation of selective refresh which keeps an activation counter table in the memory controller to avoid per-row counters in the memory. Experimental results show that the proposed method can significantly (by average 26.6% with an 8K entry table in case of read disturb error rate of 10-12) reduce the overhead of STT-RAM writes compared with the existing method that performs write-back on every row activation. The proposed method is expected to enable an important trade-off between the cost of cell reliability enhancement and write overhead in the STT-RAM design.**

Keywords; STT-RAM; main memory; read disturb error

I. INTRODUCTION

Spin-Transfer Torque RAM (STT-RAM) is considered to be a strong candidate to replace DRAM in the future since DRAM is facing scaling limits in sub-20nm technology. For mass production, however, STT-RAM has still technical challenges in material, circuit and architecture. Especially, write current reduction (for the cell) and write capability enhancement (for the cell access transistor) is critical for low power and high performance operation [1]. Reduced write current incurs a new problem called read disturb where repeated read operations flip data bits incurring bit errors [2].

A key question is how much error rate the read operation can incur. According to a recent study [2], the read disturb error rate at the worst case STT-RAM cells can reach up to 10-4 at 22nm technology. In this paper, we present a low-cost solution to resolve the read disturb error problem. Our basic idea is to selectively refresh frequently accessed (i.e., read) data. By doing that, we can avoid the accumulation of read disturb errors thereby avoiding the overhead of alternative solutions, e.g., high area/performance overhead in the strong error correction-based methods, or too frequent write-backs in the conservative methods [3].

II. READ DISTURB ERROR REDUCTION SCHEME

A. Read Disturb Problem

Like the conventional DRAM, we assume that STT-RAM consists of banks. Each bank has rows and a row buffer (sense amplifiers and latches).

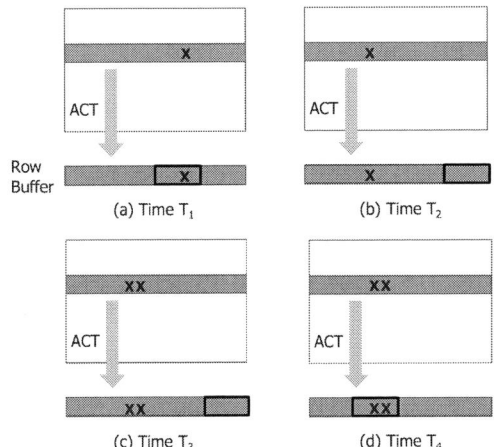

Figure 1 Read disturb problem

When reading data from the STT-RAM, an activation (ACT) a read (RD) and a precharge (PRE) commands are used to get a data. Note that, in case of write, the write operation from the row buffer to the original row in the bank is performed just before the precharge is performed. Thus, there are two types of precharge: simple precharge (PRE) and precharge after write-back (PRE_WB).

Figure 1 shows a scenario where an STT-RAM row is activated multiple times to serve read requests. We assume that STT-RAM is equipped with 1b ECC as in the conventional DRAM subsystem. In the case that read disturb errors occur on the data which are not accessed by RD/WR commands, bit errors can accumulate themselves over time finally yielding un-correctable errors as Figure 1 (b)-(d) shows. At Time T2 (Figure 1 (b)), assume that a read disturb error occurs at a location which is not accessed by the RD command. The error remains in the row after precharge since the PRE command does not change the row contents, but precharges bitlines. At time T3 (Figure 1 (c)), another read disturb error occurs, but it is not corrected since its location is not accessed in the row

978-1-5090-3220-4/16 $31.00 © 2016 IEEE   315   ISOCC 2016

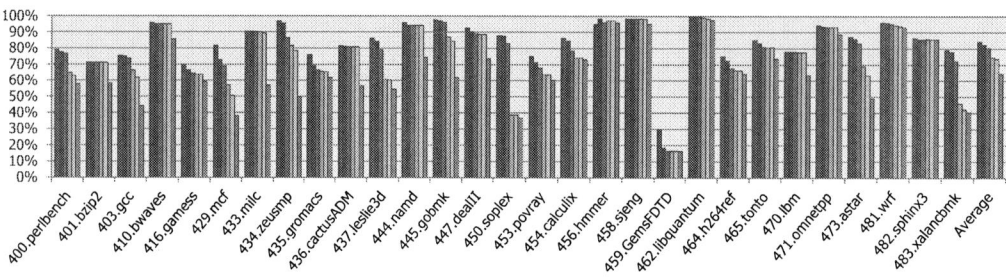

Figure 3 STT-RAM writes (normalized to the baseline): 512B row, read disturb error rate of 10^{-12}

buffer. Finally, at time T4 (Figure 1 (d)), when the data with two bit errors are accessed, those errors are uncorrectable since we can correct only one bit error.

B. Selective Refreshes

Our basic idea is to perform refresh when the row reaches its limit of row activations, e.g., 4 row activations for the 512B row. The refresh represents correcting errors, if any, in the data and writing them back to the associated row. We assume that 1b ECC is implemented on the STT-RAM as in the case of ECC-equipped DRAMs. For the refresh of currently activated row, we use the command of precharge with write-back (PRE_WB).

Figure 2 shows how the selective refresh scheme works. On each row activation, the memory controller invokes the function SelectiveRefresh(RA) where RA is the address of activated row. When invoked, the function first increments the row activation counter of the associated row, counter[RA]. If the counter reaches the threshold (TH) of row activations, then the memory controller performs refresh for the row. After the refresh is completed, the corresponding counter is reset..

```
SelectiveRefresh (RA)
1   Increment counter[RA]
2   If counter[RA] == TH,
3       Perform refresh for RA
4       Reset counter[RA] after the refresh is completed
```

Figure 2 Selective refresh scheme

The selective refresh scheme assumes that each row has its own counter of row activation. There are two issues with this implementation. First, the counter incurs area overhead. Second, the counter updates (incrementing and resetting the counter) incur additional write operations to the associated row. For our low cost implementation of selective refresh, the memory controller has an activation counter table (in short, AC table) which is a list of tuples, <address of activated row, row activation counter>.

III. EXPERIMENTS

We use an architecture simulation model called McSim [4]. It is based on the Pin environment and can simulate x86 cores and memory hierarchy. Table 1 shows the parameters of our architecture used in the experiments. We use SPEC2006 benchmarks. We run the simulation of 1B instructions.

We target the memory reliability requirement of 10^{-15}, and use the read disturb error rate of 10^{-12}. We vary the row size between 256B and 512B and derive the threshold of maximum

row activations. We compare our proposed method with the baseline in [3] (refresh on every row activation).

Figure 3 shows STT-RAM writes normalized to the baseline for the case of 512B row. In the case of infinite size AC table, the selective refresh reduces STT-RAM writes by average 35.6%. The AC table size of 8000 entries gives average 26.6% reduction in STT-RAM writes. In the case of *GemsFDTD*, however, the small AC table of 500 entries gives 83.5% reduction in STT-RAM writes. It is because this program has a small number of row activations. In addition, energy consumption is reduced to 18.9% on average.

IV. CONCLUSION

In this paper, we presented a concept of selective refresh to address the read disturb problem in STT-RAM. We also present a low cost implementation of selective refresh which reduces the area overhead.

V. ACKNOWLEDGEMENT

This work was supported by Research Resettlement Fund for the new faculty of Seoul National University and the IT R&D program of MKE/KEIT (No. 10041608, Embedded System Software for New Memory based Smart Devices).

REFERENCES

[1] A. Driskill-Smith, et al., "Latest Advances and Roadmap for In-Plane and Perpendicular STT-RAM," Proc. Int'l Memory Workshop, 2011.

[2] A. Raychowdhury, et al., "Design Space and Scalability Exploration of 1T-1STT MTJ Memory Arrays in the Presence of Variability and Disturbances," Proc. IEDM, Dec. 2009.

[3] R. Takemura, et al., "Highly-scalable Disruptive Reading Scheme for Gb-scale SPRAM and beyond," Proc. Int'l Memory Workshop, 2010.

[4] Manycore simulation infrastructure, available at http://cal.snu.ac.kr/mediawiki/index.php/McSim.

[5] C. W. Smullen, et al., "The STeTSiMS STT-RAM Simulation and Modeling System," Proc. ICCAD, Nov. 2011.

Table 1 Architectural parameters

Core	x86, 2GHz, in-order
L1 cache	32KB/32KB I/D cache, 4-way, 1 cycles
L2 cache	1MB, I/D shared, 16-way, 6 cycles, 64B line
STT-RAM	2Gb, DDR2-667, 32b, 8 banks, t_{RCD}=t_{CL}=t_{RP}=6 cycles, t_{RP_WB}=12 cycles E_{read}=1.27pJ/bit, E_{write}= 1.40pJ/bit [5][1]

---

[1] We use the energy parameter of 32MB memory in [5] since a bank has the capacity of 32MB in our STT-RAM.

Design of eMMC Controller with Multiple Channels

Chulhoon Kim
Department of Electronic Engineering
Soongsil University
Seoul, Korea
ch.kim@ssu.ac.kr

Chanho Lee
Department of Electronic Engineering
Soongsil University
Seoul, Korea
chlee@ssu.ac.kr

Abstract— **Embedded multimedia card (eMMC) is expected to replace secure digital (SD) card which is widely used for external memory and to be used widely in the embedded systems due to the improved performance and package. In this paper, we propose architecture of eMMC controller with multiple channels. It is connected to a host system using an AXI master interface for data transfer and an APB slave interface for writing command and reading responses and status. The interface for eMMC devices has multiple channels for multiple devices and each channel can be enabled so that multiple processors can request memory access. An eMMC controller is designed based on the proposed architecture using Verilog-HDL and is implemented using an FPGA.**

Keywords – eMMC; Multi-Channel; AXI; APB; Verilog-HDL; DMA Controller; SoC

I. INTRODUCTION

Embedded multimedia card (eMMC) [1] is a high performance non-volatile memory which includes NAND flash memory devices and a controller, and is expected to replace secure digital (SD) cards. Multiple channels can increase the data bandwidth since the data width of an eMMC device is limited to 8 bits. The simultaneous operation of multiple channels is not effective for multiple requests of multiple processors although it increases the data bandwidth.

In this paper, we propose an eMMC controller architecture with multiple channels for multiple requests of multiple processors. It consists of an application adaptor and eMMC adaptors. The application adaptor includes an AMBA APB slave interface [2] and an AMBA AXI master interface [3]. The APB interface is used for configuring the controller, writing eMMC commands and reading eMMC responses. The AXI interface is used for the data transfer. The application adaptor enables the eMMC adaptors. The eMMC adaptor generates commands, controls data transfer, and analyzing responses. The proposed eMMC controller architecture is designed using Verilog-HDL, and is implemented using an FPGA.

II. ARCHITECUTE OF PROPOSED CONTROLLER

A. Application adaptor

The APB slave interface of the application adaptor is used by a host system to configure the controller, writing eMMC commands and reading eMMC responses. It also includes command and argument queue so as to issue the next command without latency when the controller receives a response. The AXI master interface includes a direct memory access controller (DMAC) which reads and writes data in a host system. It sends enable signals to eMMC adaptors, and divides and combines data when more than one channels are enabled. It can distinguish the commands and the responses of the multiple channels. Other functions are similar to that of the controller with a single channel [4] except for the added functions.

B. eMMC adaptor

The eMMC adaptor is the core component of the controller. A channel controller is connected to one eMMC device to transfer data. The channel controller can be enabled or disabled by a host system through the APB interface. It issues commands at the right timing and receives responses to analyzed and to deliver them to the APB interface. It has a buffer to hold a block of data which is the minimum transfer unit between an eMMC adaptor and a device. Since the DMAC also has a buffer, we can increase the performance using double buffering [5]. The basic structure is similar those of the controller with a single channel [4]. Fig. 1 shows the block diagram of the proposed eMMC controller with multiple channels.

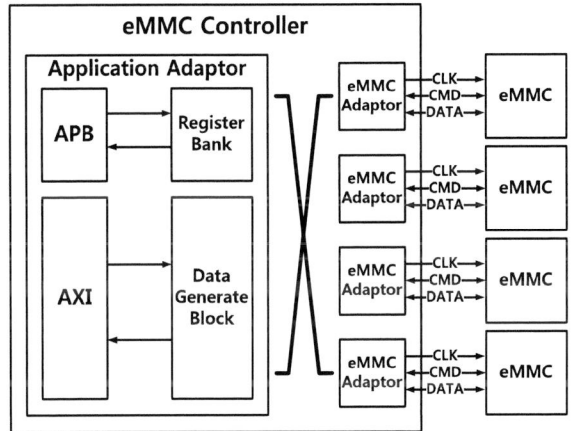

Figure 1. Block diagram of eMMC controller with multiple channels.

III. OPERATION OF CONTROLLER

A processor should configure the eMMC controller with multiple channels to determine which channel is enabled before

---

This work was supported by the National Research Foundation of Korea (NRF) grant funded by the Korea government (MOE) (NRF-2016R1D1A1B01008846) and the Industrial Core Technology Development Program (10049192, Development of a smart automotive ADAS SW-SoC for a self-driving car) funded by the Ministry of Trade, industry & Energy. The EDA tools were supported by IDEC.

978-1-5090-3220-4/16 $31.00 © 2016 IEEE

it requests a memory access. If all the channels are enabled and a memory access is requested, the application adaptor divides the data by 8 bits and distributes them to all the channels for writing request, or collects 8 bit data form each channel and combines them. The data bandwidth increases by the number of enabled channels. Multiple eMMC devices act as a single device with the increased transfer rate.

Processors can write data to specific eMMC devices since the controller transfers data to eMMC devices which are connected to enabled channels only. We can assign a channel or channels to a processor so that the processor can have a dedicated eMMC device. A processor enables its channel before it transfers data normally, and enables another channel when data sharing is necessary. We can give variable data bandwidth to each processor by assigning different number of channels. Therefore, the controller can process multiple memory access requests from multiple processors by enabling each channel independently.

IV. SIMULATION RESULT AND IMPLEMENTATION

A eMMC controller with 4 channels is designed using Verilog-HDL based on the proposed architecture and is verified by simulation using an eMMC simulation model and ModelSim. Fig. 2 shows the simulation environment. The AHB master behaves processor operations and the test slave stores data in the host system.

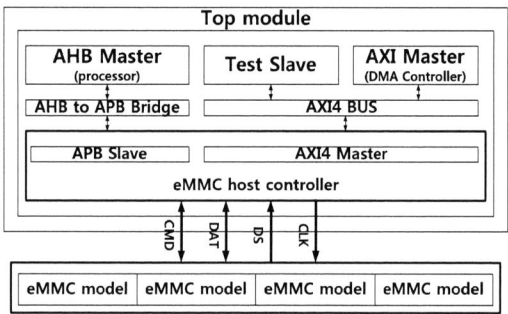

Figure 2. Simulation environment.

Fig. 3 shows the simulation results of the write data operation when all the channels are enabled. The eMMc devices operate in HS400 mode using 8bit data bus [6]. The AHB master writes 2KB data which are divided into four 512B blocks. Each data block is written to the corresponding eMMC device simultaneously. The data bandwidth increases by 4 times compared with that of a single channel operation.

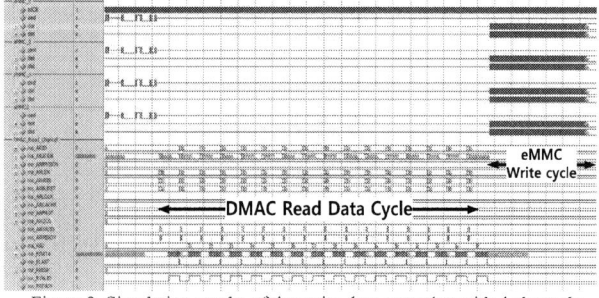

Figure 3. Simulation results of the write data operation with 4 channels enabled.

Fig. 4 shows the simulation results of reading 512B data by 4 processors with one channel enabled at a time. The simulation environment is the same as that of Fig. 3. It shows that data are read through only one channel. Each processor read data from its own eMMC device.

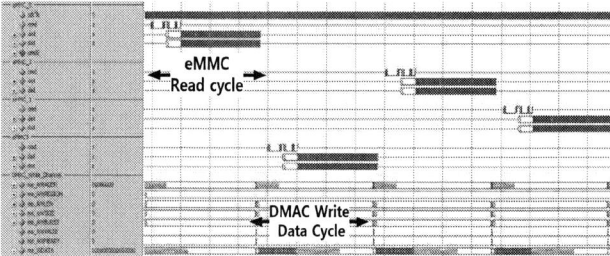

Figure 4. Simulation result of the read data operation with one channel enabled at a time.

Table 1 shows the synthesis result of the eMMC controller using Synopsys Design Compiler and a 0.18um CMOS standard library.

TABLE I.   SYNTHESIS RESULT WITH EMMC DEVICES

Maximum Frequency	285MHz
Logic Gate Count	42,085
SRAM size in buffers	12.2KB

V. CONCLUSIONS

We propose an eMMC memory control architecture and design a controller with multiple channels. Each channel can be enabled before a request of memory access and variable configurations are possible. The controller can operate with increased data bandwidth or process multiple requests of multiple processors for the dedicated devices. The operation is verified by simulation for various requests. The maximum operation frequency is 285MHz and the area for logic is 42,085 gates with a 0.18um CMOS standard library.

REFERENCES

[1] JEDEC, Embedded Multi-Media Card (eMMC) Electrical Standard(5.1), JEDEC JESD84-B51, February 2015.

[2] ARM, AMBA APB Protocol Specification., ARM IHI 0024C, April 2010.

[3] ARM, AMBA AXI and ACE Protocol Specification., ARM IHI 0022E, Febuary 2013, pp.22-122.

[4] Chulhoon Kim and Chanho Lee, "Design of programmable high performance eMMC controller," IEIE SoC Conference, Seoul, Rep. of Korae, May 2016.

[5] Manil Dev Gomony, Benny Akesson, and Kees Goossens, "Architecture and optimal configuration of a real-time multi-channel memory controller," Design, Automation & Test in Europe Conference & Exhibition (DATE), Grenoble, France, pp. 1307–1312, March 2013.

[6] Ning Fu, Yuanheng Li, Baowen Liu, Hongwei Xu and Yigang Zhang, "Realization of controlling eMMC 5.0 device based on FPGA for automation test system," IEEE AUTOTESTCON, National Harbor, MD, pp. 251–255, November 2015.

Energy-Based Iterative Cost Aggregation in Depth Estimation with a Stereo Camera

Nguyen Xuan Truong
Electrical and Computer Engineering
Seoul National University
Seoul, Korea
truongnx@capp.snu.ac.kr

Huyk-Jae Lee
Electrical and Computer Engineering
Seoul National University
Seoul, Korea
hjlee@capp.snu.ac.kr

Abstract— **This paper presents a novel algorithm for performing an efficient cost aggregation in stereo vision. The cost aggregation is re-formulated under an iterative framework with a perspective of an energy model. The convergence of global energy is exploited to calculate the number of the iterations in cost aggregation. Experimental results show that the proposed method improves the quality of the disparity.**

I. INTRODUCTION

DEPTH estimation from a pair of stereo images has been one of the most important problems in computer vision ([1]). In general, stereo matching algorithms are classified into global and local ones according to the strategies used for estimation. In general, local algorithms are much faster and more compatible to practical applications than global ones. However, leading local algorithms which generate high-quality disparity maps still have high complexity. In this paper, we explore the convergence of an iterative cost aggregation stereo matching algorithm; then propose a novel method to perform an efficient cost aggregation.

The procedure of iterative local approaches is as follows. When a truncated absolute difference (TAD) is used to estimate a left disparity map, a per-pixel cost $C^0(p,d)$ for disparity d is first estimated by using the left and "f_p" -shifted right images, where I_l and I_r are left and right images, respectively. Aggregated cost $C^k(p,d)$ of pixel p at iteration k is then recursively computed via an adaptive summation of the cost at the previous iteration on the supportive window $N(p)$. Finally, the Winner-Takes-All (WTA) technique is performed for seeking the best one among all the disparity hypotheses.

$$C^0(p,d) = \min(|I_l(x,y) - I_r(x-d,y)|, \sigma)$$
$$for\ k = 1:T$$
$$C^k(p,d) = \sum_{q \in N(p)} w(p,q)C^{k-1}(q,d) \qquad (1)$$
$$d(p) = \arg\min_d C^T(p,d)$$

The per-pixel cost is truncated with a threshold σ to limit the influence of outliers to the dissimilarity measure. When the number of iteration is one, the algorithm is exactly same as the common non-iterative stereo matching algorithm.

II. ENERGY-BASED ITERATIVE COST AGGREGATION

In the aggregation step (1), the weighting function can play an important role for gathering the information of neighboring pixels where disparity values are likely to be similar. Yoon and Kweon [2] proposed an adaptive (soft) weight approach which leverages the color and spatial similarity measures with the corresponding color images. The weighting function (or correlation) between pixels p and q is defined as follows:

$$w(p,q) = exp\left(-\sqrt{(I_p - I_q)^2}/\sigma_I - \sqrt{(p-q)^2}/\sigma_S\right) \quad (2)$$

Where σ_I and σ_S are color and spatial regularization constants. As the color similarity is measured by using a corresponding color image, it can be interpreted as a variant of joint bilateral filtering [3]. Note that a bilateral filter is an edge-preserving and noise-reduction smoothing filtering for images which preserves sharp edges by systematically looping through each pixel and adjusting weights to the adjacent pixels accordingly. Then the aggregated cost, which is iteratively computed by the filter, still preserves edges while reducing the noise.

The problem is how to define the number of cost aggregation steps in the iterative framework. In this paper, an energy-based method is used to define the number. Remind that general stereo matching algorithms can be defined as the energy minimization problem [4]. Let P be the set of pixels in an image and D be a finite set of disparities. The problem is to find a labeling that assigns a label $d \in D$ to each pixel $p \in P$. The subject is to minimize the labeling cost given by an energy function:

$$E = \sum_{p \in P} C(p,d) + \sum_{(p,q) \in N} V(d(p) - d(q)) \qquad (3)$$

where N are the (undirected) edges in the four-connected image grid graph. $C(p,d)$ is the cost of assigning disparity d to pixel p, and is referred to as the data cost. $V(d(p) - d(q))$ measures the cost of assigning disparities $d(p)$ and $d(q)$ to two neighboring pixels, and is often referred to as the discontinuity cost. If the discontinuity term is ignored, the energy function only includes the data term. It is obvious that

978-1-5090-3220-4/16 $31.00 © 2016 IEEE

the simple local solution gives an optimal solution in such case. This observation implies that an aggregation strategy can be actually interpreted into an iterative energy minimization problem in which the energy is reduced iteratively. The proposed energy-based iterative cost aggregation is based on this observation.

The procedure of the algorithm is presented in Figure 1. At first, the data cost is computed by the per-pixel cost, followed by the computation of a disparity map and its corresponding energy. As the weights only depend on the stereo images, they can be computed before the main loop to reduce the redundant computation. At each iteration the aggregated cost is computed by the variant of the bilateral filtering followed by the disparity and energy computation. The loop is terminated as the energy change is smaller than a threshold.

$$C^0(p,d) = \min(|I_l(x,y) - I_r(x-d,y)|, \sigma)$$
$$d^0(p) = \arg\min_d C^0(p,d)$$

$$E^0 = \sum_{p \in P} C^0(p,d) + \sum_{(p,q) \in N} V(d^0(p) - d^0(q))$$

$$\text{Compute } w(p,q) = \exp\left(-\sqrt{(I_p - I_q)^2}/\sigma_I - \sqrt{(p-q)^2}/\sigma_S\right)$$

$Repeat$

$$k = k + 1$$
$$C^k(p,d) = \sum_{q \in N(p)} w(p,q) C^{k-1}(q,d)$$
$$d^k(p) = \arg\min_d C^k(p,d)$$
$$E^k = \sum_{p \in P} C^0(p,d) + \sum_{(p,q) \in N} V(d^k(p) - d^k(q))$$

$Until\ |(E^k - E^{k-1})/E^{k-1}| < th$

Figure 1: Pseudocode of the proposed algorithm

III. EXPERIMENTAL RESULTS

We have implemented the proposed method and evaluated with the Middlebury test bed: 'Cones', 'Teddy', 'Venus' and Sawtooth stereo images [5]. The parameters are set as follows $\sigma = 15.0, \sigma_I = 5.0, \sigma_S = RAD + 0.5, \lambda = 0.07$ and number of disparities is fixed by 64. The loop is terminated if the change of energy is less than 1% ($th = 0.01$). Table I shows the detail results. The first column is the radius of aggregation window (RAD) which is varying among 2, 3, 4, 8, 12, 16, and 32. For each dataset, results of the energy after the 1st iteration, the final energy, the number of iterations and the running time are reported. Column 2, 6, 10 and 14 shows the energy values after the 1st iteration for 'Cones', 'Teddy', 'Venus' and 'Sawtooth' image pairs, respectively. As the aggregation

window increases, the energy decreases. The window 5x5 gives the largest initial energy, while the one 65x65 gives the smallest initial energy. Meanwhile, the window 17x17 often gives the smallest final energy as shown in the column 3, 7, 11 and 15.

The number of iterations are reported in column 4, 8, 12 and 16 while running time values are indicated in column 5, 9, 13 and 17. For each aggregation window, the running time is proportional to the number of iterations. For example, if RAD is 2 (or window is 5x5), each iterations of 'Cones' takes about 0.165s. On average, each iteration takes 0.158, 0.204, 0.300, 0.548, 0.874, 1.150 and 2.273 seconds when RAD is 2, 3, 4, 8, 12, 16 and 32, respectively. The time is proportional to RAD as we use two-pass approximation of the bilateral filter in each iteration. The horizontal aggregation followed by vertical ones is performed to make the complexity of the filter linear to the window size. It implies that the proposed algorithm is robust as a fast and simple cost aggregation method can output a high-quality disparity map (Figure 2).

ACKNOWLEDGMENTS

This research was supported by the MSIP (Ministry of Science, ICT and Future Planning), Korea, under the ITRC(Information Technology Research Center) support program (IITP-2016-H8601-16-1008) supervised by the IITP(Institute for Information & communications Technology Promotion)

REFERENCES

[1] Scharstein, D. and R. Szeliski (2002). "A Taxonomy and Evaluation of Dense Two-Frame Stereo Correspondence Algorithms." International Journal of Computer Vision 47(1): 7-42.
[2] Kuk-Jin, Y. and K. In So (2006). "Adaptive support-weight approach for correspondence search." IEEE Transactions on Pattern Analysis and Machine Intelligence 28(4): 650-656.
[3] Kopf, J., et al. (2007). Joint bilateral upsampling. ACM SIGGRAPH 2007 papers. San Diego, California, ACM: 96.
[4] Boykov, Y., et al. (2001). "Fast approximate energy minimization via graph cuts." IEEE Transactions on Pattern Analysis and Machine Intelligence 23(11): 1222-1239.
[5] http://vision.middlebury.edu/stereo/data/

TABLE I: RESULTS OF 'CONES', 'TEDDY', 'VENUS' AND 'SAWTOOTH' FOR TH = 0.01

RAD	Cones				Teddy				Venus				Sawtooth			
	1st	final	iters	time (s)	1st	final	iters	time (s)	1st	final	iters	time (s)	1st	final	iters	time (s)
2	10.074	1.488	41	6.78	7.546	1.257	38	6.24	6.142	0.640	27	4.11	4.697	0.562	28	4.21
3	8.245	1.342	30	6.31	5.918	1.155	27	5.62	4.039	0.317	43	8.64	3.078	0.401	32	6.29
4	7.020	1.261	25	6.57	4.974	0.940	32	8.02	2.964	0.268	39	9.22	2.350	0.417	23	5.72
8	4.669	**1.236**	15	8.45	3.455	0.864	20	11.08	1.458	**0.268**	20	10.65	1.435	**0.386**	13	7.02
12	3.784	1.253	12	10.71	2.873	**0.775**	18	15.45	1.099	0.272	13	11.21	1.135	0.400	9	7.96
16	3.341	1.265	12	13.82	2.531	0.779	16	18.26	0.929	0.285	10	11.56	1.014	0.423	7	8.05
32	2.911	1.485	7	15.62	2.122	0.848	10	21.99	0.665	0.386	6	13.71	0.805	0.510	4	9.51

Implementation of Low Complexity Inter Prediction for IoT Systems

Jaehyuk So, Junwon Mun, Kyungmook Oh, Jaeseok Kim
Department of Electrical and Electronic Engineering
Yonsei University
Seoul, Korea
{sojh0124, , mjw5554, okm806, jaekim}@yonsei.ac.kr

Abstract—In this paper, we presents efficient hardware design of inter picture prediction in the slim – high efficient video coding (HEVC). Compared with fully implemented HM10.0, our compression performance of inter prediction hardware block is decreased due to simplification. However our target is Real-time Encoder suitable for IoT, so our inter prediction block is small and fast. Also the verification of the inter prediction design is conducted using the ZYNQ and Virtex7.

Keywords—video coding, HEVC, inter prediction, VLSI, H.264/AVC

I. INTRODUCTION

In recent years, digital video services over network are steadily growing and the overall amount of video data rate is growing at a very fast pace. To use such video data in real-world applications, we need a video codec to compress the video data efficiently. The main goal of HEVC is to achieve 50% bit-rate reduction with similar video quality compared to the previous video coding standard, H.264/AVC. To satisfy these needs, HEVC is developed with several new coding techniques for high coding performance.

Inter prediction is one of the main parts that adapted the new coding techniques. An inter-prediction block merging technique significantly simplified the block-wise motion data signaling by inferring all motion data from already decoded blocks. However, the new techniques increase the computational complexity. Thus the efficient design of inter prediction hardware is required. In this paper, we propose a new hardware design of HEVC inter prediction. To design the simple inter prediction using Verilog, we removed some techniques in HM10.0, and simplified techniques. We perform the verification model, ZYNQ and Xilinx FPGA(Virtex-2000t) are used for hardware verification.

II. ANALYSIS OF INTER PREDICTION IN HEVC

Inter prediction makes use of the temporal correlation between pictures in order to derive a motion-compensated prediction for a block of image samples. Inter prediction mode takes a significant role in HEVC and many added functions bring largely increasing of coding efficiency. In others words, motion estimator is the important block in the HEVC. To provide a compression ratio higher than the previous standards,

the inter prediction of HEVC uses the basic unit size of 64x64, the recursive quad-tree coding unit structure, the asymmetric motion-partitioning (AMP) mode, improved sub-pixel interpolation, advanced motion vector prediction (AMVP), and merge mode. A large basic unit size, recursive quad tree coding unit structure, and AMP mode provide more flexible predictability in size partitioning than previous standards do. New features provide more flexible predictability in size partitioning than previous standards do, but they make it difficult to design the motion estimator. Basically, major two blocks in inter prediction are motion estimation and motion compensation.

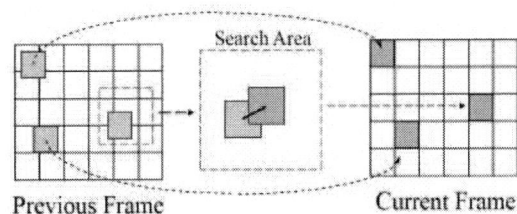

Figure 1. Motion estimation algorithm

III. PROPOSED ARCHITECTURE

To realize the real-time operation system, the efficiency parallel structure and motion vector predictor are used. We remove some functions in inter prediction and conduct basic prediction of it. First of all, there is no interpolation filter to improve accuracy of motion vectors by using fractional sub-pixel. And merge mode is non-existing so quantity of transmitting data for decoder such as motion vector difference(MVD) and partition index is slightly increased. Instead of omitting merge mode, we use only AMVP. The set of candidates can include a temporal candidate from a pre-defined reference picture and spatial candidates enable more flexibility in selection of the co-located reference. AMVP uses motion vectors on neighbor PUs and as the outcome; it gives high compression performance but results data dependency on each PU. It requires a high operation frequency when the hardware is implemented. To process the motion estimation of Pus in same CTU at the same time, the motion vector set of AMVP must be modified. The modified motion vector set of AMVP is placed outside the CTU. Figure 3 shows an example of the modified motion vector set of AMVP.

978-1-5090-3220-4/16 $31.00 © 2016 IEEE

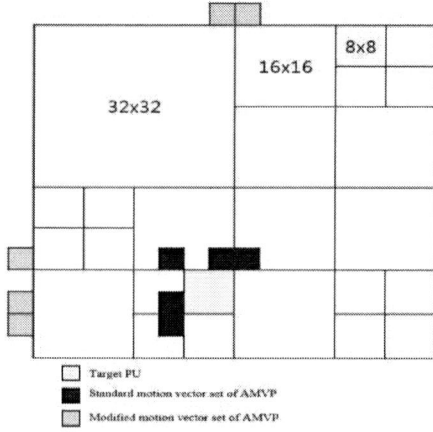

Figure 2. Example of the modified motion vector set of AMVP

The hardware architecture consists of controller, motion vector memories, search area memories, current block memory, 256 processing elements, SAD summation blocks, comparison block and AMBA bus wrapper. Motion vector memories, search area memories and current block memory save necessary data. Processing elements solves 4x4 SAD values. A SAD summation blocks solves various size SAD values by using 4x4 SAD values. A comparison block solves cost value of each prediction unit size. Controller generates appropriate control signals depending on the data flow and the stage of each block. AMBA bus transfers signals of motion estimator to tha of AHB standard.

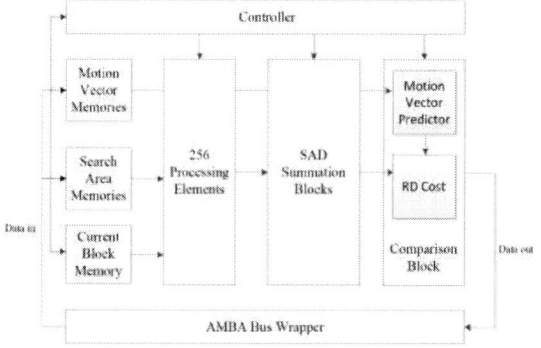

Figure 3. Block diagram of proposed inter prediction

IV. FPGA TEST OF INTER PREDICTION BLOCK

The designed hardware is verified with the hardware verification model. The result of designed logic is compared with that of HM 10.0 which is HEVC reference software. The input data is the test image and the output data is the computation result of the motion estimator. The output data of HM reference software is same with designed systems.

For the FPGA verification, the FPGA simulation results is used as the results of the hardware architecture in the verification model. FPGA environment consisted of Advanced Microcontroller Bus Architecture (AMBA).

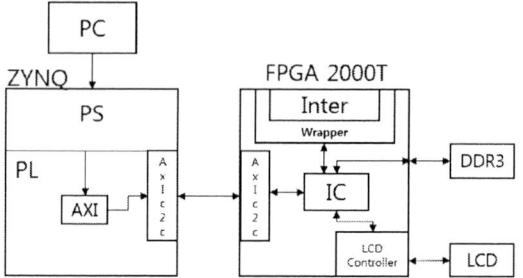

Figure 5. FPGA constitution for verification

Figure 5 shows the FPGA constitution. Xilinx ZYNQ was used as processor. FPGA is used for interface between processor, external DDR3 memory and proposed design. Xilinx virtex7 XC7V2000T FPGA is used. DDR3 memory is used for input image and output data.

V. CONCLUSION AND FUTURE WORKS

In recent years, demand for ultra-high definition resolution and high quality video services has increased. To support UHD resolution, the high performance video coding is required, therefore HEVC standard is established. Our purpose is designing the slim-HEVC for IoT system. The proposed motion estimator uses the modified motion vector set in AMVP. It gives parallel processing of PUs which is placed in one available CTU. To reduce the complexity, we omitted some blocks. However the proposed hardware blocks achieved enough throughput in targeting Full-HD(1920x1080p). The proposed motion estimator systems are implemented in Verilog HDL and verified in FPGA environments. The gate count is 700k and the internal SRAM is 20Kbytes. We will integrated our blocks (intra prediction, inter prediction, CABAC) for slim-HEVC encoder. Also, we will design the HEVC decoder using the blocks in encoder.

ACKNOWLEDGMENT

This work was supported by the industrial Core Technology Development Program(10049009, Development of Main IPs for IoT and Image Based Security Low-Power SoC) funded by the Ministry of Trade, Industry & Energy. This work was also supported by IDEC(IPC, EDA Tool, MPW).

REFERENCES

[1] Jaehyuk So, Kyungmook Oh, Jaeseok Kim, "Design and verification of intra prediction hardware for video streaming in IoT systems," International SoC Design Conference. Gyungju, pp. 283-284, Nov 2015.

[2] HEVC Test Model 10 "HM 10.0", https://hevc.hhi.fraunhofer.de/svn/svn_HEVCSofwa rc/tags/HM-10.0

[3] B. Bross et al., "High Efficiency Video Coding (HEVC) text specification draft 10 (for FDIS & Last Call)", presented at JCTVC-L1003, JCT-VC of ISO/IEC and ITU-T, Geneva, CH, Jan. 2013.

[4] ITU-T and ISO/IEC JTC 1, "Advanced video coding for generic audiovisual services", ITU-T Recommendation H.264 and ISO/IEC 14496-10 (MPEG4-AVC), 4th ed., Sept. 2008.

Halo Effect Suppression
for Single Image Haze Removal Method

Geun-Jun Kim
Department of Electronic Engineering
Dong-A University
Busan, Korea
firstaccel@gmail.com

Bongsoon Kang[*]
Department of Electronic Engineering
Dong-A University
Busan, Korea
bongsoon@dau.ac.kr

Abstract—**Haze is mainly occurred by atmospheric phenomena. Recently, many researchers in haze removal algorithm area are using single image. At the single image, we can't use depth information. To estimate the thickness of haze without depth information is not easy. As a result, single image haze removal method includes halo effect. In this paper, we propose halo effect suppression for single image haze removal method. Conventional halo removal method required a lot of time, but the proposed method is to lower the computational complexity.**

Keywords; defog, dehaze, haze removal, single image

I. INTRODUCTION

Haze removal is consumer electronics field. Due to the prevalence of the camera, we can take a picture anytime, anywhere. Many pictures have been taken are outdoors. For this reason, many image signal processing methods are developed and improved. One of them is also the haze removal algorithm. Haze is one of meteorological phenomenon. This phenomenon causes degradation of the quality of the picture. Recently, there has been studied haze removal using a single image [1-2]. Using single image means we don't know about the scene depth. Scene depth is key information for haze remove. Equation (1) is the hazy image model in the computer vision [3].

$$L(x, y) = L_0(x, y)e^{-kd(x,y)} + L_s(1 - e^{-kd(x,y)}) \qquad (1)$$

Where L is the luminance of the observed image, L_0 is the scene radiance luminance, L_s is the luminance of the sky, $e^{-kd(x,y)}$ is the transmission and k is the atmosphere coefficient and $d(x, y)$ is the distance between camera and object and means depth information.

II. PROPOSED METHOD

A. The reason for halo effect occurs

Conventional haze removal method is shown Fig. 1. Estimation of atmospheric light is finding L_s at the (1), estimation of transmission is finding $e^{-kd(x,y)}$ at the (1) and recovery of scene radiance is finding L_0 at the (1). Halo effect in haze removal method occurs at estimation of transmission. Haze is additive component of scene radiance. For this reason, when we can't use depth information, we use windowing operation for estimation of transmission [1-2].

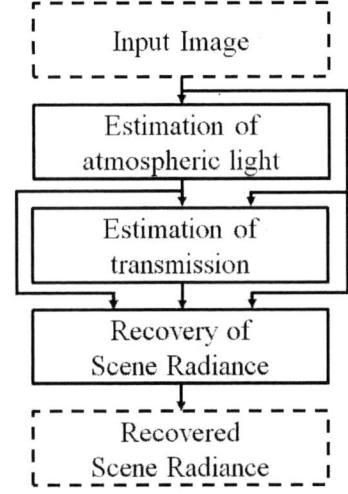

Figure 1. Conventional haze removal method

It is quite a good way to use the windowing operation to estimate the haze. But windowing operation occur halo effect at the boundary of area. Fig. 2 shows halo effect at the boundary area. The halo effect is largely generated in the abutting portion of the sky and degrades the quality of the image.

Figure 2. Halo effect in haze removal method[2]

---

[*] Corresponding Author : Bongsoon Kang

This paper was supported by Basic Science Research Program through the National Research Foundation of Korea(NRF) funded by the Ministry of Science, ICT & Future Planning(NRF-2015R1D1A1A01060427).

B. Halo effect suppression

Conventional method [2] uses additional algorithm for halo effect such as matting laplacian [4]. These methods require a large amount of computation. We propose modify estimation of transmission. The reason of using windowing operation is to estimate the component of the haze. We can see at (1), observed image is sum of scene radiance and haze component. It means hazy is always less than observed image. It is not accurate value. When we can't use depth information, other methods will also estimate the haze. Equation (2) is expression of approximate haze.

$$h(x, y) \leq \min_{c \in \{r,g,b\}} (L^c(x, y)) \qquad (2)$$

(a) Original hazy image

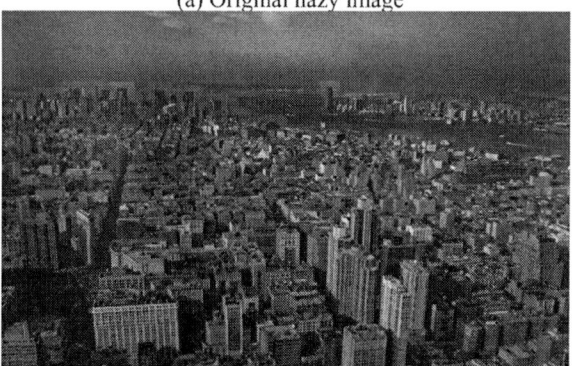

(b) Conventional scene radiance [2]

(c) The proposed scene radiance

Figure 3. Experimental result

Where h is the approximated haze, L^c is the color channel of the observed image. In addition, the haze will have a value less than the minimum value of the observed image. Equation (2) can be rewritten as (3).

$$h(x, y) \cong w * \min_{c \in \{r,g,b\}} (L^c(x, y)) \qquad (2)$$

Where w is transfer coefficient. In this paper, we set the w as 0.9

III. EXPERIMENTAL RESULT

Fig. 3 shows an experimental result. (a) is the original hazy image. And (b) is the conventional scene radiance without matting laplacian [4]. (c) is the proposed scene radiance. Comparing (b) and (c), (b) has halo effect along river. (c) has problem that is the difference in color when compared with the original hazy image.

IV. CONCLUSION

The proposed method is halo effect suppression for single image haze removal algorithm. Conventionally, counteract for halo effect require a large amount of computation. The proposed method is not requiring additional algorithm for halo effect. It is simple and efficient method for halo effect suppression. But proposed method includes color tone distortion. Future works will require an overcome the color tone distortion.

REFERENCES

[1] J. P. Tarel, N and Hautiere, "Fast visibility restoration from a single color or gray level image," IEEE International Conference on computer Vision, pp. 2201-2208, 2009.

[2] K. He at al, "Single Image Haze Removal Using Dark Channel Prior," IEEE International Conference on Computer Vision and Pattern Recognition, pp. 1956-1963, 2009.

[3] H. Koschmieder, "Theorie der horizontalen sichtweite," Beitr. Zur Phys. D. freien Atm. 1924.

[4] A. Levin, D. Lischinski, and Y. Weiss. "A closed form solution to natural image matting," IEEE International Conference on Computer Vision and Pattern Recognition, pp. 61-68, 2006.

A Design of Real Time Detection IP with Color Detection for Surveillance

Chang-Hee Park
University of Ulsan
SoC Lab.
Ulsan, Republic of Korea
pc0pc@naver.com

Hyun-Tae Kim
University of Ulsan
SoC Lab.
Ulsan, Republic of Korea
gusxo4479@naver.com

Young-Min Jang
University of Ulsan
SoC Lab.
Ulsan, Republic of Korea
min-s2@nate.com

Sang-Bock Cho
University of Ulsan
SoC Lab.
Ulsan, Republic of Korea
sbcho@ulsan.ac.kr

Abstract—**In this paper, in order to implement the IP of edge detection algorithm for surveillance in image using the camera in CCTV or vehicle black box, we designed pre-processing step that is the edge extraction algorithm. First, after input image converts into the input signals of R, G, and B, three inputs are combined, and converted to gray scale. The data and blue color value are entered separated from each other and subtraction operations. Then appropriate portion by threshold extracts only white and converts the rest to black. A proposed algorithm was implemented using Matlab program. It was verified through a RTL-level simulation of ISE14.3.**

Keywords-Internet of things(IoT), Color Detection, FPGA, IP, RGB

I. INTRODUCTION

Recently, human life are becoming more and more convenient, thanks to the development ICT(Information Communication Technology), new technology that combines the ICT in a variety of industries are being developed. Especially among technologies such as ITS(Intelligent Transportation System) and IOT(Internet of Things) is actively conducted research around the image processing system, using image processing and image information are investing a lot of manpower and cost to the IP core development. This image processing system is a pre-treatment step of the process is important to provide an accurate and clear image information.

In this paper, we designed an color detection algorithm in order to implement a preprocessing step for obtaining the image for security in IP. A proposed algorithm was implemented using Matlab program. This is converted and confirmed to Verilog through Xilinx ISE.

II. R,G,B COLOR MODEL

RGB color model[1] is the most basic model. This model is to consider the image as a combination of three components of the Red, Green, Blue. It is expressed, in the RGB model, black is R = G = B = 0, white is R = G = B = 255, red is R = 255, G = B = 0 and yellow is R = G = 255, B = 0. If R = G = B is the achromatic color in Gray color. Because R, G, B, each have a

value from 0 to 255[2]. Figure 1 is expressed RGB color model Graph in Digital.

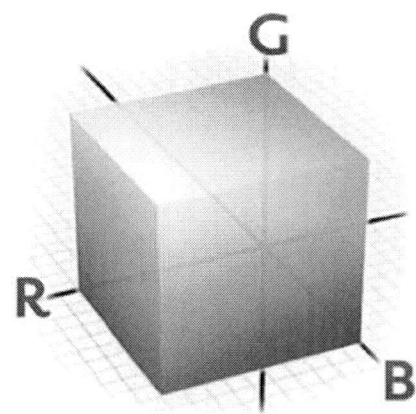

Figure 1. RGB model graph in Digital.

III. COLOR DETECTION

The method used in this paper was used for the basic data of R, G, B in the color detection method. This method is faster, without the process of transformation in the Y, cb, cr, or HSV method[2] was applied to detect a color close to the primary colors. Figure 2 is a block diagram color detection algorithm.

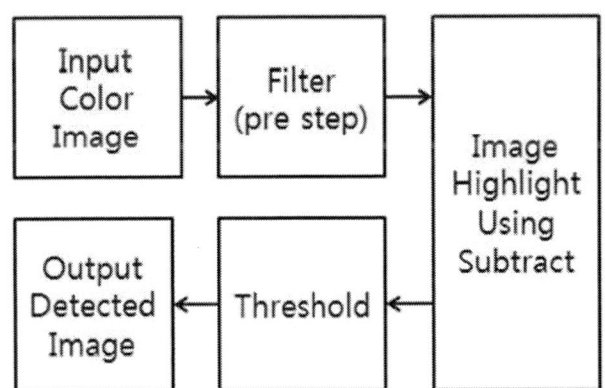

Figure 2. The block diagram of color detection algorithm.

A. Input

The original image is converted to grayscale and Deteriorated for noise cancellation to RGB input of Gaussian filter.

B. Filter

Degrade the signal passes through the Gaussian filter. This value is a Gray scale conversion by the Equation 1 [3][4].

$$Gray = (Blue + Green + Red)/3 \qquad (1)$$

C. Image Highlight

Calculates the received input image through Equation 2 [3][4] and the color highlighted.

$$Data = Blue - Gray \qquad (2)$$

The reason for using such equation is pure Blue comes close to the value of 255 the blue color value of a particular position. On the other hand, the Gray is a relatively small value if a specific location is close to the value blue. The reason is added by three values are reduced by 1/3.

D. Threshold

In this stage, we find the appropriate threshold. Then appropriate portion by threshold extracts only white and converts the rest to black.

E. Output

Finally it outputs the result to the image through the line buffer.

IV. EXPERMETAL RESULT

Figure 3 shows the results of a RTL-level verification in ISE 14.3.

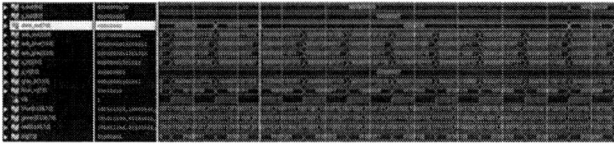

Figure 3. The simulation result of RTL-level.

The results shown confirm that over a certain portion of the difference between blue and gray out the binary value of 255 is the other part of the value 0.

Figure 4 is possible to determine the original image and the color detected result.

Figure 4. The original image and result image.

In the original image, only the blue part of the helmet can be seen that the search in white. Table 1 is Device utilization.

TABLE I. DEVICE UTILIZATION

Resources	Used	Available	Utilatio
Slice LUTs	19	53,200	1%
Occupied Slices	6	13,300	1%
ILogice	22	200	11%
OLogice	43	200	21%
Averaged Running Time 1.664ns			

V. CONCLUSION

In this paper, it designed to implement to IP Color detection algorithm to process at a faster of the pre-treatment step to obtain an image for security. Input image converted into the input signals of R, G, and B, three inputs are combined, and converted to gray scale. Then Subtraction operating between Gray value and Blue value for Blue highlight.

As a result, the detection of a color close to the primaries was faster and easier. However, it can be seen the need of the detection algorithm with only the color values in the color except for gamma value as possible in order to detect with a color complex. It plans to supplement accordingly.

The algorithm was validated by RTL-level through the ISE 14.3.

ACKNOWLEDGMENT

This work was supported by the Industrial Core Technology Development Program (10049009, Development of Main IPs for IoT and Image-Based Security Low-Power SoC) funded by the Ministry of Trade, Industry & Energy.

REFERENCES

[1] Noor A. Ibraheem, 2 Mokhtar M. Hasan, 3Rafiqul Z. Khan, 4 Pramod K. Mishra, "Understanding Color Models: A Review",ARPN. vol.2, no.3, pp.265-275, April 2012.

[2] Rafael C.Gonzalez, Richard E.woods, Steven L. Edins, Digital Image Processing using MATLAB, Prentice-Hall, 2004.

[3] Yu Wang, "IMAGE ENHANCEMENT BASED ON EQUAL AREA DUALISTIC SUB-IMAGE HISTOGRAM EQUALIZATION METHOD", IEEE Transactions on Consumer Electronics, vol. 45, no. 1, pp. 68-75, February 1999.

[4] Hyun Tae Kim, Young Min Jang, Sang Bock Cho, Fire Extinguisher Recognition using Robust Color Detection and Gamma Correction,"ITC-CSCC, p1049_P3-15,2016

Non-Photorealistic Rendering from Real Video Sequences with Discontinuity Reduction Using Fast Video Segmentation

Lu Xiao, Xiao-Xuan Huang, Yi-Chang Lu
Graduate Institute of Electronics Engineering and Graduate Institute of Communication Engineering
National Taiwan University, Taipei, Taiwan, 10617
Email: {r02943152, xxhuang, yiclu}@ntu.edu.tw

Abstract— **Non-Photorealistic Rendering (NPR) for video sequences can be regarded as an extension of NPR for images. However, if the NPR sequence is obtained from a real video using an image processing pipeline, discontinuity problems have to be addressed, especially for the background region. In this paper, we propose to adopt Fast Video Segmentation algorithm to extract the common background in the video before applying NPR. With the help of the moving object mask, the background and moving objects in the scene can be rendered respectively, and combined together to generate final NPR sequences with better qualities.**

Non-photorealistic Rensering; Video Segmentation;

(a)   (b)

Fig. 1. An example of NPR: (a) a captured image; (b) NPR from (a).

I. INTRODUCTION

Non-photorealistic rendering (NPR) and physically-based rendering (PBR) are two distinct rendering technologies. NPR techniques are used to show a certain artistic expression, while physically-based rendering is to provide a vivid viewing experience. One of the fast NPR techniques is to produce an NPR sequence directly from a real video through a series of image processing steps. Fig. 1 shows a zoomed-in example of an original video frame and a frame after applying NPR. However, if we apply NPR to each frame separately, discontinuity between frames can be easily observed. The problem mainly results from small differences between original frames being magnified after NPR processes. There are several studies to diminish NPR discontinuity problems. Winnemöller et al. [1] chose to use a bilateral filter to smooth images. Bousseau et al. [2] proposed to adopt a time-domain morphology filter. These researches attained good temporal coherence for NPR sequences. However, they did not distinguish moving objects from the background which humans often perceive differently. In this paper, we propose to combine NPR with Fast Video Segmentation [3], so that we can efficiently generate NPR video and diminish the discontinuity problem to provide better results.

II. NPR FOR VIDEO SEQUENCES

The proposed NPR flow for video sequences is illustrated as Fig. 2. We first use Fast Video Segmentation [3] to partition video contents into two parts, the common background and moving objects. In the common background region, we can apply NPR directly. As for the moving objects, we add an image

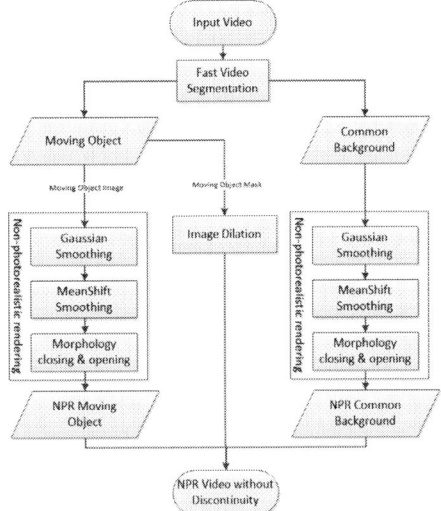

Fig. 2. The flow chart of the proposed NPR for video sequences.

dilation procedure before NPR to emphasize moving objects for better visual experiences. Finally, we combine the two parts to obtain our NPR video with discontinuity problem suppressed. The details of the flow are discussed in the following paragraphs.

In the original Fast Video Segmentation algorithm [3], the background and moving objects of a frame are estimated by the difference between the current frame and the background obtained in the previous adjacent frame:

$$BD(x, y, t) = |I(x, y, t) - BG(x, y, t-1)|, \qquad (1)$$

This work is partially supported by the Ministry of Science and Technology, Taiwan, under Grant numbers MOST 104-2221-E-002-098 and 105-2218-E-002-024.

Fig. 3. Results of the proposed NPR flow using the 70th frame as an example: (a) the 70th frame of the input video; (b) the extracted common background of the input video; (c) the calculated moving object mask of the 70th frame; (d) the differences between the 70th and 71st NPR frames without Fast Video Segmentation; (e) the 70th frame of the proposed NRP video; (f) the common NPR background; (g) the moving object of the 70th NPR frame; (h) the differences between the 70th and 71st NPR frames with the proposed method.

where BD is the background difference of a frame at time t, I is the input frame, BG is the background at time $t-1$, and x, y represent the position of the target pixel. With the help of background difference, the moving object mask can be acquired by setting a proper threshold, th:

$$MOM(x,y,t) = \begin{cases} 1, & BD(x,y,t) \geq th \\ 0, & BD(x,y,t) < th \end{cases} \quad (2)$$

As mentioned earlier, in an NPR video, subtle changes between frames can be magnified and cause discontinuity problems. Therefore, we revise the background difference equation as follows:

$$BD_{rev}(x,y,t) = \left| I(x,y,t) - BG\left(x,y,t_{final}\right) \right|, \quad (3)$$

where $BG(x,y,t_{final})$ is the final common background. Instead of using the background of the previous adjacent frame, the new background difference refers to the final common background. Since all frames refer to the same background, the moving object mask of the video is more stable. Also, to make sure that the rendered edges of moving objects are smooth, we intentionally dilate the moving object mask for later use. In our NPR design, we first apply Gaussian filter to remove high frequency components in the images to mimic human painting effects. Then we use Mean Shift algorithm to flatten the surface [4]. Since moving objects generally draw more attentions, we choose to preserve more details of moving objects by selecting smaller bandwidth parameters in Mean Shift, while the Mean Shift parameters used for the background are larger. We also apply morphological filters as suggested in [2], so the small areas can be faded out. At the final step, we combine the common NPR background and the moving objects of the NPR frame. Since every frame shares the same background, the NPR video are now temporal coherent and viewers can get a video without random blinks in the background.

An example of our experiment is shown in Fig. 3. In the video, a person is walking from left to right with a tree and buildings in the background. As expected, the calculated moving objects include the person, and leaves/grasses affected by the wind. The effectiveness of discontinuity reduction is demonstrated in Fig. 3(d) and (h). Fig. 3(d) shows the differences of two adjacent frames if we apply NPR to each frame separately. There are many noisy pixels in the background, which lead to blinking in the video. On the other hand, Fig. 3(h) shows the differences only related to actual moving objects, which offers preferable viewing experiences.

III. CONCLUSION

In this paper, we propose to use Fast Video Segmentation to identify the common background in the video and use it to diminish discontinuity in NPR videos. We can use different NPR parameters for the background and moving objects to obtain better results. With the proposed flow, NPR sequences can be efficiently generated from captured videos to provide viewers with different visual experiences.

REFERENCES

[1] H. Winnemöller, S. C. Olsen, and B. Gooch, "Real-time video abstrac- tion," in *ACM Transactions on Graphics (TOG)*, vol. 25, no. 3. ACM, 2006, pp. 1221–1226.

[2] A Bousseau, F. Neyret, J. Thollot, and D. Salesin, "Video watercol-orization using bidirectional texture advection," in *ACM Transactions on Graphics (ToG)*, vol. 26, no. 3. ACM, 2007, p. 104.

[3] S.-Y. Chien, Y.-W. Huang, B.-Y. Hsieh, S.-Y. Ma, and L.-G. Chen, "Fast video segmentation algorithm with shadow cancellation, global motion compensation, and adaptive threshold techniques," *IEEE Transactions on Multimedia*, vol. 6, no. 5, pp. 732–748, 2004.

[4] D. Comaniciu and P. Meer, "Mean shift: A robust approach toward feature space analysis," *IEEE Transactions on pattern analysis and machine intelligence*, vol. 24, no. 5, pp. 603–619, 2002.

An H.265/HEVC 4K UHD Slim Codec Design with Shared Prediction Unit Architecture

Sukho Lee and Hyunmi Kim

Department of Intelligent SoC Research, ETRI, Daejeon Korea,
{shlee99, chaos0218}@etri.re.kr

Abstract— **H.265/High Efficiency Video Coding (HEVC) is the latest next generation video compression standard posterior to H.264/AVC. However, despite its superior coding efficiency to the previous video coding standards, the complexity to implement it is an obstacle to overcome. Especially, combining the separate encoder and decoder has a disadvantage on the aspect of the size and power consumption. To solve these problems, we design an encoder based H.265/HEVC 4K slim codec. The decoder within this codec shares the prediction unit of encoder except an entropy decoder. The proposed shared prediction unit architecture saves the total size by 40 % compared to an independent codec with encoder and encoder separately. The size of logic is 2.8 M gates with 120 kB internal SRAM and the power consumption of this slim codec is within a level of encoder. The function of slim codec is verified on our designed the Xilinx Virtex-7 platform and the 4K UHD codec chip operating at 600 MHz is going to implement on a 28 nm CMOS process in this year.**

Keywords; H.265, HEVC, UHD, codec, Prediction Unit

I. INTRODUCTION

The new video compression standard H.265/HEVC has a bit-rate reduction of 50 % with similar video quality compared to H.264/AVC [1]. The major role that gives an effect on higher coding efficiency is enlargement of the coding unit size and transform size. In the advanced coding tools of the new standard, large coding unit (LCU), which is like macroblock (MB) of 16×16 pixels in H.264/AVC, increases up to 64×64 pixels and two-dimensional transform of various sizes specifies from 4×4 to 32×32. The large size of the coding and transform unit contributes to compression efficiency, but it makes complexity much heavier and causes manufacturing costs expensive. Although smart-phones today are capable to accomplish video encoding or decoding for H.265/HEVC, 4K video processing still incurs an unacceptable energy cost that can cause complete battery drain within a few minutes [2]. The high technology processor to deal with H.265/HEVC such as newly GPU and CPU has a grossly inefficient power problem so it is not suitable for 4K Ultra-HD low power applications, for example, mobile, IoT, VR and wearable devices [3]. To solve these drawbacks, we design a H.265/HEVC codec with the encoder resources themselves. With an idea that encoder has a decoder function within it, slim codec reuse prediction unit within the encoder when decoding mode. Because a intra/inter prediction unit includes not only prediction function but also compensation function to restore a decided mode, the decoder need not additional

compensation units if this codec decodes only a bit-stream with a limited encode mode on our designed encoder [4]. The architecture of the proposed codec has encoder-decoder shared prediction units and this structure makes the size of codec slim down.

II. SHARED PREDICTION UNIT

A. Inter Prediction

Inter prediction has three steps in encoder side, which are integer motion estimation (IME) whose search range (SR) is from -24 to +24 in *X* direction and from -16 to +16 in *Y* direction, fine motion estimation (FME) and motion compensation (MC). After IME process, coding unit (CU) size is decided at first and mode decision unit with rate distortion optimization (RDO) chooses FME or MERGE mode finally. Next, to restore predicted images, MC restores predicted pixels with an estimated motion vector (MV). But decoder need not process IME and FME, it does only MC process. The decoder in the proposed slim codec shares this MC unit as it is without any additional MC unit. To do this, we design additional path on MC unit for decoding. When decoding mode, entropy decoder parser transfers motion vectors to MC unit and MC makes compensated pixels for a reconstructed image. The inter reconstructed pixels enter into transform and quantization unit.

B. Intra Prediction

Intra prediction of slim codec supports twelve directional modes including dc and planar and processes two steps in encoder mode [5]. They are intra prediction and intra compensation (IC). The first step is to choose the best one mode with RDO and IC makes predicted pixels with a decided mode as like MC in inter prediction in the next step. Encoder stores predicted pixels of all modes to internal SRAMs to save processing cycles before mode decision. But decode is not necessary to do that because decoder only receives an intra prediction mode form entropy decoder parser. To share IC unit with encoder and decoder, we design addition control path for that. The intra reconstructed pixels enter into transform and quantization unit as same as inter prediction unit does.

C. Transform and Quntization

Transform size of slim codec supports from 8×8 to 32×32 and is same with CU size. The entire transform unit consists of one 32×32, two 16×16, two 8×8 DCT and one

---

This work was supported by the IT R&D Program of MSIP/IITP. [R7117-16-0153, Development of Intelligent Semiconductor Common Platform Tech. for Smart Devices]

shared 32 × 32 inverse transform unit to deal with parallel processing at RDO mode decision step [5]. Encoder needs both forward transform and inverse DCT for restore pixels but decoder only need inverse transform unit. In decoding mode, slim codec only shares inverse transform and de-quantization of transform units to make reconstructed pixels.

III. SLIME CODEC ARCHITECTURE

The proposed slim codec architecture is based on the encoder that is presented our prior work [4] and is as shown in Fig. 1.

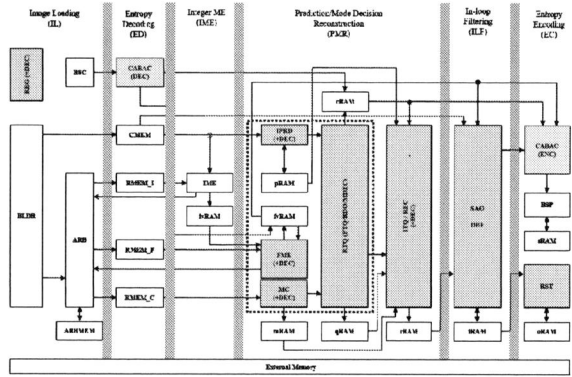

Figure 1. The proposed slim codec architecture.

Encoding mode compresses input images according to five-stage LCU level pipeline process, which are image loading (IL), integer motion estimation (IME), PMR, in-loop filtering (ILF) and entropy encoding (EC). PMR stage includes intra prediction, fine motion estimation, motion compensation, mode decision and reconstruction. PMR stage has to be in a single pipeline stage because intra prediction need reconstructed neighboring pixels immediately at next coding unit. Decoding mode decompress a bit-stream with three stage LCU level pipeline process. Entropy decoder (ED) parses a compressed stream and transfers coefficient data and additional information to PMR stage. PMR stage restore pixel images with de-quantization and inverse transform. ILF processes de-blocking filtering and sample adaptive offset (SAO) filtering and restores images for display. Because P slice has the mixed intra/inter coding units (CU) and intra compensation has to use the boundary pixels of the intra or inter CU, the switching interface for this is necessary within PMR unit.

IV. EXPERIMENTAL RESULTS

Table I shows the performance comparison of the proposed slime codec with HEVC Test Model (HM-13.0) in Low Delay P mode. The average Bjøntegaard Delta (BD)-rate, which means bit rate, increase by 34.6% and ΔY-PSNR loss is 0.5 dB [3].

V. FPGA VERIFICATION

The designed H.265/HEVC slim codec is implemented on our designed Xilinx Virtex-7 verification platform as shown Fig. 2.

Table I. Performance of Comparison with HM-13.0

Test sequences Class A (1920x1080)	BD-rate	ΔY-PSNR
Kimono	24.1 %	-0.4 dB
Park Scene	30.2 %	-0.4 dB
Cactus	44.9 %	-0.6 dB
BasketballDrive	34.9 %	-0.5 dB
BQTerrace	39.1 %	-0.5 dB
Total Average	34.6 %	-0.5 dB

The logic gate count is 2.8M and internal SRAM size is 120 kB. The total usage of LUTs on FPGA is 39 %. When encoding mode, real-time image is transmitted via HDMI Rx and the encoder compresses this image to store SD card on board. After that, decoder de-compresses a bit-stream in SD card and displays a monitor.

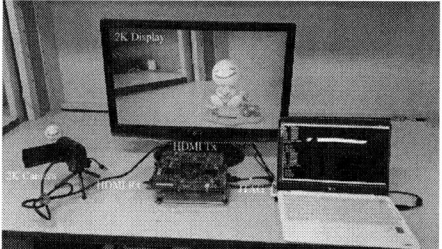

Figure 2. H.265/HEVC slim codec verification board

VI. CONCLUSIONS

We design an H.265/HEVC slim codec and it is much lighter than a conventional codec because it does not need any additional decoding unit. It has only an additional entropy decoder in the entire hardware. The proposed architecture shares prediction units and transform unit at encoding and decoding mode. This saves the total size by 40 % compared to an independent codec with encoder and encoder each. The power consumption of this slim codec is within a level of encoder. The shared type slim codec's function is verified on our designed the Xilinx Virtex-7 platform and the 4K UHD codec chip operating at 600 MHz is going to implement on a 28 nm CMOS process in this year.

REFERENCES

[1] V. Sze, M, Budagavi, and G.J. Sullivan, "High efficiency video coding: algorithms and architectures," *Springer*, 2014.

[2] C. Ju et al., "A 0.5nJ/Pixel 4K H.265/HEVC codec LSI for multi-format smartphone applications," IEEE J. Solid-State Circuits, vol. 51, no. 1, Jan. 2016, pp.56-67.

[3] S. Lee and H. Kim, "Reduced complexity single core based HEVC video codec processor for mobile 4K-UHD applications," IEEE Int. Conf. on Consumer Electronics-Berilin, Sep. 2016, To be published.

[4] S. Lee, S. Cho, H. Kim, and J. Lee, "A study on the Full-HD HEVC encoder IP design," Journal of The Institute of Electronics and Information Engineers, vol.52, no. 12, Korea, Dec. 2015, pp. 2229-2235

[5] S. Lee, H. Kim, K. Byun, and N. Eum, "On FHD 300MHz@60fps, intra/inter CU mode decision hardware architecture for the Hypernova H.265 encoder," Proc. IEEE Int. SoC Design Conference, Nov. 2014, pp.254-255.

Fast CU Size Decision Method for HEVC Using CU Split Information of Adjacent Frames

Young Ho Kim, Tae Sun Kim and
Myung Hoon Sunwoo
School of Electrical and Computer Engineering,
Ajou University
San 5, World cup-ro, Yeongtong-gu, Suwon, 16499, Korea
k8064129@ajou.ac.kr, zzktszz@ajou.ac.kr,
sunwoo@ajou.ac.kr

Jae Heon Jeong
Advanced SW Development Team,
Hyundai Mobis
Mabuk-ro 240beon-gil, Giheung-gu, Yongin-si,
Gyeonggi-do, Korea
jjh4708@gmail.com

Abstract— **High Efficiency Video Coding (HEVC) adopts a new technique, called quad-tree based coding for inter prediction. However, deciding the optimal depth of Coding Unit (CU), namely the basic unit of the quad-tree structure, is a very time-consuming task. This paper presents a fast CU size decision algorithm to reduce the complexity and encoding time for inter prediction. The proposed algorithm shows a reasonable performance in terms of saving the total encoding time up to 40% increasing 1.4% Bjonteggrad Delta Bit Rate (BDBR) compared to the HEVC reference model (HM).**

Keywords; HEVC; FCDS; RPD; collocated; candidate

I. INTRODUCTION

HEVC [1] has achieved the enhanced video quality and data compression. However, the huge computational complexity of HEVC makes it difficult to be used in resource constrained environments [2]. Hence, various CU decision methods [3]-[5] have been proposed to reduce the complexity and time of the CU decision process.

The existing algorithms [3]-[5] cannot flexibly change the candidate depths according to the CU depths of the previous frames. However, the proposed algorithm uses the information of the temporal neighboring CU. Unlike the existing algorithms, the proposed one does not strictly fix the candidate depth range. Therefore, it can be useful when there is a big difference between the depths of the collocated CUs.

II. PROPOSED ALGORITHM

A. Fast CU Depth Selection Stage (FCDS)

As the frame rate of the video is becoming higher, the texture information between adjacent frames is becoming more similar [6]. The Fast CU Depth Selection Stage (FCDS) uses this temporal similarity, which selects the CU depth by employing the collocated CU depth information of the previous frame, $Depth_{colloc}$. In FCDS, instead of computing the Rate Distortion (RD) cost of the all possible depths, only some candidate depths are evaluated using a 2N x 2N Prediction Unit (PU).

Table I shows the proportion of the selected PU type when the encoding process is finished. Experimental results show that among the all kinds of PU, 2N x 2N PU is mostly selected. It

means that the RD cost of 2N x 2N PU can be the optimal RD cost of the current CU.

TABLE I. PROPORTION OF THE SELECTED PU TYPE

sequence	Proportion of the selected PU type (%)				
	2Nx2N	2NxN	Nx2N	NxN	AMP
BQSquare	75.54	4.74	4.89	12.20	2.64
Blowing Bubbles	75.13	3.16	5.10	11.61	5
Basketball Pass	78.97	3.27	5.34	8.61	3.81
Party Scene	73.40	4.56	7.25	10.83	3.95
Basketball Drill	80.99	2.69	3.10	9.38	3.83
Four People	83.38	0.72	1.98	12.14	1.78
BQTerrace	77.70	2.30	2.78	14.96	2.26
Cactus	81.65	2.47	3.21	8.64	4.03
Kimono	88.50	2.75	2.90	0.16	5.69

TABLE II. CU DEPTH TO BE EAVLUATED

$Depth_{colloc}$	Candidate depths
0	0,1
1	0,1,2
2	1,2,3
3	2,3

In addition, FCDS computes the RD cost only when the depth of the current CU is equal to $Depth_{colloc} \pm 1$. By limiting the number of the CU depth to 2 or 3, FCDS can greatly reduce the CU decision time. In summary, Table II shows the candidate depths for the current CU.

B. Refinement PU Decision Stage (RPD)

After FCDS is completed, the RPD process refines the result within the selected CU depth. In RPD, symmetric motion partition (SMP) and asymmetric motion partition (AMP) are used for finding the best prediction mode having a minimum RD cost. Then, the CU decision process is finished. However, the CU depth can be rapidly changed in the temporally collocated neighboring CUs. To cover the exception, the proposed algorithm includes some additional CU depth checking methods.

This work was partly supported by the Mid-Career Researcher Program through an NRF grant funded by the MSIP (2014R1A2A2A01002952)

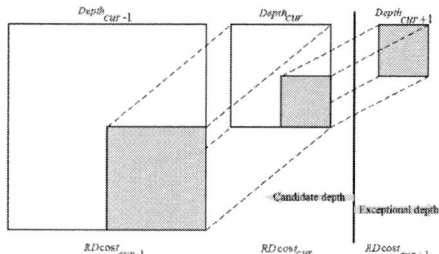

Fig. 1. CU checking method in deeper depth.

The drastic change between adjacent frames can be classified into two cases.

Case I: $Depth_{cur} > Depth_{colloc}$. For example, $Depth_{colloc}$ is 0 and $Depth_{cur}$ is equal to 3.

Case II: $Depth_{cur} < Depth_{colloc}$. For example, $Depth_{colloc}$ is 3 and $Depth_{cur}$ is equal to 0.

To cover Case I, the proposed algorithm uses $RDcost_{cur-1}$ for encoding $Depth_{cur+1}$ as shown in Fig. 1. If $Depth_{cur+1}$ is not a candidate depth and (1) is satisfied, then $Depth_{cur+1}$ will be evaluated.

$$0.25 \times RD\cos t_{cur-1} > RD\cos t_{cur} \qquad (1)$$

To cover Case II, the proposed algorithm uses the flag for encoding $Depth_{cur-1}$ as shown in Fig. 2. If $Depth_{cur-1}$ is not a candidate depth and $RDcost_{cur}$ is smaller than $RDcost_{cur+1}$, up signal flag will be set. It can be expressed as (2) and (3). If more than two up signal flag 's are set, $Depth_{cur-1}$ will be evaluated.

$$RD\cos t_{cur+1} = \sum_{i=1}^{4} RD\cos t_{up_i} \qquad (2)$$

$$\begin{cases} up\ signal\ flag = 1\ (RDcost_{cur} < RDcost_{cur+1}) \\ up\ signal\ flag = 0,\ (otherwise) \end{cases} \qquad (3)$$

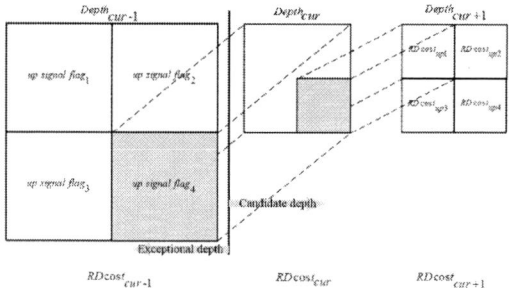

Fig. 2. CU checking method in lower depth.

III. EXPERIMENTAL RESULTS

The proposed algorithm is implemented on the HEVC software, HM14.0. The simulation is performed on Windows 7 platform with Intel core i5@ 3.40 GHz and 16 GB RAM. Each sequence was encoded using four quantization parameter (QP):

22, 27, 32 and 37. Table III shows the BDBR and time saving (TS) comparisons among the existing algorithms [4]-[5] and the proposed algorithm for the low delay profile. The proposed algorithm shows about 3.3% reduction in encoding time with respect to the algorithm in [4]. In addition, the proposed algorithm outperforms [4] with 0.3% less BDBR loss. Compared to the algorithm in [5], the proposed algorithm shows 0.4% less BDBR loss and 7.6% more encoding time saving.

TABLE III.　COMPARISONS OF DIFFERNET ALGORITHMS

sequence	[4]		[5]		proposed	
	BDBR	TS	BDBR	TS	BDBR	TS
BQSquare	1.1 %	24.2 %	0.4 %	44.3 %	1.2 %	39.8 %
Blowing Bubbles	2.1 %	32.1 %	2.4 %	22.0 %	1.0 %	32.9 %
Basketball Pass	1.8 %	41.1 %	1.6 %	14.8 %	1.3 %	35.7 %
Party Scene	2.0 %	42.7 %	3.3 %	25.1 %	2.5 %	38.2 %
Basketball Drill	3.9 %	51.9 %	1.3 %	38.5 %	2.8 %	41.4 %
Four People	-	-	2.4 %	47.8 %	0.9 %	46.2 %
BQTerrace	0.6 %	51.5 %	1.8 %	32.2 %	0.8 %	41.9 %
Cactus	2.1 %	42.2 %	3.1 %	36.4 %	1.4 %	39.4 %
Kimono	1.6 %	42.2 %	0.3 %	27.8 %	1.5 %	38.5 %
Average	1.7 %	36.4 %	1.8 %	32.1 %	1.4 %	39.7 %

IV. CONCLUSION

This paper proposed the fast CU size decision algorithm using the information of the collocated CUs to reduce the computational complexity of the quad-tree computation in HEVC. The proposed algorithm can flexibly change the candidate depth range according to the CU depth of the previous frame. Thus, it can be used when the depth of the CU is rapidly changed. The experimental results show that the proposed fast CU size decision algorithm can reduce the total encoding time about 39.7% compared to that of the original HM14.0. In addition, the average bit-rate does not increase more than 0.87%. To further enhance the accuracy of CU size decision process, future work can be done by referring to the size of the spatially neighboring CUs.

REFERENCES

[1] M.T. Pourazad, C. Doutre, M. Azimi, and P. Nasiopoulos, "HEVC: The new gold standard for video compression: How does HEVC compare with H.264/AVC," IEEE Consumer Electron. Mag., vol. 1, no. 3, pp. 36-36, Jul. 2012.

[2] James Nightingale, Qi Wang, and Christos Grecos, "Priority based methods for reducing the impact of packet loss on HEVC encoded video streams," in Proc. IS&T/SPIE Real Time Imaging 2013.

[3] H. Wei-Jhe, and H. Hsueh-Ming, "Fast coding unit decision algorithm for HEVC," in Proc. Asia-Pacific Signal Inf. Process. Assoc. Annu. Summit Conf., Oct./Nov. 2013, pp. 1–5.

[4] Jian Xiong, Hongliang Li, Qingbo Wu, and Fanman Meng, "A fast HEVC inter CU selection method based on pyramid motion divergence," IEEE Trans. Multimedia, vol. 16, no. 2, pp. 559-564, 2014.

[5] Liquan Shen, Zhi Liu, Xinpeng Zhang, Wenqiang Zhao, and Zhaoyang Zhang, "An effective CU size decision method for HEVC encoders," IEEE Trans. Multimedia, vol. 15, no. 2, pp. 465-470, Feb. 2013.

[6] G. Zhong, X. He, L. Qing, and Y. Li, "A fast inter-prediction algorithm for HEVC based on temporal and spatial correlation," Multimedia Tools and Applications, pp. 1-21, 2014.

The parallelization of convolution on a CNN using a SIMT based GPGPU

Heekyeong Jeon, Kwanho Lee, Seonghyung Han, Kwangyeob Lee
Department of Computer engineering
Seokyeong university
Seoul, Korea

Abstract— **This paper proposes a method to accelerate convolutional neural network(CNN) by utilizing GPGPU. The convolutional layer of the conventional CNN required a large number of multiplication operations. This paper seeks to reduce the number of multiplication operations through Winograd convolution operation and perform parallel processing of the convolution operation by utilizing SIMT structure of GPGPU. The experiment was conducted using ModelSim and TestDrive, and the experimental results showed that the processing time was improved by about 7%, compared to the conventional convolution operation.**

Keywords; neural network, gpu, parallelism, convolutional neural network

I. INTRODUCTION

In recent years, there has been a growing interest in machine learning used as in Google's AlphaGo[1]. Machine learning is a field that develops algorithms and technologies to allow computers to learn. In particular, variables as in image processing have been utilized in a variety of fields. Among them, the convolutional neural network is a neural network specialized in processing large amounts of data including images. However, since the convolution operation, which is the most basic operation of the convolutional neural network, consists of a number of multiplication operations, it requires a lot of processing time. In order to combat this disadvantage, a Winograd convolution algorithm is used in this paper[2][3]. The Winograd convolution improves an algorithm from a number of multiplication operations to a number of addition operations with a relatively short operation time. In addition, SIMT structure of GPGPU is used to perform the parallel processing of the convolution operation, and thus to improve the processing time.

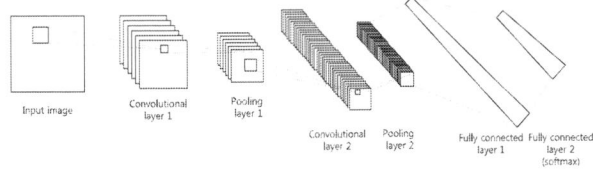

Figure 1. The architecture of the conventional convolutional neural network

II. PROPOSED CONVOLUTIONAL LAYER

A. Conventional CNN

The general convolutional neural network is composed of convolutional layer, pooling layer and fully connected layer as shown in Figure 1[4]. The convolutional layer detects the characteristics required for classification by filtering the input image data or the previous feature map data with the convolution operation. The pooling layer integrates the data with the strongest characteristics among those detected from the convolutional layer. This integrated data is classified into the specified class through the fully connected layer. This paper seeks to improve the processing speed by improving the convolutional layer with the large amounts of operations. The convolution operation is performed with a number of multiplication and additions as shown in numerical formula (1). The values obtained by multiplying the input data x by the filter value w are added and stored in the next layer data y.

$$y_{i,k} = \sum_{c=1}^{C} w_{i,c*}x_{k,c} \tag{1}$$

B. Winograd small convoltuion algorithm

The Winograd small convolution algorithm reduces the number of multiplication operations by using a Chinese remainder theorem[5]. For example, if the convolution operation that has two outputs as shown in the following formula (2) proceeds, a value can be obtained with 4 multiplications as shown in numerical formula (3), not with 6 multiplications, with respect to the output m.

$$F(2,3) = \begin{bmatrix} d_0\,d_1\,d_2 \\ d_1\,d_2\,d_3 \end{bmatrix} \begin{bmatrix} g_0 \\ g_1 \\ g_2 \end{bmatrix} = \begin{bmatrix} m_1+m_2+m_3 \\ m_2-m_3-m_4 \end{bmatrix} \tag{2}$$

$$m_1 = (d_0 - d_2)g_0 \qquad m_2 = (d_1 + d_2)\frac{g_0 + g_1 + g_2}{2}$$

$$m_3 = (d_1 - d_3)g_2 \qquad m_3 = (d_2 - d_1)\frac{g_0 - g_1 + g_2}{2} \tag{3}$$

C. SIMT based GPGPU

This paper proposes a method for parallel processing of the improved convolution algorithm in the SIMT-based GPGPU. Some Common Mistakes

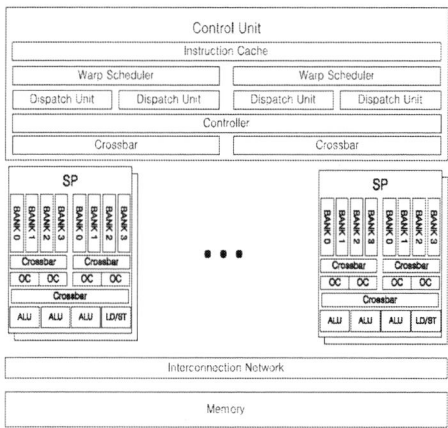

Figure 2. The architecture of GPGPU

As shown in Figure 2, the SIMT-based GPGPU has the concept of warp, which is a bundle of threads and consists of 16 warps and 16 threads. Each thread has arithmetic unit and register sets[6].

In this paper, the parallel processing was performed by taking charge of one component operation of the output matrix for each thread. If parallelization is done based on the operation for the convolution of a filter of size 3x3 with size 4x4 as an input, the existing convolution operation is made with 9 multiplications and 8 additions in a single thread. In the case of the Winograd small convolution algorithm, the convolution can be done with 2 multiplications and 12 additions in a single thread.

In this paper, the code data was stored in LUT (Look-Up Table) to perform parallel processing of the convolution operation. If the subtraction operation is substituted for the addition operation with the use of the stored code data, the same addition operation is repeated at a certain number of times. This is suitable for the SIMT structure that processes a lot of data with one command and thus can effectively perform parallel processing of the convolution operation.

III. EXPERIMENT

In the experiment, ModelSim and TestDrive[7] were used for simulation, and the operating frequency was 100MHz. The convolution was accelerated using a filter of size 3x3 with a 4x4 matrix as an input. A comparison between the existing convolution operation and the method proposed in this paper was conducted through parallel processing by component of each matrix. The experimental results are summarized in Table 1, and the results confirmed that the number of multiplications that the proposed method requires was 2, which suggests that the number of multiplications was further reduced compared to the conventional method.

TABLE I.   THE COMPARISON OF THE CONVENTIONAL METHOD AND PROPOSED METHOD

	Conventional method	Proposed method
Processing time(ns)	29770	27690
The number of multiplication	9	2
The number of addition	8	24

In addition, a comparison of the processing time revealed that the proposed method was about 7% faster than the conventional method.

IV. CONCOLUSION

This paper sought to improve the processing time and the amount of convolution operation which accounts for the largest amount of operation in the conventional neural network. With the use of Winograd small convolution algorithm, a large number of multiplications were processed into a large number of additions in the existing convolution. In addition, the algorithm was accelerated using the SIMT-based GPGPU that operates with 16 warps and16 threads. In the experiment, the results of the conventional convolution operation and the improved convolution operation were compared, and the results confirmed that there was about 7% improvement. Therefore, if the overall convolutional network is configured by improving the processing time of the convolutional layer with the largest amount of operation in the convolutional neural network, a faster processing speed than that of the conventional convolutional neural network can be expected.

ACKNOWLEDGMENT

This work was supported by the Industrial Core Technology Development Program (10049192, Development of a smart automotive ADAS SW-SoC for a self-driving car) funded By the Ministry of Trade, industry & Energy

REFERENCES

[1] David Silver, Aja Huang, Chris J. Maddison, Arthur Guez, et al., "Mastering the game of Go with deep neural networks and tree search," . Nature, 529, pp. 484-489, 28 January 2016.

[2] ShmuelWinograd, " Arithmetic complexity of computations," volume 33. Siam, 1980.

[3] Lavin, Andrew. "Fast algorithms for convolutional neural networks." arXiv preprint arXiv:1509.09308 2015.

[4] Krizhevsky, Alex, Ilya Sutskever, and Geoffrey E. Hinton. "Imagenet classification with deep convolutional neural networks." Advances in neural information processing systems. 2012.

[5] Agarwal, R., and J. Cooley. "New algorithms for digital convolution." IEEE Transactions on Acoustics, Speech, and Signal Processing 25.5 (1977): 392-410.

[6] Yunseop Hwang, Kwang yeob Lee, Junmo Jeong, "Design of SIMT Architecture-based Reconfigurable Image Signal Processor," International conference on future information & communication engineering, 25 June 2015.

[7] https://sourceforge.net/projects/test-drive/

978-1-5090-3220-4/16 $31.00 © 2016 IEEE

Transmission Timing Configuraiton for Control and Non-Payload Communication of Unmanned Aerial Vehicle

Tae Chul Hong, Kunseok Kang, Kwangjae Lim, Jae Young Ahn

Unmanned Aircraft System ICT Research Section
Electronics and Telecommunications Research Institute
Daejeon, Korea
taechori@etri.re.kr

Abstract— **For the safe control of Unmanned Aerial Vehicle (UAV), a reliable communication link is very important, so a Control and Non-Payload Communication (CNPC) standard is progressing in Radio Technical Commission for Aeronautics (RTCA) [1][2]. Currently, CNPC standard only includes the physical layer specification, so the upper layer protocol is required for implementing a CNPC transceiver. In this paper, we assume that the LTE based upper layer protocol is implemented in Digital Signal Processing (DSP) board, and analyze the configuration of timing ticks for reducing transmission delay. The application of half subframe tick can decrease 38.4% of transmission delay compared with the subframe tick.**

Keywords; UAV, CNPC, transmission timing, tick cofiguration

I. INTRODUCTION

Unmanned Aircraft System (UAS) includes UAV, Ground Radio Station (GRS), and Ground Control Station (GCS). A GRS transmits and receives radio signals, and a GCS controls an UAV. The CNPC standard of RTCA is about the control information communication between an UAV and a GCS, and the CNPC standard considers medium and large sized UAVs (above 150 kg). The scope of the CNPC standard only includes Radio Frequency (RF) and physical layer techniques, so the standard does not define upper layer protocols yet. Upper layer protocols, which are located between the physical layer and Internet Protocol (IP), are required for the implementation of a CNPC transceiver. National Space Agency (NASA) considers IEEE 802.16 based upper layer protocols, but we consider the application of LTE based upper layer protocols [3][4].

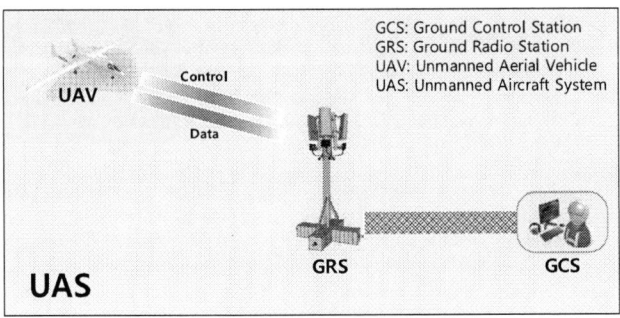

Figure 1. Unmanned Aircraft System

II. CNPC TRANSCEIVER

We consider the architecture of a CNPC transceiver, which is shown in Fig.2. It is composed of a RF/IF subsystem, a physical layer subsystem, and an upper layer subsystem. An upper layer subsystem is implemented by embedded software in a TI-TMS320C6678 DSP board. There are two type of interfaces between a physical layer subsystem and an upper layer subsystem. An upper layer subsystem receives timing ticks through General Purpose Input Output (GPIO), the data exchange between a physical layer subsystem and an upper layer subsystem is carried out through Serial Rapid Input Output (SRIO).

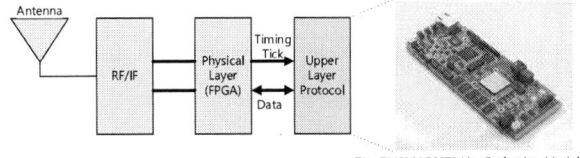

Figure 2. CNPC transceiver architecture

III. TANSMISSION TIMING ANALYSIS

The frame structure of CNPC is shown in Fig.3. CNPC uses Time Division Duplex (TDD), so one frame is consist of one tx subframe and one rx subframe. The transmission duration of a GRS is the receiving duration of an UAV. Under TDD frame structure, continuous transmissions are not possible, so the transmission delay is larger than that of the Frequency Division Duplex (FDD) frame structure. For the safe operation of UAV, the transmission delay of CNPC should be minimized. Therefore, we try to minimize the transmission delay through the transmission timing configuration.

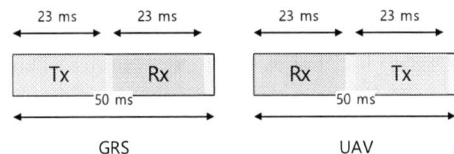

Figure 3. CNPC frame structure

Medium Access Control (MAC), which is the one of upper layer protocols and implemented through the software, needs to know the timings of processing a subframe and transferring a subframe between a FPGA board and a DSP board. Therefore, the board of a physical layer provides ticks through GPIOs. According to the ticks from the FPGA board, the MAC layer starts a subframe processing for the transmission or receiving.

It is intuitive to think of the application of subframe ticks. When the subframe tick is applied, the transfer timing according to protocol layers is shown in Fig. 4. In Fig.4, T_1 means the first subframe tick, and T_2 means the second subframe tick. At the T_1 tick, the tx subframe of a GRS and the rx subframe of a UAV are started. In Fig 4, the MAC layer of a GRS starts a subframe processing at the first tick of Nth frame, and the MAC layer transfers the subframe to physical layer (PHY) at the second tick of Nth frame. After the physical layer processing, the subframe is transmitted to an UAV at the first tick of N+1th frame. The physical layer of an UAV receives the subframe at the second tick of N+1th frame, and the MAC layer receives the subframe at the first tick of N+2th frame. In brief, the subframe of a GRS, which is transmitted at the first tick of Nth frame, is received by an UAV at the N+2th frame. Therefore, the transmission delay is about 100 ms in the best case, and 150 ms in the worst case. We think that the transmission delay according to the application of the subframe tick is not proper to the control of an UAV.

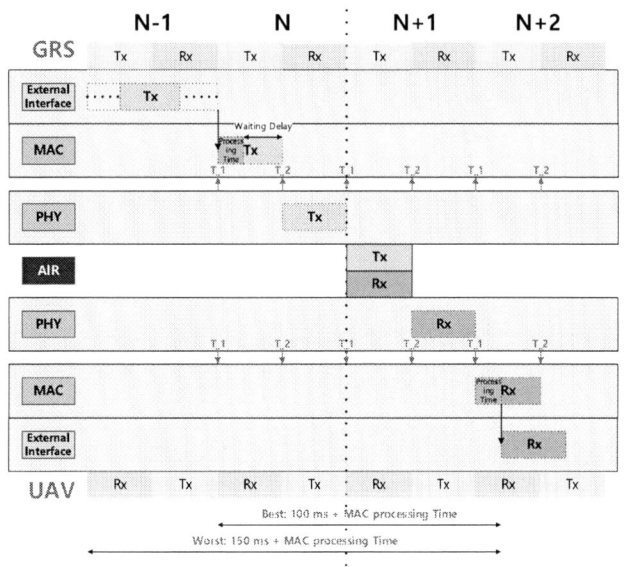

Figure 4. Timing with subframe tick

For the reduction of the transmission delay of CNPC, we consider the half subframe tick. When the half subframe tick is applied, the transmission timing is shown in Fig.5. There are 4 ticks per a frame, and a subframe processing of the MAC layer can be carried out during the half subframe time. The processing start time of MAC layer for the transmission at N+1th frame can be postponed for one subframe duration comparing to the case of the subframe tick. Therefore, the MAC layer of a GRS starts a subframe processing at the third tick of Nth frame, and the MAC layer transfers the subframe to

the physical layer at the fourth tick of Nth frame. The pipeline action of a RF/IF subsystem can allow to delay the processing start time of physical layer. Therefore, the physical layer of an UAV receives the subframe at the third tick of N+1th frame, and the MAC layer receives the subframe at the fourth tick of N+1th frame. In brief, the subframe of a GRS, which is transmitted at third tick of Nth frame, is received by an UAV at the N+1th frame. The transmission delay is about 61.5 ms in the best case and 111.5 ms in the worst case. The application of the half subframe tick decreases 38.4% of the transmission delay compared with the subframe tick.

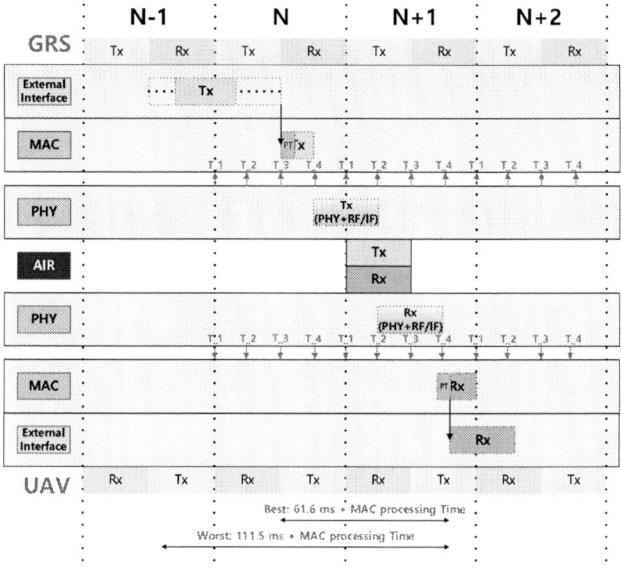

Figure 5. Timing with half subframe tick

IV. CONCLUSION

In this paper, we analyze the transmission timing of a CNPC transceiver, and propose the configuration of timing ticks for minimizing the transmission delay.

ACKNOWLEDGMENT

This work was supported by the ICT R&D program of MSIP/IITP. [R0126-16-1005, Development of High Reliable Communications and Security SW for Various Unmanned Vehicles]

REFERENCES

[1] RTCA, Inc., Minimum Operational Performance Standards (MOPS) For Unmanned Aircraft Systems (UAS) Control and Non-Payload Communications Terrestrial Link Radio Systems, Version 1.0, 2016.

[2] Kerczewski, Robert J., and James H. Griner. "Control and Non-Payload Communications Links for Integrated Unmanned Aircraft Operations.", NASA, 2012.

[3] Griner, J. "Unmanned aircraft systems (UAS) integration in the National Airspace System (NAS) project: UAS Control and Non-Payload Communication (CNPC) System Development and Testing." 2014 Integrated Communications, Navigation and Surveillance Conference (ICNS) Conference Proceedings. IEEE, 2014. pp.1-24.

[4] S. Sesia, I. Toufik, and M. Baker, LTE, The UMTS Long Term Evolution: From Theory to Practice, John Wiley & Sons Ltd., chichester, 2009.

A System-level Design of MapReduce-based Embedded Multiprocessor System-on-Chips

Huajuan Zhang[1], Hao Xiao[1a)], and Ning Wu[1]

[1] College of Electronic and Information Engineering, Nanjing University of
Aeronautics and Astronautics, Nanjing 210016, China
a) xiaohao@nuaa.edu.cn

Abstract—**With the advent of the Internet of Things, collection and processing of large datasets on embedded systems become increasingly important. Therefore, to enable embedded processors with more data processing capabilities, this paper applies the MapReduce parallel programming model to embedded multi-processor system-on-chip (MPSoC). We implement the proposed MPSoC in system-level SystemC and evaluate its performance using commercial ESL tools. Experimental results show that the proposed MPSoC can achieve up to 2.1x performance improvement compared with Phoenix on general-purpose embedded platform.**

Keywords—Multiprocessor system-on-chip; MapReduce; multiprocessor programming; embedded systems

I. INTRODUCTION

With the advent of the Internet of Things (IoT), it tends to be increasingly important to collect and process large datasets on embedded system. Currently, most IoT systems adopt centralized storage and processing architecture. To relieve the computation pressure of the cloud and to minimize the data-transfer between the mobile node and the cloud, the concept of localized pre-processing and computation have been proposed. It changes the calculating node model from a server or cloud to an embedded device. However, due to the difficulties of parallel programming, managing the concurrency of multiprocessor platforms is still not easy.

MapReduce is a successful programming framework created by Google [1], which can greatly simplify parallel programming and facilitate processing large datasets on data centers. This work aims to ease the development of dataset processing engines by applying the MapReduce model to customized programmable multiprocessor system-on-chip (MPSoC) platforms. This kind of MPSoC architecture is more suitable for data flow of the MapReduce compared with general multi-core processors or other commercial heterogeneous multi-core processors [2]. Moreover, it is programmable and easy-developed compared with FPGA [3]. The proposed MPSoC is modeled using SystemC at system-level. Then full-system simulation is carried out at system-level using Synopsys system-level design tools.

This work is supported in part by National Natural Science Foundation of China 61504059, Natural Science Foundation of Jiangsu Province BK20140834, and Fundamental Research Funds for the Central Universities NS2015043.

The rest of this brief is organized as follows. Section 2 presents the proposed MPSoC architecture, which is followed by the experimental result in Section 3. Finally, we conclude the paper in section 4.

II. HARDWARE ARCHITECTURE

Fig. 1 shows an architectural overview of the proposed MPSoC, which consists of a cluster of PEs connected by a hierarchical interconnection fabric. The architecture is conceptually coincidence with the MapReduce processing flow, and consists of three main components: the front-end for splitting input data, multiple Map-Reduce block-pipes (MRBP) for data processing, and the back-end for merging the MapReduce results. For achieving fast block data transfer and easy implementation, the proposed architecture uses multiple AHB bus fabrics to connect the three components. In order to relieve the data traffic on the bus, we separate the inter-processor control messages from the MapReduce data streams, and use another lightweight inter-processor connection to signal the control token among the PEs.

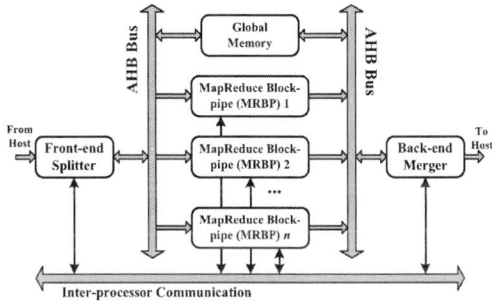

Fig. 1. Architecture overview of proposed MPSoC.

The detailed architecture and data flow will be showed in the Fig 3. The Splitter works as a DMA, as well as a central scheduler to control the MRBPs and Merge. As described in the blue dash lines, the front-end splitter is responsible for dispatching the input raw data from the host system to the global memory. Then, depending on the architectural configurations (e.g. input data size and number of block pipes), the splitter further dispatches small data blocks to each data buffer of the MRBP. Finally, as described in the red dash lines, when the data buffer is ready, the splitter directly triggers the MRBRs to start. MRBP is the kernel of the architecture, which consists of a map-reduce pair connected by a shared AHB bus. Based to the MapReduce model, the mapper PE processes the raw data to intermediate <key, value> pairs, and the reduce PE further polishes the intermediate pairs to <key, values> pairs. In order to facilitate the exchange of intermediate <key, value> pairs, we adopt a ping-pong buffer structure, whereby the mapper and reducer can work alternately in a block-pipe manner. Finally, after getting the results from each MRBP, the back-end subsystem start to collect the separated <key, values> pairs to be a single entity and returns it to the host system. In this subsystem architecture, we use two-layered merge mechanism to improve the throughput of merge processing, which is implemented by using the multilayer AHB fabric.

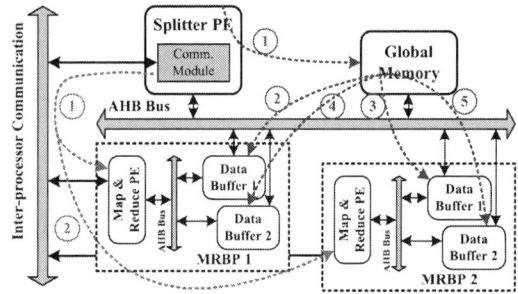

Fig. 2. Front-end subsystem: architecture and data flow.

III. EXPERIMENTAL RESULT

The proposed MPSoC is modeled in system-level SystemC and the simulation is carried out using Synopsys Platform Architect. We evaluate four benchmarks, including Word Count, String Match, Linear Regression and Histogram, which have been implemented using the MapReduce programming model. For comparison, we also implement these benchmarks on Xilinx ZYNQ platform (ARM A9 dual-core processor) using Phoenix MapReduce framework. In addition, a single-thread implementation is also evaluated as a baseline reference. Table 1 lists the detailed architecture configurations.

TABLE I. PROCESSOR ARCHITECTURE CHARACTERISTICS

	ZYNQ	Proposed
Processor	ARM A9	Customized MPSoC
Num. of Cores	2	4 MRBPs
L1 Cache	32KB(I), 32KB(D)	None
L2 Cache	512KB	None
Frequency	667MHz	667MHz

Fig. 3 shows the performance results of the benchmarks with different data sizes in terms of execution time. As shown, using the proposed MPSoC leads to a better performance than the other two architectures. Moreover, as the data size increases, the gained performance achieved by the proposed MPSoC also increases, which shows its potential capability in dealing with big datasets. Fig. 4 further shows the normalized execution time of 1GB dataset. As shown, the proposed MPSoC is 1.7x to 3.0x faster than the single-thread implementation, while it is 1.5x to 2.1x faster than the dual-core phoenix implementation.

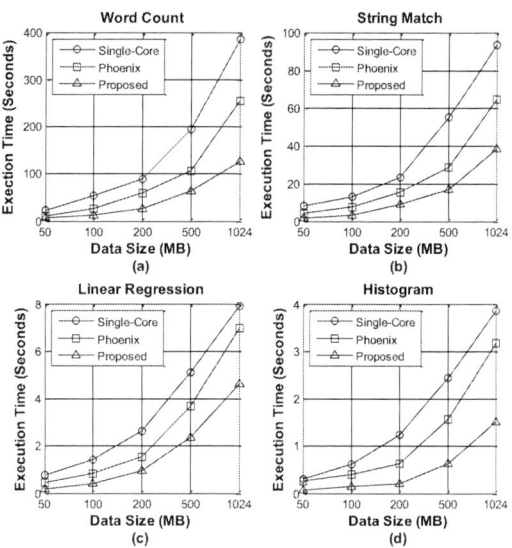

Fig. 3. Execution time for different applications.

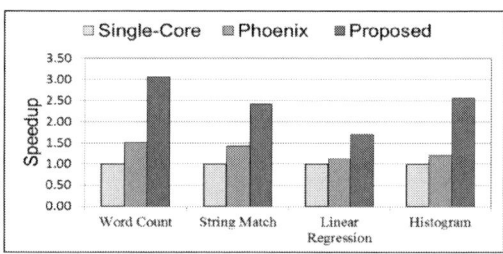

Fig. 4. Speedup for different applications using 1GB datasets.

IV. CONCLUTION

This paper presents a customized MPSoC architecture for the IoT application. It uses multiple MRBPs to increase the efficiency of the parallel computation. The hybrid shared-memory facilitates the data transfer between PEs. While the global communication network achieves fast inter-processor communication and relieves the traffic contention of the shared AHB bus. Experimental result shows that our architecture can achieve up to 2.1x speedup compared with general-purpose embedded multicore processor. In our future work, we will design these PEs using our application specific instruction-set processor (ASIP) methodology [4] that enables exploiting instruction-level acceleration of map/reduce tasks.

REFERENCES

[1] J. Dean and S. Ghemawat: Commun. ACM 51 (2008) 107.

[2] W. Fang, B. He, Q. Luo, and N. K. Govindaraju: IEEE Trans. Parallel. Distrib. Syst. 22 (2011) 608.

[3] Y. Shan, B. Wang, J. Yan, Y. Wang, N. Xu, and H. Yang: ACM SIGDA Int. Symp. FPGA (2010) 93.

[4] H. Xiao, T. Isshiki, D. Li, H. Kunieda, Y. Nakase, and S. Kimura: IPSJ Trans. Syst. LSI Des. Methodol. 5 (2012) 118.

Radio-Frequency Energy-Harvesting IC with DC-DC Converter

Donghoon Seong, Kichang Jang, Wonjoon Hwang, Hyeondeok Jeon, and Joongho Choi

Department of Electrical and Computer Engineering
University of Seoul, Seoul, Korea
Email: jchoi@uos.ac.kr

Abstract—**This paper presents an energy harvesting system that uses a radio frequency (RF) signal. It consists of an RF voltage multiplier-rectifier and DC-DC converter. A Dickson voltage multiplier is used as the RF voltage rectifier. A buck-boost converter is used as the DC combiner. The input frequency of the RF signal is 2.4 GHz, and the output DC voltage is programmable. The proposed harvesting system is implemented in a 0.18-um CMOS process.**

Keywords — Energy Harvesting System, RF Rectifier, DC-DC buck-boost converter

I. INTRODUCTION

There have been significant developments for energy harvesting technologies that can extract power from sunlight, mechanical vibration, thermal gradients, RF waves, and so on [1]. One of the most promising energy harvesting methods should be to utilize propagating radio-frequency signals [2]. The way of extracting energy from RF signals can be one of the most efficient way to drive IoT-based hardware with minimum usage of battery or even without battery.

II. RF ENERGY HARVESTING SYSTEM DESIGN TECHNIQUE

The proposed RF energy harvesting system consists of an RF voltage rectifier and DC-DC converter. Because the RF signal is an AC signal with very small magnitude, a voltage multiplier circuit is implemented with rectifier one. A following DC-DC converter is used for providing the proper form of supply voltage that can charge the battery and drive the electronic circuitry load.

A. RF Rectifier Optimization

A charge pump circuit, known as the Dickson voltage multiplier, is used to rectify the input AC signal [3]. The circuit schematic of a conventional Dickson voltage multiplier is shown in Fig. 1. The voltage multiplier consists of multiple stages of rectifiers. Each stage consists of a peak rectifier formed by D_{Vn} and C_{Vn} and voltage clamp formed by D_{Hn} and C_{Hn}. Energy conversion efficiency varies according to the stage

and voltage multiplier capacitor. Performance optimization of the voltage multiplier is required to maximize energy conversion efficiency.

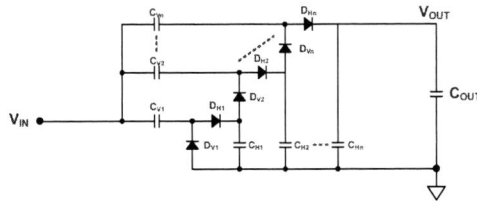

Figure 1.   Circuit schematic of RF voltage multiplier-rectifier

Simulation results of the RF rectifier circuit at a 60 kΩ output load are shown in Fig. 2. Schottky diodes are used for rectification devices. In our simulation conditions, the energy conversion efficiency decreases as the number of voltage multiplier stages increases due to input impedance mismatch of the multiplier. On the other hand, the energy conversion efficiency increases with capacitor values because of loss reduction. In our system, maximum energy conversion efficiency of 24% occurs at $C_N = 5$ pF for 3-stage multiplier.

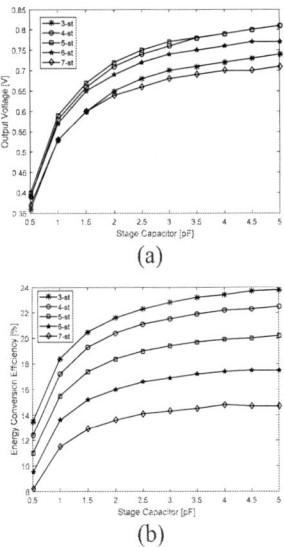

(a)

(b)

Figure 2.   Simulation results of the RF voltage multiplier. (a) Output voltage (b) Conversion efficiency.

B. DC-DC Converter

As shown in Fig. 2, the output conditions of the RF energy harvesting block might vary according to the number of stage and considerations on efficiency as well as the incoming signal magnitude. A buck-boost converter, shown in Fig. 3, is included as the DC-DC converter. In addition, the output of the converter can be a battery (or super capacitor) to be charged or other power management unit such as low-dropout regulator. A buck-boost converter topology is chosen to process various input and output conditions.

The proposed buck-boost converter consists of on-switch control block, bias and reference block, 100kHz ring oscillator circuit, and burst-mode control block. The internal bias circuit and oscillator circuit generate the main clock, which controls operations of buck-boost converter. Since the load current is small, burst-mode timing scheme is included for better efficiency. The off-switch devices are implemented with diodes in order to alleviate the burden of driving circuits.

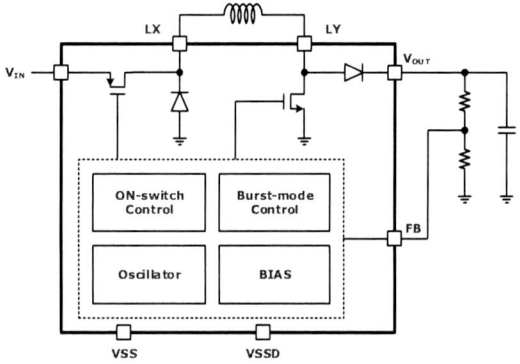

Figure 3.   Block diagram of the buck-boost converter.

The typical operation of a buck-boost converter is shown in Fig. 4. Although input voltage of buck-boost converter is changed, output voltage of buck-boost converter is kept constant. By burst-mode timing scheme, the buck-boost converter is switching operation only as the output voltage is smaller than lower limit reference voltage. If output voltage is greater than upper limit reference voltage, buck-boost converter is not switching operation until output voltage is decreased to lower limit reference voltage.

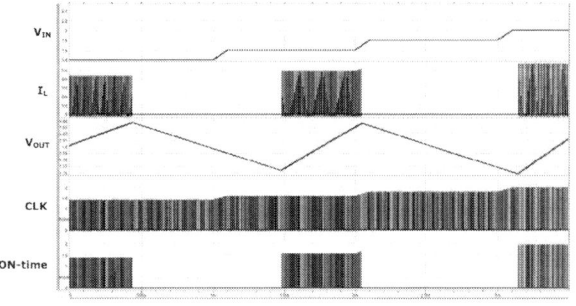

Figure 4.   Operation of the buck-boost converter

III.   DESIGN IMPLEMENTATION

The proposed RF energy harvesting IC is designed and fabricated in a 0.18-μm CMOS process. The input RF frequency band is 2.4 GHz for processing most available WIFI applications. The output voltage can be variable with external scaling resistors. The layout of the RF energy harvest system is as shown in Fig. 5, where the area is 0.98 mm². Three types of RF voltage multiplier and 3-channel DC-DC converter are included for prototype IC.

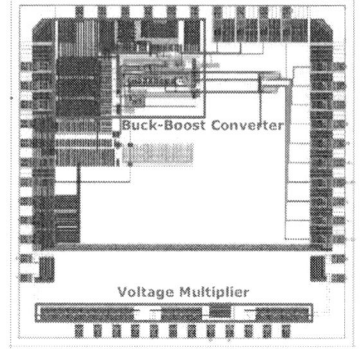

Figure 5.   Layout of the proposed RF energy harvest

IV.   CONCLUSION

In this paper, we present an RF energy harvesting IC in a 0.18-um CMOS process. The proposed RF energy harvesting IC is suitable for a Wi-Fi band at 2.4 GHz. The designed system consists of a RF voltage rectifier and DC-DC converter. Dickson voltage multiplier is used as the voltage rectifier and a buck-boost converter is employed as the DC-DC converter.

ACKNOWLEDGMENT

This research was supported by the MSIP(Ministry of Science, ICT and Future Planning), Korea, under the ITRC(Information Technology Research Center) support program (IITP-2016-H8501-16-1010) supervised by the IITP(Institute for Information & communications Technology Promotion)

REFERENCES

[1] M. Marzencki, M. Defosseux, and S. Skandar, "MEMS Vibration Energy Harvesting Devices With Passive Resonance Frequency Adaptation Capability," *Journal of Microelectromechanical Systems*, vol. 18, no. 6, pp. 1444–1453, Dec. 2009.

[2] K. Finkenzeller, *RFID Handbook: Fundamentals and Applications in Contactless Smart Cards and Identification*, 2nd ed. Chicester, Sussex, U.K.: Wiley, 2003.

[3] T. Le, K. Mayaram and T. Fiez, "Efficient Far-Field Radio Frequency Energy Harvesting for Passively Powered Sensor Networks," *IEEE Journal of Solid-State Circuits*, vol 43, no. 5, pp. 1287-1302, May 2008.

[4] R. W. Erickson and D. Maksimovic, *Fundamentals of Power Electronics*. Norwell, MA, USA: Kluwer, 2011.

[5] U. Karthaus and M. Fischer, "Fully integrated passive UHF RFID transponder IC with 16.7-μW minimum RF input power," *IEEE J. Solid-State Circuits*, vol. 38, no. 10, pp. 1602-1608, Oct. 2003.

Design and Verification of sensorless BLDC motor start-up Logic with FPGA

Hyun-Young Lee, Byeong-Chan Jeon, Won-ki Park, and Sung-Chul Lee

SoC Platform Research Center
Korea Electronics Technology Institute
Seongnam, Korea
sapho@keti.re.kr

Abstract— **This paper shows the Design and implementation of sensorless BLDC motor drive logic with FPGA. In this paper BEMF method was used to detect rotator position and start-up algorithm and implementation were suggested. This system was operated with the supply voltage range up to 24 V at1.5A ,120mNm load.**

Keywords; Sensorless; BLDC Motor; BEMF; FPGA;

I. INTRODUCTION

In Sensored BLDC motor, Hall sensors are required to obtain the information of Rotor position. However, Sensorless motor without hall sensors is preferred because costs and reliability issues.

A sensorless BLDC motor drive is shown in Fig.1. When motor rotates, each winding generates BEMF which opposes the main voltage supplied to the windings according to the Lenz's law.

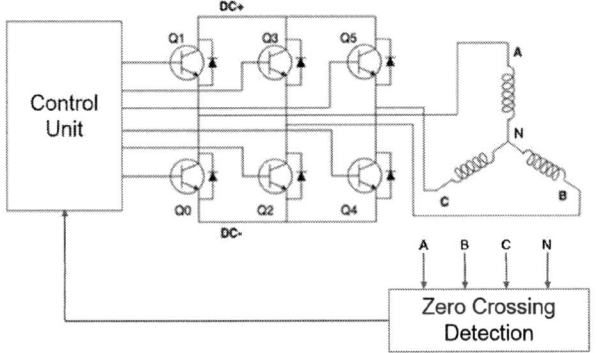

Figure 1. System schematic of sensorless BLDC motor drive

In this paper, Sensorless BLDC motor drive algorithm based on BEMF detection methods was designed and verified with FPGA.

II. DESIGN AND IMPLEMENTATION OF SENSORLESS MOTOR DRIVE ALGORITHM

A. Principle of sensorless motor drive algorithm

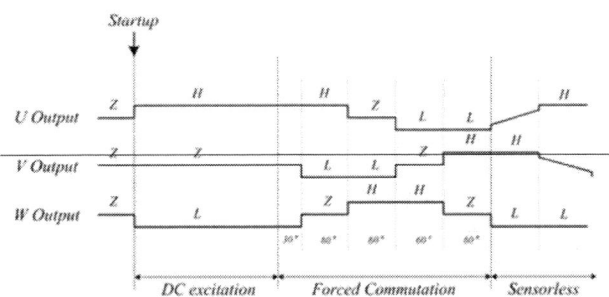

Figure 2. Sensorless Start-Up

It is hard to detect zero crossing point when motor speed is slow or start-up operation. Therefore, start-up algorithm and logic is critical to drive sensorless BLDC motor. Start-up algorithm consists of three areas.

In DC excitation area , Rotor position is aligned through forced DC excitation. After that, rotators speed is pulled up until BEMF is detected in Forced commutation area. As BEMF is detected, motor is controlled by the zero-crossing-point detection in Sensorless mode.

B. Algorithm of sensorless BLDC motor

Fig. 2. shows the algorithm of sensorless BLDC motor drive . When 'BRAKE' signal is 'H', Motor is in stationary state (3 upper switches are off and 3 lower switches are on). Motor state goes to "WAIT STATE" mode when BRAKE signal is 'L'. In "WAIT STATE" mode, upper and lower switches are all off. In "WAIT STATE" mode, motor goes to align mode. In align mode, upper switch at U phase and lower switch at W phase are turned on. PWM duty of upper switch at U phase is fixed at 50 %, which means current is limited to a half of maximum current. If 'art_int_d 'signal is "H", motor enters 'FORCED COMMUTATION' mode which also limits the current through the PWM duty as does the align mode.

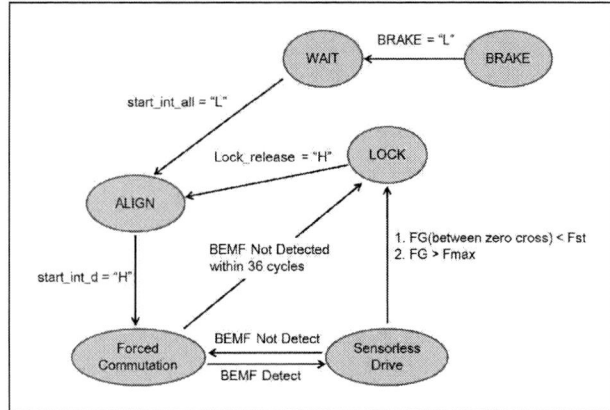

Figure 2. Drive Mode Sequence

Commutation is carried out with initial start up frequency by FST<1:0> registers. After the BEMF is detected more than 6 consecutive times, motor state goes to "SENSORLESS DRIVE" mode. Its state returns to the Forced Commutation mode if BEMF isn't detected in SENSORLESS DRIVE MODE. In FORCED COMMUTATION mode, after failing of BEMF detection in 36 cycles, motor enters "LOCK" state. If the motor's speed is slow (FG_OUT <FST) or higher than FMAX, it goes into LOCK mode and all upper and lower drivers go to off state during LOCK TIME. After LOCK TIME, Lock_release = "H" and enter into align mode again.

III. IMPLEMENTATION AND VERIFICATION

A. System Implementation with FPGA

To verify sensorless BLDC motor drive algorithm, its system was implemented using FPGA. Figure. 3 and Figure. 4 each show the implemented FPGA board Block diagram and system board.

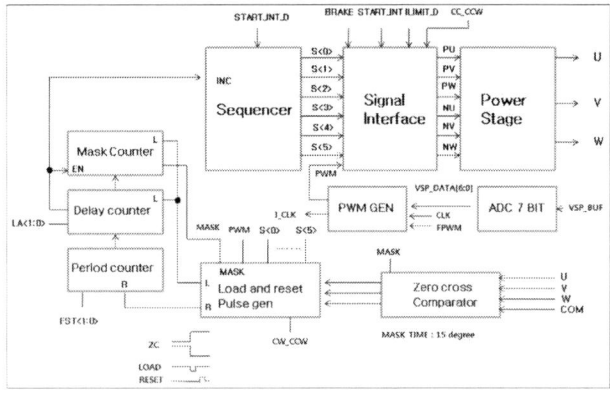

Figure 3. FPGA Board Block diagram

 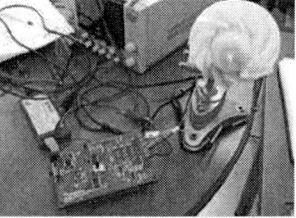

(a) Test Board   (b) Test Setup

Figure 4. FPGA Test Board

B. Verification.

1) Start-Up Test : FIG.5. shows the test results of the Start-up algorithm when motor input voltage is 12V, 18V. In DC excitation period, U phase sets to H , W phase sets to L and V phase sets to Z (High impedance) during initial 250ms, After that , motor state goes to sensorless mode through Forced commutation state.

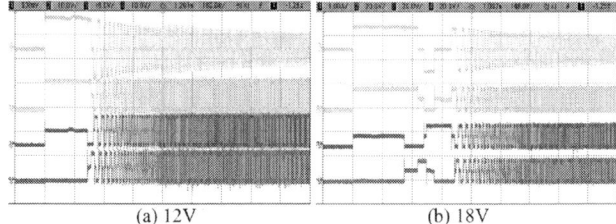

(a) 12V   (b) 18V

Figure 5. Start-Up Test

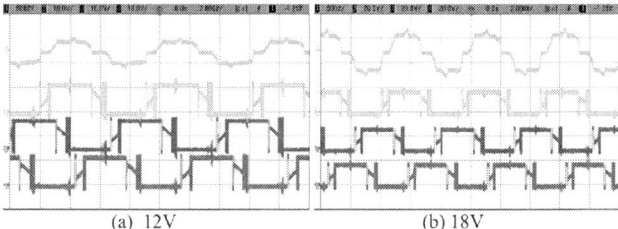

(a) 12V   (b) 18V

Figure 6. Normal Operation

2) Normal Operation : Fig.6. shows the output phase current and phase voltage according to the input voltage during normal operation in SENSORLESS DRIVE mode. Fig.6 shows that It's operation works well.

IV. CONCULSION

In this paper, an algorithm to detect the BEMF Zero-Crossing Point at startup condition was introduced. To verify its algorithm, Sensorless BLDC motor control logic was implemented with FPGA (Xilinx Spartan-6). . This algorithm was verified at Start-up and Normal operation modes with various conditions. The result shows its algorithm works well

REFERENCES

[1] T. H. Kim and M. Ehasani, "Sensorless control of the BLDC motors from near-zero to high speeds," IEEE Trans. on Power Electronics, vol. 19, no. 6, pp. 1635-1645, 2004.

[2] T. S. Kim, B. G. Park, D. M. Lee, J. S. Ryu, and D.S. Hyun, "A new approach to sensorless control method for brushless DC motors," International Journal of Control, Automation, and Systems, vol. 6, no. 4, pp.477-487, Oct. 2008.

A Dimmable and Power-compensated AC Direct LED Driver with High Efficiency

Donglie Gu, Shengpeng Tang, Jianxiong Xi, and
Lenian He

Institute of VLSI Design, Zhejiang University
Hangzhou, China, 310027
Email: gudl@vlsi.zju.edu.cn

Kexu Sun

Department of Electrical Engineering
Southern Methodist University
Dallas, Texas, USA
Email: sunkexu@gmail.com

Abstract—A dimmable and power-compensated AC direct LED driver with high efficiency for lighting applications is presented. A Source-Coupled Pair (SCP) is adopted to realize the soft self-commutating. There is no additional sensing or control circuitry, which lower the system power dissipation. A pulse-width modulation (PWM) dimming technique is applied in regulating the LED's brightness. Furthermore, the driver includes a Power Compensation module which enables the output power of LEDs approximately constant under various line voltage. The LED driver IC is designed and implemented in CSMC's 700 V 1 um BCD process. The post-layout simulation results show that under 220 V|rms AC voltage, the driver has a high efficiency of 90.28% and realizes dimming with a 90% duty ratio, 10 KHz dimming signal. The output power compensated only changes 4% under a 10% variation in input voltage.

Keywords—*soft self-commutating; power compensation; dimming; SCP; high efficiency*

I. INTRODUCTION

As a new type of semiconductor solid-state light source, Light-emitting diode (LED) has been widely used in lighting system for its long life and environmental friendliness [1]. Many self-commutating methods using minimum control circuitry for multiple-string LED drivers have been proposed [2]-[6]. Those works all operate under the fixed input line voltage. However, in practical applications, the input voltage can jitter in complicated grid environment. As depicted in Fig. 1, when the amplitude of line voltage raises, the more current would flow through LEDs in the same period of time [2]. Similarly, the current decreases when the amplitude of line voltage falls. Hence, the average power and the luminous intensity of LEDs cannot stay constant, which would damage human vision. To overcome these

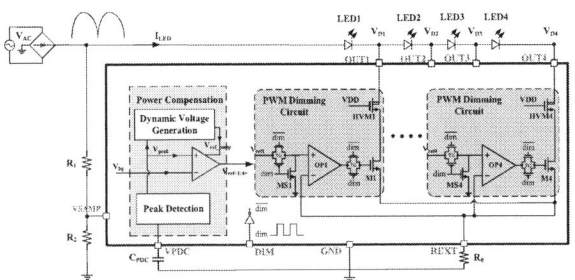

Figure 2. Proposed dimmable LED driver with Power Compensation

problems, a dimmable and power-compensated LED driver applied under various line voltage is presented.

II. PRINCIPLE OF OPERATION

A. Proposed LED Driver

Fig. 2 depicts the proposed LED driver which includes a bridge rectifier, four LED strings, sampling resistors R_1 and R_2, a capacitor C_{PDC} and a controller IC. The IC is comprised of a Power Compensation module and four corresponding PWM dimming circuits.

B. Principle for soft self-commutation and dimming control

As shown in Fig. 2, the high-voltage NMOSFETs HVMs with gate connected to VDD are introduced to withhold the line voltage and supply the adequate current capacity for SCPs. Hence, the SCPs can be implemented by the low-voltage NMOSFETs Ms whose gates are controlled by OPs for the sake of fast dynamic response. When the signal dim is at low level, the transmission-gates TGs are on and the MOS switches MSs

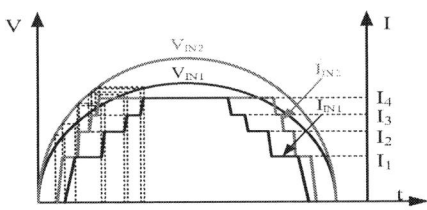

Figure 1. Timming diagram of LED current I_{IN1}, I_{IN2} in a 4-string LED driver under normal line voltage V_{IN1} and increased line voltage V_{IN2}

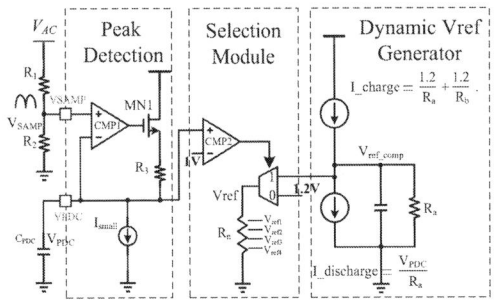

Figure 3. The circuit of power compensation module

978-1-5090-3220-4/16 $31.00 © 2016 IEEE   343   ISOCC 2016

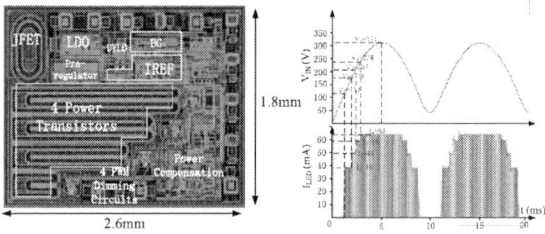

Figure 4.   (a) The chip layout of the proposed LED driver

are off, which transmit the reference voltages $V_{ref1} \sim V_{ref4}$ to the corresponding non-inverting inputs of OPs. When the drain voltages $V_{D1} \sim V_{Di}$ $(1 \leq i \leq 4)$ are available at the same time, the gate highest-biased MOSFET would carry all the current equal to V_{refi}/R_0 because of the negative feedback clamping by OPi and Mi. Therefore, the combination of the SCP with incremental gate voltages and the LED strings in series enables soft self-commutation without any additional sensing and control circuitry [2], which can achieve considerable efficiency.

Conversely, when the signal dim is at high level, the transmission-gates TGs are off and the MOS switches MSs are on. Each non-inverting input of OPs is grounded and all Ms are off. There is no current flowing through LED strings.

III.   PRINCIPLE OF POWER COMPENSATION

The circuit of Power Compensation presented in Fig. 3 includes three parts, Peak Detection, Selection Module and Dynamic Voltage Generator. In Peak Detection, there exists a fixed discharge current I_{small} for C_{PDC}. When the voltage V_{PDC} is lower than the sampled line voltage V_{SAMP}, the comparator CMP1 outputs high level and the MOS transistor MN1 is turned on to charge the capacitor C_{PDC} with a large current. Both the constant small discharge current and the conditioned large charge current make V_{PDC} approximately equal to the peak value of V_{SAMP}. Next, the Selection Module receives V_{PDC} and determines reference voltages V_{ref1-4} which are the input of the OP1-OP4 as shown in Fig. 2. When V_{PDC} is lower than 1 V, the mux would select 1.2 V as a new reference voltage. Or the mux would choose V_{ref_comp} which is generated by Dynamic Vref Generator. This value is determined by charge current I_charge and discharge current $I\_discharge$. The final reference voltage can be formulated as follows:

$$V_{ref} = \begin{cases} 1.2 & V_{PDC} \leq 1 \\ 1.2 + (1.2 - V_{PDC})\frac{R_a}{R_b} & V_{PDC} > 1 \end{cases} \qquad (1)$$

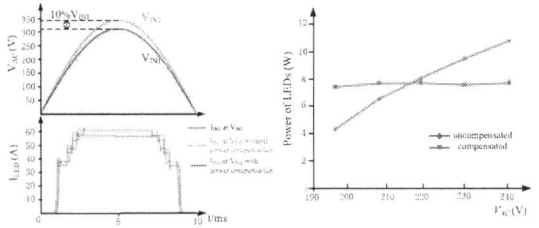

Figure 5.   (a) Timming diagram of V_{IN1}, V_{IN2}, I_{IN1}, I_{IN2} and I_{IN3}
(b) Curves of output power with and without power compensation

TABLE I.   COMPARISON OF PUBLISHED RESULTS AND THIS WORK

Ref.	[5]	[6]	This Work
PF	95%	94%	97.74%
THD	30%	20%	16.33%
Efficiency	80%	75%	90.28%
Output Power	6.8W	8.1W	7.65W

V_{ref1-4} are generated according to V_{ref} by the resistance divider R_n. Operating with these dynamic reference voltages, the current flowing through LED strings can be regulated to an approximately constant value once the line voltage fluctuates.

IV.   SIMULATION RESULTS

The controller IC is implemented by CSMC's 700 V 1 um BCD process. The chip layout is showed in Fig. 4(a). Fig. 4(b) presents the post-layout simulation waveforms of the dimmed LED strings current under the 220V|rms 50Hz AC input and 10 KHz, 90% duty ratio dimming signal condition. The comparison between the compensated and non-compensated LED current is depicted in Fig. 5(a). Fig. 5(b) indicates that there is only 4 percent of variance in output power when the line voltage changes normally 10 percent. As shown in Table I, the driver proposed operates better with PF, THD and Efficiency.

V.   CONCLUSION

A dimmable and power-compensated AC direct LED Driver with high efficiency is addressed in this paper. Based on the SCP soft self-commutating structure, a PWM dimming method is adopted to dim the LED lamp by controlling the on-off time of the SCP. The IC includes the Power Compensation Module, which keeps the output power of LEDs approximately constant under various input voltage. According to simulation results, the proposed driver achieves a high efficiency (90.28%) and realizes dimming in 10 KHz, 0.9 duty ratio under the 220V|rms 50Hz AC condition. The corrected output power only changes 4% under a 10% variation in the input voltage.

REFERENCES

[1]   M. O. Holcomb et al., "The LED lightbulb: are we there yet? progress and challenges for solid state illumination," *Lasers and Electro-Optics, 2003. CLEO '03. Conference on*, Baltimore, MD, USA, 2003, pp. 4 pp.-.

[2]   J. Kim, J. Lee and S. Park, "A soft self-commutating method using minimum control circuitry for multiple-string LED drivers," *2013 IEEE International Solid-State Circuits Conference Digest of Technical Papers*, San Francisco, CA, 2013, pp. 376-377.

[3]   Y. Yeh, M. Chen, X. Li, H. Shinohara and T. Yoshihara, "AC direct multiple-string LED driver with low THD and minimum components," 2015 International SoC Design Conference (ISOCC), Gyungju, 2015, pp. 117-118.doi: 10.1109/ISOCC.2015.7401680

[4]   Si Fu; Minjie Chen; Xutao Lee; Yoshihara, T., "A high efficiency multichannel LED driver based on converter-free technique and load adaptive method," SoC Design Conference (ISOCC), ISOCC.2014

[5]   Kang E, Kim J, Oh D, et al. A 6.8-W purely-resistive AC light-emitting diode driver circuit with 95% power factor[C]//Power Electronics and ECCE Asia (ICPE & ECCE), 2011 IEEE 8th International Conference on. IEEE, 2011: 778-781.

[6]   Dayal R, Modepalli K, Parsa L. A direct AC LED driver with high power factor without the use of passive components[C]//Energy Conversion Congress and Exposition (ECCE), 2012 IEEE. IEEE, 2012.

HV Switch Using Differential Voltage Shaping Driver for 13 Series Li-ion Battery Cells BMS

Tzung-Je Lee, *Member, IEEE*
Department of Computer Science and Information Engineering
Cheng Shiu University, Kaohsiung, Taiwan 83347
Email: tjlee@gcloud.csu.edu.tw

Abstract—**This paper presents a HV switch for the 13 series Li-ion battery cells. In order to operate at the large current of 1.0 A with the battery voltage of 46.68 V, the Differential Voltage Shaping Driver is employed to improve the slew rate of the driving voltage by 77.9%. Besides, the isolation is improved from -46.182 dB to -52.254 dB. The proposed design is implemented using a typical 0.25 um 1P3M 60V BCD process.**

Keywords—Lithium-ion, battery, BMS, HV switch

I. INTRODUCTION

Lithium-ion battery cells operate in series and in parallel to provide the required loading voltage and current in various electronics products [1]. For 13 series battery cells with typical voltage of 3.6 V and 2.2 Ah capacity [2], it requires a HV switch controlled by the BMS (Battery Monitoring System) to deal with the large voltage of 46.8 V and the large current of 1.0 A, as shown in Fig. 1. However, the operation voltage and current in the state-of-the-art designs are too small [3], [4]. Thus, this paper proposes a HV (High Voltage) switch by using a Differential Voltage Shaping Driver to improve the driving voltage for the applications of 46.8 V and 1.0 A.

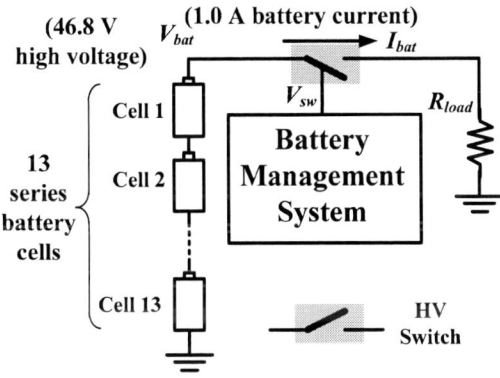

Fig. 1. The HV switch for the 13 series Li-ion battery cells.

II. HV SWITCH WITH DIFFERENTIAL VOLTAGE SHAPING DRIVER

Fig. 2 (a) reveals the schematic of the traditional HV switch. The HV PMOS transistors, M_{P1} and M_{P2}, are connected in series as the switching transistors. The driving voltage is determined by the control signal V_{SW}. When M_{n1} is turned on, V_{drive0} is coupled to $V_{bat,n-1}$ to turn on M_{P1} and M_{P2}. When M_{n1} is off, V_{drive0} is at $V_{bat,n}$ to turn off the switching transistors. The size of M_{P1} and M_{P2} are very large to provide the large current

of 1.0 A. Due to the large gate capacitive loads, the transient response is limited by the slew rate of V_{drive0}.

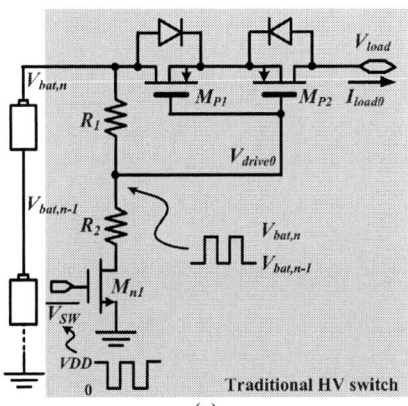

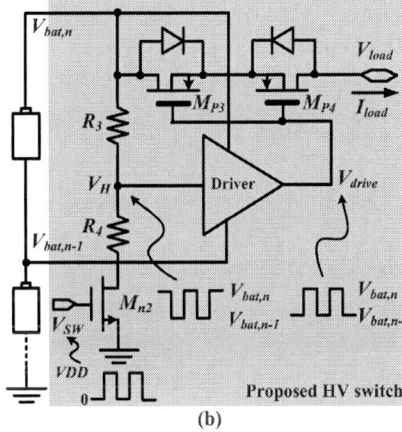

Fig. 2. Schematic of (a) the traditional and (b) the proposed HV switch.

Fig. 2 (b) shows the schematic of the proposed HV switch. The Differential Voltage Shaping Driver is employed to improve the slew rate of V_{drive}. The Differential Voltage Shaping Driver is composed of the differential signal generator (Diff Gen), the differential signal shaper (Diff Shaper), and the differential-to-single converter (DtoS Conv), as shown in Fig. 3. Diff Gen generates the inverse signal of V_H. Thus, V_H and V_{H2} become the differential signals and are amplified by the Diff Shaper. The amplified differential signals, V_{a1} and V_{a2}, are then converted into the single voltage, V_{drive}, by DtoS Conv. Notably, M311-M314 are the cross-coupled structures, which can improve the transient response. M319 is self-biased by M317.

978-1-5090-3220-4/16 $31.00 © 2016 IEEE   345   ISOCC 2016

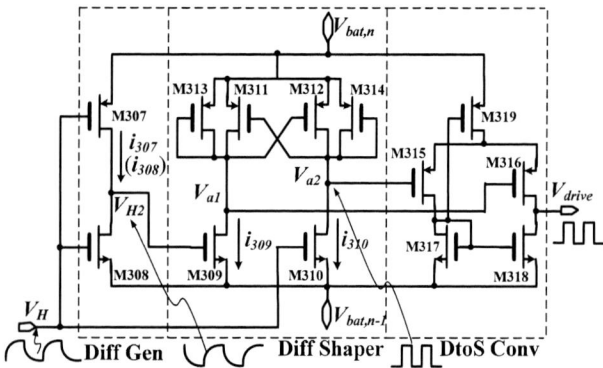

Fig. 3. Schematic of the Differential Voltage Shaping Driver.

III. IMPLEMENTATION AND SIMULATION RESULTS

The proposed HV switch is implemented using a typical 0.25 um 1P3M 60 V BCD process. Fig. 4 reveals the simulated waveforms of the Differential Voltage Shaping Driver at 100 KHz. Because the gate capacitive load of the M307, M308 and M310 is quite small, the differential signals, V_H and V_{H2}, could be driven from 43.2 V to 46.8 V easily. The slew rate of V_{a1} and V_{a2} is then improved by Diff Shaper. The single driving voltage, V_{drive}, is finally obtained. Fig. 5 reveals the waveforms compared to the traditional HV switch, as shown in Fig. 2 (a). The slew rate of the driving signal is improved by 77.9% (from 1.8237 V/us to 8.2568 V/us) at the worst case. Besides, the isolation is improved from -46.182 dB to -52.254 dB. Table I reveals the comparison of the specifications with several prior works. The proposed design possesses the best performance on the operation voltage, current, and the turned-on resistance.

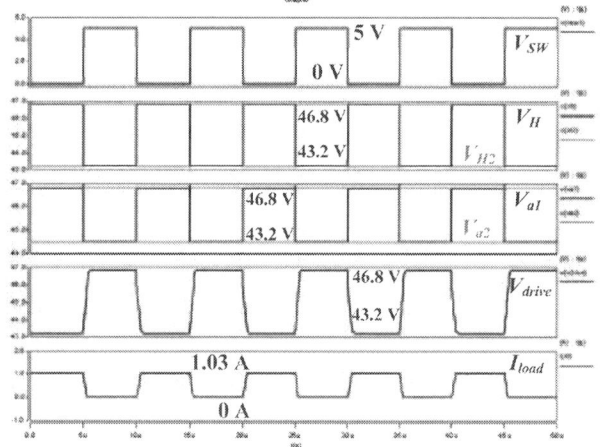

Fig. 4. Simulated waveforms of the proposed HV Differential Voltage Shaping Driver.

IV. CONCLUSION

This paper proposes the HV switch for the 13 series battery cells BMS. By using the Differential Voltage Shaping Driver, the proposed HV switch is adequate to the large current of 1.0 A and large voltage of 46.8 V with the isolation of -52.25 dB and the turned-on resistance of 1.81 Ω.

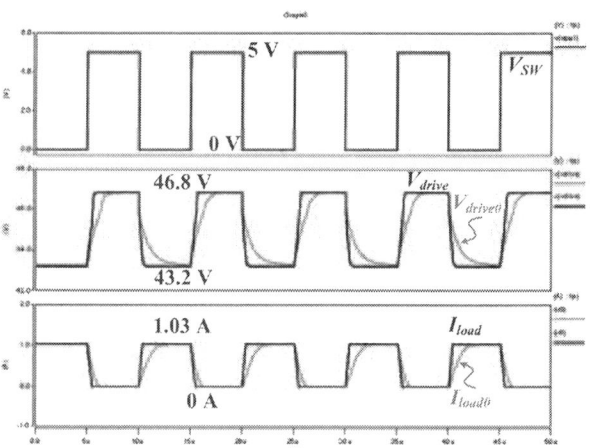

Fig. 5. Simulated waveforms of the proposed HV switch and the traditional HV switch.

TABLE I.   COMPARISON WITH SEVERAL PRIOR WORKS

	This work	Traditional*	[3]
Year	2016	N/A	2012
Publication	ISOCC	N/A	ICICDT
Process (um)	0.25	0.25	0.25
Supply voltage (V)	5.0	5.0	N/A
Battery voltage (V)	46.8	46.8	40
Battery current (A)	1.0	1.0	N/A
On resistance (Ω)	1.81	1.91	16.8
Efficiency (%)	91.29	84.76	N/A
Isoaltion (dB)	-52.25 @1MHz	-46.18 @1MHz	-79.4 @10MHz

Note: * The traditional design is simulated by this work at the same conditions except for the driver.

ACKNOWLEDGMENT

This research was partially supported by the Ministry of Science and Technology under grant no. MOST 105-2221-E-230-013 and Cheng Shiu University under grant 105C18. Moreover, the authors would like to express their deepest gratefulness to CIC (Chip Implementation Center) of NARL (National Applied Research Laboratories), Taiwan, for their thoughtful chip fabrication service.

REFERENCES

[1] C.-H. Kim, M.-Y. Kim, Y.-D. Kim, and G.-W. Moon, "A modularized charge equalizer using battery monitoring IC for series connected Li-Ion battery strings in an electric vehicle," *IEEE Transactions on Power Electronics*, vol. 28, no. 8 pp. 3779-3787, Aug., 2013.

[2] *bq24105 Datasheet*, Synchronous switch mode, Li-Ion and Li-Polymer charge-management IC with integrated power FETs, Texas Instruments, 2010.

[3] C.-L. Chen, Y. Hu, W. Luo, C.-Y. Juan, and C.-C. Wang, "A high voltageanalog multiplexer with digital calibration for battery management systems," in Proc. *IEEE Inter. Conf. IC Des. Technol.*, May 2012, pp. 1-4.

[4] M. Du and H. Lee, "An Integrated speed and accuracy-enhanced CMOScurrent sensor with dynamically biased shunt feedback for current-mode buck regulators," *IEEE Trans. on Circuits Systems-I, Reg. Papers*, vol. 57, no. 10, pp. 2804-2814, Oct. 2010.

A CMOS buck converter
with PFM / Hysteretic mode

Tae-Heon Lee, Jong-Gu Kim, and Kwang-Sub Yoon
Department of Electronic Engineering
Inha University
Incheon, Korea
xogjs9503@naver.com, ksyoon@inha.ac.kr

Abstract— **This paper proposes design of a CMOS DC-DC buck converter with Fixed Hysteretic mode and PFM(Pulse Frequency Modulation) mode. The inherent problems of a slow transient time from heavy load to light load that the conventional Dual DC-DC buck converters with PWM/PFM mode has faced have been resolved by using the proposed DC-DC buck converter which employed Hysteretic mode in heavy load. It improves transient time in load regulation. This also can maintain range of wide load current, advantage of Dual mode buck converter. The proposed Dual Buck Converter is fabricated in 180nm CMOS 1-poly 6-metal process and occupies a core effective area of 1.35 mm^2. Measurement environment are input voltage range of 2.7~3.3V, output voltage 1.2V and load current range from 10mA to 500mA. And measurement result shows that the maximum efficiency is 90% and ripple voltage is less 1.32mV.**

Keywords; DC-DC Buck Converter, Dual, FVC, mode controller, mobile applications

I. INTRODUCTION

Recently, the number of mobile device is increasing by encountering easily wearable device and IoT(internet of Things). Mobile device is required by designing System held DPSS(Dynamic Partial Shutdown Strategy) Mode and Awake Scheduling in wide load range in order to use effectively battery. Thus researches on dual mode PMIC (Power Management Integrated Chip) are actively conducting. [1, 2]

In various feedback modes, hysteretic mode has EMI noise due to the variation of switching frequency and the noise makes efficiency degraded. And output ripple voltage of the hysteretic mode is not small. PFM (Pulse Frequency Modulation) mode generates switching frequency depending on load current. The mode has high efficiency in light load because of low switching frequency. But the mode becomes low efficiency due to switching loss. PWM (Pulse Width Modulation) mode has high efficiency in heavy load and small output ripple. But it is difficult to make compensation in the PWM. The mode has lower efficiency in light load because many sub-circuits consume powers. The proposed circuit employed fixed switching frequency hysteretic Mode with PLL (phase locked loop) architecture [3] in heavy load and PFM mode in light load. The hysteretic mode is capable of removing EMI noise conventional that hysteretic converter suffered from because switching frequency is fixed. By using PFM mode in

light load, the proposed circuit with a fixed switching frequency can minimize switching loss caused by high switching frequency in light load. Thus the proposed circuit can maintain high efficiency in wide range load. It also can obtain faster transient time than conventional Dual buck converter because the proposed buck converter contains hysteretic mode.

The rest of this paper is organized as follows. Section II describes the proposed architecture and design of the main block. The simulation results are presented and discussed to compare the performances of the proposed circuit with those of the conventional circuits in section III. Finally, the conclusions are drawn in section IV.

II. THE PROPOSED ARCHITECTURE

The overall architecture proposed in section II is shown in Fig. 1. It consists of the conventional hysteretic buck converter with ramp generator, PLL architecture, PFM buck, PFM & Hysteretic mode controller. PFM & Hysteretic mode controller

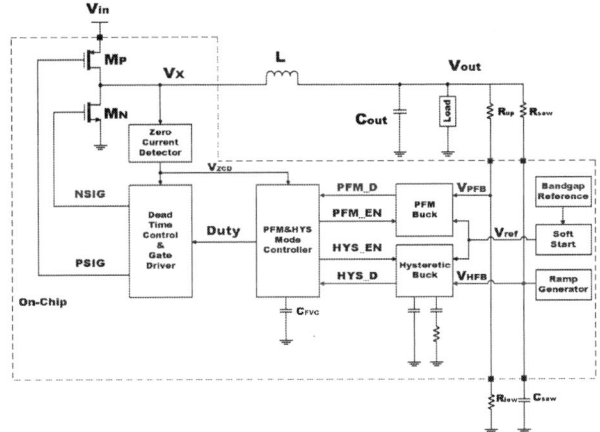

Fig. 1 Proposed Dual buck converter Block Diagram

are connected with PFM duty node and Hysteretic duty node to control depending on load current. PFM & HYS Mode controller is composed of FVC (Frequency to Voltage

978-1-5090-3220-4/16 $31.00 © 2016 IEEE   347   ISOCC 2016

Converter) and digital circuits. FVC receives the output signal (V$_{ZCD}$) of ZCD (Zero Current Detector) and operates mode controller to classify either heavy load or light load. The FVC is presented in Fig. 2. The FVC with C$_{FVC}$ receives signals (V$_{ZCD}$ and HYS_D). The multiplexer (MUX) in FVC receives PFM_EN to control amplitude of the V$_{FVC}$ according to each mode. The comparator compares Vref with V$_{SAW}$, the ramp-wave generated by V$_{FVC}$. V$_{FVC}$ is described as (1).

,where V$_{SAW}$ is the ramp wave made by pulse wave of the V$_{FVC}$. I$_{mode}$ and T$_{mode}$ are the current and the time of mode, respectively, depending on load current. Output voltage of the comparator decides D and Q of the D-latch, which determines mode of the proposed circuit. PFM buck is connected between the V$_{PFB}$ node and mode controller. Hysteretic buck is connected with the output node of ramp generator to use small inductor.

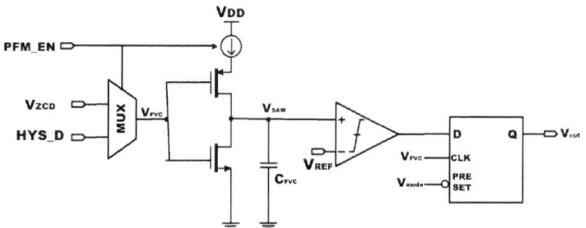

Fig. 2 FVC for PFM & HYS mode controller

III. SIMULATION RESULTS

As the load current changes from 150mA to 30mA, the ramp waveform, V$_{SAW}$ and output voltage, V$_{ctrl}$ of FVC mode controller are shown in Fig. 3 (a). The waveforms of V$_{SAW}$, Vref and V$_{ctrl}$ with load current changing from 30mA to 150mA are presented in Fig. 3 (b). The waveform shown in Fig. 3(c) demonstrates the hysteretic mode of the proposed circuit with fixed switching frequency of 1.97MHz. As the load current goes below 30mA, the buck converter switches to PFM mode through waveform of inductor current. Switching frequency becomes hundreds of hertz in PFM mode.

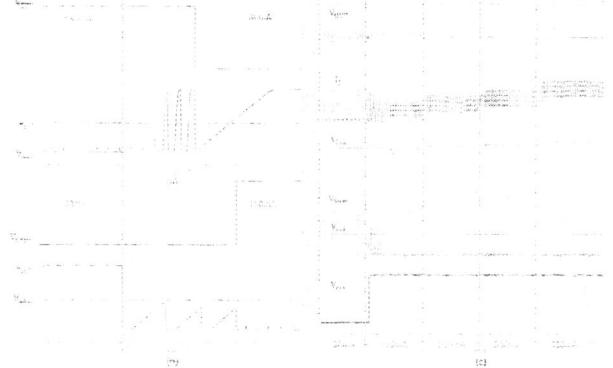

Fig. 3 Simulated waveform : PFM/HYS MODE Controller

The performance comparison was made in table between the conventional works and the proposed circuit. The proposed one achieved not only wider load current range (10-500mA),

TABLE I.   **PERFORMANCE COMPARISON**

	[4]	[5]	This work
Mode	Hysteretic	Dual Mode	Dual Mode
Freq(light load)	50kHz	100~600kHz	9.7~250kHz
Freq(heavy load)	1.38MHz		1.97MHz
Inductor	10uH	10uH	1.2uH
Capacitor	20uF	10uF	22uF
V$_{IN}$/V$_{OUT}$	2.7~4.2V/2V	2.7~5V/1V	2.7~3.3V/1.2V
Ripple voltage	5.73mV	20~36mV	1.37~10mV
Load range	50~500mA	≤ 460mA	10~500mA
Transient Response/overshoot	8.26us/79mV	15us/68mV	3.7us/2.5mV
Efficiency	89~95%	94~95%	84~90%

but also a lower ripple output voltage with a faster transient time(3.7us) and a similar efficiency(84-90%).

IV. CONCLUSIONS

This paper describes the proposed dual mode (PFM for light load and hysteretic for heavy load) buck converter for mobile applications. The proposed circuit was able to achieve high efficiency and fast transient time in all load current. The proposed circuit employs the dual mode included hysteretic mode with fixed frequency of 1.97MHz and PFM mode in light load to obtain fast transient time and high efficiency.

ACKNOWLEDGMENT

This research was supported by the MSIP(Ministry of Science, ICT and Future Planning), Korea, under the ITRC(Information Technology Research Center) support program (IITP-2016-H8501-16-1010) supervised by the IITP(Institute for Information & communications Technology Promotion). This research was also supported by Basic Science Research Program through the National Research Foundation of Korea(NRF) funded by the Ministry of Education(2010-0020163). This work was supported by the IDEC.

REFERENCES

[1] C. Schurgers, V. Tsiatsis, S. Ganeriwal and M. Srivastava "Optimizing Sensor Networks in the Energy-Latency-Density Design Space" IEEE Transactions on Mobile Computing (Volume:1 , Issue: 1), pp 1536-1233, Jan-Mar 2002

[2] O. Ocakoglu and O. Ercetin "Energy Efficient Random Sleep-Awake Schedule Design" IEEE Communications Letters (Volume:10 , Issue: 7), pp 528-530, July 2006

[3] Tae-Jin Jeong, Kwang S. Yoon, "A CMOS Hysteretic DC-DC Buck Converter with a constant switching frequency", Journal of Semiconductor Technology and Science (JSTS), VOL. 15, NO.4, AUGUST, pp471~476, 2015.

[4] C.J. Chuang and H.P. Chou, "An Efficient Fast Response Hysteretic Buck Converter with Adaptive Synthetic Ripple Modulator", 8th International Conference Power Electronics-ECCE Asia, pp 625-627, May 2011.

[5] Wan-Rone Liou, Member, IEEE, Mei-Ling Yeh, and Yueh Lung Kuo "A High Efficiency Dual-Mode Buck Converter IC For Portable Applications," Power Electronics, IEEE Transactions on. pp.667-677, 2008.

A Single Inductor Multiple Output(SIMO) Buck/Boost DC-DC Converter with Output Error-Driven Random Control

Hyunbin Park and Shiho Kim

School of Integrated Technology, and YICT
Yonsei University, Seoul, Korea
{ bin9000, shiho }@yonsei.ac.kr

Abstract— **We propose an output error-driven random controlled Single Inductor Multiple Output (SIMO) buck/boost DC-DC converter operated in the Continuous Conduction Mode (CCM). Random selection of output switches based on maximum instantaneous output error enables us a novel control scheme which leads to a fast transient response with mitigating the cross-regulation problem of buck-boost converters. A freewheeling period for CCM control is not necessary in the proposed output error-driven control scheme thanks to the instantaneous control of output switches, hence, we can achieve high power conversion efficiency as well as stability of operation. Both simulated and experimental results show the effectiveness of the proposed control scheme. The measured power conversion efficiency of the converter with 3 buck/boost output nodes is approximately 91.0%.**

Keywords; Single Inductor Multiple Output(SIMO) converter, buck/boost, error-driven random control, cross regulation, continuous conduction mode

I. INTRODUCTION (HEADING 1)

The SIMO DC-DC converter regulates multiple DC voltage levels with high power conversion efficiency which is requirement of Power Management Integrated Circuits (PMICs) for battery-powered devices. Outputs of the SIMO converter share single off-chip inductor and power switches to achieve low chip area consumption. However, the share causes cross regulation problem. The SIMO converter operated in Discontinuous Conduction Mode (DCM)/Pseudo-Continuous Conduction Mode (PCCM) suppress the cross regulation problem by employing idle time or freewheeling stage [1, 2]. Error-based control rapidly responds to load change since the conduction sequence of output switches is determined by load conditions [3]. Step up/down converter using Time Multiplexing (TM) control is proposed where unbalance loading problem is resolved [4]. As far as the authors know, error-based control scheme was only applied to buck SIMO converter [3], a buck-boost SIMO converter with error-driven random control is not reported yet. The purpose of this presentation is providing a buck/boost SIMO DC-DC converter employing error-driven random control.

This paper is a result of a research project supported by SK Hynix Inc.

II. PROPOSED ERROR-BASED AUTO-BUCK-BOOST CONTROL TECHNIQUE

A. The charging and discharging phases of inductor current

TM control is adopted in the proposed technique where only one charging phase and one discharging phase of inductor current exist in one period, which is controlled by Digital Pulse Width Modulation (DPWM) with fixed frequency. In the charging phase, power switches S_1 and S_3 are closed and rest of switches are opened in Fig. 1 (a). In the discharging phase, power switch S_2 and one of the output power switch selected are closed.

B. Control Strategy

The analog subtractor block in Fig. 1(b) provides voltage level named Verror, subtracting scaled output voltage level from predefined reference level. The control circuits determine the output for charge to be distributed in the discharging phase that presents the highest Verror level. The analog adder block provides voltage level named Vsum, adding all Verror voltage levels. The analog adder and subtractor block consist of an op-amp and resistors. Average inductor current level is adjusted by

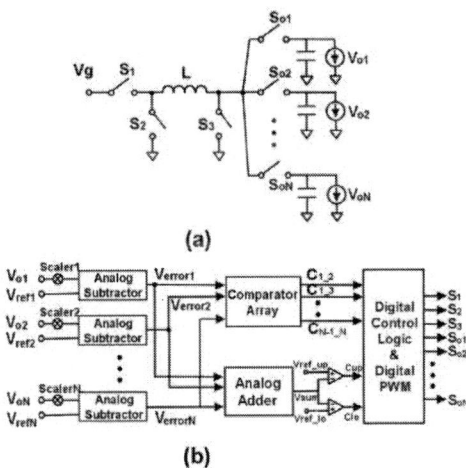

(a)

(b)

Figure 1. Circuit diagram of the proposed SIMO converter (a) power stage with N outputs; (b) block diagram of the control architecture

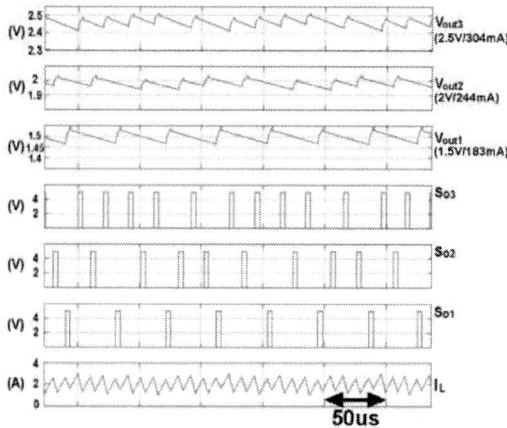

Figure 2. Simulated steady state waveform of output voltages, voltages applied on the gates of output switches, and inductor current where Vg is 2V.

controlling duty cycle of DPWM so that output voltages are regulated even if the load condition is changed. The Vsum voltage level is compared with two reference voltages, Vref_up and Vref_lo. If Vsum is higher than Vref_up, then digital part of control circuits increases duty cycle. On the contrary, if Vsum is lower than Vref_lo, the duty cycle is decreased. If Vsum is between the two reference voltage levels, the duty cycle is maintained.

III. SIMULATION RESULTS

The proposed converter with three outputs is simulated with MATLAB. Input voltage (Vg) and three outputs are set to 2V, 2.5V, 2V, and 1.5V respectively. Inductance of the inductor is 5.8uH, capacitance of output capacitors is 110uF, and Frequency is 98kHz.

Fig. 2 shows the steady state waveforms with load currents 304mA(Iout3), 244mA(Iout2), and 183mA(Iout1). The third output is the most frequently selected since the load current is the largest. Fig. 3 shows the load transient responses. Load current of the second output changes from 20mA to 200mA. To support the load change, duty ratio is increased to raise inductor current level.

IV. EXPERIMENTAL RESULTS

A prototype of the proposed SIMO converter with three

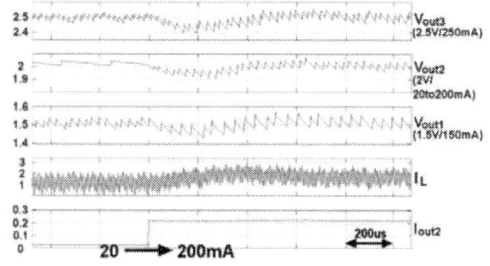

Figure 3. Simulated load transient responses, where I_{out2} is changed from 20mA to 200mA, I_{out3} = 250mA, and I_{out1} = 150mA.

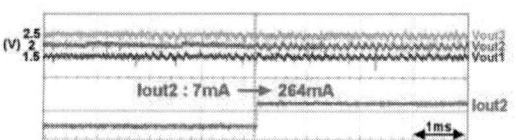

Figure 4. Measured load transient response, where I_{out2} is changed from 7mA to 264mA, I_{out3} = 250mA, and I_{out1} = 150mA. Vout3,

outputs is implemented with discrete components and FPGA with following parameters: Vg = 2V, Vout3 = 2.5V, Vout2 = 2V, Vout1 = 1.5V, L = 5.8uH, Cout = 110uF, and fsw = 98kHz. Fig. 4 shows both buck and boost outputs are regulated and cross regulation effect is not observed noticeably with variation of 257mA in second load current where cross regulation is 0.43mV/mA. Input voltage of the proposed converter ranges from 0.8V to 5V. Measured power efficiency of the converter is 91.02% at total output power 1.31W. The proposed converter regulates in wide range of input voltages with low cross regulation with very simple architecture and achieves high efficiency at heavy loads over 1W with CCM control.

V. CONCLUSIONS

We proposed an output error-driven random controlled SIMO buck/boost DC-DC converter having three different output voltage levels and loads. The Error-based direct Random control on output switches in buck-boost converter leads to Fast response with respect to load change, and therefore to mitigate cross-regulation problem of SIMO converters. High power conversion efficiency is achieved since CCM control is implemented without employing freewheeling period. Simulation and experimental results show the improvement both in the efficiency and drawback of the proposed control.

ACKNOWLEDGMENT

This work was supported by MSIP as a part of the "IT Consilience Creative Program (IITP-2015-R0346-15-1008)" supervised by IITP (Institute for Information & Communications Technology Promotion). EDA tools for circuit design and simulation environment were supported by IDEC.

REFERENCES

[1] D. Ma et al., "Single-Inductor Multiple-Output Switching Converters With Time-Multiplexing Control in Discontinuous Conduction Mode", IEEE J. Solid-state Circuits, Vol. 38, No. 1, 2003.

[2] D. Ma, W. Ki, and C. Tsui, "A Pseudo-CCM/DCM SIMO Switching Converter With Freewheel Switching", IEEE J. Solid-state Circuits, Vol. 38, No. 6, 2003.

[3] M. Jung et al., "An Error-Based Controlled Single-Inductor 10-Output DC-DC Buck Converter With High Efficiency Under Light Load Using Adaptive Pulse Modulation", IEEE J. Solid-state Circuits, Vol. 50, No. 12, 2015.

[4] Y. Zheng et al., "A Single-Inductor Multiple-Output Auto-Buck-Boost DC-DC Converter With Autophase Allocation", IEEE Trans. on Power Electronics, Vol. 31, No3, 2016.

A Synchronous Boost Converter with High Speed and High Accuracy Peak Current Control Unit

Shengpeng Tang, Xianzhi Meng, Donglie Gu
Jianxiong xi and Lenian He
Institute of VLSI Design, Zhejiang University
Hangzhou, China, 310027
Email: tangsp@zju.edu.cn

Kexu Sun
Department of Electrical Engineering
Southern Methodist University
Dallas, Texas, USA
Email: sunkexu@gmail.com

Abstract— **A 3 MHz, 48W Boost converter with high speed and high accuracy Peak Current Control Unit (PCCU) is presented. The Boost controller IC adopts a novel PCCU consisting of a fully differential open-loop operational transconductance amplifier (OTA) and a trans-impedance amplifier (TIA), which can minimize the delay and error of the whole control loop. In the PCCU, the compensated output of error amplifier is moved forward to improve the control accuracy further. This IC is designed in CSMC 0.8 um 60 V BCD process. The simulation results show the Boost converter operates at 3 MHz and the peak current control error is 1.3% under the worst situation. The delay of peak current trigger point to the low side gate driver (LDRV) output is 25.14 ns and the delay of the compensated output of error amplifier to LDRV output is 37.56 ns.**

Keywords—Peak Current Control Unit (PCCU); high speed; high acurracy; Boost converter; fully differential open-loop OTA, TIA; delay; controller IC; low side gate driver (LDRV) output

I. INTRODUCTION

As fundamental power management units, Boost converters are widely applied to Battery-Powered Systems, HEV/EV Powertrain and Automotive Power Systems. Among most common Boost control schemes, Peak Current Mode (PCM) has advantage of fast response on variation of IN/OUT, simple loop frequency compensation and inherent cycle-by-cycle current limit [1]. However, the conventional PCM Boost converter [2]-[3] can hardly achieve both high switch frequency (Multi-MHz) and high accuracy due to low bandwidth of the conventional peak current sensing and control circuitry [4]. To solve the problem, the converter adopting a novel PCCU with high speed and high accuracy is proposed.

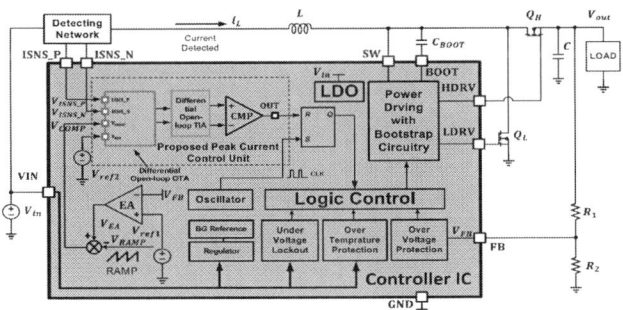

Figure 1. The Boost Controller IC application diagram with PCCU

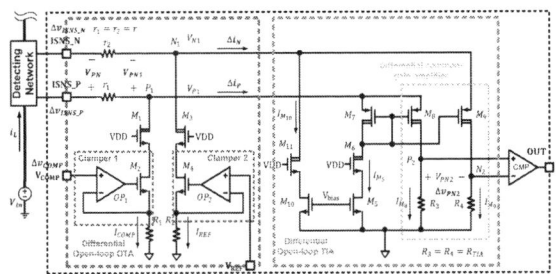

Figure 2. The circuit of PCCU

II. PROPOSED BOOST CONVERTER WITH PCCU

Fig. 1 depicts a synchronous rectifier Boost converter with a controller IC. Based on peak current mode, the control loop of this IC is the same as conventional Boost controller ICs, except for the proposed PCCU which determines the majority of the control loop delay.

Within the blue dashed wireframe, this PCCU includes a Differential Open-loop OTA, a Differential Open-loop TIA and a comparator CMP. The OTA receives two pairs of differential voltage signals, V_{ISNS_P}, V_{ISNS_N} and V_{COMP}, V_{ref2}. Then, it transforms the two pairs of differential voltage signals into one pair of differential current signals and transmits them to the TIA. The TIA amplifies the differential current signals and outputs differential voltage signals which are directly connected to CMP.

Instead of conventional closed-loop circuitry, the proposed PCCU employs fully differential open-loop amplifiers for high bandwidth. To improve the control accuracy further, the compensated output V_{COMP} of EA is moved forward to compare with sampled peak-current voltage signals rather than processed at the last stage analog circuitry CMP [2]-[4].

III. PRINCIPLE OF PROPOSED PCCU

The circuit of proposed PCCU is presented in Fig. 2. As depicted in red dashed wireframe, the Differential Open-loop OTA includes resistors r_1, r_2 and two branches whose current is determined by corresponding operational amplifier clamper. As shown in green dashed wireframe, the differential open-loop TIA mainly includes a differential common-gate amplifier and the branch of current I_{M_5} supplies the amplifier with quiescent current bias. In order to eliminate quiescent voltage difference of the TIA input, the branch carrying I_{M10} is introduced to

978-1-5090-3220-4/16 $31.00 © 2016 IEEE   351   ISOCC 2016

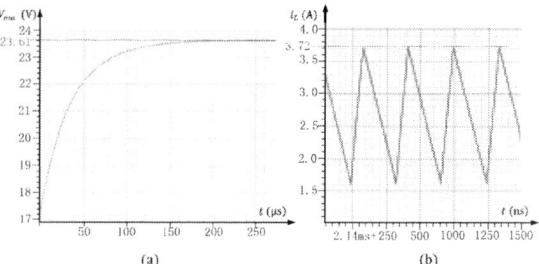

Figure 3. Under shortest switch on-time D=0.25 condition (a) the simulation waveform of V_{out} (b) the simulation waveform of i_L

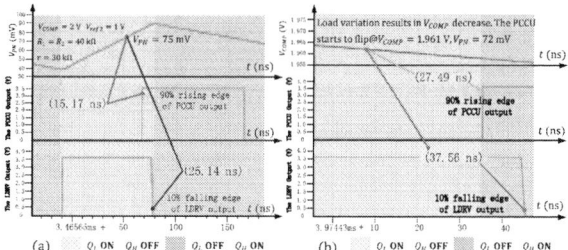

Figure 4. (a) the time delay from V_{PN} to PCUU output and LDRV output (b) the time delay from V_{COMP} to PCUU output and LDRV output

TABLE I. COMPARISON OF PUBLISHED RESULTS AND THIS WORK

Reference	[3]	[5]	This work
Delay of V_{PN} to control output (ns)	N/A	N/A	15.17
Delay of V_{PN} to LDRV output(ns)	N/A	30	25.14
Delay of V_{COMP} to control output (ns)	N/A	N/A	27.49
Delay of V_{COMP} to LDRV output(ns)	N/A	25	37.56
Switch frequency (MHz)	0.5	2.2	3
Maximum propagation delay (ns)	>400	<100	<70
Error of peak current control	≤ 4%	N/A	≤ 1.4%
Output power (W)	3	20	48
Efficiency@full load	N/A	91.3%	90.3%

compensate the N_1 potential drop brought by branch carrying I_{M5}. M_1, M_3, M_{11} and $M_6 \sim M_9$ are high voltage MOSFET, which improve current source performance and withstand high voltage from Boost converter input. When the voltage difference V_{PN} between port **ISNS_P** and **ISNS_N** equals zero, the quiescent voltage difference V_{PN1_Q} of V_{PN1} can be derived as follow

$$V_{PN1\_Q} = -(I_{COMP} - I_{REF})r = -\left(\frac{r}{R_1}V_{COMP} - \frac{r}{R_2}V_{REF}\right) \quad (1)$$

where V_{PN1_Q} is normally set below zero. While the inductor current i_L increases with Q_L ON and Q_H OFF, V_{PN} and V_{PN1} increases. When V_{PN} exceeds $-V_{PN1\_Q}$, the CMP of the PCCU flips and output high level. Afterwards, the Power Driving module of the IC would output low level at LDRV and high level bootstrapped at HDRV, which turns off the Q_L and turns on the Q_H. Hence, the inductor current starts to decrease. When the current drops below the peak current set, the PCCU flips to low level again and the Q_L keeps OFF until next CLK high level arrives at the SET of RS flip-flop. In this way, the proposed converter set the inductor peak current trigger point $V_{tr} = -V_{PN1\_Q}$ and realizes the cycle-by-cycle current control with the procedures illustrated above.

In the proposed PCCU, the compensated EA output V_{COMP} connected to OP_1 stays nearly constant during a switch period without load variation. Although the current I_{COMP} is generated by the closed-loop OPA clamper, the whole module can be also regarded as a "pure" open-loop high-speed system. In addition, there exists two key differential poles at $P_1 N_1$ and $P_2 N_2$, but they can be easily pushed extremely far for adequate bandwidth by optimizing the parameters of circuit.

IV. SIMULATION RESULTS

The Boost controller IC with the proposed PCCU is designed by CSMC 0.5um 60V BCD process. The oscillator frequency of this IC is 3 MHz and the proposed PCCU trigger voltage V_{tr} is 75 mV. The converter parameters are defined, $V_{in} = 18$ V, $R_{LOAD} = 12 \Omega$, $L = 749.3$ nH, $C = 10 \mu$F, $R_1 = 22.8$ MΩ, $R_2 = 1.2$ MΩ, $R_{SENSE} = 24.06$ mΩ, where R_{SENSE} is one of various implementation for Detecting Network. The ideal peak current can be obtained below

$$i_{peak} = \frac{V_{tr}}{R_{SENSE}} + \frac{V_{in}}{L} t_{delay\_tr\_to\_LDRVoutput} \quad (2)$$

Fig. 3 shows the simulation waveform of V_{out} and i_L with 3 MHz switch frequency and the smallest duty cycle 0.25. Fig. 4

shows the time delay of peak current set point V_{PN} and compensated EA output V_{COMP} to corresponding output. The i_{peak} is set as 3.67 A, while the simulation value is 3.72 A. There is only 1.4% peak current control error during the shortest conducting time. As presented in Table I, the proposed Boost converter with the PCCU shows better performance in the delay of V_{PN} to corresponding output, maximum propagation delay, switch frequency and peak current control error.

V. CONCLUSION

A synchronous Boost converter with a high speed and high accuracy PCCU is addressed. Due to open-loop fully differential amplifiers (OTA, TIA) and moving forward of EA compensated output signal, the proposed PCCU can easily realize pretty high speed and accuracy. With the high speed and high accuracy PCCU, a Boost controller IC for 3 MHz and 48W application can be reasonably designed in CSMC 0.5um 60V BCD process. The simulation results show that the converter operates with rather small control delay and error. The delay of V_{PN} to LDRV is 25.14 ns and there is at most 1.4% percent error on peak current control under different duty cycle configurations.

REFERENCES

[1] R. B. Ridley, "A new continuous-time model for current-mode control," in Proc. Power Conversion Intell. Motion Conf., 1989, pp. 455–464.

[2] Yeong-Seuk Kim; Bo-Mi sNo; Jun-Sik Min; Said Al-Sarawi; Derek Abbott,"Onchip current sensing circuit for Currentlimited minimum off-time PFM boost converter" ISOCC.2009

[3] WeiweiHuang; XiaoYang; ChaodongLing"A novel current sensing circuit for boost DC-DC converter", ACSI. 2012

[4] M. Du, H. Lee and J. Liu, "A 5-MHz 91% peak-power-efficiency buck regulator with auto-selectable peak- and valley-current control," in *IEEE Journal of Solid-State Circuits*, vol. 46, no. 8, pp. 1928-1939, Aug. 2011.

[5] Texas Instruments, "LM5022-Q1 2.2MHz, 60V low-side controller for boost and sepic", SNVSAG9-MARCH 2016

A Digital Low-Dropout(DLDO) Regulator with 14dB Power Supply Rejection Enhancement

Byung Gun Joung[*], Yangho Seo[*] and Chulwoo Kim[**]

Department of Nano-Semiconductor Engineering*, Department of Electrical Engineering**
Korea University, Seoul, Republic of Korea
jbg@kilby.korea.ac.kr

Abstract— **Digital LDO (low-dropout) which has 0.45V V_OUT and a good PSR characteristic is presented for NTV operation block. It consists of comparator, counter, power switch array, and proposed adaptive VSS driver. The proposed power MOSFET driver is composed with one inverter and two additional transistors and it leads 14dB PSR improvement for overall frequency range. Minimum resolution is 1mV at 400kHz and the peak current efficiency is 99%.**

Keywords : Near-Threshold Voltage (NTV), Digital Low-Dropout (DLDO) regulator, Power Supply Rejection (PSR).

I. INTRODUCTION

Recently, there are a various research that makes a circuit operate with low supply voltage to minimize its power consumption. Called NTV (near-threshold voltage) operation, it has a supply voltage as low as the threshold voltage of MOSFET. So in order to operate properly, many analog circuits are replaced into digital one.

Although NTV operation is usually composed of digital circuit, supply voltage variation might be critical. Of course digital circuit has an immunity for supply voltage variation, but in the case of NTV operation, gate delay is highly dependent on the supply voltage. So little change in supply voltage can cause the timing violation.

To avoid designing a circuit with a large timing margin, digital LDO with PSR improvement technique is proposed. In the next session, the concept of the proposed digital LDO regulator and circuit implementation are discussed in detail. Then, simulation results are reported to verify the theory.

II. PROPOSED CIRCUIT DESCRIPTION

A. Conventional Digital LDO Operation

An analog LDO has an error amplifier in the feedback loop. And it makes the LDO hard to regulate low supply voltage, because the gain of the amplifier will be degraded. So, an analog error amplifier is replaced with digital control logic due to its strong functionality regardless with the supply voltage level.

However digital LDO has a poor power supply rejection characteristic. Because it can control turn on and off, the over-

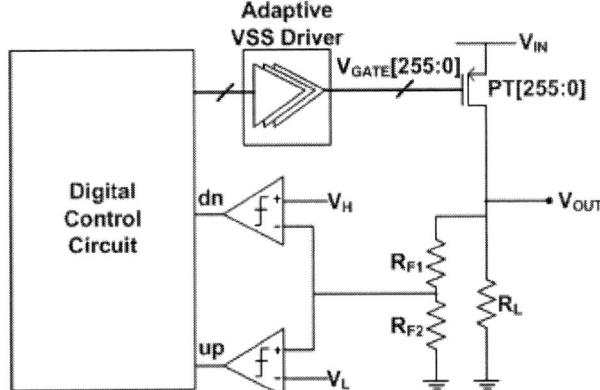

Figure 1 Top block diagram of proposed digital LDO

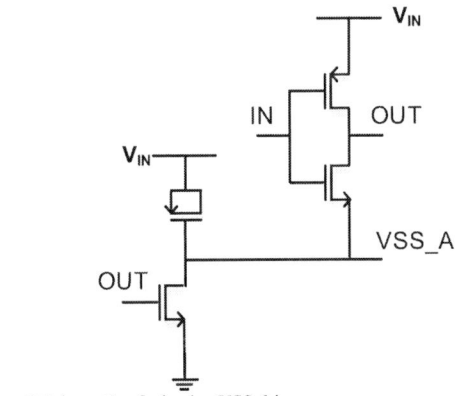

Figure 2 Schematic of adaptive VSS driver

drive voltage of the power transistor changes with the same amount of the change on the input. And it causes critical output voltage change.

B. Proposed Digital LDO Operation

Figure 1 shows top block diagram of the proposed digital LDO with adaptive VSS driver. It consists of two comparators for hysteresis window control and digital control block. The hysteresis window control can detect the steady state and hold the state, so the no ripple on the output is presented. The digital control block determines the number of power transistor turn on and off. It acts like a shifter with binary gain.

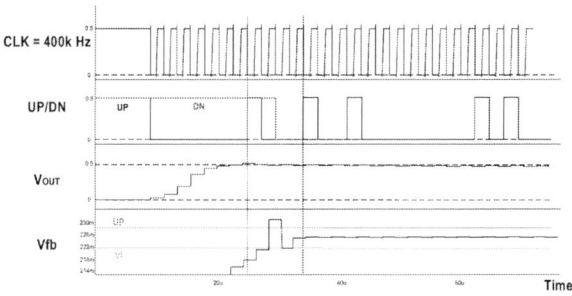

Figure 3 Top simulation result

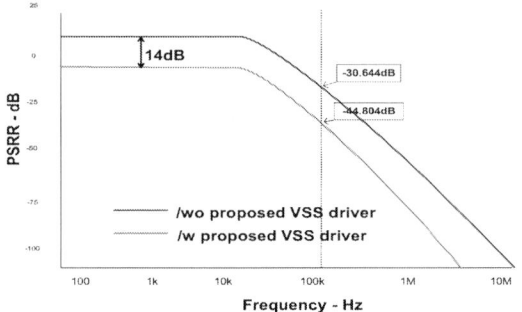

Figure 4 Comparison of power supply rejection

So when it detects the transient case, the gain of the shifter is increased, thus the transient response time is reduced.

C. Adaptive VSS Driver

To prevent the ripple on the input of LDO, adaptive VSS driver is used. With two additional MOSFET shown in figure 2, it acts like high pass filter. This filter allows to pass only the AC component of V_{IN}, thus ripple on V_{IN} node is copied to VSS_A node. The overdrive voltage of the power transistor is:

$$V_{overdrive} = V_{sg,PT} - V_{th}$$
$$= V_{IN} - VSS\_A - V_{th} \qquad (1)$$

And it remains constant, so the ripple on V_{IN} doesn't affect the output node V_{OUT}.

III. SIMUALTION RESULTS

Top simulation result is shown in figure 3. The proposed DLDO generates 0.45V regulated output voltage from 0.5V input voltage. It is designed to have maximum load current under 1mA and output capacitance is 1nF.

In figure 4, the PSR simulation result is shown. The proposed one has more than 14 dB PSR improvement

throughout whole frequency range. The performance summary is given in table 1.

Table 1. Performance summary and comparison

Spec	[2]	[3]	[4]	This work
Process	0.13μm	90nm	0.13μm	0.18μm
VIN [V]	0.5-1.2	0.5	0.65	0.5
VOUT [V]	0.45-1.14	0.45	0.95	0.45
Quiescent Current [A]	11.2u	N/A	12.7u	10u
PSR @low frequency	N/A	N/A	N/A	-3.95dB
Current Efficiency	98.3%	99.6%	99.9%	99%

IV. CONCLUSION

In this paper, Digital LDO regulator with adaptive VSS driver is proposed. It is fabricated with a 0.18μm CMOS process. It regulates 0.45V output voltage for NTV supply with 0.5V input voltage. The operating frequency is 400 kHz-10 MHz and it has 14dB PSR improvement for overall frequency range.

ACKNOWLEDGMENT

This work was supported by the IT R&D program of MOTIE/KEIT. [10052716, Design technology development of ultra-low voltage operating circuit and IP for smart sensor SoC]

REFERENCES

[1] Bangda Yang, Brian Drost, Sachin Rao, and Pavan Kumar Hanumolu. "A High-PSR LDO using a Feedforward Supply-Noise Cancellation Technique." *Custom Integrated Circuits Conference(CICC). IEEE*, 2011.

[2] Chan, Yunsheng, and Yingchieh Ho. "Design of a near-threshold digital LDO with fast transient response." *SoC Design Conference (ISOCC), International. IEEE*, 2014.

[3] Bin Nasir, Saad, Samantak Gangopadhyay, and Arijit Raychowdhury. "5.6 A 0.13μm fully digital low-dropout regulator with adaptive control and reduced dynamic stability for ultra-wide dynamic range." *Solid-State Circuits Conference-(ISSCC). IEEE International. IEEE*, 2015.Bin Nasir, Saad, Samantak Gangopadhyay, and Arijit Raychowdhury. "5.6 A 0.13μm fully digital low-dropout regulator with adaptive control and reduced dynamic stability for ultra-wide dynamic range." *Solid-State Circuits Conference-(ISSCC). IEEE International. IEEE*, 2015.

[4] W. J. Huang, *et al*, "Sub-1V capacitor-free low-dropout regulator," *IET Electron. Lett*, 2006, pp. 1395-1396.

FPGA Power Estimation Simulator for Dynamic Input Data

Taehee You, Jeongbin Kim, Minyoung Im and Eui-Young Chung
School of Electrical and Electronic Engineering
Yonsei University
Seoul, Republic of Korea
Email: {xoqhd1212, jbkim404, minyoung}@dtl.yonsei.ac.kr, eychung@yonsei.ac.kr

Abstract — As the availability of field-programmable gate arrays (FPGAs) increases, the importance of their power management has become crucial. For an efficient power management scheme, an accurate power estimation is required. The power consumption of FPGAs differs depending on the input, and previous power estimation methods have limitations which make it difficult to predict the input patterns which affect the power consumption of FPGA. Therefore, we propose a simulator which is able to estimate the power in consideration of input data. It estimates the power consumption more accurately at a minute level. From the result of experiment, we identify that there is a great gap on power estimation between previous methods and the proposed one.

Keywords—field-programmable gate array(FPGA); power; estimation; simulator;

I. INTRODUCTION

Field-programmable gate arrays (FPGAs) are reconfigurable logic and are available to use in programming of various applications. However, they usually consume more power compared to application specific integrated circuits (ASICs) since FPGAs consist of many logic blocks, wires, as well as switch boxes which are required to make them reconfigurable [1]. As they are also used in power hungry area such as internet of things (IoT), smartphones and hardware accelerators, the importance of managing the power of FPGA is increasing and thus more accurate power estimation schemes are required.

There exist a few simulation-based and probabilistic-based power estimation methods. Simulation-based ones provide higher accuracy but as they consume more time, i.e. in a day level, as the size of the application programmed in FPGAs increases, the cost of power estimation is quickly becoming a great burden. Probabilistic-based ones predict the activity of each component consisting an FPGA and then calculates its dynamic and static power. The studies are dominated by the ones that enhance the accuracy of power models and probabilistic activity on each component [2]. However, there exist critical issues in these methods. First of all, it is difficult to predict the pattern of input data as applications mapped in FPGAs become complicated are spread in various fields. That is, the power estimation based on the activity prediction of a FPGA component has become less accurate. Secondly, the power

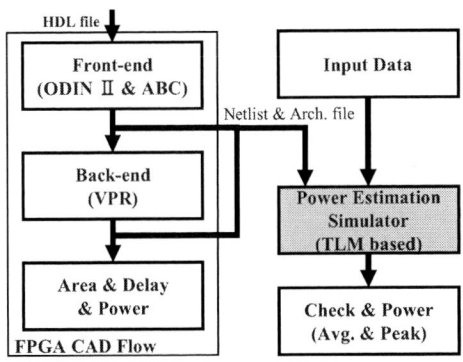

Fig. 1. Proposed Power Estimation flow

consumption of each component differs depending on the input data. For instance, the variability of power consumption can be as high as 69% with 4 input LUT depending on the input data [3]. Finally, it is impossible to estimate the peak power because the previous power estimations only target to measure the average power. The estimation of the peak power is essential as recent FPGAs are integrated in other power-critical systems.

In this paper, we design a power estimation simulator based on transaction-level modeling (TLM). It constructs the circuit of target applications using the information provided by FPGA computer-aided-design (CAD) tools and calculates the dynamic and static power every clock through the comparison of the current data and previous data. As a result, it estimates average and peak power in a minute level.

II. POWER ESTIMATION SIMULATOR

A. Power Estimation Methodolgy

A FPGA CAD consists of front and back-end as shown fig. 1. We use ODIN II[4], ABC[5], VPR[6] which are popular academic FPGA CAD tools. Although conventional FPGA CAD tools support power estimation, they do not consider the input data. We design a simulator that enables an accurate power estimation considering power variation by input data in a minute level. The simulator, which utilizes the information from CAD tools and FPGA architecture, constructs connections between components in FPGA. Then, it calculates

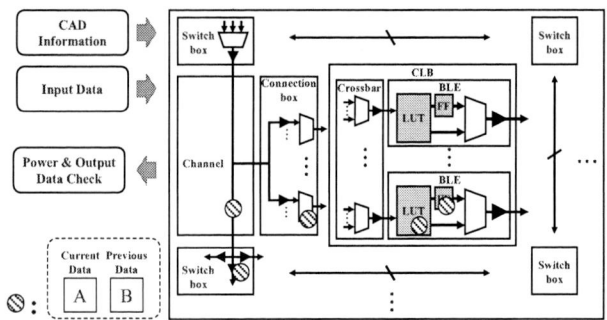

Fig. 2.Proposed Power Estimation constructer

the dynamic and static power of each component according to the input data in order to calculate the average and peak power of the overall system.

As shown fig. 2, the components of FPGA are composed of switch boxes, channels, connection boxes and configurable logic blocks (CLBs) which consist of look-up tables (LUTs), flip-flops and a crossbar logic. These components except the LUTs and the flip-flops are composed of MUXs, buffers and wires. In order to estimate these power components, our simulator stores the current data and the previous data. The power model of each components is provided by VPR CAD tool. The power of each component is divided into dynamic and static power, they are represented by $P_{dynamic} = 1/2CV^2f$ and $P_{static} = VI$, respectively. And the capacitance of each component is determined by FPGA architecture. The power of LUTs and flip-flops are determined according to the in/out data combinations. As for the wires and buffers, the dynamic power consumption occurs upon data toggle (e.g. $0 \rightarrow 1$, $1 \rightarrow 0$). For MUXs, the dynamic power is calculated when data toggle of in/out pin occurs and static power is calculated for unselected pins. Finally, total power can be measured by combining these calculated dynamic and static power. This power modeling method yields a simulator which can measure the average and the peak power consumption of each application and the accuracy would be similar to the real FPGA environment.

To validate our simulator, we compare the output data of proposed simulator with the output data from ModelSim simulation, since the proposed simulator estimates the power depending on changes of input data, the correct functionality is crucial for an accurate power estimation.

III. EXPERIMENT

We identify the applicability of the proposed power estimation simulator by comparing the result of VPR simulation in various applications. Table.1 shows the result of VPR and the proposed simulator using ten thousand input data set. We identify that the maximum difference value between power consumption of VPR and that of dynamic power estimation simulator is 13.13 times. In case of VPR, it estimate the power regardless of input data since it is calculated using toggle probability of each component. However, our proposed simulator has different estimation results depending on input data which enables the peak power estimation.

IV. CONCLUSION

In this paper we proposed a simulator which is able to estimate power which depend on input data of applications. Also, through the comparison of the power estimation result of the simulator to the result of VPR, we identify that there were large differences in power estimation. As a future work, we shall apply a high-accuracy power modeling of components to the simulator for a higher accurate power estimation.

ACKNOWLEDGMENT

This work was supported by the National Research Foundation of Korea (NRF) grant funded by the Korea government (MSIP) (2016R1A2B4011799) and by SK-Hynix Semiconductor Inc.

REFERENCES

[1] Kuon, Ian, Russell Tessier, and Jonathan Rose, "FPGA architecture: Survey and challenges," Foundations and Trends in Electronic Design Automation 2.2, 2008, pp. 135-253.

[2] Tang et al., "FPGA-SPICE: A simulation-based power estimation framework for FPGAs," In: Computer Design (ICCD), 2015 33rd IEEE International Conference on. IEEE, 2015, pp. 696-703.

[3] F. Li et al., "Power Modeling and Characteristics of Field Programmable Gate Arrays," IEEE TCAD, Vol. 24, No. 11, 2005, pp. 1712-1724.

[4] Jamieson, Peter, et al. "Odin II-an open-source verilog HDL synthesis tool for CAD research," FCCM, 2010 18th IEEE Annual International Symposium on. IEEE, 2010. pp. 149-156.

[5] Brayton, Robert, and Alan Mishchenko, "ABC: An academic industrial-strength verification tool," Computer Aided Verification. Springer Berlin Heidelberg, 2010. pp. 24-40.

[6] Betz, Vaughn, and Jonathan Rose. "VPR: A new packing, placement and routing tool for FPGA research," Field-Programmable Logic and Applications. Springer Berlin Heidelberg, 1997, pp.213-22.

TABLE I.   RESULT OF POWER ESTIMATION

Application	Power (uW)		
	VPR	Proposed Simulator	
		Avg.	Peak.
Dag3_mod	1798	889.53	1261.41
4_bit_shift_register	1061	383.26	638.08
Logic_w_Dff2	708.4	80.80	122.84
If_common	1167	622.26	1074.95
Stmt_all_mod	2306	718.05	956.66
Stmt_compare_padding	935.1	351.94	604.06

Full System Verification of Compatible Microprocessors with a Dual Physical Core Verification Platform

Jyun-Yan Li and Ing-Jer Huang
Department of Computer Science and Engineering
National Sun Yat-sen University
Kaohsiung, Taiwan
jyli@esl.cse.nsysu.edu.tw, ijhuang@cse.nsysu.edu.tw

Abstract— **This paper presents a dual physical core verification (DPCV) platform that verifies the device under verification (DUV) microprocessor with an existing compatible one at run time. Working as fault tolerance, both cores execute the same program and compare each data transaction. The DPCV synchronizes both microprocessors when data write or external interrupt occurs for data consistency and buffers the I/O data read for the DUV. If the data transactions are inconsistent, the DPCV suspends both cores after the fault for verification.**

Keywords: fault-tolerant, functional verification, compatible microprocessor

I. INTRODUCTION

In recent decades, the same instruction set of microprocessors are developed with different constraints such as performance, power, area and so on. At the verification stage, the functionalities of the compatible microprocessor are verified by the software-based self-test (SBST) [1] in the microprocessor core [2], cache memory [3] and cache controller [4]. However, the functional design faults still existing in the microprocessor may cause system crash when executing complex programs such as the OS in the real environment. The program traces of the golden microprocessor which is the existing complete microprocessor can help to develop the correctness of the device under verification (DUV) microprocessor. But the program traces of the OS may be different because the external interrupt and I/O accesses are difficult to reproduce.

The fault-tolerant system redundantly executes the same operation to detect the transient or permanent fault. In the Chip-level Redundant Threading (CRT) [5], two cores execute the same program simultaneously. The input replicator feeds the data to both cores and the output comparator compares the operating results. If the operation result is inconsistent, the system would rollback to the last checkpoint. However, this methodology assumes the cache is protected by the error correcting code (ECC) and the fault occurs in the processor core. This is not enough to verify the microprocessor if the faults are occurred in the cache RAM or cache controller.

This paper proposes a dual physical core verification (DPCV) platform to compare the data transaction at run time

with the fault tolerance concept. The error-free golden core verifies the DUV core which is under development with the same instruction set architecture (ISA) at run time. We use the data from the load/store instructions as the signature. The signature represents the operating result of that code section. The DPCV compares the data of both cores to verify the correctness of that code section. When the data are inconsistent, the platform knows that the fault occurs between the last two load/store instructions and both cores are suspended immediately for further debugging.

II. DUAL CORE VERIFICATION PLATFORM

A. Overview

The CRT is adopted to construct the DPCV platform. In Fig. 1, a compatible microprocessor (DUV) is connected to an existed environment which a golden core is constructed on an on-chip bus and this bus connects the on-board bus to access the on-board I/Os. The verification engine, the verification controller, I/O read queue, interrupt synchronization unit and write sink as marked grey in Fig. 1, are used for monitoring and verifying both data transaction, synchronizing both core, and passing the I/O data. The functionality of those modules is described in subsequent subsections.

The first major difference of this proposal from

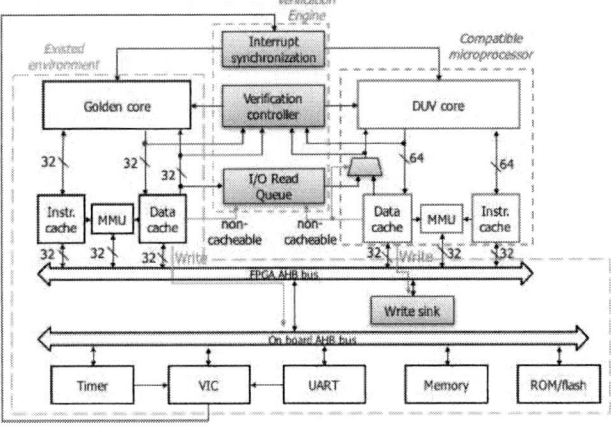

Figure 1. The Dual Physical Core Verification (DPCV) platform overview

conventional CRT [5] is that both cores can execute the program independently without lock-step at runtime. The second is that the organization of the DUV can be different with golden microprocessor.

B. Verification controller

The verification controller simultaneously monitors the data address and the load/store data between the core and the data cache. Since the load/store instruction occupies an average of 30% of the program, each data transaction is like a signature that is monitored and compared with the golden core. The advantage of this methodology is to ignore the complex operation of the core and the data cache since the verification controller is only concerned when the data is ready after a data request. When the data are inconsistent, the verification controller suspends both cores after the fault.

Since both cores are executing the same program and accessing the same virtual address, the memory data will incur data inconsistency. The verification controller also controls both cores for synchronization when one of the cores wants to write data.

To improve the performance, the verification controller allocates a queue for each core to buffer the load data and the core is suspended when the buffer is full. The buffer size is a trade-off between performance and fault detect latency. Large buffer size can decrease performance overhead, but it increases the fault detect latency.

C. I/O read queue and write sink

Reading data from I/O may vary at different times, and the verification controller may misjudge the correctness of the data when the cores access the I/O. For this data inconsistency, the I/O read queue is responsible for buffering the I/O load data of the golden core and passing data to the DUT core. In most cases, the I/O access is a non-cacheable operation and also enforces the cache to access data from the bus. To identify I/O access, some information about the data cache is required to determine a non-cacheable read. We extract the non-cacheable bit, which is judged by the cache controller, to determine I/O access.

Some I/O can only be written once, such as a display via UART. When both cores write the same data, the characters are displayed twice. The write sink can prevent this situation. Only the golden core can write to memory or to I/O and the DUT writes to write sink.

D. Interrupt synchronization unit

The I/O sends a signal to interrupt the core for I/O service. The core interrupts current process and jumps to the interrupt service routine (ISR). In the DPCV, two cores executing at different speeds with different interrupted addresses and backup data may incur misjudgment in the verification controller. The interrupt synchronization unit synchronizes both cores that are interrupted at the same instruction.

The counter records which core is faster. The counter is increased by one when the golden core is jumping and is decreased by one when the DUV is jumping. When the counter is zero it implies that both cores are in the same iteration.

III. CASE STUDY

The two academic ARM7-like [6] microprocessors are used to construct the DPCV platform on the Socle MDK-3D FPGA development board [7]. The version of U-boot is 1.1.4 and Linux kernel is 2.6.29. The purpose is to demonstrate the verification engine that the run-control and comparison of mechanisms are worked correctly.

The write sink is turned off to make the DUV's memory access explicitly observable. Fig. 2 shows that both cores have successfully passed through the U-boot, the Linux kernel booting process and the command execution (i.e. ls). Note that characters are duplicated since both cores write to the same terminal and the verification controller synchronizes them properly. The I/O read queue and the interrupt synchronization unit are also working correctly at the Linux booting.

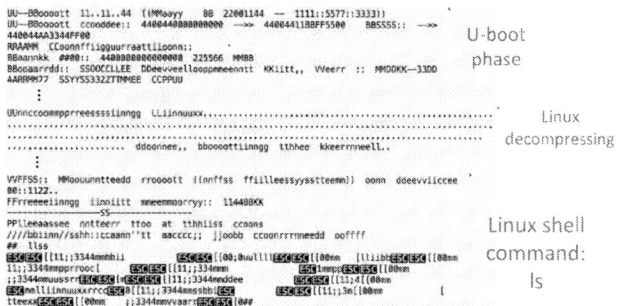

Figure 2. Trace of the golden and DUV core booting Linux kernel

IV. SUMMARY AND CONCLUSIONS

We have proposed a dual physical core verification (DPCV) platform for the verification of instruction-set compatible microprocessors. This platform adopts based fault-tolerant system in which memory accesses, I/O transactions and external interrupts are properly buffered, synchronized and compared. If there is any inconsistency, the platform halts at the exact memory/IO/interrupt events for effective bug identification. We will investigate the issues of variations of processor pipeline structures and caches in the future.

REFERENCES

[1] M. Psarakis, D. Gizopoulos, E. Sanchez and M. Sonza Reorda, "Microprocessor Software-Based Self-Testing," IEEE Design & Test of Computers, Vol. 27, 2010, pp. 4-19.

[2] A. Paschalis and D. Gizopoulos, "Effective Software-Based Self-Test Strategies for On-Line Periodic Testing of Embedded Processors," The Design, Automation and Test in Europe Conference and Exhibition, Vol. 1, 2004, pp. 578-583.

[3] G. Theodorou, N. Kranitis, A. Paschalis and D. Gizopoulos, "Software-Based Self-Test for Small Caches in Microprocessors," IEEE Transactions on Computer-Aided Design of Integrated Circuits and Systems, Vol. 33, 2014, pp. 1991-2004.

[4] Y. Lin, Y.-Y. Tsai, K.-J. Lee, C.-W. Yen and C.-H. Chen, "A Software-Based Test Methodology for Direct-Mapped Data Cache," The 17th Asian Test Symposium, 2008, pp. 363-368.

[5] S. S. Mukherjee, M. Kontz and S. K. Reinhardt, "Detailed Design and Evaluation of Redundant Multithreading Alternatives," The 29th Annual International Symposium on Computer Architecture, 2002, pp. 99-110.

[6] ARM7TDMI Data Sheet, ARM Ltd., 1995.

[7] 3D Multimedia Development kit, Socle technology Corp.

Memory ECC Architecutre Utilizing Memory Column Spares

Jong Hyuk Park[*][+]

[*]Memory Division,
Samsung Electronics
Hwasung, Korea
jh99.park@skku.edu

Joon-Sung Yang[+]

[+]Department of Semiconductor and Display Engineering
Sungkyunkwan University
Suwon, Korea
js.yang@skku.edu

Abstract— **Advanced process technology allows high memory density. However, the memory cell shrinkage introduces more memory defects and this causes a memory yield problem. To overcome the issue, memory ECC has become a critical solution. This paper proposes a hardware architecture to support memory ECC utilizing memory spares. Overheads imposed by the proposed architecture are analyzed and compared against conventional non-ECC memory architecture.**

Keywords; Memory ECC, Hardware Architecture, Memory Spares

I. Introduction

With an advancement of manufacturing technology, the feature size becomes smaller. With the same area, memories can have more transistors and this allows a memory density increase. The need for lower power operation keeps increasing and the supply voltage goes lower. Hence, the memory cells have less margins and this causes operation instability. Memories have become vulnerable to soft errors. The soft errors are transient behaviors which are not easily detectable from manufacturing tests.

Memory ECC (error correcting code) is used to deal with the occurrence of transient errors. The capability of memory ECC is determined by its generation for single-error-correcting code, double-error-correcting code, double-adjacent-error-correcting code and etc [1 - 6]. The longer codewords in ECC would give the better error correcting performance. It is important to have longer check bits. However, the longer code words require a larger memory array. The larger memory array has a higher probability of having more transient errors. References [1] and [2] introduced an ECC using memory spare columns to store additional check bits. These approaches allow to reduce a memory array size while keeping the same ECC capability.

Reference [2] used a linear block code. In a (n, k) linear block code, k data bits are encoded by n-bit codewords. The number of check bits is $r=(n-k)$. The $(r \times n)$ parity-check matrix (H-matrix) completely defines the code. C is a codeword of the code if and only if :

$$H \cdot C^T = 0$$

where C^T is the transpose of the codeword C. Let each

element in the error vector E is a 1 if the corresponding bit is in error and a 0 if the bit is error-free, then an erroneous message can be represented as $V_{error} = V \oplus E$. The syndrome, S, is defined as :

$$S = H \cdot V_{error} = H \cdot (V \oplus E) = H \cdot V \oplus H \cdot E = H \cdot E$$

In H-matrix, the additional row can be added by [1, 2]. Fig. 1(a) shows an example of (7, 3) SEC-DED Hsiao code. Fig. 1(b) illustrates the H-matrix with an additional row and this significantly enhances the ECC capability, miscorrection probability.

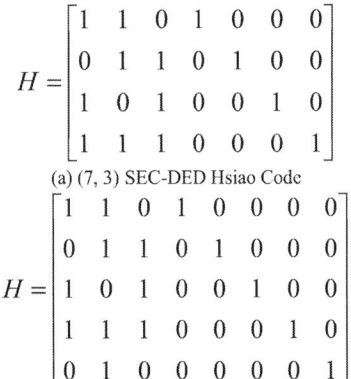

(a) (7, 3) SEC-DED Hsiao Code

(b) Adding One Row in (7, 3) SEC-DED Hsiao Code

Fig. 1. Example of (7, 3) Hsiao Code and Its Extension

II. Memory Ecc by Utilizing Memory Spares

During manufacturing test process, defective memory cells are replaced either by memory spares (repair row or repair column). In [2], repair columns are used to store additional check bits. If there is an available spare column, it is used to tell the repair information. Fig. 2 shows an example with four repair columns [2]. Two defective cells are replaced by SP2 and SP3. Since there is two unused spare columns, one (SP4) of two is used to store the repair information. 0 tells that no defective cell is located on the row and 1 indicates that there would be one or more defective cells in the corresponding row.

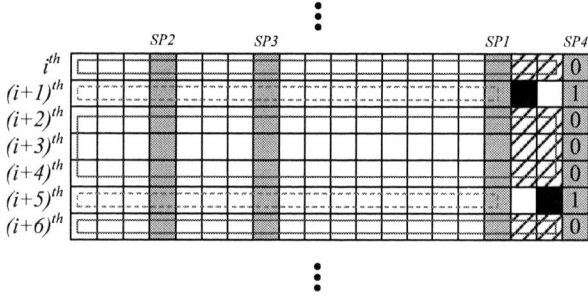

Fig. 2. Example of Repair Process by Spare Columns [2]

III. PROPOSED ARCHITESCTURE AND CONCLUSIONS

A. ECC Architecture

To support the ECC operation in [2], this paper proposes a hardware architecture for ECC using memory spare columns. Fig. 3 shows the proposed architecture. One of unused spare column stores defect information and the information is fed into Defect Info logic. The Defect Info logic decodes the value and generates inputs to a syndrome generator. In front of syndrome generator, this architecture requires an additional AND gate and this may add a read data delay. However, it should be noted that ECC operation requires extra hardware such as syndrome generator and correction logic. The AND gate can be optimized with these logic to minimize the read data delay. The syndrome generator output is used to flag an error detection and this information is used to correct the dataword. This architecture supports tripe error miscorrection

probability reduction and achieves a reliable memory operation.

B. Conclusion

It is critical to secure the reliable memory operation to guarantee the correct system behavior. As the system evolves as a multi-core system, unreliable data can cause a serious system down time. Hence, ECC is considered as a crucial part of a memory.

In this paper, the architecture to support memory ECC utilizing memory spare columns in [2] is proposed. The proposed method requires less hardware overheads such as *Defect Info* logic and one AND gate in front of syndrome generator. It should be noted that AND gate delay in data read can be optimized with syndrome generator and correction logic.

REFERENCES

[1] Datta, R., and N. A. Touba, "Exploiting Unused Spare Columns to Improve Memory ECC," VLSI Test Symposium, 2009. VTS '09. 27th IEEE, pp. 47-52, May, 2009.

[2] J.-S. Yang, "Improving Memory ECC via Defective Memory Columns", IEEE/ACM Intl. SoC Design Conference, pp. 63-64, Nov 2015.

[3] Dutta, A., "Low Cost Adjacent Double Error Correcting Code with Complete Elimination of Miscorrection within a Dispersion Window for Multiple Bit Upset Tolerant Memory," VLSI and System-on-Chip, 2012 IEEE/IFIP 20th International Conference on, pp. 287-290, Oct., 2012.

[4] Hamming, R. W., "Error Correcting and Error Detecting Codes," Bell Sys. Tech. Journal, Vol. 29, pp.147-160, Apr., 1950.

[5] Hsiao, M. Y., "A Class of Optimal Minimum Oddweight-column SECDED codes," IBM Journal of Research and Development, Vol. 14, pp. 395-401. Jul., 1970

[6] Hung, L., H. Irie, M. Goshima, and S. Sakai, "Utilization of SECDED for soft error and variation-induced defect tolerance in caches," *Proc. Of Design, Automation & Test in Europe Conference*, pp. 1 – 6, 2007.

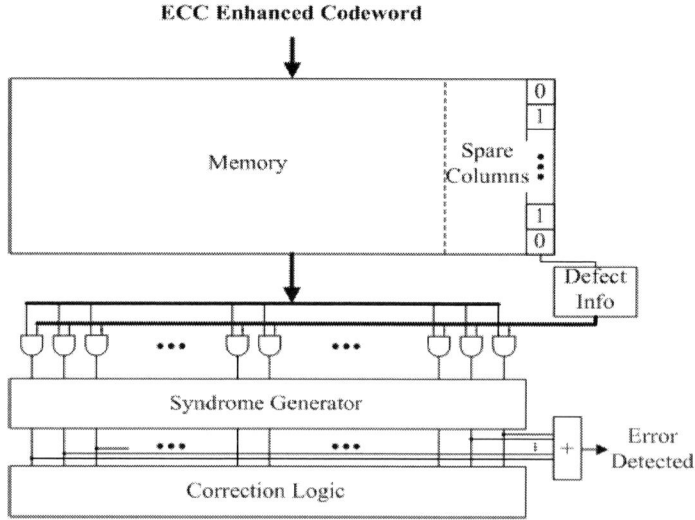

Fig. 3. Proposed Memory ECC Architecture

IEEE
445 Hoes Lane
Piscataway, NJ 08854-4141

ISBN 978-1-5090-3220-4